Continuous Distributions

Beta

$$f(x) = \frac{\Gamma(\alpha + \beta)}{\Gamma(\alpha)\Gamma(\beta)} x^{\alpha-1}(1-x)^{\beta-1}, \qquad 0 < x < 1$$

$$\mu = \frac{\alpha}{\alpha + \beta}, \quad \sigma^2 = \frac{\alpha\beta}{(\alpha + \beta + 1)(\alpha + \beta)^2}$$

Chi-Square
$\chi^2(r)$

$$f(x) = \frac{1}{\Gamma(r/2)2^{r/2}} x^{r/2-1}e^{-x/2}, \qquad 0 \le x < \infty$$

$$M(t) = \frac{1}{(1 - 2t)^{r/2}}, \qquad t < \frac{1}{2}$$

$$\mu = r, \ \sigma^2 = 2r$$

GW00322850

Exponential

$$f(x) = \frac{1}{\theta} e^{-x/\theta}, \qquad 0 \le x < \infty$$

$$M(t) = \frac{1}{1 - \theta t}, \qquad t < 1/\theta$$

$$\mu = \theta, \ \sigma^2 = \theta^2$$

Gamma

$$f(x) = \frac{1}{\Gamma(\alpha)\theta^\alpha} x^{\alpha-1}e^{-x/\theta}, \qquad 0 \le x < \infty$$

$$M(t) = \frac{1}{(1 - \theta t)^\alpha}, \qquad t < 1/\theta$$

$$\mu = \alpha\theta, \ \sigma^2 = \alpha\theta^2$$

Normal
$N(\mu, \sigma^2)$

$$f(x) = \frac{1}{\sigma\sqrt{2\pi}} e^{-(x-\mu)^2/(2\sigma^2)}, \qquad -\infty < x < \infty$$

$$M(t) = e^{\mu t + \sigma^2 t^2/2}$$

$$E(X) = \mu, \ \text{Var}(X) = \sigma^2$$

Uniform
$U(a, b)$

$$f(x) = \frac{1}{b - a}, \qquad a \le x \le b$$

$$M(t) = \frac{e^{tb} - e^{ta}}{t(b - a)}, \qquad t \ne 0; \ M(0) = 1$$

$$\mu = \frac{a + b}{2}, \ \sigma^2 = \frac{(b - a)^2}{12}$$

Fifth Edition

Probability and Statistical Inference

Robert V. Hogg

University of Iowa

Elliot A. Tanis

Hope College

 PRENTICE HALL, Upper Saddle River, NJ 07458

Library of Congress Cataloging-in-Publication Data

HOGG, ROBERT V.
 Probability and statistical inference / Robert V. Hogg, Elliot A. Tanis
   p.   cm.
 Includes bibliographical references and index.
 ISBN 0–13–254608–6
 1. Probabilities. 2. Mathematical statistics. I. Tanis, Elliot.
 II. Title.
QA273.H694 1997   96–31778
519.2—dc20   CIP

Acquisitions Editor: *Ann Heath*
Editorial Assistant: *Mindy Ince*
Editorial Director: *Tim Bozik*
Editor-in-Chief: *Jerome Grant*
Assistant Vice President of Production and Manufacturing: *David W. Riccardi*
Editorial/Production Supervision: *Robert C. Walters*
Managing Editor: *Linda Mihatov Behrens*
Executive Managing Editor: *Kathleen Schiaparelli*
Manufacturing Buyer: *Alan Fischer*
Manufacturing Manager: *Trudy Pisciotti*
Marketing Manager: *John Tweeddale*
Marketing Assistant: *Diana Penha*
Creative Director: *Paula Maylahn*
Cover Director: *Jayne Conte*

 © 1997 by Prentice-Hall, Inc.
Simon & Schuster/A Viacom Company Company
Upper Saddle River, New Jersey 07458

Previous editions copyright © 1993, 1988, 1983, 1977 by Macmillan Publishing
Company, a division of Macmillan, Inc.

All rights reserved. No part of this book may
be reproduced, in any form or by any means,
without permission in writing from the publisher.

Printed in the United States of America

10 9 8 7 6 5 4 3

ISBN 0-13-254608-6

Prentice-Hall International (UK) Limited, *London*
Prentice-Hall of Australia Pty. Limited, *Sydney*
Prentice-Hall Canada Inc., *Toronto*
Prentice-Hall Hispanoamericana, S.A., *Mexico*
Prentice-Hall of India Private Limited, *New Delhi*
Prentice-Hall of Japan, Inc., *Tokyo*
Simon & Schuster Asia Pte. Ltd., *Singapore*
Editora Prentice-Hall do Brasil, Ltda., *Rio de Janeiro*

Contents

List of Figures vi
List of Tables ix
Preface xi

1 Summary and Display of Data 1
  1.1 Basic Concepts 1
  1.2 Numerical Characteristics 15
  1.3 Exploratory Data Analysis 22
  1.4 Graphical Comparisons of Data Sets 32
  1.5 Time Sequences 41
  1.6 Scatter Plots, Least Squares, and Correlation 48

2 Probability 62
  2.1 Properties of Probability 62
  2.2 Methods of Enumeration 71
  2.3 Conditional Probability 82
  2.4 Independent Events 93
  2.5 Bayes' Theorem 101

3 Discrete Distributions 105
  3.1 Random Variables of the Discrete Type 105
  3.2 Mathematical Expectation 112
  3.3 The Mean, Variance, and Standard Deviation 118
  3.4 The Discrete Uniform and Hypergeometric Distributions 126
  3.5 Bernoulli Trials and the Binomial Distribution 135
  3.6 Geometric and Negative Binomial Distributions 147
  3.7 The Poisson Distribution 157
  3.8 Generating Functions 168

4 Continuous Distributions 180
  4.1 Random Variables of the Continuous Type 180
  4.2 The Uniform and Exponential Distributions 191
  4.3 The Gamma and Chi-Square Distributions 201
  4.4 The Normal Distribution 209
  4.5 Distributions of Functions of a Random Variable 222
  4.6 Mixed Distributions and Censoring 230

5   Sampling Distribution Theory   **237**

5.1   Independent Random Variables   237
5.2   Distributions of Sums of Independent Random Variables   245
5.3   Random Functions Associated with Normal Distributions   254
5.4   The Central Limit Theorem   262
5.5   Approximations for Discrete Distributions   270
5.6   Limiting Moment-Generating Functions   277

6   Estimation   **282**

6.1   Point Estimation   282
6.2   Confidence Intervals for Means   294
6.3   Confidence Intervals for Difference of Two Means   305
6.4   Confidence Intervals for Variances   312
6.5   Confidence Intervals for Proportions   320
6.6   Sample Size   326

7   Tests of Statistical Hypotheses   **336**

7.1   Tests about Proportions   336
7.2   Power and Sample Size   346
7.3   Tests about One Mean and One Variance   358
7.4   Tests of the Equality of Two Normal Distributions   372
7.5   Chi-Square Goodness-of-Fit Tests   385
7.6   Contingency Tables   399

8   Linear Models   **414**

8.1   Tests of the Equality of Several Means   414
8.2   Two-Factor Analysis of Variance   425
8.3   Simple Regression Problem   438
8.4   More Regression   450

9   Quality Improvement Through Statistical Methods   **462**

9.1   Shewhart Control Charts   462
9.2   General Factorial and 2^k Factorial Designs   475
9.3   More on Design of Experiments   483
9.4   Quality Improvement by Understanding Variability   488

10   Nonparametric Methods   **493**

10.1   Order Statistics   493
10.2   Confidence Intervals for Percentiles   504
10.3   Binomial Tests for Percentiles   512
10.4   The Wilcoxon Test   519
10.5   Two-Sample Distribution-Free Tests   527
10.6   Run Test and Test for Randomness   536
10.7   Kolmogorov-Smirnov Goodness-of-Fit Test   544

11 Multivariate Distributions **552**

11.1 Multivariate Distributions 552
11.2 The Correlation Coefficient 565
11.3 Conditional Distributions 572
11.4 The Bivariate Normal Distribution 583
11.5 Correlation Analysis 590
11.6 Transformations of Random Variables 596

12 Theory of Statistical Inference **605**

12.1 Sufficient Statistics 605
12.2 Bayesian Estimation 609
12.3 Asymptotic Distributions of Maximum Likelihood Estimators 615
12.4 Chebyshev's Inequality and Convergence in Probability 620
12.5 Best Critical Regions 625
12.6 Likelihood Ratio Tests 633

A REVIEW OF SELECTED MATHEMATICAL TECHNIQUES **641**

A.1 Algebra of Sets 641
A.2 Mathematical Tools for the Hypergeometric Distribution 645
A.3 Limits 648
A.4 Infinite Series 650
A.5 Integration 654
A.6 Multivariate Calculus 656

B REFERENCES **661**

C TABLES **663**

Table I   Binomial Coefficients 664
Table II   The Binomial Distribution 665
Table III   The Poisson Distribution 672
Table IV   The Chi-Square Distribution 675
Table Va   The Standard Normal Distribution Function 676
Table Vb   The Standard Normal Right Tail Probabilities 677
Table VI   The t Distribution 678
Table VII   The F Distribution 679
Table VIII   The Kolmogrov-Smirnov Acceptance Limits 684
Table IX   Random Numbers on the Interval $(0, 1)$ 685
Table X   Divisors for the Confidence Interval for σ
              of Minimum Length 686
Table XI   Distribution Function of the Correlation Coefficient R 687

Answers to Odd-Numbered Exercises **688**

Index **717**

List of Figures

1.1   Number of children per family   6
1.2   Weights of candy bars   9
1.3   Losses due to wind-related catastrophes   10
1.4   Box plot of scores from 50 statistics examinations   27
1.5   Box plot for 40 wind-related catastrophes   28
1.6   Box plot with fences for 40 wind-related catastrophes   29
1.7   Box plots of the hardness of parts from two furnaces   33
1.8   Box plots for deflections of three beams   34
1.9 q-q plot for misfeeding lead wires   36
1.10 Box plots of misfeeding lead wires   36
1.11 Sales of a company over time   42
1.12 Value of a share of stock plotted biweekly   43
1.13 Output of a worker for each month   44
1.14 Calculus scores over time   45
1.15 Prewash concentrate fill weights   46
1.16 Scatter plot of final exam score (y) vs. preliminary exam score (x) 49
1.17 Scatter plot of fuel efficiency versus weight   53
1.18 Four scatter plots and linear fits   57
1.19 Plot of 25 quadratically related plots   58

2.1   Algebra of sets   63
2.2   Fraction of experiments having at least one match   64
2.3   Tree diagram   72
2.4   Conditional Probability   84
2.5   Dice example   85

3.1   Bar graph and probability histogram   108
3.2   Sum of two tetrahedral dice   110
3.3   Hypergeometric probability histograms   129
3.4   Binomial probability histograms   138
3.5   Presidential approval histogram   140
3.6   Number of heads when five coins are flipped   143
3.7   Binomial (solid) and hypergeometric (dashed) histograms   144
3.8   Negative binomial probability histograms   149
3.9   Poisson probability histograms   160
3.10 Barium histogram   161
3.11 Poisson (solid) and binomial (dashed) probability histograms   164
3.12 Operating characteristic curve   165

4.1   Lengths between bad records on tape   181
4.2   Continuous distribution p.d.f. and c.d.f.   184
4.3   Illustration of percentiles π_{32} and π_{92} 187
4.4   Uniform p.d.f. and distribution function   191
4.5 q-q plot and distribution functions for the uniform distribution   193
4.6   Box plot of the sample from the uniform distribution   193
4.7   Exponential p.d.f. and distribution function   196
4.8   Gamma p.d.f.'s: $\theta = 4$ with $\alpha = 2, 3, 4$; $\alpha = 4$ with $\theta = 2, 3, 4$ 203
4.9   Chi-square p.d.f.'s with $r = 2, 3, 5, 8$ 204
4.10 Chi-square tails, $r = 5, \alpha = 0.10$ 206
4.11 Standard normal p.d.f.   211
4.12 Upper 100α percent point, z_α 213
4.13 q-q plot, carrot weights vs. $N(0, 1)$ quantiles   218
4.14 Lognormal distributions   223
4.15 Exponential q-q plot and box plot   227
4.16 Mixed distribution function for Example 4.6.2   232
4.17 Mixed distribution function for Example 4.6.3   232

5.1   Normal sample mean p.d.f.'s   255
5.2   P.d.f.'s of sums of uniform random variables   266
5.3   Sums of chi-square random variables   267
5.4   Sum of triangular random variables   268
5.5   Sum of U-shaped random variables   268
5.6   Probability histogram for $b(4, 1/4)$ 271
5.7   Normal approximation for the binomial distribution   272
5.8   Normal approximation of Poisson, $\lambda = 20$ 274

6.1   Confidence intervals using z and t 296
6.2 t distribution p.d.f.'s and right tail probability   299

7.1   P.d.f of $\overline{X}$ under H_0 and H_1 347
7.2   Power function: $K(p) = P(Y \le 6;\ p)$, where Y is $b(20, p)$ 349
7.3   Power function: $K(\mu) = 1 - \Phi([62 - \mu]/2)$ 350
7.4   Illustration of p-value   353
7.5   Test about mean of tumor growths   360
7.6   Test about variance of IQ scores   364
7.7   Critical region and p-value for pea stem growths   373
7.8   Box plots for pea stem growths   374
7.9   Box plots for fill weights   376
7.10 Box plots for gum thicknesses   378
7.11 Test of variances for pea stem growths   380
7.12 U-shaped p.d.f.   394

8.1   Box plots for pull-out tests   422
8.2   Scatter plot and least squares regression line   439
8.3   A 95% confidence interval for $\mu(x)$ 457
8.4   A 95% prediction band for Y 457

9.1   Typical control chart   464
9.2 $\bar{x}$-chart   466
9.3 s-chart   466
9.4   Plot of 50 console opening times   467
9.5 R-chart   469
9.6 p-chart   471
9.7 q-q plot of normal percentiles vs. estimated effects   482
9.8   Goodness response to cake-baking experiment   484
9.9   Results of cake-baking experiments   485

10.1 The Wilcoxon distribution for $n = 4$ and $n = 10$ 523
10.2 Two distribution functions with $F(z) \geq G(z)$, $m_X < m_Y$ 528
10.3 q-q plot of Cauchy vs. $N(0, 1)$ 533
10.4 H_0: X is $U(0, 1)$ 546
10.5 H_0: X has a gamma distribution   547
10.6 A 90% confidence band for an unknown distribution function   548

11.1 Discrete joint p.d.f.   554
11.2 Joint p.d.f. $f(x, y) = e^{-x-y}$ 558
11.3 Trinomial distribution   569
11.4 Joint and marginal p.d.f.'s   574
11.5 Conditional p.d.f.'s of X, given y 574
11.6 Conditional p.d.f.'s of Y, given x 574
11.7 Conditional p.d.f. of Y, given $x = 5, 10, 15$ 585
11.8 Bivariate normal p.d.f.   587
11.9 Contours for the bivariate normal distribution   587
11.10 R p.d.f.'s when $n = 8$ and $n = 14$ 593
11.11 Beta distribution p.d.f.'s   598
11.12 F distribution p.d.f.'s   602

12.1 The likelihood ratio for testing H_0: $\mu = 160$ 635

A.1   Sample space for two rolls of a four-sided die   642
A.2   Venn diagrams   644
A.3   Venn diagrams illustrating De Morgan's Laws   645
A.4 $A = \{(x, y): \ 0 \leq x \leq 1, \ 0 \leq y \leq x\}$ 658

List of Tables

1.1	Number of children per family	5
1.2	Frequency table of number of children	6
1.3	Candy bar weights	8
1.4	Frequency table of candy bar weights	8
1.5	Scores from 50 statistics examinations	22
1.6	Stem-and-leaf display of statistics examinations	23
1.7	Ordered stem-and-leaf display of statistics examinations	23
1.8	Ordered stem-and-leaf display of statistics examinations	24
1.9	Ordered stem-and-leaf display of 60 ACT scores	25
1.10	The hardness of parts from two furnaces	33
1.11	Finding corresponding quantiles for samples of different sizes	37
1.12	Fuel efficiency versus weight for 10 different cars	52
1.13	Fuel efficiency/weight statistics for 10 different cars	54
2.1	Number of samples of r out of n objects	79
3.1	Sum of two tetrahedral dice	109
3.2	Barium-133	161
4.1	Random exponential observations	226
4.2	Exponential quantiles	226
7.1	Test of hypotheses for one proportion	340
7.2	Test of hypotheses for two proportions	341
7.3	Values of the power function	350
7.4	Tests of hypotheses about one mean, variance known	358
7.5	Tests of hypotheses for one mean, variance unknown	359
7.6	Tests about the variance	364
7.7	Tests of hypotheses for equality of two means	374
7.8	Tests of hypotheses of the equality of variances	379
7.9	Grouped Geiger counter data	391
7.10	Summary of U-shaped data	393
7.11	Grouped values of w_n	394
7.12	Undergraduates at the University of Iowa	405

8.1   One-factor random samples   415
8.2   Analysis-of-variance table   418
8.3   Illustrative data   419
8.4   ANOVA table for illustrative data   419
8.5   Pull-out test data   421
8.6   ANOVA table for pull-out tests   421
8.7   Two-way ANOVA table, one observation per call   428
8.8   Gas mileage data   429
8.9   ANOVA table for gas mileage data   430
8.10 Two-way ANOVA table, c observation per cell   433
8.11 Illustration of arrays for numbers of digits   433
8.12 Cell, row, and column means when adding numbers   434
8.13 ANOVA table when adding numbers   434
8.14 Food consumption by rats   436
8.15 Mileage performance   437
8.16 Data   437
8.17 Calculations for test score data   442
8.18 Tests about the slope of the regression line   450
8.19 ANOVA table for testing $H_0: \beta = 0$ 452

9.1   Some constants used with control charts   463
9.2   Console opening times   465
9.3   Data on yellow candies   470
9.4   ANOVA table   476
9.5 2^2 design   477
9.6 2^3 design   478
9.7 2^4 design   478
9.8 2^3 design decomposition   479
9.9   Seven measures order   481

10.1 Information for confidence intervals for m 506
10.2 Body fat percentages, beginning (x) and end (y) of semester   514
10.3 Ordered absolute differences from 160   523
10.4 Ordered absolute values, changes in percentage of body fat   525

Preface

We are pleased with the reception that was given to the first four editions of *Probability and Statistical Inference*. The fifth edition is still designed for use in a course having from three to six semester hours of credit. No previous study of statistics is assumed, and a standard course in calculus provides an adequate mathematical background. Starred sections are optional in that they are not needed in subsequent sections. This, however, does not mean that they are unimportant, and we hope many of you will study them.

Although we still view this book as the basis of a junior or senior level course in the mathematics of probability and statistics that is taught by many departments of mathematics and/or statistics, we have tried to respond to statisticians' recent view of what is good statistical education. In particular, many believe in a data-oriented approach and, if time permits, will assign projects involving the collection of data by teams composed of from four to eight students. Accordingly, Chapter 1 includes good descriptive statistics, exploratory data analysis, the method of least squares, and correlation. Such things as stem-and-leaf displays, box-and-whisker diagrams, quantile-quantile plots, time-sequences, and scatter plots are included. These graphical tools are then used throughout the rest of the text.

Expanded in This Edition

Chapter 2 provides some basic concepts in probability, much of which is illustrated by the traditional counting problems, although some of this material can be omitted if the instructor so desires. Chapters 3 and 4 contain, respectively, basic discrete and continuous distributions, including their generating functions. Much of the material in the first four chapters is similar to that in the fourth edition. We did, however, move certain considerations of probability from the first chapter to the second chapter. Moreover, we included probability-generating functions as well as moment-generating functions in Chapter 3 on discrete distributions. Almost all of the material related to quality improvement by statistical methods that was scattered throughout the fourth edition, including the first chapter, is collected together in a new Chapter 9.

The start of Chapter 5 has been greatly simplified. We found that it was a major hurdle for students to jump from the case of one random variable to that of many random variables. Thus, in our presentation we have introduced only the essential material about several random variables that is needed for the consideration of basic sampling distribution theory, including the central limit theorem. This background allows us to consider an introduction to estimation and tests of statistical hypotheses found in Chapters 6 and 7, respectively. In this edition, the t- and F-distributions are not introduced until needed in Chapter 6. Because much of the material on multivariate distributions has been delayed until Chapter 11, many topics from the theory of statistics are not given until Chapters 11 and 12.

Flexible Organization

These changes allow the instructor a great deal of flexibility in teaching from this textbook. For example, in a two-semester sequence, selections from the first five chapters may be used for a probability course. The second semester on statistics would include Chapter 6, covering point and interval estimation and Chapter 7, tests of statistical hypotheses, which now contains chi-square tests. For a more applied course, selections may be made from Chapter 8, analysis of variance and regression, Chapter 9, involving quality improvement by statistics methods, and Chapter 10, concerning nonparametric methods. For more theoretical courses, Chapters 6 and 7 may be followed by Chapter 11 dealing with multivariate distributions and Chapter 12 considering a number of topics in the theory of statistical inference. Obviously, an instructor may select a mixture of the applied and theoretical offerings.

In a four-semester hour course or a two-quarter sequence, most of the material in Chapters 1–7 can be covered. This would give the student a good introduction to probability and statistical inference.

Appendix A contains helpful notions concerning algebra of sets and certain summations, as well as suggestions about limits, series, integration, and multivariate calculus. Basic mathematical notions are included here for those who feel that their mathematics is a little "rusty."

Features

Throughout the book, figures and many new real applications will help the student understand statistics and what statistical methods can accomplish. For some exercises, it is assumed that calculators or computers are available; thus, the solutions will not always involve "nice" numbers. Solutions using a computer are given if a complicated data set is involved. The data sets for these problems are available on a data disk that accompanies the text.

Ancilliaries

A **Solutions Manual** (ISBN 0–13–254616–7), containing worked out solutions to the even-numbered exercises in the text, is available to instructors from the publisher. The **Data Disk** (ISBN 0–13–254624–8), is available to instructors for their own use and for distribution to students. Upon request, it is available packaged with the text. Many of the numerical exercises in the text were solved using *Maple*.

For additional exercises that involve simulations, a separate manual *Probability and Statistics Explorations with MAPLE* by Zaven Karian and Elliot Tanis (ISBN 0–13–436759–6) is available for purchase. Several exercises in that manual also make use of the power of *Maple* as a computer algebra system.

Acknowledgments

We are indebted to the *Biometrika* Trustees for permission to include Tables IV and VII, which are abridgements and adaptions of tables published in *Biometrika Tables for Statisticians*. We are also grateful to the Literary Executor of the late Sir Ronald A. Fisher, F.R.S., to Dr. Frank Yates, F.R. S., and to Longman Group Ltd, London, for permission to use Table III from their book *Statistical Tables for Biological, Agricultural, and Medical Research* (6th ed., 1974), reproduced as our Table VI.

We wish to thank our colleagues, students, and friends for many suggestions and for their generosity in supplying data for exercises and examples. In particular we would like to acknowledge the helpful reviews of Jack Brown, Carole Bernett, Richard M. Dudley, M. Eugene Meyer, and Steven Hillmer. Moreover, our good friend, Tim Robertson, made several suggestions that we used in this edition. Also, we thank John Krueger, Joshua Levy, and Sueann Keen for their help with the preparation of the manuscript, and Zaven Karian for his supplementary *Maple* procedures that were used for solving exercises and drawing figures. We also thank the University of Iowa and Hope College for providing time and encouragement. Finally, our families, through five editions, have been most understanding during the preparation of all of this material; we truly appreciate their patience and needed their love.

R.V.H.
E.A.T.

1

SUMMARY AND DISPLAY
OF DATA

1.1 BASIC CONCEPTS

It is usually difficult to explain to the general public what statisticians do. Many think of us as "math nerds" who seem to enjoy dealing with numbers. And there is some truth to that concept. But if we consider the bigger picture, many recognize that statisticians can be extremely useful in many investigations.

Consider the following.

1. There is some problem or situation that needs to be considered; so statisticians are often asked to work with investigators or research scientists.
2. Suppose that some measure (or measures) are needed to help us understand the situation better. The measurement problem is often extremely difficult, and creating good measures is a valuable skill. For illustration, in higher education, how do we measure good teaching? This is a question to which we have not found a satisfactory answer, although several measures, such as student evaluations, have been used in the past.
3. After the measuring instrument has been developed, we must collect data through observation, possibly the results of a survey or an experiment.
4. Using these data, statisticians summarize the results, often using descriptive statistics and graphical methods.
5. These summaries are then used to analyze the situation. Here it is possible that statisticians make what are called statistical inferences.
6. Finally, a report is presented, along with some recommendations that are based upon the data and the analysis of them. Frequently, such a recommendation might be to perform the survey or experiment again, possibly changing some of the questions or factors involved. This is how statistics is used in what is referred

to as the scientific method, because often the analysis of the data suggests other experiments. Accordingly, the scientist must consider different possibilities in his or her search for an answer and thus performs similar experiments over and over again.

So the discipline of statistics deals with the *collection* and *analysis of data*. When measurements are taken, even seemingly under the same conditions, the results usually vary. Despite this variability, a statistician tries to find a pattern; yet due to the "noise," not all of the points lie on the pattern. In the face of this variability, he or she must still determine the best way to describe the pattern. Accordingly, statisticians know that mistakes will be made in data analysis, and they try to minimize those errors as much as possible and then give bounds on the possible errors. By considering these bounds, decision makers can decide how much confidence they want to place on these data and the analysis of them. If the bounds are wide, perhaps more data should be collected. If they are small, however, the person involved in the study might want to make a decision and proceed accordingly.

Variability is a fact of life, and proper statistical methods can help us understand data collected under inherent variability. Because of this variability, many decisions have to be made that involve uncertainties. In medical research, interest may center on the effectiveness of a new vaccine for mumps; an agronomist must decide if an increase in yield can be attributed to a new strain of wheat; a meteorologist is interested in predicting the probability of rain; the state legislature must decide whether increasing speed limits will result in more accidents; the admissions officer of a college must predict the college performance of an incoming freshman; a biologist is interested in estimating the clutch size for a particular type of bird; an economist desires to estimate the unemployment rate; an environmentalist tests whether new controls have resulted in a reduction in pollution.

In reviewing the preceding, relatively short list of possible areas of applications of statistics, the reader should recognize that good statistics is closely associated with careful thinking in many investigations. For illustration, students should appreciate how statistics is used in the endless cycle of the scientific method. We observe Nature and ask questions, we run experiments and collect data that shed light on these questions, we analyze the data and compare the results of the analysis to what we thought previously, we raise new questions, and on and on. Or if you like, statistics is clearly part of the important "plan–do–study–act" cycle: Questions are raised and investigations planned and carried out. The resulting data are studied and analyzed and then acted upon, often raising new questions.

There are many aspects of statistics. Some people get interested in the subject by collecting data and trying to make sense out of these observations. In some cases the answers are obvious and little training in statistical methods is necessary. But if a person goes very far in many investigations, he or she soon realizes that there is a need for some theory to help describe the error structure associated with the various estimates of the patterns. That is, at some point appropriate probability and mathematical models are required to make sense out of complicated data sets.

Statistics and the probabilistic foundation on which statistical methods are based can provide the models to help people make decisions such as these. So in this book, we are more concerned about the mathematical, rather than the applied, aspects of statistics, although we give enough real examples so that the reader can get a good sense of a number of important applications of statistical methods.

In the study of statistics we consider experiments for which the outcome cannot be predicted with certainty. Such experiments are called **random experiments**. Each experiment ends in an outcome that cannot be determined with certainty before the experiment is performed. However, the experiment is such that the collection of every possible outcome can be described and perhaps listed. This collection of all outcomes is called the **outcome space** or, more simply, the **space** S. The following examples will help illustrate what we mean by random experiments, outcomes, and their associated spaces. In this section we look at *observed* outcomes or measurements of experiments, giving some graphical methods for visualizing these data.

Example 1.1.1

Two dice are cast and the number of spots on the sides that are "up" are counted. The outcome space is $S = \{2, 3, 4, 5, 6, 7, 8, 9, 10, 11, 12\}$.   ◇

Example 1.1.2

Each of six students selects an integer at random from the first 52 positive integers. We are interested in whether at least two of these six integers match (M) or whether all are different (D). Thus $S = \{M, D\}$.   ◇

Example 1.1.3

A fair coin is flipped successively at random until the first head is observed. If we let x denote the number of flips of the coin that are required, then $S = \{x : x = 1, 2, 3, 4, \ldots\}$, which consists of an infinite, but countable, number of outcomes.   ◇

Example 1.1.4

A box of breakfast cereal contains one of four different prizes. The purchase of one box of cereal yields one of the prizes as the outcome, and the sample space is the set of four different prizes.   ◇

Example 1.1.5

In Example 1.1.4, assume that the prizes are put into the boxes randomly. A family continues to buy this cereal until they obtain a complete set of the four different prizes. The number of boxes of cereal that must be purchased is one of the outcomes in $S = \{b : b = 4, 5, 6, \ldots, \}$.   ◇

Example 1.1.6

A fair coin is flipped successively at random until **heads** is observed on **two** successive flips. If we let y denote the number of flips of the coin that are required, then $S = \{y : y = 2, 3, 4, \ldots\}$.   ◇

Example 1.1.7

To determine the percentage of body fat for a person, one measurement that is made is a person's weight under water. If w denotes this weight in kilograms, then the sample space could be $S = \{w : 0 < w < 7\}$, as we know from past experience that this weight does not exceed 7 kilograms.   ◇

Example 1.1.8

An ornithologist is interested in the clutch size (number of eggs in a nest) for gallinules, a bird that lives in a marsh. If we let c equal the clutch size, then a possible sample space would be $S = \{c : c = 0, 1, 2, \ldots, 15\}$, as 15 is the largest known clutch size.   ◇

Note that the outcomes of a random experiment can be numerical, as in Examples 1.1.3, 1.1.6, 1.1.7, and 1.1.8, but they do not have to be, as shown by Examples 1.1.2 and 1.1.4. Often, we "mathematize" those latter outcomes by assigning numbers to them. For instance, in Example 1.1.2, we could denote the outcome $\{D\}$ by the number zero and the outcome $\{M\}$ by the number one. In general, measurements on outcomes associated with random experiments are called **random variables** and these are usually denoted by some capital letter toward the end of the alphabet, such as X, Y, and Z.

Consider the numbers of outcomes in the sample spaces in these examples. In Examples 1.1.1, 1.1.4, and 1.1.8 the number of outcomes is finite. In Examples 1.1.3 and 1.1.6 the number of possible outcomes is infinite but countable. That is, there are as many outcomes as there are counting numbers (positive integers). The space for Example 1.1.7 is different from the other examples in that the set of possible outcomes is an interval of numbers. Theoretically, the weight could be any one of an infinite number of possible weights; here the number of possible outcomes is not countable. However, from a practical point of view, reported weights are selected from a finite number of possibilities because we can read and record the answer to an accuracy determined only by our scale. Many times, however, it is better to conceptualize the space as an interval of outcomes, and Example 1.1.7 is an example of a space of the continuous type.

If we consider a random experiment and its space, we note that under repeated performances of the experiment, some outcomes occur more frequently than others. For illustration, in Example 1.1.3, if this coin-flipping experiment is repeated over and over, the first head is observed on the first flip more often than on the second flip. If we can somehow, by theory or observations, determine the fractions of times a random experiment ends in the respective outcomes, we have described a *distribution of the random variable* (sometimes called a *population*). Often, we cannot determine this distribution through theoretical reasoning but must actually perform the random experiment a number of times to obtain guesses or *estimates* of these fractions. The collection of the observations that are obtained from these repeated trials is often called a *sample*. The making of a conjecture about a distribution of a random variable based on the sample is called a *statistical inference*. That is, in statistics, we try to argue from the sample to the population. To understand the background behind statistical inferences that are made from the sample, we need a knowledge of some probability, basic distributions, and sampling distribution theory; these topics are considered in the early part of this book.

Given a sample or set of measurements, we would like to determine methods for describing the data. We begin with some graphical techniques, first for discrete data and then for continuous data. We will show the relationship between certain sets of data and probability models that are given later.

Suppose that we have some **counting** or **discrete data**. For example, you record the number of children in the family of each of your classmates. Or perhaps your state, on a regular basis, selects six integers out of the first 49 integers for the state lottery. You could count the number of those six integers that are odd.

In general, suppose that we repeat a random experiment a number of times, say n. Suppose that a certain outcome has occurred f times in these n trials; the number f is called the **frequency** of the outcome. The ratio f/n is called the **relative frequency** of the outcome. A relative frequency is usually very unstable for small values of n, but it tends to stabilize about some number, say p, as n increases. The number p is called the **probability** of the outcome and this concept is discussed in Chapter 2.

To develop an understanding of a particular set of discrete data, you can summarize the data in a frequency table and then construct a histogram of the data. A **frequency table** provides the number of occurrences of each possible outcome. A **histogram** presents graphically the tallied data as illustrated in the following example.

Example 1.1.9

In Table 1.1 the number of children in each family of 100 students in two statistics classes is recorded. We shall construct a frequency table in which we tally the number of times that each outcome was observed. We also add a column that gives the relative frequency of each outcome.

TABLE 1.1 Number of children per family

2	2	5	3	4	4	3	3	6	4	3	4	4	4	4	2	5	9	2	3
1	3	5	2	4	4	4	3	3	2	2	4	2	2	6	6	1	3	3	3
3	2	3	4	7	3	3	3	2	2	2	2	3	2	3	2	3	2	5	2
3	2	2	2	4	3	3	2	3	2	4	3	3	3	4	2	4	1	2	2
2	4	3	3	3	5	2	3	3	2	2	3	3	4	2	2	2	7	2	3

In a table of numbers such as Table 1.1 it is often difficult to detect features of the measurements such as the minimum, the maximum, or the measurement that occurs most often (called the typical value). Frequently, this typical value is called the **mode**. The frequency table, Table 1.2, helps to point out these features. A more visual way to present these data is with a **histogram**. To construct a **frequency histogram**, center a rectangle with a base of length one at each observed integer value and make

TABLE 1.2   Frequency table of number of children

Number of Children	Tabulation	Frequency	Relative Frequency
1	‖‖	3	0.03
2	ℕ ℕ ℕ ℕ ℕ ℕ ℕ ‖‖‖	34	0.34
3	ℕ ℕ ℕ ℕ ℕ ℕ ℕ ‖‖‖	34	0.34
4	ℕ ℕ ℕ ‖‖	18	0.18
5	ℕ	5	0.05
6	‖‖	3	0.03
7	‖	2	0.02
8		0	0.00
9	‖	1	0.01
		100	1.00

the height equal to the frequency of this integer. To construct a **relative frequency histogram** or **density histogram**, make the height of each rectangle equal to the relative frequency of the outcome. Note that the total area of a relative frequency histogram is equal to one. This is illustrated in Figure 1.1.   ◇

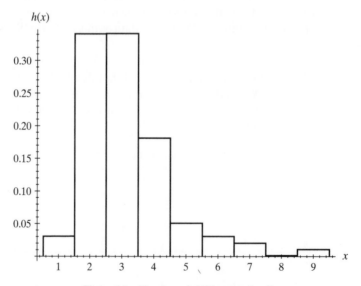

Figure 1.1   Number of children per family

One reason for constructing histograms is to picture the data more easily. Also, a histogram helps us to use the observed data to say something about the source of the data or the population from which the data were selected, provided that the data were selected randomly. For example, these 100 observations of number of children could perhaps help us say something about the number of children in families having a student at Hope College. If we would like to say something about the number of children in families in the United States, we could select a representative sample of many families and look at the number of children in these families. Note that 0 is a possible outcome for the latter experiment while that is not true of families sending at least one of their children to Hope College.

Many experiments or observations of random phenomena do not have integers as outcomes but instead are measurements selected from an interval of numbers. For example, you could find the length of time that it takes when waiting in line to buy frozen yogurt. Or, the weight of a "1-pound" package of hot dogs could perhaps be any number between 0.94 and 1.25 pounds. The weight of a miniature Baby Ruth candy bar could be any number between 20 and 27 grams. Even though such times and weights could be selected from an interval of values, times and weights are generally rounded off so that the data often look like discrete data. If conceptually the measurements could come from an interval of possible outcomes, we call it data from a continuous-type population or, more simply, **continuous-type data**.

Given a set of continuous-type data, we shall group the data into classes and then construct a histogram of the grouped data. This will help us better visualize the data. The following guidelines and terminology will be used to group continuous-type data into classes of equal length. These guidelines can also be used for sets of discrete data that have a large range.

1. Determine the largest (maximum) and smallest (minimum) observations. The **range** is the difference, $R = \text{maximum} - \text{minimum}$.
2. Generally select from $k = 5$ to $k = 20$ classes which are usually nonoverlapping intervals of equal length, so that these classes cover the interval from the minimum to the maximum.
3. Each interval begins and ends halfway between two possible values of the measurements which have been rounded off to a given number of decimal places.
4. The first interval should begin about as much below the smallest value as the last interval ends above the largest.
5. The intervals are called **class intervals** and the boundaries are called **class boundaries**. We shall denote these k class intervals by
$$(c_0, c_1), (c_1, c_2), \ldots, (c_{k-1}, c_k).$$
6. The **class limits** are the smallest and the largest possible observed (recorded) values in a class.
7. The **class mark** is the midpoint of a class.

A frequency table is constructed that lists the class intervals, the class limits, a tabulation of the measurements in the various classes, the frequency f_i of each class, the class marks, and (optionally) a column that is used to construct a relative frequency (density) histogram. A frequency histogram is constructed by drawing a rectangle for each class having as its base the class interval and height equal to the frequency of the class. For the relative frequency histogram, each rectangle has an **area** equal to the relative frequency f_i/n of the observations for the class. That is, the function defined by

$$h(x) = \frac{f_i}{(n)(c_i - c_{i-1})} \quad \text{for} \quad c_{i-1} < x \le c_i, \quad i = 1, 2, \ldots, k,$$

is called a **relative frequency histogram**, where f_i is the frequency of the ith class and n is the total number of observations. Clearly, if the class intervals are of equal length, the relative frequency histogram, $h(x)$, is proportional to the **frequency histogram** f_i, for $c_{i-1} < x \le c_i, i = 1, 2, \ldots, k$. The frequency histogram should be used only in those situations in which the class intervals are of equal length.

Example 1.1.10

The weights in grams of 40 miniature Baby Ruth candy bars, with the weights ordered, are given in Table 1.3. We shall group these data and then construct a histogram to visualize the distribution of weights. The range of the data is

$$R = 26.7 - 20.5 = 6.2.$$

TABLE 1.3   Candy bar weights

20.5	20.7	20.8	21.0	21.0	21.4	21.5	22.0	22.1	22.5
22.6	22.6	22.7	22.7	22.9	22.9	23.1	23.3	23.4	23.5
23.6	23.6	23.6	23.9	24.1	24.3	24.5	24.5	24.8	24.8
24.9	24.9	25.1	25.1	25.2	25.6	25.8	25.9	26.1	26.7

The interval, $(20.5, 26.7)$, could be covered with $k = 8$ classes of width 0.8, $k = 9$ classes of width 0.7, and there are other possibilities. We shall use $k = 7$ classes of width 0.9. The first class interval will be $(20.45, 21.35)$ and the last class interval will be $(25.85, 26.75)$. The data are grouped in Table 1.4.

TABLE 1.4   Frequency table of candy bar weights

Class Interval	Class Limits	Tabulation	Frequency (f_i)	$h(x)$	Class Mark
$(20.45, 21.35)$	20.5–21.3	卌	5	5/36	20.9
$(21.35, 22.25)$	21.4–22.2	‖‖	4	4/36	21.8
$(22.25, 23.15)$	22.3–23.1	卌 ‖‖	8	8/36	22.7
$(23.15, 24.05)$	23.2–24.0	卌 ‖	7	7/36	23.6
$(24.05, 24.95)$	24.1–24.9	卌 ‖‖	8	8/36	24.5
$(24.95, 25.85)$	25.0–25.8	卌	5	5/36	25.4
$(25.85, 26.75)$	25.9–26.7	‖‖	3	3/36	26.3

A relative frequency histogram of these data is given in Figure 1.2. Note that the total area of this histogram is equal to one. We could also construct a frequency histogram in which the heights of the rectangles would be equal to the frequencies of the classes. The shape of the two histograms is the same. Later we will see the reason for preferring the relative frequency histogram. In particular, we will be superimposing on the relative frequency histogram the graph of a continuous function, called a probability density function, that corresponds to a probability model.   ◇

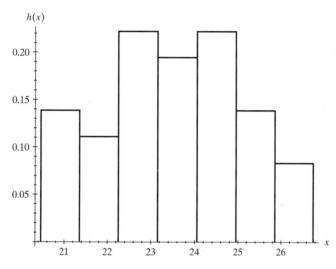

Figure 1.2   Weights of candy bars

In some instances it is not always desirable to use class intervals of equal widths in the construction of the frequency distribution and histogram. This is particularly true if the data are skewed with a very long tail. This is illustrated in Example 1.1.11.

Example 1.1.11

The following 40 losses, due to wind-related catastrophes, were recorded to the nearest $1 million. These data include only those losses of $2 million or more; and, for convenience, they have been ordered and recorded in millions.

$$
\begin{array}{cccccccccc}
2 & 2 & 2 & 2 & 2 & 2 & 2 & 2 & 2 & 2 \\
2 & 2 & 3 & 3 & 3 & 3 & 4 & 4 & 4 & 5 \\
5 & 5 & 5 & 6 & 6 & 6 & 6 & 8 & 8 & 9 \\
15 & 17 & 22 & 23 & 24 & 24 & 25 & 27 & 32 & 43
\end{array}
$$

The selection of class boundaries is more subjective in this case. It makes sense to let $c_0 = 1.5$ and $c_1 = 2.5$ because only values of $2 million or more are recorded and there are 12 observations equal to 2. We could let $c_2 = 6.5$, $c_3 = 29.5$, and

$c_4 = 49.5$. This would yield the following relative frequency histogram:

$$h(x) = \begin{cases} \dfrac{12}{40}, & 1.5 < x \le 2.5, \\[2ex] \dfrac{15}{(40)(4)}, & 2.5 < x \le 6.5, \\[2ex] \dfrac{11}{(40)(23)}, & 6.5 < x \le 29.5, \\[2ex] \dfrac{2}{(40)(26)}, & 29.5 < x \le 49.5. \end{cases}$$

It is displayed in Figure 1.3. It takes some experience before a person can display a relative frequency histogram that is most meaningful.

Figure 1.3 Losses due to wind-related catastrophes

The areas of the four rectangles, 0.300, 0.375, 0.275, 0.050, are the respective relative frequencies. It is important to note in the case of unequal lengths among class intervals that the areas, not the heights, of the rectangles are proportional to the frequencies. In particular, the first and second classes have frequencies $f_1 = 12$ and $f_2 = 15$, and yet the height of the first is greater than the height of the second while here $f_1 < f_2$. If we have equal lengths among the class intervals, then the heights are proportional to the frequencies. ◇

For continuous-type data, the interval with the largest class height is called the **modal class** and the respective class mark is called the **mode**. Hence in the last example, $x = 2$ is the mode and (1.5, 2.5) the modal class.

EXERCISES

1.1.1 Describe the outcome space for each of the following experiments:

    (a) A student is selected at random from a statistics class, and the student's ACT score in mathematics is determined. HINT: ACT test scores in mathematics are integers between 1 and 36, inclusive.

    (b) A candy bar with a 20.4-gram label weight is selected at random from a production line and is weighed.

    (c) A coin is tossed three times, and the sequence of heads and tails is observed.

1.1.2 Describe the outcome space for each of the following experiments:

    (a) Consider families that have three children each and select one such family at random. Describe the sample space S as 3-tuples, agreeing for example, that "gbb" would indicate that the youngest is a girl and the two oldest are boys.

    (b) A rat is selected at random from a cage, and its sex is determined.

    (c) A state selects a three-digit integer at random for one of its lottery games.

1.1.3 The *National Geographic* reported the following litter sizes for 20 lions:

$$2 \quad 2 \quad 1 \quad 3 \quad 2 \quad 2 \quad 3 \quad 1 \quad 4 \quad 3$$
$$2 \quad 1 \quad 3 \quad 2 \quad 2 \quad 2 \quad 2 \quad 1 \quad 2 \quad 1$$

    (a) Construct a frequency table for these data.

    (b) Draw a frequency histogram.

    (c) Draw a relative frequency histogram.

    (d) Is there a typical litter size?

1.1.4 Some ornithologists were interested in the clutch size of the common gallinule. They observed the number of eggs in each of 117 nests, yielding the following data:

```
 7   5  13   7   7   8   9   9    9   8   8   9   9   7   7
 5   9   7   7   4   9   8   8   10   9   7   8   8   8   7
 9   7   7  10   8   7   9   7   10   8   9   7  11  10   9
 9   4   8   6   8   9   9   9    8   8   5   8   8   9   9
14  10   8   9   9   9   8   7    9   7   9  10  10   7   6
11   7   7   6   9   7   7   6    8   9   4   6   9   8   9
 7   9   9   9   9   8   8   8    9   9   9   8  10   9   9
 8   5   7   8   7   6   7   7    7   6   5   9
```

    (a) Construct a frequency table for these data.

    (b) Draw a histogram.

    (c) What is the mode (the typical clutch size)?

1.1.5 The ornithologists in the preceding exercise noticed that some of the gallinules had a second brood during the summer. They were interested in comparing the clutch sizes for the second broods with those for the first brood. They were able to collect the following clutch sizes for second broods.

$$
\begin{array}{cccccccccccccccc}
4 & 4 & 6 & 5 & 5 & 9 & 6 & 5 & 6 & 9 & 7 & 9 & 4 & 8 & 6 & 5 & 9 \\
8 & 8 & 7 & 6 & 8 & 8 & 9 & 10 & 9 & 9 & 7 & 4 & 6 & 8 & 7 & 5 &
\end{array}
$$

(a) Construct a frequency table for these data.

(b) Draw a histogram.

(c) Is there a typical clutch size for second broods?

1.1.6 Before buying enough cereal to obtain a set of four prizes (Example 1.1.4), a family decided to estimate the number of boxes they would have to buy using simulation. They rolled a four-sided die until they had observed each face at least once. They repeated this 100 times and recorded the numbers of rolls needed as follows:

$$
\begin{array}{cccccccccccccccc}
8 & 6 & 6 & 4 & 13 & 11 & 19 & 13 & 4 & 15 & 8 & 13 & 5 & 8 & 9 \\
16 & 9 & 5 & 12 & 6 & 4 & 8 & 6 & 6 & 6 & 11 & 6 & 5 & 10 & 5 \\
6 & 5 & 8 & 4 & 5 & 14 & 5 & 7 & 5 & 7 & 5 & 8 & 4 & 9 & 8 \\
9 & 13 & 8 & 14 & 5 & 24 & 4 & 5 & 6 & 7 & 5 & 4 & 8 & 7 & 6 \\
11 & 4 & 5 & 6 & 5 & 10 & 10 & 4 & 5 & 14 & 10 & 15 & 6 & 9 & 8 \\
7 & 10 & 14 & 10 & 8 & 8 & 10 & 7 & 9 & 10 & 8 & 10 & 9 & 7 & 5 \\
12 & 7 & 16 & 6 & 5 & 11 & 4 & 11 & 5 & 5 & & & & &
\end{array}
$$

(a) Construct a frequency table for these data.

(b) Draw a histogram.

(c) Could a statistician tell this family how many boxes of cereal they should expect to purchase to obtain all four prizes? Note that she cannot tell the family exactly, but she might give it some type of average.

1.1.7 The costs of 28 "300-level" textbooks for the spring were as follows:

$$
\begin{array}{ccccccc}
14.00 & 5.99 & 26.35 & 35.95 & 15.95 & 24.95 & 19.95 \\
32.95 & 59.00 & 9.95 & 69.95 & 61.35 & 14.95 & 12.95 \\
16.95 & 10.95 & 57.35 & 29.95 & 5.95 & 41.95 & 66.95 \\
19.85 & 11.95 & 15.95 & 50.25 & 74.65 & 68.00 & 69.95
\end{array}
$$

(a) Group these data into seven classes using as class intervals $(5.455, 15.455)$, $(15.455, 25.455)$, and so on.

(b) Construct a histogram and interpret the histogram.

1.1.8 In the Old Kent River Bank Run (a 25-kilometer race) the times are listed separately for males and females and also by age groupings. A 53-year-old male runner was interested in the distribution of race times for males 50–54 years old. There were 125 runners in this category and their race times are listed in order in minutes and seconds.

96:21	98:24	100:26	101:23	101:29	101:38	103:11	103:58	104:28
105:02	105:14	106:00	106:14	108:49	109:28	109:32	109:42	110:49
110:54	111:48	111:54	112:17	112:33	112:48	112:54	112:55	113:09
113:10	113:15	113:43	114:30	114:36	115:28	115:39	115:44	116:11
116:34	116:56	117:45	118:10	118:11	118:12	118:28	118:30	118:44
119:10	119:27	119:53	120:17	120:19	121:12	122:15	122:17	122:46
122:57	122:59	123:46	124:04	124:06	124:49	124:55	125:14	125:15
125:16	126:09	126:12	126:29	126:43	127:21	127:25	127:57	128:02
128:23	129:13	129:21	129:39	130:01	130:07	130:29	130:34	131:07
131:22	131:58	132:27	132:51	132:58	133:51	134:04	134:40	134:53
135:05	135:41	136:27	136:51	136:52	136:53	137:09	137:13	137:31
137:48	138:01	138:13	138:44	139:34	139:51	140:03	140:05	141:20
142:28	143:08	143:46	144:10	144:35	145:40	147:05	148:12	151:18
151:39	152:00	156:13	160:43	161:12	162:29	166:42	169:54	

(a) Group these data into 10 classes with a class width of 8 minutes.

(b) Draw a histogram of the grouped data.

(c) Describe the shape of the histogram. Is it what you would have expected? (Similar data can be found following most road races.)

1.1.9 In Example 1.1.7 the sample space for underwater weights was given. These weights for 82 male students were

3.7	3.6	4.0	4.3	3.8	3.4	4.1	4.0	3.7	3.4	3.5	3.8	3.7	4.9
3.5	3.8	3.3	4.8	3.4	4.6	3.5	5.3	4.4	4.2	2.5	3.1	5.2	3.8
3.3	3.4	4.1	4.6	4.0	1.4	4.3	3.8	4.7	4.4	5.0	3.2	3.1	4.2
4.9	4.5	3.8	4.2	2.7	3.8	3.8	2.0	3.4	4.9	3.3	4.3	5.6	3.2
4.7	4.5	5.2	5.0	5.0	4.0	3.8	5.3	4.5	3.8	3.8	3.4	3.6	3.3
4.2	5.1	4.0	4.7	6.5	4.4	3.6	4.7	4.5	2.3	4.0	3.7		

(a) Group these data into 11 classes with a class width of 0.5 kilogram, in which the class marks are 1.5, 2.0, 2.5, . . . , 6.5, respectively.

(b) Draw a histogram of the grouped data.

(c) Describe the shape of the histogram. Is it what you would have expected?

1.1.10 A small part for an automobile rear-view mirror was produced on two different punch presses. In order to describe the distribution of the weights of those parts, a random sample was selected, and each piece was weighed in grams, resulting in the following data set:

3.968	3.534	4.032	3.912	3.572	4.014	3.682	3.608
3.669	3.705	4.023	3.588	3.945	3.871	3.744	3.711
3.645	3.977	3.888	3.948	3.551	3.796	3.657	3.667
3.799	4.010	3.704	3.642	3.681	3.554	4.025	4.079
3.621	3.575	3.714	4.017	4.082	3.660	3.692	3.905
3.977	3.961	3.948	3.994	3.958	3.860	3.965	3.592
3.681	3.861	3.662	3.995	4.010	3.999	3.993	4.004
3.700	4.008	3.627	3.970	3.647	3.847	3.628	3.646
3.674	3.601	4.029	3.603	3.619	4.009	4.015	3.615
3.672	3.898	3.959	3.607	3.707	3.978	3.656	4.027
3.645	3.643	3.898	3.635	3.865	3.631	3.929	3.635
3.511	3.539	3.830	3.925	3.971	3.646	3.669	3.931
4.028	3.665	3.681	3.984	3.664	3.893	3.606	3.699
3.997	3.936	3.976	3.627	3.536	3.695	3.981	3.587
3.680	3.888	3.921	3.953	3.847	3.645	4.042	3.692
3.910	3.672	3.957	3.961	3.950	3.904	3.928	3.984
3.721	3.927	3.621	4.038	4.047	3.627	3.774	3.983
3.658	4.034	3.778					

(a) Construct a frequency distribution using about 10 (say, 8 to 12) classes.

(b) Draw a histogram of the data.

(c) Describe the shape of this distribution.

1.1.11 An insurance company experienced the following mobile home losses in 10,000's of dollars for 50 catastrophic events:

1	2	2	3	3	4	4	5	5	5
5	6	7	7	9	9	9	10	11	12
22	24	28	29	31	33	36	38	38	38
39	41	48	49	53	55	74	82	117	134
192	207	224	225	236	280	301	308	351	527

(a) Group these data into six to eight classes. As in Example 1.1.11 the classes do not have the same width.

(b) Construct a relative frequency histogram.

(c) Describe the distribution of losses.

1.1.12 In Example 1.1.3, a fair coin is flipped at random until the first head is observed. Let A_x equal the event that exactly x flips are required for $x = 1, 2, 3, 4, \ldots$. By performing this experiment or simulating it on a computer, to what does the relative frequency of x, say f_x/n, $x = 1, 2, 3, 4, 5$, get close?

1.2 NUMERICAL CHARACTERISTICS

In Section 1.1 we looked at graphical techniques for describing a set of measurements. We now consider some numerical characteristics of a set of data. Suppose that a random experiment is repeated n times with observations $x_1, x_2, \ldots, x_n$. Often, the collection of these n values, $x_1, x_2, \ldots, x_n$ is called a **sample**. There are many characteristics associated with these data. A measure of the center of the data is called the mean of the sample, or more simply, the **sample mean**, and it is defined by

$$\bar{x} = \frac{1}{n} \sum_{i=1}^{n} x_i.$$

The variance of the sample, or more simply, the **sample variance**, is defined by

$$s^2 = \frac{1}{n-1} \sum_{i=1}^{n} (x_i - \bar{x})^2.$$

The **sample standard deviation**, $s = \sqrt{s^2} > 0$, gives a measure of how dispersed the data are from the sample mean. At this stage of the study of statistics it is difficult to get a good understanding or meaning of the standard deviation s, but you can roughly think of it as the average distance of the values $x_1, x_2, \ldots, x_n$ from the mean $\bar{x}$. This is not true exactly for, in general,

$$s \neq \frac{1}{n} \sum_{i=1}^{n} |x_i - \bar{x}|,$$

but it is fair to write, to help your understanding, that

$$s \approx \frac{1}{n} \sum_{i=1}^{n} |x_i - \bar{x}|.$$

For example, rolling a fair eight-sided die five times could result in the sample of $n = 5$ observations

$$x_1 = 3, \quad x_2 = 7, \quad x_3 = 2, \quad x_4 = 5, \quad x_5 = 3.$$

In this case

$$\bar{x} = \frac{3 + 7 + 2 + 5 + 3}{5} = 4$$

and

$$s^2 = \frac{(3-4)^2 + (7-4)^2 + (2-4)^2 + (5-4)^2 + (3-4)^2}{4} = \frac{16}{4} = 4.$$

It follows that $s = \sqrt{4} = 2$. Roughly, s can be thought of as the average distance that the x-values are away from the sample mean $\bar{x}$. Clearly, this is not exact because in this example the distances from $\bar{x} = 4$ are 1, 3, 2, 1, 1, with an average of 1.6. In general, s will be somewhat larger than this average distance, but this approximation should give the reader some idea about the meaning of a standard deviation.

There is an alternative way of computing s^2 because

$$\sum_{i=1}^{n}(x_i - \bar{x})^2 = \sum_{i=1}^{n}(x_i^2 - 2\bar{x}x_i + \bar{x}^2) = \sum_{i=1}^{n}x_i^2 - 2\bar{x}\sum_{i=1}^{n}x_i + \sum_{i=1}^{n}\bar{x}^2$$

$$= \sum_{i=1}^{n}x_i^2 - 2\bar{x}(n\bar{x}) + n\bar{x}^2 = \sum_{i=1}^{n}x_i^2 - n\bar{x}^2.$$

Thus

$$s^2 = \frac{\displaystyle\sum_{i=1}^{n}x_i^2 - n\bar{x}^2}{n-1} = \frac{\sum_{i=1}^{n}x_i^2 - \dfrac{1}{n}\left(\displaystyle\sum_{i=1}^{n}x_i\right)^2}{n-1}.$$

In our simple example,

$$s^2 = \frac{3^2 + 7^2 + 2^2 + 5^2 + 3^2 - (5)(4)^2}{4} = \frac{16}{4} = 4.$$

Of course, in this computation, $5(4)^2$ could be replaced by $(20)^2/5 = 80$.

There are some sets of data, especially when the data are discrete, for which certain measurements are repeated several times. In these cases it is useful to first group the data before finding the sample mean and sample variance, using the formulas that are given after the next example.

Example 1.2.1

Suppose that a six-sided die is rolled $n = 100$ times resulting in the 100 numbers:

```
3 6 5 1 4 5 4 3 5 6 5 4 4 5 5 2 4 6 2 2
4 3 2 3 4 6 1 5 6 5 1 4 3 4 6 4 2 6 4 3
6 4 4 4 5 5 1 5 3 1 1 4 6 4 6 1 5 1 3 5
1 1 5 2 4 2 3 1 2 3 3 2 4 6 3 5 5 4 2 4
2 2 2 2 3 4 6 6 5 2 1 6 2 2 1 5 6 4 2 2
```

For these data,

$$\sum_{i=1}^{100}x_i = 358 \qquad \text{and} \qquad \sum_{i=1}^{100}x_i^2 = 1548.$$

Thus

$$\bar{x} = 358/100 = 3.58 \qquad \text{and} \qquad s^2 = \frac{1548 - (358)^2/100}{99} = 2.6905.$$

It follows that $s = \sqrt{2.6905} = 1.64$. Since there are only six possible outcomes, $\{1, 2, 3, 4, 5, 6\}$, each occurring with a certain frequency, we can modify our computations of the mean and the variance by taking advantage of repeated measurements. Denote the outcomes by u_i and the frequencies by f_i. Then we can calculate the following sums.

Outcome (u_i)	Frequency (f_i)	$f_i u_i$	$f_i u_i^2$	$f_i(u_i - \overline{u})^2$
1	13	13	13	86.5332
2	19	38	76	47.4316
3	13	39	117	4.3732
4	22	88	352	3.8808
5	18	90	450	36.2952
6	15	90	540	87.8460
	100	358	1548	266.3600

Here $\sum_{i=1}^{100} x_i = \sum_{i=1}^{6} f_i u_i = 358$, and $\sum_{i=1}^{100} x_i^2 = \sum_{i=1}^{6} f_i u_i^2 = 1548$, so

$$\overline{x} = \frac{358}{100} = 3.58 \quad \text{and} \quad s^2 = \frac{266.3600}{99} = \frac{1548 - (358)^2/100}{99} = 2.6905.$$

$\diamond$

The preceding example can be generalized when the data $x_1, x_2, \ldots, x_n$ have k possible outcomes $u_1, u_2, \ldots, u_k$ with respective frequencies $f_1, f_2, \ldots, f_k$. The sample mean is, with $n = \sum_{i=1}^{k} f_i$,

$$\overline{x} = \frac{1}{n} \sum_{i=1}^{n} x_i = \frac{1}{n} \sum_{i=1}^{k} f_i u_i = \overline{u}$$

and the sample variance is

$$s_x^2 = \frac{1}{n-1} \sum_{i=1}^{n} (x_i - \overline{x})^2 = \frac{1}{n-1} \sum_{i=1}^{k} f_i(u_i - \overline{u})^2 = s_u^2.$$

The alternative formula for s^2 is given by

$$s_x^2 = \frac{\sum_{i=1}^{n} x_i^2 - \frac{1}{n}\left(\sum_{i=1}^{n} x_i\right)^2}{n-1} = \frac{\sum_{i=1}^{k} f_i u_i^2 - \frac{1}{n}\left(\sum_{i=1}^{k} f_i u_i\right)^2}{n-1} = s_u^2.$$

Remark   Many hand-held calculators can find the sample mean and the sample standard deviation for ungrouped as well as for grouped data. **Learn how to use your calculator.** And, of course, for large data sets, take advantage of statistical packages on computers.

Now suppose that continuous-type data for an experiment have been grouped in a frequency table with k classes. In such a case we could assume that the possible outcomes are approximately equal to the midpoints of the classes, say $u_1, u_2, \ldots, u_k$. Recall that these midpoints are called the **class marks**. Denote the respective frequencies as $f_1, f_2, \ldots, f_k$ and let $n = \sum_{i=1}^{k} f_i$. In these cases we have k points $u_1, u_2, \ldots, u_k$ with respective frequencies $f_1, f_2, \ldots, f_k$. The sample mean and the sample variance of the u values of these grouped data are, respectively,

$$\bar{u} = \frac{1}{n} \sum_{i=1}^{k} f_i u_i$$

and

$$s_u^2 = \frac{1}{n-1} \sum_{i=1}^{k} f_i (u_i - \bar{u})^2 = \frac{\sum_{i=1}^{k} f_i u_i^2 - \frac{1}{n}\left(\sum_{i=1}^{k} f_i u_i\right)^2}{n-1}$$

$$= \frac{\sum_{i=1}^{k} f_i u_i^2 - n\bar{u}^2}{n-1}.$$

For grouped "continuous data," these statistics can be used as respective approximations of $\bar{x}$ and s_x^2 which are computed before grouping. But, in the discrete case, we have noted that $\bar{x} = \bar{u}$ and $s_x^2 = s_u^2$.

Example 1.2.2

We construct the following table to find the sample mean and sample variance of the grouped data in Example 1.1.10.

Class Mark (u_i)	Frequency (f_i)	$f_i u_i$	$f_i u_i^2$	$f_i (u_i - \bar{u})^2$
20.9	5	104.5	2,184.05	34.0605
21.8	4	87.2	1,900.96	11.6964
22.7	8	181.6	4,122.32	5.2488
23.6	7	165.2	3,898.72	0.0567
24.5	8	196.0	4,802.00	7.8408
25.4	5	127.0	3,225.80	17.8605
26.3	3	78.9	2,075.07	23.3523
	40	940.4	22,208.92	100.1160

The mean and variance of these grouped data are

$$\bar{u} = \frac{940.4}{40} = 23.51 \quad \text{and} \quad s_u^2 = \frac{100.1160}{39} = 2.5671.$$

Note that the variance could have been found using the formula

$$s_u^2 = \frac{22{,}208.92 - (40)(23.51)^2}{39} = \frac{100.1160}{39} = 2.5671.$$

Here $s_u = \sqrt{2.5671} = 1.60$ is the standard deviation of the grouped data.

Note that $\bar{u}$ and s_u are approximations for $\bar{x}$ and s_x, the sample mean and sample standard deviation of the ungrouped data. Usually, the approximation is quite good and in this particular example, $\bar{x} = 23.505$ and $s_x = 1.64$, so the approximations are very good. ◇

Given a set of measurements, the sample mean is the center of the data such that the deviations from that center sum to zero; that is, $\sum_{i=1}^{n}(x_i - \bar{x}) = 0$. The sample standard deviation gives a measure of how spread out the data are from the sample mean. If the histogram is "mound-shaped" or "bell-shaped," the following empirical rule gives rough approximations to the percentages of the data that fall between certain points.

Empirical Rule: Let $x_1, x_2, \ldots, x_n$ have a sample mean $\bar{x}$ and sample standard deviation s. If the histogram of these data is "bell-shaped," then

- Approximately 68% of the data are in the interval $(\bar{x} - s, \bar{x} + s)$.
- Approximately 95% of the data are in the interval $(\bar{x} - 2s, \bar{x} + 2s)$.
- Approximately 99.7% of the data are in the interval $(\bar{x} - 3s, \bar{x} + 3s)$.

For the data in Examples 1.1.10 and 1.2.2, the number of weights that fall within one standard deviation of the mean, $(23.505 - 1.64, 23.505 + 1.64)$, is 27 or 67.5%. For these particular weights, 100% fall within two standard deviations of $\bar{x}$. Thus the histogram is missing part of the "bell" in the tails for the empirical rule to hold. The theory needed to explain why this approximation does or does not work is considered in Chapter 4. When you draw a histogram, it is useful to indicate on the histogram the location of $\bar{x}$ as well as the points $\bar{x} \pm s$ and $\bar{x} \pm 2s$.

EXERCISES

1.2.1 One characteristic of a car's storage console that is checked by the manufacturer is the time in seconds that it takes for the lower storage compartment door to open completely. A random sample of size $n = 5$ yielded the following times:

$$1.1 \quad 0.9 \quad 1.4 \quad 1.1 \quad 1.0$$

(a) Find the sample mean, $\bar{x}$.

(b) Find the sample variance, s^2.

(c) Find the sample standard deviation, s.

1.2.2 The grip strengths at the end of a health dynamics course for 15 male students were

$$58 \quad 52 \quad 46 \quad 57 \quad 52 \quad 45 \quad 65 \quad 71$$
$$57 \quad 54 \quad 48 \quad 58 \quad 35 \quad 44 \quad 53$$

Find the mean, the variance, and the standard deviation of these 15 grip strengths.

1.2.3 The *National Geographic* reported the following litter sizes for 20 lions (Exercise 1.1.3):

$$2 \quad 2 \quad 1 \quad 3 \quad 2 \quad 2 \quad 3 \quad 1 \quad 4 \quad 3$$
$$2 \quad 1 \quad 3 \quad 2 \quad 2 \quad 2 \quad 2 \quad 1 \quad 2 \quad 1$$

(a) Here we have only four possible outcomes; so proceed as in Example 1.2.1 and determine the frequency of each. Which value is the mode?

(b) Find the mean, the variance, and the standard deviation of these 20 litter sizes.

1.2.4 Some ornithologists were interested in the clutch size of the common gallinule. They observed the number of eggs in 117 nests that are recorded in Exercise 1.1.4.

(a) Find the sample mean of these data.

(b) Find the sample standard deviation.

(c) Find the mode.

(d) If you constructed a histogram of these data in Exercise 1.1.4, locate on your histogram $\bar{x}, \bar{x} \pm s, \bar{x} \pm 2s$.

1.2.5 In Exercise 1.1.6 you were given data for the number of rolls of a four-sided die that are needed to observe each face at least once. Use these data to help a family estimate the number of boxes of cereal that they must expect to purchase to obtain each of four different prizes, assuming that the prizes are put into the boxes of cereal randomly. In particular, compute the mean of the 100 values.

1.2.6 Denote a sample of n numbers by $x_1, x_2, \ldots, x_n$. Suppose that we transform each observation x_i into a new observation z_i using the transformation $z_i = ax_i + b, i = 1, 2, \ldots, n$. Show that the relationships between the means and the variances of $x_1, x_2, \ldots, x_n$ and $z_1, z_2, \ldots, z_n$ are

$$\bar{z} = a\bar{x} + b \qquad \text{and} \qquad s_z^2 = a^2 s_x^2.$$

1.2.7 Using the following sample: 1.0, 2.3, 3.0, 4.2, and 1.5, and the results of Exercise 1.2.6, find five observations with a mean three units larger and a variance four times as large as these.

1.2.8 The costs of 28 "300-level" textbooks for the spring were given in Exercise 1.1.7.

(a) In that exercise, these data were grouped into seven classes using as class intervals (5.455, 15.455), (15.455, 25.455), and so on. Compute $\bar{u}$ and s_u^2, where $u_i, i = 1, 2, \ldots, 7$, are the class marks.

(b) Calculate the sample mean $\bar{x}$ and sample standard deviation s_x using the original x values.

1.2.9 Before Halloween two bags of miniature Clark bars were purchased and each bar was weighed (in grams), 19 in the first bag and 23 in the second bag, yielding the following weights:

19.8	20.3	20.2	19.6	20.0	19.4	15.4	19.9	20.5	20.5	19.8
15.5	21.8	20.0	15.4	20.6	15.7	20.4	21.3	16.7	15.6	14.2
15.9	16.8	15.2	15.0	16.0	16.4	15.4	16.0	16.1	17.0	16.0
15.1	14.1	15.8	15.7	16.6	15.2	15.5	14.9	15.3		

(a) Group these weights using as class boundaries 13.95–14.95, 14.95–15.95, and so on, and construct a histogram.

(b) Calculate the values of the sample mean (either $\bar{x}$ or $\bar{u}$) and sample standard deviation (either s_x or s_u).

(c) Locate on your histogram $\bar{x}, \bar{x} \pm s_x, \bar{x} \pm 2s_x$ (or $\bar{u}, \bar{u} \pm s_u, \bar{u} \pm 2s_u$).

(d) Give an interpretation.

1.2.10 Decks have become very popular items for American homeowners. Lumberyards sell many styles of decks during the summer months, but they all use the same type of nail. The question that always arises is how many nails to buy. If 800 nails are needed, how many grams of nails should be purchased? The weights (in grams) of the following 50 nails should help to answer the question:

```
9.42  8.69  8.93  8.27  8.82  8.66  8.90  8.31  9.15  9.63
9.41  8.56  8.82  8.58  8.43  8.05  8.56  8.55  8.88  8.73
8.29  8.79  8.51  8.85  9.34  9.21  8.38  8.51  8.41  8.98
8.58  9.21  8.27  8.76  9.26  8.59  8.36  8.71  8.51  8.88
9.20  8.24  8.57  8.85  8.69  8.85  9.08  9.40  9.25  8.79
```

(a) Group these data into seven classes using as class intervals (7.995, 8.245), (8.245, 8.495), and so on.

(b) Construct a histogram.

(c) Calculate the sample mean and sample standard deviation for the grouped data.

(d) How many grams of nails would you recommend?

1.2.11 In the casino game roulette, if a player bets $1 on red, the probability of winning $1 is 18/38 and the probability of losing $1 is 20/38. Let X equal the number of successive $1 bets that a player makes before losing $5. One hundred observations of X were simulated on a computer, yielding the following data:

```
23  127  877   65  101   45   61    95   21   43
53   49   89    9   75   93   71    39   25   91
15  131   63   63   41    7   37    13   19  413
65   43   35   23  135  703   83     7   17   65
49  177   61   21    9   27  507     7    5   87
13  213   85   83   75   95  247  1815    7   13
71   67   19  615   11   15    7   131   47   25
25    5  471   11    5   13   75    19  307   33
57   65    9   57   35   19    9    33   11   51
27    9   19   63  109  515  443    11   63    9
```

(a) Find the sample mean and sample standard deviation of these data.

(b) Construct a relative frequency histogram using about 10 classes. These do not need to be of the same length.

(c) Locate on your histogram $\bar{x}, \bar{x} \pm s, \bar{x} \pm 2s, \bar{x} \pm 3s$.

(d) If a friend asked you how many bets could typically be placed before losing $5, would $\bar{x}$ be a good answer? Why?

1.3 EXPLORATORY DATA ANALYSIS

In Section 1.1 we studied methods, such as the histogram, of describing data. In recent years statisticians have become more concerned about better descriptive statistical methods. One such person is John W. Tukey. For more details, see the books by Tukey[1] (1977) and Velleman and Hoaglin[2] (1981).

Possibly the easiest way to begin is with an example to which all of us can relate. Say that we have the following 50 test scores on a statistics examination:

93	77	67	72	52	83	66	84	59	63
75	97	84	73	81	42	61	51	91	87
34	54	71	47	79	70	65	57	90	83
58	69	82	76	71	60	38	81	74	69
68	76	85	58	45	73	75	42	93	65

Noting that 34 and 97 are the smallest and largest scores, respectively, we can make a conventional tally such as that in Table 1.5. Along with the corresponding histogram, this tabulation describes the sample fairly well. However, it is important to note that in this summary the original test scores are lost.

TABLE 1.5 Scores from 50 statistics examinations

Class Interval	Tabulation	Frequency (f_i)
(29.5, 39.5)	‖	2
(39.5, 49.5)	‖‖	4
(49.5, 59.5)	ⅢⅡ	7
(59.5, 69.5)	ⅢⅢ	10
(69.5, 79.5)	ⅢⅢⅢ	13
(79.5, 89.5)	ⅢⅢ	9
(89.5, 99.5)	Ⅲ	5

We can do much the same thing, but keep those original values, through a **stem-and-leaf display**. For this particular data set, we could use the following procedure. The first number in the set, 93, is recorded as follows: the 9 (in the "tens" place) is treated as the stem, and the 3 (in the "units" place) is the corresponding leaf. Note

---

[1] Tukey, John W., *Exploratory Data Analysis*, Addison-Wesley Publishing Company, Reading, Mass., 1977

[2] Velleman, Paul F. and Hoaglin, David C., *Applications, Basics, and Computing of Exploratory Data Analysis*, Duxbury Press, Boston, Mass., 1981

that this leaf of 3 is the first digit after the stem of 9 in Table 1.6. The second number, 77, is given by the leaf of 7 after the stem of 7; the third number, 67, by the leaf of 7 after the stem of 6; the fourth number, 72, as the leaf of 2 after the stem of 7 (note that this is the second leaf on the 7 stem); and so on. Table 1.6 is called a stem-and-leaf display. If the leaves are carefully aligned vertically, this table has the same effect as a histogram, but the original numbers are not lost.

TABLE 1.6   Stem-and-leaf display of scores from 50 statistics examinations

Stem	Leaves	Frequency
3	4 8	2
4	2 7 5 2	4
5	2 9 1 4 7 8 8	7
6	7 6 3 1 5 9 0 9 8 5	10
7	7 2 5 3 1 9 0 6 1 4 6 3 5	13
8	3 4 4 1 7 3 2 1 5	9
9	3 7 1 0 3	5

There are times when it is helpful to modify the stem-and-leaf display by ordering the leaves in each row from smallest to largest. The resulting stem-and-leaf diagram is called an **ordered stem-and-leaf display**. Table 1.7 gives an ordered stem-and-leaf display using the data in Table 1.6.

TABLE 1.7   Ordered stem-and-leaf display of statistics examinations

Stem	Leaves	Frequency
3	4 8	2
4	2 2 5 7	4
5	1 2 4 7 8 8 9	7
6	0 1 3 5 5 6 7 8 9 9	10
7	0 1 1 2 3 3 4 5 5 6 6 7 9	13
8	1 1 2 3 3 4 4 5 7	9
9	0 1 3 3 7	5

There is another modification that can also be helpful. Suppose that we want two rows of leaves with each original stem. We can do this by recording leaves 0, 1, 2, 3, and 4 with a stem adjoined with an asterisk (∗) and leaves 5, 6, 7, 8, and 9 with a stem adjoined with a dot (●). Of course, in our example, by going from seven original

TABLE 1.8   Ordered stem-and-leaf display of statistics examinations

Stem	Leaves	Frequency
3∗	4	1
3•	8	1
4∗	2 2	2
4•	5 7	2
5∗	1 2 4	3
5•	7 8 8 9	4
6∗	0 1 3	3
6•	5 5 6 7 8 9 9	7
7∗	0 1 1 2 3 3 4	7
7•	5 5 6 6 7 9	6
8∗	1 1 2 3 3 4 4	7
8•	5 7	2
9∗	0 1 3 3	4
9•	7	1

classes to 14 classes, we lose a certain amount of smoothness with this particular data set, as illustrated in Table 1.8, which is also ordered.

Clearly, some imagination must be used if class intervals of different lengths are required. Tukey suggested the scheme used in the next example.

Example 1.3.1

The following numbers represent ACT composite scores for 60 entering freshmen at a certain college:

26	19	22	28	31	29	25	23	20	33	23	26
30	27	26	29	20	23	18	24	29	27	32	24
25	26	22	29	21	24	20	28	23	26	30	19
27	21	32	28	29	23	25	21	28	22	25	24
19	24	35	26	25	20	31	27	23	26	30	29

An ordered stem-and-leaf display is given in Table 1.9. Here leaves are recorded as zeros and ones with a stem adjoined with an asterisk (∗), twos and threes with a stem adjoined with t, fours and fives with a stem adjoined with f, sixes and sevens with a stem adjoined with s, and eights and nines with a stem adjoined with a dot (•).   ◇

There is a reason for constructing ordered stem-and-leaf diagrams. For a sample of n observations, $x_1, x_2, \ldots, x_n$, when the observations are ordered from small to large, the resulting ordered data are called the **order statistics** of the sample. For years statisticians have found that order statistics and certain functions of the order

TABLE 1.9   Ordered stem-and-leaf display
of 60 ACT scores

Stem	Leaves	Frequency
1●	8 9 9 9	4
2*	0 0 0 0 1 1 1	7
2t	2 2 2 3 3 3 3 3 3	9
2f	4 4 4 4 4 5 5 5 5 5	10
2s	6 6 6 6 6 6 6 7 7 7 7	11
2●	8 8 8 8 9 9 9 9 9	5
3*	0 0 0 1 1	3
3t	2 2 3	3
3f	5	1

statistics are extremely valuable. It is very easy to determine the values of the sample in order from an ordered stem-and-leaf display. For illustration, consider the values in Table 1.7 or Table 1.8. The order statistics of the 50 test scores are

$$
\begin{array}{cccccccccc}
34 & 38 & 42 & 42 & 45 & 47 & 51 & 52 & 54 & 57 \\
58 & 58 & 59 & 60 & 61 & 63 & 65 & 65 & 66 & 67 \\
68 & 69 & 69 & 70 & 71 & 71 & 72 & 73 & 73 & 74 \\
75 & 75 & 76 & 76 & 77 & 79 & 81 & 81 & 82 & 83 \\
83 & 84 & 84 & 85 & 87 & 90 & 91 & 93 & 93 & 97
\end{array}
$$

Sometimes we give ranks to these order statistics and use the rank as the subscript on y. The first order statistic $y_1 = 34$ has rank 1; the second order statistic $y_2 = 38$ has rank 2; the third order statistic $y_3 = 42$ has rank 3; the fourth order statistic $y_4 = 42$ has rank 4, ...; and the 50th order statistic $y_{50} = 97$ has rank 50. It is also about as easy to determine these values from the ordered stem-and-leaf display. We see that $y_1 \leq y_2 \leq \cdots \leq y_{50}$.

From either these order statistics or the corresponding ordered stem-and-leaf display, it is rather easy to find the **sample percentiles**. If $0 < p < 1$, the $(100p)$th sample percentile has *approximately* np sample observations less than it and also $n(1-p)$ sample observations greater than it. One way of achieving this is to take the $(100p)$th sample percentile as the $(n+1)p$th order statistic, provided that $(n+1)p$ is an integer. If $(n+1)p$ is not an integer but is equal to r plus some proper fraction, say a/b, use a weighted average of the rth and the $(r+1)$st order statistics. That is, define the $(100p)$th sample percentile as

$$ \tilde{\pi}_p = y_r + (a/b)(y_{r+1} - y_r) = (1 - a/b)y_r + (a/b)y_{r+1}. $$

Note that this is simply a linear interpolation between y_r and y_{r+1}. [If $p < 1/(n+1)$ or $p > n/(n+1)$, that sample percentile is not defined.]

For illustration, consider the 50 ordered test scores. With $p = 1/2$, we find the 50th percentile by averaging the 25th and 26th order statistics, since $(n+1)p =$

$(51)(1/2) = 25.5$. Thus the 50th percentile is

$$\tilde{\pi}_{0.50} = (1/2)y_{25} + (1/2)y_{26} = (71 + 71)/2 = 71.$$

With $p = 1/4$, we have $(n+1)p = (51)(1/4) = 12.75$; and thus the 25th sample percentile is

$$\tilde{\pi}_{0.25} = (1 - 0.75)y_{12} + (0.75)y_{13} = (0.25)(58) + (0.75)(59) = 58.75.$$

With $p = 3/4$, so that $(n + 1)p = (51)(3/4) = 38.25$, the 75th sample percentile is

$$\tilde{\pi}_{0.75} = (1 - 0.25)y_{38} + (0.25)y_{39} = (0.75)(81) + (0.25)(82) = 81.25.$$

Note that *approximately* 50%, 25%, and 75% of the sample observations are less than 71, 58.75, and 81.25, respectively.

Special names are given to certain percentiles. The 50th percentile is the **median** of the sample. The 25th, 50th, and 75th percentiles are the **first, second**, and **third quartiles** of the sample. For notation we let $\tilde{q}_1 = \tilde{\pi}_{0.25}$, $\tilde{q}_2 = \tilde{m} = \tilde{\pi}_{0.50}$, and $\tilde{q}_3 = \tilde{\pi}_{0.75}$. The 10th, 20th, ..., and 90th percentiles are the **deciles** of the sample. So note that the 50th percentile is also the median, the second quartile, and the fifth decile. By using the set of 50 test scores, since $(51)(2/10) = 10.2$ and $(51)(9/10) = 45.9$, the second and ninth deciles are

$$\tilde{\pi}_{0.20} = (0.8)y_{10} + (0.2)y_{11} = (0.8)(57) + (0.2)(58) = 57.2$$

and

$$\tilde{\pi}_{0.90} = (0.1)y_{45} + (0.9)y_{46} = (0.1)(87) + (0.9)(90) = 89.7,$$

commonly called the 20th and 90th percentiles, respectively.

Let us consider, along with the three quartiles, the extremes of the sample, namely the first and nth order statistics. (In the test scores data, they are 34 and 97, respectively.) These provide a **five-number summary** of a set of data consisting of the minimum, the first quartile, the median, the third quartile, and the maximum, written in that order. Furthermore, the difference between the third and first quartiles is called the **interquartile range, IQR**. That is, $IQR = \tilde{q}_3 - \tilde{q}_1$, which equals $81.25 - 58.75 = 22.50$ with the test scores data.

There is a graphical means for displaying the five-number summary of a set of data that is called a **box-and-whisker diagram**. To construct a horizontal box-and-whisker diagram, or more simply, a **box plot**, draw a horizontal axis that is scaled to the data. Above the axis draw a rectangular box with the left and right sides drawn at $\tilde{q}_1$ and $\tilde{q}_3$ with a vertical line segment drawn at the median, $\tilde{q}_2 = \tilde{m}$. A left whisker is drawn as a horizontal line segment from the minimum to the midpoint of the left side of the box, and a right whisker is drawn as a horizontal line segment from the midpoint of the right side of the box to the maximum. Note that the length of the box is equal to the IQR. The left and right whiskers contain the first and fourth quarters of the data, while the two middle quarters of the data are contained, respectively, in the two sections of the box, one to the left and one to the right of the median line.

Example 1.3.2

Using the test score data, we have that the five-number summary is given by

$$y_1 = 34, \quad \tilde{q}_1 = 58.75, \quad \tilde{q}_2 = \tilde{m} = 71, \quad \tilde{q}_3 = 81.25, \quad y_{50} = 97.$$

The box-and-whisker diagram or box plot of these data is given in Figure 1.4. The facts that the long whisker is to the left and the box from 58.75 to 71 is slightly longer than that from 71 to 81.25 lead us to say that these data are slightly *skewed to the left*. ◇

Figure 1.4   Box plot of scores from 50 statistics examinations

The next example illustrates how the box plot depicts data that are *skewed to the right*.

Example 1.3.3

Using the 40 losses due to wind damage that are given in Example 1.1.11, the minimum and maximum are 2 and 43. Since $(40 + 1)(1/4) = 10.25$, $(40 + 1)(2/4) = 20.5$, and $(40 + 1)(3/4) = 30.75$, we have that

$$\tilde{q}_1 = y_{10} + 0.25(y_{11} - y_{10}) = 2,$$
$$\tilde{m} = (y_{20} + y_{21})/2 = 5,$$
$$\tilde{q}_3 = y_{30} + 0.75(y_{31} - y_{30}) = 13.5.$$

Thus the five-number summary is given by

$$y_1 = 2, \quad \tilde{q}_1 = 2, \quad \tilde{q}_2 = \tilde{m} = 5, \quad \tilde{q}_3 = 13.5, \quad y_{40} = 43.$$

The box plot associated with these data is given in Figure 1.5. Note that the box plot indicates that the data are very skewed to the right. In fact, because $y_1 = \tilde{q}_1 = 2$, the length of the left whisker equals zero. ◇

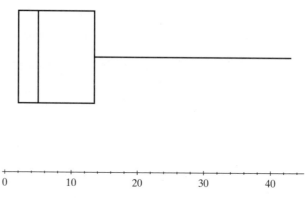

Figure 1.5   Box plot for 40 wind-related catastrophes

Sometimes we are interested in picking out observations that seem to be much larger or much smaller than most of the other observations. That is, we are looking for outliers. Tukey suggested a method for defining outliers that is resistant to the effect of one or two extreme values and makes use of the IQR. In a box-and-whisker diagram, construct **inner fences** to the left and right of the box at a distance of 1.5 times the IQR. **Outer fences** are constructed in the same way at a distance of 3 times the IQR. Observations that lie between the inner and outer fences are called **suspected outliers**. Observations that lie beyond the outer fences are called **outliers**. When you are analyzing a set of data, suspected outliers deserve a closer look and outliers should be looked at very carefully. It does not follow that suspected outliers should be removed from the data unless some error (such as a recording error) has been made. Moreover, it sometimes helps to determine the cause of extreme values, because outliers can often provide useful insights to the situation under consideration (such as a better way of doing things).

Example 1.3.4

Continuing with Example 1.3.3, we find that the interquartile range is $IQR = 13.5-2 = 11.5$. Thus the inner fences would be constructed at a distance of $1.5(11.5) = 17.25$ to the left and right of the box, and the outer fences would be constructed at a distance of $3(11.5) = 34.5$ to the left and right of the box. Figure 1.6 shows a box plot with the fences. Of course, since no data fall to the left of the box, there are no fences to the left. From this box plot we see that there are two suspected outliers. However, because of the nature of the data, they are consistent with the type of losses that the insurance industry can expect to see.   ◇

Some functions of two or more order statistics are quite important in modern statistics. We mention and illustrate two more, along with the IQR, using the 50 test scores:

(a) **Midrange** $=$ average of the extremes

$$= \frac{y_1 + y_n}{2} = \frac{34 + 97}{2} = 65.5.$$

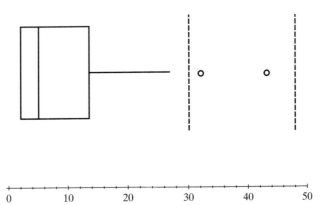

Figure 1.6   Box plot with fences for 40 wind-related catastrophes

(b) **Range** = difference of the extremes

$$= y_n - y_1 = 97 - 34 = 63.$$

(c) **Interquartile range** = difference of third and first quartiles

$$= \tilde{q}_3 - \tilde{q}_1 = 81.25 - 58.75 = 22.5.$$

Thus we see that the mean, the median, and the midrange are measures of the middle of the sample. In some sense, the standard deviation, the range, and the interquartile range provide measures of spread of the sample.

EXERCISES

1.3.1 Let X denote the concentration of $CaCO_3$ in milligrams per liter. Twenty observations of X are:

130.8	129.9	131.5	131.2	129.5
132.7	131.5	127.8	133.7	132.2
134.8	131.7	133.9	129.8	131.4
128.8	132.7	132.8	131.4	131.3

(a) Construct an ordered stem-and-leaf display, using stems of 127, 128, ..., 134.

(b) Find the midrange, range, interquartile range, median, sample mean, and sample variance.

(c) Draw a box-and-whisker diagram.

1.3.2 A company manufactures mints that have a label weight of 20.4 grams. It regularly samples from the production line and weighs the selected mints. During two mornings of production it sampled 81 mints with the following weights:

21.8	21.7	21.7	21.6	21.3	21.6	21.5	21.3	21.2
21.0	21.6	21.6	21.6	21.5	21.4	21.8	21.7	21.6
21.6	21.3	21.9	21.9	21.6	21.0	20.7	21.8	21.7
21.7	21.4	20.9	22.0	21.3	21.2	21.0	21.0	21.9
21.7	21.5	21.5	21.1	21.3	21.3	21.2	21.0	20.8
21.6	21.6	21.5	21.5	21.2	21.5	21.4	21.4	21.3
21.2	21.8	21.7	21.7	21.6	20.5	21.8	21.7	21.5
21.4	21.4	21.9	21.8	21.7	21.4	21.3	20.9	21.9
20.7	21.1	20.8	20.6	20.6	22.0	22.0	21.7	21.6

(a) Construct an ordered stem-and-leaf display using stems of $20f$, $20s$, $20\bullet$, $21*$, ..., $22*$.

(b) Find (i) the three quartiles, (ii) the 60th percentile, and (iii) the 15th percentile.

(c) Construct a box-and-whisker display.

1.3.3 *PC Magazine* of February 1994 includes "Going Mainstream" by John R. Quain, in which tests of 26 CD-ROM drives are reported. One of the tests was a random read of a 4K (4,000 byte) block of data. The read rates, recorded in kilobytes (K) per second, were:

29.21	14.50	25.10	22.90	30.80	30.62	27.88	20.65	30.90
30.60	6.12	19.94	19.61	21.51	28.80	21.72	21.95	18.89
15.60	29.38	25.45	30.41	30.11	30.51	28.17	21.42	

(a) Construct an ordered stem-and-leaf display using three-digit leaves. That is, 29.21 would have the stem of 2 and leaf of (921). Would much be lost if, before constructing the stem-and-leaf display, each number was rounded to the nearest integer?

(b) Find the five-number summary of the data.

(c) Construct a box-and-whisker display.

(d) What type of skewness is indicated by your displays?

1.3.4 For the roulette data in Exercise 1.2.11:

(a) Determine the order statistics.

(b) Find the five-number summary of the data.

(c) Draw a box-and-whisker diagram.

(d) Find the locations of the inner and outer fences and draw a box plot that shows the fences, the suspected outliers, and the outliers.

(e) Find $\bar{x}$, the sample mean.

(f) In your opinion does the median or sample mean give a better measure of the center of these data?

1.3.5 In the casino game roulette, if a player bets \$1 on red (or on black or on odd or on even), the probability of winning \$1 is 18/38 and the probability of losing \$1 is 20/38. Suppose that a player begins with \$5 and makes successive \$1 bets. Let Y equal the player's maximum capital before losing the \$5. One hundred observations of Y were simulated on a computer, yielding the following data:

25	9	5	5	5	9	6	5	15	45
55	6	5	6	24	21	16	5	8	7
7	5	5	35	13	9	5	18	6	10
19	16	21	8	13	5	9	10	10	6
23	8	5	10	15	7	5	5	24	9
11	34	12	11	17	11	16	5	15	5
12	6	5	5	7	6	17	20	7	8
8	6	10	11	6	7	5	12	11	18
6	21	6	5	24	7	16	21	23	15
11	8	6	8	14	11	6	9	6	10

(a) Construct an ordered stem-and-leaf display.

(b) Find the five-number summary of the data and draw a box-and-whisker diagram.

(c) Calculate the IQR and the locations of the inner and outer fences.

(d) Draw a box plot that shows the fences, suspected outliers, and outliers.

(e) Find the 90th percentile.

1.3.6 The weights (in grams) of 25 indicator housings used on gauges are as follows:

102.0	106.3	106.6	108.8	107.7
106.1	105.9	106.7	106.8	110.2
101.7	106.6	106.3	110.2	109.9
102.0	105.8	109.1	106.7	107.3
102.0	106.8	110.0	107.9	109.3

(a) Construct an ordered stem-and-leaf display using integers as the stems and tenths as the leaves.

(b) Find the five-number summary of the data and draw a box plot.

(c) Are there any suspected outliers or outliers?

1.3.7 For the insurance data in Exercise 1.1.11:

(a) Find the five-number summary of the data and draw a box-and-whisker diagram.

(b) Calculate the IQR and the locations of the inner and outer fences.

(c) Draw a box plot that shows the fences, suspected outliers, and outliers.

1.3.8 Using the costs of the "300-level" textbooks in Exercise 1.1.7:

(a) Construct an ordered stem-and-leaf display using as stems 1∗, 1•, 2∗, 2•, and so on, and three-digit leaves. (A leaf would consist of the ones place and two decimal places.) Would much be lost if each of the original prices was rounded to the nearest integer?

(b) Calculate the five-number summary of the data and draw a box plot.

(c) Are there any suspected outliers?

1.4 GRAPHICAL COMPARISONS OF DATA SETS

Graphical methods are extremely valuable in showing how two or more data sets are related to each other. In this section we extend two graphical techniques from exploratory data analysis, namely, stem-and-leaf displays and box-and-whisker diagrams, as well as give a third graphical technique. We begin with an extension of ordered stem-and-leaf diagrams, which are then used to help us find the order statistics for the other graphical techniques.

To compare the characteristics of two treatments (two populations), take a random sample of size n from the first, $x_1, x_2, \ldots, x_n$, and a random sample of size m from the second, $y_1, y_2, \ldots, y_m$. Assuming that these measurements are about the same order of magnitude, we construct a **back-to-back stem-and-leaf display** or a **two-sided stem-and-leaf display**. In doing this, place the stems that are appropriate for both sets of data down the center of the page, placing the leaves of the x-values on the left and those for the y-values on the right. It is helpful, for further applications, to order the leaves from small to large, beginning at the stem and proceeding outward.

Example 1.4.1

At a heat-treating company, iron castings and steel forgings are heat-treated to achieve desired mechanical properties and machinability. One steel forging is "annealed" to soften the part for easy machining. Two lots of this part, made of 1020 steel, are heat-treated in two different furnaces. The parts are heat-treated to a hardness specification, in this case the Rockwell G scale. The specification for this part is 36–66 (Rockwell G), and it is desirable for the parts to be close to the middle of the hardness specification. The hardness results for furnace 10 are

$$
\begin{array}{cccccccccccc}
49 & 47 & 48 & 49 & 51 & 48 & 47 & 48 & 49 & 51 & 47 & 46 & 48 \\
49 & 50 & 46 & 46 & 47 & 48 & 49 & 46 & 46 & 47 & 48 & 47 &
\end{array}
$$

The hardness results for furnace 14 are

$$
\begin{array}{cccccccccccc}
43 & 43 & 47 & 48 & 36 & 43 & 43 & 47 & 47 & 51 & 41 & 43 & 46 \\
43 & 49 & 41 & 38 & 44 & 40 & 46 & 39 & 38 & 41 & 38 & 41 &
\end{array}
$$

The back-to-back stem-and-leaf display in Table 1.10 gives an excellent comparison of the results of the two furnaces.

This back-to-back stem-and-leaf display shows quite dramatically that the two furnaces are not operating similarly and that furnace number 10 has not only less variability, but also produces hardness values much closer to the middle of the hardness specification.   ◇

Recall that the box-and-whisker diagram was useful in showing where the middle half of the observations falls and whether there are any outliers, and also gave an

TABLE 1.10 Hardness of parts from two furnaces

Frequency	Furnace 10 Leaves	Stem	Furnace 14 Leaves	Frequency
0		3s	6	1
0		3●	8 8 8 9	4
0		4*	0 1 1 1 1	4
0		4t	3 3 3 3 3 3	6
0		4f	4	1
11	7 7 7 7 7 7 6 6 6 6 6	4s	6 6 7 7 7	5
11	9 9 9 9 9 8 8 8 8 8 8	4●	8 9	2
3	1 1 0	5*	1	1

indication of the skewness of the data. Box plots can also be used to compare two or more sets of data by drawing the box plots on the same graph. The box plots may be drawn either horizontally or vertically.

We continue with the preceding example.

Example 1.4.2

If we use the hardness results for furnaces 10 and 14 in the preceding example, since Table 1.10 gives an ordered back-to-back stem-and-leaf display, the quartiles can be found easily. The locations of the first and third quartiles are the averages of the 6th and 7th ordered observations and the 19th and 20th ordered observations, respectively, and the median is the 13th observation. So for furnace 10 (F10), the five-number summary is 46, 47, 48, 49, and 51. The five-number summary for furnace 14 (F14) is 36, 40.5, 43, 46.5, and 51. The box plots for the two furnaces are given in Figure 1.7. Note that this figure confirms what we had noted from the back-to-back stem-and-leaf display. ◇

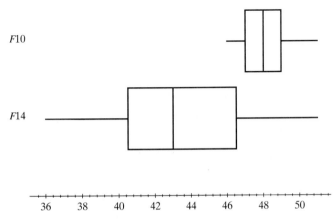

Figure 1.7 Box plots of the hardness of parts from two furnaces

Clearly, more than two box-and-whisker displays can be used to compare three or more samples and thus their corresponding distributions. More formal methods of testing the equality of several means are considered in Section 8.1.

Example 1.4.3

Hogg and Ledolter [see R. V. Hogg and J. Ledolter, *Applied Statistics for Engineers and Physical Scientists*, 2nd ed. (New York: Macmillan, 1992)] report that a civil engineer wishes to compare the strengths of three different types of beams, one (A) made of steel and two (B and C) made of different and more expensive alloys. A certain deflection (in units of 0.001 inch) was measured for each beam when submitted to a given force; thus a small deflection would indicate a beam of great strength. The order statistics for the three samples of sizes $n_1 = 8$, $n_2 = 6$, and $n_3 = 6$ are the following:

$$
\begin{array}{llllllll}
\text{A:} & 79 & 82 & 83 & 84 & 85 & 86 & 86 & 87 \\
\text{B:} & 74 & 75 & 76 & 77 & 78 & 82 \\
\text{C:} & 77 & 78 & 79 & 79 & 79 & 82 \\
\end{array}
$$

The corresponding box-and-whisker displays are given in Figure 1.8. The deflection seems to be greatest in A, while B and C are about the same. However, with such small sample sizes, it is difficult to say anything for certain. ◇

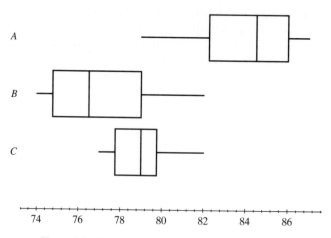

Figure 1.8   Box plots for deflections of three beams

We now introduce a third graphical technique for comparing two sets of data. To begin, suppose that each of two sets of data contains n numbers. Furthermore, order each set so that $x_1 \leq x_2 \leq x_3 \leq \cdots \leq x_n$ and $y_1 \leq y_2 \leq y_3 \leq \cdots \leq y_n$. Thus we have the order statistics for each random sample. Each of x_r and y_r is called the **quantile of order** $r/(n+1)$ for the respective samples. Note that x_r and y_r are also called the $100[r/(n+1)]$th percentiles.

In a **quantile-quantile plot** or a q-q **plot**, the quantiles of one sample are plotted against the corresponding quantiles of the other sample. That is, with equal sample sizes, we graph the points (x_1, y_1), (x_2, y_2), ..., (x_n, y_n). If both samples were exactly the same, each pair of corresponding quantiles would plot on a straight line with slope 1 and intercept 0. If, on the other hand, the mean of the first sample (which is plotted as the first coordinate) is shifted over, say d units, but otherwise the samples are the same, pairs would plot on a straight line with slope 1 but the intercept is now $-d$. A slope less than 1 would indicate that the variability of the first sample (plotted as the x-coordinate of the pair) is greater than that of the second (plotted as the y-coordinate of the pair). Of course, a slope greater than one would also indicate that the variability of the second sample is greater than that of the first. With a little thought, other graphs suggest other situations. For illustration, suppose that the q-q plot is on a curve whose slope increases with higher quantiles. This suggests that the variability of the first sample decreases relative to that of the second with higher values of each. For instance, the first sample might be skewed to the left and the second skewed to the right.

Usually with real data sets, the q-q plot is not exactly on a straight line. Let us consider a case study by J. H. Sheesley. [See R. D. Snee, L. B. Hare, and J. R. Trout, *Experiments in Industry* (Milwaukee, WI: American Society for Quality Control, 1985).]

Example 1.4.4

In Sheesley's study, the performance of two different types of lead wires was compared in the production of ordinary household light bulbs. Due to the fact that some of the lead wires made by the old method would misfeed, the company changed the way it produced lead wires in hopes of improving the situation. To compare the old method with the new method, the results of 12 production runs with the old method were compared with 12 runs with the new. The two sets of order statistics of the average hourly number of misfeeding leads were the following:

Old:	10.8	17.6	18.0	18.3	19.2	19.6
	19.9	21.4	22.7	23.2	23.7	39.4
New:	7.8	11.2	11.5	12.4	16.7	16.8
	18.9	21.5	23.7	25.6	26.9	28.1

Since the ith order statistics of each sample is the $[i/(n + 1)]$th quantile of the sample, $i = 1, 2, \ldots, 12$, the 12 pairs (quantile from old, quantile from new) are plotted in Figure 1.9. From the figure we note, except for the extreme quantiles, the sample values of the old wires are closer together than those of the new wires. There are more points below the 45° line through the origin; this would suggest that the old distribution might have a larger mean than that of the new distribution, but the improvement has not been outstanding.

It is interesting to note that two box-and-whisker diagrams alongside each other would have provided about the same information. Since $(0.25)(13) = 3.25$, $(0.5)(13) = 6.5$, and $(0.75)(13) = 9.75$, the 25th, 50th, and 75th percentiles are the

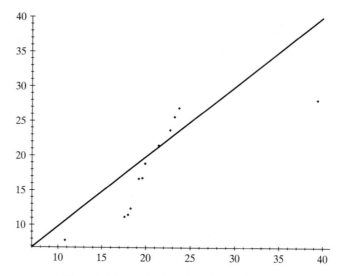

Figure 1.9 q-q plot for misfeeding lead wires

respective weighted averages of the 3rd and 4th, 6th and 7th, and the 9th and 10th order statistics. The two box-and-whisker diagrams are depicted in Figure 1.10. Thus we see that the interquartile range is much smaller for the old than the new and the whiskers longer for the old method. In addition, the comparison of the two box-and-whisker diagrams indicates that, based on these data, the new method produces somewhat fewer misfeeds than the old. It seems to us, however, as if even more improvements could be made and then more data collected.   ◇

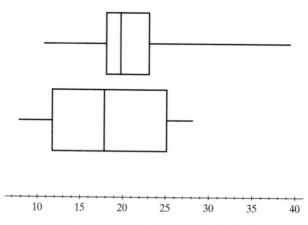

Figure 1.10   Box plots of misfeeding lead wires

We will not always be so fortunate as to have equal sample sizes. If the sample sizes are small but unequal, the order statistics of the smaller sample can be used to determine which quantiles are to be used.

In general, suppose that we would like to find the $(100p)$th percentile (pth quantile) of a sample. If $r = (n + 1)p$ is an integer, then x_r is that percentile; for example, if $n = 9$ and $p = 0.30$, then $r = (9 + 1)(0.30) = 3$, so x_3 is the 30th percentile or quantile of order $p = 3/10$. If $(n+1)p$ is not an integer, we use a weighted average of the two adjacent order statistics, x_r and x_{r+1}, where r is the greatest integer $[(n+1)p]$ in $(n+1)p$. For example, if $n = 11$ and we want to find the 30th percentile, we calculate $r = [(11 + 1)(0.30)] = [3.6] = 3$. We then use the 3rd order statistic plus 0.6 of the difference between the 4th and 3rd order statistics or, equivalently, a weighted average of the 3rd and 4th order statistics:

$$x_3 + (0.6)(x_4 - x_3) = (0.4)x_3 + (0.6)x_4.$$

Suppose that we want to make a q-q plot using $n = 11$ observations of X and $m = 13$ observations of Y. We shall use the quantiles of the smaller sample to select the corresponding quantiles of the larger sample. That is, for the smaller sample the $n = 11$ order statistics $x_1 < x_2 < \cdots < x_{11}$ are the respective $(100p)$th percentiles (pth quantiles) of the sample for $p = 1/12, 2/12, 3/12, \ldots, 11/12$. To find the corresponding percentiles (quantiles) of the larger sample of size $m = 13$ using the order statistics $y_1 \le y_2 \le \cdots \le y_{13}$, we can use Table 1.11 (see Exercise 1.4.4).

TABLE 1.11   Finding corresponding quantiles for samples of different sizes

p	$p(n+1)$	pth Quantile of x's	$p(m+1)$	pth Quantile of y's
1/12	1	x_1	1.17	$(0.83)y_1 + (0.17)y_2$
2/12	2	x_2	2.33	$(0.67)y_2 + (0.33)y_3$
3/12	3	x_3	3.50	$(0.50)y_3 + (0.50)y_4$
4/12	4	x_4	4.67	$(0.33)y_4 + (0.67)y_5$
5/12	5	x_5	5.83	$(0.17)y_5 + (0.83)y_6$
6/12	6	x_6	7.00	y_7
7/12	7	x_7	8.17	$(0.83)y_8 + (0.17)y_9$
8/12	8	x_8	9.33	$(0.67)y_9 + (0.33)y_{10}$
9/12	9	x_9	10.50	$(0.50)y_{10} + (0.50)y_{11}$
10/12	10	x_{10}	11.67	$(0.33)y_{11} + (0.67)y_{12}$
11/12	11	x_{11}	12.83	$(0.17)y_{12} + (0.83)y_{13}$

EXERCISES

1.4.1 Construct an ordered back-to-back stem-and-leaf display for the data in Example 1.4.4. Does this confirm the conclusions given in that example?

1.4.2 At the beginning of the semester in a health dynamics course, some students were weighed under water in order to determine their percentage of body fat. The weights in kilograms for 25 females were

1.2	1.5	1.2	1.5	1.6	2.0	1.1	2.3	1.0
2.0	1.8	1.5	2.1	1.9	2.1	1.9	1.9	1.4
1.8	2.7	0.8	1.6	1.4	1.7	0.9		

The weights in kilograms for 25 males were

5.2	3.7	5.2	2.7	4.7	5.2	3.8	4.2	4.0
4.1	4.5	3.7	4.2	4.6	4.3	3.6	4.2	3.2
3.4	3.1	4.2	2.8	3.7	4.9	3.9		

(a) Construct an ordered back-to-back stem-and-leaf display.
(b) Find the five-number summary for each set of weights.
(c) Draw box-and-whisker diagrams on the same graph.
(d) Use the paired order statistics to draw a q-q plot.
(e) What conclusions can you draw from your graphical displays?

1.4.3 At the beginning of the semester in the health dynamics course, the forced vital capacity (FVC) (the volume of air a person can expel from the lungs) was one measurement taken for some students. The FVC in liters for 25 females were

2.9	2.8	3.4	3.2	3.7	3.8	4.1	3.1	4.2
3.8	3.7	3.4	3.5	3.5	2.7	4.2	3.4	4.0
3.5	3.7	3.4	3.4	3.1	3.1	3.6		

The FVC in liters for 25 males were

5.1	5.4	5.8	4.3	6.0	5.4	5.7	5.6	5.3
4.7	6.1	6.3	4.6	5.5	6.7	5.3	4.3	4.1
4.8	4.5	5.8	4.1	4.8	5.3	6.7		

(a) Construct an ordered back-to-back stem-and-leaf display.
(b) Find the five-number summary for each set of volumes.
(c) Draw box-and-whisker diagrams on the same graph.
(d) Use the paired order statistics to draw a q-q plot.
(e) What conclusions can you draw from your graphical displays?

1.4.4 A botanist is interested in comparing the growth response of dwarf pea stems to two different levels of the hormone indoleacetic acid (IAA). Using 16-day-old pea plants, the botanist obtains 5-millimeter sections and floats these sections on solutions with different hormone concentrations to observe the effects of the hormone. Let X and Y denote, respectively, the growth in millimeters that can be attributed to the hormone during the first 26 hours after sectioning for $(0.5)10^{-4}$ and 10^{-4} levels of concentration of IAA. Given $n = 11$ observations of X,

0.8 1.8 1.0 0.1 0.9 1.7 1.0 1.4 0.9 1.2 0.5,

and $m = 13$ observations of Y,

1.0 0.8 1.6 2.6 1.3 1.1 2.4 1.8 2.5 1.4 1.9 2.0 1.2.

    (a) Construct an ordered back-to-back stem-and-leaf display.

    (b) Find the five-number summary for each set of lengths.

    (c) Draw box-and-whisker diagrams on the same graph.

    (d) To construct a q-q plot, use Table 1.11 to find the quantiles of the observations of Y.

    (e) What conclusions can you draw from your graphical displays?

1.4.5 In order to test a new golf ball, each of 50 professional golfers hit the new ball three times off the tee and then hit a competing brand three times off the tee. Of course, half of the golfers hit the competing brand first, and half hit the new ball first. The total distances (in yards) for the new brand are

785	680	810	740	697	787	830	845	866	782
773	754	685	753	807	835	742	766	709	724
746	783	786	796	816	739	781	823	789	764
775	828	821	821	808	732	636	705	764	692
703	740	714	746	804	690	740	751	807	773

The respective total distances for the competing brand are

700	643	803	706	654	728	769	783	817	766
741	741	681	760	765	783	717	714	668	728
729	699	713	780	805	699	714	788	715	728
703	804	784	783	782	747	656	688	703	668
721	720	695	721	774	680	726	739	798	746

    (a) Construct an ordered back-to-back stem-and-leaf display.

    (b) Find the five-number summary for each set of distances.

    (c) Draw box-and-whisker diagrams on the same graph.

    (d) Draw a q-q plot. (You may use quantiles of order $i/17, i = 1, 2, \ldots, 16$.)

    (e) Find the sample means and sample standard deviations.

    (f) Write a comparison of the two brands of golf balls.

1.4.6 A process puts a coating on a piece of glass so that it is sensitive to touch. The resistance measurement is taken at a particular location for parts coated on December 7. The resistance measurement is taken at the same location for parts coated on December 22. A random sample of 15 observations taken on December 7 yielded the following measurements. (The data are ordered for your convenience.)

    166 175 175 176 176 177 177 178 178 182 184 185 186 187 189

A random sample of 15 observations taken on December 22 yielded the following measurements:

    155 158 159 160 160 161 161 161 162 162 162 163 163 170 175

    (a) Construct a back-to-back stem-and-leaf display.

    (b) Give the five-number summary for each set of data and draw box plots side by side.

    (c) Write a comparison of the production process on these two days.

1.4.7 An ecology laboratory studied tree dispersion patterns for the sugar maple, whose seeds are dispersed by the wind, and the American beech, whose seeds are dispersed by mammals. In a plot 50 meters by 50 meters, they measured distances between like trees yielding the following distances in meters for 19 American beech trees:

5.00	5.00	6.50	4.25	4.25	8.80	6.50	7.15	6.15	2.70
2.70	11.40	9.70	6.10	9.35	2.85	4.50	4.50	6.50	

and the following distances for 19 sugar maple trees:

6.00	4.00	6.00	6.45	5.00	5.00	5.50	2.35	2.35	3.90
3.90	5.35	3.15	2.10	4.80	3.10	5.15	3.10	6.25	

(a) Find the five-number summary for each set of distances.

(b) Construct box plots on the same figure.

(c) Construct a back-to-back stem-and-leaf diagram.

(d) Interpret your output.

1.4.8 Data were collected during a step–direction experiment in the biomechanics lab at Hope College. The goal of the study is to establish differences in stepping responses between healthy young and healthy old adults. In one part of the experiment the subjects are told in what direction they should take a step. Then, when given a signal, the subject takes a step in that direction and the time it takes for the person to lift his or her foot to take the step is measured. The direction is repeated a few times throughout the testing, so for each subject, a mean of all the liftoff times in a certain direction is calculated. The mean liftoff times for the anterior direction, ordered for your convenience, are

Young subjects:	397	433	450	468	485	488	498	504	561
	565	569	576	577	579	581	586	696	
Old subjects:	463	538	549	573	588	590	594	626	627
	653	674	728	818	835	863	888	936	

(a) Construct a back-to-back stem-and-leaf diagram.

(b) Give the five-number summary for each set of times and construct box plots on the same figure.

(c) Write a comparison of these times.

1.4.9 The precipitation of platinum sulfide by three different methods led to the following results (already ordered in magnitude):

Method 1:	10.28	10.29	10.31	10.33	10.33	10.33	10.34
Method 2:	10.22	10.23	10.23	10.24	10.27	10.28	10.29
Method 3:	10.22	10.25	10.30	10.33	10.36	10.38	10.38

(a) Draw three box-and-whisker diagrams on the same figure.

(b) Construct three q-q plots: method 1 against method 2, method 2 against method 3, method 3 against method 1.

(c) Make some appropriate comments about the graphical displays.

1.4.10 For an aerosol product, there are three weights: the tare weight (container weight), the concentrate weight, and the propellant weight. Let X_1, X_2, and X_3 denote the weights in grams of the propellant for products produced on three different days. Nine observations of each random variable were as follows:

$$
\begin{array}{llllll}
X_1: & 43.06 & 43.32 & 42.63 & 42.86 & 43.05 \\
     & 42.87 & 42.94 & 42.80 & 42.36 & \\
X_2: & 42.33 & 42.81 & 42.13 & 42.41 & 42.39 \\
     & 42.10 & 42.42 & 41.42 & 42.52 & \\
X_3: & 42.83 & 42.57 & 42.96 & 43.16 & 42.25 \\
     & 42.24 & 42.20 & 41.97 & 42.61 &
\end{array}
$$

(a) Find the five-number summary for each set of weights.

(b) Construct three box-and-whisker diagrams on the same graph.

(c) What conclusions can you draw?

1.5 TIME SEQUENCES

Thus far we have emphasized the importance of good summaries of data sets, in particular visual displays such as histograms, stem-and-leaf displays, q-q plots, and box-and-whisker diagrams. Here we want to consider another important variable, namely time, that often contributes to variability. In a **time sequence**, observations are recorded in the order in which they were collected as ordered pairs where the x-coordinate denotes the time (in order of collection, which could be days, weeks, years, etc.) and the y-coordinate records the observation. If measurements are taken sequentially and plotted in this time sequence, trends, cycles, or major changes in process (like government interventions) are often observed. For illustration, consider the three time sequences in Figure 1.11.

Figure 1.11(a) might be a recording of sales of a company for the last 12 years. Despite occasional drops in sales from one year to the next, there has been an upward trend. Suppose, however, that only the last two points had been recorded for the stockholders. Since sales went down, they might believe that there are some difficulties and that changes should be made. If they look at the entire sequence, however, they clearly see the long-term improvement and only variability about that upward trend accounts for the last drop. On the other hand, if several successive drops appear, then some action might be needed.

The points in Figure 1.11(b) might also represent sales, but now plotted by quarters. Clearly, in this business under consideration, there is a cyclic effect with increases from the first to second quarters and from second to third but decreases from the third to the fourth and the fourth to the first quarter of the next year. Again we note that there seems to be an overall upward trend.

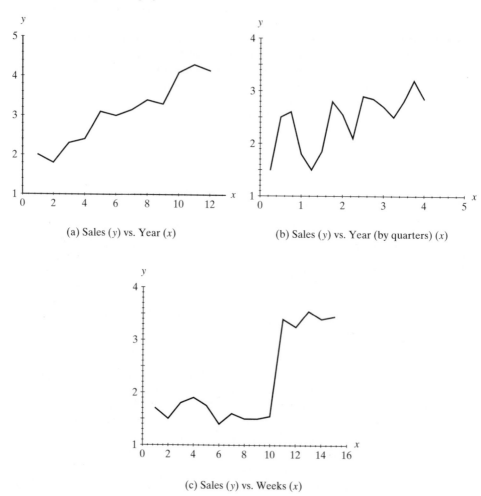

(a) Sales (y) vs. Year (x)

(b) Sales (y) vs. Year (by quarters) (x)

(c) Sales (y) vs. Weeks (x)

Figure 1.11   Sales of a company over time

Figure 1.11(c) could be the plot of some measure of success (possibly sales) of the company. Clearly, something happened after the tenth week. Perhaps management did not like the past level of performance and instituted a new program (possibly one based on statistical quality control). On the other hand, there might be some outside intervention (like war) that increased the sales of this particular product.

We have listed three possible scenarios associated with these three time sequences. There might be others; but, in any case, we clearly see that plotting the points in a time sequence often helps explore the variability of the outcomes associated with a particular system. That is, time is often an important factor in dealing with variability, possibly when purchasing stock. The following is a simple example based on one student's experience.

Example 1.5.1

A student monitored some stock that she purchased in March for $5 per share. Every two weeks she recorded the value of a share yielding the following paired data, (x, y), with x equal to the time in weeks after purchase and y the value of a share at that time.

$(0, 5.00)$	$(2, 5.00)$	$(4, 5.25)$	$(6, 5.25)$	$(8, 5.50)$
$(10, 5.50)$	$(12, 5.25)$	$(14, 5.75)$	$(16, 6.25)$	$(18, 6.50)$
$(20, 6.25)$	$(22, 7.25)$	$(24, 7.50)$		

A time sequence of these data is shown in Figure 1.12. ◇

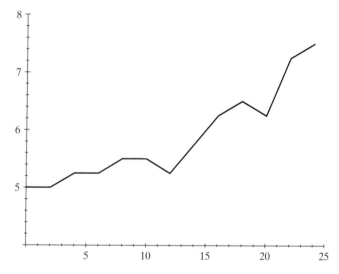

Figure 1.12   Value of a share of stock plotted biweekly

Sometimes, however, time is not an important factor in the response variable and yet time sequence plots still help us understand the concept of variation. Suppose that a process or individual performs at some average level, sometimes doing better and sometimes worse than that average. For example, suppose that a certain worker's output is plotted by the month in Figure 1.13.

Let us say the manager of the company decides to do the following. On good months (like the second month) the worker is given a "pat on the back" (maybe a bonus or even being named "worker of the month"). What happens? Usually the performance goes down the next month. Recall that the worker is averaging below the performance of that high month, that is, she was performing above average when she was made worker of the month. So there is a tendency to do worse the month following a very good performance. (Statisticians sometimes call this "regression toward the mean.")

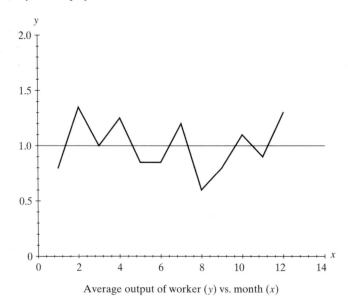

Average output of worker (y) vs. month (x)

Figure 1.13   Output of a worker for each month

This explains what is often called the "sophomore jinx" in sports. A player might have an unusually good first year, actually one far above his average. So he tends to do worse the next year. Fans will say it's the sophomore jinx; he probably got overconfident and didn't work as hard. Although there might be some truth to this, it could also be that he worked just as hard as before but still went down because his first year was over his average. So, often the "rookie of the year" will do worse the second year because not many players win that award with an average year. That is, he was performing far above his average when he won the award.

Back to the worker described by Figure 1.13. On bad months (like the eighth month), the manager could call in the worker and have a serious talk with her or, in some way, penalize her. What happens? Usually, she does better the next month (again regression toward the mean). So the boss gets the idea that it is better to get tough because the workers seem to do better after these little talks. Of course, like the "pat on the back" technique, this does not really work because the workers keep varying about their average. They usually do better after a very poor performance and worse after an excellent performance.

If the manager does not like the the average level of performance, then he or she must do something to change the system. Most of the time, the individual workers can do very little to change the process; major decisions for improvements must be made by the manager or even in the boardroom. After these decisions, the workers must be given "road maps directing them to use these improvements." Most workers (as well as students) want to do better and would like to take pride in their performance. Managers (as well as teachers) should provide for the opportunity to improve the self-esteem of those working for them. The following example could help some of you in analyzing performances in events that are repeated over time.

Example 1.5.2

Consider the following 20 quiz scores of a student in a two-semester calculus course. Each quiz was graded on a 20-point scale and they are given in the order in which they were taken.

$$
\begin{array}{cccccccccc}
18 & 20 & 16 & 12 & 15 & 15 & 10 & 13 & 14 & 9 \\
12 & 10 & 8 & 11 & 12 & 11 & 10 & 11 & 8 & 6
\end{array}
$$

We could calculate the average of all 20 quiz scores, which is 12.05. We could calculate the average of the first 10 and the second 10. A graphic display of what has happened to this student over the course of the two semesters is shown in Figure 1.14. Since the time history is indicated in a time sequence, it is clear that the student was not making a great effort throughout the year in this calculus class. If interested, students can correct trends like this, but they must make changes in their behavior to do so. By continuing the past pattern nothing will change and the downward trend will continue. What improvements could be made to change this? Each student must answer this in his or her own way.   ◇

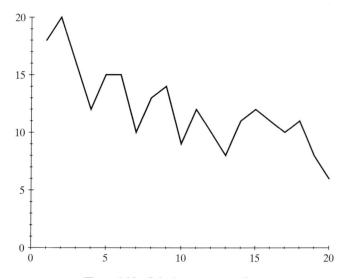

Figure 1.14   Calculus scores over time

The next example illustrates how, in a production process, tracking something over time can be beneficial.

Example 1.5.3

Containers of a prewash concentrate are selected regularly from the production line and the concentrate is weighed. The target weight is 330.5–336.5 grams. We now construct a time-sequence plot for the following 60 weights that were observed in the order left to right, line by line.

337.9	338.1	337.7	338.3	337.3	338.1	337.3	338.0	337.9	338.9
337.6	338.2	338.1	338.6	338.0	338.4	334.4	334.3	334.5	333.9
335.1	334.0	335.6	334.1	335.0	334.8	335.2	335.2	335.2	334.0
334.5	334.6	335.4	335.9	335.6	335.8	334.5	335.4	335.0	335.5
335.1	335.0	334.1	334.6	334.4	334.7	333.8	335.0	334.0	334.3
334.0	333.7	334.7	333.5	334.8	333.9	335.7	335.1	335.1	334.6

It is pretty clear that during the beginning of production, the fill weights were slightly above the target weight (Figure 1.15). This is good news for the customer as long as the containers do not overflow. However, if the trend had continued, this would have resulted in a loss of profit for the company. After adjustments were made, the process seemed to come under control, with the weights reflecting the normal variability for a process like this.   ◇

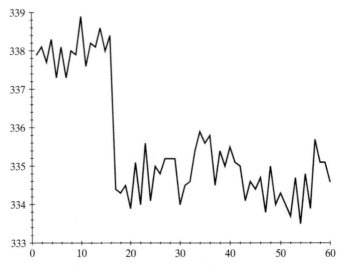

Figure 1.15   Prewash concentrate fill weights

EXERCISES

1.5.1 The following data give the percentage of solids in a product. The specifications call for the percentage to be between 60.0% and 72.0%. The data were observed in the order left to right, line by line.

69.8	71.3	65.6	66.3	70.1	71.9	69.6	71.9	71.1	71.7
71.9	69.8	66.8	68.3	64.4	64.2	65.1	63.7	66.2	61.9
66.1	62.9	66.9	67.3	63.3	63.4	67.2	67.4	65.5	66.2
67.5	67.3	66.9	66.5	65.5	63.9	64.6	62.3	66.2	67.2
66.0	69.8	69.7	71.0	69.8	66.0	70.3	65.5	67.0	66.8
67.6	68.6	66.5	66.2	70.4	68.1	64.3	65.2	68.0	65.1

(a) Construct a time sequence of these percentages.

(b) Interpret your output.

1.5.2 Twenty-four "3-pound" bags of apples were filled and weighed automatically in the following time order by scale number 5, yielding the following weights:

3.26	3.62	3.39	3.12	3.53	3.30	3.10	3.26
3.19	3.22	3.14	3.39	3.31	3.21	3.49	3.41
3.02	3.17	3.20	3.12	3.42	3.36	3.21	3.26

(a) Construct and interpret a time sequence of these weights.

(b) Compare scale number 5 with scale number 6 in the next exercise.

1.5.3 Twenty-four "3-pound" bags of apples were filled and weighed automatically in the following time order by scale number 6, yielding the following weights:

3.22	2.97	3.00	3.08	3.04	3.09	3.04	3.21
3.00	3.05	3.11	3.00	3.10	3.13	3.08	3.22
3.12	3.04	3.03	3.02	3.10	3.09	3.18	3.00

(a) Construct and interpret a time sequence of these weights.

(b) Compare scale number 6 with scale number 5 in the preceding exercise.

1.5.4 The year, x, and the birthrate, y (number of births per 1000 population), for 1960–1987 were as follows:

(1960,23.7)	(1961,23.3)	(1962,22.4)	(1963,21.7)
(1964,21.1)	(1965,19.4)	(1966,18.4)	(1967,17.8)
(1968,17.6)	(1969,17.9)	(1970,18.4)	(1971,17.2)
(1972,15.6)	(1973,14.8)	(1974,14.8)	(1975,14.6)
(1976,14.6)	(1977,15.1)	(1978,15.0)	(1979,15.6)
(1980,15.9)	(1981,15.8)	(1982,15.9)	(1983,15.5)
(1984,15.5)	(1985,15.8)	(1986,15.6)	(1987,15.7)

(a) Construct a time sequence of these birthrates.

(b) Interpret your output.

1.5.5 Dish drop viscosity has a specification of 120–230 Hz. One hundred observations were selected in the following time order and measured, yielding the following viscosities (read across, line by line):

158	147	158	159	169	151	166	151	143	169
153	174	151	164	185	168	140	180	176	154
160	187	145	164	158	169	153	149	144	157
156	183	157	140	162	158	160	180	154	160
164	168	154	158	164	159	153	170	158	170
150	161	169	166	154	157	138	155	134	165
161	172	156	145	153	143	152	152	156	163
179	157	135	172	143	154	165	145	152	145
171	189	144	154	147	187	147	159	167	151
153	168	148	188	152	165	155	140	157	176

(a) Construct a time sequence of these viscosities.

(b) If there seems to be no significant time factor, construct a histogram of these viscosities.

(c) Use the two graphs to interpret these measurements.

1.6 SCATTER PLOTS, LEAST SQUARES, AND CORRELATION

In Section 1.5 we plotted a response variable, such as sales, against another variable, which in that section was time. By doing so, we often obtain valuable insights about the situation under consideration. Moreover, we can make similar bivariate plots when one of the variables is not time but some other type of explanatory variable. For example, a student's ACT or SAT score can be of some help in predicting that student's grade-point average in college. Also, the weight of a car frequently suggests the number of gallons of gasoline needed to drive that car 100 miles. In each of these examples, the role of the time variable is taken by another explanatory variable: the scholastic aptitude score and the weight of the car, respectively. We denote this "independent" variable by x and the response variable by y. In our examples, y equals college GPA and number of gallons needed to drive the car 100 miles, respectively.

Suppose that we observe n points with coordinates (x_i, y_i), $i = 1, 2, \ldots, n$, where x_i is a preliminary exam score and y_i is a final exam score. These might plot as in Figure 1.16; here $n = 10$. Such a diagram is called a **scatter plot**. Here it looks as if a straight line $y = \alpha + \beta x$ would fit these data fairly well. Clearly, not all the points lie on any straight line, but one line could serve as the pattern about which the points fall. One such line is also depicted in Figure 1.16.

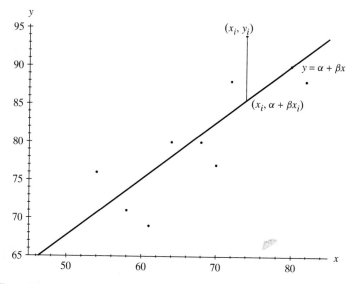

Figure 1.16   Scatter plot of final exam score (y) vs. preliminary exam score (x)

How do we select values of the parameters, α and β, of the line so that, in some sense, the line "best" fits the data? One way of doing this is as follows. Look at a typical point, which we have denoted by (x_i, y_i). The point on the line that is either directly below or above that point (x_i, y_i) is the point $(x_i, \alpha + \beta x_i)$. Accordingly, the vertical distance between these two points is $|y_i - \alpha - \beta x_i|$. So, in some way, we want to fit the line to minimize these distances $|y_i - \alpha - \beta x_i|$, $i = 1, 2, \ldots, n$; that is, we want the line $y = \alpha + \beta x$ to be as close as possible to the observed points. One popular way to do this is to square each of the n distances and to add these squared values together to obtain

$$H(\alpha, \beta) = \sum_{i=1}^{n} (y_i - \alpha - \beta x_i)^2.$$

We then want to select α and β to minimize $H(\alpha, \beta)$. Since we are *minimizing* the *sum of squared values*, we call this the **method of least squares**. We now have a calculus problem: We want to find the values of α and β that minimize the function $H(\alpha, \beta)$.

Remark   If you have not studied multivariate calculus, we simply note here that the rule for finding a minimum is about the same as that in the calculus of a single variable. Recall that there one takes the first derivative and equates it to zero. For functions of two variables, we take two first derivatives: Treat β as a constant and take the derivative with respect to α, and then treat α as a constant and take the derivative with respect to β. These two derivatives are denoted, respectively, by

$$\frac{\partial H(\alpha, \beta)}{\partial \alpha} \quad \text{and} \quad \frac{\partial H(\alpha, \beta)}{\partial \beta}$$

and are called first partial derivatives. These are equated to 0 in our search for the values of α and β that minimize $H(\alpha, \beta)$. Just as in one-variable calculus, there are certain properties that the second derivatives must satisfy to be certain that we have obtained a relative minimum, but we will not worry about these here for, in this case, the solution does in fact provide the absolute minimum.

We set the two first partial derivatives equal to zero, namely

$$\frac{\partial H(\alpha, \beta)}{\partial \alpha} = \sum_{i=1}^{n} 2(y_i - \alpha - \beta x_i)(-1) = 0,$$

$$\frac{\partial H(\alpha, \beta)}{\partial \beta} = \sum_{i=1}^{n} 2(y_i - \alpha - \beta x_i)(-x_i) = 0.$$

Rewriting these derivatives, we obtain (because $\sum_{i=1}^{n} \alpha = n\alpha$)

$$n\alpha + \left(\sum_{i=1}^{n} x_i \right) \beta = \sum_{i=1}^{n} y_i,$$

$$\left(\sum_{i=1}^{n} x_i \right) \alpha + \left(\sum_{i=1}^{n} x_i^2 \right) \beta = \sum_{i=1}^{n} x_i y_i.$$

If the first of these equations is multiplied by $\left(\sum_{i=1}^{n} x_i \right)$ and the second is multiplied by n, the difference of the two equations eliminates the α term and gives us

$$\left[n \sum_{i=1}^{n} x_i^2 - \left(\sum_{i=1}^{n} x_i \right)^2 \right] \beta = n \sum_{i=1}^{n} x_i y_i - \left(\sum_{i=1}^{n} x_i \right) \left(\sum_{i=1}^{n} y_i \right).$$

That is, if we denote the β value of the solution by $\widehat{\beta}$, it is given by

$$\widehat{\beta} = \frac{n \sum_{i=1}^{n} x_i y_i - \left(\sum_{i=1}^{n} x_i \right) \left(\sum_{i=1}^{n} y_i \right)}{n \sum_{i=1}^{n} x_i^2 - \left(\sum_{i=1}^{n} x_i \right)^2} = \frac{\sum_{i=1}^{n} x_i y_i - \frac{1}{n} \left(\sum_{i=1}^{n} x_i \right) \left(\sum_{i=1}^{n} y_i \right)}{\sum_{i=1}^{n} x_i^2 - \frac{1}{n} \left(\sum_{i=1}^{n} x_i \right)^2}.$$

Substituting $\widehat{\beta}$ in an earlier equation, we see that the α value of the solution is

$$\widehat{\alpha} = \frac{\sum_{i=1}^{n} y_i - \left(\sum_{i=1}^{n} x_i \right) \widehat{\beta}}{n} = \bar{y} - \widehat{\beta} \bar{x},$$

where $\bar{x} = (\sum_{i=1}^{n} x_i)/n$ and $\bar{y} = (\sum_{i=1}^{n} y_i)/n$. That is, the best-fitting straight line, in the sense of least squares, is

$$\widehat{y} = \widehat{\alpha} + \widehat{\beta} x = \bar{y} - \widehat{\beta} \bar{x} + \widehat{\beta} x = \bar{y} + \widehat{\beta}(x - \bar{x}).$$

If we study the solution $\widehat{\beta}$, its denominator equals $\sum_{i=1}^{n}(x_i - \overline{x})^2$ because

$$\sum_{i=1}^{n}(x_i - \overline{x})^2 = \sum_{i=1}^{n} x_i^2 - 2\overline{x}\sum_{i=1}^{n} x_i + n\overline{x}^2 = \sum_{i=1}^{n} x_i^2 - \frac{1}{n}\left(\sum_{i=1}^{n} x_i\right)^2.$$

Moreover, the numerator is equal to $\sum_{i=1}^{n}(x_i - \overline{x})(y_i - \overline{y})$ because

$$\sum_{i=1}^{n}(x_i - \overline{x})(y_i - \overline{y}) = \sum_{i=1}^{n}(x_i y_i - \overline{x} y_i) - \overline{y}\sum_{i=1}^{n}(x_i - \overline{x})$$

$$= \sum_{i=1}^{n} x_i y_i - \sum_{i=1}^{n} \overline{x} y_i$$

$$= \sum_{i=1}^{n} x_i y_i - \frac{1}{n}\left(\sum_{i=1}^{n} x_i\right)\left(\sum_{i=1}^{n} y_i\right).$$

That is,

$$\widehat{\beta} = \frac{\displaystyle\sum_{i=1}^{n}(x_i - \overline{x})(y_i - \overline{y})}{\displaystyle\sum_{i=1}^{n}(x_i - \overline{x})^2}.$$

Recall that $s_x^2 = [1/(n-1)]\sum_{i=1}^{n}(x_i - \overline{x})^2$ and $s_y^2 = [1/(n-1)]\sum_{i=1}^{n}(y_i - \overline{y})^2$ are called the sample variances of the x-values and the y-values, respectively. In this last expression for $\widehat{\beta}$, the numerator divided by $n - 1$, is called the **covariance** of x and y and is given by

$$c_{xy} = \frac{1}{n-1}\sum_{i=1}^{n}(x_i - \overline{x})(y_i - \overline{y}).$$

If we consider the covariance of the *standardized* x-values and y-values, namely $(x_i - \overline{x})/s_x$ and $(y_i - \overline{y})/s_y$, $i = 1, 2, \ldots, n$, we obtain another popular statistic called the **correlation coefficient**,

$$r = \frac{1}{n-1}\sum_{i=1}^{n}\left(\frac{x_i - \overline{x}}{s_x}\right)\left(\frac{y_i - \overline{y}}{s_y}\right) = \frac{\dfrac{1}{n-1}\displaystyle\sum_{i=1}^{n}(x_i - \overline{x})(y_i - \overline{y})}{s_x s_y} = \frac{c_{xy}}{s_x s_y}.$$

Thus we now see that

$$\widehat{\beta} = \frac{c_{xy}}{s_x^2} = \left(\frac{c_{xy}}{s_x s_y}\right)\left(\frac{s_y}{s_x}\right) = r\frac{s_y}{s_x}$$

and the best-fitting straight line, sometimes called the **least squares regression line**, is

$$\widehat{y} = \overline{y} + r\frac{s_y}{s_x}(x - \overline{x}).$$

Again it must be emphasized that all the points (x_i, y_i), $i = 1, 2, \ldots, n$, do *not* lie on this line, but will be reasonably close to it.

To calculate the value of r, it is sometimes useful to note that

$$r = \frac{\dfrac{1}{n-1} \displaystyle\sum_{i=1}^{n} (x_i - \bar{x})(y_i - \bar{y})}{\sqrt{\dfrac{1}{n-1} \displaystyle\sum_{i=1}^{n} (x_i - \bar{x})^2} \sqrt{\dfrac{1}{n-1} \displaystyle\sum_{i=1}^{n} (y_i - \bar{y})^2}}$$

$$= \frac{\displaystyle\sum_{i=1}^{n} x_i y_i - \dfrac{1}{n} \left(\displaystyle\sum_{i=1}^{n} x_i \right)\left(\displaystyle\sum_{i=1}^{n} y_i \right)}{\sqrt{\displaystyle\sum_{i=1}^{n} x_i^2 - \dfrac{1}{n}\left(\displaystyle\sum_{i=1}^{n} x_i \right)^2} \sqrt{\displaystyle\sum_{i=1}^{n} y_i^2 - \dfrac{1}{n}\left(\displaystyle\sum_{i=1}^{n} y_i \right)^2}} .$$

An example using data from H. V. Henderson and P. F. Velleman, "Building Multiple Regression Models Interactively" [*Biometrics* **37**:391–411 (1981)], will help to illustrate these characteristics.

Example 1.6.1

Consider Table 1.12, in which x equals the weight of a car in thousands of pounds and y equals the number of gallons of gasoline needed to drive 100 miles. The scatter plot of these $n = 10$ data points is given in Figure 1.17. We find that the means are $\bar{x} = 29.0/10 = 2.90$, $\bar{y} = 43.9/10 = 4.39$, and the standard deviations are $s_x = \sqrt{5.18/9} = 0.759$, $s_y = \sqrt{14.589/9} = 1.273$. Since the covariance equals

TABLE 1.12 Fuel efficiency vs. weight for 10 different cars

Car	x	y	$(x - \bar{x})(y - \bar{y})$	$(x - \bar{x})^2$	$(y - \bar{y})^2$
AMC Concord	3.4	5.5	0.555	0.25	1.2321
Chevrolet Caprice	3.8	5.9	1.359	0.81	2.2801
Ford Country Squire Wagon	4.1	6.5	2.532	1.44	4.4521
Chevrolet Chevette	2.2	3.3	0.763	0.49	1.1881
Toyota Corona	2.6	3.6	0.237	0.09	0.6241
Ford Mustang Ghia	2.9	4.6	0.000	0.00	0.0441
Mazda GLC	2.0	2.9	1.341	0.81	2.2201
AMC Sprint	2.7	3.6	0.158	0.04	0.6241
VW Rabbit	1.9	3.1	1.290	1.00	1.6641
Buick Century	3.4	4.9	0.255	0.25	0.2601
	29.0	43.9	8.490	5.18	14.5890

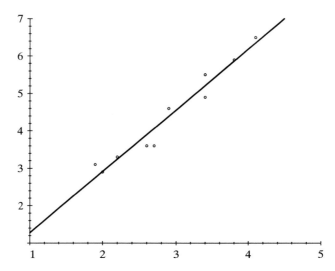

Figure 1.17   Scatter plot of fuel efficiency vs. weight

$$c_{xy} = \frac{8.490}{9} = 0.943,$$

we find that

$$r = \frac{8.490/9}{\sqrt{5.18/9}\,\sqrt{14.589/9}} = 0.977.$$

The best-fitting straight line is

$$\widehat{y} = 4.39 + (0.977)\frac{1.273}{0.759}(x - 2.90) = 1.639x - 0.363,$$

and this is actually the line depicted in Figure 1.17. Note that we could also have found the slope of the line using

$$\widehat{\beta} = \frac{\displaystyle\sum_{i=1}^{n}(x_i - \overline{x})(y_i - \overline{y})}{\displaystyle\sum_{i=1}^{n}(x_i - \overline{x})^2} = \frac{8.490}{5.18} = 1.639.$$

The values of r and $\widehat{\beta}$ can sometimes be found more easily by constructing a table like Table 1.13.

TABLE 1.13   Fuel efficiency/weight
statistics for 10 different cars

x	y	x^2	xy	y^2
3.4	5.5	11.56	18.70	30.25
3.8	5.9	14.44	22.42	34.81
4.1	6.5	16.81	26.65	42.25
2.2	3.3	4.84	7.26	10.89
2.6	3.6	6.76	9.36	12.96
2.9	4.6	8.41	13.34	21.16
2.0	2.9	4.00	5.80	8.41
2.7	3.6	7.29	9.72	12.96
1.9	3.1	3.61	5.89	9.61
3.4	4.9	11.56	16.66	24.01
29.0	43.9	89.28	135.80	207.31

Using the sums in Table 1.13, we obtain

$$
r = \frac{\sum_{i=1}^{n} x_i y_i - \frac{1}{n}\left(\sum_{i=1}^{n} x_i\right)\left(\sum_{i=1}^{n} y_i\right)}{\sqrt{\sum_{i=1}^{n} x_i^2 - \frac{1}{n}\left(\sum_{i=1}^{n} x_i\right)^2}\sqrt{\sum_{i=1}^{n} y_i^2 - \frac{1}{n}\left(\sum_{i=1}^{n} y_i\right)^2}}
$$

$$
= \frac{135.80 - (29.0)(43.9)/10}{\sqrt{89.28 - (29.0)^2/10}\,\sqrt{207.31 - (43.9)^2/10}}
$$

$$
= \frac{8.49}{\sqrt{5.18}\,\sqrt{14.589}} = 0.977
$$

and

$$
\widehat{\beta} = \frac{\sum_{i=1}^{n} x_i y_i - \frac{1}{n}\left(\sum_{i=1}^{n} x_i\right)\left(\sum_{i=1}^{n} y_i\right)}{\sum_{i=1}^{n} x_i^2 - \frac{1}{n}\left(\sum_{i=1}^{n} x_i\right)^2}
$$

$$
= \frac{135.80 - (29.0)(43.9)/10}{89.28 - (29.0)^2/10}
$$

$$
= \frac{8.490}{5.180} = 1.639.
$$

Of course, $\widehat{\alpha} = \bar{y} - \widehat{\beta}\bar{x} = 4.39 - (1.639)(2.90) = -0.363.$ ◇

A few remarks should be made about the correlation coefficient r.

(a) If we think of the horizontal and vertical lines through the point $(\bar{x}, \bar{y})$ as dividing the plane into four quadrants, note that points in the first and third quadrants give positive products $(x_i - \bar{x})(y_i - \bar{y})$ and those in the second and fourth quadrants produce negative products. Thus, if there are more and larger deviations from $(\bar{x}, \bar{y})$ in the first and third quadrants than in the second and fourth, the correlation coefficient is positive. That is, if larger x-values tend to produce larger y-values and smaller x-values tend to produce smaller y-values, then $r > 0$. On the other hand, $r < 0$ if the points tend to favor the second and fourth quadrants; this happens when large x-values tend to produce small y-values and small x-values tend to produce large y-values.

(b) Note that

$$\sum_{i=1}^{n} [(y_i - \bar{y}) - t(x_i - \bar{x})]^2 \geq 0$$

for every real t. That is, the quadratic expression in t,

$$\left[\sum_{i=1}^{n} (x_i - \bar{x})^2\right] t^2 - 2\left[\sum_{i=1}^{n} (x_i - \bar{x})(y_i - \bar{y})\right] t + \left[\sum_{i=1}^{n} (y_i - \bar{y})^2\right],$$

is greater than or equal to zero for all real t. Hence its discriminant, $b^2 - 4ac$, must be nonpositive; that is,

$$\left[2\sum_{i=1}^{n} (x_i - \bar{x})(y_i - \bar{y})\right]^2 - 4\left[\sum_{i=1}^{n} (x_i - \bar{x})^2\right]\left[\sum_{i=1}^{n} (y_i - \bar{y})^2\right] \leq 0.$$

Thus

$$\frac{\left[\sum_{i=1}^{n} (x_i - \bar{x})(y_i - \bar{y})\right]^2}{\left[\sum_{i=1}^{n} (x_i - \bar{x})^2\right]\left[\sum_{i=1}^{n} (y_i - \bar{y})^2\right]} = r^2 \leq 1.$$

It follows that $-1 \leq r \leq 1$, since $r^2 \leq 1$.

(c) The correlation coefficient r quantifies the *linear* association between x and y. If $r = 1$, then all the points are on a straight line with positive slope, s_y/s_x. If, on the other hand, $r = -1$, then all the points are on a straight line with negative slope, $-s_y/s_x$. Moreover, r could equal zero, indicating a weak linear relationship between x and y; yet there could be some other strong nonlinear relationship, like a quadratic one (see Exercise 1.6.10). In any case, one statistic, here r, might miss certain aspects of the data that are discernible to the eye. That is the reason we encourage you to plot the data in looking for the patterns.

From Example 1.6.1, with $r = 0.977$, we see that there is a strong linear relationship. This certainly agrees with our observation of the scatter plot. Thus, for these data, the weight of the car is highly correlated with the number of gallons needed to travel a specified distance. In this case, weight is the major explanation of mileage performance. However, two variables could be highly correlated, and yet one does not cause the other. For example, before air conditioning and modern hospitals, it was true that infant mortality was highly correlated to the softness of asphalt streets. Clearly, one was not causing the other, but a third variable (temperature) made both of the other variables increase or decrease together.

To visualize better the relationship between r and a scatter plot of n observed points $(x_1, y_1), (x_2, y_2), \ldots, (x_n, y_n)$, we have generated four different sets of 50 pairs of observations. In the next example we list the corresponding values of $\bar{x}, \bar{y}, s_x^2, s_y^2, r$, and the observed best-fitting line. Each set of points and corresponding line are plotted on the same graph.

Example 1.6.2

Consider four samples of $n = 50$ pairs of points $(x_1, y_1), (x_2, y_2), \ldots, (x_{50}, y_{50})$. The corresponding sample characteristics are

(a) $\bar{x} = 9.7145$, $s_x^2 = 14.2151$, $\bar{y} = 10.1156$, $s_y^2 = 13.0909$, and $r = 0.9276$, so the best-fitting line is

$$\hat{y} = 10.1156 + 0.9276 \sqrt{\frac{13.0909}{14.2151}} (x - 9.7145) = 0.8902x + 1.4681.$$

(b) $\bar{x} = 11.7587$, $s_x^2 = 21.3780$, $\bar{y} = 10.1313$, $s_y^2 = 13.0362$, and $r = 0.4246$, so the best-fitting line is

$$\hat{y} = 10.1313 + 0.4246 \sqrt{\frac{13.0362}{21.3780}} (x - 11.7587) = 0.3316x + 6.2325.$$

(c) $\bar{x} = 10.1178$, $s_x^2 = 15.5668$, $\bar{y} = 10.0817$, $s_y^2 = 14.7901$, and $r = -0.1946$, so the best-fitting line is

$$\hat{y} = 10.0817 - 0.1946 \sqrt{\frac{14.7901}{15.5668}} (x - 10.1178) = -0.1897x + 12.0009.$$

(d) $\bar{x} = 10.3652$, $s_x^2 = 16.8695$, $\bar{y} = 9.2948$, $s_y^2 = 18.6522$, and $r = -0.7207$, so the best-fitting line is

$$\hat{y} = 9.2948 - 0.7207 \sqrt{\frac{18.6522}{16.8695}} (x - 10.3652) = -0.7578x + 17.1498.$$

In Figure 1.18, these respective lines and the corresponding sample points are plotted. Note the effect that the value of r has on the slope of the line and the variability of the points about that line.   ◇

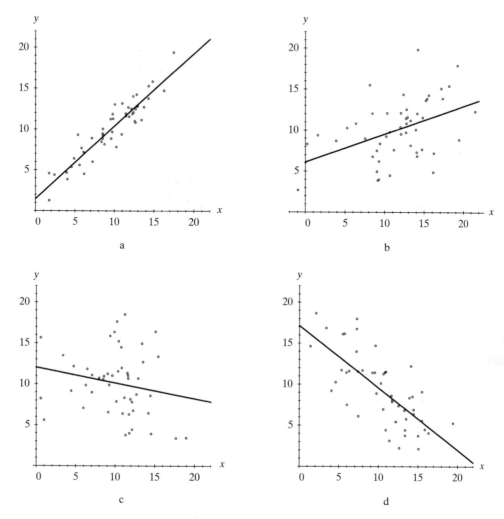

Figure 1.18 Four scatter plots and linear fits

The next example shows that two paired variables may be clearly related (dependent), yet have a correlation coefficient r close to zero. This, however, is not unexpected, since we recall that r does, in a sense, measure the *linear* relationship between two random variables. That is, the linear relationship between the variables could be zero, whereas higher order ones could be quite strong.

Example 1.6.3

Twenty-five observations of paired variables were simulated, yielding the following data:

(6.91, 17.52)	(4.32, 22.69)	(2.38, 17.61)	(7.98, 14.29)
(8.26, 10.77)	(2.00, 12.87)	(3.10, 18.63)	(7.69, 16.77)
(2.21, 14.97)	(3.42, 19.16)	(8.18, 11.15)	(5.39, 22.41)
(1.19, 7.50)	(3.21, 19.06)	(5.47, 23.89)	(7.35, 16.63)
(2.32, 15.09)	(7.54, 14.75)	(1.27, 10.75)	(7.33,17.42)
(8.41, 9.40)	(8.72, 9.83)	(6.09, 22.33)	(5.30, 21.37)
(7.30, 17.36)			

For this set of observations $\bar{x} = 5.33$, $\bar{y} = 16.17$, $s_x^2 = 6.521$, $s_y^2 = 20.865$, and $r = -0.06$. Note that r is very close to zero even though the variables seem very dependent; that is, it seems that a quadratic expression would fit the data very well. In Exercise 1.6.7 the reader is asked to fit $y = a + bx + cx^2$ to these 25 points by the method of least squares. See Figure 1.19 for a plot of these 25 points.   ◇

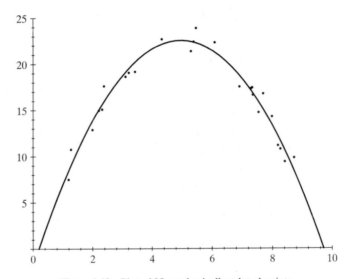

Figure 1.19   Plot of 25 quadratically related points

EXERCISES

1.6.1 A reader survey for a birding magazine asked respondents to give the number of binoculars (x) and the number of bird reference books (y) that they owned. The following data were returned for 10 respondents:

(1, 0)	(3, 3)	(6, 7)	(2, 6)	(2, 1)
(5, 9)	(1, 1)	(3, 1)	(3, 10)	(0, 1)

(a) Find the value of the correlation coefficient r.
(b) Find the equation of the least squares regression line.
(c) Plot the data and the least squares regression line on the same graph.

1.6.2 There was an interest in discovering whether some college students learned to use the library more effectively during the academic year. A library knowledge test was given at the beginning and the end of the semester with the scores denoted by x and y, respectively. The scores for $n = 14$ students were as follows:

$$(15, 21) \quad (19, 21) \quad (22, 25) \quad (17, 16) \quad (13, 19)$$
$$(24, 23) \quad (20, 23) \quad (18, 22) \quad (22, 22) \quad (19, 19)$$
$$(13, 18) \quad (17, 12) \quad (15, 20) \quad (17, 21)$$

(a) Find the value of the correlation coefficient r.

(b) Find the equation of the least squares regression line.

(c) Plot the data and the least squares regression line on the same graph.

1.6.3 The respective high school and college GPAs for 20 college seniors as ordered pairs (x, y) are

$$(3.75, 3.19) \quad (3.45, 3.34) \quad (2.87, 2.23) \quad (3.60, 3.46)$$
$$(3.42, 2.97) \quad (4.00, 3.79) \quad (2.65, 2.55) \quad (3.10, 2.50)$$
$$(3.47, 3.15) \quad (2.60, 2.26) \quad (4.00, 3.76) \quad (2.30, 2.11)$$
$$(2.47, 2.11) \quad (3.36, 3.01) \quad (3.60, 2.92) \quad (3.65, 3.09)$$
$$(3.30, 3.05) \quad (2.58, 2.63) \quad (3.80, 3.22) \quad (3.79, 3.27)$$

(a) Verify that $\bar{x} = 3.29$, $\bar{y} = 2.93$, $s_x^2 = 0.283$, $s_y^2 = 0.260$, and $r = 0.92$.

(b) Find the equation of the best-fitting line.

(c) Plot the 20 points and the line on the same graph.

1.6.4 The following data give the ACT math and ACT verbal scores, say (x, y), for 15 students:

$$(16, 19) \quad (18, 17) \quad (22, 18) \quad (20, 23) \quad (17, 20)$$
$$(25, 21) \quad (21, 24) \quad (23, 18) \quad (24, 18) \quad (31, 25)$$
$$(27, 29) \quad (28, 24) \quad (30, 24) \quad (27, 23) \quad (28, 24)$$

(a) Verify that $\bar{x} = 23.8$, $\bar{y} = 21.8$, $s_x^2 = 22.457$, $s_y^2 = 11.600$, and $r = 0.626$.

(b) Find the equation of the best-fitting line.

(c) Plot the 15 points and the line on the same graph.

1.6.5 Each of 16 professional golfers hits off the tee a brand A golf ball and a brand B golf ball, eight hitting ball A before ball B and eight hitting ball B before ball A. Let x and y equal the distances traveled in yards for ball A and ball B, respectively. The following data, (x, y), were observed:

$$(265, 252) \quad (272, 276) \quad (246, 243) \quad (260, 246)$$
$$(274, 275) \quad (263, 246) \quad (255, 244) \quad (258, 245)$$
$$(276, 259) \quad (274, 267) \quad (274, 260) \quad (269, 267)$$
$$(244, 251) \quad (212, 222) \quad (254, 255) \quad (224, 231)$$

(a) Verify that $\bar{x} = 257.50$, $\bar{y} = 252.44$, $s_x^2 = 341.333$, $s_y^2 = 218.796$, and $r = 0.867$.

(b) Find the equation of the best-fitting line.

(c) Plot the 16 points and the line on the same graph.

1.6.6 Fourteen pairs of gallinules were captured and weighed. Let x equal the male weight and y the female weight. The following weights in grams were observed:

$$(405, 321) \quad (396, 378) \quad (457, 351) \quad (450, 320) \quad (415, 365)$$
$$(403, 328) \quad (370, 372) \quad (435, 314) \quad (425, 375) \quad (425, 355)$$
$$(415, 355) \quad (400, 340) \quad (425, 398) \quad (420, 330)$$

(a) Verify that $\bar{x} = 417.2$, $s_x = 22.36$, $\bar{y} = 350.1$, $s_y = 25.56$, and $r = -0.252$.

(b) Find the equation of the best-fitting line.

(c) Plot the 14 points and the line on the same graph.

1.6.7 We would like to fit the quadratic curve $y = a+bx+cx^2$ to a set of points (x_1, y_1), (x_2, y_2), ..., (x_n, y_n) by the method of least squares. To do this, let

$$h(a, b, c) = \sum_{i=1}^{n} (y_i - a - bx_i - cx_i^2)^2.$$

(a) By setting the three first partial derivatives of h with respect to a, b, and c equal to 0, show that a, b, and , and c satisfy the following set of equations (called normal equations), all sums going from 1 to n:

$$an + b \sum x_i + c \sum x_i^2 = \sum y_i;$$
$$a \sum x_i + b \sum x_i^2 + c \sum x_i^3 = \sum x_i y_i;$$
$$a \sum x_i^2 + b \sum x_i^3 + c \sum x_i^4 = \sum x_i^2 y_i.$$

(b) For the data given in Example 1.6.3, $\sum x_i = 133.34$, $\sum x_i^2 = 867.75$, $\sum x_i^3 = 6197.21$, $\sum x_i^4 = 46{,}318.88$, $\sum y_i = 404.22$, $\sum x_i y_i = 2138.38$, $\sum x_i^2 y_i = 13{,}380.30$ with $n = 25$. Show that $a = -1.88$, $b = 9.86$, and $c = -0.995$.

(c) Plot the points and this least squares quadratic regression curve on the same graph.

1.6.8 After keeping records for six years, a telephone company is interested in predicting the number of telephones that will be in service in year 7. The following data are available. The number of telephones in service is given in thousands.

Year	Number of Telephones
1	91
2	93
3	95
4	99
5	102
6	105

(a) Find the quadratic curve of best fit, $y = a+bx+cx^2$, that could be used for this prediction.

(b) Plot the points and the curve on the same graph.

(c) Predict the number of telephones that will be in service in year 7.

Remark   Predicting a future value can often be difficult, tricky, and sometimes dangerous. For illustration, there might be some intervention (possibly one by a government or nature) that changes the situation completely. If, however, the situation is rather stable, the method given here is reasonable for a prediction in the relative near future. If one needed to make many predictions, he or she should study techniques of time series in greater detail.

1.6.9 For male freshmen in a health fitness program, let x equal a participant's percentage of body fat at the beginning of the semester and let y equal the change in this percentage (percentage at the end of the semester minus percentage at the beginning of the semester, so that a negative y indicates a loss). Twelve observations of (x, y) are

$$\begin{array}{cccc}
(13.1, 1.1) & (16.8, 0.5) & (17.9, -1.3) & (10.6, -2.2) \\
(8.2, -1.1) & (10.4, -0.2) & (17.4, -2.0) & (10.5, -1.4) \\
(5.4, 0.5) & (14.3, -4.6) & (11.1, 1.0) & (5.3, 1.7)
\end{array}$$

(a) Verify that $\bar{x} = 11.750$, $s_x = 4.298$, $\bar{y} = -0.667$, $s_y = 1.788$, and $r = -0.3945$.

(b) Find the equation of the best-fitting line (i.e., the least squares regression line).

(c) Plot the points and the line on the same graph.

1.6.10 [The following information comes from the Westview Blueberry Farm and National Oceanic and Atmospheric Administration (NOAA) Reports.] For the paired data, (x, y), x gives the Holland rainfall for June and y gives the blueberry production in thousands of pounds from the Westview Blueberry Farm. The data come from the years 1971 to 1989 for those years in which the last frost occurred May 10 or earlier.

$$\begin{array}{cccc}
(4.11, 56.2) & (5.49, 45.3) & (5.35, 31.0) & (6.53, 30.1) \\
(5.18, 40.0) & (4.89, 38.5) & (2.09, 50.0) & (1.40, 45.8) \\
(4.52, 45.9) & (1.11, 32.4) & (0.60, 18.2) & (3.80, 56.1)
\end{array}$$

(a) Find the correlation coefficient for these data.

(b) Find the least squares regression line.

(c) Make a scatter plot of the data with the least squares regression line on the plot.

(d) Find the least squares quadratic regression curve.

(e) Give a short interpretation.

1.6.11 *Parade* (March 13, 1994) stated that "Heart surgeries, including bypass and angioplasty, totaled 196,000 in 1980 and reached 839,000 in 1991." (*Source: National Hospital Discharge Survey,* 1991.) The data for the years 1980 to 1991 in thousands are

$$\begin{array}{cccccc}
(1980, 196) & (1981, 217) & (1982, 243) & (1983, 275) & (1984, 314) & (1985, 379) \\
(1986, 490) & (1987, 588) & (1988, 674) & (1989, 719) & (1990, 781) & (1991, 839)
\end{array}$$

(a) Make a scatter plot (here a time sequence) of the data with the least squares regression line superimposed.

(b) Find and interpret the correlation coefficient.

(c) If possible, find a polynomial that provides a better fit of these data. Construct a scatter plot of the data with this polynomial superimposed.

2

PROBABILITY

2.1 PROPERTIES OF PROBABILITY

In Section 1.1 the collection of all possible outcomes (the *universal set*) of a random experiment is denoted by S and is called the *outcome space*. Given an outcome space S, let A be a part of the collection of outcomes in S, that is, $A \subset S$. Then A is called an **event**. When the random experiment is performed and the outcome of the experiment is in A, we say that **event A has occurred**.

Since, in studying probability, the words *set* and *event* are interchangeable, the reader might want to review **algebra of sets**, provided in the Appendix. For convenience, however, we remind the reader of a little of that terminology:

- $\emptyset$ denotes the **null** or **empty** set.
- $A \subset B$ means that A is a **subset** of B.
- $A \cup B$ is the **union** of A and B.
- $A \cap B$ is the **intersection** of A and B.
- A' is the **complement** of A (i.e., all elements in S that are not in A).

Some of these sets are depicted in Figure 2.1.

Special terminology associated with events that is often used by statisticians includes the following:

1. $A_1, A_2, \ldots, A_k$ are **mutually exclusive events** means that $A_i \cap A_j = \emptyset, i \neq j$, that is, $A_1, A_2, \ldots, A_k$ are disjoint sets.
2. $A_1, A_2, \ldots, A_k$ are **exhaustive events** means that $A_1 \cup A_2 \cup \cdots \cup A_k = S$.

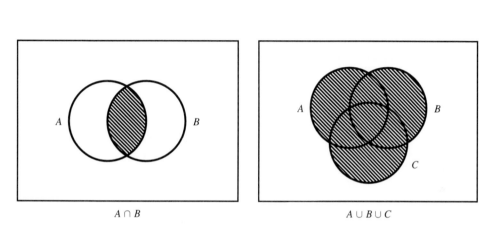

Figure 2.1   Algebra of sets

So if $A_1, A_2, \ldots, A_k$ are **mutually exclusive and exhaustive** events, we know that $A_i \cap A_j = \emptyset, i \neq j$, and $A_1 \cup A_2 \cup \cdots \cup A_k = S$.

We are interested in defining what is meant by the probability of A, denoted by $P(A)$, and often called the chance of A occurring. To help us understand what is meant by the probability of A, $P(A)$, consider repeating the experiment a number of times, say n times. Count the number of times that event A actually occurred throughout these n performances; this number is called the frequency of event A and is denoted by $\mathcal{N}(A)$. The ratio $\mathcal{N}(A)/n$ is called the **relative frequency** of event A in these n repetitions of the experiment. A relative frequency is usually very unstable for small values of n, but it tends to stabilize as n increases. This suggests that we associate with event A a number, say p, that is equal to or approximately equal to the number about which the relative frequency tends to stabilize. This number p can then be taken as the number that the relative frequency of event A will be near in future performances of the experiment. Thus, although we cannot predict the outcome of a random experiment with certainty, we can, for a large value of n, predict fairly accurately the relative frequency associated with event A. The number p assigned to event A is called the **probability** of event A, and is denoted by $P(A)$. That is, $P(A)$ represents the proportion of outcomes of a random experiment that terminate in the event A as the number of trials of that experiment increases without bound.

The following example will help to illustrate some of the ideas just presented.

Example 2.1.1

A fair six-sided die is rolled six times. If the face numbered k is the outcome on roll k for $k = 1, 2, \ldots, 6$, we say that a match has occurred. The experiment is called a success if at least one match occurs during the six trials. Otherwise, the experiment is called a failure. The sample space is $S = \{\text{success}, \text{failure}\}$. Let $A = \{\text{success}\}$. We would like to assign a value to $P(A)$. Accordingly, this experiment was simulated 500 times on a computer. Figure 2.2 depicts the results of this simulation, and the following table summarizes a few of the results:

n	$\mathcal{N}(A)$	$\mathcal{N}(A)/n$
50	33	0.660
100	59	0.590
250	161	0.644
500	329	0.658

Figure 2.2   Fraction of experiments having at least one match

The probability of event A is not intuitively obvious, but it will be shown in Example 2.4.6 that $P(A) = 1 - (1 - 1/6)^6 = 0.665$. This assignment is certainly supported by the simulation (although not proved by it).   ◇

Example 2.1.1 shows that at times intuition cannot be used to assign probabilities although simulation can perhaps help to assign a probability empirically. The following example gives an illustration where intuition can help in assigning a probability to an event.

Example 2.1.2

A disk 2 inches in diameter is thrown at random on a tiled floor, where each tile is a square with sides 4 inches in length. Let C be the event that the disk will land entirely on one tile. In order to assign a value to $P(C)$, consider the center of the disk. In what region must the center lie to assure that the disk lies entirely on one tile? If you draw a picture, it should be clear that the center must lie within a square having sides of length 2 and lying in the center of a tile. Since the area of this square is 4 and the area of a tile is 16, it makes sense to let $P(C) = 4/16$.   ◇

Sometimes the nature of an experiment is such that the probability of A can be assigned easily. For example, when a state lottery randomly selects a three-digit integer, we would expect each of the 1000 possible three-digit numbers to have the same chance of being selected, namely $1/1000$. If we let $A = \{233, 323, 332\}$, then it makes sense to let $P(A) = 3/1000$. Or if we let $B = \{234, 243, 324, 342, 423, 432\}$, then we would let $P(B) = 6/1000$, the probability of the event B. Probabilities of events associated with many random experiments are perhaps not quite as obvious and straightforward as was seen in Example 2.1.1.

So we wish to associate with A a number $P(A)$ about which the relative frequency $\mathcal{N}(A)/n$ of the event A tends to stabilize with large n. A function such as $P(A)$, which is evaluated for a set A, is called a **set function**. In this section we consider the probability set function $P(A)$ and discuss some of its properties. In succeeding sections we describe how the probability set function is defined for particular experiments.

To help decide what properties the probability set function should satisfy, consider properties possessed by the relative frequency $\mathcal{N}(A)/n$. For example, $\mathcal{N}(A)/n$ is always nonnegative. If $A = S$, the sample space, then the outcome of the experiment will always belong to S, and thus $\mathcal{N}(S)/n = 1$. Also, if A and B are two mutually exclusive events, then $\mathcal{N}(A \cup B)/n = \mathcal{N}(A)/n + \mathcal{N}(B)/n$. Hopefully, these remarks will help to motivate the following definition.

Definition 2.1.1.   Probability is a set function P that assigns to each event A in the sample space S a number $P(A)$, called the probability of the event A, such that the following properties are satisfied:

  (a) $P(A) \geq 0$.
  (b) $P(S) = 1$.

(c) If $A_1, A_2, A_3, \ldots$ are events and $A_i \cap A_j = \emptyset, i \neq j$, then

$$P(A_1 \cup A_2 \cup \cdots \cup A_k) = P(A_1) + P(A_2) + \cdots + P(A_k)$$

for each positive integer k, and

$$P(A_1 \cup A_2 \cup A_3 \cup \cdots) = P(A_1) + P(A_2) + P(A_3) + \cdots$$

for an infinite, but countable, number of events.

The following theorems give some other important properties of the probability set function. When one considers these theorems, it is important to understand the theoretical concepts and proofs. However, if the reader keeps the relative frequency concept in mind, the theorems should also have some intuitive appeal.

Theorem 2.1.1.   For each event A,

$$P(A) = 1 - P(A').$$

Proof.   We have

$$S = A \cup A' \quad \text{and} \quad A \cap A' = \emptyset.$$

Thus, from properties (b) and (c), it follows that

$$1 = P(A) + P(A').$$

Hence

$$P(A) = 1 - P(A'). \qquad \square$$

Example 2.1.3

A fair coin is flipped successively until the same face is observed on successive flips. Let $A = \{x : x = 3, 4, 5, \ldots\}$; that is, A is the event that it will take three or more flips of the coin to observe the same face on two consecutive flips. To find $P(A)$, we first find the probability of $A' = \{x : x = 2\}$, the complement of A. In two flips of a coin, the possible outcomes are $\{HH, HT, TH, TT\}$, and we assume that each of these four points has the same chance of being observed. Thus

$$P(A') = P(\{HH, TT\}) = \frac{2}{4}.$$

It follows from Theorem 2.1.1 that

$$P(A) = 1 - P(A') = 1 - \frac{2}{4} = \frac{2}{4}. \qquad \diamond$$

Theorem 2.1.2. $P(\emptyset) = 0.$

Proof.   In Theorem 2.1.1, take $A = \emptyset$ so that $A' = S$. Thus

$$P(\emptyset) = 1 - P(S) = 1 - 1 = 0. \qquad \square$$

Theorem 2.1.3.   If events A and B are such that $A \subset B$, then $P(A) \le P(B)$.

Proof.   Now

$$B = A \cup (B \cap A') \quad \text{and} \quad A \cap (B \cap A') = \emptyset.$$

Hence, from property (c),

$$P(B) = P(A) + P(B \cap A') \ge P(A)$$

because from property (a),

$$P(B \cap A') \ge 0. \qquad \square$$

Theorem 2.1.4.   For each event A, $P(A) \le 1$.

Proof.   Since $A \subset S$, we have by Theorem 2.1.3 and property (b) that

$$P(A) \le P(S) = 1,$$

which gives the desired result. $\qquad \square$

Property (a) along with Theorem 2.1.4 shows that for each event A,

$$0 \le P(A) \le 1.$$

Theorem 2.1.5.   If A and B are any two events, then

$$P(A \cup B) = P(A) + P(B) - P(A \cap B).$$

Proof.   The event $A \cup B$ can be represented as a union of mutually exclusive events, namely,

$$A \cup B = A \cup (A' \cap B).$$

Hence, by property (c),

$$P(A \cup B) = P(A) + P(A' \cap B). \qquad (2.1)$$

However,

$$B = (A \cap B) \cup (A' \cap B),$$

which is a union of mutually exclusive events. Thus

$$P(B) = P(A \cap B) + P(A' \cap B)$$

and

$$P(A' \cap B) = P(B) - P(A \cap B).$$

If this result is substituted in Equation 2.1 we obtain

$$P(A \cup B) = P(A) + P(B) - P(A \cap B),$$

which is the desired result. $\qquad \square$

Example 2.1.4

A faculty leader was meeting two students in Paris, one arriving by train from Amsterdam and the other arriving by train from Brussels at approximately the same time. Let A and B be the events that the trains are on time, respectively. Suppose from past experience we know that $P(A) = 0.93$, $P(B) = 0.89$, and $P(A \cap B) = 0.87$. Then

$$P(A \cup B) = P(A) + P(B) - P(A \cap B)$$
$$= 0.93 + 0.89 - 0.87 = 0.95$$

is the probability that at least one train is on time.   ◇

Theorem 2.1.6.   If A, B, and C are any three events, then

$$P(A \cup B \cup C) = P(A) + P(B) + P(C) - P(A \cap B)$$
$$- P(A \cap C) - P(B \cap C) + P(A \cap B \cap C).$$

Proof.   Write

$$A \cup B \cup C = A \cup (B \cup C)$$

and apply Theorem 2.1.5. The details are left as an exercise.   □

Example 2.1.5

Continuing with Example 2.1.4, say that a third student is arriving from Cologne. Let C be the event that this train is on time with $P(C) = 0.91$, $P(B \cap C) = 0.85$, $P(A \cap C) = 0.86$, and $P(A \cap B \cap C) = 0.81$. Then

$$P(A \cup B \cup C) = P(A) + P(B) + P(C) - P(A \cap B) - P(A \cap C)$$
$$- P(B \cap C) + P(A \cap B \cap C)$$
$$= 0.93 + 0.89 + 0.91 - 0.87 - 0.85 - 0.86 + 0.81$$
$$= 0.96$$

is the probability that at least one of these trains is on time.   ◇

Let a probability set function be defined on a sample space S. Let $S = \{e_1, e_2, \ldots, e_m\}$, where each e_i is a possible outcome of the experiment. The integer m is called the total number of ways in which the random experiment can terminate. If each of these outcomes has the same probability of occurring, we say that the m outcomes are **equally likely**. That is,

$$P(\{e_j\}) = \frac{1}{m}, \qquad i = 1, 2, \ldots, m.$$

If the number of outcomes in an event A is h, the integer h is called the number of ways that are favorable to the event A. In this case $P(A)$ is equal to the number of ways favorable to the event A divided by the total number of ways in which the experiment can terminate. That is, under this assumption of equally likely outcomes, we have that

$$P(A) = \frac{h}{m} = \frac{N(A)}{N(S)},$$

where $h = N(A)$ is the number of ways that A can occur and $m = N(S)$ is the number of ways that S can occur. Exercise 2.1.17 considers this assignment of probability in a more theoretical manner.

It should be emphasized that in order to assign the probability h/m to the event A, we must assume that each of the outcomes $e_1, e_2, \ldots, e_m$ has the same probability $1/m$. This assumption is then an important part of our probability model; if it is not realistic in an application, the probability of the event A cannot be computed this way. Actually, we have used this result in the simple case given in Example 2.1.3 because it seemed realistic to assume that each of the possible outcomes, HH, HT, TH, TT, had the same chance of being observed.

Example 2.1.6

Let a card be drawn at random from an ordinary deck of 52 playing cards. The sample space S is the set of $m = 52$ different cards, and it is reasonable to assume that each of these cards has the same probability for selection, $1/52$. Accordingly, if A is the set of outcomes that are kings, $P(A) = 4/52 = 1/13$ because there are $h = 4$ kings in the deck. That is, $1/13$ is the probability of drawing a card that is a king provided that each of the 52 cards has the same probability. ◇

In Example 2.1.6, the computations are very easy because there is no difficulty in the determination of the appropriate values of h and m. However, instead of drawing only one card, suppose that 13 are taken at random and without replacement. We can think of each possible 13-card hand as being an outcome in a sample space, and it is reasonable to assume that each of these outcomes has the same probability. To use the foregoing method to assign the probability of a hand, consisting of seven spades and six hearts, for illustration, we must be able to count the number h of all such hands as well as the number m of possible 13-card hands. In these more complicated situations, we need better methods of determining h and m. We discuss some of these counting techniques in Section 2.2.

EXERCISES

2.1.1 Draw one card at random from a standard deck of cards. The sample space S is the collection of the 52 cards. Assume that the probability set function assigns $1/52$ to each of these 52 outcomes. Let

$$A = \{x : x \text{ is a jack, queen, or king}\},$$

$$B = \{x : x \text{ is a 9, 10, or jack and } x \text{ is red}\},$$

$$C = \{x : x \text{ is a club}\},$$

$$D = \{x : x \text{ is a diamond, a heart, or a spade}\}.$$

Find (a) $P(A)$, (b) $P(A \cap B)$, (c) $P(A \cup B)$, (d) $P(C \cup D)$, and (e) $P(C \cap D)$.

2.1.2 A coin is tossed four times, and the sequence of heads and tails is observed.

(a) List each of the 16 sequences in the sample space S.

(b) Let events A, B, C, and D be given by $A = \{$at least 3 heads$\}$, $B = \{$at most 2 heads$\}$, $C = \{$heads on the third toss$\}$, and $D = \{$1 head and 3 tails$\}$. If the probability set function assigns 1/16 to each outcome in the sample space, find (i) $P(A)$, (ii) $P(A \cap B)$, (iii) $P(B)$, (iv) $P(A \cap C)$, (v) $P(D)$, (vi) $P(A \cup C)$, and (vii) $P(B \cap D)$.

2.1.3 A field of beans is planted with three seeds per hill. For each hill of beans, let A_i be the event that i seeds germinate, $i = 0, 1, 2, 3$. Suppose that $P(A_0) = 1/64$, $P(A_1) = 9/64$, and $P(A_2) = 27/64$. Give the value of $P(A_3)$.

2.1.4 Consider the trial on which a 3 is first observed in successive rolls of a four-sided die. Let A be the event that 3 is observed on the first trial. Let B be the event that at least two trials are required to observe a 3. Assuming that each side has probability 1/4, find (a) $P(A)$, (b) $P(B)$, and (c) $P(A \cup B)$.

2.1.5 A fair eight-sided die is rolled once. Let $A = \{2, 4, 6, 8\}$, $B = \{3, 6\}$, $C = \{2, 5, 7\}$, and $D = \{1, 3, 5, 7\}$. Assume that each face has the same probability.

    (a) Give the values of (i) $P(A)$, (ii) $P(B)$, (iii) $P(C)$, and (iv) $P(D)$.

    (b) Give the values of (i) $P(A \cap B)$, (ii) $P(B \cap C)$, and (iii) $P(C \cap D)$.

    (c) Give the values of (i) $P(A \cup B)$, (ii) $P(B \cup C)$, and (iii) $P(C \cup D)$ using Theorem 2.1.5.

2.1.6 If $P(A) = 0.4$, $P(B) = 0.5$, and $P(A \cap B) = 0.3$, find (a) $P(A \cup B)$, (b) $P(A \cap B')$, and (c) $P(A' \cup B')$.

2.1.7 If $S = A \cup B$, $P(A) = 0.7$, and $P(B) = 0.9$, find $P(A \cap B)$.

2.1.8 If $P(A) = 0.4$, $P(B) = 0.5$, and $P(A \cup B) = 0.7$, find (a) $P(A \cap B)$ and (b) $P(A' \cup B')$.

2.1.9 Roll a fair six-sided die three times. Let $A_1 = \{$1 or 2 on the first roll$\}$, $A_2 = \{$3 or 4 on the second roll$\}$, and $A_3 = \{$5 or 6 on the third roll$\}$. It is given that $P(A_i) = 1/3$, $i = 1, 2, 3$; $P(A_i \cap A_j) = (1/3)^2$, $i \neq j$; and $P(A_1 \cap A_2 \cap A_3) = (1/3)^3$.

    (a) Use Theorem 2.1.6 to find $P(A_1 \cup A_2 \cup A_3)$.

    (b) Show that $P(A_1 \cup A_2 \cup A_3) = 1 - (1 - 1/3)^3$.

2.1.10 Prove Theorem 2.1.6.

2.1.11 For each positive integer n, let $P(\{n\}) = (1/2)^n$. Consider the events $A = \{n : 1 \leq n \leq 10\}$, $B = \{n : 1 \leq n \leq 20\}$ and $C = \{n : 11 \leq n \leq 20\}$. Find (a) $P(A)$, (b) $P(B)$, (c) $P(A \cup B)$, (d) $P(A \cap B)$, (e) $P(C)$, and (f) $P(B')$.

2.1.12 Let x equal a number that is selected randomly from the closed interval from zero to one, that is, $[0, 1]$. Use your intuition to assign values to

    (a) $P(\{x : 0 \leq x \leq 1/3\})$.

    (b) $P(\{x : 1/3 \leq x \leq 1\})$.

    (c) $P(\{x : x = 1/3\})$.

    (d) $P(\{x : 1/2 < x < 5\})$.

2.1.13 A typical roulette wheel used in a casino has 38 slots that are numbered $1, 2, 3, \ldots, 36, 0, 00$, respectively. The 0 and 00 slots are colored green. Half of the remaining slots are red and half are black. Also, half of the integers between 1 and 36 inclusive are odd, half are even, and 0 and 00 are defined to be neither odd nor even. A ball is rolled around the wheel and ends up in one of the slots; we assume that each slot has equal probability of 1/38 and we are interested in the number of the slot in which the ball falls.

(a) Define the sample space S.

(b) Let $A = \{0, 00\}$. Give the value of $P(A)$.

(c) Let $B = \{14, 15, 17, 18\}$. Give the value of $P(B)$.

(d) Let $D = \{\text{odd}\}$. Give the value of $P(D)$.

2.1.14 The five numbers 1, 2, 3, 4, and 5 are written, respectively, on five disks of the same size and placed in a hat. Two disks are drawn without replacement from the hat, and the numbers written on them are observed.

(a) List the 10 possible outcomes for this experiment as pairs of numbers (order not important).

(b) If each of the 10 outcomes has probability 1/10, assign a value to the probability that the sum of the two numbers drawn is (i) 3 and (ii) between 6 and 8 inclusive.

2.1.15 Divide a line segment into two parts by selecting a point at random. Use your intuition to assign a probability to the event that the larger segment is at least two times longer than the shorter segment.

2.1.16 Let the interval $[-r, r]$ be the base of a semicircle. If a point is selected at random from this interval, assign a probability to the event that the length of the perpendicular segment from this point to the semicircle is less than $r/2$.

2.1.17 Let $S = A_1 \cup A_2 \cup \cdots \cup A_m$, where events $A_1, A_2, \ldots, A_m$ are mutually exclusive and exhaustive.

(a) If $P(A_1) = P(A_2) = \cdots = P(A_m)$, show that $P(A_i) = 1/m$, $i = 1, 2, \ldots, m$.

(b) If $A = A_1 \cup A_2 \cup \cdots \cup A_h$, where $h < m$, and part (a) holds, prove that $P(A) = h/m$.

2.2 METHODS OF ENUMERATION

In this section we develop counting techniques that are useful in determining the number of outcomes associated with the events of certain random experiments. We begin with a consideration of the multiplication principle.

Multiplication Principle: Suppose that an experiment (or procedure) E_1 has n_1 outcomes and for each of these possible outcomes an experiment (procedure) E_2 has n_2 possible outcomes. The composite experiment (procedure) $E_1 E_2$ that consists of performing first E_1 and then E_2 has $n_1 n_2$ possible outcomes.

Example 2.2.1

Let E_1 denote the selection of a rat from a cage containing one female (F) rat and one male (M) rat. Let E_2 denote the administering of either drug A (A), drug B (B), or a placebo (P) to the selected rat. The outcome for the composite experiment can be denoted by an ordered pair, such as (F, P). In fact, the set of all possible outcomes, namely $(2)(3) = 6$ of them, can be denoted by the following rectangular array:

(F, A)	(F, B)	(F, P)
(M, A)	(M, B)	(M, P)

◇

Another way of illustrating the multiplication principle is with a tree diagram like that in Figure 2.3. The diagram shows that there are $n_1 = 2$ possibilities (branches) for the sex of the rat and that for each of these outcomes there are $n_2 = 3$ possibilities (branches) for the drug.

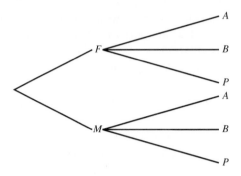

Figure 2.3   Tree diagram

Clearly, the multiplication principle can be extended to a sequence of more than two experiments or procedures. Suppose that the experiment E_i has n_i, $i = 1, 2, \ldots, m$, possible outcomes after previous experiments have been performed. The composite experiment $E_1 E_2 \cdots E_m$, which consists of performing E_1, then E_2, $\ldots$, and finally E_m, has $n_1 n_2 \cdots n_m$ possible outcomes.

Example 2.2.2

A certain food service gives the following choices for dinner: E_1, soup or tomato juice; E_2, steak or shrimp; E_3, french fried potatoes, mashed potatoes, or a baked potato; E_4, corn or peas; E_5, Jello, tossed salad, cottage cheese, or cole slaw; E_6, cake, cookies, pudding, brownie, vanilla ice cream, chocolate ice cream, or orange sherbet; E_7, coffee, tea, milk, or punch. How many different dinner selections are possible if one of the listed choices is made for each of E_1, E_2, $\ldots$, E_7? By the multiplication principle there are

$$(2)(2)(3)(2)(4)(7)(4) = 2688$$

different combinations.   ◇

Although the multiplication principle is fairly simple and easy to understand, it will be extremely useful as we now develop various counting techniques.

Suppose that n positions are to be filled with n different objects. There are n choices for filling the first position, $n - 1$ for the second, $\ldots$, 1 choice for the last position. So by the multiplication principle there are

$$n(n - 1) \cdots (2)(1) = n!$$

possible arrangements. The symbol $n!$ is read n factorial. By definition, we take $0! = 1$; that is, we say that zero positions can be filled with zero objects in one way.

Definition 2.2.1.   Each of the $n!$ arrangements (in a row) of n different objects is called a **permutation** of the n objects.

Example 2.2.3

The number of permutations of the four letters a, b, c, and d is clearly $4! = 24$. However, the number of possible four-letter code words using the four letters a, b, c, and d if letters may be repeated is $4^4 = 256$, because in this case each selection can be performed in four ways.   ◇

If only r positions are to be filled with objects selected from n different objects, $r \leq n$, the number of possible ordered arrangements is

$$_nP_r = n(n-1)(n-2)\cdots(n-r+1).$$

That is, there are n ways to fill the first position, $(n-1)$ ways to fill the second, and so on, until there are $[n-(r-1)] = (n-r+1)$ ways to fill the rth position. In terms of factorials we have that

$$_nP_r = \frac{n(n-1)\cdots(n-r+1)(n-r)\cdots(3)(2)(1)}{(n-r)\cdots(3)(2)(1)} = \frac{n!}{(n-r)!}.$$

Definition 2.2.2.   Each of the $_nP_r$ arrangements is called a **permutation of n objects taken r at a time**.

Example 2.2.4

The number of possible four-letter code words, selecting from the 26 letters in the alphabet, in which all four letters are different is

$$_{26}P_4 = (26)(25)(24)(23) = \frac{26!}{22!} = 358{,}800.$$ ◇

Example 2.2.5

The number of ways of selecting a president, a vice president, a secretary, and a treasurer in a club consisting of 10 persons is

$$_{10}P_4 = 10 \cdot 9 \cdot 8 \cdot 7 = \frac{10!}{6!} = 5040.$$ ◇

Suppose that a set contains n objects. Consider the problem of drawing r objects from this set. The order in which the objects are drawn may or may not be important. In addition, it is possible that a drawn object is replaced before the next object is drawn. Accordingly, we give some definitions and show how the multiplication principle can be used to count the number of possibilities.

Definition 2.2.3.   If r objects are selected from a set of n objects, and if the order of selection is noted, the set of r objects selected is called an **ordered sample of size r**.

Definition 2.2.4.   Sampling with replacement occurs when an object is selected and then replaced before the next object is selected.

By the multiplication principle, the number of possible ordered samples of size r taken from a set of n objects is n^r when sampling with replacement.

Example 2.2.6

A die is rolled five times. The number of possible ordered samples is $6^5 = 7776$. Note that rolling a die is equivalent to sampling with replacement.   ◇

Example 2.2.7

An urn contains 10 balls numbered 0, 1, 2, ..., 9. If four balls are selected, one at a time and with replacement, the number of possible ordered samples is $10^4 = 10,000$. Note that this is the number of four-digit integers between 0000 and 9999, inclusive.   ◇

Definition 2.2.5.   Sampling without replacement occurs when an object is not replaced after it has been selected.

By the multiplication principle, the number of possible ordered samples of size r taken from a set of n objects, sampling without replacement, is

$$n(n-1)\cdots(n-r+1) = \frac{n!}{(n-r)!}$$

which is equivalent to $_nP_r$, the number of permutations of n objects taken r at a time.

Example 2.2.8

The number of ordered samples of five cards that can be drawn without replacement from a standard deck of 52 playing cards is

$$(52)(51)(50)(49)(48) = \frac{52!}{47!} = 311,875,200.\qquad ◇$$

Often the order of selection is not important and interest centers only on the selected set of r objects. That is, we are interested in the number of subsets of size r that can be selected from a set of n different objects. In order to find the number of (unordered) subsets of size r, we count, in two different ways, the number of ordered subsets of size r that can be taken from the n distinguishable objects. By equating these two answers, we are able to count the number of (unordered) subsets of size r.

Let C denote the number of (unordered) subsets of size r that can be selected from n different objects. We can obtain each of the $_nP_r$ ordered subsets by first selecting one of the C unordered subsets of r objects and then ordering these r objects. Since the latter can be carried out in $r!$ ways, the multiplication principle yields $(C)(r!)$ ordered subsets; so $(C)(r!)$ must equal $_nP_r$. Thus we have

$$(C)(r!) = \frac{n!}{(n-r)!}$$

or

$$C = \frac{n!}{r!\,(n-r)!}.$$

We denote this answer by either $_nC_r$ or $\binom{n}{r}$; that is,

$$_nC_r = \binom{n}{r} = \frac{n!}{r!\,(n-r)!}.$$

Accordingly, a set of n different objects possesses

$$\binom{n}{r} = \frac{n!}{r!\,(n-r)!}$$

unordered subsets of size $r \le n$.

We could also say that the number of ways in which r objects can be selected without replacement from n objects, when the order of selection is disregarded, is $\binom{n}{r} = {}_nC_r$, and this can be read as "n choose r." This motivates the following definition.

Definition 2.2.6.   Each of the $_nC_r$ unordered subsets is called a **combination of n objects taken r at a time**, where

$$_nC_r = \binom{n}{r} = \frac{n!}{r!\,(n-r)!}.$$

Example 2.2.9

The number of possible five-card hands (hands in five-card poker) drawn from a deck of 52 playing cards is

$$_{52}C_5 = \binom{52}{5} = \frac{52!}{5!\,47!} = 2{,}598{,}960. \qquad \diamond$$

Example 2.2.10

The number of possible 13-card hands (hands in bridge) that can be selected from a deck of 52 playing cards is

$$_{52}C_{13} = \binom{52}{13} = \frac{52!}{13!\,39!} = 635{,}013{,}559{,}600. \qquad \diamond$$

The numbers $\binom{n}{r}$ are frequently called **binomial coefficients**, since they arise in the expansion of a binomial. We illustrate this by giving a justification of the binomial expansion

$$(a + b)^n = \sum_{r=0}^{n} \binom{n}{r} b^r a^{n-r}. \qquad (2.2)$$

In the expansion of

$$(a + b)^n = (a + b)(a + b) \cdots (a + b),$$

either an a or a b is selected from each of the n factors. One possible product is then $b^r a^{n-r}$; this occurs when b is selected from each of r factors and a from each of the remaining $n - r$ factors. But the latter operation can be completed in $\binom{n}{r}$ ways, which then must be the coefficient of $b^r a^{n-r}$, as shown in Equation 2.2.

The binomial coefficients are given in Table I in the Appendix for selected values of n and r. Note that for some combinations of n and r, the table uses the fact that

$$\binom{n}{r} = \frac{n!}{r! \, (n - r)!} = \frac{n!}{(n - r)! \, r!} = \binom{n}{n - r}.$$

That is, the number of ways in which r objects can be selected out of n objects is equal to the number of ways in which $n - r$ objects can be selected out of n objects.

Example 2.2.11

Assume that each of the $\binom{52}{5} = 2,598,960$ five-card hands drawn from a deck of 52 playing cards has the same probability for being selected. The number of possible five-card hands that are all spades (event A) is

$$N(A) = \binom{13}{5}\binom{39}{0}$$

because the 5 spades can be selected from the 13 spades in $\binom{13}{5}$ ways, after which zero nonspades can be selected in $\binom{39}{0} = 1$ way. We have

$$\binom{13}{5} = \frac{13!}{5! \, 8!} = 1287$$

from Table I. Thus the probability of an all-spade five-card hand is

$$P(A) = \frac{N(A)}{N(S)} = \frac{1287}{2,598,960} = 0.000495.$$

Suppose now that the event B is the set of outcomes in which exactly three cards are kings and exactly two cards are queens. We can select the three kings in any one of $\binom{4}{3}$ ways and the two queens in any one of $\binom{4}{2}$ ways. By the multiplication principle, the number of outcomes in B is

$$N(B) = \binom{4}{3}\binom{4}{2}\binom{44}{0},$$

where $\binom{44}{0}$ gives the number of ways in which 0 cards are selected out of the non-kings and non-queens and of course is equal to 1. Thus

$$P(B) = \frac{N(B)}{N(S)} = \frac{\binom{4}{3}\binom{4}{2}\binom{44}{0}}{\binom{52}{5}} = \frac{24}{2,598,960} = 0.0000092.$$

Finally, let C be the set of outcomes in which there are exactly two kings, two queens, and one jack. Then

$$P(C) = \frac{N(C)}{N(S)} = \frac{\binom{4}{2}\binom{4}{2}\binom{4}{1}\binom{40}{0}}{\binom{52}{5}} = \frac{144}{2,598,960} = 0.000055$$

because the numerator of this fraction is the number of outcomes in C.   ◇

Now suppose that a set contains n objects of two types, r of one type and $n - r$ of the other type. The number of permutations of n different objects is $n!$. However, in this case, the objects are not all distinguishable. To count the number of distinguishable arrangements, first select r out of the n positions for the objects of the first type. This can be done in $\binom{n}{r}$ ways. Then fill in the remaining positions with the objects of the second type. Thus the number of distinguishable arrangements is

$$_nC_r = \binom{n}{r} = \frac{n!}{r!\,(n-r)!}.$$

Definition 2.2.7.   Each of the $_nC_r$ permutations of n objects, r of one type and $n-r$ of another type, is called a **distinguishable permutation**.

Example 2.2.12

A coin is flipped 10 times and the sequence of heads and tails is observed. The number of possible 10-tuplets that result in four heads and six tails is

$$\binom{10}{4} = \frac{10!}{4!\,6!} = \frac{10!}{6!\,4!} = \binom{10}{6} = 210.$$ ◇

Example 2.2.13

Students on a boat send signals back to shore by arranging seven colored flags on a vertical flagpole. If they have four orange and three blue flags, they can send

$$\binom{7}{4} = \frac{7!}{4!\,3!} = 35$$

different signals. Note that if they had seven flags of different colors, they could send $7! = 5040$ different signals. ◇

The foregoing results can be extended. Suppose that in a set of n objects, n_1 are similar, n_2 are similar, ..., n_s are similar, where $n_1 + n_2 + \cdots + n_s = n$. The number of distinguishable permutations of the n objects is (see Exercise 2.2.16)

$$\binom{n}{n_1, n_2, \ldots, n_s} = \frac{n!}{n_1!\,n_2! \cdots n_s!}. \tag{2.3}$$

Example 2.2.14

If the students on the boat have three red flags, four yellow flags, and two blue flags to arrange on a vertical pole, the number of possible signals is

$$\binom{9}{3, 4, 2} = \frac{9!}{3!\,4!\,2!} = 1260.$$ ◇

The argument used in determining the binomial coefficients in the expansion of $(a + b)^n$ can be extended to find the expansion of $(a_1 + a_2 + \cdots + a_s)^n$. The coefficient of $a_1^{n_1} a_2^{n_2} \cdots a_s^{n_s}$ is

$$\binom{n}{n_1, n_2, \ldots, n_s} = \frac{n!}{n_1!\,n_2! \cdots n_s!}.$$

This is sometimes called a **multinomial coefficient**.

When r objects are selected out of n objects, we are often interested in the number of possible outcomes. We have seen that for ordered samples, there are n^r possible outcomes when sampling with replacement and $_nP_r$ outcomes when sampling without replacement. For unordered samples, there are $_nC_r$ outcomes when sampling without replacement. Each of the outcomes above is equally likely provided that the experiment is performed in a fair manner.

Although not needed in the study of probability it is interesting to count the number of possible samples of size r that can be selected out of n objects when the order is irrelevant and when sampling with replacement. For example, if a six-sided die is rolled 10 times (or 10 six-sided dice are rolled once), how many possible unordered outcomes are there? To count the number of possible outcomes, think of listing r 0's for the r objects that are to be selected. Then insert $(n - 1)$ |'s to partition the r objects into n sets, the first set giving objects of the first kind, and so on. So if $n = 6$ and $r = 10$ in the die illustration, a possible outcome is

$$0\ 0\ |\ |\ 0\ 0\ 0\ |\ 0\ |\ 0\ 0\ 0\ |\ 0,$$

which says there are two 1's, zero 2's, three 3's, one 4, three 5's, and one 6. In general, each outcome is a permutation of r 0's and $(n-1)$ |'s. Each distinguishable permutation is equivalent to an unordered sample. The number of distinguishable permutations, and hence the number of unordered samples of size r that can be selected out of n objects when sampling with replacement, is

$$_{n-1+r}C_r = \frac{(n-1+r)!}{r!\,(n-1)!}.$$

Example 2.2.15

Roll a pair of six-sided dice. The number of distinguishable outcomes is

$$\frac{(6-1+2)!}{2!\,(6-1)!} = \frac{7!}{2!\,5!} = 21.$$

You should be able to list these 21 outcomes. Note that these outcomes are not equally likely under the usual probabilistic assumptions about dice.   ◇

The number of samples of size r that can be selected out of n objects is summarized in Table 2.1.

TABLE 2.1   Number of samples of r out of n objects

	With Replacement	Without Replacement
Ordered	n^r	$\dfrac{n!}{(n-r)!}$
Unordered	$\dfrac{(n-1+r)!}{(n-1)!\,r!}$	$\dfrac{n!}{r!\,(n-r)!}$

EXERCISES

2.2.1 A boy found a bicycle lock for which the combination was unknown. The correct combination is a four-digit number, $d_1d_2d_3d_4$ where d_i, $i = 1, 2, 3, 4$, is selected from 1, 2, 3, 4, 5, 6, 7, 8. How many different lock combinations are possible with such a lock?

2.2.2 How many different signals can be made using four flags of different colors on a vertical flagpole if exactly three flags are used for each signal?

2.2.3 How many different license plates are possible if a state uses

(a) Two letters followed by a four-digit integer (leading zeros permissible and the letters and digits can be repeated)?

(b) Three letters followed by a three-digit integer?

2.2.4 From a collection of nine paintings, four are to be selected to hang side by side on a gallery wall in positions 1, 2, 3, and 4. In how many ways can this be done?

2.2.5 How many four-letter code words are possible using the letters in HOPE if

  (a) The letters may not be repeated?

  (b) The letters may be repeated?

2.2.6 Some albatrosses return to the world's only mainland colony of royal albatrosses on Otago Peninsula near Dunedin, New Zealand, every two years to nest and raise their young. In order to learn more about the albatross, colored plastic bands are placed on their legs so that they can be identified from a distance. Suppose that three bands are placed on one leg, selecting from the colors red, yellow, green, white, and blue. Find the number of different color codes that are possible for banding an albatross if

  (a) The three bands are different colors.

  (b) Repeated colors are permissible.

2.2.7 In a state lottery four digits are drawn at random one at a time with replacement from 0 to 9. Suppose that you win if any permutation of your selected integers is drawn. Give the probability of winning if you select

  (a) 6, 7, 8, 9.

  (b) 6, 7, 8, 8.

  (c) 7, 7, 8, 8.

  (d) 7, 8, 8, 8.

2.2.8 Suppose that Chicago and Toronto are playing in a preliminary round of the Stanley Cup hockey playoffs in which they play until one team wins three games. That is, it is a best-of-five series that ends when one team has three victories. Considering the possible orderings for the winning team, in how many ways could this series end?

2.2.9 The World Series in baseball continues until either the American League team or the National League team wins four games. How many different orders are possible if the series goes

  (a) Four games?

  (b) Five games?

  (c) Six games?

  (d) Seven games?

2.2.10 How many different varieties of pizza can be made if you have the following choices: size—small, medium, or large; crust—thin 'n crispy, hand-tossed, or pan; toppings—(cheese is automatic) there are 12 toppings from which you may select from 0 to 12?

2.2.11 A cafe lets you order a deli sandwich your way. There are 6 choices for bread, 4 choices for meat, 4 choices for cheese, and 12 different garnishes (condiments). How many different sandwich possibilities are there if you choose

  (a) One bread, one meat, and one cheese?

  (b) One bread, one meat, one cheese, and from 0 to 12 garnishes?

  (c) One bread; 0, 1, or 2 meats; 0, 1, or 2 cheeses; and from 0 to 12 garnishes?

2.2.12 Pascal's triangle gives a method for calculating the binomial coefficients; it begins as follows:

$$
\begin{array}{ccccccccc}
 & & & & 1 & & & & \\
 & & & 1 & & 1 & & & \\
 & & 1 & & 2 & & 1 & & \\
 & 1 & & 3 & & 3 & & 1 & \\
1 & & 4 & & 6 & & 4 & & 1 \\
\end{array}
$$

1   5   10   10   5   1

⋮   ⋮   ⋮   ⋮   ⋮

The nth row of this triangle gives the coefficients for $(a+b)^{n-1}$. To find an entry in the table other than a 1 on the boundary, add the two nearest numbers in the row directly above. The equation

$$\binom{n}{r} = \binom{n-1}{r} + \binom{n-1}{r-1},$$

called **Pascal's equation**, explains why Pascal's triangle works. Prove that this equation is correct.

2.2.13 Among nine presidential candidates at a debate, three are Republicans and six are Democrats.
  (a) In how many different ways can the nine candidates be lined up?
  (b) How many lineups by party are possible if each candidate is labeled either R or D?
  (c) For each of the nine candidates, you are to decide whether the candidate did a good job or a poor job; that is, give each of the nine candidates a grade of G or P. How many different "score cards" are possible?

2.2.14 Prove:

$$\sum_{r=0}^{n}(-1)^r\binom{n}{r} = 0 \quad \text{and} \quad \sum_{r=0}^{n}\binom{n}{r} = 2^n.$$

HINT: Consider $(1-1)^n$ and $(1+1)^n$ or use Pascal's equation and proof by induction.

2.2.15 A poker hand is defined as drawing five cards at random without replacement from a deck of 52 playing cards. Find the probability of each of the following poker hands:

  (a) Four of a kind (four cards of equal face value and one card of a different value).
  (b) Full house (one pair and one triple of cards with equal face value).
  (c) Three of a kind (three equal face values plus two cards of different values).
  (d) Two pairs (two pairs of equal face value plus one card of a different value).
  (e) One pair (one pair of equal face value plus three cards of different values).

2.2.16 Prove Equation 2.3. HINT: First select n_1 positions in $\binom{n}{n_1}$ ways, after which select n_2 from the remaining $n - n_1$ positions in $\binom{n-n_1}{n_2}$ ways, and so on. Finally, use the multiplication rule.

2.2.17 There are three teams in a cross-country race. Team A has 5 runners, team B has 6 runners, and team C has 7 runners. In how many ways can the runners cross the finish line if we are only interested in the team for which they run? That is, what is the number of distinguishable permutations of 5 A's, 6 B's, and 7 C's? (Note that for scoring purposes, only the scores of the first 5 runners for each team count.)

2.2.18 A box of candy hearts contains 52 hearts of which 19 are white, 10 are tan, 7 are pink, 3 are purple, 5 are yellow, 2 are orange, and 6 are green. If you select 9 pieces of candy randomly from the box, without replacement, give the probability that

    (a) Three of the hearts are white.

    (b) Three are white, 2 are tan, 1 is pink, 1 is yellow, and 2 are green.

2.2.19 An office furniture manufacturer makes modular storage files. It offers its customers two choices for the base and four choices for the top, and the modular storage files come in five different heights. The customer may choose any combination of the five different-sized modules so that the finished file has a base, a top, and 1, 2, 3, 4, 5, or 6 storage modules.

    (a) How many choices does the customer have if the completed file has four storage modules, a top, and a base? The order in which the four storage modules are stacked is irrelevant.

    (b) The manufacturer would like to use in its advertising the number of different files that are possible—selecting one of the two bases, one of the four tops, and then either 1 or 2 or 3 or 4 or 5 or 6 storage modules, selecting any combination of the five different sizes with the order of stacking irrelevant. What is the number of possibilities?

2.2.20 The "eating club" is hosting a make-your-own sundae at which the following are provided:

Ice Cream Flavors	Toppings
Chocolate	Caramel
Cookies-n-cream	Hot fudge
Strawberry	Marshmallow
Vanilla	m&m's
	Nuts
	Strawberries

    (a) How many sundaes are possible using 1 flavor of ice cream and 3 different toppings?

    (b) How many sundaes are possible using 1 flavor of ice cream and from 0 to 6 toppings?

    (c) How many different combinations of flavors of 3 scoops of ice cream are possible if it is permissible to make all 3 scoops the same flavor?

2.3 CONDITIONAL PROBABILITY

We introduce the idea of conditional probability by means of an example.

Example 2.3.1

Suppose that we are given 20 tulip bulbs that are very similar in appearance and told that 8 tulips will bloom early, 12 will bloom late, 13 will be red, and 7 will be yellow, in accordance with the following combinations:

	Early (E)	Late (L)	Totals
Red (R)	5	8	13
Yellow (Y)	3	4	7
	8	12	20

If one bulb is selected at random, the probability that it will produce a red tulip (R) is given by $P(R) = 13/20$, under the assumption that each bulb is "equally likely." Suppose, however, that close examination of the bulb will reveal whether it will bloom early (E) or late (L). If we consider an outcome only if it results in a tulip bulb that will bloom early, only eight outcomes in the sample space are now of interest. Thus it is natural to assign, under this limitation, the probability of 5/8 to R; that is, $P(R \mid E) = 5/8$, where $P(R \mid E)$ is read as the probability of R given that E has occurred. Note that

$$P(R \mid E) = \frac{5}{8} = \frac{N(R \cap E)}{N(E)} = \frac{N(R \cap E)/20}{N(E)/20} = \frac{P(R \cap E)}{P(E)},$$

where $N(R \cap E)$ and $N(E)$ are the numbers of outcomes in events $R \cap E$ and E, respectively.   ◇

Example 2.3.1 is illustrative of a number of common situations. That is, in some random experiments, we are interested only in those outcomes that are elements of a subset B of the sample space S. This means, for our purposes, that the sample space is effectively the subset B. We are now confronted with the problem of defining a probability set function with B as the "new" sample space. That is, for a given event A we want to define $P(A \mid B)$, the probability of A considering only those outcomes of the random experiment that are elements of B. The preceding example gives us the clue to that definition. That is, for experiments in which each outcome is equally likely, it makes sense to define $P(A \mid B)$ by

$$P(A \mid B) = \frac{N(A \cap B)}{N(B)},$$

where $N(A \cap B)$ and $N(B)$ are the numbers of outcomes in $A \cap B$ and B, respectively. If we divide the numerator and the denominator of this fraction by $N(S)$, the number of outcomes in the sample space, we have

$$P(A \mid B) = \frac{N(A \cap B)/N(S)}{N(B)/N(S)} = \frac{P(A \cap B)}{P(B)}.$$

We are thus led to the following definition.

Definition 2.3.1. The **conditional probability** of an event A given that event B has occurred is defined by

$$P(A \mid B) = \frac{P(A \cap B)}{P(B)},$$

provided that $P(B) > 0$.

A formal use of the definition is given in the following example.

Example 2.3.2

If $P(A) = 0.4$, $P(B) = 0.5$, and $P(A \cap B) = 0.3$, then $P(A \mid B) = 0.3/0.5 = 0.6$ and $P(B \mid A) = P(A \cap B)/P(A) = 0.3/0.4 = 0.75$.   ◇

We can think of the "given B" as specifying the new sample space for which we now want to calculate the probability of that part of A that is contained in B to determine $P(A \mid B)$. The following two examples illustrate this idea.

Example 2.3.3

Suppose that $P(A) = 0.7$, $P(B) = 0.3$, and $P(A \cap B) = 0.2$. These probabilities are listed on the Venn diagram in Figure 2.4. Given that the outcome of the experiment belongs to B, what is the probability of A? We are effectively restricting the sample space to B; of the probability $P(B) = 0.3$, 0.2 corresponds to $P(A \cap B)$ and hence to A. That is, 0.2/0.3 = 2/3 of the probability of B corresponds to A. Of course, formally by definition, we also obtain

$$P(A \mid B) = \frac{P(A \cap B)}{P(B)} = \frac{0.2}{0.3} = \frac{2}{3}.$$ ◇

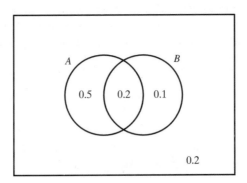

Figure 2.4   Conditional probability

Example 2.3.4

A pair of four-sided dice is rolled and the sum is determined. Let A be the event that a sum of 3 is rolled and let B be the event that a sum of 3 or a sum of 5 is rolled. In a sequence of rolls the probability that a sum of 3 is rolled before a sum of 5 is rolled can be thought of as the conditional probability of a sum of 3 given that a sum of 3 or 5 has occurred; that is, the conditional probability of A given B,

$$P(A \mid B) = \frac{P(A \cap B)}{P(B)} = \frac{P(A)}{P(B)} = \frac{2/16}{6/16} = \frac{2}{6}.$$

Note that for this example, the only outcomes of interest are those having a sum of 3 or a sum of 5, and of these six equally likely outcomes, two have a sum of 3 (see Figure 2.5; see also Exercise 2.3.13). $\diamond$

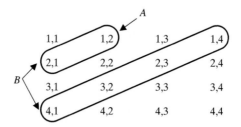

Figure 2.5   Dice example

It is interesting to note that conditional probability satisfies the axioms for a probability function, namely, with $P(B) > 0$,

(a) $P(A \mid B) \geq 0$.

(b) $P(B \mid B) = 1$.

(c) If $A_1, A_2, A_3, \ldots$ are mutually exclusive events, then

$$P(A_1 \cup A_2 \cup \cdots \cup A_k \mid B) = P(A_1 \mid B) + P(A_2 \mid B) + \cdots + P(A_k \mid B),$$

for each positive integer k, and

$$P(A_1 \cup A_2 \cup \cdots \mid B) = P(A_1 \mid B) + P(A_2 \mid B) + \cdots$$

for an infinite, but countable, number of events.

Properties (a) and (b) are evident because

$$P(A \mid B) = \frac{P(A \cap B)}{P(B)} \geq 0,$$

since $P(A \cap B) \geq 0$ and $P(B) > 0$, and

$$P(B \mid B) = \frac{P(B \cap B)}{P(B)} = \frac{P(B)}{P(B)} = 1.$$

Property (c) holds because, for the second part of property (c),

$$P(A_1 \cup A_2 \cup \cdots \mid B) = \frac{P[(A_1 \cup A_2 \cup \cdots) \cap B]}{P(B)}$$

$$= \frac{P[(A_1 \cap B) \cup (A_2 \cap B) \cup \cdots]}{P(B)}.$$

But $(A_1 \cap B), (A_2 \cap B), \ldots$ are also mutually exclusive events; so

$$
\begin{aligned}
P(A_1 \cup A_2 \cup \cdots \mid B) &= \frac{P(A_1 \cap B) + (A_2 \cap B) + \cdots}{P(B)} \\
&= \frac{P(A_1 \cap B)}{P(B)} + \frac{P(A_2 \cap B)}{P(B)} + \cdots \\
&= P(A_1 \mid B) + P(A_2 \mid B) + \cdots.
\end{aligned}
$$

The first part of property (c) is proved in a similar manner.

Many times the conditional probability of an event is clear because of the nature of an experiment. The following example illustrates this.

Example 2.3.5

At a fair a vendor has 25 helium balloons on strings: 10 balloons are yellow, 8 are red, and 7 are green. A balloon is selected at random and sold. Given that the balloon sold is yellow, what is the probability that the next balloon selected at random is also yellow? Of the 24 remaining balloons, 9 are yellow; so a natural value to assign to this conditional probability is 9/24.   ◇

In Example 2.3.5, let A be the event that the first balloon selected is yellow, and let B be the event that the second balloon selected is yellow. Suppose that we are interested in the probability that both selected balloons are yellow. That is, we are interested in finding $P(A \cap B)$. We noted in Example 2.3.5 that

$$
P(B \mid A) = \frac{P(A \cap B)}{P(A)} = \frac{9}{24}.
$$

Thus, multiplying through by $P(A)$, we have

$$
P(A \cap B) = P(A)P(B \mid A) = P(A)\left(\frac{9}{24}\right) \tag{2.4}
$$

or

$$
P(A \cap B) = \left(\frac{10}{25}\right)\left(\frac{9}{24}\right).
$$

That is, Equation 2.4 gives us a general rule for the probability of the intersection of two events once we know the conditional probability $P(B \mid A)$.

Definition 2.3.2. The probability that two events, A and B, both occur is given by the **multiplication rule**,

$$
P(A \cap B) = P(A)P(B \mid A)
$$

or by

$$
P(A \cap B) = P(B)P(A \mid B).
$$

Sometimes, after considering the nature of the random experiment, one can make reasonable assumptions so that it is easier to assign $P(B)$ and $P(A \mid B)$ rather than $P(A \cap B)$. Then $P(A \cap B)$ can be computed with these assignments. This will be illustrated in Examples 2.3.6 and 2.3.7.

Example 2.3.6

A bowl contains seven blue chips and three red chips. Two chips are to be drawn successively at random and without replacement. We want to compute the probability that the first draw results in a red chip (A) and the second draw results in a blue chip (B). It is reasonable to assign the following probabilities:

$$P(A) = \frac{3}{10} \quad \text{and} \quad P(B \mid A) = \frac{7}{9}.$$

The probability of red on the first draw and blue on the second draw is

$$P(A \cap B) = \frac{3}{10} \cdot \frac{7}{9} = \frac{7}{30}. \qquad \diamond$$

It should be noted that in many instances, it is possible to compute a probability by two seemingly different methods. For illustration, consider Example 2.3.6 but find the probability of drawing a red chip on each of the two draws. Following that example, it is

$$\frac{3}{10} \cdot \frac{2}{9} = \frac{1}{15}.$$

However, we can also find this probability using combinations as follows:

$$\frac{\binom{3}{2}\binom{7}{0}}{\binom{10}{2}} = \frac{\frac{(3)(2)}{(1)(2)}}{\frac{(10)(9)}{(1)(2)}} = \frac{1}{15}.$$

Thus we obtain the same answer, as we should, provided that our reasoning is consistent with the underlying assumptions.

Example 2.3.7

From an ordinary deck of playing cards, cards are to be drawn successively at random and without replacement. The probability that the third spade appears on the sixth draw is computed as follows: Let A be the event of two spades in the first five cards drawn, and let B be the event of a spade on the sixth draw. Thus the probability that we wish to compute is $P(A \cap B)$. It is reasonable to take

$$P(A) = \frac{\binom{13}{2}\binom{39}{3}}{\binom{52}{5}} = 0.274 \quad \text{and} \quad P(B \mid A) = \frac{11}{47} = 0.234.$$

The desired probability $P(A \cap B)$ is the product of these numbers,

$$P(A \cap B) = (0.274)(0.234) = 0.064.$$ ◇

Example 2.3.8

Continuing with Example 2.3.4 in which a pair of four-sided dice is rolled, the probability for rolling a sum of 3 on the first roll and then, continuing the sequence of rolls, rolling a sum of 3 before rolling a sum of 5 is

$$\frac{2}{16} \cdot \frac{2}{6} = \frac{4}{96} = \frac{1}{24}.$$ ◇

The multiplication rule can be extended to three or more events. In the case of three events we have, by using the multiplication rule for two events,

$$P(A \cap B \cap C) = P[(A \cap B) \cap C]$$
$$= P(A \cap B)P(C \mid A \cap B).$$

But

$$P(A \cap B) = P(A)P(B \mid A).$$

Hence

$$P(A \cap B \cap C) = P(A)P(B \mid A)P(C \mid A \cap B).$$

This type of argument can be used to extend the multiplication rule to more than three events, and the general formula for k events can be officially proved by mathematical induction.

Example 2.3.9

Four cards are to be dealt successively at random and without replacement from an ordinary deck of playing cards. The probability of receiving in order a spade, a heart, a diamond, and a club is

$$\frac{13}{52} \cdot \frac{13}{51} \cdot \frac{13}{50} \cdot \frac{13}{49},$$

a result that follows from the extension of the multiplication rule and reasonable assignments to the probabilities involved. ◇

We close this section with a different type of example.

Example 2.3.10

A grade-school boy has five blue and four white marbles in his left pocket and four blue and five white marbles in his right pocket. If he transfers one marble at random from his left to his right pocket, what is the probability of his then drawing a blue marble from his right pocket? For notation let BL, BR, and WL denote drawing blue from left pocket, blue from right pocket, and white from left pocket, respectively. Then

$$
\begin{aligned}
P(BR) &= P(BL \cap BR) + P(WL \cap BR) \\
&= P(BL)P(BR \mid BL) + P(WL)P(BR \mid WL) \\
&= \frac{5}{9} \cdot \frac{5}{10} + \frac{4}{9} \cdot \frac{4}{10} = \frac{41}{90}
\end{aligned}
$$

is the desired probability. $\diamond$

EXERCISES

2.3.1 A common test for AIDS is called the ELISA test. Among 1,000,000 people who are given the ELISA test, we can expect results similar to those given in the following table:

	Carry AIDS Virus (B_1)	Do Not Carry Aids Virus (B_2)	Total
Test positive (A_1)	4,885	73,630	78,515
Test negative (A_2)	115	921,370	921,485
	5,000	995,000	1,000,000

If one of these 1,000,000 people is selected randomly, find the following probabilities: (a) $P(B_1)$; (b) $P(A_1)$; (c) $P(A_1 \mid B_2)$; (d) $P(B_1 \mid A_1)$. (e) In words, what do parts (c) and (d) say?

2.3.2 The following table classifies 1456 people by their sex and by whether or not they favor a gun law:

	Male (S_1)	Female (S_2)	Total
Favor (A_1)	392	649	1041
Oppose (A_2)	241	174	415
	633	823	1456

Compute the following probabilities if one of these 1456 persons is selected randomly: (a) $P(A_1)$; (b) $P(A_1 \mid S_1)$; (c) $P(A_1 \mid S_2)$. (d) Interpret your answers to parts (b) and (c).

2.3.3 Let A_1 and A_2 be the events that a person is left eye dominant or right eye dominant, respectively. When a person folds their hands, let B_1 and B_2 be the events that their left thumb and right thumb, respectively, are on top. A survey in one statistics class yielded the following table:

	B_1	B_2	Total
A_1	5	7	12
A_2	14	9	23
	19	16	35

If a student is selected randomly, find the probabilities: (a) $P(A_1 \cap B_1)$; (b) $P(A_1 \cup B_1)$; (c) $P(A_1 \mid B_1)$; (d) $P(B_2 \mid A_2)$. (e) If the students had their hands folded and you hoped to select a right-eye-dominant student, would you select a "right thumb on top" or a "left thumb on top" student? Why?

2.3.4 Two cards are drawn successively and without replacement from an ordinary deck of playing cards. Compute the probability of drawing

(a) Two hearts.

(b) A heart on the first draw, a club on the second draw.

(c) A heart on the first draw, an ace on the second draw. HINT: Note that a heart can be drawn by getting the ace of hearts or one of the other 12 hearts.

2.3.5 Suppose that $P(A) = 0.7$, $P(B) = 0.5$, and $P([A \cup B]') = 0.1$.

(a) Find $P(A \cap B)$.

(b) Give $P(A \mid B)$.

(c) Give $P(B \mid A)$.

2.3.6 A hand of 13 cards is to be dealt at random and without replacement from an ordinary deck of playing cards. Given that the hand contains at least two kings, find the conditional probability that there are at least three kings in the hand.

2.3.7 Suppose that the genes for eye color for a certain male fruit fly are (R, W) and the genes for eye color for the mating female fruit fly are (R, W), where R and W represent red and white, respectively. Their offspring receive one gene for eye color from each parent.

(a) Define the sample space for the genes for eye color for the offspring.

(b) Assume that each of the four possible outcomes has equal probability. If an offspring ends up with either two red genes or one red and one white gene for eye color, its eyes will look red. Given that an offspring's eyes look red, what is the conditional probability that it has two red genes for eye color?

2.3.8 Suppose that there are 14 songs on a compact disk (CD) and you like 8 of them. When using the random button selector on a CD player, each of the 14 songs is played once in a random order. Find the probability that among the first 2 songs that are played,

(a) You like both of them.

(b) You like neither of them.

(c) You like exactly one of them.

2.3.9 Suppose that there are 12 songs on a compact disk (CD) of which two are your favorites. When using the random button selector on a CD player, each of the 12 selections is played once in a random order. Find the probability that the second of your two favorites (one has already been played) is

(a) The third song that is played.

(b) The sixth song that is played.

2.3.10 An urn contains 17 balls marked LOSE and 3 balls marked WIN. You and an opponent take turns selecting at random a single ball from the urn without replacement. The person who selects the third WIN ball wins the game. It does not matter who selected the first two WIN balls.

(a) If you draw first, find the probability that you win the game on your fourth draw.

(b) If you draw first, what is the probability that you win?

HINT: You could win on your second, third, fourth, . . . , or tenth draw, not on your first.

2.3.11 In a string of 12 Christmas tree light bulbs, 3 are defective. The bulbs are selected at random and tested, one at a time, until the third defective bulb is found. Compute the probability that the third defective bulb is the

(a) Third bulb tested.

(b) Fifth bulb tested.

(c) Tenth bulb tested.

2.3.12 A small grocery store had l0 cartons of milk, 2 of which were sour. If you are going to buy the sixth carton of milk sold that day at random, compute the probability of selecting a carton of sour milk.

2.3.13 In the gambling game Craps, a pair of dice is rolled and the outcome of the experiment is the sum of the points on the upsides of the six-sided dice. The bettor wins on the first roll if the sum is 7 or 11. The bettor loses on the first roll if the sum is 2, 3, or 12. If the sum is 4, 5, 6, 8, 9, or l0, that number is called the bettor's "point." Once the point is established, the rule is: If the bettor rolls a 7 before the "point," the bettor loses; but if the "point" is rolled before a 7, the bettor wins.

(a) List the 36 outcomes in the sample space for the roll of a pair of dice. Assume that each of them has a probability of 1/36.

(b) Find the probability that the bettor wins on the first roll. That is, find the probability of rolling a 7 or 11, $P(7 \text{ or } 11)$.

(c) Given that 8 is the outcome on the first roll, find the probability that the bettor now rolls the point 8 before rolling a 7 and thus wins. Note that at this stage in the game the only outcomes of interest are 7 and 8. Thus find $P(8 \mid 7 \text{ or } 8)$.

(d) The probability that a bettor rolls an 8 on the first roll and then wins is given by $P(8)P(8 \mid 7 \text{ or } 8)$. Show that this probability is (5/36)(5/11).

(e) Show that the total probability that a bettor wins in the game of Craps is 0.49293.

HINT: Note that the bettor can win in one of several mutually exclusive ways: by rolling a 7 or 11 on the first roll or by establishing one of the points 4, 5, 6, 8, 9, or 10 on the first roll and then obtaining that point before a 7 on successive rolls.

2.3.14 A single card is drawn at random from each of six well-shuffled decks of playing cards. Let A be the event that all six cards drawn are different.

(a) Find $P(A)$.

(b) Find the probability that at least two of the drawn cards match.

2.3.15 Consider the birthdays of the students in a class of size r. Assume that the year consists of 365 days.

(a) How many different ordered samples of birthdays are possible (r in sample) allowing repetitions (with replacement)?

(b) Repeat part (a) except require that all the students have different birthdays (without replacement).

(c) If we can assume that each ordered outcome in part (a) has the same probability, what is the probability that at least two students have the same birthday?

(d) For what value of r is the probability in part (c) about equal to 1/2? Is this number surprisingly small? HINT: Use a calculator or computer to find r.

2.3.16 You are a member of a class of 18 students. A bowl contains 18 chips, 1 blue and 17 red. Each student is to take 1 chip from the bowl without replacement. The student who draws the blue chip is guaranteed an A for the course.

(a) If you have a choice of drawing first, fifth, or last, which position would you choose? Justify your choice using probability.

(b) Suppose that the bowl contains 2 blue and 16 red chips. What position would you now choose?

2.3.17 A drawer contains four black, six brown, and eight olive socks. Two socks are selected at random from the drawer.

(a) Compute the probability that both socks are the same color.

(b) Compute the probability that both socks are olive if it is known that they are the same color.

2.3.18 Bowl A contains three red and two white chips, and bowl B contains four red and three white chips. A chip is drawn at random from bowl A and transferred to bowl B. Compute the probability of then drawing a red chip from bowl B.

2.3.19 An urn contains four balls numbered 1 through 4. The balls are selected one at a time without replacement. A match occurs if ball numbered m is the mth ball selected. Let the event A_i denote a match on the ith draw, $i = 1, 2, 3, 4$. Show that

(a) $P(A_i) = \dfrac{3!}{4!}$.

(b) $P(A_i \cap A_j) = \dfrac{2!}{4!}$.

(c) $P(A_i \cap A_j \cap A_k) = \dfrac{1!}{4!}$.

(d) The probability of at least one match is

$$P(A_1 \cup A_2 \cup A_3 \cup A_4) = 1 - \frac{1}{2!} + \frac{1}{3!} - \frac{1}{4!}.$$

(e) Extend this exercise so that there are n balls in the urn. Show that the probability of at least one match is

$$P(A_1 \cup A_2 \cup \cdots \cup A_n) = 1 - \frac{1}{2!} + \frac{1}{3!} - \frac{1}{4!} + \cdots + \frac{(-1)^{n+1}}{n!}$$
$$= 1 - \left(1 - \frac{1}{1!} + \frac{1}{2!} - \frac{1}{3!} + \cdots + \frac{(-1)^n}{n!}\right).$$

(f) What is the limit of this probability as n increases without bound?

2.4 INDEPENDENT EVENTS

For certain pairs of events, the occurrence of one of them may or may not change the probability of the occurrence of the other. In the latter case they are said to be **independent events**. However, before giving the formal definition of independence, let us consider an example.

Example 2.4.1

Flip a coin twice and observe the sequence of heads and tails. The sample space is then

$$S = \{HH, HT, TH, TT\}.$$

It is reasonable to assign a probability of 1/4 to each of these four outcomes. Let

$$A = \{\text{heads on the first flip}\} = \{HH, HT\},$$
$$B = \{\text{tails on the second flip}\} = \{HT, TT\},$$
$$C = \{\text{tails on both flips}\} = \{TT\}.$$

Now $P(B) = 2/4 = 1/2$. However, if we are given that C has occurred, then $P(B \mid C) = 1$ because $C \subset B$. That is, the knowledge of the occurrence of C has changed the probability of B. On the other hand, if we are given that A has occurred,

$$P(B \mid A) = \frac{P(A \cap B)}{P(A)} = \frac{1/4}{2/4} = \frac{1}{2} = P(B).$$

So the occurrence of A has not changed the probability of B. Hence the probability of B does not depend upon knowledge about event A, so we say that A and B are independent events. That is, events A and B are independent if the occurrence of one of them does not affect the probability of the occurrence of the other. A more mathematical way of saying this is

$$P(B \mid A) = P(B) \quad \text{or} \quad P(A \mid B) = P(A),$$

provided that $P(A) > 0$ or, in the latter case, $P(B) > 0$. With the first of these equalities and the multiplication rule (Definition 2.3.2), we have

$$P(A \cap B) = P(A)P(B \mid A) = P(A)P(B).$$

The second of these equalities, namely $P(A \mid B) = P(A)$, gives us the same result:

$$P(A \cap B) = P(B)P(A \mid B) = P(B)P(A). \qquad \diamond$$

This example motivates the following definition of independent events.

Definition 2.4.1. Events A and B are **independent** if and only if $P(A \cap B) = P(A)P(B)$. Otherwise, A and B are called **dependent** events.

Events that are independent are sometimes called **statistically independent**, **stochastically independent**, or **independent in a probability sense**, but in most instances we use independent without a modifier if there is no possibility of misunderstanding. It is interesting to note that the definition always holds if $P(A) = 0$ or $P(B) = 0$ because then $P(A \cap B) = 0$, since $(A \cap B) \subset A$ and $(A \cap B) \subset B$. Thus the left- and right-hand members of $P(A \cap B) = P(A)P(B)$ are both equal to zero and thus are equal to each other.

Example 2.4.2

A red die and a white die are rolled. Let event $A = \{4$ on the red die$\}$ and event $B = \{$sum of dice is odd$\}$. Of the 36 equally likely outcomes, 6 are favorable to A, 18 are favorable to B, and 3 are favorable to $A \cap B$. Thus

$$P(A)P(B) = \frac{6}{36} \cdot \frac{18}{36} = \frac{3}{36} = P(A \cap B).$$

Hence A and B are independent by Definition 2.4.1. $\qquad \diamond$

Example 2.4.3

A red die and a white die are rolled. Let event $C = \{5$ on red die$\}$ and event $D = \{$sum of dice is 11$\}$. Of the 36 equally likely outcomes, 6 are favorable to C, 2 are favorable to D, and 1 is favorable to $C \cap D$. Thus

$$P(C)P(D) = \frac{6}{36} \cdot \frac{2}{36} = \frac{1}{108} \neq \frac{1}{36} = P(C \cap D).$$

Hence C and D are dependent events by Definition 2.4.1. $\qquad \diamond$

Theorem 2.4.1. If A and B are independent events, then the following pairs of events are also independent:

(a) A and B'.

(b) A' and B.

(c) A' and B'.

Proof.   We know that conditional probability satisfies the axioms for a probability function. Hence, if $P(A) > 0$, then $P(B' \,|\, A) = 1 - P(B \,|\, A)$. Thus

$$P(A \cap B') = P(A)P(B' \,|\, A) = P(A)[1 - P(B \,|\, A)]$$
$$= P(A)[1 - P(B)]$$
$$= P(A)P(B'),$$

since $P(B \,|\, A) = P(B)$ by hypothesis. Thus A and B' are independent events. The proofs for parts (b) and (c) are left as exercises.   □

Before extending the definition of independent events to more than two events, we present the following example.

Example 2.4.4

An urn contains four balls numbered 1, 2, 3, and 4. One ball is to be drawn at random from the urn. Let the events A, B, and C be defined by $A = \{1, 2\}$, $B = \{1, 3\}$, $C = \{1, 4\}$. Then $P(A) = P(B) = P(C) = 1/2$. Furthermore,

$$P(A \cap B) = \frac{1}{4} = P(A)P(B),$$

$$P(A \cap C) = \frac{1}{4} = P(A)P(C),$$

$$P(B \cap C) = \frac{1}{4} = P(B)P(C),$$

which implies that A, B, and C are independent in pairs (called **pairwise independence**). However, since $A \cap B \cap C = \{1\}$, we have

$$P(A \cap B \cap C) = \frac{1}{4} \neq \frac{1}{8} = P(A)P(B)P(C).$$

That is, something seems to be lacking for the complete independence of A, B, and C.   ◇

This example illustrates the reason for the second condition in Definition 2.4.2.

Definition 2.4.2.   Events A, B, and C are **mutually independent** if and only if the following two conditions hold:

(a) They are pairwise independent; that is,

$$P(A \cap B) = P(A)P(B), \qquad P(A \cap C) = P(A)P(C),$$

and

$$P(B \cap C) = P(B)P(C).$$

(b) $P(A \cap B \cap C) = P(A)P(B)P(C).$

Definition 2.4.2 can be extended to mutual independence of four or more events. In this extension, each pair, triple, quartet, and so on, must satisfy this type of multiplication rule. If there is no possibility of misunderstanding, *independent* is often used without the modifier *mutually* when considering several events.

Example 2.4.5

A rocket has a built-in redundant system. In this system, if component K_1 fails, it is bypassed and component K_2 is used. If component K_2 fails, it is bypassed and component K_3 is used. (An example of such components is computer systems.) Suppose that the probability of failure of any one of these components is 0.15 and assume that the failures of these components are mutually independent events. Let A_i denote the event that component K_i fails for $i = 1, 2, 3$. Because the system fails if K_1 fails and K_2 fails and K_3 fails, the probability that the system does not fail is given by

$$
\begin{aligned}
P[(A_1 \cap A_2 \cap A_3)'] &= 1 - P(A_1 \cap A_2 \cap A_3) \\
&= 1 - P(A_1)P(A_2)P(A_3) \\
&= 1 - (0.15)^3 \\
&= 0.9966.
\end{aligned}
$$

One way to increase the reliability of such a system is to add more components (realizing that this also adds weight and takes up space). For example, if a fourth component K_4 were added to this system, the probability that the system does not fail is

$$ P[(A_1 \cap A_2 \cap A_3 \cap A_4)'] = 1 - (0.15)^4 = 0.9995. \qquad \diamond $$

The proof and illustration of the following results are left as exercises. If A, B, and C are mutually independent events, then the following events are also independent:

(a) A and $(B \cap C)$

(b) A and $(B \cup C)$

(c) A' and $(B \cap C')$

In addition, A', B', and C' are mutually independent.

Many experiments consist of a sequence of n trials that are mutually independent. If the outcomes of the trials, in fact, do not have anything to do with one another, then events, such that each is associated with a different trial, should be independent in the probability sense. That is, if the event A_i is associated with the ith trial, $i = 1, 2, \ldots, n$, then

$$ P(A_1 \cap A_2 \cap \cdots \cap A_n) = P(A_1)P(A_2) \cdots P(A_n). $$

Example 2.4.6

A fair six-sided die is rolled 6 independent times. Let A_i be the event that side i is observed on the ith roll, called a match on the ith trial, $i = 1, 2, \ldots, 6$. Thus $P(A_i) = 1/6$, and $P(A_i') = 1 - 1/6 = 5/6$. If we let B denote the event that at least one match occurs, then B' is the event that no matches occur. Thus

$$P(B) = 1 - P(B') = 1 - P(A'_1 \cap A'_2 \cap \cdots \cap A'_6)$$

$$= 1 - \frac{5}{6} \cdot \frac{5}{6} \cdot \frac{5}{6} \cdot \frac{5}{6} \cdot \frac{5}{6} \cdot \frac{5}{6} = 1 - \left(\frac{5}{6}\right)^6$$

is the probability of B.   ◇

The sample space for an experiment of n trials is a set of n-tuples, where the ith component denotes the outcome on the ith trial. For example, if a six-sided die is rolled five times,

$$S = \{(O_1, O_2, O_3, O_4, O_5) : O_i = 1, 2, 3, 4, 5, \text{ or } 6, \text{ for } i = 1, 2, 3, 4, 5\}.$$

That is, S is a set of five-tuples, where each component is one of the first six positive integers.

If a coin is tossed two times,

$$S = \{(O_1, O_2) : O_i = H \text{ or } T, i = 1, 2\}.$$

We often drop the commas and parentheses and let, for illustration, $(H, T) = HT$, as in Example 2.4.1.

Example 2.4.7

Urn I contains two red balls and three white balls; urn II contains two red balls and one white ball; urn III contains one red ball and three white balls. At the ith trial a ball is drawn from urn i, $i = $ I, II, III. Assume that the trials are independent. Under reasonable assumptions, the probabilities of some of the possible outcomes are (in an obvious notation)

$$P(\{R, R, R\}) = P(RRR) = \frac{2}{5} \cdot \frac{2}{3} \cdot \frac{1}{4},$$

$$P(\{W, R, W\}) = P(WRW) = \frac{3}{5} \cdot \frac{2}{3} \cdot \frac{3}{4},$$

$$P(\{R, W, W\}) = P(RWW) = \frac{2}{5} \cdot \frac{1}{3} \cdot \frac{3}{4}.$$ ◇

Example 2.4.8

An urn contains three red, two white, and four yellow balls. An ordered sample of size 3 is drawn from the urn. If the balls are drawn with replacement so that one outcome does not change the probabilities of the others, the trials are independent. Under reasonable assumptions, the probabilities of the two given outcomes (in an obvious notation) are

$$P(RWY) = \frac{3}{9} \cdot \frac{2}{9} \cdot \frac{4}{9} = \frac{8}{243}$$

and

$$P(\{YYR \text{ or } RWW\}) = \frac{4}{9} \cdot \frac{4}{9} \cdot \frac{3}{9} + \frac{3}{9} \cdot \frac{2}{9} \cdot \frac{2}{9} = \frac{20}{243}.$$

If the balls are drawn without replacement, the trials are dependent. The probabilities, again under reasonable assumptions, of the two outcomes are

$$P(RWY) = \frac{3}{9} \cdot \frac{2}{8} \cdot \frac{4}{7} = \frac{1}{21}$$

and

$$P(\{YYR \text{ or } RWW\}) = \frac{4}{9} \cdot \frac{4}{8} \cdot \frac{3}{7} + \frac{3}{9} \cdot \frac{2}{8} \cdot \frac{2}{7} = \frac{7}{84}. \qquad \diamond$$

Example 2.4.9

Suppose that on five consecutive days an "instant winner" lottery ticket is purchased and the probability of winning is 1/5 on each day. Assuming independent trials,

$$P(WWLLL) = \left(\frac{1}{5}\right)^2 \left(\frac{4}{5}\right)^3$$

and

$$P(LWLWL) = \frac{4}{5} \cdot \frac{1}{5} \cdot \frac{4}{5} \cdot \frac{1}{5} \cdot \frac{4}{5} = \left(\frac{1}{5}\right)^2 \left(\frac{4}{5}\right)^3.$$

In general, the probability of purchasing two winning tickets and three losing tickets is

$$\binom{5}{2}\left(\frac{1}{5}\right)^2 \left(\frac{4}{5}\right)^3 = \frac{5!}{2!3!}\left(\frac{1}{5}\right)^2 \left(\frac{4}{5}\right)^3 = 0.2048$$

because there are $\binom{5}{2}$ ways to select the positions (or the days) for the winning tickets and each of these $\binom{5}{2}$ ways has the probability $(1/5)^2(4/5)^3$. $\qquad \diamond$

EXERCISES

2.4.1 Let A and B be independent events with $P(A) = 0.7$ and $P(B) = 0.2$. Compute (a) $P(A \cap B)$, (b) $P(A \cup B)$, and (c) $P(A' \cup B')$.

2.4.2 Let $P(A) = 0.3$ and $P(B) = 0.6$.

   (a) Find $P(A \cup B)$ when A and B are independent.

   (b) Find $P(A \mid B)$ when A and B are mutually exclusive.

2.4.3 Let A and B be independent events with $P(A) = 1/4$ and $P(B) = 2/3$. Compute (a) $P(A \cap B)$, (b) $P(A \cap B')$, (c) $P(A' \cap B')$, (d) $P[(A \cup B)']$, and (e) $P(A' \cap B)$.

2.4.4 Prove parts (b) and (c) of Theorem 2.4.1.

2.4.5 If $P(A) = 0.8$, $P(B) = 0.5$, and $P(A \cup B) = 0.9$, are A and B independent events? Why?

2.4.6 If A, B, and C are mutually independent, show that the following pairs of events are independent: A and $(B \cap C)$, A and $(B \cup C)$, A' and $(B \cap C')$. Also show that A', B' and C' are mutually independent.

2.4.7 Each of three persons fires one shot at a target. Let A_i denote the event that the target is hit by person i, $i = 1, 2, 3$. If we assume that A_1, A_2, A_3 are mutually independent and if $P(A_1) = 0.7$, $P(A_2) = 0.9$, $P(A_3) = 0.8$, compute the probability that exactly two people hit the target (i.e., one person misses).

2.4.8 Die A has orange on one face and blue on five faces, die B has orange on two faces and blue on four faces, and die C has orange on three faces and blue on three faces. These are unbiased dice. If the three dice are rolled, find the probability that exactly two of the three dice come up orange.

2.4.9 Suppose that A, B, and C are mutually independent events and that $P(A) = 0.5$, $P(B) = 0.8$, and $P(C) = 0.9$. Find the probabilities that

(a) All three events occur.

(b) Exactly two of the three events occur.

(c) None of the events occur.

2.4.10 Let D_1, D_2, D_3 be three four-sided dice whose sides have been labeled as follows:

$$D_1 : 0\,3\,3\,3 \qquad D_2 : 2\,2\,2\,5 \qquad D_3 : 1\,1\,4\,6$$

These dice are rolled at random. Let A, B, and C be the events that the outcome on die D_1 is larger than the outcome on D_2, the outcome on D_2 is larger than the outcome on D_3, the outcome on D_3 is larger than the outcome on D_1, respectively. Show that (a) $P(A) = 9/16$, (b) $P(B) = 9/16$, and (c) $P(C) = 10/16$. Do you find it interesting that each of the probabilities that D_1 "beats" D_2, D_2 "beats" D_3, and D_3 "beats" D_1 is greater than 1/2? Thus it is difficult to determine the "best" die.

2.4.11 Let A and B be two events.

(a) If the events A and B are mutually exclusive, are A and B always independent? If the answer is no, can they ever be independent? Explain.

(b) If $A \subset B$, can A and B ever be independent events? Explain.

2.4.12 Flip an unbiased coin five independent times. Compute the probability of

(a) $HHTHT$.

(b) $THHHT$.

(c) $HTHTH$.

(d) Three heads occurring in the five trials.

2.4.13 An urn contains two red balls and four white balls. Sample successively five times at random and with replacement so that the trials are independent. Compute the probability of each of the two sequences $WWRWR$ and $RWWWR$.

2.4.14 In Example 2.4.5, suppose that the probability of failure of a component is $p = 0.4$. Find the probability that the system does not fail if the number of redundant components is (a) 3 and (b) 8.

2.4.15 An urn contains 10 red and 10 white balls. The balls are drawn from the urn at random, one at a time. Find the probability that the fourth white ball is the sixth ball drawn if the sampling is done

(a) With replacement.

(b) Without replacement.

(c) In the World Series the American League (red) and National League (white) teams play until one team wins four games. Do you think that this urn model could be used to describe the probabilities of a 4-, 5-, 6-, or 7-game series? If your answer is yes, would you choose sampling with or without replacement in your model? (For your information, the numbers of 4-, 5-, 6-, and 7-game series, up to 1995, were 15, 20, 19, 33. In 1903 and 1919–1921, winners had to take 5 out of 9 games. Three of those series went eight, one went seven. The 1994 series was cancelled.)

2.4.16 An urn contains five balls, one marked WIN and four marked LOSE. You and another player take turns selecting a ball from the urn, one at a time. The first person to select the WIN ball is the winner. If you draw first, find the probability that you will win if the sampling is done

(a) With replacement.

(b) Without replacement.

2.4.17 For a time if you bought a Lotto ticket for the Michigan lottery, you could say "yes" to Zinger. The word "Yes" then appeared next to a six-digit integer (selected randomly from 000000 to 999999, inclusive) on your Lotto ticket. To win something playing Zinger, your numbers had to match, in exact order drawn from left to right, the first 2, 3, 4, 5, or 6 digits of the Zinger number selected. Find the probability of matching

(a) The first 2 but not the first 3 digits.

(b) The first 3 but not the first 4 digits.

(c) The first 4 but not the first 5 digits.

(d) The first 5 but not the first 6 digits.

(e) All 6 digits.

2.4.18 Extend Example 2.4.6 to an n-sided die. That is, suppose that a fair n-sided die is rolled n independent times. A match occurs if side i is observed on the ith trial, $i = 1, 2, \ldots, n$.

(a) Show that the probability of at least one match is

$$1 - \left(\frac{n-1}{n}\right)^n = 1 - \left(1 - \frac{1}{n}\right)^n.$$

(b) Find the limit of this probability as n increases without bound.

2.4.19 An urn contains n balls numbered from 1 through n. A random sample of n balls is selected from the urn, one at a time. A match occurs if ball numbered i is selected on the ith draw. For $n = 1$ to 15, find the probability of at least one match if the sampling is done

(a) With replacement (see Exercise 2.4.18).

(b) Without replacement (see Exercise 2.3.19).

(c) How much does n affect these probabilities?

(d) How does sampling with and without replacement affect these probabilities?

(e) Illustrate these probabilities empirically, either performing the experiments physically or simulating them on a computer.

*2.5 BAYES' THEOREM

In this section we consider the well-known Bayes' theorem. We begin with an example.

Example 2.5.1

Bowl B_1 contains two red and four white chips; bowl B_2 contains one red and two white chips; and bowl B_3 contains five red and four white chips. Say that the probabilities for selecting the bowls are not the same but are given by $P(B_1) = 1/3$, $P(B_2) = 1/6$, and $P(B_3) = 1/2$, where B_1, B_2, and B_3 are the events that bowls B_1, B_2, and B_3 are chosen, respectively. The experiment consists of selecting a bowl with these probabilities and then drawing a chip at random from that bowl. Let us compute the probability of event R, drawing a red chip, say $P(R)$. Note that $P(R)$ is dependent first of all on which bowl is selected and then on the probability of drawing a red chip from the selected bowl. That is, the event R is the union of the mutually exclusive events $B_1 \cap R$, $B_2 \cap R$, and $B_3 \cap R$. Thus

$$P(R) = P(B_1 \cap R) + P(B_2 \cap R) + P(B_3 \cap R)$$
$$= P(B_1)P(R \mid B_1) + P(B_2)P(R \mid B_2) + P(B_3)P(R \mid B_3)$$
$$= \frac{1}{3} \cdot \frac{2}{6} + \frac{1}{6} \cdot \frac{1}{3} + \frac{1}{2} \cdot \frac{5}{9} = \frac{4}{9}.$$

Suppose now that the outcome of the experiment is a red chip, but we do not know from which bowl it was drawn. Accordingly, we compute the conditional probability that the chip was drawn from bowl B_1, namely, $P(B_1 \mid R)$. From the definition of conditional probability and the result above, we have that

$$P(B_1 \mid R) = \frac{P(B_1 \cap R)}{P(R)}$$
$$= \frac{P(B_1)P(R \mid B_1)}{P(B_1)P(R \mid B_1) + P(B_2)P(R \mid B_2) + P(B_3)P(R \mid B_3)}$$
$$= \frac{(1/3)(2/6)}{(1/3)(2/6) + (1/6)(1/3) + (1/2)(5/9)} = \frac{2}{8}.$$

Similarly, we have that

$$P(B_2 \mid R) = \frac{P(B_2 \cap R)}{P(R)} = \frac{(1/6)(1/3)}{4/9} = \frac{1}{8}$$

and

$$P(B_3 \mid R) = \frac{P(B_3 \cap R)}{P(R)} = \frac{(1/2)(5/9)}{4/9} = \frac{5}{8}.$$

Note that the conditional probabilities $P(B_1 \mid R)$, $P(B_2 \mid R)$, and $P(B_3 \mid R)$ have changed from the original probabilities $P(B_1)$, $P(B_2)$, and $P(B_3)$ in a way that agrees with your intuition. Namely, once the red chip has been observed, the probability concerning B_3 seems more favorable than originally because B_3 has a larger percentage of red chips than do B_1 and B_2. The conditional probabilities of B_1 and B_2 decrease from their original ones once the red chip is observed. Frequently, the original probabilities are called *prior probabilities*, and the conditional probabilities are the *posterior probabilities*.   ◇

We generalize the result of Example 2.5.1. Let $B_1, B_2, \ldots, B_m$ constitute a *partition* of the sample space S. That is,

$$S = B_1 \cup B_2 \cup \cdots \cup B_m \quad \text{and} \quad B_i \cap B_j = \emptyset, \quad i \neq j.$$

Of course, the events $B_1, B_2, \ldots, B_m$ are mutually exclusive and exhaustive (since the union of the disjoint sets equals the sample space S). Furthermore, suppose that the **prior probability** of the event B_i is positive; that is, $P(B_i) > 0, i = 1, \ldots, m$. If A is an event, then A is the union of m mutually exclusive events, namely,

$$A = (B_1 \cap A) \cup (B_2 \cap A) \cup \cdots \cup (B_m \cap A).$$

Thus

$$P(A) = \sum_{i=1}^{m} P(B_i \cap A)$$

$$= \sum_{i=1}^{m} P(B_i) P(A \mid B_i). \tag{2.5}$$

If $P(A) > 0$, we have that

$$P(B_k \mid A) = \frac{P(B_k \cap A)}{P(A)}, \qquad k = 1, 2, \ldots, m. \tag{2.6}$$

Using Equation 2.5 and replacing $P(A)$ in Equation 2.6, we have **Bayes' theorem**:

$$P(B_k \mid A) = \frac{P(B_k) P(A \mid B_k)}{\sum_{i=1}^{m} P(B_i) P(A \mid B_i)}, \qquad k = 1, 2, \ldots, m.$$

The conditional probability $P(B_k \mid A)$ is often called the **posterior probability** of B_k. The following example illustrates one application of Bayes' result.

Example 2.5.2

In a certain factory, machines I, II, and III are all producing springs of the same length. Of their production, machines I, II, and III produce 2%, 1%, and 3% defective springs, respectively. Of the total production of springs in the factory, machine I produces 35%, machine II produces 25%, and machine III produces 40%. If one spring is selected at random from the total springs produced in a day, the probability that it is defective, in an obvious notation, equals

$$P(D) = P(I)P(D \mid I) + P(II)P(D \mid II) + P(III)P(D \mid III)$$

$$= \left(\frac{35}{100}\right)\left(\frac{2}{100}\right) + \left(\frac{25}{100}\right)\left(\frac{1}{100}\right) + \left(\frac{40}{100}\right)\left(\frac{3}{100}\right) = \frac{215}{10,000}.$$

If the selected spring is defective, the conditional probability that it was produced by machine III is, by Bayes' formula,

$$P(III \mid D) = \frac{P(III)P(D \mid III)}{P(D)} = \frac{(40/100)(3/100)}{215/10,000} = \frac{120}{215}.$$

Note how the posterior probability of III increased from the prior probability of III after the defective spring was observed because III produces a larger percentage of defectives than do I and II. ◇

EXERCISES

2.5.1 Bowl B_1 contains 2 white chips, bowl B_2 contains 2 red chips, bowl B_3 contains 2 white and 2 red chips, and bowl B_4 contains 3 white chips and 1 red chip. The probabilities of selecting bowl B_1, B_2, B_3, or B_4 are 1/2, 1/4, 1/8, and 1/8, respectively. A bowl is selected using these probabilities, and a chip is then drawn at random. Find

(a) $P(W)$, the probability of drawing a white chip.

(b) $P(B_1 \mid W)$, the conditional probability that bowl B_1 had been selected, given that a white chip was drawn.

2.5.2 Bean seeds from supplier A have an 85% germination rate and those from supplier B have a 75% germination rate. A seed packaging company purchases 40% of their bean seeds from supplier A and 60% from supplier B and mixes these seeds together.

(a) Find the probability that a seed selected at random from the mixed seeds will germinate, say $P(G)$.

(b) Given that a seed germinates, find the probability that the seed was purchased from supplier A.

2.5.3 The Belgium 20-franc coin ($B20$), the Italian 500-lire coin ($I500$), and the Hong Kong 5-dollar coin ($HK5$) are approximately the same size. Coin purse 1 ($C1$) contains 6 of each of these coins. Coin purse 2 ($C2$) contains 9 $B20$'s, 6 $I500$'s, and 3 $HK5$'S. A fair four-sided die is rolled. If the outcome is $\{1\}$, a coin is selected randomly from $C1$. If the outcome belongs to $\{2, 3, 4\}$, a coin is selected randomly from $C2$. Find

(a) $P(B20)$, the probability of selecting a Belgian coin.

(b) $P(C1 \mid B20)$, the probability that the coin was selected from $C1$, given that it was a Belgian coin.

2.5.4 Bowl A contains two red chips; bowl B contains two white chips; and bowl C contains one red chip and one white chip. A bowl is selected at random (with equal probabilities), and one chip is taken at random from that bowl.

(a) Compute the probability of selecting a white chip, say $P(W)$.

(b) If the chip selected is white, compute the conditional probability that the other chip in the bowl is red.

HINT: Compute $P(C \mid W)$.

2.5.5 A package, say A, of 24 crocus bulbs contains 8 yellow, 8 white, and 8 purple crocus bulbs. A package, say B, of 24 crocus bulbs contains 6 yellow, 6 white, and 12 purple crocus bulbs. One of the two packages is selected at random.

    (a) If 3 bulbs from this package were planted and all 3 yielded purple flowers, compute the conditional probability that package B was selected.

    (b) If the 3 bulbs yielded 1 yellow flower, 1 white flower, and 1 purple flower, compute the conditional probability that package A was selected.

2.5.6 Each bag in a large box contains 25 tulip bulbs. Three-fourths of the bags contain bulbs for 5 red and 20 yellow tulips; one-fourth of the bags contain bulbs for 15 red and 10 yellow tulips. A bag is selected at random and one bulb is selected randomly and planted. Give the probability that it will produce a

    (a) Red tulip, say $P(R)$.

    (b) Yellow tulip, say $P(Y)$.

    (c) If the tulip is red, find the conditional probability that a bag having 15 red and 10 yellow tulips was selected.

2.5.7 Bowl C contains 6 red chips and 4 blue chips. Five of these 10 chips are selected at random and without replacement and put in bowl D, which was originally empty. One chip is then drawn at random from bowl D. Given that this chip is blue, find the conditional probability that 2 red chips and 3 blue chips were transferred from bowl C to bowl D.

3

DISCRETE DISTRIBUTIONS

3.1 RANDOM VARIABLES OF THE DISCRETE TYPE

An outcome space S may be difficult to describe if the elements of S are not numbers. We shall now discuss how we can use a rule by which each outcome of a random experiment, an element s of S, may be associated with a real number x. We begin the discussion with an example.

Example 3.1.1

A rat is selected at random from a cage and its sex is determined. The set of possible outcomes is female and male. Thus the outcome space is $S = \{\text{female, male}\} = \{\text{F, M}\}$. Let X be a function defined on S such that $X(\text{F}) = 0$ and $X(\text{M}) = 1$. Thus X is a real-valued function that has the outcome space S as its domain and the set of real numbers $\{x : x = 0, 1\}$ as its range. We call X a random variable and, in this example, the space associated with X is $\{x : x = 0, 1\}$. ◇

We now formulate the definition of a random variable.

Definition 3.1.1. Given a random experiment with an outcome space S, a function X that assigns to each element s in S one and only one real number $X(s) = x$ is called a **random variable**. The **space** of X is the set of real numbers $\{x : X(s) = x, s \in S\}$, where $s \in S$ means that the element s belongs to the set S.

Remark As we give examples of random variables and their probability distributions, the reader soon recognizes that when observing a random experiment the experimenter must take some type of measurement (or measurements). This measurement can be thought of as the outcome of a random variable. We would simply like to know the probability of a measurement ending in A, a subset of the space of X. If this is known for all subsets A, then we know the probability distribution of the random variable. Obviously, in practice, we do not very often know this distribution exactly. Hence statisticians make conjectures about these distributions; that is, we construct probabilistic models for random variables. The ability of a statistician to model a real situation appropriately is a valuable trait; some of us are better than others. In this chapter we introduce some probability models in which the spaces of the random variables consist of sets of integers.

It may be that the set S has elements that are themselves real numbers. In such an instance we could write $X(s) = s$ so that X is the identity function and the space of X is also S. This is illustrated in Example 3.1.2.

Example 3.1.2

Let the random experiment be the cast of a die, observing the number of spots on the side facing up. The outcome space associated with this experiment is $S = \{1, 2, 3, 4, 5, 6\}$. For each $s \in S$, let $X(s) = s$. The space of the random variable X is then $\{1, 2, 3, 4, 5, 6\}$.

If we associate a probability of 1/6 with each outcome, then $P(X = 5) = 1/6$, $P(2 \leq X \leq 5) = 4/6$, and $P(X \leq 2) = 2/6$, for example, seem to be reasonable assignments, where $\{2 \leq X \leq 5\}$ means $\{X = 2, 3, 4, \text{ or } 5\}$ and $\{X \leq 2\}$ means $\{X = 1 \text{ or } 2\}$, in this example.   ◇

The student will no doubt recognize two major difficulties here:

1. In many practical situations the probabilities assigned to the events are unknown.
2. Since there are many ways of defining a function X on S, which function do we want to use?

As a matter of fact, the solutions to these problems in particular cases are major concerns in applied statistics. In considering question 2, statisticians try to determine what *measurement* (or measurements) should be taken on an outcome; that is, how best do we "mathematize" the outcome (which, for the anthropologist, might be a skull)? These measurement problems are most difficult and can be answered only by getting involved in a practical project. For problem 1, we often need, through repeated observations (called sampling), to estimate these probabilities or percentages. For example, what percentage of newborn girls in the University of Iowa Hospital weigh less than 7 pounds? Here a newborn baby girl is the outcome, and we have measured her one way (by weight); but obviously there are many other ways of measuring her. If we let X be the weight in pounds, we are interested in the probability $P(X < 7)$, and we can estimate this only by repeated observations. One obvious way of estimating this is by use of the relative frequency of $\{X < 7\}$ after a number of observations. If additional assumptions can be made, we will study, in this text, other ways of estimating this probability. It is the latter aspect with which mathematical statistics is concerned. That is, if we assume certain models, we find that the theory of statistics can explain how best to draw conclusions or make predictions.

In many instances, it is clear exactly what function X the experimenter wants to define on the outcome space. For example, the caster in the dice game Craps is concerned about the sum of the spots, say X, that are up on the pair of dice. Hence we go directly to the space of X, which is usually denoted by R. After all, in the dice game the caster is directly concerned only with the probabilities associated with X. Hence for convenience, the reader can, in many instances, think of the space of X as being the outcome space.

Let X denote a random variable with space R. Suppose that we know how the probability is distributed over the various subsets A of R; that is, we can compute $P(X \in A)$. In this sense, we speak of the distribution of the random variable X, meaning, of course, the distribution of probability associated with the space R of X.

Let X denote a random variable with one-dimensional space R, a subset of the real numbers. Suppose that the space R contains a countable number of points; that is, R contains either a finite number of points or the points of R can be put into a one-to-one correspondence with the positive integers. Such a set R is called a set of discrete points or simply a discrete outcome space. Furthermore, the random variable X is called a random variable of the **discrete type**, and X is said to have a distribution of the discrete type.

For a random variable X of the discrete type, the probability $P(X = x)$ is frequently denoted by $f(x)$, and this function $f(x)$ is called the **probability density function**. Note that some authors refer to $f(x)$ as the probability function, the frequency function, or the probability mass function. We prefer "probability density function," and it is hereafter abbreviated p.d.f.

Let $f(x)$ be the p.d.f. of the random variable X of the discrete type, and let R be the space of X. Since $f(x) = P(X = x)$, $x \in R$, $f(x)$ must be positive for $x \in R$ and we want all these probabilities to add to 1 because each $P(X = x)$ represents the fraction of times x can be expected to occur. Moreover, to determine the probability associated with the event $A \in R$, we would sum the probabilities of the x values in A. This leads us to the following definition.

Definition 3.1.2.   The p.d.f. $f(x)$ of a discrete random variable X is a function that satisfies the following properties:

(a) $f(x) > 0$, $x \in R$.

(b) $\sum_{x \in R} f(x) = 1$.

(c) $P(X \in A) = \sum_{x \in A} f(x)$,   where $A \subset R$.

We usually let $f(x) = 0$ when $x \notin R$ and thus the domain of $f(x)$ is the set of real numbers. When we define the p.d.f. $f(x)$ and do not say zero elsewhere, then we tacitly mean that $f(x)$ has been defined at all x's in the space R, and it is assumed that $f(x) = 0$ elsewhere, namely, $f(x) = 0$, $x \notin R$. Since the probability $P(X = x) = f(x) > 0$ when $x \in R$ and since R contains all the probability associated with X, R is sometimes referred to as the **support** of X as well as the space of X.

Example 3.1.3

Roll a four-sided die twice and let X equal the larger of the two outcomes if they are different and the common value if they are the same. The outcome space for this experiment is $S = \{(d_1, d_2): d_1 = 1, 2, 3, 4; d_2 = 1, 2, 3, 4\}$, where we assume that each of these 16 points has probability 1/16. Then $P(X = 1) = P[(1, 1)] = 1/16$, $P(X = 2) = P[\{(1, 2), (2, 1), (2, 2)\}] = 3/16$, and similarly, $P(X = 3) = 5/16$ and $P(X = 4) = 7/16$. That is, the p.d.f. of X can be written simply as

$$f(x) = P(X = x) = \frac{2x - 1}{16}, \qquad x = 1, 2, 3, 4. \tag{3.1}$$

We could add that $f(x) = 0$ elsewhere; but if we do not, the reader should take $f(x)$ to equal zero when $x \notin R$.   ◇

A better understanding of a particular probability distribution can often be obtained with a graph that depicts the p.d.f. of X. Note that the graph of the p.d.f., when $f(x) > 0$, would be simply the set of points $\{[x, f(x)]: x \in R\}$, where R is the space of X. Two types of graphs can be used to give a better visual appreciation of the p.d.f., namely, a bar graph and a probability histogram. A **bar graph** of the p.d.f. $f(x)$ of the random variable X is a graph having a vertical line segment drawn from $(x, 0)$ to $[x, f(x)]$ at each x in R, the space of X. If X can only assume integer values, a **probability histogram** of the p.d.f. $f(x)$ is a graphical representation that has a rectangle of height $f(x)$ and a base of length 1, centered at x for each $x \in R$, the space of X. Figure 3.1 displays a bar graph and a probability histogram for the p.d.f. $f(x)$ defined in Equation 3.1.

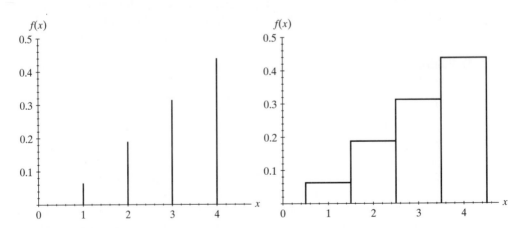

Figure 3.1   Bar graph and probability histogram

In Section 2.1 we discussed the relationship between the probability $P(A)$ of an event A and the relative frequency $\mathcal{N}(A)/n$ of occurrences of event A in n repetitions of an experiment. We now extend those ideas. Suppose that a random experiment is repeated n independent times. Let $A = \{X = x\}$, the event that x is the outcome of the experiment. Then we would expect the relative frequency $\mathcal{N}(A)/n$ to be close to $f(x)$. The following example illustrates this.

Example 3.1.4

A tetrahedron (four-sided die with outcomes 1, 2, 3, 4) is rolled twice. Let X equal the sum of the two outcomes. Then the possible values that X can equal are 2, 3, 4, 5, 6, 7, and 8. The following argument suggests that the p.d.f. of X is given by $f(x) = (4 - |x - 5|)/16$, for $x = 2, 3, 4, 5, 6, 7, 8$. That is, $f(2) = 1/16$, $f(3) = 2/16$, $f(4) = 3/16$, $f(5) = 4/16$, $f(6) = 3/16$, $f(7) = 2/16$, and $f(8) = 1/16$. Intuitively, these probabilities seem correct if we think of the 16 points (result on first roll, result on second roll), and assume that each has probability 1/16. Then note that $X = 2$ only for the point $(1, 1)$; $X = 3$ for the two points $(2, 1)$ and $(1, 2)$; and so on. This experiment was simulated 1000 times on a computer. Table 3.1 lists the results and compares the relative frequencies with the corresponding probabilities.

TABLE 3.1   Sum of two tetrahedral dice

x	Number of Observations of x	Relative Frequency of x	Probability of $\{X = x\}$, $f(x)$
2	71	0.071	0.0625
3	124	0.124	0.1250
4	194	0.194	0.1875
5	258	0.258	0.2500
6	177	0.177	0.1875
7	122	0.122	0.1250
8	54	0.054	0.0625

A graph can be used to display the results given in Table 3.1. The probability histogram of the p.d.f. $f(x)$ of X is given in Figure 3.2. The heights of the dashed lines represent the observed relative frequencies of the corresponding x values. The histogram associated with the dashed lines is the relative frequency histogram. For random experiments of the discrete type, this relative frequency histogram of a set of data gives an estimate of the probability histogram of the associated random variable when the latter is unknown. Estimation is considered in detail later in this book.   ◇

Figure 3.2   Sum of two tetrahedral dice

EXERCISES

3.1.1 Let the p.d.f. of X be defined by $f(x) = x/9$, $x = 2, 3, 4$.

(a) Draw a bar graph for this p.d.f.

(b) Draw a probability histogram for this p.d.f.

3.1.2 Let a chip be taken at random from a bowl that contains six white chips, three red chips, and one blue chip. Let the random variable $X = 1$ if the outcome is a white chip; let $X = 5$ if the outcome is a red chip; and let $X = 10$ if the outcome is a blue chip.

(a) Find the p.d.f. of X.

(b) Graph the p.d.f. as a bar graph.

3.1.3 For each of the following, determine the constant c so that $f(x)$ satisfies the conditions of being a p.d.f. for a random variable X. Depict each p.d.f. as a bar graph.

(a) $f(x) = x/c$, $x = 1, 2, 3, 4$.

(b) $f(x) = cx$, $x = 1, 2, 3, \ldots, 10$.

(c) $f(x) = c(1/4)^x$, $x = 1, 2, 3, \ldots$.

(d) $f(x) = c(x + 1)^2$, $x = 0, 1, 2, 3$.

(e) $f(x) = x/c$, $x = 1, 2, 3, \ldots, n$.

3.1.4 The state of Michigan generates a three-digit number at random six days a week for their daily lottery. The numbers are generated one digit at a time. Consider the following set of 50 three-digit numbers as 150 one-digit integers that were generated at random:

169	938	506	757	594	656	444	809	321	545
732	146	713	448	861	612	881	782	209	752
571	701	852	924	766	633	696	023	601	789
137	098	534	826	642	750	827	689	979	000
933	451	945	464	876	866	236	617	418	988

Let X denote the outcome when a single digit is generated.

(a) With true random numbers, what is the p.d.f. of X? Draw the probability histogram.

(b) For the 150 observations, determine the relative frequencies of 0, 1, 2, 3, 4, 5, 6, 7, 8, and 9, respectively.

(c) Draw the relative frequency histogram of the observations on the same graph paper as that of the probability histogram. Use a colored or dashed line for the relative frequency histogram.

3.1.5 The p.d.f. of X is $f(x) = (5 - x)/10$, $x = 1, 2, 3, 4$.

(a) Graph the p.d.f. as a bar graph.

(b) Use the following independent observations of X, simulated on a computer, to construct a table like Table 3.1:

```
3 1 2 2 3 2 2 2 1 3 3 2 3 2 4 4 2 1 1 3
3 1 2 2 1 1 4 2 3 1 1 1 2 1 3 1 1 3 3 1
1 1 1 1 1 4 1 3 1 2 4 1 1 2 3 4 3 1 4 2
2 1 3 2 1 4 1 1 1 2 1 3 4 3 2 1 4 4 1 3
2 2 2 1 2 3 1 1 4 2 1 4 2 1 2 3 1 4 2 3
```

(c) Construct a probability histogram and a relative frequency histogram like Figure 3.2.

3.1.6 Let a random experiment be the cast of a pair of unbiased dice, each having six faces, and let the random variable X denote the sum of the dice.

(a) With reasonable assumptions, determine the p.d.f. $f(x)$ of X. HINT: Picture the sample space consisting of the 36 points, (result on first die, result on second die), and assume that each has probability 1/36. Find the probability for each possible outcome of X, namely, $x = 2, 3, 4, \ldots, 12$.

(b) Draw the bar graph for $f(x)$.

3.1.7 Let a random experiment be the cast of a pair of unbiased six-sided dice and let X equal the smaller of the outcomes if they are different and the common value if they are equal.

(a) With reasonable assumptions, find the p.d.f. of X.

(b) Draw a probability histogram.

3.1.8 Let the p.d.f. of X be defined by $f(x) = (1 + |x - 3|)/11$, $x = 1, 2, 3, 4, 5$. Graph the p.d.f. of X as a bar graph.

3.1.9 Five cards are selected at random without replacement from a standard, thoroughly shuffled 52-card deck of playing cards. Let X equal the number of face cards (kings, queens, jacks) in the hand. Forty observations of X yielded the following data:

```
2 1 2 1 0 0 1 0 1 1 0 2 0 2 3 0 1 1 0 3
1 2 0 2 0 2 0 1 0 1 1 2 1 0 1 1 2 1 1 0
```

(a) Argue that the p.d.f. of X is

$$f(x) = \frac{\binom{12}{x}\binom{40}{5-x}}{\binom{52}{5}}, \qquad x = 0, 1, 2, 3, 4, 5$$

and thus $f(0) = 2109/8330$, $f(1) = 703/1666$, $f(2) = 209/833$, $f(3) = 55/833$, $f(4) = 165/21{,}658$, $f(5) = 33/108{,}290$.

(b) Draw a probability histogram for this distribution.

(c) Determine the relative frequencies of 0, 1, 2, 3, and superimpose the relative frequency histogram on your probability histogram.

3.2 MATHEMATICAL EXPECTATION

An extremely important concept in summarizing important characteristics of distributions of probability is that of mathematical expectation, which is introduced by an example.

Example 3.2.1

An enterprising young man who needs a little extra money devises a game of chance in which some of his friends might wish to participate. The game that he proposes is to let the participant cast an unbiased die and then receive a payment according to the following schedule: If the event $A = \{1, 2, 3\}$ occurs, he receives 1¢; if $B = \{4, 5\}$ occurs, he receives 5¢; and if $C = \{6\}$ occurs, he receives 35¢. The probabilities of the respective events are assumed to be 3/6, 2/6, and 1/6, since the die is unbiased. The problem that now faces the young man is the determination of the amount that should be charged for the opportunity of playing the game. He reasons, correctly, that if the game is played a large number of times about 3/6 of the trials will require a payment of 1¢, about 2/6 of them will require one of 5¢, and about 1/6 of them will require one of 35¢. Thus the approximate average payment is

$$(1)\left(\frac{3}{6}\right) + (5)\left(\frac{2}{6}\right) + (35)\left(\frac{1}{6}\right) = 8.$$

That is, he expects to pay 8¢ "on the average." Note that he never pays exactly 8¢; the payment is either 1¢, 5¢, or 35¢. However, the "weighted average" of 1, 5, and 35, in which the weights are the respective probabilities 3/6, 2/6, and 1/6, equals eight. Such a weighted average is called the mathematical expectation of payment. Thus, if the young man decides to charge 10¢ per play, he would make 2¢ per play "on the average." Since the most that a player would lose at the charge of 10¢ per play is 9¢, the young man might find that several players are attracted by the possible gain of 25¢. ◇

A more mathematical way of formulating the preceding example would be to let X be the random variable defined by the outcome of the cast of the die. Thus the p.d.f. of X is

$$f(x) = \frac{1}{6}, \qquad x = 1, 2, 3, 4, 5, 6.$$

In terms of the observed value x, the payment is given by the function

$$u(x) = \begin{cases} 1, & x = 1, 2, 3, \\ 5, & x = 4, 5, \\ 35, & x = 6. \end{cases}$$

The mathematical expectation of payment is then equal to

$$\sum_{x=1}^{6} u(x) f(x) = (1)\left(\frac{1}{6}\right) + (1)\left(\frac{1}{6}\right) + (1)\left(\frac{1}{6}\right) + (5)\left(\frac{1}{6}\right) + (5)\left(\frac{1}{6}\right) + (35)\left(\frac{1}{6}\right)$$

$$= (1)\left(\frac{3}{6}\right) + (5)\left(\frac{2}{6}\right) + (35)\left(\frac{1}{6}\right)$$

$$= 8.$$

This discussion suggests the more general definition of mathematical expectation of a function of X.

Definition 3.2.1.   If $f(x)$ is the p.d.f. of the random variable X of the discrete type with space R and if the summation

$$\sum_{x \in R} u(x) f(x), \qquad \text{which is sometimes written} \qquad \sum_{R} u(x) f(x),$$

exists, then the sum is called the **mathematical expectation** or the **expected value** of the function $u(X)$, and it is denoted by $E[u(X)]$. That is,

$$E[u(X)] = \sum_{x \in R} u(x) f(x).$$

We can think of the expected value $E[u(X)]$ as a weighted mean of $u(x)$, $x \in R$, where the weights are the probabilities $f(x) = P(X = x)$, $x \in R$.

Remark   The usual definition of mathematical expectation of $u(X)$ requires that the sum converge absolutely; that is,

$$\sum_{x \in R} |u(x)| f(x)$$

converges and is finite. However, in this book, each $u(x)$ is such that the convergence is absolute, and we do not burden the student with this additional requirement. Moreover, sometimes $E[u(X)]$ is called, more simply, the expectation of $u(X)$.

There is another important observation that must be made about the consistency of this definition. Certainly, this function $u(X)$ of the random variable X is itself a random variable, say Y. Suppose that we find the p.d.f. of Y to be $g(y)$ on the support R_1. Then $E(Y)$ is given by the summation

$$\sum_{y \in R_1} y\, g(y).$$

In general, it is true that

$$\sum_{x \in R} u(x) f(x) = \sum_{y \in R_1} y\, g(y);$$

that is, the same expectation is obtained by either method. We do not prove this general result but only illustrate it in the following example.

Example 3.2.2

Let the random variable X have the p.d.f.

$$f(x) = \frac{1}{3}, \qquad x \in R.$$

where $R = \{-1, 0, 1\}$. Let $u(X) = X^2$. Then

$$\sum_{x \in R} x^2 f(x) = (-1)^2 \left(\frac{1}{3}\right) + (0)^2 \left(\frac{1}{3}\right) + (1)^2 \left(\frac{1}{3}\right) = \frac{2}{3}.$$

However, the support of the random variable $Y = X^2$ is $R_1 = \{0, 1\}$ and

$$P(Y = 0) = P(X = 0) = \frac{1}{3},$$

$$P(Y = 1) = P(X = -1) + P(X = 1) = \frac{1}{3} + \frac{1}{3} = \frac{2}{3}.$$

That is,

$$g(y) = \begin{cases} \dfrac{1}{3}, & y = 0, \\[2mm] \dfrac{2}{3}, & y = 1; \end{cases}$$

and $R_1 = \{0, 1\}$. Hence

$$\sum_{y \in R_1} y\, g(y) = (0) \left(\frac{1}{3}\right) + (1) \left(\frac{2}{3}\right) = \frac{2}{3},$$

which illustrates the preceding observation. ◇

Before presenting additional examples, we list some useful facts about mathematical expectation in the following theorem.

Theorem 3.2.1.   When it exists, mathematical expectation E satisfies the following properties:

(a) If c is a constant, $E(c) = c$.

(b) If c is a constant and u is a function,

$$E[c\,u(X)] = cE[u(X)].$$

(c) If c_1 and c_2 are constants and u_1 and u_2 are functions, then

$$E[c_1u_1(X) + c_2u_2(X)] = c_1E[u_1(X)] + c_2E[u_2(X)].$$

Proof.   First, we have for the proof of property (a) that

$$E(c) = \sum_{x \in R} cf(x) = c\sum_{x \in R} f(x) = c$$

because

$$\sum_{x \in R} f(x) = 1.$$

Next, to prove property (b), we see that

$$E[c\,u(X)] = \sum_{x \in R} c\,u(x)f(x)$$

$$= c\sum_{x \in R} u(x)f(x)$$

$$= c\,E[u(X)].$$

Finally, the proof of property (c) is given by

$$E[c_1u_1(X) + c_2u_2(X)] = \sum_{x \in R}[c_1u_1(x) + c_2u_2(x)]f(x)$$

$$= \sum_{x \in R} c_1u_1(x)f(x) + \sum_{x \in R} c_2u_2(x)f(x).$$

By applying property (b), we obtain

$$E[c_1u_1(X) + c_2u_2(X)] = c_1E[u_1(X)] + c_2E[u_2(X)]. \qquad \square$$

Property (c) can be extended to more than two terms by mathematical induction; that is, we have

$$(c)' \qquad E\left[\sum_{i=1}^{k} c_i\,u_i(X))\right] = \sum_{i=1}^{k} c_i\,E[u_i(X)].$$

Because of property (c)', mathematical expectation E is often called a **linear** or **distributive** operator.

Example 3.2.3

Let X have the p.d.f.

$$f(x) = \frac{x}{10}, \qquad x = 1, 2, 3, 4.$$

Then

$$E(X) = \sum_{x=1}^{4} x\left(\frac{x}{10}\right)$$

$$= (1)\left(\frac{1}{10}\right) + (2)\left(\frac{2}{10}\right) + (3)\left(\frac{3}{10}\right) + (4)\left(\frac{4}{10}\right) = 3,$$

$$E(X^2) = \sum_{x=1}^{4} x^2\left(\frac{x}{10}\right)$$

$$= (1)^2\left(\frac{1}{10}\right) + (2)^2\left(\frac{2}{10}\right) + (3)^2\left(\frac{3}{10}\right) + (4)^2\left(\frac{4}{10}\right) = 10,$$

and

$$E[X(5 - X)] = 5E(X) - E(X^2) = (5)(3) - 10 = 5. \qquad \diamond$$

Example 3.2.4

Let $u(x) = (x - b)^2$, where b is not a function of X, and suppose $E[(X - b)^2]$ exists. To find that value of b for which $E[(X - b)^2]$ is a minimum, we write

$$g(b) = E[(X - b)^2] = E[X^2 - 2bX + b^2]$$
$$= E(X^2) - 2bE(X) + b^2$$

because $E(b^2) = b^2$. To find the minimum, we differentiate $g(b)$ with respect to b, set $g'(b) = 0$ and solve for b as follows:

$$g'(b) = -2E(X) + 2b = 0,$$
$$b = E(X).$$

Because $g''(b) = 2 > 0$, $E(X)$ is the value of b that minimizes $E[(X - b)^2]$. In the next section we learn that $E(X)$ and $E[(X - b)^2]$, when $b = E(X)$, have special names. $\qquad \diamond$

EXERCISES

3.2.1 The number 3 appears on each of two chips in a bowl and the number 9 on a third chip in the bowl. Consider the following game. A player draws one chip at random from the bowl and receives either \$3 or \$9, depending on the number on that chip. If he or she plays this game a large number of times, how many dollars can the player expect to receive on the average per play if we assume that each chip has a probability of 1/3?

3.2.2 Let X be a number selected at random from the first 10 positive integers. Assuming equal probabilities on these 10 integers, compute $E[X(11 - X)]$.

3.2.3 Find $E(X)$ for each of the distributions given in Exercise 3.1.3. HINT: For part (c), note that the difference $E(X) - (1/4)E(X)$ is equal to the sum of a geometric series.

3.2.4 Let the random variable X have the p.d.f.

$$f(x) = \frac{(|x| + 1)^2}{9}, \qquad x = -1, 0, 1.$$

Compute $E(X)$, $E(X^2)$, and $E(3X^2 - 2X + 4)$.

3.2.5 In a particular lottery 3,000,000 tickets are sold each week for 50¢ apiece. Out of the 3,000,000 tickets, 12,006 are drawn at random and without replacement and awarded prizes: twelve thousand $25 prizes, four $10,000 prizes, one $50,000 prize, and one $200,000 prize. If you purchased a single ticket each week, what is the expected value of this game to you?

3.2.6 In a state lottery a three-digit integer (000–999) is selected at random. If a player bets $1 on a particular number, the payoff, if that number is selected, is $500 minus the $1 paid for the ticket. Let X equal the payoff to the bettor, namely $-\$1$ or $\$499$, and find $E(X)$.

3.2.7 In the gambling game Chuck-a-Luck, for a $1 bet it is possible to win $1, $2, or $3 with respective probabilities 75/216, 15/216, and 1/216. One dollar is lost with probability 125/216. Let X equal the payoff for this game and find $E(X)$. Note that when a bet is won, the $1 that was bet is returned to the bettor in addition to the $1, $2, or $3 that is won.

3.2.8 Let the p.d.f. of X be defined by $f(x) = 6/(\pi^2 x^2)$, $x = 1, 2, 3, \ldots$. Show that $E(X)$ is not finite in this case.

3.2.9 Let us select at random a number from the first n positive integers. If the payment is equal to the reciprocal of the number, find an expression for the expected payment. Evaluate this number when $n = 5$. Approximate this number when $n = 100$. HINT: Use a modification of the integral test for testing convergence of series.

3.2.10 Let X be a random variable with support $\{1, 2, 3, 5, 15, 25, 50\}$, each point of which has the same probability $1/7$. Argue that $c = 5$ is the value of c that minimizes $h(c) = E(|X - c|)$. Compare this to the value of b that minimizes $g(b) = E[(X - b)^2]$.

3.2.11 At one time if you purchased a Lotto ticket for the Michigan lottery, you could say "yes" to Zinger. The word "Yes" then appeared next to a six-digit integer on your Lotto ticket. To win something playing Zinger, your number had to match, in exact order drawn from left to right, the first 2, 3, 4, 5, or 6 digits of the Zinger number selected. If you matched the first 2 digits, you won $20. Matching the first 3 digits, you won $100; matching the first 4 digits, you won $500; matching the first 5 digits, you won $5000; and matching all 6 digits, you won $100,000. Note that you could win at most one of these prizes at a time.

(a) What was the expected payoff of this game?

(b) If it cost $1 to play, what was your expected loss on each play?

3.2.12 A roulette wheel used in a U.S. casino has 38 slots, of which 18 are red, 18 are black, and 2 are green. A roulette wheel used in a French casino has 37 slots, of which 18 are red, 18 are black, and 1 is green. A ball is rolled around the wheel and ends up in one of the slots with equal probability. Suppose that a player bets on red. If a $1 bet is placed, the player wins $1 if the ball ends up in a red slot (the player's $1 bet is returned). If the ball ends up in a black or green slot, the player loses $1. Find the expected value of this game to the player in

(a) The United States.

(b) France.

3.2.13 In the gambling game Craps (see Exercise 2.3.13), the player wins $1 with probability 0.49293 and loses $1 with probability 0.50707 for each $1 bet. What is the expected value of the game to the player?

3.2.14 In the casino game High-Low, there are three possible bets. Assume that $1 is the size of the bet. A pair of fair six-sided dice is rolled and their sum is calculated. If you bet low, you win $1 if the sum of the dice is {2, 3, 4, 5, 6}. If you bet high, you win $1 if the sum of the dice is {8, 9, 10, 11, 12}. If you bet on {7}, you win $4 if a sum of 7 is rolled. Find the expected value of the game to the bettor for each of these bets.

3.3 THE MEAN, VARIANCE, AND STANDARD DEVIATION

Certain mathematical expectations are so important that they have special names. In this section we consider two of them: the mean and the variance.

If X is a random variable of the discrete type with p.d.f. $f(x)$ and space

$$R = \{b_1, b_2, b_3, \ldots\},$$

then

$$E(X) = \sum_{x \in R} x f(x)$$

$$= b_1 f(b_1) + b_2 f(b_2) + b_3 f(b_3) + \cdots$$

is the weighted average of the numbers belonging to R, where the weights are given by the p.d.f. $f(x)$. We call $E(X)$ the **mean** of X (or the mean of the distribution) and denote it by μ. That is, $\mu = E(X)$.

Remark   In mechanics, the weighted average of the points $b_1, b_2, b_3, \ldots$ in one-dimensional space is called the centroid of the system. Those readers without a mechanics background can think of the centroid as being the point of balance for the system in which the weights $f(b_1), f(b_2), f(b_3), \ldots$ are placed upon the points $b_1, b_2, b_3, \ldots$.

Example 3.3.1

Let X have the p.d.f.

$$f(x) = \begin{cases} \dfrac{1}{8}, & x = 0, 3, \\ \dfrac{3}{8}, & x = 1, 2. \end{cases}$$

The mean of X is

$$\mu = E(X) = 0\left(\frac{1}{8}\right) + 1\left(\frac{3}{8}\right) + 2\left(\frac{3}{8}\right) + 3\left(\frac{1}{8}\right) = \frac{3}{2}. \qquad \diamond$$

The next example shows that if the outcomes of X are equally likely (i.e., each of the outcomes has the same probability), then the mean of X is the arithmetic average of these outcomes.

Example 3.3.2

Roll a fair die and let X denote the outcome. Thus X has the p.d.f. $f(x) = 1/6$, $x = 1, 2, 3, 4, 5, 6$. The mean of X is

$$\mu = E(X) = \sum_{x=1}^{6} x\left(\frac{1}{6}\right) = \frac{1 + 2 + 3 + 4 + 5 + 6}{6} = \frac{7}{2},$$

which is the arithmetic average of the first six positive integers.   ◇

We have noted that the mean $\mu = E(X)$ is the centroid of a system of weights or a measure of the central location of the probability distribution of X. A measure of the dispersion or spread of a distribution is defined as follows. If $u(x) = (x - \mu)^2$ and $E[(X - \mu)^2]$ is finite, the **variance**, frequently denoted by σ^2 or Var(X), of a random variable X of the discrete type (or variance of the distribution) is defined by

$$\sigma^2 = E[(X - \mu)^2] = \sum_{x \in R}(x - \mu)^2 f(x).$$

The positive square root of the variance is called the **standard deviation** of X and is denoted by

$$\sigma = \sqrt{\text{Var}(X)} = \sqrt{E[(X - \mu)^2]}.$$

Example 3.3.3

Let X equal the outcome when rolling a fair die. From Example 3.3.2 we know that $\mu = 7/2 = 3.5$. Thus

$$\sigma^2 = E[(X - 3.5)^2] = \sum_{x=1}^{6}(x - 3.5)^2\left(\frac{1}{6}\right)$$

$$= \left[(1 - 3.5)^2 + (2 - 3.5)^2 + \cdots + (6 - 3.5)^2\right]\left(\frac{1}{6}\right)$$

$$= \frac{35}{12}.$$

The standard deviation of X is $\sigma = \sqrt{35/12} = 1.708$, approximately.   ◇

It is worthwhile to note that the variance can be computed in another manner. We have

$$\sigma^2 = E[(X - \mu)^2] = E(X^2 - 2\mu X + \mu^2),$$

which, by the distributive property of E, is

$$\sigma^2 = E(X^2) - 2\mu E(X) + \mu^2 = E(X^2) - 2\mu^2 + \mu^2 = E(X^2) - \mu^2.$$

Sometimes $\sigma^2 = E(X^2) - \mu^2$ provides an easier way of computing Var(X) than does $\sigma^2 = E[(X - \mu)^2]$. Thus, in Example 3.3.3 we could have first computed

$$E(X^2) = \sum_{x=1}^{6} x^2 \left(\frac{1}{6}\right) = \frac{1^2 + 2^2 + 3^2 + 4^2 + 5^2 + 6^2}{6} = \frac{91}{6}$$

and then

$$\sigma^2 = \mathrm{Var}(X) = \frac{91}{6} - \left(\frac{7}{2}\right)^2 = \frac{35}{12}.$$

Example 3.3.4

Let the p.d.f. of X be defined by $f(x) = x/6$, $x = 1, 2, 3$. The mean of X is

$$\mu = E(X) = 1\left(\frac{1}{6}\right) + 2\left(\frac{2}{6}\right) + 3\left(\frac{3}{6}\right) = \frac{7}{3}.$$

To find the variance and standard deviation of X we first find

$$E(X^2) = 1^2\left(\frac{1}{6}\right) + 2^2\left(\frac{2}{6}\right) + 3^2\left(\frac{3}{6}\right) = \frac{36}{6} = 6.$$

Thus the variance of X is

$$\sigma^2 = E(X^2) - \mu^2 = 6 - \left(\frac{7}{3}\right)^2 = \frac{5}{9}$$

and the standard deviation of X is

$$\sigma = \sqrt{5/9} = 0.745. \qquad \diamond$$

Although most students understand that $\mu = E(X)$ is, in some sense, a measure of the middle of the distribution of X, it is difficult to get much of a feeling for the variance and the standard deviation. The following example illustrates that the standard deviation is a measure of dispersion or spread of the points belonging to the space R.

Example 3.3.5

Let X have the p.d.f. $f(x) = 1/3$, $x = -1, 0, 1$. Here the mean is

$$\mu = \sum_{x=-1}^{1} xf(x) = (-1)\left(\frac{1}{3}\right) + (0)\left(\frac{1}{3}\right) + (1)\left(\frac{1}{3}\right) = 0.$$

Accordingly, the variance, denoted by σ_X^2, is

$$\sigma_X^2 = E[(X - 0)^2]$$

$$= \sum_{x=-1}^{1} x^2 f(x)$$

$$= (-1)^2 \left(\frac{1}{3}\right) + (0)^2 \left(\frac{1}{3}\right) + (1)^2 \left(\frac{1}{3}\right)$$

$$= \frac{2}{3},$$

so the standard deviation is $\sigma_X = \sqrt{2/3}$. Next let another random variable Y have the p.d.f. $g(y) = 1/3$, $y = -2, 0, 2$. Its mean is also zero, and it is easy to show that $\text{Var}(Y) = 8/3$, so the standard deviation of Y is $\sigma_Y = 2\sqrt{2/3}$. Here the standard deviation of Y is twice that of the standard deviation of X, reflecting the fact that the probability of Y is spread out twice as much as that of X.   ◇

Now let X be a random variable with mean μ_X and variance σ_X^2. Of course, $Y = aX + b$, where a and b are constants, is a random variable, too. The mean of Y is

$$\mu_Y = E(Y) = E(aX + b) = aE(X) + b = a\mu_X + b.$$

Moreover, the variance of Y is

$$\sigma_Y^2 = E[(Y - \mu_Y)^2] = E[(aX + b - a\mu_X - b)^2] = E[a^2(X - \mu_X)^2] = a^2\sigma_X^2.$$

Thus $\sigma_Y = |a|\sigma_X$. For illustration, in Example 3.3.5 we note that the relationship between the two distributions could be explained by $Y = 2X$ so that $\sigma_Y^2 = 4\sigma_X^2$ and thus $\sigma_Y = 2\sigma_X$, which we had observed there. In addition, we see that adding or subtracting a constant from X does not change the variance. For illustration, $\text{Var}(X - 1) = \text{Var}(X)$, because $a = 1$ and $b = -1$.

Let r be a positive integer. If

$$E(X^r) = \sum_{x \in R} x^r f(x)$$

is finite, it is called the rth **moment** of the distribution about the origin. The expression **moment** has its origin in the study of mechanics. In addition, the expectation

$$E[(X - b)^r] = \sum_{x \in R} (x - b)^r f(x)$$

is called the rth moment of the distribution about b.

For a given positive integer r,

$$E[(X)_r] = E[X(X - 1)(X - 2) \cdots (X - r + 1)]$$

is called the rth factorial moment. We note that the second factorial moment is equal to the difference of the second and first moments:

$$E[X(X - 1)] = E(X^2) - E(X).$$

There is another formula that can be used for computing the variance that uses the second factorial moment and sometimes simplifies the calculations. First find the values of $E(X)$ and $E[X(X-1)]$. Then

$$\sigma^2 = E[X(X-1)] + E(X) - [E(X)]^2,$$

since, using the distributive property of E, this becomes

$$\sigma^2 = E(X^2) - E(X) + E(X) - [E(X)]^2 = E(X^2) - \mu^2.$$

Example 3.3.6

Continuing with Example 3.3.4, we find that

$$E[X(X-1)] = (1)(0)\left(\frac{1}{6}\right) + (2)(1)\left(\frac{2}{6}\right) + (3)(2)\left(\frac{3}{6}\right) = \frac{22}{6}.$$

So

$$\sigma^2 = E[X(X-1)] + E[X] - [E(X)]^2$$
$$= \frac{22}{6} + \frac{7}{3} - \left(\frac{7}{3}\right)^2 = \frac{5}{9}$$

is the variance of X.   ◇

Remark   In Section 1.2, we called $x_1, x_2, \ldots, x_n$ a sample if these were n observations resulting from n independent replications of a random experiment. We create the **empirical distribution** if we place the weight (probability) of $1/n$ on each of n observations $x_1, x_2, \ldots, x_n$. Then the mean of this empirical distribution is

$$\sum_{i=1}^{n} x_i \frac{1}{n} = \frac{\displaystyle\sum_{i=1}^{n} x_i}{n} = \overline{x}.$$

The sample mean $\overline{x}$ is the **mean of the empirical distribution**. We shall see that $\overline{x}$ is usually close in value to $\mu = E(X)$; thus, when μ is unknown, $\overline{x}$ can be used to estimate μ.

Similarly, the **variance of the empirical distribution** can be computed. Let v denote this variance so that it is equal to

$$v = \sum_{i=1}^{n} (x_i - \overline{x})^2 \frac{1}{n} = \sum_{i=1}^{n} x_i^2 \frac{1}{n} - \overline{x}^2 = \frac{1}{n}\sum_{i=1}^{n} x_i^2 - \overline{x}^2.$$

This last statement is true because, in general,

$$\sigma^2 = E(X^2) - \mu^2.$$

Note that there is a relationship between the sample variance s^2 and variance v of the empirical distribution, namely $s^2 = nv/(n-1)$. Of course, with large n, the difference between s^2 and v is very small. Usually, we use s^2 to estimate σ^2 when σ^2 is unknown. The reason for this is given later in the book.

EXERCISES

3.3.1 Let the p.d.f. of X be given by $f(0) = 3/10$, $f(1) = 3/10$, $f(2) = 1/10$, and $f(3) = 3/10$. Compute the mean, variance, and standard deviation of X.

3.3.2 If the p.d.f. of X is given by $f(x)$, (i) depict the p.d.f. as a probability histogram, and find the values of (ii) μ, (iii) σ^2, and (iv) σ. (v) Mark the five points μ, $\mu \pm \sigma$, $\mu \pm 2\sigma$ on the graph.

(a) $f(x) = \dfrac{2 - |x - 4|}{4}$, $x = 3, 4, 5.$

(b) $f(x) = \dfrac{4 - |x - 4|}{16}$, $x = 1, 2, 3, 4, 5, 6, 7.$

3.3.3 Find the mean and variance for the following discrete distributions:

(a) $f(x) = \dfrac{1}{5}$, $x = 5, 10, 15, 20, 25.$

(b) $f(x) = 1$, $x = 5.$

(c) $f(x) = \dfrac{4 - x}{6}$, $x = 1, 2, 3.$

3.3.4 Suppose that a school has 20 classes: 16 with 25 students in each, three with 100 students in each, and one with 300 students, for a total of 1000 students.

(a) What is the average class size?

(b) Select a student randomly out of the 1000 students. Let the random variable X equal the size of the class to which this student belongs and define the p.d.f. of X.

(c) Find the mean of X. Does this answer surprise you?

3.3.5 For the following distributions, find $\mu = E(X)$ and then find $\sigma^2 = E[(X - \mu)^2]$:

(a) $f(x) = \dfrac{1 + |x - 3|}{11}$, $x = 1, 2, 3, 4, 5.$

(b) $f(0) = 8/27$, $f(1) = 12/27$, $f(2) = 6/27$, $f(3) = 1/27$.

3.3.6 For the following distributions, find $\mu = E(X)$, $E[X(X - 1)]$, and $\sigma^2 = E[X(X - 1)] + E(X) - \mu^2$:

(a) $f(x) = \dfrac{3!}{x!(3 - x)!} \left(\dfrac{1}{4}\right)^x \left(\dfrac{3}{4}\right)^{3-x}$, $x = 0, 1, 2, 3.$

(b) $f(x) = \dfrac{4!}{x!(4 - x)!} \left(\dfrac{1}{2}\right)^4$, $x = 0, 1, 2, 3, 4.$

3.3.7 Given $E(X + 4) = 10$ and $E[(X + 4)^2] = 116$, determine (a) $\mathrm{Var}(X + 4)$, (b) μ, and (c) σ^2.

3.3.8 Let μ and σ^2 denote the mean and variance of the random variable X. Find $E[(X - \mu)/\sigma]$ and $E\{[(X - \mu)/\sigma]^2\}$.

3.3.9 A measure of **skewness** is defined by

$$\frac{E[(X - \mu)^3]}{\{E[(X - \mu)^2]\}^{3/2}} = \frac{E[(X - \mu)^3]}{(\sigma^2)^{3/2}} = \frac{E[(X - \mu)^3]}{\sigma^3}.$$

When a distribution is symmetrical about the mean, the skewness is equal to zero. If the probability histogram has a longer "tail" to the right than to the left, the measure of skewness is positive, and we say that the distribution is skewed positively or to the right. If the probability histogram has a longer tail to the left than to the right, the measure of skewness is negative, and we say that the distribution is skewed negatively or to the left. If the p.d.f. of X is given by $f(x)$, (i) depict the p.d.f. as a probability histogram, and find the values of (ii) the mean, (iii) the standard deviation, and (iv) skewness.

(a)

$$f(x) = \begin{cases} \dfrac{2^{6-x}}{64}, & x = 1, 2, 3, 4, 5, 6, \\ \dfrac{1}{64}, & x = 7. \end{cases}$$

(b)

$$f(x) = \begin{cases} \dfrac{1}{64}, & x = 1, \\ \dfrac{2^{x-2}}{64}, & x = 2, 3, 4, 5, 6, 7. \end{cases}$$

3.3.10 Let the p.d.f. of X be defined by $f(x) = (2 - |x - 2|)/4$, $x = 1, 2, 3$.

(a) Calculate (i) μ, (ii) σ^2, and (iii) σ.

(b) Depict the p.d.f. as a probability histogram.

(c) Use the 50 following simulated observations of X and calculate the following characteristics of the empirical distribution: (i) the mean $\overline{x}$; (ii) the sample variance s^2; (iii) the sample standard deviation s.

```
2  2  2  3  2  2  2  2  2  3
1  2  1  2  3  1  1  2  2  3
1  1  2  3  2  1  2  2  3  3
3  2  2  2  2  2  3  3  1  3
2  2  2  2  2  3  2  3  2  2
```

(d) Superimpose the relative frequency histogram of these observations on the probability histogram.

3.3.11 Let X equal the larger outcome when a pair of four-sided dice is rolled. The p.d.f. of X is

$$f(x) = \frac{2x - 1}{16}, \qquad x = 1, 2, 3, 4.$$

(a) Find the mean, variance, and standard deviation of X.

(b) Calculate the sample mean, sample variance, and sample standard deviation of the following 100 simulated observations of X and compare these answers to those in part (a):

```
4  4  4  4  2  2  2  3  1  4  3  3  2  3  2  4  4  2  3  4
4  3  4  3  4  3  3  2  4  4  3  2  3  3  3  2  4  4  3  4
1  4  3  4  4  4  3  2  4  4  4  3  1  3  2  4  4  4  4  1
3  4  3  2  4  4  3  3  1  3  3  3  3  2  2  2  3  4  3  3
2  4  2  3  3  2  4  4  3  4  4  4  4  3  4  4  4  4  4  4
```

(c) Draw the graphs of the probability histogram and the relative frequency histogram on the same figure.

3.3.12 Let X be the smaller outcome when a pair of four-sided dice is rolled. The p.d.f. of X is

$$f(x) = \frac{9 - 2x}{16}, \qquad x = 1, 2, 3, 4.$$

(a) Find the mean, variance, and standard deviation of X.

(b) Calculate the sample mean, sample variance, and sample standard deviation of the following 100 simulated observations of X:

```
1 2 3 1 2 2 2 2 4 1 1 2 2 3 1 2 1 3 1 1
1 1 2 3 1 2 1 2 2 2 3 1 3 1 1 1 3 2 1 2
3 2 1 1 2 1 2 1 1 1 1 1 3 2 1 4 2 3 1 3
1 2 3 1 1 1 3 2 1 1 1 1 2 2 2 3 1 3 1 1
1 1 3 2 1 2 1 1 1 4 1 1 4 1 2 3 2 2 4 1
```

(c) Draw the graphs of the probability histogram and the relative frequency histogram on the same figure.

3.3.13 A Bingo card has 25 squares with numbers on 24 of them, the center being a free square. The integers that are placed on the Bingo card are selected randomly and without replacement from 1 to 75, inclusive. When a game called Cover-up is played, balls numbered from 1 to 75, inclusive, are selected randomly and without replacement until a player covers each of the numbers on a card. Let X equal the number of balls that must be drawn to cover all the numbers on a single card.

(a) Argue that the p.d.f. of X, for $x = 24, 25, \ldots, 75$, is

$$f(x) = \frac{\binom{24}{23}\binom{51}{x-24}}{\binom{75}{x-1}} \cdot \frac{1}{75-(x-1)} = \frac{\binom{24}{24}\binom{51}{x-24}}{\binom{75}{x}} \cdot \frac{24}{x}.$$

(b) What value of X is most likely to occur? In other words, what is the mode of this distribution?

(c) To show that the mean of X is $(24)(76)/25 = 72.96$ use the combinatorial identity

$$\binom{k+n+1}{k+1} = \sum_{x=k}^{n+k}\binom{x}{k}.$$

(d) Show that $E[X(X+1)] = \dfrac{24 \cdot 77 \cdot 76}{26} = 5401.8462.$

(e) Calculate $\mathrm{Var}(X) = E[X(X+1)] - E[X] - [E(X)]^2 = \dfrac{46{,}512}{8125} = 5.7246$ and $\sigma = 2.39.$

(f) The following 100 observations of X were simulated on the computer.

```
75 73 74 74 73 75 75 71 71  74 75 74 74 73 75
75 75 71 68 75 74 72 73 75  74 67 73 71 74 74
73 75 71 73 71 68 74 75 66  75 75 74 75 71 75
72 75 74 75 75 72 75 75 74  73 75 62 64 75 72
74 74 75 73 75 75 73 74 75  68 75 69 74 75 61
73 73 73 72 74 68 74 71 73  75 73 74 72 74 73
75 73 71 70 62 74 74 72 75  74
```

Compare these data with the probability model. In particular, make the following comparisons: (i) $\bar{x}$ with μ; (ii) s^2 with σ^2; (iii) s with σ; (iv) $(1/100) \sum_{i=1}^{100} x_i(x_i + 1)$ with $E[X(X+1)]$.

(g) Compare the probability histogram with the relative frequency histogram.

3.4 THE DISCRETE UNIFORM AND HYPERGEOMETRIC DISTRIBUTIONS

We now begin the task of developing discrete probability models that are appropriate for a variety of random experiments. For each of them we will give some examples and applications as well as the p.d.f., mean, variance, and standard deviation. Note that we usually use letters at the end of the alphabet, such as W, Y, Z as well as X, as random variables. Also note that the distributions that we are about to develop are summarized on the front endpapers as well as in the Appendix.

Perhaps the easiest probability model is the one that puts equal probability on each of the points in its space. We begin with an example.

Example 3.4.1

Several states select a three- or four-digit integer randomly for their daily lottery game. Each of these games depends on its ability to select the digits from 0 to 9, inclusive, randomly. That is, we are interested in a random variable Y that is equal to an integer selected randomly from 0 to 9, inclusive. It seems appropriate to define the p.d.f. of Y by

$$g(y) = P(Y = y) = \frac{1}{10}, \qquad y = 0, 1, 2, 3, 4, 5, 6, 7, 8, 9. \qquad \diamond$$

Generalizing this example, let X equal an integer that is selected randomly from the first m positive integers. We say that X has a **discrete uniform distribution on the integers** $1, 2, \ldots, m$. The p.d.f. of X is defined by

$$f(x) = \frac{1}{m}, \qquad x = 1, 2, \ldots, m.$$

The mean of X is given by

$$\mu = E(X) = \sum_{x=1}^{m} x\left(\frac{1}{m}\right) = \frac{1}{m} \sum_{x=1}^{m} x$$

$$= \left(\frac{1}{m}\right) \frac{m(m+1)}{2} = \frac{m+1}{2}.$$

To find the variance of X, we first find

$$E(X^2) = \sum_{x=1}^{m} x^2\left(\frac{1}{m}\right) = \frac{1}{m} \sum_{x=1}^{m} x^2$$

$$= \left(\frac{1}{m}\right) \frac{m(m+1)(2m+1)}{6} = \frac{(m+1)(2m+1)}{6}.$$

Thus the variance of X is

$$\sigma^2 = \text{Var}(X) = E[(X - \mu)^2]$$

$$= E(X^2) - \mu^2 = \frac{(m+1)(2m+1)}{6} - \left(\frac{m+1}{2}\right)^2$$

$$= \frac{m^2 - 1}{12}.$$

For example, we find that if X equals the outcome when rolling a fair six-sided die, the p.d.f. of X is

$$f(x) = \frac{1}{6}, \qquad x = 1, 2, 3, 4, 5, 6;$$

the respective mean and variance of X (see Examples 3.3.2 and 3.3.3) are

$$\mu = \frac{6+1}{2} = 3.5 \quad \text{and} \quad \sigma^2 = \frac{6^2 - 1}{12} = \frac{35}{12}.$$

If we let $Y = X - 1$, then Y is equal to an integer selected randomly from the first m nonnegative integers: $0, 1, 2, \ldots, m - 1$. The p.d.f. of Y is

$$g(y) = P(Y = y) = P(X - 1 = y) = P(X = y + 1) = \frac{1}{m}, \quad y = 0, 1, \ldots, m-1.$$

Furthermore, the mean and variance of Y are

$$\mu_Y = E(Y) = E(X - 1) = E(X) - 1 = \frac{m+1}{2} - 1 = \frac{m-1}{2};$$

$$\sigma_Y^2 = \text{Var}(Y) = \text{Var}(X - 1) = \text{Var}(X) = \frac{m^2 - 1}{12}.$$

Thus, for Example 3.4.1, the mean and variance of Y are

$$\mu_Y = \frac{10 - 1}{2} = \frac{9}{2}$$

and

$$\sigma_Y^2 = \text{Var}(Y) = \frac{10^2 - 1}{12} = \frac{99}{12}.$$

Our next probability model uses the material in Section 2.2. Consider a collection of $N = N_1 + N_2$ similar objects, N_1 of them belonging to one of two dichotomous classes (red chips, say) and N_2 of them belonging to the second class (blue chips, say). A collection of n objects is selected from these N objects at random and without replacement. Find the probability that exactly x (where the integer x satisfies $x \leq n, x \leq N_1$, and $n - x \leq N_2$) of these n objects are red (i.e., x belong to the first class and $n - x$ belong to the second). Of course, we can select x red chips in any one of $\binom{N_1}{x}$ ways and $n - x$ blue chips in any one of $\binom{N_2}{n-x}$ ways. By the multiplication principle, the product $\binom{N_1}{x}\binom{N_2}{n-x}$ equals the number of ways the joint operation can be performed. If we assume that each of the $\binom{N}{n}$ ways of selecting n objects from $N = N_1 + N_2$ objects has the same probability, we have that the probability of selecting exactly x red chips is

$$P(X = x) = \frac{\binom{N_1}{x}\binom{N_2}{n-x}}{\binom{N}{n}},$$

where $x \leq n$, $x \leq N_1$, and $n - x \leq N_2$. We say that the random variable X has a **hypergeometric distribution**.

Example 3.4.2

In a small pond there are 50 fish, 10 of which have been tagged. If a fisherman's catch consists of 7 fish, selected at random and without replacement, and X denotes the number of tagged fish, the probability that exactly 2 tagged fish are caught is

$$P(X = 2) = \frac{\binom{10}{2}\binom{40}{5}}{\binom{50}{7}} = \frac{(45)(658,008)}{99,884,400} = \frac{246,753}{832,370} = 0.2964,$$

approximately.   ◇

Example 3.4.3

A lot, consisting of 100 fuses, is inspected by the following procedure. Five fuses are chosen at random and tested; if all 5 blow at the correct amperage, the lot is accepted. Suppose that the lot contains 20 defective fuses. If X is a random variable equal to the number of defective fuses in the sample of 5, the probability of accepting the lot is

$$P(X = 0) = \frac{\binom{20}{0}\binom{80}{5}}{\binom{100}{5}} = \frac{19,513}{61,110} = 0.3193,$$

approximately. More generally, if x is one of the numbers 0, 1, 2, 3, 4, or 5, then

$$P(X = x) = \frac{\binom{20}{x}\binom{80}{5-x}}{\binom{100}{5}}, \qquad x = 0, 1, 2, 3, 4, 5.$$ ◇

Example 3.4.4

Some examples of hypergeometric probability histograms are given in Figure 3.3. The values of N_1, N_2, and n are given with each figure.   ◇

Figure 3.3   Hypergeometric probability histograms

It should be true that $\sum_{x=0}^{n} f(x) = 1$. To show this, and also to find the mean and variance for a hypergeometric random variable, we need some mathematical tools that are given in the Appendix. For example, Theorem A.1 that is proved there is

$$\binom{N}{n} = \sum_{x=0}^{n} \binom{N_1}{x}\binom{N_2}{n-x}, \tag{3.2}$$

where $N = N_1 + N_2$ and it is understood that $\binom{k}{j} = 0$ if $j > k$. From this result,

dividing both sides of the equation by $\binom{N}{n}$, it follows that $\sum_{x=0}^{n} f(x) = 1$.

It requires a little tricky calculation to show that

$$\mu = n\left(\frac{N_1}{N}\right) = np$$

where $p = N_1/N$, and

$$\sigma^2 = n\left(\frac{N_1}{N}\right)\left(\frac{N_2}{N}\right)\left(\frac{N - n}{N - 1}\right) = np(1 - p)\left(\frac{N - n}{N - 1}\right).$$

Hints for these calculations are given in Exercises 3.4.16 and 3.4.17.

Example 3.4.5

For Example 3.4.2, $N_1 = 10$, $N_2 = 40$, $n = 7$. The mean of X is

$$\mu = n\left(\frac{N_1}{N}\right) = 7\left(\frac{10}{50}\right) = 1.4,$$

the variance of X is

$$\sigma^2 = n\left(\frac{N_1}{N}\right)\left(\frac{N_2}{N}\right)\left(\frac{N - n}{N - 1}\right)$$

$$= 7\left(\frac{10}{50}\right)\left(\frac{40}{50}\right)\left(\frac{43}{49}\right) = \frac{172}{175},$$

and the standard deviation of X is

$$\sigma = \sqrt{\sigma^2} = 0.9914. \qquad \diamond$$

Example 3.4.6

A supplier ships parts to another company in lots of 25 parts, some of which could be defective. The receiving company has an *acceptance sampling plan* which states that $n = 5$ parts are to be taken at random and without replacement from each lot. If there are zero defectives among those 5 parts, the entire lot is accepted; otherwise, the lot is rejected. That is, if one or more defectives are found among the 5 in the sample, the lot is rejected. If X is the number of defectives among the 5 sampled parts, X must have a hypergeometric distribution; but here we do not know the values of N_1, the number of defectives in the lot, and N_2, the number of good parts in the lot, but only that $N_1 + N_2 = 25$. Clearly, we want the probability of accepting the lot, namely $P(X = 0)$, to be large if N_1 is very small. This probability is usually called the *operating characteristic curve* when treated as a function of N_1 or equivalently, the fraction defective, $p = N_1/25$, in the lot. That is, the operating characteristic curve is

$$OC(p) = P(X = 0) = \frac{\binom{N_1}{0}\binom{25 - N_1}{5}}{\binom{25}{5}},$$

where $p = N_1/25$. Letting $N_1 = 0, 1, 2, \ldots$ so that $p = 0.00, 0.04, 0.08, \ldots$, we find using Table I in the Appendix that

$$OC(0.00) = \frac{\binom{0}{0}\binom{25}{5}}{\binom{25}{5}} = 1.00,$$

$$OC(0.04) = \frac{\binom{1}{0}\binom{24}{5}}{\binom{25}{5}} = 0.800,$$

$$OC(0.08) = \frac{\binom{2}{0}\binom{23}{5}}{\binom{25}{5}} = 0.633,$$

$$OC(0.12) = \frac{\binom{3}{0}\binom{22}{5}}{\binom{25}{5}} = 0.496,$$

$$OC(0.16) = \frac{\binom{4}{0}\binom{21}{5}}{\binom{25}{5}} = 0.383,$$

and so on. Upon observing these probabilities, the company might find this acceptance sampling plan unsatisfactory because possibly, with $N_1 = (25)(0.04) = 1$, $OC(0.04) = P(X = 0) = 0.8$ is too low and, with $N_1 = (25)(0.16) = 4$, $OC(0.16) = P(X = 0) = 0.383$ is too high. Accordingly, if this is so, the plan must be changed. In practice in industry, usually lot sizes are much larger than 25, sometimes running into the thousands, and the sample sizes might be in the hundreds rather than $n = 5$. Here we kept the numbers small so we could use Table I, but we illustrate what is actually done in practice with those larger values in Section 3.7.   ◇

EXERCISES

3.4.1 In a lottery, a three-digit integer is selected at random from 000 to 999, inclusive. Let X equal the integer that is selected on a particular day. Find

(a) The p.d.f. of X.

(b) The mean of X.

(c) The variance of X.

(d) The standard deviation of X.

3.4.2 Let X equal the outcome when rolling a fair 12-sided die. Give the p.d.f. of X and find the mean, variance, and standard deviation of X.

3.4.3 Let X equal the number of the month of the year in which a person is born; that is, record 1 for January, 2 for February, and so on. Does X have the same distribution, approximately, as the outcome when rolling a fair 12-sided die in Exercise 3.4.2? Use the following birth months of 60 students of one of the authors to help answer this question.

```
1   1  11   4  2  11  5  10  8  3   4   1  10   7   8
9   7   1   9  1  11  4   8  1  7   8  12   2   3  12
7   5   4   6  8  12  6  12  9  9   4  10   7   1  10
9  11   8  12  7   1  4   6  5  5   6   4   5  10  10
```

In particular, find

(a) The sample mean, $\bar{x}$.

(b) The sample variance, s^2.

(c) The sample standard deviation, s.

Compare these sample characteristics with the distribution characteristics from Exercise 3.4.2. Also,

(d) Construct a relative frequency histogram and superimpose the probability histogram.

(e) What is your conclusion using these data?

3.4.4 Let X have a uniform distribution on the integers from 1 to 365, inclusive.

(a) What is the p.d.f. of X?

(b) Find the mean, variance, and standard deviation of X.

(c) Let Y equal the day of the year on which a person is born. Does Y have the same distribution as X, approximately? To help answer this question, use the following data which give the day of the year on which 50 beginning math students were born.

```
230  283  273   47  340    8  183   27   82  174
277   51  193   39  157  280  301  243  317  148
160  206  352  163   56  139  295   28  120  193
313  145  221   55   92  268  193  319   61   98
335  126  104  360   34   69  220    4  298  120
```

For these data, find (i) the sample mean, (ii) the sample variance, and (iii) the sample standard deviation, comparing these sample characteristics of Y with the distribution characteristics of X. What is your conclusion with these limited data?

3.4.5 Let a and b be integers with $a < b$. Let Y have a discrete uniform distribution on the integers between a and b, inclusive.

(a) Define the p.d.f. of Y.

(b) What are the mean, variance, and standard deviation of Y?

3.4.6 A bag contains 24 pieces of candy, of which 12 are peppermints and 12 are butterscotch. Let X equal the number of peppermints in a sample of five pieces of candy selected at random and without replacement from the bag. Find

(a) $P(X = 2)$.

(b) $P(X \le 2)$.

(c) $\mu = E(X)$.

(d) $\sigma^2 = \text{Var}(X)$.

3.4.7 Suppose that there are 3 defective items in a lot of 50 items. A sample of size 10 is taken at random and without replacement. Let X denote the number of defective items in the sample. Find

(a) The probability that the sample contains exactly 1 defective item.

(b) The probability that the sample contains at most 1 defective item.

(c) The mean and variance of X.

3.4.8 A jar contains 25 pieces of candy, of which 11 are yogurt-covered nuts and 14 are yogurt-covered raisins. Let X equal the number of nuts in a random sample of 7 pieces of candy that are selected without replacement. Find

(a) $P(X = 3)$.

(b) $P(X = 6)$.

(c) The mean and variance of X.

3.4.9 In a lot of 100 light bulbs, there are 5 bad bulbs. An inspector inspects 10 bulbs selected at random. Find the probability of finding at least one defective bulb. HINT: First compute the probability of finding no defectives in the sample.

3.4.10 In Lotto 49, Michigan's lottery game, a player selects 6 integers out of the first 49 positive integers. The state then randomly selects 6 out of the first 49 integers. Cash prizes are given to a player who matches 4, 5, or 6 integers. Let X equal the number of integers selected by a player that match integers selected by the state.

(a) State the p.d.f. of X.

(b) Calculate the mean, variance, and standard deviation of X.

(c) What value of X is most likely to occur?

(d) On February 25, 1995, the jackpot was worth $45,000,000. When the prize is this large, a lot of bets are placed. Out of the 25,000,000 bets that were placed, 3 people matched all 6 numbers, with each winning $15,000,000 (most of this paid by losers during preceding games), 390 matched 5 numbers to win $2500, and 22,187 matched 4 numbers to win $100. Are these numbers of winners consistent with the probability model?

(e) A mathematics professor convinced some colleagues to pool their lotto bets so that they were able to purchase 138 tickets together. They let the state computer randomly select the numbers on which they placed their bets. Among their 138 bets, 65 matched 0 of the winning Lotto numbers, $\{3, 20, 33, 34, 43, 46\}$, 55 matched 1, 16 matched 2, and 2 matched 3 numbers. How do these results compare with what they could have expected?

3.4.11 In Michigan's lottery game Keno, the player picks 10 integers between 1 and 80, inclusive. The state selects 22 integers randomly from 1 to 80, inclusive. Let X equal the number of integers selected by the player that match integers selected by the state. If $X = 10, 9, 8, 7, 6$, the player wins \$250,000, \$2500, \$250, \$25, and \$7, respectively. If $X = 0$, the player receives a free \$1 Instant Ticket.

    (a) State the p.d.f. of X.

    (b) Calculate the mean, variance, and standard deviation of X.

    (c) How many standard deviations above the mean are 6, 7, 8, 9, and 10, respectively?

    (d) How many standard deviations below the mean is 0?

    (e) What is the expected value of this game to the player, assuming that the prize for 0 matches is worth \$1 and the cost for one ticket is \$1?

    (f) Of course, in part (e), a \$1 ticket is not really worth \$1, but rather the expected value of the game, say E. Hence we have the equation

$$E = (E)P(X = 0) + (7)P(X = 6) + (25)P(X = 7) + (250)P(X = 8)$$
$$+ (2500)P(X = 9) + (250{,}000)P(X = 10).$$

    Solve for E.

3.4.12 A bridge hand is defined as 13 cards selected at random without replacement from a deck of 52 playing cards.

    (a) What is the probability of being dealt at random a bridge hand that does not contain a spade?

    (b) What is the probability that a bridge hand contains exactly five hearts?

    (c) Find the probability that a bridge hand contains at most one ace.

3.4.13 For the distributions depicted in Figure 3.3, find μ, σ^2, and σ when

    (a) $N_1 = 8$, $N_2 = 32$, and $n = 16$.

    (b) $N_1 = 16$, $N_2 = 16$, and $n = 16$.

    (c) $N_1 = 16$, $N_2 = 64$, and $n = 16$.

    (d) $N_1 = 40$, $N_2 = 40$, and $n = 16$.

3.4.14 A lot of $N_1 + N_2 = 25$ items is accepted if the number of defectives, X, among $n = 5$ items taken at random and without replacement from the lot is less than or equal to 1. The operating characteristic curve is

$$OC(p) = P(X \le 1) = P(X = 0) + P(X = 1),$$

where $p = N_1/25$. Determine the probabilities $OC(0.04)$, $OC(0.08)$, $OC(0.12)$, and $OC(0.16)$ and compare them to those in Example 3.4.6.

3.4.15 A lot of $N_1 + N_2 = 100$ items is accepted if the number of defectives, X, among $n = 10$ items taken at random and without replacement from the lot is equal to zero. The operating characteristic curve is $OC(p) = P(X = 0)$, where $p = N_1/100$. Evaluate $OC(p)$ where $p = 0.00, 0.01, 0.02$, and 0.05. Here Table I cannot be used, but these probabilities are not difficult to evaluate with a small calculator.

3.4.16 To find the mean of a hypergeometric random variable,

(a) First note that

$$\binom{N}{n} = \binom{N}{n}\binom{N-1}{n-1}.$$

(b) In the summation for the mean, make the change of variables $k = x - 1$ and replace the denominator using the expression in part (a).

(c) In the resulting expression, use Equation 3.2 with N and n replaced with $N-1$ and $n-1$ respectively, to show that, with $p = N_1/N$,

$$\mu = n\left(\frac{N_1}{N}\right) = np.$$

3.4.17 To find the variance of a hypergeometric random variable:

(a) First show that

$$E[X(X-1)] = \frac{N_1(N_1-1)(n)(n-1)}{N(N-1)}$$

making a change of variables $k = x - 2$ and noting that

$$\binom{N}{n} = \frac{N(N-1)}{n(n-1)}\binom{N-2}{n-2}.$$

(b) Now show that, with $p = N_1/N$,

$$\sigma^2 = n\left(\frac{N_1}{N}\right)\left(\frac{N_2}{N}\right)\left(\frac{N-n}{N-1}\right) = np(1-p)\left(\frac{N-n}{N-1}\right).$$

3.5 BERNOULLI TRIALS AND THE BINOMIAL DISTRIBUTION

The probability models for random experiments described in this section occur frequently in applications.

A **Bernoulli experiment** is a random experiment, the outcome of which can be classified in but one of two mutually exclusive and exhaustive ways, say, success or failure (e.g., female or male, life or death, nondefective or defective). A sequence of **Bernoulli trials** occurs when a Bernoulli experiment is performed several *independent* times so that the probability of success, say p, remains the *same* from trial to trial. That is, in such a sequence we let p denote the probability of success on each trial. In addition, we shall frequently let $q = 1 - p$ denote the probability of failure; that is, we shall use q and $1 - p$ interchangeably.

Example 3.5.1

Suppose that the probability of germination of a beet seed is 0.8 and the germination of a seed is called a success. If we plant 10 seeds and can assume that the germination of one seed is independent of the germination of another seed, this would correspond to 10 Bernoulli trials with $p = 0.8$.   ◇

Example 3.5.2

In the Michigan daily lottery the probability of winning when placing a six-way boxed bet is 0.006. A bet placed on each of 12 successive days would correspond to 12 Bernoulli trials with $p = 0.006$. ◇

Let X be a random variable associated with a Bernoulli trial by defining it as follows:

$$X(\text{success}) = 1 \quad \text{and} \quad X(\text{failure}) = 0.$$

That is, the two outcomes, success and failure, are denoted by one and zero, respectively. The p.d.f. of X can be written as

$$f(x) = p^x (1 - p)^{1-x}, \quad x = 0, 1,$$

and we say that X has a **Bernoulli distribution**. The expected value of X is

$$\mu = E(X) = \sum_{x=0}^{1} x \, p^x (1 - p)^{1-x} = (0)(1 - p) + (1)(p) = p,$$

and the variance of X is

$$\sigma^2 = \text{Var}(X) = \sum_{x=0}^{1} (x - p)^2 p^x (1 - p)^{1-x}$$

$$= (0 - p)^2 (1 - p) + (1 - p)^2 p = p(1 - p) = pq.$$

It follows that the standard deviation of X is

$$\sigma = \sqrt{p(1 - p)} = \sqrt{pq}.$$

In a sequence of n Bernoulli trials, we shall let X_i denote the Bernoulli random variable associated with the ith trial. An observed sequence of n Bernoulli trials will then be an n-tuple of zeros and ones, and we often call this collection a **random sample** of size n from a Bernoulli distribution.

Example 3.5.3

Out of millions of instant lottery tickets, suppose that 20% are winners. If five such tickets are purchased, $(0, 0, 0, 1, 0)$ is a possible observed sequence in which the fourth ticket is a winner and the other four are losers. Assuming independence among winning and losing tickets, the probability of this outcome is

$$(0.8)(0.8)(0.8)(0.2)(0.8) = (0.2)(0.8)^4. \qquad ◇$$

Example 3.5.4

If five beet seeds are planted in a row, a possible observed sequence would be $(1, 0, 1, 0, 1)$, in which the first, third, and fifth seeds germinated and the other two did not. If the probability of germination is $p = 0.8$, the probability of this outcome is, assuming independence,

$$(0.8)(0.2)(0.8)(0.2)(0.8) = (0.8)^3(0.2)^2. \qquad \diamond$$

In a sequence of Bernoulli trials, we are often interested in the total number of successes and not in the order of their occurrence. If we let the random variable X equal the number of observed successes in n Bernoulli trials, the possible values of X are $0, 1, 2, \ldots, n$. If x successes occur, where $x = 0, 1, 2, \ldots, n$, then $n - x$ failures occur. The number of ways of selecting x positions for the x successes in the n trials is

$$\binom{n}{x} = \frac{n!}{x!(n-x)!}.$$

Since the trials are independent and since the probabilities of success and failure on each trial are, respectively, p and $q = 1 - p$, the probability of each of these ways is $p^x(1-p)^{n-x}$. Thus the p.d.f. of X, say $f(x)$, is the sum of the probabilities of these $\binom{n}{x}$ mutually exclusive events; that is,

$$f(x) = \binom{n}{x} p^x(1-p)^{n-x}, \qquad x = 0, 1, 2, \ldots, n.$$

These probabilities are called binomial probabilities, and the random variable X is said to have a **binomial distribution**.

Summarizing, a binomial experiment satisfies the following properties:

1. A Bernoulli (success–failure) experiment is performed n times.
2. The trials are independent.
3. The probability of success on each trial is a constant p; the probability of failure is $q = 1 - p$.
4. The random variable X equals the number of successes in the n trials.

A binomial distribution will be denoted by the symbol $b(n, p)$ and we say that the distribution of X is $b(n, p)$. The constants n and p are called the **parameters** of the binomial distribution; they correspond to the number n of independent trials and the probability p of success on each trial. Thus, if we say that the distribution of X is $b(12, 1/4)$, we mean that X is the number of successes in a random sample of size $n = 12$ from a Bernoulli distribution with $p = 1/4$.

Example 3.5.5

In the instant lottery with 20% winning tickets, if X is equal to the number of winning tickets among $n = 8$ that are purchased, the probability of purchasing 2 winning tickets is

$$f(2) = P(X = 2) = \binom{8}{2}(0.2)^2(0.8)^6 = 0.2936.$$

The distribution of the random variable X is $b(8, 0.2)$. $\qquad \diamond$

Example 3.5.6

In order to obtain a better feeling for the effect of the parameters n and p on the distribution of probabilities, four probability histograms are displayed in Figure 3.4.   ◇

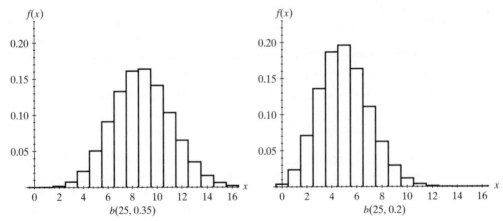

Figure 3.4   Binomial probability histograms

Example 3.5.7

In Example 3.5.1, the number X of seeds that germinate in $n = 10$ independent trials is $b(10, 0.8)$: that is,

$$f(x) = \binom{10}{x}(0.8)^x(0.2)^{10-x}, \qquad x = 0, 1, 2, \ldots, 10.$$

In particular,

$$P(X \le 8) = 1 - P(X = 9) - P(X = 10)$$
$$= 1 - 10(0.8)^9(0.2) - (0.8)^{10} = 0.6242.$$

Also, we could find, with a little more work,

$$P(X \le 6) = \sum_{x=0}^{6} \binom{10}{x}(0.8)^x(0.2)^{10-x}. \qquad \diamond$$

Cumulative probabilities like those in the preceding example are often of interest. We call the function defined by

$$F(x) = P(X \le x)$$

the **cumulative distribution function**, or more simply, the **distribution function** of the random variable X. Values of the distribution function of a random variable X that is $b(n, p)$ are given in Table II in the Appendix for selected values of n and p. The use of this table is illustrated in the next example.

Example 3.5.8

Leghorn chickens are raised for laying eggs. Let $p = 0.5$ be the probability of a female chick hatching. Assuming independence, let X equal the number of female chicks out of 10 newly hatched chicks selected at random. Then the distribution of X is $b(10, 0.5)$. The probability of 5 or fewer female chicks is, using Table II,

$$P(X \le 5) = 0.6230.$$

The probability of exactly 6 female chicks is

$$P(X = 6) = \binom{10}{6}\left(\frac{1}{2}\right)^6\left(\frac{1}{2}\right)^4$$
$$= P(X \le 6) - P(X \le 5)$$
$$= 0.8281 - 0.6230 = 0.2051,$$

since $P(X \le 6) = 0.8281$. The probability of at least 6 female chicks is

$$P(X \ge 6) = 1 - P(X \le 5) = 1 - 0.6230 = 0.3770. \qquad \diamond$$

Although probabilities for the binomial distribution $b(n, p)$ are given in Table II for selected values of p that are less than or equal to 0.5, the next example demonstrates that this table can also be used for values of p that are greater than 0.5. And in later sections we learn how to approximate certain binomial probabilities with those of other distributions.

Example 3.5.9

Suppose that we are in one of those rare times when 65% of the American public approve of the way the president of the United States is handling his job. Take a random sample of $n = 8$ Americans and let Y equal the number who give approval. Then the distribution of Y is $b(8, 0.65)$. To find $P(Y \geq 6)$, note that

$$P(Y \geq 6) = P(8 - Y \leq 8 - 6) = P(X \leq 2),$$

where $X = 8 - Y$ counts the number who disapprove. Since $q = 1 - p = 0.35$ equals the probability of disapproval by each person selected, the distribution of X is $b(8, 0.35)$ (see Figure 3.5). From Table II, since $P(X \leq 2) = 0.4278$, it follows that $P(Y \geq 6) = 0.4278$. Similarly,

$$\begin{aligned} P(Y \leq 5) &= P(8 - Y \geq 8 - 5) \\ &= P(X \geq 3) = 1 - P(X \leq 2) \\ &= 1 - 0.4278 = 0.5722 \end{aligned}$$

and

$$\begin{aligned} P(Y = 5) &= P(8 - Y = 8 - 5) \\ &= P(X = 3) = P(X \leq 3) - P(X \leq 2) \\ &= 0.7064 - 0.4278 = 0.2786. \end{aligned}$$ ◇

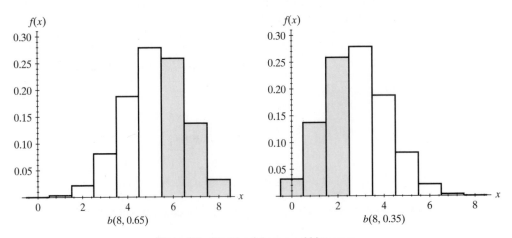

Figure 3.5 Presidential approval histogram

Recall that if n is a positive integer, then

$$(a + b)^n = \sum_{x=0}^{n} \binom{n}{x} b^x a^{n-x}.$$

Thus the sum of the binomial probabilities, if we use the binomial expansion above with $b = p$ and $a = 1 - p$, is

$$\sum_{x=0}^{n} \binom{n}{x} p^x (1-p)^{n-x} = [(1-p) + p]^n = 1,$$

a result that had to follow from the fact that $f(x)$ is a p.d.f.

We use the binomial expansion to find the mean and the variance of a binomial random variable X that is $b(n, p)$. The mean is given by

$$\mu = E(X) = \sum_{x=0}^{n} x \, \frac{n!}{x!(n-x)!} \, p^x (1-p)^{n-x}.$$

Since the first term of this sum is equal to zero, this can be written as

$$\mu = \sum_{x=1}^{n} \frac{n!}{(x-1)!(n-x)!} \, p^x (1-p)^{n-x}$$

because $x/x! = 1/(x-1)!$ when $x > 0$.

If we let $k = x - 1$ in the latter sum, we obtain

$$\mu = \sum_{k=0}^{n-1} \frac{n!}{k!(n-k-1)!} \, p^{k+1} (1-p)^{n-k-1}$$

$$= np \sum_{k=0}^{n-1} \frac{(n-1)!}{k!(n-1-k)!} \, p^k (1-p)^{n-1-k} = np$$

because the last summand is that of the binomial p.d.f. $b(n-1, p)$.

To find the variance, we first find the second factorial moment $E[X(X-1)]$:

$$E[X(X-1)] = \sum_{x=0}^{n} x(x-1) \, \frac{n!}{x!(n-x)!} \, p^x (1-p)^{n-x}.$$

The first two terms in this summation equal zero; thus we find that

$$E[X(X-1)] = \sum_{x=2}^{n} \frac{n!}{(x-2)!(n-x)!} \, p^x (1-p)^{n-x}$$

after observing that $x(x-1)/x! = 1/(x-2)!$ when $x > 1$. Letting $k = x - 2$, we obtain

$$E[X(X-1)] = \sum_{k=0}^{n-2} \frac{n!}{k!(n-k-2)!} \, p^{k+2} (1-p)^{n-k-2}$$

$$= n(n-1)p^2 \sum_{k=0}^{n-2} \frac{(n-2)!}{k!(n-2-k)!} \, p^k (1-p)^{n-2-k}.$$

Since the last summand is that of the binomial p.d.f. $b(n - 2, p)$, we obtain

$$E[X(X - 1)] = n(n - 1)p^2.$$

Thus

$$\sigma^2 = \text{Var}(X) = E(X^2) - [E(X)]^2 = E[X(X - 1)] + E(X) - [E(X)]^2$$
$$= n(n - 1)p^2 + np - (np)^2 = -np^2 + np = np(1 - p).$$

Summarizing, if X is $b(n, p)$, then

$$\mu = np, \quad \sigma^2 = np(1 - p) = npq, \quad \text{and} \quad \sigma = \sqrt{np(1 - p)}.$$

Note that when p is the probability of success on each trial, the expected number of successes in n trials is np, a result that agrees with most of our intuitions.

Example 3.5.10

Suppose that observation over a long period of time has disclosed that, on the average, one out of 10 items produced by a process is defective. Select five items independently from the production line and test them. Let X denote the number of defective items among the $n = 5$ items. Then X is $b(5, 0.1)$. Furthermore,

$$E(X) = 5(0.1) = 0.5, \quad \text{Var}(X) = 5(0.1)(0.9) = 0.45.$$

For example, the probability of observing at most one defective item is

$$P(X \le 1) = \binom{5}{0}(0.1)^0(0.9)^5 + \binom{5}{1}(0.1)^1(0.9)^4 = 0.9185. \qquad \diamond$$

The next example shows the relationship between the binomial probability model and a set of observed data. Replications of this example would of course yield different results, in which the fits could be better or worse than the given one.

Example 3.5.11

Consider the simple experiment of flipping a fair coin five independent times. If X equals the number of heads that is observed, then X is $b(5, 0.5)$, $\mu = 2.5$, $\sigma^2 = 1.25$, and $\sigma = 1.118$. This experiment was simulated 100 times, yielding the following data:

```
2 3 2 4 1 2 1 1 4 2  4 2 0 4 4 2 4 4 3 4
2 2 4 4 1 1 3 3 1 4  2 3 1 2 4 1 2 5 3 2
4 3 2 2 2 3 5 2 0 3  2 1 3 4 2 2 4 0 2 1
3 3 2 3 2 1 3 2 2 2  1 1 3 3 1 1 4 2 1 5
3 2 3 0 3 5 3 2 4 3  3 5 2 3 3 1 3 2 1 1
```

For these data $\bar{x} = 2.47$, $s^2 = 1.5243$, and $s = 1.235$. In Figure 3.6 the probability histogram and the relative frequency histogram (as dashed lines) are given. $\qquad \diamond$

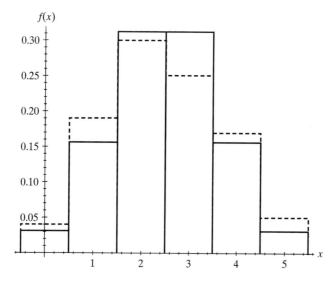

Figure 3.6   Number of heads when five coins are flipped

Suppose that an urn contains N_1 success balls and N_2 failure balls and we let $p = N_1/(N_1 + N_2)$. Let X equal the number of success balls in a random sample of size n that is taken from this urn. If the sampling is done one at a time with replacement, the distribution of X is $b(n, p)$; and if the sampling is done without replacement, X has a hypergeometric distribution with p.d.f.

$$f(x) = \frac{\binom{N_1}{x}\binom{N_2}{n - x}}{\binom{N_1 + N_2}{n}}, \qquad x = 0, 1, \ldots, N_1.$$

When $N_1 + N_2$ is large and n is relatively small, it makes little difference if the sampling is done with or without replacement. In Figure 3.7 the probability histograms are compared for different combinations of n, N_1, and N_2. You are asked to compute some of these probabilities in Exercise 3.5.19.

EXERCISES

3.5.1 An urn contains 7 red and 11 white balls. Draw one ball at random from the urn. Let $X = 1$ if a red ball is drawn, and let $X = 0$ if a white ball is drawn. Give the p.d.f., mean, and variance of X.

3.5.2 Suppose that in Exercise 3.5.1, $X = 1$ if a red ball is drawn, and $X = -1$ if a white ball is drawn. Give the p.d.f., mean, and variance of X.

3.5.3 On a six-question multiple-choice test there are five possible answers, of which one is correct (C) and four are incorrect (I). If a student guesses randomly and independently, find the probability of

$b(8, 0.2)$; Hypergeometric, $N_1 = 8$, $N_2 = 32$, $n = 8$

$b(8, 0.2)$; Hypergeometric, $N_1 = 16$, $N_2 = 64$, $n = 8$

$b(16, 0.2)$; Hypergeometric, $N_1 = 16$, $N_2 = 64$, $n = 16$

$b(16, 0.2)$; Hypergeometric, $N_1 = 32$, $N_2 = 128$, $n = 16$

Figure 3.7 Binomial (solid) and hypergeometric (dashed) probability histograms

(a) Being correct only on questions 1 and 4 (i.e., scoring C, I, I, C, I, I).

(b) Being correct on two questions.

3.5.4 Suppose that 40% of American homes have a microwave oven. Let X equal the number of American homes in a random sample of $n = 25$ homes that have a microwave oven. Find the probability that

(a) X is at most 11.

(b) X is at least 7.

(c) X is equal to 8.

(d) Give the mean, variance, and standard deviation of X.

3.5.5 Suppose that 70% of all residents of the Wellsburg, West Virginia, area are at risk for heart disease (*Time*, December 12, 1988). Let X equal the number out of $n = 16$ of these residents who are selected randomly that are at risk for heart disease. Find the probability that

(a) X is at least 13.

(b) X is at most 11.

(c) X is equal to 12.

(d) Give the mean, variance, and standard deviation of X.

3.5.6 It is claimed that 15% of the ducks in a particular region have patent schistosome infection. Suppose that seven ducks are selected at random. Let X equal the number of ducks that are infected.

(a) Assuming independence, how is X distributed?

(b) Find (i) $P(X \geq 2)$, (ii) $P(X = 1)$, and (iii) $P(X \leq 3)$.

3.5.7 Suppose that 2000 points are selected independently and at random from the unit square $S = \{(x, y) : 0 \leq x < 1, 0 \leq y < 1\}$. Let W equal the number of points that fall in $A = \{(x, y) : x^2 + y^2 < 1\}$.

(a) How is W distributed?

(b) Give the mean, variance, and standard deviation of W.

(c) What is the expected value of $W/500$?

(d) By selecting n pairs of random numbers from Table IX in the Appendix (or using a random number generator on a computer), generate n observations of W and use these to give an estimate for π. (Of course, we know the real value of π, and more will be said about estimation later.)

(e) By using n triplets of random numbers, could you extend part (d) to estimate the volume of a ball of radius 1 in 3-space, namely $(4/3)\pi$?

(f) Using a computer, extend these techniques to estimate the "volume" of a ball of radius 1 in n-space.

3.5.8 It is believed that 20% of Americans do not have any health insurance. Suppose that this is true and let X equal the number with no health insurance in a random sample of $n = 15$ Americans.

(a) How is X distributed?

(b) Give the mean, variance, and standard deviation of X.

(c) Find $P(X \geq 5)$.

3.5.9 Let p equal the proportion of all college and university students who would answer "yes" to the question, "Would you drink from the same glass as your friend if you suspected that this friend were an AIDS virus carrier?" Assume that $p = 0.10$. Let X equal the number of students out of a random sample of size $n = 9$ who would answer "yes" to this question.

(a) How is X distributed?

(b) Give the values of the mean, variance, and standard deviation of X.

(c) Find (i) $P(X = 2)$ and (ii) $P(X \geq 2)$.

3.5.10 Suppose that, in manufacturing, the probability of a certain item being defective is $p = 0.05$. Suppose further that the quality of an item is independent of the quality of the other manufactured items. An inspector selects six items at random. Let X equal the number of defective items in the sample.

(a) How is X distributed?

(b) Give the values of $E(X)$ and $\text{Var}(X)$.

(c) Find (i) $P(X = 0)$, (ii) $P(X \leq 1)$, and (iii) $P(X \geq 2)$.

3.5.11 According to a *Des Moines Register* poll, 40% of Iowa farmers support an independent Palestinian state. Let X equal the number of Iowa farmers out of a random sample of $n = 20$ who support an independent Palestinian state. Find

(a) $P(X \leq 10)$.

(b) $P(X \geq 8)$.

(c) $P(X = 9)$.

3.5.12 One of the 12 courses in a Chinese vegetarian banquet at a Buddhist temple was a platter of mushrooms. When attempting to pick up a round, slippery mushroom with chopsticks, an American tourist was successful 55% of the time on the first try. Let X equal the number of times that the tourist was successful on the first try at picking up 14 mushrooms. Assume that the "first tries" are independent trials.

(a) Give the mean and variance of X.

(b) Find $P(X < 8)$ and $P(X > 6)$.

3.5.13 A random variable X has a binomial distribution with mean 6 and variance 3.6. Find $P(X = 4)$.

3.5.14 A certain type of mint has a label weight of 20.4 grams. Suppose that the probability is 0.90 that a mint weighs more than 20.7 grams. Let X equal the number of mints that weigh more than 20.7 grams in a sample of 8 mints selected at random.

(a) How is X distributed if we assume independence?

(b) Find (i) $P(X = 8)$, (ii) $P(X \leq 6)$, and (iii) $P(X \geq 6)$.

3.5.15 It is given that the probability of germination of each beet seed is 0.8. Thus if X denotes the number of seeds that germinate in a set of nine seeds, then X is $b(9, 0.8)$, provided that the assumption of independence is valid.

(a) Give the values of μ, σ^2, and σ.

(b) Draw the probability histogram for the p.d.f. of X.

(c) If 100 sets of nine beet seeds were planted, calculate $\bar{x}$, s^2, and s for the following 100 observations.

```
7 9 7 7 9 7 9 8 5   8 8 6 8 7 8 8 9 7   8 8
8 7 8 7 7 7 6 7 9   7 6 7 6 7 6 5 7 8   7 4
8 8 7 6 9 7 8 7 7   6 7 9 6 6 8 8 8 7   9 5
7 4 7 8 8 9 6 7 6   8 7 5 7 7 9 7 7 8   7 9
7 5 9 8 7 7 8 9 8   5 9 9 8 6 9 8 8 8   9 7
```

(d) Superimpose the relative frequency histogram of these observations on the graph of the probability histogram.

3.5.16 In the casino game Chuck-a-Luck, three unbiased six-sided dice are rolled. One possible bet is $1 on fives, and the payoff is equal to $1 for each five on that roll. In addition, the dollar bet is returned if at least one five is rolled. The dollar that was bet is lost only if no fives are rolled. Let X denote the payoff for this game. Then X can equal $-1, 1, 2$, or 3.

(a) Determine the p.d.f. $f(x)$.

(b) Calculate μ, σ^2, and σ.

(c) Depict the p.d.f. as a probability histogram.

(d) Use the 100 following simulated observations of X to calculate $\bar{x}$, s^2, and s:

−1	1	−1	−1	1	−1	−1	−1	−1	1
−1	1	−1	2	1	−1	−1	−1	1	−1
1	−1	−1	1	1	−1	−1	−1	−1	1
−1	−1	2	1	1	−1	−1	1	−1	1
−1	−1	1	2	−1	−1	1	−1	−1	−1
−1	1	−1	−1	−1	1	−1	1	−1	2
−1	1	−1	1	−1	2	1	−1	−1	1
−1	1	1	−1	1	1	−1	3	1	−1
−1	−1	1	1	1	1	−1	−1	2	−1
1	−1	1	−1	1	2	−1	1	1	−1

(e) Superimpose the relative frequency histogram of these observations on the probability histogram.

3.5.17 It is claimed that for a particular lottery, 1/10 of the 50,000,000 tickets will win a prize. What is the probability of winning at least one prize if you purchase (a) 10 tickets and (b) 15 tickets?

3.5.18 For the lottery described in Exercise 3.5.17, find the smallest number of tickets that must be purchased so that the probability of winning at least one prize is greater than (a) 0.50 and (b) 0.95.

3.5.19 Construct a table that gives the probabilities for the $b(8, 0.2)$ distribution, the hypergeometric distribution with $N_1 = 8$, $N_2 = 32$, and $n = 8$, and the hypergeometric distribution with $N_1 = 16$, $N_2 = 64$, and $n = 8$ (see Figure 3.7).

3.6 GEOMETRIC AND NEGATIVE BINOMIAL DISTRIBUTIONS

To obtain a binomial random variable, we observed a sequence of n Bernoulli trials and counted the number of successes. Suppose now that we do not fix the number of Bernoulli trials in advance but instead continue to observe the sequence of Bernoulli trials until a certain number, say r, of successes occurs. The random variable of interest is the number of trials needed to observe the rth success.

We first discuss this problem when $r = 1$. That is, consider a sequence of Bernoulli trials with probability p of success. This sequence is observed until the first success occurs. Let X denote the trial number on which this first success occurs. For example, if F and S represent failure and success, respectively, and the sequence starts with $F, F, F, S, \ldots$, then $X = 4$. Moreover, because the trials are independent, the probability of such a sequence is

$$P(X = 4) = (q)(q)(q)(p) = q^3 p = (1 - p)^3 p.$$

In general, the p.d.f., $f(x) = P(X = x)$, of X is given by

$$f(x) = (1 - p)^{x-1} p, \qquad x = 1, 2, \ldots$$

because there must be $x - 1$ failures before the first success that occurs on trial x. We say that X has a **geometric distribution**.

Recall that for a geometric series (see the Appendix for a review), the sum is given by

$$\sum_{k=0}^{\infty} ar^k = \sum_{k=1}^{\infty} ar^{k-1} = \frac{a}{1-r}$$

when $|r| < 1$. Thus

$$\sum_{x=1}^{\infty} f(x) = \sum_{x=1}^{\infty} (1-p)^{x-1} p = \frac{p}{1-(1-p)} = 1,$$

so that $f(x)$ does satisfy the properties of a p.d.f.

From the sum of a geometric series we also note that when k is an integer,

$$P(X > k) = \sum_{x=k+1}^{\infty} (1-p)^{x-1} p = \frac{(1-p)^k p}{1-(1-p)} = (1-p)^k = q^k,$$

and thus the value of the distribution function at a positive integer k is

$$P(X \le k) = \sum_{x=1}^{k} (1-p)^{x-1} p = 1 - P(X > k) = 1 - (1-p)^k = 1 - q^k.$$

Example 3.6.1

Some biology students were checking the eye color for a large number of fruit flies. For the individual fly, suppose that the probability of white eyes is 1/4 and the probability of red eyes is 3/4, and that we may treat these observations as having independent Bernoulli trials. The probability that at least four flies have to be checked for eye color to observe a white-eyed fly is given by

$$P(X \ge 4) = P(X > 3) = q^3 = \left(\frac{3}{4}\right)^3 = \frac{27}{64} = 0.4219.$$

The probability that at most four flies have to be checked for eye color to observe a white-eyed fly is given by

$$P(X \le 4) = 1 - q^4 = 1 - \left(\frac{3}{4}\right)^4 = \frac{175}{256} = 0.6836.$$

The probability that the first fly with white eyes is the fourth fly that is checked is

$$P(X = 4) = q^{4-1} p = \left(\frac{3}{4}\right)^3 \left(\frac{1}{4}\right) = \frac{27}{256} = 0.1055.$$

It is also true that

$$P(X = 4) = P(X \le 4) - P(X \le 3)$$
$$= [1 - (3/4)^4] - [1 - (3/4)^3]$$
$$= \left(\frac{3}{4}\right)^3 \left(\frac{1}{4}\right).$$

In general,

$$f(x) = P(X = x) = \left(\frac{3}{4}\right)^{x-1} \left(\frac{1}{4}\right), \qquad x = 1, 2, 3, \ldots.$$

A probability histogram is shown in Figure 3.8 with $r = 1$, $p = 0.25$.   ◇

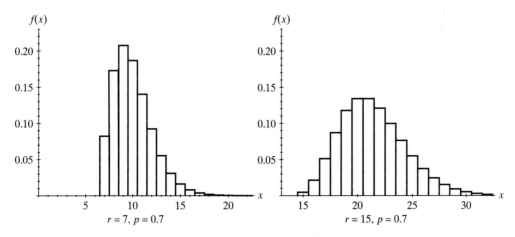

Figure 3.8   Negative binomial probability histograms

To find the mean and the variance for the geometric distribution, we will use the following results about the sum and the first and second derivatives of a geometric series. For $-1 < w < 1$, let

$$g(w) = \sum_{k=0}^{\infty} a\, w^k = \frac{a}{1-w}. \tag{3.3}$$

Then

$$g'(w) = \sum_{k=1}^{\infty} a\,k\,w^{k-1} = \frac{a}{(1-w)^2} \tag{3.4}$$

and

$$g''(w) = \sum_{k=2}^{\infty} a\,k\,(k-1)w^{k-2} = \frac{2a}{(1-w)^3}. \tag{3.5}$$

If X has a geometric distribution and $0 < p < 1$, then the mean of X is given by

$$E(X) = \sum_{k=1}^{\infty} x\,q^{x-1}p = \frac{p}{(1-q)^2} = \frac{1}{p}$$

using Equation 3.4 with $a = p$ and $w = q$. Note, for example, that if $p = 1/4$ is the probability of success, then $E(X) = 1/(1/4) = 4$ trials are needed on the average to observe a success.

To find the variance of X, we first shall find the second factorial moment $E[X(X-1)]$. We have

$$E[X(X-1)] = \sum_{x=1}^{\infty} x(x-1)q^{x-1}p = \sum_{x=2}^{\infty} p\,q\,x(x-1)q^{x-2} = \frac{2pq}{(1-q)^3} = \frac{2q}{p^2}$$

using Equation 3.5 with $a = pq$ and $w = q$. Thus the variance of X is

$$\text{Var}(X) = E(X^2) - [E(X)]^2 = \{E[X(X-1)] + E(X)\} - [E(X)]^2$$

$$= \frac{2q}{p^2} + \frac{1}{p} - \frac{1}{p^2} = \frac{2q+p-1}{p^2} = \frac{q}{p^2} = \frac{1-p}{p^2}.$$

The standard deviation of X is

$$\sigma = \sqrt{(1-p)/p^2}.$$

Example 3.6.2

Continuing with Example 3.6.1, with $p = 1/4$, we obtain

$$\mu = \frac{1}{1/4} = 4,$$

$$\sigma^2 = \frac{3/4}{(1/4)^2} = 12,$$

and

$$\sigma = \sqrt{12} = 3.4641.$$

Locate μ, $\mu \pm \sigma$, and $\mu \pm 2\sigma$ on the graph in Figure 3.8 ($r = 1$, $p = 0.25$). Note that

$$P(X < 4) = 1 - \frac{27}{64} = 0.5781.$$

The probability that X is more than one standard deviation below the mean is

$$P(X < 4 - 3.4641) = P(X \leq 0) = 0.$$

However, the probability that X is more than one standard deviation above the mean is

$$P(X > 4 + 3.4641) = P(X > 7) = (0.75)^7 = 0.1335. \qquad \diamond$$

Example 3.6.3

Let X equal the number of people that you have to ask, selecting randomly, to find someone who was born in the same month as you were. Assuming that each month is equally likely, the p.d.f. of X is $f(x) = (11/12)^{x-1}(1/12)$, $x = 1, 2, 3, \ldots$. The mean, variance, and standard deviation of X are

$$\mu = \frac{1}{1/12} = 12, \quad \sigma^2 = \frac{11/12}{(1/12)^2} = (11)(12) = 132, \quad \text{and} \quad \sigma = 11.489.$$

Also,

$$P(X > 23) = (11/12)^{23} = 0.1352,$$
$$P(X < 10) = 1 - (11/12)^9 = 0.5430. \qquad \diamond$$

Consider the problem of obtaining a random ordering of the first n positive integers. For example, suppose that we would like a random ordering (permutation) of the first six positive integers. We could obtain this random ordering by rolling a fair six-sided die. The first cast of the die would give the first outcome in the random ordering. To obtain the second number in the ordering, only five of the six possible outcomes are eligible. After the first $k - 1$ positions have been filled with unique integers, the number of candidates for position k is $6 - k + 1$ for $k = 1, 2, 3, 4, 5, 6$. The probability of selecting one of these eligible integers is $p_k = (6 - k + 1)/6$. If X_k denotes the number of trials (casts of the die) needed to observe the first success (an integer that has not yet been selected), then X_k has a geometric distribution with $p = p_k$.

An interesting problem is to determine the average number of casts of the die that are needed to obtain a random ordering of 1, 2, 3, 4, 5, 6. Let X_k equal the number of casts required to fill position k. If we let $W = X_1 + X_2 + \cdots + X_6$, then W denotes the total number of casts required. So, intuitively, the average number of casts required is

$$
\begin{aligned}
E(W) &= E(X_1 + X_2 + \cdots + X_6) \\
&= E(X_1) + E(X_2) + \cdots + E(X_6) \\
&= 1 + \frac{1}{5/6} + \frac{1}{4/6} + \cdots + \frac{1}{1/6} = 14.7.
\end{aligned}
$$

Remark   The fact that $E(X_1 + X_2 + \cdots + X_6) = E(X_1) + E(X_2) + \cdots + E(X_6)$ is considered in Section 5.2.

Example 3.6.4

A computer science student was interested in simulating on the computer the shuffling of a deck of 52 playing cards. Essentially, this problem requires the random ordering of the first 52 positive integers. A random number generator can be used to obtain this ordering. If we think of this random number generator as a 52-sided die, what is the average or expected number of times that this "die" would have to be cast to obtain the random ordering? Letting W denote the total number of casts, we obtain

$$E(W) = \sum_{k=1}^{52} \frac{1}{(52 - k + 1)/52} = \sum_{k=1}^{52} \frac{52}{53 - k} = 235.978. \qquad \diamond$$

We turn now to the more general problem of observing a sequence of Bernoulli trials until exactly r successes occur, where r is a fixed positive integer. Let the random variable X denote the number of trials needed to observe the rth success. That is, X is the trial number on which the rth success is observed. By the multiplication rule of probabilities, the p.d.f. of X, say $g(x)$, equals the product of the probability

$$\binom{x-1}{r-1} p^{r-1}(1-p)^{x-r} = \binom{x-1}{r-1} p^{r-1} q^{x-r}$$

of obtaining exactly $r - 1$ successes in the first $x - 1$ trials and the probability p of a success on the rth trial. Thus the p.d.f. of X is

$$g(x) = \binom{x-1}{r-1} p^{r}(1-p)^{x-r} = \binom{x-1}{r-1} p^{r} q^{x-r}, \qquad x = r, r+1, \ldots.$$

We say that X has a **negative binomial distribution**.

Example 3.6.5

Suppose that the biology students in Example 3.6.1 check the eye color of fruit flies until the third white-eyed fruit fly is observed. Then, with $p = 1/4$, the probability of observing the third white-eyed fruit fly on the thirteenth trial is

$$\binom{13-1}{3-1} \left(\frac{1}{4}\right)^3 \left(\frac{3}{4}\right)^{13-3} = 0.0581.$$

The probability that at most 12 flies have to be checked to find the third white-eyed fly is

$$\sum_{x=3}^{12} \binom{x-1}{3-1} \left(\frac{1}{4}\right)^3 \left(\frac{3}{4}\right)^{x-3} = 0.6093. \qquad \diamond$$

The reason for calling this the negative binomial distribution is the following. Consider $h(w) = (1 - w)^{-r}$, the binomial $(1 - w)$ with the negative exponent $-r$. Using Maclaurin's series expansion (see Section A.4), we have

$$(1 - w)^{-r} = \sum_{k=0}^{\infty} \frac{h^{(k)}(0)}{k!} w^k = \sum_{k=0}^{\infty} \binom{r + k - 1}{r - 1} w^k. \tag{3.6}$$

If we let $x = k + r$ in the summation, then $k = x - r$ and

$$(1 - w)^{-r} = \sum_{x=r}^{\infty} \binom{r + x - r - 1}{r - 1} w^{x-r} = \sum_{x=r}^{\infty} \binom{x - 1}{r - 1} w^{x-r},$$

the summand of which is, except for the factor p^r, the negative binomial probability when $w = q$. In particular, we see that the sum of the probabilities for the negative binomial distribution is 1 because

$$\sum_{x=r}^{\infty} g(x) = \sum_{x=r}^{\infty} \binom{x - 1}{r - 1} p^r q^{x-r} = p^r (1 - q)^{-r} = 1.$$

In Exercise 3.6.21 the reader is asked to show that the mean of X is

$$\mu = E(X) = \frac{r}{p}.$$

Note that this mean is r times the mean of a geometric random variable.

In Exercise 3.6.22 the reader is asked to show that the variance of X is

$$\sigma^2 = \frac{rq}{p^2} = \frac{r(1 - p)}{p^2}.$$

Example 3.6.6

Suppose that during practice, a basketball player can make a free throw 80% of the time. Furthermore, assume that a sequence of free-throw shooting can be thought of as independent Bernoulli trials. Let X equal the minimum number of free throws that this player must attempt to make a total of 10 shots. The p.d.f. of X is

$$g(x) = \binom{x - 1}{10 - 1} (0.80)^{10} (0.20)^{x-10}, \qquad x = 10, 11, 12, \ldots.$$

The mean, variance, and standard deviation of X are

$$\mu = 10 \left(\frac{1}{0.80} \right) = 12.5, \qquad \sigma^2 = \frac{10(0.20)}{0.80^2} = 3.125, \qquad \text{and} \qquad \sigma = 1.768.$$

And we have, for example,

$$P(X = 12) = g(12) = \binom{11}{9} (0.80)^{10} (0.20)^2 = 0.2362. \qquad \diamond$$

Example 3.6.7

In order to consider the effect of p and r on the negative binomial distribution, Figure 3.8 gives the probability histograms for four combinations of p and r. Note that since $r = 1$ in the first of these, it represents a geometric p.d.f.   ◇

EXERCISES

3.6.1 If a student selects true or false at random on an examination, assuming independence among answers, determine the probability that

(a) The first correct answer is that of question 3.

(b) At most three questions must be answered to obtain the first correct answer.

3.6.2 Let X equal the number of people selected at random that you must ask in order to find someone with the same birthday as yours. Assuming each day of the year is equally likely (and ignoring February 29),

(a) What is the p.d.f. of X?

(b) Give the values of the mean, variance, and standard deviation of X.

(c) Find $P(X > 400)$ and $P(X < 300)$.

3.6.3 For each question on a multiple-choice test, there are five possible answers of which exactly one is correct. If a student selects answers at random, give the probability that the first question answered correctly is question 4.

3.6.4 The probability that a machine produces a defective item is 0.01. Each item is checked as it is produced. Assume that these are independent trials and compute the probability that at least 100 items must be checked to find one that is defective.

3.6.5 Apples are packaged automatically in "3-pound" bags. Suppose that 4% of the time the bag of apples weighs less than 3 pounds. If you select bags randomly and weigh them in order to find one underweight bag of apples, find the probability that the number of bags that must be selected is

(a) At least 20.

(b) At most 20.

(c) Exactly 20.

3.6.6 Suppose that the probability of manufacturing a defective rear-view mirror for a car is 0.03. Suppose further that the quality of any one mirror is independent of the quality of any other mirror. If an inspector selects mirrors at random from the production line, give the probability that the first defective mirror is the eighth mirror selected.

3.6.7 According to a representative for an automobile manufacturer, the company uses 3000 lock-and-key combinations on its vehicles. Suppose that you find a key for one of these cars.

(a) Give the expected number of vehicles that you would have to check to find one that your key fits.

(b) Give the probability that you would have to check at least 3000 vehicles to find one that your key fits.

(c) Give the probability that at most 2000 vehicles would have to be checked to find one that your key fits.

3.6.8 For a particular instant lottery the bettor can determine immediately whether a purchased ticket is a winning ticket. Suppose that the probability of a winning ticket is $p = 1/10$.

  (a) What is the expected number of tickets that must be purchased in order to find a winning ticket?

  (b) Give the probability that more than 10 tickets must be purchased to find a winning ticket.

  (c) Give the probability that exactly 20 tickets must be purchased to find two winning tickets.

3.6.9 Flip an unbiased coin in a sequence of independent trials. Compute the conditional probability that the first head is observed on the fifth trial, given that tails are observed on each of the first three trials.

3.6.10 Let X have a geometric distribution. Show that

$$P(X > k + j \mid X > k) = P(X > j),$$

where k and j are nonnegative integers. NOTE: We sometimes say that in this situation there has been loss of memory.

3.6.11 Let X equal the number of flips of a fair coin that are required to observe the same face on consecutive flips.

  (a) Find the p.d.f. of X.

  (b) Give the values of the mean, variance, and standard deviation of X.

  (c) Find the values of (i) $P(X \le 3)$, (ii) $P(X \ge 5)$, and (iii) $P(X = 3)$.

3.6.12 Let X equal the number of flips of a fair coin that are required to observe heads on consecutive flips. Let f_n equal the nth Fibonacci number where $f_1 = 1$, $f_2 = 1$, and $f_n = f_{n-1} + f_{n-2}$, $n = 3, 4, 5, \ldots$.

  (a) Show that the p.d.f. of X is

$$f(x) = \frac{f_{x-1}}{2^x}, \qquad x = 2, 3, 4, \ldots.$$

  (b) Use the fact that

$$f_x = \frac{1}{\sqrt{5}} \left[\left(\frac{1 + \sqrt{5}}{2} \right)^x - \left(\frac{1 - \sqrt{5}}{2} \right)^x \right]$$

  to show that $\sum_{x=2}^{\infty} f(x) = 1$.

  (c) Show that $\mu = E(X) = 6$.

  (d) Show that $E[X(X-1)] = 52$ so that the variance of X is $\sigma^2 = 22$ and the standard deviation is $\sigma = 4.690$.

  (e) Either simulate this experiment on the computer or physically toss a coin to support the theoretical characteristics of this distribution.

3.6.13 Let X equal the number of flips of a fair coin that are required to observe heads-tails on consecutive flips.

  (a) Find the p.d.f. of X.

  (b) Determine the values of the mean, variance, and standard deviation of X.

  (c) Find the values of (i) $P(X \le 3)$, (ii) $P(X \ge 5)$, and (iii) $P(X = 3)$.

3.6.14 **(a)** A fair six-sided die is rolled until a 3 is observed. What is the expected number of rolls needed?

(b) A pair of fair six-sided dice is rolled. What is the expected number of rolls needed to observe at least one 3 on at least one of the two dice?

(c) What is the smallest number of dice required so that the expected number of rolls needed to observe at least one 3 does not exceed 2?

3.6.15 One of four different prizes was randomly put into each box of a cereal. If a family decided to buy this cereal until it obtained at least one of each of the four different prizes, what is the expected number of boxes of cereal that must be purchased?

3.6.16 A tea distributor randomly placed one of 20 English porcelain miniature animals in a 100-bag box.

(a) On the average, how many boxes of tea must be purchased by a customer to obtain a complete collection consisting of 20 different animals?

(b) If the customer uses one tea bag per day, how long will it take, approximately, to obtain a complete collection?

3.6.17 An urn contains 11 red balls. A ball is drawn at random from the urn and replaced by a black ball. What is the expected number of draws needed so that all of the balls in the urn are black? NOTE: This exercise is based on an experiment in biology in which the students studied how food was "eaten" by paramecia.

3.6.18 Show that 63/512 is the probability that the fifth head is observed on the tenth independent flip of an unbiased coin.

3.6.19 A CD player has a magazine that holds six CDs. The machine is capable of randomly selecting a CD at random and then selecting a song randomly from that CD. Suppose that five CDs are albums by Paul McCartney and one is by Billy Joel. The player selects songs until a song by Billy Joel is played after which the machine is turned off. Give the probability that the machine is turned off after

(a) The sixth song.

(b) At least five songs have been played.

(c) Suppose now that the songs continue to be played until the second song by Billy Joel has been played. Find the probability that at most four songs are played.

3.6.20 Let X have a negative binomial distribution. As illustrated in Figure 3.8, verify that

(a) $P(X = 5) = P(X = 6)$ when $r = 4$ and $p = 0.6$.

(b) $P(X = 20) = P(X = 21)$ when $r = 15$ and $p = 0.7$.

3.6.21 To find the mean of a negative binomial random variable, X, use Maclaurin's series expansion for $h(w) = (1 - w)^{-r}$, Equation 3.6.

(a) Show that

$$h'(w) = r(1 - w)^{-r-1} = \sum_{k=1}^{\infty} \binom{r + k - 1}{r - 1} k w^{k-1}.$$

(b) Using the series expansion for $h'(w)$, show that $E(X - r) = p^r q r (1 - q)^{-r-1}$.

(c) Show that $E(X) = r/p$.

3.6.22 To find the variance of a negative binomial random variable, X, use the series expansion for $h'(w)$ in Exercise 3.6.21(a).

(a) Show that

$$h''(w) = r(r+1)(1-w)^{-r-2} = \sum_{k=2}^{\infty} \binom{r+k-1}{r-1} k(k-1) w^{k-2}.$$

(b) Show that $E[(X-r)(X-r-1)] = (q^2/p^2)(r)(r+1)$.

(c) Use Exercise 3.6.21(b) and part (b) of this exercise to show that the variance of $X - r$ is $\text{Var}(X-r) = rq/p^2$.

(d) Find the variance of X.

3.6.23 Suppose that a basketball player, different from the one in Example 3.6.6, can make a free throw 60% of the time. Let X equal the minimum number of free throws that this player must attempt to make a total of 10 shots.

(a) Give the mean, variance, and standard deviation of X.

(b) Find $P(X = 16)$.

3.7 THE POISSON DISTRIBUTION

Some experiments result in counting the number of times particular events occur in given times or on given physical objects. For example, we could count the number of phone calls arriving at a switchboard between 9 and 10 A.M., the number of flaws in 100 feet of wire, the number of customers that arrive at a ticket window between 12 noon and 2 P.M., or the number of defects in a 100-foot roll of aluminum screen that is 2 feet wide. Each count can be looked upon as a random variable associated with an approximate Poisson process provided the conditions in Definition 3.7.1 are satisfied.

Definition 3.7.1.   Let the number of changes that occur in a given continuous interval be counted. We have an **approximate Poisson process** with parameter $\lambda > 0$ if the following are satisfied:

(a) The numbers of changes occurring in nonoverlapping intervals are independent.

(b) The probability of exactly one change in a sufficiently short interval of length h is approximately λh.

(c) The probability of two or more changes in a sufficiently short interval is essentially zero.

Remark   In this definition, we have modified the usual requirements of a Poisson process by using the words *approximately* and *essentially* in conditions (b) and (c). We do this to avoid some advanced mathematics. Thus we refer to this as the *approximate* Poisson process.

Suppose that an experiment satisfies the three points of an approximate Poisson process. Let X denote the number of changes in an interval of "length 1" (where "length 1" represents one unit of the quantity under consideration). We would like to find an approximation for $P(X = x)$, where x is a nonnegative integer. To achieve this, we partition the unit interval into n subintervals of equal length $1/n$. If n is sufficiently large (i.e., much larger than x), we shall approximate the probability that x changes occur in this unit interval by finding the probability that one change occurs in each of exactly x of these n subintervals. The probability of one change occurring in any one subinterval of length $1/n$ is approximately $\lambda(1/n)$ by condition (b). The probability of two or more changes in any one subinterval is essentially zero by condition (c). So for each subinterval, exactly one change occurs with a probability of approximately $\lambda(1/n)$. Consider the occurrence or nonoccurrence of a change in each subinterval as a Bernoulli trial. By condition (a) we have a sequence of n Bernoulli trials with probability p approximately equal to $\lambda(1/n)$. Thus an approximation for $P(X = x)$ is given by the binomial probability

$$\frac{n!}{x!\,(n-x)!} \left(\frac{\lambda}{n}\right)^x \left(1 - \frac{\lambda}{n}\right)^{n-x}.$$

If n increases without bound, we have that

$$\lim_{n \to \infty} \frac{n!}{x!\,(n-x)!} \left(\frac{\lambda}{n}\right)^x \left(1 - \frac{\lambda}{n}\right)^{n-x}$$

$$= \lim_{n \to \infty} \frac{n(n-1)\cdots(n-x+1)}{n^x} \frac{\lambda^x}{x!} \left(1 - \frac{\lambda}{n}\right)^n \left(1 - \frac{\lambda}{n}\right)^{-x}.$$

Now, for fixed x, we have (see Section A.3)

$$\lim_{n \to \infty} \frac{n(n-1)\cdots(n-x+1)}{n^x} = \lim_{n \to \infty} \left[(1)\left(1 - \frac{1}{n}\right)\cdots\left(1 - \frac{x-1}{n}\right)\right] = 1,$$

$$\lim_{n \to \infty} \left(1 - \frac{\lambda}{n}\right)^n = e^{-\lambda},$$

$$\lim_{n \to \infty} \left(1 - \frac{\lambda}{n}\right)^{-x} = 1.$$

Thus

$$\lim_{n \to \infty} \frac{n!}{x!\,(n-x)!} \left(\frac{\lambda}{n}\right)^x \left(1 - \frac{\lambda}{n}\right)^{n-x} = \frac{\lambda^x e^{-\lambda}}{x!} = P(X = x).$$

The distribution of probability associated with this process has a special name. We say that the random variable X has a **Poisson distribution** if its p.d.f. is of the form

$$f(x) = \frac{\lambda^x e^{-\lambda}}{x!}, \qquad x = 0, 1, 2, \ldots,$$

where $\lambda > 0$.

It is easy to see that $f(x)$ enjoys the properties of a p.d.f. because clearly $f(x) \geq 0$ and, from the Maclaurin's series expansion of e^λ (see Section A.4), we have

$$\sum_{x=0}^{\infty} \frac{\lambda^x e^{-\lambda}}{x!} = e^{-\lambda} \sum_{x=0}^{\infty} \frac{\lambda^x}{x!} = e^{-\lambda} e^\lambda = 1.$$

To discover the exact role of the parameter $\lambda > 0$, let us find some of the characteristics of the Poisson distribution.

The mean for the Poisson distribution is given by

$$E(X) = \sum_{x=0}^{\infty} x \frac{\lambda^x e^{-\lambda}}{x!} = e^{-\lambda} \sum_{x=1}^{\infty} \frac{\lambda^x}{(x-1)!}$$

because $(0) f(0) = 0$ and $x/x! = 1/(x-1)!$ when $x > 0$. If we let $k = x - 1$, then

$$E(X) = e^{-\lambda} \sum_{k=0}^{\infty} \frac{\lambda^{k+1}}{k!} = \lambda e^{-\lambda} \sum_{k=0}^{\infty} \frac{\lambda^k}{k!} = \lambda e^{-\lambda} e^\lambda = \lambda.$$

That is, the parameter λ is the mean of the Poisson distribution.

We first determine the second factorial moment $E[X(X-1)]$ in order to find the variance. We have

$$E[X(X-1)] = \sum_{x=0}^{\infty} x(x-1) \frac{\lambda^x e^{-\lambda}}{x!} = e^{-\lambda} \sum_{x=2}^{\infty} \frac{\lambda^x}{(x-2)!}$$

because $(0)(0-1)f(0) = 0$, $(1)(1-1)f(1) = 0$, and $x(x-1)/x! = 1/(x-2)!$ when $x > 1$ If we let $k = x - 2$, then

$$E[X(X-1)] = e^{-\lambda} \sum_{k=0}^{\infty} \frac{\lambda^{k+2}}{k!} = \lambda^2 e^{-\lambda} \sum_{k=0}^{\infty} \frac{\lambda^k}{k!} = \lambda^2 e^{-\lambda} e^\lambda = \lambda^2.$$

Thus

$$\text{Var}(X) = E(X^2) - [E(X)]^2 = E[X(X-1)] + E(X) - [E(X)]^2$$
$$= \lambda^2 + \lambda - \lambda^2 = \lambda.$$

That is, for the Poisson distribution, $\mu = \sigma^2 = \lambda$.

Table III in the Appendix gives values of the distribution function of a Poisson random variable for selected values of λ. This table is illustrated in the next example.

Example 3.7.1

Let X have a Poisson distribution with a mean of $\lambda = 5$. Then using Table III,

$$P(X \leq 6) = \sum_{x=0}^{6} \frac{5^x e^{-5}}{x!} = 0.762,$$

$$P(X > 5) = 1 - P(X \leq 5) = 1 - 0.616 = 0.384,$$

and

$$P(X = 6) = P(X \le 6) - P(X \le 5) = 0.762 - 0.616 = 0.146. \qquad \diamond$$

Example 3.7.2

In order to consider the effect of λ on the p.d.f. $f(x)$ of X, Figure 3.9 shows the probability histograms of $f(x)$ for four different values of λ. $\qquad \diamond$

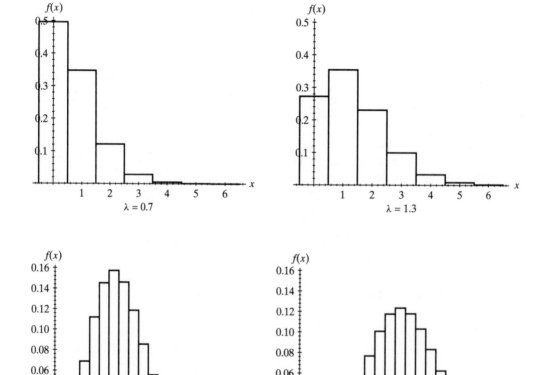

Figure 3.9   Poisson probability histograms

Example 3.7.3

Let X denote the number of alpha particles emitted by barium–133 in 1/10 of a second and counted by a Geiger counter. One hundred observations of X produced the data in Table 3.2. For these data, $\bar{x} = 559/100 = 5.59$, and

$s^2 = 3,619/99 - (559)^2/9900 = 4.992$. Thus we see that the sample mean and sample variance are fairly close to each other. In Figure 3.10 the probability histogram for a Poisson distribution with $\lambda = 5.59$ is drawn. The dashed lines represent the relative frequency histogram of the data. Accordingly, it looks as if the Poisson distribution provides a reasonable probability model for this random variable X. Later in the book, formal tests of the goodness of fit of various models are given.   ◇

TABLE 3.2   Barium–133

Outcome (x)	Frequency (f)	fx	fx^2
1	1	1	1
2	4	8	16
3	13	39	117
4	19	76	304
5	16	80	400
6	15	90	540
7	9	63	441
8	12	96	768
9	7	63	567
10	2	20	200
11	1	11	121
12	1	12	144
	100	559	3619

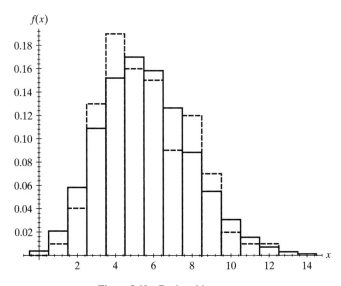

Figure 3.10   Barium histograms

If events in a Poisson process occur at a mean rate of λ per unit, then the expected number of occurrences in an interval of length t is λt. For example, if phone calls arrive at a switchboard following a Poisson process at a mean rate of three per minute, then the expected number of phone calls in a 5-minute period is $(3)(5) = 15$. Or if calls arrive at a mean rate of 22 in a 5-minute period, the expected number of calls per minute is $\lambda = 22(1/5) = 4.4$. Moreover, the number of occurrences, say X, in the interval of length t has the Poisson p.d.f.

$$f(x) = \frac{(\lambda t)^x e^{-\lambda t}}{x!}, \qquad x = 0, 1, 2, \ldots .$$

This follows by treating the interval of length t as if it were the "unit interval" with mean λt instead of λ.

Example 3.7.4

Flaws (bad records) on a used computer tape occur on the average of one flaw per 1200 feet. If one assumes a Poisson distribution, what is the distribution of X, the number of flaws in a 4800-foot roll? The expected number of flaws in $4800 = 4(1200)$ feet is 4; that is, $E(X) = 4$. Thus the p.d.f. of X is

$$f(x) = \frac{4^x e^{-4}}{x!}, \qquad x = 0, 1, 2, \ldots ,$$

and, in particular,

$$P(X = 0) = \frac{4^0 e^{-4}}{0!} = e^{-4} = 0.018,$$

$$P(X \leq 4) = 0.629,$$

by Table III.   ◇

Example 3.7.5

Telephone calls enter a college switchboard on the average of two every 3 minutes. If one assumes an approximate Poisson process, what is the probability of five or more calls arriving in a 9-minute period? Let X denote the number of calls in a 9-minute period. We see that $E(X) = 6$; that is, on the average, six calls will arrive during a 9-minute period. Thus,

$$P(X \geq 5) = 1 - P(X \leq 4) = 1 - \sum_{x=0}^{4} \frac{6^x e^{-6}}{x!}$$

$$= 1 - 0.285 = 0.715,$$

using Table III.   ◇

Not only is the Poisson distribution important in its own right, but it can also be used to approximate probabilities for a binomial distribution. If X has a Poisson distribution with parameter λ, we saw that with n large,

$$P(X = x) \approx \binom{n}{x}\left(\frac{\lambda}{n}\right)^x\left(1 - \frac{\lambda}{n}\right)^{n-x},$$

where $p = \lambda/n$ so that $\lambda = np$ in the binomial probability above. That is, if X has the binomial distribution $b(n, p)$ with large n and small p, then

$$\frac{(np)^x e^{-np}}{x!} \approx \binom{n}{x}p^x(1 - p)^{n-x}.$$

This approximation is reasonably good if n is large. But since λ was a fixed constant in that earlier argument, p should be small since $np = \lambda$. In particular, the approximation is quite accurate if $n \geq 20$ and $p \leq 0.05$ or if $n \geq 100$ and $p \leq 0.10$, but it is not bad in other situations violating these bounds somewhat, like $n = 50$ and $p = 0.12$.

Example 3.7.6

A manufacturer of Christmas tree light bulbs knows that 2% of its bulbs are defective. Assuming independence, we have a binomial distribution with parameters $p = 0.02$ and $n = 100$. To approximate the probability that a box of 100 of these bulbs contains at most three defective bulbs, we use the Poisson distribution with $\lambda = 100(0.02) = 2$, which gives

$$\sum_{x=0}^{3} \frac{2^x e^{-2}}{x!} = 0.857$$

from Table III. Using the binomial distribution, we obtain, after some tedious calculations,

$$\sum_{x=0}^{3} \binom{100}{x}(0.02)^x(0.98)^{100-x} = 0.859.$$

Hence, in this case, the Poisson approximation is extremely close to the true value, but much easier to find.   ◇

Remark   With the availability of better and faster computing power, it is often very easy to find binomial probabilities. So do not use the Poisson approximation if you are able to find the probability exactly.

Example 3.7.7

In Figure 3.11 Poisson probability histograms have been drawn along with dashed lines representing the respective binomial probabilities so that we can see whether or not these are close to each other. If the distribution of X is $b(n, p)$, the approximating Poisson distribution has a mean of $\lambda = np$. Note that the approximation is not good when p is large (e.g., $p = 0.30$).   ◇

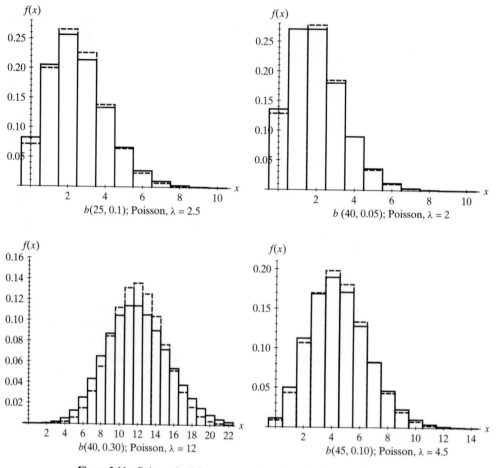

Figure 3.11   Poisson (solid) and binomial (dashed) probability histograms

Example 3.7.8

A lot of 1000 parts is shipped to a company. A sampling plan dictates that $n = 100$ parts are to be taken at random and without replacement and the lot accepted if no more than two of these 100 parts are defective. Here $Ac = 2$ is usually called the **acceptance number**. The operating characteristic curve

$$OC(p) = P(X \le 2),$$

where p is the fraction defective in the lot, is really the sum of the three hypergeometric probabilities

$$P(X = x) = \frac{\binom{N_1}{x}\binom{1000 - N_1}{100 - x}}{\binom{1000}{100}}, \qquad x = 0, 1, 2,$$

where $N_1 = (1000)(p)$. However, we have seen that the hypergeometric distribution can be approximated by the binomial distribution, which in turn can be approximated by the Poisson distribution when n is large and p is small. This is exactly our situation since $n = 100$ and since we are interested in values of p in the range 0.00 to 0.10. Thus

$$OC(p) = P(X \leq 2) \approx \sum_{x=0}^{2} \frac{(100p)^x e^{-100p}}{x!}.$$

Using Table III, we find that some approximate values of the operating characteristic curve are $OC(0.01) = 0.920$, $OC(0.02) = 0.677$, $OC(0.03) = 0.423$, $OC(0.05) = 0.125$, and $OC(0.10) = 0.003$ and it is plotted in Figure 3.12. If this is not a satisfactory operating characteristic curve for the type of parts involved, the sample size n and the acceptance number Ac must be changed.   ◇

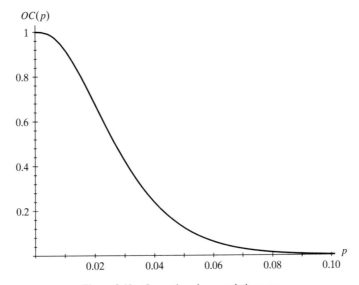

Figure 3.12   Operating characteristic curve

EXERCISES

3.7.1 Let X have a Poisson distribution with a mean of 4. Find

(a) $P(2 \leq X \leq 5)$.

(b) $P(X \geq 3)$.

(c) $P(X \leq 3)$.

3.7.2 Let X have a Poisson distribution with a variance of 3. Find $P(X = 2)$.

3.7.3 Customers arrive at a travel agency at a mean rate of 11 per hour. Assuming that the number of arrivals per hour has a Poisson distribution, give the probability that more than 10 customers arrive in a given hour.

3.7.4 If X has a Poisson distribution so that $3P(X = 1) = P(X = 2)$, find $P(X = 4)$.

3.7.5 Flaws in a certain type of drapery material appear on the average of one in 150 square feet. If we assume the Poisson distribution, find the probability of at most one flaw in 225 square feet.

3.7.6 A certain type of aluminum screen that is 2 feet wide has on the average one flaw in a 100-foot roll. Find the probability that a 50-foot roll has no flaws.

3.7.7 With probability 0.001, a prize of $499 is won in the Michigan daily lottery when a $1 straight bet is placed. Let Y equal the number of $499 prizes won by a gambler after placing n straight bets. Note that Y is $b(n, 0.001)$. After placing $n = 2000$ $1 bets, a gambler is behind if $\{Y \leq 4\}$. Use the Poisson distribution to approximate $P(Y \leq 4)$ when $n = 2000$.

3.7.8 Suppose that the probability of suffering a side effect from a certain flu vaccine is 0.005. If 1000 persons are inoculated, find approximately the probability that

   (a) At most 1 person suffers.

   (b) 4, 5, or 6 persons suffer.

3.7.9 A roll of a biased die results in a two only 1/10 of the time. Let X denote the number of two's in 100 rolls of this die. Approximate

   (a) $P(3 \leq X \leq 7)$.

   (b) $P(X \geq 5)$.

3.7.10 The mean of a Poisson random variable X is $\mu = 9$. Compute

$$P(\mu - 2\sigma < X < \mu + 2\sigma).$$

3.7.11 A Geiger counter was set up in the physics laboratory to record the number of alpha particle emissions of carbon–14 in 0.5 second. The following are 50 observations:

4	6	6	12	11	11	10	5	10	7
9	6	9	11	9	6	4	9	11	8
10	11	7	4	5	8	6	5	8	7
6	3	4	6	4	12	7	14	5	9
7	10	9	6	6	12	10	7	12	13

   (a) Calculate the sample mean and sample variance for these data. Are these two measurements close to each other?

   (b) On the same graph depict the probability histogram for a Poisson distribution with $\lambda = 8$ (we use 8 only because it is the closest value of λ to $\bar{x}$ in Table III), with the relative frequency histogram of the data superimposed.

3.7.12 For determining the half-lives of radioactive isotopes, it is important to know what the background radiation is in a given detector over a period of time. Data taken in a γ-ray detection experiment over 300 one-second intervals yielded the following data:

```
0 2 4 6 6 1 7 4 6 1 1 2 3 6 4 2 7 4 4 2
2 5 4 4 4 1 2 4 3 2 2 5 0 3 1 1 0 0 5 2
7 1 3 3 3 2 3 1 4 1 3 5 3 5 1 3 3 0 3 2
6 1 1 4 6 3 6 4 4 2 2 4 3 3 6 1 6 2 5 0
6 3 4 3 1 1 4 6 1 5 1 1 4 1 4 1 1 1 3 3
4 3 3 2 5 2 1 3 5 3 2 7 0 4 2 3 3 5 6 1
4 2 6 4 2 0 4 4 7 3 5 2 2 3 1 3 1 3 6 5
4 8 2 2 4 2 2 1 4 7 5 2 1 1 4 1 4 3 6 2
1 1 2 2 2 2 3 5 4 3 2 2 3 3 2 4 4 3 2 2
3 6 1 1 3 3 2 1 4 5 5 1 2 3 3 1 3 7 2 5
4 2 0 6 2 3 2 3 0 4 4 5 2 5 3 0 4 6 2 2
2 2 2 5 2 2 3 4 2 3 7 1 1 7 1 3 6 0 5 3
0 0 3 3 0 2 4 3 1 2 3 3 3 4 3 2 2 7 5 3
5 1 1 2 2 6 1 3 1 4 4 2 3 4 5 1 3 4 3 1
0 3 7 4 0 5 2 5 4 4 2 2 3 2 4 6 5 5 3 4
```

Do these look like observations of a Poisson random variable with mean $\lambda = 3$? To help answer this question,

(a) Find the frequencies of $0, 1, 2, \ldots, 8$.

(b) Calculate the sample mean and sample variance. Are they approximately equal to each other?

(c) Construct a probability histogram with $\lambda = 3$ and a relative frequency histogram on the same graph.

(d) What is your conclusion? (Later in the book we consider tests of hypotheses which help give a statistical answer.)

3.7.13 Let X equal the number of telephone calls that are received at the college switchboard from on-campus during a 15-minute period. The following numbers of calls were received during each of 26 time periods:

```
4  8  5  3  1  3  2  5  6  7  4  4  5
2  3  6  4  1  2  5  6  7  5  7  5  1
```

(a) Calculate the sample mean and sample variance for these data. Are they approximately equal to each other?

(b) Assume that $\lambda = 4.2$. Compare $P(X \leq 3)$ with the proportion of observations that are less than or equal to 3.

(c) Compare $P(X > 5)$ with the proportion of observations that are greater than 5.

(d) Does it look like the Poisson distribution with $\lambda = 4.2$ could be the correct probability model based on these limited data?

3.7.14 Let X equal the number of green peanut m&m's in packages of size 22. Forty-five observations of X yielded the following frequencies for the possible outcomes of X:

Outcome (x):	0	1	2	3	4	5	6	7	8	9
Frequency:	0	2	4	5	7	9	8	5	3	2

(a) Use these data to construct a relative frequency histogram and superimpose over it a Poisson probability histogram with a mean of $\lambda = 5$.

(b) Do these data appear to be observations of a Poisson random variable? (In a later chapter we describe a statistical test used to determine whether the answer is yes or no.)

3.7.15 Let X equal the number of chocolate chips in a chocolate-chip cookie. Sixty-two observations of X yielded the following frequencies for the possible outcomes of X:

Outcome (x):	0	1	2	3	4	5	6	7	8	9	10
Frequency:	0	0	2	8	7	13	13	10	4	4	1

(a) Use these data to graph the relative frequency histogram and the Poisson probability histogram with $\lambda = 5.6$ on the same figure.

(b) Do these data seem to be observations of a Poisson random variable with a mean of $\lambda = 5.6$? (In a later chapter we describe a statistical test used to help answer this and similar questions.)

3.7.16 While testing a used computer tape for bad records, a computer operator counted the number of flaws per 100 feet of tape. Let X equal this random variable. Given that 40 observations of X yielded 5 zeros, 7 ones, 12 twos, 9 threes, 5 fours, 1 five, and 1 six,

(a) Construct a relative frequency histogram and superimpose a Poisson p.d.f. with $\lambda = 2.2$.

(b) Does the Poisson distribution seem to be a possible probability model for this experiment?

3.7.17 A student trainer worked in the training room in the physical education building. The number of students coming in to be taped or for treatment of other injuries was recorded each half-hour yielding the following data:

$$
\begin{array}{cccccccccccc}
3 & 3 & 2 & 0 & 4 & 5 & 6 & 4 & 4 & 3 & 2 & 1 & 2 & 3 & 0 \\
5 & 5 & 3 & 2 & 3 & 5 & 4 & 1 & 2 & 0 & 3 & 2 & 4 & 2 & 6
\end{array}
$$

(a) Find the sample mean, $\bar{x}$.

(b) Find the sample variance, s^2.

(c) Find the sample standard deviation, s.

(d) Draw a relative frequency histogram of these data with an appropriate Poisson probability histogram superimposed.

3.7.18 A lot of 10,000 items is accepted if a sample of $n = 400$ has no more than $Ac = 3$ defective items. Determine the approximate values of $OC(0.002)$, $OC(0.004)$, $OC(0.006)$, $OC(0.01)$, and $OC(0.02)$ and plot the operating characteristic curve.

3.7.19 Consider a lot of 5000 items. Design a sampling plan (i.e., find n and Ac) so that $OC(0.02) = 0.95$ and $OC(0.08) = 0.10$ approximately. HINT: Start with $n = 50$ and $Ac = 2$ and modify as necessary, noting that increasing n creates a steeper OC curve.

3.8 GENERATING FUNCTIONS

In the preceding sections we have seen the importance of the mean, variance, and standard deviation of a random variable X. For some distributions it was fairly difficult to find directly $E(X)$ and $E(X^2)$, the first and second moments, or $E[X(X-1)]$, the second factorial moment, which was sometimes used in finding the variance. In this section we define two functions of a real variable t, the first of which can help us find the factorial moments of X and the second of which can be used to find moments like $E(X)$ and $E(X^2)$. While these generating characteristics of these two functions are extremely important, the most important aspect of each of the two functions is a certain uniqueness property that is described later in this section.

Definition 3.8.1.   Let X be a random variable of the discrete type with p.d.f. $f(x)$ and space R. If

$$\eta(t) = E(t^X) = \sum_{x \in R} t^x f(x)$$

exists and is finite for t values in some interval including $t = 0$ and $t = 1$, the function $\eta(t)$ is called the **probability-generating function of X**.

In this definition we have taken a function of x, namely $f(x)$, and created a function $\eta(t)$ of t. The following example illustrates how we can find $\eta(t)$ for a particular distribution.

Example 3.8.1

Consider a random variable X that has the geometric p.d.f.

$$f(x) = (1 - p)^{x-1}p, \qquad x = 1, 2, 3, \ldots,$$

where $0 < p < 1$. Then

$$\eta(t) = \sum_{x=1}^{\infty} t^x (1 - p)^{x-1} p = pt \sum_{x=1}^{\infty} [(1 - p)t]^{x-1}$$

$$= \frac{pt}{1 - (1 - p)t}, \qquad \text{provided that} \quad -1 < (1 - p)t < 1.$$

That is, if $-1/(1 - p) < t < 1/(1 - p)$, $\eta(t)$ exists and that interval of t includes $t = 0$ and $t = 1$ since $0 < p < 1$. $\diamond$

To see why $\eta(t)$ is called the probability-generating function, let us consider the case in which the space R consists of a collection of nonnegative integers. This restriction is true for all of the major discrete distributions in this chapter, namely the uniform, hypergeometric, Bernoulli, binomial, geometric, negative binomial, and Poisson; thus it is not a very limiting restriction and yet it is convenient for the following explanation. Initially, suppose there are only a finite number of points in R, which is true for the binomial distribution for illustration. That is, here,

$$\eta(t) = \sum_{x=0}^{n} t^x f(x)$$

where

$$f(x) = \binom{n}{x} p^x (1 - p)^{n-x}, \qquad x = 0, 1, 2, \ldots, n.$$

We see, writing out the summation, that

$$\eta(t) = f(0) + f(1)t + f(2)t^2 + \cdots + f(n)t^n$$

is simply a polynomial of nth degree, the coefficients of which are the respective probabilities of the events $\{X = 0\}, \{X = 1\}, \{X = 2\}, \ldots, \{X = n\}$. That is, the probability $P(X = r)$ is the coefficient of t^r, $r = 0, 1, 2, \ldots, n$, and, in particular, $\eta(0) = P(X = 0)$.

If we take the first derivative,

$$\eta'(t) = f(1) + 2f(2)t + 3f(3)t^2 + \cdots + nf(n)t^{n-1},$$

and set $t = 0$, we obtain

$$\eta'(0) = f(1) = P(X = 1).$$

The second derivative,

$$\eta''(t) = 2f(2) + (3)(2)f(3)t + \cdots + (n)(n-1)f(n)t^{n-2},$$

evaluated at $t = 0$ gives

$$\eta''(0) = 2f(2) = 2P(X = 2)$$

or, equivalently,

$$P(X = 2) = \frac{\eta''(0)}{2!}.$$

Continuing this, we see that

$$P(X = 3) = \frac{\eta'''(0)}{3!}$$

and, in general,

$$P(X = r) = \frac{\eta^{(r)}(0)}{r!}, \qquad r = 0, 1, 2, \ldots, n.$$

That is, we can generate the probabilities $P(X = x)$, $x \in R$, from $\eta(t)$ and its derivatives.

To emphasize the usefulness of this function $\eta(t)$, let us actually sum that polynomial in the binomial case. That is, here,

$$\eta(t) = \sum_{x=0}^{n} t^x \binom{n}{x} p^x (1-p)^{n-x}$$

$$= \sum_{x=0}^{n} \binom{n}{x} (pt)^x (1-p)^{n-x}$$

$$= [(1-p) + pt]^n, \qquad -\infty < t < \infty,$$

from the general formula of the binomial expansion when $a = 1 - p$ and $b = pt$. Note that

$$\eta'(t) = n[(1-p) + pt]^{n-1}(p)$$

and
$$\eta''(t) = (n)(n-1)[(1-p)+pt]^{n-2}p^2.$$
At $t=0$, the function $\eta(t)$ and these two derivatives equal
$$\eta(0) = (1-p)^n = P(X=0),$$
$$\eta'(0) = n(1-p)^{n-1}p = P(X=1),$$
$$\eta''(0) = n(n-1)(1-p)^{n-2}p^2 = (2!)P(X=2).$$
Of course, we could take additional derivatives and generate all the probabilities, $P(X=r), r = 0, 1, 2, \ldots, n$.

However, there is another use for $\eta(t)$ and its derivatives. Suppose that we evaluate these functions at $t=1$ rather than at $t=0$. We obtain, in the binomial case with $\eta(t)$ and its first two derivatives, that
$$\eta(1) = [(1-p)+p]^n = 1$$
$$\eta'(1) = n[(1-p)+p]^{n-1}p = np$$
$$\eta''(1) = n(n-1)[(1-p)+p]^{n-2}p^2 = n(n-1)p^2.$$
Note, by comparing these results with those of Section 3.5, that
$$\eta'(1) = E(X) \quad \text{and} \quad \eta''(1) = E[X(X-1)].$$
Do these results hold in general, and if so, why? Consider
$$\eta(t) = \sum_{x \in R} t^x f(x).$$
If we interchange the order of differentiation and summation, which we can do provided that the resulting summations exist, we obtain
$$\eta'(t) = \sum_{x \in R} xt^{x-1}f(x),$$
$$\eta''(t) = \sum_{x \in R} x(x-1)t^{x-2}f(x),$$
and for each positive integer r,
$$\eta^{(r)}(t) = \sum_{x \in R} x(x-1)\cdots(x-r+1)t^{x-r}f(x).$$
Setting $t=1$ gives us
$$\eta(1) = \sum_{x \in R} 1f(x) = 1,$$
$$\eta'(1) = \sum_{x \in R} xf(x) = E(X),$$
$$\eta''(1) = \sum_{x \in R} x(x-1)f(x) = E[X(X-1)],$$

and for each positive integer r,

$$\eta^{(r)}(1) = \sum_{x \in R} x(x-1)\cdots(x-r+1)f(x)$$

$$= E[X(X-1)\cdots(X-r+1)] = E[(X)_r],$$

the rth factorial moment of X.

Since $E[X(X-1)\cdots(X-r+1)] = E[(X)_r]$ is often called a factorial moment, the probability-generating function $\eta(t)$ is often called the **factorial moment-generating function**. We also note that

$$\eta'(1) = E(X) = \mu$$

and

$$\eta''(1) + \eta'(1) - [\eta'(1)]^2 = E[X(X-1)] + E(X) - [E(X)]^2$$

$$= E(X^2) - [E(X)]^2 = \sigma^2.$$

Let us use these results with the probability-generating function we obtained Example 3.8.1.

Example 3.8.2

For the geometric distribution, the probability-generating function is

$$\eta(t) = \frac{pt}{[1-(1-p)t]}, \qquad -\frac{1}{1-p} < t < \frac{1}{1-p}.$$

Here we have that

$$\eta'(t) = \frac{[1-(1-p)t]p + pt(1-p)}{[1-(1-p)t]^2}$$

$$= \frac{p}{[1-(1-p)t]^2},$$

$$\eta''(t) = \frac{p(-2)[-(1-p)]}{[1-(1-p)t]^3}.$$

The mean of the geometric distribution is

$$\mu = E(X) = \eta'(1) = \frac{p}{[1-(1-p)]^2} = \frac{1}{p}.$$

The variance of the geometric distribution is

$$\sigma^2 = \eta''(1) + \eta'(1) - [\eta'(1)]^2$$

$$= \frac{2p(1-p)}{[1-(1-p)]^3} + \frac{1}{p} - \frac{1}{p^2}$$

$$= \frac{1-p}{p^2} = \frac{q}{p^2}.$$

Of course, we first found the moments of the geometric distribution in Section 3.6.

So we have seen that the probability-generating function $\eta(t)$ and its derivatives are useful for generating probabilities and factorial moments. That is, once we sum $E(t^X)$ to obtain $\eta(t)$ we can find, in the discrete case, probabilities and factorial moments from this function $\eta(t)$ and its derivatives.

Generating these characteristics is a very important use of this function $\eta(t)$, but there is another important aspect of it that we now observe. Suppose that there are two random variables, say X and Y, that have exactly the same probability-generating function $\eta(t)$. Then their respective probabilities must be the same; that is, $P(X = r) = P(Y = r)$ since both of these probabilities can be generated from this common probability-generating function. So X and Y have the same distribution and we say that the probability-generating function, $\eta(t)$, uniquely determines the distribution of a random variable. Thus having $\eta(t)$ is as worthwhile as having the p.d.f. $f(x)$. The reader will appreciate this fact more when we find that we can determine a distribution of a random variable by first finding its probability-generating function—or, equivalently, its moment-generating function, a function which we now define.

Definition 3.8.2.   Let X be a random variable of the discrete type with p.d.f. $f(x)$ and space R. If there is a positive number h such that

$$E(e^{tX}) = \sum_{x \in R} e^{tx} f(x)$$

exists and is finite for $-h < t < h$, then the function of t defined by

$$M(t) = E(e^{tX})$$

is called the **moment-generating function of** X.

In this definition we have taken a function of x, namely $f(x)$, and created a function $M(t)$ of t. We again use the geometric distribution for illustration.

Example 3.8.3

Consider the random variable X that has the geometric p.d.f.

$$f(x) = (1 - p)^{x-1}p, \qquad x = 1, 2, 3, \dots .$$

Here

$$M(t) = \sum_{x=1}^{\infty} e^{tx}(1 - p)^{x-1}p = p\,e^t \sum_{x=1}^{\infty} [(1 - p)e^t]^{x-1}.$$

The summation is the sum of a geometric series which exists provided that $(1-p)e^t < 1$ or, equivalently, $t < -\ln(1 - p)$. That is,

$$M(t) = \frac{p\,e^t}{1 - (1 - p)e^t}, \qquad t < -\ln(1 - p).$$

Note we could have taken $\eta(t)$ found in Example 3.8.1 and replaced t by e^t to obtain this $M(t)$.   ◇

From the theory of mathematical analysis, it can be shown that the existence of $M(t)$, for $-h < t < h$, implies that derivatives of $M(t)$ of all orders exist at $t = 0$; moreover, it is permissible to interchange differentiation and summation. Thus

$$M'(t) = \sum_{x \in R} x e^{tx} f(x),$$

$$M''(t) = \sum_{x \in R} x^2 e^{tx} f(x),$$

and for each positive integer r,

$$M^{(r)}(t) = \sum_{x \in R} x^r e^{tx} f(x).$$

Setting $t = 0$, we see that

$$M'(0) = \sum_{x \in R} x f(x) = E(X),$$

$$M''(0) = \sum_{x \in R} x^2 f(x) = E(X^2),$$

and, in general,

$$M^{(r)}(0) = \sum_{x \in R} x^r f(x) = E(X^r).$$

In particular, if the moment-generating function exists,

$$\mu = M'(0) \qquad \text{and} \qquad \sigma^2 = M''(0) - [M'(0)]^2.$$

The argument above shows that we can find the moments of X by differentiating $M(t)$. It must be emphasized that in use, we generally first evaluate the summation representing $M(t)$, obtaining a closed-form solution, and then differentiate that solution to obtain the moments of X. The following examples illustrate the use of the moment-generating function for finding the first and second moments and then the mean and variance.

Example 3.8.4

Continuing with Example 3.8.3, suppose that we take two derivatives of the moment-generating function for the geometric distribution. We have

$$M'(t) = \frac{[1 - (1-p)e^t]pe^t - [-(1-p)e^t]pe^t}{[1 - (1-p)e^t]^2} = \frac{pe^t}{[1 - (1-p)e^t]^2}$$

and

$$M''(t) = \frac{pe^t + p(1-p)e^{2t}}{[1 - (1-p)e^t]^3}.$$

Thus

$$\mu = E(X) = M'(0) = \frac{1}{p}$$

and

$$\sigma^2 = \text{Var}(X) = M''(0) - [M'(0)]^2$$
$$= \frac{p + p(1-p)}{p^3} - \left(\frac{1}{p}\right)^2 = \frac{1-p}{p^2}.$$

◇

Example 3.8.5

Let X have a binomial distribution $b(n, p)$ with p.d.f.

$$f(x) = \binom{n}{x} p^x (1-p)^{n-x}, \qquad x = 0, 1, 2, \ldots, n.$$

The moment-generating function of X is

$$M(t) = E(e^{tX}) = \sum_{x=0}^{n} e^{tx} \binom{n}{x} p^x (1-p)^{n-x}$$

$$= \sum_{x=0}^{n} \binom{n}{x} (pe^t)^x (1-p)^{n-x}.$$

Using the formula for the binomial expansion with $a = 1 - p$ and $b = pe^t$, we see that

$$M(t) = [(1-p) + pe^t]^n, \qquad -\infty < t < \infty.$$

Again we note that we could have obtained $M(t)$ by taking $\eta(t)$ found earlier and replaced t by e^t. The first two derivatives of $M(t)$ are

$$M'(t) = n[(1-p) + pe^t]^{n-1}(pe^t)$$

and

$$M''(t) = n(n-1)[(1-p) + pe^t]^{n-2}(pe^t)^2 + n[(1-p) + pe^t]^{n-1}(pe^t).$$

Thus

$$\mu = E(X) = M'(0) = np$$

and

$$\sigma^2 = E(X^2) - [E(X)]^2 = M''(0) - [M'(0)]^2$$
$$= n(n-1)p^2 + np - (np)^2 = np(1-p),$$

as was shown in Section 3.5.

In the special case when $n = 1$, X has a Bernoulli distribution and

$$M(t) = (1-p) + pe^t$$

for all real values of t, $\mu = p$, and $\sigma^2 = p(1-p)$.

◇

Example 3.8.6

Let X have a Poisson distribution with p.d.f.

$$f(x) = \frac{\lambda^x e^{-\lambda}}{x!}, \qquad x = 0, 1, 2, \ldots.$$

The moment-generating function of X is

$$M(t) = E(e^{tX}) = \sum_{x=0}^{\infty} e^{tx} \frac{\lambda^x e^{-\lambda}}{x!} = e^{-\lambda} \sum_{x=0}^{\infty} \frac{(\lambda e^t)^x}{x!}.$$

From the series representation of the exponential function, we have that

$$M(t) = e^{-\lambda} e^{\lambda e^t} = e^{\lambda(e^t - 1)}$$

for all real values of t. Now

$$M'(t) = \lambda e^t e^{\lambda(e^t - 1)}$$

and

$$M''(t) = (\lambda e^t)^2 e^{\lambda(e^t - 1)} + \lambda e^t e^{\lambda(e^t - 1)}.$$

The values of the mean and variance of X are

$$\mu = M'(0) = \lambda$$

and

$$\sigma^2 = M''(0) - [M'(0)]^2 = (\lambda^2 + \lambda) - \lambda^2 = \lambda. \qquad \diamondsuit$$

We have noted the close association between $\eta(t)$ and $M(t)$, namely

$$M(t) = \eta(e^t)$$

or, equivalently,

$$\eta(t) = M(\ln t).$$

This relationship implies that $M(t)$ also uniquely determines the distribution. As statisticians tend to use the moment-generating function more than the probability-generating function, we do so in this book. However, we do not want the reader to forget their close association; that is, $\eta(t)$ and $M(t) = \eta(e^t)$ are equivalent and both uniquely determine the distribution. As a matter of fact, we now observe how $M(t)$ can be used to find $P(X = r)$ in the discrete case.

If X has a p.d.f. $f(x)$ with support $\{b_1, b_2, \ldots\}$, then

$$M(t) = \sum_{x \in R} e^{tx} f(x)$$

$$= f(b_1) e^{tb_1} + f(b_2) e^{tb_2} + \cdots.$$

Hence the coefficient of e^{tb_i} is $f(b_i) = P(X = b_i)$. That is, if we write a moment-generating function of a discrete-type random variable X in the form above, the probability of any value of X, say b_i, is the coefficient of e^{tb_i}.

Example 3.8.7

Let the moment-generating function of X be defined by

$$M(t) = \frac{1}{15} e^t + \frac{2}{15} e^{2t} + \frac{3}{15} e^{3t} + \frac{4}{15} e^{4t} + \frac{5}{15} e^{5t}.$$

Then, for example, the coefficient of e^{2t} is 2/15. Thus $f(2) = P(X = 2) = 2/15$. In general, we see that the p.d.f. of X is

$$f(x) = \frac{x}{15}, \qquad x = 1, 2, 3, 4, 5.$$ ◇

When the moment-generating function exists, derivatives of all orders exist at $t = 0$. Thus it is possible to represent $M(t)$ as a Maclaurin's series, namely,

$$M(t) = M(0) + M'(0) \left(\frac{t}{1!} \right) + M''(0) \left(\frac{t^2}{2!} \right) + M'''(0) \left(\frac{t^3}{3!} \right) + \cdots.$$

If the Maclaurin's series expansion of $M(t)$ exists and the moments are given, we can frequently sum the Maclaurin's series to obtain the closed form of $M(t)$. This is illustrated in the next example.

Example 3.8.8

Let the moments of X be defined by

$$E(X^r) = 0.8, \qquad r = 1, 2, 3, \ldots.$$

The moment-generating function of X is then

$$M(t) = M(0) + \sum_{r=1}^{\infty} 0.8 \left(\frac{t^r}{r!} \right) = 1 + 0.8 \sum_{r=1}^{\infty} \frac{t^r}{r!}$$

$$= 0.2 + 0.8 \sum_{r=0}^{\infty} \frac{t^r}{r!} = 0.2 e^{0t} + 0.8 e^{1t}.$$

Thus

$$P(X = 0) = 0.2 \qquad \text{and} \qquad P(X = 1) = 0.8.$$ ◇

Let us summarize this section by finding the probability-generating and moment-generating functions of a negative binomial distribution.

Example 3.8.9

Let X have a negative binomial distribution with p.d.f.

$$f(x) = \binom{x - 1}{r - 1} p^r (1 - p)^{x-r}, \qquad x = r, r + 1, r + 2, \ldots.$$

The probability-generating function of X is

$$\eta(t) = \sum_{x=r}^{\infty} t^x \binom{x-1}{r-1} p^r (1-p)^{x-r}$$

$$= (pt)^r \sum_{x=r}^{\infty} \binom{x-1}{r-1} [(1-p)t]^{x-r}$$

$$= \frac{(pt)^r}{[1-(1-p)t]^r}, \qquad |t| < \frac{1}{1-p}.$$

The moment-generating function of X is

$$M(t) = \eta(e^t)$$

$$= \frac{(pe^t)^r}{[1-(1-p)e^t]^r}, \qquad t < -\ln(1-p).$$

Either of these generating functions could be used to show that

$$\mu = \frac{r}{p}$$

and

$$\sigma^2 = \frac{r(1-p)}{p^2}.$$

However, the calculations get a little messy and an easier way to find these characteristics is given in Exercise 3.8.9.   ◇

EXERCISES

3.8.1 Find the probability-generating function when the p.d.f. of X is defined by

  (a) $f(x) = \dfrac{1}{3}, \qquad x = 1, 2, 3.$

  (b) $f(x) = 1, \qquad x = 5.$

  (c) $f(x) = \dfrac{5-x}{10}, \qquad x = 1, 2, 3, 4.$

  (d) $f(x) = (0.3)^x (0.7)^{1-x}, \qquad x = 0, 1.$

3.8.2 Define the p.d.f. and give the values of μ, σ^2, and σ when the moment-generating function of X is defined by

  (a) $M(t) = 1/3 + (2/3)e^t.$

  (b) $M(t) = (0.25 + 0.75e^t)^{12}.$

  (c) $M(t) = e^{4.6(e^t - 1)}.$

3.8.3 Let X have a Poisson distribution with mean λ. Show that the probability-generating function of X is $\eta(t) = e^{\lambda(t-1)}$ directly by finding $E(t^X)$.

3.8.4 Find the moment-generating function, mean, and variance of X if the p.d.f. of X is $f(x) = (1/2)(2/3)^x, x = 1, 2, 3, 4, \ldots.$

3.8.5 If the moment-generating function of X is

$$M(t) = \frac{2}{5} e^t + \frac{1}{5} e^{2t} + \frac{2}{5} e^{3t},$$

find the mean, variance, p.d.f. of X, and the probability-generating function.

3.8.6 (i) Give the name of the distribution of X (if it has a name), (ii) find the values of μ and σ^2, and (iii) calculate $P(1 \le X \le 2)$ when the moment-generating function of X is given by

(a) $M(t) = (0.3 + 0.7e^t)^5$.

(b) $M(t) = e^{4(e^t - 1)}$.

(c) $M(t) = \dfrac{0.3e^t}{1 - 0.7e^t}$, $t < -\ln(0.7)$.

(d) $M(t) = 0.45 + 0.55e^t$.

(e) $M(t) = 0.3e^t + 0.4e^{2t} + 0.2e^{3t} + 0.1e^{4t}$.

(f) $M(t) = (0.6e^t)^2(1 - 0.4e^t)^{-2}$, $t < -\ln(0.4)$.

(g) $M(t) = \sum_{x=1}^{10} (0.1)e^{tx}$.

3.8.7 Find $P(X = 3)$ if the probability-generating function of X is

(a) $\eta(t) = (0.65 + 0.35t)^{14}$.

(b) $\eta(t) = e^{5.6(t-1)}$.

(c) $\eta(t) = 0.25t + 0.35t^3 + 0.40t^5$.

(d) $\eta(t) = \dfrac{0.35t}{1 - 0.65t}$, $-1/0.65 < t < 1/0.65$.

(e) $\eta(t) = \sum_{x=1}^{15} (1/15)t^x$.

3.8.8 If $E(X^r) = 5^r, r = 1, 2, 3, \ldots$, find $M(t)$, the moment-generating function of X, and the p.d.f. of X.

3.8.9 Let the moment-generating function $M(t)$ of X exist for $-h < t < h$. Consider the function $R(t) = \ln M(t)$. The first two derivatives of $R(t)$ are, respectively,

$$R'(t) = \frac{M'(t)}{M(t)} \quad \text{and} \quad R''(t) = \frac{M(t)M''(t) - [M'(t)]^2}{[M(t)]^2}.$$

Setting $t = 0$, show that

(a) $\mu = R'(0)$.

(b) $\sigma^2 = R''(0)$.

3.8.10 Use the result of Exercise 3.8.9 to find the mean and variance of

(a) The Bernoulli distribution.

(b) The binomial distribution.

(c) The Poisson distribution.

(d) The geometric distribution.

(e) The negative binomial distribution.

4

CONTINUOUS DISTRIBUTIONS

4.1 RANDOM VARIABLES OF THE CONTINUOUS TYPE

Random variables whose spaces are not composed of a countable number of points but are intervals or a union of intervals are said to be of the **continuous type**. Recall that the relative frequency (or density) histogram $h(x)$ associated with n observations of a random variable of that type is a nonnegative function defined so that the total area between its graph and the x axis equals one. In addition, $h(x)$ is constructed so that the integral

$$\int_a^b h(x)\,dx$$

is approximately equal to the relative frequency of the interval $\{x : a < x < b\}$ and hence can be thought of as an estimate of the probability $P(a < X < b)$.

Let us now consider what happens to the function $h(x)$ in the limit, as n increases without bound and as the lengths of the class intervals decrease to zero. It is to be hoped that $h(x)$ will become closer and closer to some function, say $f(x)$, that gives the true probabilities, such as $P(a < X < b)$, through the integral

$$P(a < X < b) = \int_a^b f(x)\,dx.$$

That is, $f(x)$ should be a nonnegative function such that the total area between its graph and the x axis equals one. Moreover, the probability $P(a < X < b)$ is the area bounded by the graph of $f(x)$, the x axis, and the lines $x = a$ and $x = b$. Thus we say that the **probability density function (p.d.f.)** of a random variable X of the **continuous type**, with space R that is an interval or union of intervals, is an integrable function $f(x)$ satisfying the following conditions:

(a) $f(x) > 0, \qquad x \in R.$
(b) $\int_R f(x)\, dx = 1.$
(c) The probability of the event $X \in A$ is

$$P(X \in A) = \int_A f(x)\, dx.$$

The corresponding distribution of probability is said to be one of the continuous type.

Example 4.1.1

Let the random variable X be the distance in feet between bad records on a used computer tape. Suppose that a reasonable probability model for X is given by the p.d.f.

$$f(x) = \frac{1}{40} e^{-x/40}, \qquad 0 \le x < \infty.$$

Note that $R = \{x : 0 \le x < \infty\}$ and $f(x) > 0$ for $x \in R$. Also,

$$\begin{aligned}
\int_R f(x)\, dx &= \int_0^\infty \frac{1}{40} e^{-x/40}\, dx \\
&= \lim_{b \to \infty} \left[-e^{-x/40} \right]_0^b \\
&= 1 - \lim_{b \to \infty} e^{-b/40} = 1.
\end{aligned}$$

The probability that the distance between bad records is greater than 40 feet is

$$P(X > 40) = \int_{40}^\infty \frac{1}{40} e^{-x/40}\, dx = e^{-1} = 0.368.$$

The p.d.f. and the probability of interest are depicted in Figure 4.1.   ◇

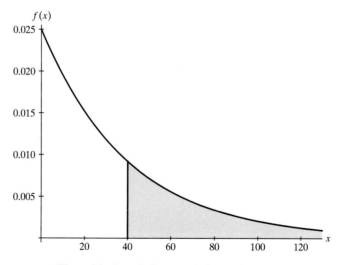

Figure 4.1   Lengths between bad records on tape

So that we can avoid repeated references to the space (or support) R of the random variable X, we shall adopt the same convention when describing probability density functions of the continuous type as we did in the discrete case. We extend the definition of the p.d.f. $f(x)$ to the entire set of real numbers by letting it equal zero when $x \notin R$. For example,

$$f(x) = \begin{cases} \dfrac{1}{40}e^{-x/40}, & 0 \leq x < \infty, \\ 0, & \text{elsewhere,} \end{cases}$$

has the properties of a p.d.f. of a continuous-type random variable X having support $\{x : 0 \leq x < \infty\}$. It will always be understood that $f(x) = 0$ when $x \notin R$, even when this is not explicitly written out.

The **distribution function** of a random variable X of the continuous type, defined in terms of the p.d.f. of X, is given by

$$F(x) = P(X \leq x) = \int_{-\infty}^{x} f(t)\,dt.$$

Here, again, $F(x)$ cumulates all of the probability less than or equal to x. From the fundamental theorem of calculus we have, for x values for which the derivative $F'(x)$ exists, that $F'(x) = f(x)$.

Example 4.1.2

Continuing with Example 4.1.1, if the p.d.f. of X is

$$f(x) = \begin{cases} 0, & -\infty < x < 0, \\ \dfrac{1}{40}e^{-x/40}, & 0 \leq x < \infty \end{cases}$$

the distribution function of X is $F(x) = 0$ for $x < 0$ and, for $x \geq 0$,

$$F(x) = \int_{-\infty}^{x} f(t)\,dt = \int_{0}^{x} \frac{1}{40}e^{-t/40}\,dt$$
$$= \left[-e^{-t/40}\right]_{0}^{x} = 1 - e^{-x/40}.$$

Note that

$$F'(x) = \begin{cases} 0, & -\infty < x < 0, \\ \dfrac{1}{40}e^{-x/40}, & 0 < x < \infty. \end{cases}$$

Also $F'(0)$ does not exist. [Sketch a graph of $y = F(x)$ to see why this is true.]   ◇

Since there are no steps or jumps in a distribution function, $F(x)$, of the continuous type, it must be true that

$$P(X = b) = 0$$

for all real values of b. This agrees with the fact that the integral $\int_b^b f(x)\,dx$ is taken to be zero in calculus. Thus we see that

$$P(a \le X \le b) = P(a < X < b) = P(a \le X < b) = P(a < X \le b) = F(b) - F(a),$$

provided that X is a random variable of the continuous type. Moreover, we can change the definition of a p.d.f. of a random variable of the continuous type at a finite (actually countable) number of points without altering the distribution of probability. For illustration,

$$f(x) = \begin{cases} 0, & -\infty < x < 0, \\ \dfrac{1}{40} e^{-x/40}, & 0 \le x < \infty, \end{cases}$$

and

$$f(x) = \begin{cases} 0, & -\infty < x \le 0, \\ \dfrac{1}{40} e^{-x/40}, & 0 < x < \infty, \end{cases}$$

are equivalent in the computation of probabilities involving this random variable.

Example 4.1.3

Let Y be a continuous random variable with p.d.f. $g(y) = 2y,\ 0 < y < 1$. The distribution function of Y is defined by

$$G(y) = \begin{cases} 0, & y < 0, \\ \displaystyle\int_0^y 2t\,dt = y^2, & 0 \le y < 1, \\ 1, & 1 \le y. \end{cases}$$

Figure 4.2 gives the graph of the p.d.f. $g(y)$ and the graph of the distribution function $G(y)$. For examples of computations of probabilities, consider

$$P\left(\frac{1}{2} < Y \le \frac{3}{4}\right) = G\left(\frac{3}{4}\right) - G\left(\frac{1}{2}\right) = \left(\frac{3}{4}\right)^2 - \left(\frac{1}{2}\right)^2 = \frac{5}{16}$$

and

$$P\left(\frac{1}{4} \le Y < 2\right) = G(2) - G\left(\frac{1}{4}\right) = 1 - \left(\frac{1}{4}\right)^2 = \frac{15}{16}. \qquad \diamond$$

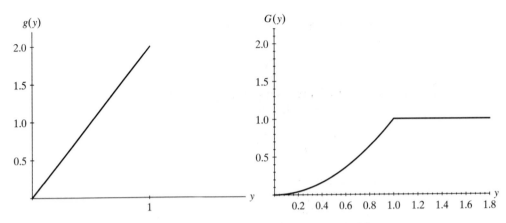

Figure 4.2   Continuous distribution p.d.f. and c.d.f.

Recall that the p.d.f. $f(x)$ of a random variable of the discrete type is bounded by one because $f(x)$ gives a probability, namely

$$f(x) = P(X = x).$$

For random variables of the continuous type, the p.d.f. does not have to be bounded [see Exercises 4.1.2(c) and 4.1.3(c)]. The restriction is that the area between the p.d.f. and the x axis must equal one. Furthermore, it should be noted that the p.d.f. of a random variable X of the continuous type does not need to be a continuous function. For example,

$$f(x) = \begin{cases} \dfrac{1}{2}, & 0 < x < 1 \text{ or } 2 < x < 3, \\ 0, & \text{elsewhere,} \end{cases}$$

enjoys the properties of a p.d.f. of a distribution of the continuous type, and yet $f(x)$ has discontinuities at $x = 0, 1, 2$, and 3. However, the distribution function associated with a distribution of the continuous type is always a continuous function.

For continuous-type random variables, the definitions associated with mathematical expectation are the same as those in the discrete case except that integrals replace summations. For illustration, let X be a continuous random variable with a p.d.f. $f(x)$.

The **expected value of** X or **mean of** X is

$$\mu = E(X) = \int_{-\infty}^{\infty} x f(x)\, dx.$$

The **variance of** X is

$$\sigma^2 = \text{Var}(X) = E[(X - \mu)^2] = \int_{-\infty}^{\infty} (x - \mu)^2 f(x)\, dx.$$

The **standard deviation of** X is

$$\sigma = \sqrt{\text{Var}(X)}.$$

The **moment-generating function**, if it exists, is

$$M(t) = \int_{-\infty}^{\infty} e^{tx} f(x)\, dx, \qquad -h < t < h.$$

Moreover, results such as $\sigma^2 = E(X^2) - \mu^2$ and $\mu = M'(0), \sigma^2 = M''(0) - [M'(0)]^2$ are still valid. Again it is important to note that the moment-generating function, if it is finite for $-h < t < h$ for some $h > 0$, completely determines the distribution.

Remark   In both the discrete and continuous cases, it should be noted that if the rth moment, $E(X^r)$, exists and is finite, then the same is true of all lower order moments, $E(X^k), k = 1, 2, \ldots, r - 1$. However the converse is not true; for illustration the first moment can exist and be finite, but the second moment is not necessarily finite (see Exercise 4.1.9). Moreover, if $E(e^{tX})$ exists and is finite for $-h < t < h$, then all moments exist and are finite, but the converse is not necessarily true.

Example 4.1.4

For the random variable Y in Example 4.1.3,

$$\mu = E(Y) = \int_0^1 y\,(2y)\, dy = \left[\left(\frac{2}{3}\right)y^3\right]_0^1 = \frac{2}{3}$$

and

$$\sigma^2 = \text{Var}(Y) = E(Y^2) - \mu^2$$

$$= \int_0^1 y^2 (2y)\, dy - \left(\frac{2}{3}\right)^2 = \left[\left(\frac{1}{2}\right)y^4\right]_0^1 - \frac{4}{9} = \frac{1}{18},$$

are the mean and variance, respectively, of Y. $\diamond$

Example 4.1.5

Let X have the p.d.f.

$$f(x) = \begin{cases} xe^{-x}, & 0 \le x < \infty, \\ 0, & \text{elsewhere.} \end{cases}$$

Then

$$M(t) = \int_0^\infty e^{tx} x e^{-x}\, dx = \lim_{b\to\infty} \int_0^b x e^{-(1-t)x}\, dx$$

$$= \lim_{b\to\infty} \left[-\frac{x e^{-(1-t)x}}{1-t} - \frac{e^{-(1-t)x}}{(1-t)^2} \right]_0^b$$

$$= \lim_{b\to\infty} \left[-\frac{b e^{-(1-t)b}}{1-t} - \frac{e^{-(1-t)b}}{(1-t)^2} \right] + \frac{1}{(1-t)^2}$$

$$= \frac{1}{(1-t)^2},$$

provided that $t < 1$. Note that $M(0) = 1$, which is true for every moment-generating function. Now

$$M'(t) = \frac{2}{(1-t)^3} \qquad \text{and} \qquad M''(t) = \frac{6}{(1-t)^4}.$$

Thus

$$\mu = M'(0) = 2$$

and

$$\sigma^2 = M''(0) - [M'(0)]^2 = 6 - 2^2 = 2. \qquad \diamond$$

Let X be a continuous-type random variable with p.d.f. $f(x)$ and distribution function $F(x)$. The $(100p)$th **percentile** is a number π_p such that the area under $f(x)$ to the left of π_p is p. That is,

$$p = \int_{-\infty}^{\pi_p} f(x)\, dx = F(\pi_p).$$

The 50th percentile is called the **median**. We let $m = \pi_{0.50}$. The 25th and 75th percentiles are called the **first** and **third quartiles**, respectively, denoted by $q_1 = \pi_{0.25}$ and $q_3 = \pi_{0.75}$. Of course, the median $m = \pi_{0.50} = q_2$ is also called the second quartile.

Example 4.1.6

Let the p.d.f. of X be $f(x) = 1 - |x - 1|, 0 \le x \le 2$. Then the distribution function of X is defined by

$$F(x) = \begin{cases} 0, & x < 0, \\[2mm] \dfrac{x^2}{2}, & 0 \le x < 1, \\[3mm] 1 - \dfrac{(2-x)^2}{2}, & 1 \le x < 2, \\[3mm] 1, & 2 \le x. \end{cases}$$

To find, for example, the 32nd percentile, $\pi_{0.32}$, we solve

$$F(\pi_{0.32}) = 0.32.$$

Since $\pi_{0.32}$ is between 0 and 1 for this distribution because $0.32 < F(1) = 1/2$, we have

$$F(\pi_{0.32}) = \frac{\pi_{0.32}^2}{2} = 0.32$$

or

$$\pi_{0.32} = \sqrt{0.64} = 0.8.$$

To find the 92nd percentile, $\pi_{0.92}$, we first note that the 92nd percentile is between 1 and 2. Thus we solve

$$F(\pi_{0.92}) = 1 - \frac{(2 - \pi_{0.92})^2}{2} = 0.92,$$

which yields

$$\pi_{0.92} = 1.6.$$

The 32nd and 92nd percentiles are shown on the graph of the p.d.f. in Figure 4.3 and on the graph of the distribution function in Figure 4.3. ◇

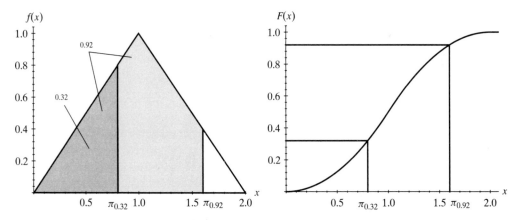

Figure 4.3   Percentiles π_{32} and π_{92}

We conclude this section by reviewing, through the next example, the important ideas.

Example 4.1.7

Let X have the p.d.f.

$$f(x) = e^{-x-1}, \qquad -1 < x < \infty.$$

Then

$$P(1 \le X) = \int_1^\infty e^{-x-1} dx = e^{-1} \left[-e^{-x} \right]_1^\infty = e^{-2} = 0.135.$$

Also, we have

$$M(t) = E(e^{tX}) = \int_{-1}^{\infty} e^{tx} e^{-x-1} dx$$

$$= e^{-1} \left[\frac{-e^{-(1-t)x}}{1-t} \right]_{-1}^{\infty}$$

$$= e^{-t}(1-t)^{-1}, \qquad t < 1.$$

Thus, since

$$M'(t) = (e^{-t})(1-t)^{-2} - e^{-t}(1-t)^{-1}$$

and

$$M''(t) = (e^{-t})(2)(1-t)^{-3} - 2e^{-t}(1-t)^{-2} + e^{-t}(1-t)^{-1},$$

we have

$$\mu = M'(0) = 0 \qquad \text{and} \qquad \sigma^2 = M''(0) - [M'(0)]^2 = 1.$$

The distribution function is

$$F(x) = \int_{-1}^{x} e^{-w-1} dw = e^{-1} \left[-e^{-w} \right]_{-1}^{x}$$

$$= e^{-1} \left[e^1 - e^{-x} \right] = 1 - e^{-x-1}, \qquad -1 < x < \infty,$$

and zero for $x \le -1$.

For illustration, the median $\pi_{0.5}$ is found by solving

$$F(\pi_{0.5}) = 0.5,$$

which is equivalent to

$$-\pi_{0.5} - 1 = \ln(0.5);$$

so

$$\pi_{0.5} = \ln 2 - 1 = -0.307. \qquad \diamond$$

EXERCISES

4.1.1 Let the random variable X have the p.d.f. $f(x) = 2(1-x), 0 \le x \le 1$, zero elsewhere.
  (a) Sketch the graph of this p.d.f.
  (b) Determine and sketch the graph of the distribution function of X.
  (c) Find (i) $P(0 \le X \le 1/2)$, (ii) $P(1/4 \le X \le 3/4)$, (iii) $P(X = 3/4)$, and (iv) $P(X \ge 3/4)$.

4.1.2 For each of the following functions, (i) find the constant c so that $f(x)$ is a p.d.f. of a random variable X, (ii) find the distribution function, $F(x) = P(X \le x)$, and (iii) sketch graphs of the p.d.f. $f(x)$ and the distribution function $F(x)$.
  (a) $f(x) = x^3/4, \qquad 0 < x < c$.
  (b) $f(x) = (3/16)x^2, \qquad -c < x < c$.
  (c) $f(x) = c/\sqrt{x}, \qquad 0 < x < 1$. Is this p.d.f. bounded?

4.1.3 For each of the following functions, (i) find the constant c so that $f(x)$ is a p.d.f. of a random variable X, (ii) find the distribution function, $F(x) = P(X \leq x)$, and (iii) sketch graphs of the p.d.f. $f(x)$ and the distribution function $F(x)$.

   (a) $f(x) = 4x^c$, $0 \leq x \leq 1$.

   (b) $f(x) = c\sqrt{x}$, $0 \leq x \leq 4$.

   (c) $f(x) = c/x^{3/4}$, $0 < x < 1$.

4.1.4 For each of the distributions in Exercise 4.1.2, find μ, σ^2, and σ.

4.1.5 For each of the distributions in Exercise 4.1.3, find μ, σ^2, and σ.

4.1.6 Let $f(x) = \cos x, 0 \leq x \leq \pi/2$, be the p.d.f. of X.

   (a) Sketch the graph of this p.d.f.

   (b) Determine and sketch the graph of the distribution function of X.

4.1.7 Let $f(x) = (1/2) \sin x, 0 \leq x \leq \pi$, be the p.d.f. of X.

   (a) Find μ and σ^2.

   (b) Sketch the graph of the p.d.f. of X.

   (c) Determine and sketch the graph of the distribution function of X.

4.1.8 The p.d.f. of X is $f(x) = c/x^2, 1 < x < \infty$, zero elsewhere.

   (a) Calculate the value of c so that $f(x)$ is a p.d.f.

   (b) Show that $E(X)$ is not finite.

4.1.9 The p.d.f. of Y is $g(y) = d/y^3, 1 < y < \infty$, zero elsewhere.

   (a) Calculate the value of d so that $g(y)$ is a p.d.f.

   (b) Find $E(Y)$.

   (c) Show that $\mathrm{Var}(Y)$ is not finite.

4.1.10 Sketch the graphs of the following probability density functions. Also find and sketch the graphs of the distribution functions associated with these distributions. Note carefully the relationship between the shape of the graph of the p.d.f. and the concavity of the graph of the distribution function.

   (a) $f(x) = \left(\dfrac{3}{2}\right)x^2$, $-1 < x < 1$.

   (b) $f(x) = \dfrac{1}{2}$, $-1 < x < 1$.

   (c) $f(x) = \begin{cases} x+1, & -1 < x < 0, \\ 1-x, & 0 \leq x < 1. \end{cases}$

4.1.11 Find the mean and variance for each of the distributions in Exercise 4.1.10.

4.1.12 Let $R(t) = \ln M(t)$, where $M(t)$ is the moment-generating function of a random variable of the continuous type. Show that

   (a) $\mu = R'(0)$.

   (b) $\sigma^2 = R''(0)$.

4.1.13 If $M(t) = (1 - t)^{-2}$, $t < 1$, use $R(t) = \ln M(t)$ and the result in Exercise 4.1.12 to find (a) μ and (b) σ^2.

4.1.14 Find the moment-generating function $M(t)$ of the distribution with p.d.f. $f(x) = (1/10)e^{-x/10}$, $0 < x < \infty$. Use $M(t)$ or $R(t) = \ln M(t)$ to determine the mean μ and the variance σ^2.

4.1.15 The logistic distribution is associated with the distribution function $F(x) = (1 + e^{-x})^{-1}$, $-\infty < x < \infty$. Find the p.d.f. of the logistic distribution and show that its graph is symmetric about $x = 0$.

4.1.16 Let $f(x) = 1/2$, $0 < x < 1$ or $2 < x < 3$, zero elsewhere, be the p.d.f. of X.

(a) Sketch the graph of this p.d.f.

(b) Define the distribution function of X and sketch its graph.

(c) Find $q_1 = \pi_{0.25}$.

(d) Find $m = \pi_{0.50}$. Is it unique?

(e) Find $q_3 = \pi_{0.75}$.

4.1.17 Let $f(x) = 1/2$, $-1 < x < 1$. Find (a) $m = \pi_{0.50}$, (b) $q_1 = \pi_{0.25}$, and (c) $\pi_{0.90}$.

4.1.18 Let $f(x) = (x + 1)/2$, $-1 < x < 1$. Find (a) $\pi_{0.64}$, (b) $q_1 = \pi_{0.25}$, and (c) $\pi_{0.81}$.

4.1.19 Let the random variable X_n have the p.d.f. $f(x_n) = n$, $0 < x_n < 1/n$.

(a) Define the distribution function of X_n, say $F_n(x_n)$.

(b) Graph the p.d.f. and the distribution function for $n = 1, 5$, and 10.

4.1.20 One hundred observations of X having a certain distribution were taken and then ordered, yielding

0.19	0.25	0.26	0.26	0.28	0.31	0.39	0.42	0.44	0.45
0.48	0.53	0.54	0.57	0.66	0.69	0.70	0.72	0.73	0.79
0.83	0.85	0.87	0.90	0.91	0.94	0.96	0.99	1.01	1.02
1.02	1.03	1.05	1.07	1.13	1.13	1.14	1.20	1.22	1.27
1.28	1.30	1.34	1.34	1.37	1.39	1.40	1.43	1.43	1.46
1.49	1.50	1.51	1.53	1.54	1.56	1.56	1.57	1.60	1.60
1.61	1.62	1.62	1.62	1.63	1.64	1.66	1.67	1.70	1.71
1.71	1.71	1.72	1.73	1.73	1.74	1.75	1.76	1.77	1.80
1.81	1.84	1.85	1.85	1.86	1.87	1.87	1.88	1.90	1.91
1.91	1.92	1.92	1.93	1.93	1.93	1.94	1.95	1.98	1.99

(a) Calculate the sample mean $\bar{x}$.

(b) Calculate the sample standard deviation s.

(c) Group these data into 10 classes: $(0.00, 0.20]$, $(0.20, 0.40]$, ..., $(1.80, 2.00]$.

(d) Construct a relative frequency histogram with the p.d.f. $f(x) = x/2$, $0 < x < 2$, superimposed. Is there a good fit?

(e) Find μ and σ, assuming that the p.d.f. of X is $f(x) = x/2$, $0 < x \leq 2$. Is $\bar{x}$ close to μ? Is s close to σ?

4.2 THE UNIFORM AND EXPONENTIAL DISTRIBUTIONS

Let the random variable X denote the outcome when a point is selected at random from an interval $[a, b]$, $-\infty < a < b < \infty$. If the experiment is performed in a fair manner, it is reasonable to assume that the probability that the point is selected from the interval $[a, x]$, $a \leq x < b$ is $(x - a)/(b - a)$. That is, the probability is proportional to the length of the interval so that the distribution function of X is

$$
F(x) = \begin{cases} 0, & x < a, \\ \dfrac{x - a}{b - a}, & a \leq x < b, \\ 1, & b \leq x, \end{cases}
$$

Because X is a continuous-type random variable, $F'(x)$ is equal to the p.d.f. of X whenever $F'(x)$ exists; thus when $a < x < b$, we have $f(x) = F'(x) = 1/(b - a)$.

The random variable X has a **uniform distribution** if its p.d.f. is equal to a constant on its support. In particular, if the support is the interval $[a, b]$, then

$$
f(x) = \frac{1}{b - a}, \qquad a \leq x \leq b.
$$

Moreover, we shall say that X is $U(a, b)$. This distribution is also referred to as **rectangular** because the graph of $f(x)$ suggests that name. See Figure 4.4 for the graph of $f(x)$ and the distribution function $F(x)$. Note that we could have taken $f(a) = 0$ or $f(b) = 0$ without altering the probabilities, since this is a continuous-type distribution, and we shall do this in some cases.

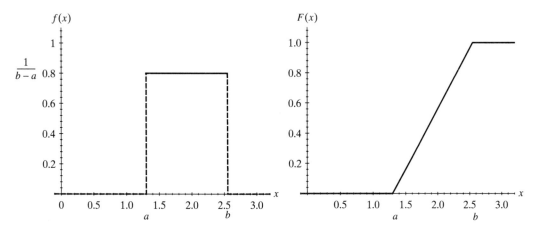

Figure 4.4   Uniform p.d.f. and distribution function

The mean, variance, and moment-generating function of X are not difficult to calculate (see Exercise 4.2.1). They are

$$\mu = \frac{a+b}{2}, \qquad \sigma^2 = \frac{(b-a)^2}{12},$$

$$M(t) = \begin{cases} \dfrac{e^{tb}-e^{ta}}{t(b-a)}, & t \neq 0, \\ 1, & t = 0. \end{cases}$$

An important uniform distribution is that for which $a=0$ and $b=1$, namely $U(0,1)$. If X is $U(0,1)$, approximate values of X can be simulated on most computers using a random number generator. In fact, it should be called a **pseudo-random number generator** because the programs that produce the random numbers are usually such that if the starting number (the seed number) is known, all subsequent numbers in the sequence may be determined by simple arithmetical operations. Yet, despite their deterministic origin, these computer-produced numbers do behave as if they were truly randomly generated, and we shall not encumber our terminology by adding *pseudo*. Examples of computer-produced random numbers are given in the Appendix in Table IX. Treat each of these numbers as a decimal; that is, divide each of them by 10^4.

The next example looks at some of the numbers in Table IX.

Example 4.2.1

In this example we use some statistical techniques that were studied earlier and that can be used to help determine whether the uniform distribution is the appropriate probability model. In particular, to help decide whether or not the data in Table IX are observations from a uniform distribution, $U(0,1)$, we could construct a histogram and empirical distribution function of all of the data, comparing these with the p.d.f. and distribution function of the uniform distribution, $U(0,1)$. Instead, the techniques we illustrate in this example use fewer numbers of observations. Suppose that we select the first 19 numbers in the third column of Table IX. They are

0.6960	0.4658	0.4930	0.7843	0.1233	0.4779	0.4871
0.0315	0.6814	0.2581	0.7244	0.9705	0.0460	0.8287
0.2055	0.2906	0.5334	0.8071	0.3384		

For these 19 numbers, the sample mean is $\bar{x} = 0.4865$ and the sample standard deviation is $s = 0.2797$. These should be compared with the mean, $\mu = 0.5000$, and standard deviation, $\sigma = \sqrt{1/12} = 0.2887$, of the uniform distribution, $U(0,1)$. Graphically, a technique that can sometimes yield insight into the data is the q-q plot of the data quantiles versus the corresponding $U(0,1)$ distribution quantiles. We first order the data as follows:

0.0315	0.0460	0.1233	0.2055	0.2581	0.2906	0.3384
0.4658	0.4779	0.4871	0.4930	0.5334	0.6814	0.6960
0.7244	0.7843	0.8071	0.8287	0.9705		

The reason for selecting a sample of size 19 is because the respective quantiles are associated with $i/(n+1), i = 1, 2, \ldots, 19$, namely $0.05, 0.10, 0.15, \ldots, 0.95$. The respective quantiles of the $U(0, 1)$ distribution are $0.05, 0.10, 0.15, \ldots, 0.95$. The q - q plot is shown in Figure 4.5. Note that the points do fall close to the line $y = x$.

Since the data are ordered, it is also quite easy to construct the empirical distribution function and draw the theoretical distribution function on the same graph. This is illustrated in Figure 4.5. In Section 10.7 a statistical test is given that is based on the greatest vertical distance between the empirical and theoretical distribution functions. This test is called the Kolmogorov-Smirnov goodness-of-fit test, but for now we simply use an "eyeball" test.

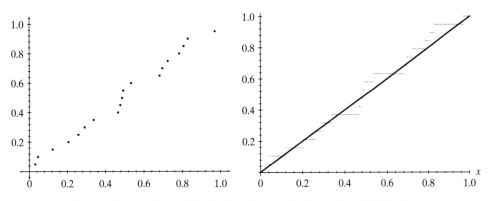

Figure 4.5 q-q plot and distribution functions for the uniform distribution

In addition, we consider a box-and-whisker diagram with the five-number summary of the minimum, 0.0315; the first quartile, $\widehat{q}_1 = 0.2581$; the median, $\widehat{m} = 0.4871$; the third quartile, $\widehat{q}_3 = 0.7244$; and the maximum, 0.9705. The box plot is shown in Figure 4.6. Note especially the symmetry of this box plot.

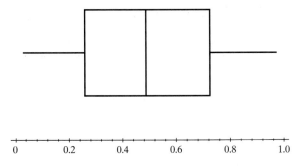

Figure 4.6   Box plot of the sample from the uniform distribution

We must point out that every set of 19 observations from Table IX or from a $U(0, 1)$ distribution will not yield characteristics and figures exactly like these, and in some cases the fit will perhaps not look very good. In addition, larger samples are preferable to samples of size 19, especially if a computer can help with the computations and the graphics.   ◇

We turn now to a continuous distribution that is related to the Poisson distribution. When previously observing a process of the (approximate) Poisson type, we counted the number of changes occurring in a given interval. This number was a discrete-type random variable with a Poisson distribution. But not only is the number of changes a random variable; the waiting times between successive changes are also random variables. However, the latter are of the continuous type, since each of them can assume any positive value. In particular, let W denote the waiting time until the first change occurs when observing a Poisson process in which the mean number of changes in the unit interval is λ. Then W is a continuous-type random variable, and we proceed to find its distribution function.

Because this waiting time is nonnegative, the distribution function $F(w) = 0$, $w < 0$. For $w \geq 0$,

$$
\begin{aligned}
F(w) &= P(W \leq w) = 1 - P(W > w) \\
&= 1 - P(\text{no changes in } [0, w]) \\
&= 1 - e^{-\lambda w},
\end{aligned}
$$

since we discovered previously that $e^{-\lambda w}$ equals the probability of no changes in an interval of length w. That is, if the mean number of changes per unit interval is λ, then the mean number of changes in an interval of length w is proportional to w, namely, λw. Thus, when $w > 0$, the p.d.f. of W is given by

$$
F'(w) = f(w) = \lambda e^{-\lambda w}.
$$

We often let $\lambda = 1/\theta$ and say that the random variable X has an **exponential distribution** if its p.d.f. is defined by

$$
f(x) = \frac{1}{\theta} e^{-x/\theta}, \qquad 0 \leq x < \infty,
$$

where the parameter $\theta > 0$. Accordingly, the waiting time W until the first change in a Poisson process has an exponential distribution with $\theta = 1/\lambda$. To determine the exact meaning of the parameter θ, we first find the moment-generating function. It is

$$
M(t) = \int_0^\infty e^{tx} \left(\frac{1}{\theta} \right) e^{-x/\theta} \, dx = \lim_{b \to \infty} \int_0^b \left(\frac{1}{\theta} \right) e^{-(1-\theta t)x/\theta} \, dx
$$

$$
= \lim_{b \to \infty} \left[-\frac{e^{-(1-\theta t)x/\theta}}{1 - \theta t} \right]_0^b = \frac{1}{1 - \theta t}, \qquad t < \frac{1}{\theta}.
$$

Thus

$$M'(t) = \frac{\theta}{(1 - \theta t)^2}$$

and

$$M''(t) = \frac{2\theta^2}{(1 - \theta t)^3}.$$

Hence, for an exponential distribution, we have

$$\mu = M'(0) = \theta \quad \text{and} \quad \sigma^2 = M''(0) - [M'(0)]^2 = \theta^2.$$

So if λ is the mean number of changes in the unit interval, then $\theta = 1/\lambda$ is the mean waiting time for the first change. In particular, suppose that $\lambda = 7$ is the mean number of changes per minute; then the mean waiting time for the first change is 1/7 of a minute, a result that agrees with our intuition.

Example 4.2.2

Let X have an exponential distribution with a mean of 40. The p.d.f. of X is

$$f(x) = \frac{1}{40} e^{-x/40}, \qquad 0 \le x < \infty.$$

See Figure 4.1 for the graph of this p.d.f. The probability that X is less than 36 is

$$P(X < 36) = \int_0^{36} \frac{1}{40} e^{-x/40} \, dx = 1 - e^{-36/40} = 0.593. \qquad \diamond$$

Let X have an exponential distribution with mean $\mu = \theta$. Then the distribution function of X is

$$F(x) = \begin{cases} 0, & -\infty < x < 0, \\ 1 - e^{-x/\theta}, & 0 \le x < \infty. \end{cases}$$

For example, the p.d.f. and distribution function are graphed in Figure 4.7 for $\theta = 5$. The median, m, is found by solving $F(m) = 0.5$. That is,

$$1 - e^{-m/\theta} = 0.5.$$

Thus

$$m = -\theta \ln(0.5) = \theta \ln(2).$$

So with $\theta = 5$, the median is $m = -5 \ln(0.5) = 3.466$. Both the median and the mean $\theta = 5$ are indicated on the graphs.

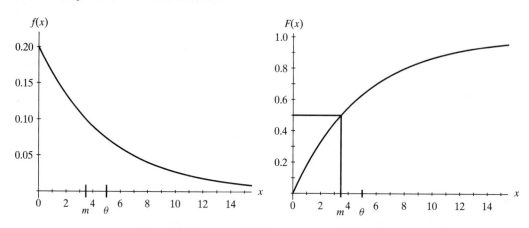

Figure 4.7 Exponential p.d.f. and distribution function

It is useful to note that for an exponential random variable, X, we have that

$$P(X > x) = 1 - F(x) = 1 - (1 - e^{-x/\theta})$$
$$= e^{-x/\theta},$$

when $x > 0$.

Example 4.2.3

Customers arrive in a certain shop according to an approximate Poisson process at a mean rate of 20 per hour. What is the probability that the shopkeeper will have to wait more than 5 minutes for the arrival of the first customer? Let X denote the waiting time in minutes until the first customer arrives and note that $\lambda = 1/3$ is the expected number of arrivals per minute. Thus

$$\theta = \frac{1}{\lambda} = 3$$

and

$$f(x) = \frac{1}{3} e^{-(1/3)x}, \qquad 0 \le x < \infty.$$

Hence

$$P(X > 5) = \int_5^\infty \frac{1}{3} e^{-(1/3)x} \, dx = e^{-5/3} = 0.1889.$$

The median time until the first arrival is

$$m = -3 \ln(0.5) = 2.0794. \qquad \diamond$$

Example 4.2.4

Suppose that the life of a certain type of electronic component has an exponential distribution with a mean life of 500 hours. If X denotes the life of this component (or the time to failure of this component), then

$$P(X > x) = \int_x^\infty \frac{1}{500} e^{-t/500} \, dt = e^{-x/500}.$$

Suppose that the component has been in operation for 300 hours. The conditional probability that it will last for another 600 hours is

$$P(X > 900 \mid X > 300) = \frac{P(X > 900)}{P(X > 300)} = \frac{e^{-900/500}}{e^{-300/500}} = e^{-6/5}.$$

It is important to note that this conditional probability is exactly equal to $P(X > 600) = e^{-6/5}$. That is, the probability that it will last an additional 600 hours, given that it has operated 300 hours, is the same as the probability that it would last 600 hours when first put into operation. Thus, for such components, an old component is as good as a new one, and we say that the failure rate is constant. Certainly, with constant failure rate, there is no advantage in replacing components that are operating satisfactorily. Obviously, this is not true in practice because most would have an increasing failure rate with time; hence the exponential distribution is probably not the best model for the probability distribution of such a life.   ◇

Remark   In Exercise 4.2.11, the result of Example 4.2.4 is generalized; namely, if the length of life has an exponential distribution, the probability that it will last a time of at least $x + y$ units, given that it has lasted at least x units, is exactly the same as the probability that it will last at least y units when first put into operation. In effect, this states that the exponential distribution has a forgetfulness (or no memory) property. It is also interesting to observe that, for continuous random variables whose support is $(0, \infty)$, the exponential distribution is the only distribution with this forgetfulness property. Recall, however, that when we considered distributions of the discrete type, we noted that the geometric distribution has this property (see Exercise 3.6.10).

EXERCISES

4.2.1 Show that the mean, variance, and moment-generating function of the uniform distribution are as given in this section.

4.2.2 Let $f(x) = 1/2$, $-1 \le x \le 1$, be the p.d.f. of X. Graph the p.d.f. and distribution function, and record the mean and variance of X.

4.2.3 Customers arrive randomly at a bank teller's window. Given that one customer arrived during a particular 10-minute period, let X equal the time within the 10 minutes that the customer arrived. If X is $U(0, 10)$, find

(a) The p.d.f. of X.

(b) $P(X \ge 8)$.

(c) $P(2 \le X < 8)$.

(d) $E(X)$.

(e) $\mathrm{Var}(X)$.

4.2.4 If the moment-generating function of X is

$$M(t) = \frac{e^{5t} - e^{4t}}{t}, \qquad t \neq 0, \qquad \text{and} \qquad M(0) = 1,$$

find (a) $E(X)$, (b) Var(X), and (c) $P(4.2 < X \leq 4.7)$.

4.2.5 Let Y have a uniform distribution $U(0, 1)$, and let

$$W = a + (b - a)Y, \qquad a < b.$$

   (a) Find the distribution function of W. HINT: Find $P[a + (b - a)Y \leq w]$.

   (b) How is W distributed?

4.2.6 Let X have an exponential distribution with a mean of $\theta = 20$. Compute

   (a) $P(10 < X < 30)$.

   (b) $P(X > 30)$.

   (c) $P(X > 40 \mid X > 10)$.

4.2.7 Let the p.d.f. of X be $f(x) = (1/2)e^{-x/2}, 0 \leq x < \infty$.

   (a) What are the mean, variance, and moment-generating function of X?

   (b) Calculate $P(X > 3)$.

   (c) Calculate $P(X > 5 \mid X > 2)$.

4.2.8 Telephone calls enter a college switchboard according to a Poisson process on the average of two every 3 minutes. Let X denote the waiting time until the first call that arrives after 10 A.M.

   (a) What is the p.d.f. of X?

   (b) Find $P(X > 2)$.

4.2.9 Determine the p.d.f., the mean, and the variance of X if the moment-generating function of X is given by

   (a) $M(t) = \dfrac{1}{1 - 3t}, \qquad t < 1/3.$

   (b) $M(t) = \dfrac{3}{3 - t}, \qquad t < 3.$

4.2.10 Let X equal the number of students who use a college card catalog every 15 minutes. Assume that X has a Poisson distribution with mean 5. Let W equal the time in minutes between two student arrivals. Then W has an exponential distribution with mean $\theta = 3$. Find

   (a) $P(W > 6)$.

   (b) $P(W > 12 \mid W > 6)$.

4.2.11 Let X have an exponential distribution with mean $\theta > 0$. Show that

$$P(X > x + y \mid X > x) = P(X > y).$$

4.2.12 Let $F(x)$ be the distribution function of the continuous-type random variable X, and assume that $F(x) = 0$ for $x \leq 0$ and $0 < F(x) < 1$ for $0 < x$. Prove that if

$$P(X > x + y \mid X > x) = P(X > y),$$

then

$$F(x) = 1 - e^{-\lambda x}, \qquad 0 < x.$$

HINT: Show that $g(x) = 1 - F(x)$ satisfies the functional equation

$$g(x + y) = g(x)g(y),$$

which implies that $g(x) = a^{cx}$.

4.2.13 Let X equal the number of bad records in each 100 feet of a used computer tape. Assume that X has a Poisson distribution with mean 2.5. Let W equal the number of feet before the first bad record is found.

(a) Give the mean number of flaws per foot.

(b) How is W distributed?

(c) Give the mean and variance of W.

(d) Find (i) $P(W \leq 20)$, (ii) $P(W > 40)$, and (iii) $P(W > 60 \,|\, W > 20)$.

4.2.14 Let W equal the number of feet between bad records on the computer tape in Exercise 4.2.13. Ten observations of W are

$$67 \quad 16 \quad 7 \quad 35 \quad 97 \quad 28 \quad 74 \quad 5 \quad 9 \quad 37$$

(a) Calculate the sample mean $\overline{w}$ and sample variance s^2 of these data.

(b) Graph the empirical distribution function of these data along with the distribution function for the exponential distribution with mean $\theta = 40$.

(c) Do these data look like observations from an exponential distribution with a mean of 40?

4.2.15 Let X equal the time (in minutes) between calls that are made over the public safety radio. On four different days (February 14, 21, and 28, and March 6) and during a period of one hour on each day, the following observations of X were made:

$$
\begin{array}{ccccccccccc}
5 & 7 & 8 & 20 & 17 & 2 & 24 & 8 & 8 & 6 & 4 \\
3 & 42 & 10 & 18 & 5 & 7 & 8 & 4 & 5 & 10 &
\end{array}
$$

If calls arrive randomly in accordance with an approximate Poisson process, then the distribution of X should be approximately exponential.

(a) Calculate the values of the sample mean and sample standard deviation. Are they close to each other in value?

(b) Construct a q-q plot of the ordered observations versus the respective quantiles of the exponential distribution with a mean of $\theta = 1$. If this is approximately linear, the exponential model is supported. Since the mean of these data is not close to 1, this line will not have slope 1, but a linear fit will still indicate an exponential model. What is your conclusion?

(c) Construct a box-and-whisker diagram. Does it indicate that the data are skewed, as would be true for observations from an exponential distribution?

4.2.16 Some biology students were interested in analyzing the amount of time that bees spend in flower patches gathering nectar. In particular, they were wondering whether it made a difference whether there was a low density or a high density of flowers. Thirty-seven bees visited the low-density flower patch and spent the following times (in seconds):

$$
\begin{array}{cccccccccc}
10 & 40 & 20 & 85 & 105 & 85 & 270 & 40 & 60 & 90 \\
360 & 30 & 520 & 210 & 45 & 16 & 6 & 420 & 90 & 515 \\
71 & 448 & 353 & 526 & 279 & 279 & 505 & 157 & 240 & 840 \\
240 & 180 & 300 & 450 & 225 & 60 & 300 & & &
\end{array}
$$

Thirty-nine bees visited the high-density flower patch and spent the following times (in seconds):

235	210	95	146	195	840	185	610	680	990
146	404	119	47	9	4	10	169	270	95
329	151	211	127	154	35	225	140	158	116
46	113	149	420	120	45	10	18	105	

(a) Calculate the sample mean and sample standard deviation for the low-density flower patch. Are they close in value to each other?

(b) Calculate the sample mean and sample standard deviation for the high-density flower patch. Are they close in value to each other?

(c) Construct back-to-back stem-and-leaf diagrams.

(d) Construct two box plots on the same diagram.

(e) How would you compare the times bees spend in the low- and high-density flower patches?

(f) Does the exponential model look like it could be used for these times?

4.2.17 There are times when a shifted exponential model is appropriate. That is, let the p.d.f. of X be

$$f(x) = \frac{1}{\theta} e^{-(x-\alpha)/\theta}, \qquad \alpha < x < \infty.$$

(a) Define the distribution function of X.

(b) Calculate the mean and variance of X.

4.2.18 A certain type of aluminum screen 2 feet in width has on the average three flaws in a 100-foot roll.

(a) What is the probability that the first 40 feet in a roll contain no flaws?

(b) What assumption did you make to solve part (a)?

4.2.19 Let X have an exponential distribution with mean θ.

(a) Find the first quartile, q_1.

(b) How far is the first quartile below the mean?

(c) Find the third quartile, q_3.

(d) How far is the third quartile above the mean?

4.2.20 Let X have a logistic distribution with p.d.f.

$$f(x) = \frac{e^{-x}}{(1 + e^{-x})^2}, \qquad -\infty < x < \infty.$$

Show that

$$Y = \frac{1}{1 + e^{-X}}$$

has a $U(0, 1)$ distribution. HINT: Find $G(y) = P(Y \leq y) = P\left(\frac{1}{1+e^{-X}} \leq y\right)$, when $0 < y < 1$.

4.3 THE GAMMA AND CHI-SQUARE DISTRIBUTIONS

In the (approximate) Poisson process with mean λ, we have seen that the waiting time until the first change has an exponential distribution. We now let W denote the waiting time until the αth change occurs and find the distribution of W.

The distribution function of W, when $w \geq 0$, is given by

$$F(w) = P(W \leq w) = 1 - P(W > w)$$
$$= 1 - P(\text{fewer than } \alpha \text{ changes occur in } [0, w])$$
$$= 1 - \sum_{k=0}^{\alpha-1} \frac{(\lambda w)^k e^{-\lambda w}}{k!} \tag{4.1}$$

since the number of changes in the interval $[0, w]$ has a Poisson distribution with mean λw. Because W is a continuous-type random variable, $F'(w)$ is equal to the p.d.f. of W whenever this derivative exists. We have, provided that $w > 0$,

$$F'(w) = \lambda e^{-\lambda w} - e^{-\lambda w} \sum_{k=1}^{\alpha-1} \left[\frac{k(\lambda w)^{k-1}\lambda}{k!} - \frac{(\lambda w)^k \lambda}{k!} \right]$$
$$= \lambda e^{-\lambda w} - e^{-\lambda w} \left[\lambda - \frac{\lambda(\lambda w)^{\alpha-1}}{(\alpha-1)!} \right]$$
$$= \frac{\lambda(\lambda w)^{\alpha-1}}{(\alpha-1)!} e^{-\lambda w}.$$

If $w < 0$, then $F(w) = 0$ and $F'(w) = 0$. A p.d.f. of this form is said to be one of the gamma type, and the random variable W is said to have a gamma distribution.

Before determining the characteristics of the gamma distribution, let us consider the gamma function for which the distribution is named. The **gamma function** is defined by

$$\Gamma(t) = \int_0^\infty y^{t-1} e^{-y} \, dy, \qquad 0 < t.$$

This integral is positive for $0 < t$ because the integrand is positive. Values of it are often given in a table of integrals. If $t > 1$, integration of the gamma function of t by parts yields

$$\Gamma(t) = \left[-y^{t-1} e^{-y} \right]_0^\infty + \int_0^\infty (t-1) y^{t-2} e^{-y} \, dy$$
$$= (t-1) \int_0^\infty y^{t-2} e^{-y} \, dy = (t-1)\Gamma(t-1).$$

For example, $\Gamma(6) = 5\Gamma(5)$ and $\Gamma(3) = 2\Gamma(2) = (2)(1)\Gamma(1)$. Whenever $t = n$, a positive integer, we have, by repeated application of $\Gamma(t) = (t-1)\Gamma(t-1)$, that

$$\Gamma(n) = (n-1)\Gamma(n-1) = (n-1)(n-2)\cdots(2)(1)\Gamma(1).$$

However,

$$\Gamma(1) = \int_0^\infty e^{-y}\, dy = 1.$$

Thus, when n is a positive integer, we have

$$\Gamma(n) = (n-1)!;$$

and, for this reason, the gamma function is called the generalized factorial. Incidentally, $\Gamma(1)$ corresponds to $0!$, and we have noted that $\Gamma(1) = 1$, which is consistent with earlier discussions.

Let us now formally define the p.d.f. of the gamma distribution and find its characteristics. The random variable X has a **gamma distribution** if its p.d.f. is defined by

$$f(x) = \frac{1}{\Gamma(\alpha)\theta^\alpha} x^{\alpha-1} e^{-x/\theta}, \qquad 0 \le x < \infty.$$

Hence, W, the waiting time until the αth change in a Poisson process, has a gamma distribution with parameters α and $\theta = 1/\lambda$. To see that $f(x)$ actually has the properties of a p.d.f., note that $f(x) \ge 0$ and

$$\int_{-\infty}^\infty f(x)\, dx = \int_0^\infty \frac{x^{\alpha-1} e^{-x/\theta}}{\Gamma(\alpha)\theta^\alpha}\, dx$$

which, by the change of variables $y = x/\theta$, equals

$$\int_0^\infty \frac{(\theta y)^{\alpha-1} e^{-y}}{\Gamma(\alpha)\theta^\alpha} \theta\, dy = \frac{1}{\Gamma(\alpha)} \int_0^\infty y^{\alpha-1} e^{-y}\, dy = \frac{\Gamma(\alpha)}{\Gamma(\alpha)} = 1.$$

The moment-generating function of X is (Exercise 4.3.3)

$$M(t) = \frac{1}{(1-\theta t)^\alpha}, \qquad t < 1/\theta.$$

The mean and variance are (Exercise 4.3.4)

$$\mu = \alpha\theta \qquad \text{and} \qquad \sigma^2 = \alpha\theta^2.$$

Example 4.3.1

Suppose that an average of 30 customers per hour arrive at a shop in accordance with a Poisson process. That is, if a minute is our unit, then $\lambda = 1/2$. What is the probability that the shopkeeper will wait more than 5 minutes before both of the first two customers arrive? If X denotes the waiting time in minutes until the second customer arrives, then X has a gamma distribution with $\alpha = 2, \theta = 1/\lambda = 2$. Hence

$$P(X > 5) = \int_5^\infty \frac{x^{2-1} e^{-x/2}}{\Gamma(2)2^2}\, dx = \int_5^\infty \frac{x e^{-x/2}}{4}\, dx$$

$$= \frac{1}{4}\left[(-2)x e^{-x/2} - 4e^{-x/2}\right]_5^\infty$$

$$= \frac{7}{2} e^{-5/2} = 0.287.$$

We could also have used Equation 4.1 with $\lambda = 1/2$ because α is an integer. From Equation 4.1 we have

$$P(X > x) = \sum_{k=0}^{\alpha-1} \frac{(x/\theta)^k \, e^{-x/\theta}}{k!}.$$

Thus, with $x = 5$, $\alpha = 2$, and $\theta = 2$, this is equal to

$$P(X > 5) = \sum_{k=0}^{2-1} \frac{(5/2)^k \, e^{-5/2}}{k!}$$

$$= e^{-5/2}\left(1 + \frac{5}{2}\right) = \left(\frac{7}{2}\right)e^{-5/2}. \qquad \diamond$$

Example 4.3.2

Telephone calls arrive at a switchboard at a mean rate of $\lambda = 2$ per minute according to a Poisson process. Let X denote the waiting time in minutes until the fifth call arrives. The p.d.f. of X, with $\alpha = 5$ and $\theta = 1/\lambda = 1/2$, is

$$f(x) = \frac{2^5 x^4}{4!} e^{-2x}, \qquad 0 \le x < \infty.$$

The mean and the variance of X are, respectively, $\mu = 5/2$ and $\sigma^2 = 5/4$. $\qquad \diamond$

In order to see the effect of the parameters on the shape of the p.d.f., several combinations of α and θ have been used for graphs that are displayed in Figure 4.8. Note that for a fixed θ, as α increases, the probability moves to the right. The same is true for increasing θ, with fixed α. Since $\theta = 1/\lambda$, as θ increases, λ decreases. That is, if $\theta_2 > \theta_1$, then $\lambda_2 = 1/\theta_2 < \lambda_1 = 1/\theta_1$. So if the mean number of changes per unit decreases, the waiting time to observe α changes can be expected to increase.

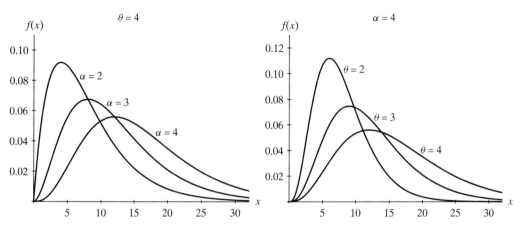

Figure 4.8   Gamma p.d.f.'s: $\theta = 4$ with $\alpha = 2, 3, 4$; $\alpha = 4$ with $\theta = 2, 3, 4$

We now consider a special case of the gamma distribution that plays an important role in statistics. Let X have a gamma distribution with $\theta = 2$ and $\alpha = r/2$, where r is a positive integer. The p.d.f. of X is

$$f(x) = \frac{1}{\Gamma(r/2)2^{r/2}}\, x^{r/2-1}e^{-x/2}, \qquad 0 \le x < \infty.$$

We say that X has a **chi-square distribution with r degrees of freedom**, which we abbreviate by saying X is $\chi^2(r)$. The mean and the variance of this chi-square distribution are

$$\mu = \alpha\theta = \left(\frac{r}{2}\right)2 = r \qquad \text{and} \qquad \sigma^2 = \alpha\theta^2 = \left(\frac{r}{2}\right)2^2 = 2r.$$

That is, the mean equals the number of degrees of freedom, and the variance equals twice the number of degrees of freedom. An explanation of "number of degrees of freedom" is given later. From the results concerning the more general gamma distribution, we see that its moment-generating function is

$$M(t) = (1 - 2t)^{-r/2}, \qquad t < \frac{1}{2}.$$

In Figure 4.9 the graphs of chi-square p.d.f.'s for $r = 2, 3, 5$, and 8 are given. Note the relationship between the mean, $\mu = r$, and the point at which the p.d.f. obtains its maximum (see Exercise 4.3.13).

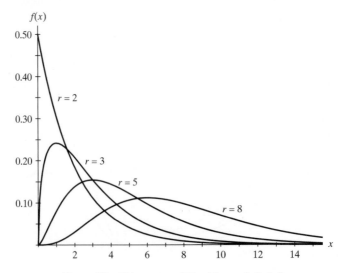

Figure 4.9   Chi-square p.d.f.'s with $r = 2, 3, 5, 8$

Because the chi-square distribution is so important in applications, tables have been prepared giving the values of the distribution function

$$F(x) = \int_0^x \frac{1}{\Gamma(r/2)2^{r/2}} w^{r/2-1} e^{-w/2} \, dw$$

for selected values of r and x. For an example, see Table IV in the Appendix.

Example 4.3.3

Let X have a chi-square distribution with $r = 5$ degrees of freedom. Then, using Table IV, we obtain

$$P(1.145 \leq X \leq 12.83) = F(12.83) - F(1.145) = 0.975 - 0.050 = 0.925$$

and

$$P(X > 15.09) = 1 - F(15.09) = 1 - 0.99 = 0.01. \qquad \diamond$$

Example 4.3.4

If X is $\chi^2(7)$, two constants, a and b, such that

$$P(a < X < b) = 0.95$$

are $a = 1.690$ and $b = 16.01$. Other constants a and b can be found, and we are restricted in our choices only by the limited table. $\qquad \diamond$

Probabilities like that of Example 4.3.4 are so important in statistical applications that we use special symbols for a and b. Let α be a positive probability (that is usually less than 0.5) and let X have a chi-square distribution with r degrees of freedom. Then $\chi_\alpha^2(r)$ is a number such that

$$P[X \geq \chi_\alpha^2(r)] = \alpha.$$

That is, $\chi_\alpha^2(r)$ is the $100(1 - \alpha)$ percentile (or upper 100α percent point) of the chi-square distribution with r degrees of freedom. Then the 100α percentile is the number $\chi_{1-\alpha}^2(r)$ such that

$$P[X \leq \chi_{1-\alpha}^2(r)] = \alpha.$$

That is, the probability to the right of $\chi_{1-\alpha}^2(r)$ is $1 - \alpha$ (see Figure 4.10).

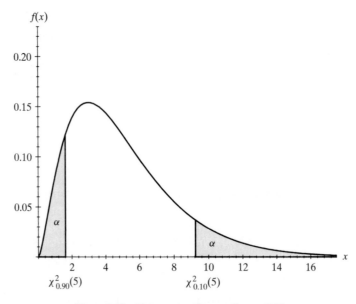

Figure 4.10   Chi-square tails, $r = 5, \alpha = 0.10$

Example 4.3.5

Let X have a chi-square distribution with five degrees of freedom. Then, using Table IV, $\chi^2_{0.10}(5) = 9.236$ and $\chi^2_{0.90}(5) = 1.610$. These are the points that are indicated on Figure 4.10.   ◇

Example 4.3.6

If customers arrive at a shop on the average of 30 per hour in accordance with a Poisson process, what is the probability that the shopkeeper will have to wait longer than 9.390 minutes for the first nine customers to arrive? Note that the mean rate of arrivals per minute is $\lambda = 1/2$. Thus, $\theta = 2$ and $\alpha = r/2 = 9$. If X denotes the waiting time until the ninth arrival, X is $\chi^2(18)$. Hence

$$P(X > 9.390) = 1 - 0.05 = 0.95.$$ ◇

Example 4.3.7

If X has an exponential distribution with a mean of 2, the p.d.f. of X is

$$f(x) = \frac{1}{2} e^{-x/2} = \frac{x^{2/2-1} e^{-x/2}}{\Gamma(2/2) 2^{2/2}}, \qquad 0 \le x < \infty.$$

That is, X is $\chi^2(2)$. Thus, for illustration,

$$P(0.051 < X < 7.378) = 0.975 - 0.025 = 0.95.$$ ◇

EXERCISES

4.3.1 Telephone calls enter a college switchboard at a mean rate of 2/3 call per minute according to a Poisson process. Let X denote the waiting time until the tenth call arrives.

(a) What is the p.d.f. of X?

(b) What are the moment-generating function, mean, and variance of X?

4.3.2 If X has a gamma distribution with $\theta = 4$ and $\alpha = 2$, find $P(X < 5)$.

4.3.3 Find the moment-generating function for the gamma distribution with parameters α and θ. HINT: In the integral representing $E(e^{tX})$, change variables by letting $y = (1 - \theta t)x/\theta$, where $1 - \theta t > 0$.

4.3.4 Use the moment-generating function of a gamma distribution to show that $E(X) = \alpha\theta$ and $\text{Var}(X) = \alpha\theta^2$.

4.3.5 If the moment-generating function of a random variable W is

$$M(t) = (1 - 7t)^{-20},$$

find the p.d.f., mean, and variance of W.

4.3.6 Let X denote the number of alpha particles emitted by barium–133 and observed by a Geiger counter in a fixed position. Assume that $\lambda = 14.7$ is the mean number of counts per second. Let W denote the waiting time to observe 100 counts. Twenty-five independent observations of W are

6.9	7.3	6.7	6.4	6.3
5.9	7.0	7.1	6.5	7.6
7.2	7.1	6.1	7.3	7.6
7.6	6.7	6.3	5.7	6.7
7.5	5.3	5.4	7.4	6.9

(a) Give the p.d.f., mean, and variance of W.

(b) Calculate the sample mean and sample variance of the 25 observations of W.

(c) Use the relative frequency of event $\{W \le 6.6\}$ to approximate $P(W \le 6.6)$.

4.3.7 Consider the computer tape described in Exercise 4.2.13 but now let W equal the number of feet before the third bad record is encountered.

(a) How is W distributed?

(b) Give the mean and variance of W.

(c) Find $P(W \le 100)$.

HINT: Use Equation 4.1 with $\lambda w = 2.5$.

4.3.8 Let X equal the number of alpha-particle emissions of carbon–14 that are counted by a Geiger counter per second. Assume that the distribution of X is Poisson with mean 16. Let W equal the time in seconds before the seventh count is made.

(a) Give the distribution of W.

(b) Find $P(W \le 0.5)$.

4.3.9 If X is $\chi^2(17)$, find
  (a) $P(X < 7.564)$.
  (b) $P(X > 27.59)$.
  (c) $P(6.408 < X < 27.59)$.
  (d) $\chi^2_{0.95}(17)$.
  (e) $\chi^2_{0.025}(17)$.

4.3.10 If X is $\chi^2(12)$, find constants a and b such that
$$P(a < X < b) = 0.90 \quad \text{and} \quad P(X < a) = 0.05.$$

4.3.11 If X is $\chi^2(23)$, find
  (a) $P(14.85 < X < 32.01)$.
  (b) Constants a and b such that $P(a < X < b) = 0.95$ and $P(X < a) = 0.025$.
  (c) The mean and variance of X.
  (d) $\chi^2_{0.05}(23)$ and $\chi^2_{0.95}(23)$.

4.3.12 If the moment-generating function of X is $M(t) = (1 - 2t)^{-12}$, $t < 1/2$, find
  (a) $E(X)$.
  (b) $\text{Var}(X)$.
  (c) $P(15.66 < X < 42.98)$.

4.3.13 Find the point at which a chi-square p.d.f. obtains its maximum when $r > 2$.

4.3.14 Cars arrive at a toll booth at a mean rate of five cars every 10 minutes according to a Poisson process. Find the probability that the toll collector will have to wait longer than 26.30 minutes before collecting the eighth toll.

4.3.15 If 15 observations are taken independently from a chi-square distribution with four degrees of freedom, find the probability that at most three of the sample items exceed 7.779.

4.3.16 If 10 observations are taken independently from a chi-square distribution with 19 degrees of freedom, find the probability that exactly 2 of the 10 sample items exceed 30.14.

4.3.17 Suppose that 100 random numbers from the interval $[0, 1)$ are generated on a computer. Then let Y_i equal the number of random numbers, x, such that $(i - 1)/10 \leq x < i/10$, $i = 1, 2, \ldots, 10$. It will be shown in Section 7.5 that
$$Q = \sum_{i=1}^{10} \frac{(Y_i - 10)^2}{10}$$
has a distribution that is approximately $\chi^2(9)$. The following data give 100 observations of Q:

12.6	6.8	4.6	9.2	12.6	17.4	7.8	12.4	5.4	1.8
14.6	7.0	2.8	11.2	4.8	12.6	9.0	8.6	18.2	12.6
5.2	17.6	6.2	19.8	9.8	4.4	8.6	6.2	4.0	17.0
7.6	6.2	15.4	2.0	11.2	12.8	5.4	10.6	12.2	5.2
1.8	16.2	7.4	9.4	10.2	13.4	14.0	6.0	9.2	9.0
10.6	16.2	5.6	7.2	14.2	7.2	9.4	11.4	6.6	15.2
7.8	10.6	2.6	12.2	7.6	9.8	18.6	15.0	6.4	10.0
8.2	7.0	4.8	7.2	5.4	5.0	10.4	10.0	14.2	11.8
3.0	23.0	8.2	7.6	4.4	10.4	6.2	14.2	4.0	6.8
4 4	12.0	8.8	4.0	11.4	4.0	12.2	3.8	13.0	10.6

(a) Construct a frequency distribution.

(b) Draw the relative frequency histogram. Superimpose the $\chi^2(9)$ p.d.f.

(c) Calculate the sample mean and sample variance. With $r = 9$ are they close to $r = 9$ and $2r = 18$, respectively?

(d) Use the data to approximate $P(Q \leq 3.325)$ and $P(Q \geq 14.68)$. Compare the approximations with the probabilities selected from Table IV for the $\chi^2(9)$ distribution.

4.4 THE NORMAL DISTRIBUTION

The normal distribution is perhaps the most important distribution in statistical applications since many measurements have (approximate) normal distributions. One explanation of this fact is the role of the normal distribution in the central limit theorem. One form of this theorem is considered in Section 5.4.

We give the definition of the p.d.f. for the normal distribution, verify that it is a p.d.f., and then justify the use of μ and σ^2 in this formula. That is, we will show that μ and σ^2 are actually the mean and the variance of this distribution. The random variable X has a **normal distribution** if its p.d.f. is defined by

$$f(x) = \frac{1}{\sigma\sqrt{2\pi}} \exp\left[-\frac{(x-\mu)^2}{2\sigma^2}\right], \qquad -\infty < x < \infty,$$

where μ and σ are parameters satisfying $-\infty < \mu < \infty$, $0 < \sigma < \infty$, and also where $\exp[v]$ means e^v. Briefly, we say that X is $N(\mu, \sigma^2)$.

Clearly, $f(x) > 0$. We now evaluate the integral

$$I = \int_{-\infty}^{\infty} \frac{1}{\sigma\sqrt{2\pi}} \exp\left[-\frac{(x-\mu)^2}{2\sigma^2}\right] dx,$$

showing that it is equal to 1. In I, change variables of integration by letting $z = (x - \mu)/\sigma$. Then

$$I = \int_{-\infty}^{\infty} \frac{1}{\sqrt{2\pi}} e^{-z^2/2} \, dz.$$

Since $I > 0$, if $I^2 = 1$, then $I = 1$. Now

$$I^2 = \frac{1}{2\pi} \left[\int_{-\infty}^{\infty} e^{-x^2/2} \, dx\right]\left[\int_{-\infty}^{\infty} e^{-y^2/2} \, dy\right],$$

or, equivalently,

$$I^2 = \frac{1}{2\pi} \int_{-\infty}^{\infty}\int_{-\infty}^{\infty} \exp\left(-\frac{x^2+y^2}{2}\right) dx \, dy.$$

Letting $x = r\cos\theta$, $y = r\sin\theta$ (i.e. using polar coordinates), we have

$$I^2 = \frac{1}{2\pi} \int_0^{2\pi}\int_0^{\infty} e^{-r^2/2} r \, dr \, d\theta$$

$$= \frac{1}{2\pi} \int_0^{2\pi} d\theta = \frac{1}{2\pi} 2\pi = 1.$$

Thus $I = 1$, and we have shown that $f(x)$ has the properties of a p.d.f. The moment-generating function of X is

$$M(t) = \int_{-\infty}^{\infty} \frac{e^{tx}}{\sigma\sqrt{2\pi}} \exp\left[-\frac{(x-\mu)^2}{2\sigma^2}\right] dx$$

$$= \int_{-\infty}^{\infty} \frac{1}{\sigma\sqrt{2\pi}} \exp\left\{-\frac{1}{2\sigma^2}[x^2 - 2(\mu + \sigma^2 t)x + \mu^2]\right\} dx.$$

To evaluate this integral, we complete the square in the exponent

$$x^2 - 2(\mu + \sigma^2 t)x + \mu^2 = [x - (\mu + \sigma^2 t)]^2 - 2\mu\sigma^2 t - \sigma^4 t^2.$$

Thus

$$M(t) = \exp\left(\frac{2\mu\sigma^2 t + \sigma^4 t^2}{2\sigma^2}\right) \int_{-\infty}^{\infty} \frac{1}{\sigma\sqrt{2\pi}} \exp\left\{-\frac{1}{2\sigma^2}[x - (\mu + \sigma^2 t)]^2\right\} dx.$$

Note that the integrand in the last integral is like the p.d.f. of a normal distribution with μ replaced by $\mu + \sigma^2 t$. However, the normal p.d.f. integrates to one for all real μ, in particular when it equals $\mu + \sigma^2 t$. Thus

$$M(t) = \exp\left(\frac{2\mu\sigma^2 t + \sigma^4 t^2}{2\sigma^2}\right) = \exp\left(\mu t + \frac{\sigma^2 t^2}{2}\right).$$

Now

$$M'(t) = (\mu + \sigma^2 t) \exp\left(\mu t + \frac{\sigma^2 t^2}{2}\right)$$

and

$$M''(t) = [(\mu + \sigma^2 t)^2 + \sigma^2] \exp\left(\mu t + \frac{\sigma^2 t^2}{2}\right).$$

Thus

$$E(X) = M'(0) = \mu,$$

$$\text{Var}(X) = M''(0) - [M'(0)]^2 = \mu^2 + \sigma^2 - \mu^2 = \sigma^2.$$

That is, the parameters μ and σ^2 in the p.d.f. of X are the mean and the variance of X.

Example 4.4.1

If the p.d.f. of X is

$$f(x) = \frac{1}{\sqrt{32\pi}} \exp\left[-\frac{(x+7)^2}{32}\right], \qquad -\infty < x < \infty,$$

then X is $N(-7, 16)$. That is, X has a normal distribution with a mean $\mu = -7$, variance $\sigma^2 = 16$, and the moment-generating function

$$M(t) = \exp(-7t + 8t^2). \qquad \diamond$$

Example 4.4.2

If the moment-generating function of X is

$$M(t) = \exp(5t + 12t^2),$$

then X is $N(5, 24)$, and its p.d.f. is

$$f(x) = \frac{1}{\sqrt{48\pi}} \exp\left[-\frac{(x-5)^2}{48}\right], \qquad -\infty < x < \infty. \qquad \diamond$$

If Z is $N(0, 1)$, we shall say that Z has a standard normal distribution. Moreover, the distribution function of Z is

$$\Phi(z) = P(Z \leq z) = \int_{-\infty}^{z} \frac{1}{\sqrt{2\pi}} e^{-w^2/2}\, dw.$$

It is not possible to evaluate this integral by finding an antiderivative that can be expressed as an elementary function. However, numerical approximations for integrals of this type have been tabulated and are given in Tables Va and Vb. The bell-shaped curved in Figure 4.11 represents the graph of the p.d.f. of Z, and the shaded area equals $\Phi(z_0)$.

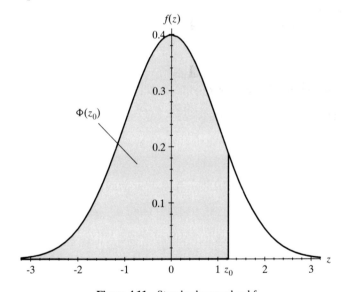

Figure 4.11   Standard normal p.d.f.

Values of $\Phi(z)$ for $z \geq 0$ are given in Table Va. Because of the symmetry of the standard normal p.d.f., it is true that $\Phi(-z) = 1 - \Phi(z)$ for all real z. Thus Table Va is sufficient. However, it is sometimes convenient to be able to read $\Phi(-z)$, for $z > 0$, directly from a table. This can be done using values in Table Vb, which lists right-tail probabilities. Again, because of the symmetry of the standard normal p.d.f., when $z > 0$, $\Phi(-z) = P(Z \leq -z) = P(Z > z)$ can be read directly from Table Vb.

Example 4.4.3

If Z is $N(0, 1)$, then, using Table Va, we obtain

$$P(Z \leq 1.24) = \Phi(1.24) = 0.8925,$$

$$P(1.24 \leq Z \leq 2.37) = \Phi(2.37) - \Phi(1.24) = 0.9911 - 0.8925 = 0.0986,$$

$$P(-2.37 \leq Z \leq -1.24) = P(1.24 \leq Z \leq 2.37) = 0.0986.$$

Using Table Vb, we find that

$$P(Z > 1.24) = 0.1075,$$

$$P(Z \leq -2.14) = P(Z \geq 2.14) = 0.0162,$$

and using both tables, we obtain

$$P(-2.14 \leq Z \leq 0.77) = P(Z \leq 0.77) - P(Z \leq -2.14)$$
$$= 0.7794 - 0.0162 = 0.7632. \qquad \diamond$$

There are times when we want to read the normal probability table in the opposite way, essentially finding the inverse of the standard normal distribution function. That is, given a probability p, find a constant a so that $P(Z \leq a) = p$. This is illustrated in the next example.

Example 4.4.4

If the distribution of Z is $N(0, 1)$, to find constants a and b such that

$$P(Z \leq a) = 0.9147 \qquad \text{and} \qquad P(Z \geq b) = 0.0526,$$

find the given probabilities in Tables Va and Vb, respectively, and read off the corresponding values of z. From Table Va we see that $a = 1.37$, and from Table Vb we see that $b = 1.62$. $\qquad \diamond$

In statistical applications we are often interested in finding a number z_α such that

$$P(Z \geq z_\alpha) = \alpha,$$

where Z is $N(0, 1)$ and α is usually less than 0.5. That is, z_α is the $100(1 - \alpha)$ percentile (sometimes called the upper 100α percent point) for the standard normal distribution (see Figure 4.12). The value of z_α is given in Table Va for selected values of α. For other values of α, z_α can be found in Table Vb.

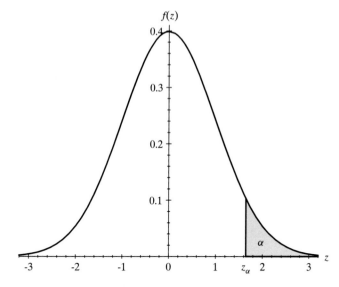

Figure 4.12   Upper 100α percent point, z_α

Because of the symmetry of the normal p.d.f.,

$$P(Z \leq -z_\alpha) = P(Z \geq z_\alpha) = \alpha.$$

Also, since the subscript of z_α is the right-tail probability,

$$z_{1-\alpha} = -z_\alpha.$$

For example,

$$z_{0.95} = z_{1-0.05} = -z_{0.05}.$$

Example 4.4.5

To find $z_{0.0125}$, note that

$$P(Z \geq z_{0.0125}) = 0.0125.$$

Thus

$$z_{0.0125} = 2.24$$

from Table Vb. Also,

$$z_{0.05} = 1.645 \qquad \text{and} \qquad z_{0.025} = 1.960$$

using the last rows of Table Va.   ◇

Remark   Recall that the $(100p)$th percentile, π_p, for a random variable X is a number such that $P(X \leq \pi_p) = p$. If Z is $N(0, 1)$, since

$$P(Z \geq z_\alpha) = \alpha,$$

it follows that

$$P(Z < z_\alpha) = 1 - \alpha.$$

Thus z_α is the $100(1-\alpha)$th percentile for the standard normal distribution, $N(0, 1)$. For example, $z_{0.05} = 1.645$ is the $100(1-0.05) = 95$th percentile and $z_{0.95} = -1.645$ is the $100(1 - 0.95) = 5$th percentile.

If X is $N(\mu, \sigma^2)$, the next theorem shows that the random variable $(X - \mu)/\sigma$ is $N(0, 1)$. Thus Tables Va and Vb can be used to find probabilities concerning X.

Theorem 4.4.1.   If X is $N(\mu, \sigma^2)$, then $Z = (X - \mu)/\sigma$ is $N(0, 1)$.

Proof.   The distribution function of Z is

$$P(Z \leq z) = P\left(\frac{X - \mu}{\sigma} \leq z\right) = P(X \leq z\sigma + \mu)$$

$$= \int_{-\infty}^{z\sigma+\mu} \frac{1}{\sigma\sqrt{2\pi}} \exp\left[-\frac{(x - \mu)^2}{2\sigma^2}\right] dx.$$

In the integral representing $P(Z \leq z)$, use the change of variable of integration given by $w = (x - \mu)/\sigma$ (i.e., $x = w\sigma + \mu$) to obtain

$$P(Z \leq z) = \int_{-\infty}^{z} \frac{1}{\sqrt{2\pi}} e^{-w^2/2} dw.$$

But this is the expression for $\Phi(z)$, the distribution function of a standardized normal random variable. Hence Z is $N(0, 1)$.   □

This theorem can be used to find probabilities about X, which is $N(\mu, \sigma^2)$, as follows:

$$P(a \leq X \leq b) = P\left(\frac{a - \mu}{\sigma} \leq \frac{X - \mu}{\sigma} \leq \frac{b - \mu}{\sigma}\right)$$

$$= \Phi\left(\frac{b - \mu}{\sigma}\right) - \Phi\left(\frac{a - \mu}{\sigma}\right),$$

since $(X - \mu)/\sigma$ is $N(0, 1)$.

Example 4.4.6

If X is $N(3, 16)$, then

$$P(4 \leq X \leq 8) = P\left(\frac{4 - 3}{4} \leq \frac{X - 3}{4} \leq \frac{8 - 3}{4}\right)$$

$$= \Phi(1.25) - \Phi(0.25) = 0.8944 - 0.5987 = 0.2957,$$

$$P(0 \leq X \leq 5) = P\left(\frac{0 - 3}{4} \leq Z \leq \frac{5 - 3}{4}\right)$$

$$= \Phi(0.5) - \Phi(-0.75) = 0.6915 - 0.2266 = 0.4649,$$

and

$$P(-2 \leq X \leq 1) = P\left(\frac{-2-3}{4} \leq Z \leq \frac{1-3}{4}\right)$$
$$= \Phi(-0.5) - \Phi(-1.25) = 0.3085 - 0.1056 = 0.2029. \diamond$$

Example 4.4.7

If X is $N(25, 36)$, we find a constant c such that

$$P(|X - 25| \leq c) = 0.9544.$$

We want

$$P\left(\frac{-c}{6} \leq \frac{X - 25}{6} \leq \frac{c}{6}\right) = 0.9544.$$

Thus

$$\Phi\left(\frac{c}{6}\right) - \left[1 - \Phi\left(\frac{c}{6}\right)\right] = 0.9544$$

and

$$\Phi\left(\frac{c}{6}\right) = 0.9772.$$

Hence $c/6 = 2$ and $c = 12$. That is, the probability that X falls within two standard deviations of its mean is the same as the probability that the standard normal variable Z falls within two units (standard deviations) of zero. $\diamond$

In the next theorem we give a relationship between the chi-square and normal distributions.

Theorem 4.4.2. If the random variable X is $N(\mu, \sigma^2)$, $\sigma^2 > 0$, then the random variable $V = (X - \mu)^2/\sigma^2 = Z^2$ is $\chi^2(1)$.

Proof. Because $V = Z^2$, where $Z = (X - \mu)/\sigma$ is $N(0, 1)$, the distribution function $G(v)$ of V is, for $v \geq 0$,

$$G(v) = P(Z^2 \leq v) = P(-\sqrt{v} \leq Z \leq \sqrt{v}).$$

That is, with $v \geq 0$,

$$G(v) = \int_{-\sqrt{v}}^{\sqrt{v}} \frac{1}{\sqrt{2\pi}} e^{-z^2/2} \, dz = 2 \int_{0}^{\sqrt{v}} \frac{1}{\sqrt{2\pi}} e^{-z^2/2} \, dz.$$

If we change the variable of integration by writing $z = \sqrt{y}$, then, since $D_y(z) = 1/(2\sqrt{y})$, we have

$$G(v) = \int_0^v \frac{1}{\sqrt{2\pi y}} e^{-y/2} \, dy, \qquad 0 \le v.$$

Of course, $G(v) = 0$, when $v < 0$. Hence the p.d.f. $g(v) = G'(v)$ of the continuous-type random variable V is, by one form of the fundamental theorem of calculus,

$$g(v) = \frac{1}{\sqrt{\pi}\sqrt{2}} v^{1/2-1} e^{-v/2}, \qquad 0 < v < \infty.$$

Since $g(v)$ is a p.d.f., it must be true that

$$\int_0^\infty \frac{1}{\sqrt{\pi}\sqrt{2}} v^{1/2-1} e^{-v/2} \, dv = 1.$$

The change of variables $x = v/2$ yields

$$1 = \frac{1}{\sqrt{\pi}} \int_0^\infty x^{1/2-1} e^{-x} \, dx = \frac{1}{\sqrt{\pi}} \Gamma\left(\frac{1}{2}\right).$$

Hence $\Gamma(1/2) = \sqrt{\pi}$, and thus V is $\chi^2(1)$. □

Example 4.4.8

If Z is $N(0, 1)$, then

$$P(|Z| < 1.96 = \sqrt{3.842}) = 0.95$$

and of course,

$$P(Z^2 < 3.842) = 0.95$$

from the chi-square table with $r = 1$. ◇

Given a set of observations of a random variable X, it is a challenge to determine the distribution of X. In particular, how can we decide whether or not X has an approximate normal distribution? If we have a large number of observations of X, a histogram of the observations can often be helpful (see Exercises 4.4.11 and 4.4.12). For small samples a q-q plot can be used to check on whether the sample arises from a normal distribution. For example, suppose that the quantiles of a sample were plotted against the corresponding quantiles of a certain normal distribution and these pairs of points were on a straight line with slope 1 and intercept 0. Of course, we would then believe that we have an ideal sample from that normal distribution with that certain mean and standard deviation. This plot, however, requires that we know the mean and the standard deviation of this normal distribution, and we usually do not. However, since the quantile, q_p, of $N(\mu, \sigma^2)$ is related to the corresponding one,

z_{1-p}, of $N(0, 1)$ by $q_p = \mu + \sigma z_{1-p}$, we can always plot the quantiles of the sample against the corresponding ones of $N(0, 1)$ and get the needed information. That is, if the sample quantiles are plotted as the x-coordinate of the pair and the $N(0, 1)$ quantiles as the y-coordinate and if the graph is almost a straight line, then it is reasonable to assume that the sample arises from a normal distribution. Moreover, the reciprocal of the slope of that straight line is a good estimate of the standard deviation σ because $z_{1-p} = (q_p - \mu)/\sigma$.

Example 4.4.9

Have you ever wondered about the weights of carrots in a prepackaged "1-pound" bag? Let X equal the weight of such a bag of carrots. Would you expect X to have a normal distribution? One of the authors frequently has his carrots weighed at the checkout. Say that $n = 12$ observations of X were

$$
\begin{array}{cccccc}
1.12 & 1.13 & 1.19 & 1.25 & 1.06 & 1.31 \\
1.12 & 1.23 & 1.29 & 1.17 & 1.20 & 1.11
\end{array}
$$

With 12 observations, the standard normal quantiles that we need correspond to $p = 1/13, 2/13, \ldots, 12/13$.

k	Weights (x)	$p = k/13$	z_{1-p}
1	1.06	0.0769	−1.43
2	1.11	0.1538	−1.02
3	1.12	0.2308	−0.74
4	1.12	0.3077	−0.50
5	1.13	0.3846	−0.29
6	1.17	0.4615	−0.10
7	1.19	0.5385	0.10
8	1.20	0.6154	0.29
9	1.23	0.6923	0.50
10	1.25	0.7692	0.74
11	1.29	0.8462	1.02
12	1.31	0.9231	1.43

A q-q plot of these data is shown in Figure 4.13. Note that the points do fall close to a straight line, so the normal probability model seems to be appropriate based on these few data.

This is an experiment that you could easily replicate to obtain more data and perhaps confirm these results or show that they are not true.   ◇

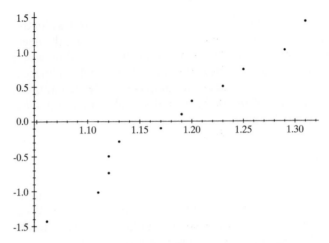

Figure 4.13 q-q plot, carrot weights vs. $N(0, 1)$ quantiles

EXERCISES

4.4.1 If Z is $N(0, 1)$, find

    (a) $P(0.53 < Z \leq 2.06)$.

    (b) $P(-0.79 \leq Z < 1.52)$.

    (c) $P(Z > -1.77)$.

    (d) $P(Z > 2.89)$.

    (e) $P(|Z| < 1.96)$.

    (f) $P(|Z| < 1)$.

    (g) $P(|Z| < 2)$.

    (h) $P(|Z| < 3)$.

4.4.2 If Z is $N(0, 1)$, find

    (a) $P(0 \leq Z \leq 0.87)$.

    (b) $P(-2.64 \leq Z \leq 0)$.

    (c) $P(-2.13 \leq Z \leq -0.56)$.

    (d) $P(|Z| > 1.39)$.

    (e) $P(Z < -1.62)$.

    (f) $P(|Z| > 1)$.

    (g) $P(|Z| > 2)$.

    (h) $P(|Z| > 3)$.

4.4.3 Find the values of (a) $z_{0.01}$, (b) $-z_{0.005}$, (c) $z_{0.0475}$, and (d) $z_{0.985}$.

4.4.4 Find the values of (a) $z_{0.10}$, (b) $-z_{0.05}$, (c) $-z_{0.0485}$, and (d) $z_{0.9656}$.

4.4.5 If Z is $N(0, 1)$, find values of c such that

(a) $P(Z \geq c) = 0.025$.

(b) $P(|Z| \leq c) = 0.95$.

(c) $P(Z > c) = 0.05$.

(d) $P(|Z| \leq c) = 0.90$.

4.4.6 If the moment-generating function of X is $M(t) = \exp(166t + 200t^2)$, find

(a) The mean of X.

(b) The variance of X.

(c) $P(170 < X < 200)$.

(d) $P(148 \leq X \leq 172)$.

4.4.7 If X is normally distributed with a mean of 6 and a variance of 25, find

(a) $P(6 \leq X \leq 12)$.

(b) $P(0 \leq X \leq 8)$.

(c) $P(-2 < X \leq 0)$.

(d) $P(X > 21)$.

(e) $P(|X - 6| < 5)$.

(f) $P(|X - 6| < 10)$.

(g) $P(|X - 6| < 15)$.

(h) $P(|X - 6| < 12.88)$.

4.4.8 If the moment-generating function of X is $M(t) = \exp(-6t + 32t^2)$, find

(a) $P(-4 \leq X < 16)$.

(b) $P(-10 < X \leq 0)$.

4.4.9 If X is $N(650, 625)$, find

(a) $P(600 \leq X < 660)$.

(b) A constant $c > 0$ such that $P(|X - 650| \leq c) = 0.9544$.

4.4.10 If X is $N(\mu, \sigma^2)$, show that $Y = aX + b$ is $N(a\mu + b, a^2\sigma^2)$, $a \neq 0$. HINT: Find the distribution function $P(Y \leq y)$ of Y and, in the resulting integral, let $w = ax + b$ or, equivalently, $x = (w - b)/a$.

4.4.11 A packaged product has a label weight of 450 grams. The company goal is to fill each package with at least 450 grams but at most 458 grams. To check these goals, a random sample of 100 packages was selected and weighed, yielding the following weights rounded to the nearest gram:

457	457	455	457	454	454	457	455	456	459
457	458	456	456	461	457	458	452	457	460
453	458	452	454	454	456	455	456	451	454
456	457	457	453	455	459	458	457	458	457
461	457	455	458	458	455	457	458	456	463
455	455	455	456	456	456	455	456	460	456
456	457	458	454	455	456	459	457	457	451
450	453	453	459	450	453	452	458	456	457
451	458	456	460	455	455	456	460	457	456
457	456	460	459	457	455	461	455	457	457

(a) Construct a frequency table and a corresponding relative frequency histogram.

(b) Do these data seem to come from a normal distribution with mean μ about equal to $\bar{x} = 456.2$ and variance about equal to $s^2 = 5.96$? Sketch the p.d.f. for the normal distribution $N(456.2, 5.96)$ on your histogram.

4.4.12 A company manufactures windows that are then inserted into an automobile. Each window has 5 studs for attaching it. A pull-out test is used to determine the force required to pull a stud out of a window. (Note that this is an example of destructive testing.) Let X equal the force required for pulling studs out of position 4. Sixty observations of X were as follows:

159	150	147	160	155	142	143	151	154	133
151	146	140	146	137	148	154	157	142	153
135	144	135	165	118	158	126	147	123	140
125	151	153	158	144	163	150	150	137	164
137	156	139	134	171	144	160	147	155	175
162	160	149	149	158	152	165	131	150	120

(a) Construct an ordered stem-and-leaf diagram using as stems 11•, 12*, 12•, and so on.

(b) Construct a q-q plot using about every 5th observation in the ordered array and the corresponding quantiles of $N(0, 1)$.

(c) Does it look like X has a normal distribution?

4.4.13 When you fly, the usual snack is a bag of peanuts. Let X equal the weight of peanuts in a bag that has a 14-gram label weight. Sixteen observations of X, that have been ordered, are

13.9	14.4	14.6	14.7	14.7	15.2	15.2	15.2
15.3	15.4	15.4	15.5	15.6	15.6	15.9	16.4

Determine whether X could have a normal distribution by constructing a q-q plot. What is your conclusion?

4.4.14 A candymaker produces mints that have a label weight of 20.4 grams. Assume that the distribution of the weights of these mints is $N(21.37, 0.16)$.

(a) Let X denote the weight of a single mint selected at random from the production line. Find $P(X > 22.07)$.

(b) Suppose that 15 mints are selected independently and weighed. Let Y equal the number of these mints that weigh less than 20.857 grams. Then find $P(Y \le 2)$.

4.4.15 Let X equal the birth weight (in grams) of babies in the United States. Assuming that the distribution of X is $N(3315, 575^2)$, find

(a) $P(2584.75 \le X \le 4390.25)$.

(b) $P(2619.25 \le X \le 3642.75)$.

(c) $P(3119.50 \le X \le 3579.50)$.

(d) Let Y equal the number of babies that weigh less than 2719 grams at birth among 25 of these babies selected independently. Find $P(Y \le 4)$.

4.4.16 Let X equal the birth weight (in grams) of babies in Singapore. Assuming that the distribution of X is $N(3135, 459^2)$, find

(a) $P(2546.56 \le X \le 3723.44)$.

(b) $P(2379.94 \le X \le 3890.06)$.

(c) $P(2067.37 \le X \le 4202.63)$.

(d) Let Y equal the number of babies that weigh less than 2546.56 grams at birth among 20 of these babies selected independently. Find $P(Y \le 3)$.

4.4.17 A chemistry major weighed 19 plain m&m's (in grams) on a ± 0.0001 scale. The ordered weights are

0.7938	0.8032	0.8089	0.8222	0.8268	0.8383	0.8442
0.8490	0.8528	0.8572	0.8674	0.8734	0.8786	
0.8850	0.8873	0.8920	0.9069	0.9150	0.9243	

(a) Construct a stem-and-leaf diagram using three-digit leaves, with stems 0.7•, 0.8∗, 0.8t, and so on.

(b) Construct a q-q plot of these data and the corresponding quantiles for the standard normal distribution, $N(0, 1)$.

(c) Do these data look like observations from a normal distribution? Why?

4.4.18 Let the distribution of X be $N(\mu, \sigma^2)$. Show that the points of inflection of the graph of the p.d.f. of X occur at $x = \mu \pm \sigma$.

4.4.19 If X is $N(7, 4)$, find $P[15.364 \le (X - 7)^2 \le 20.096]$.

4.4.20 If the moment-generating function of X is $M(t) = e^{500t + 5000t^2}$, then find $P[27,060 \le (X - 500)^2 \le 50,240]$.

4.4.21 The graphs of the moment-generating functions of three normal distributions, $N(0, 1)$, $N(-1, 1)$, and $N(2, 1)$ are given in the figure. Identify them.

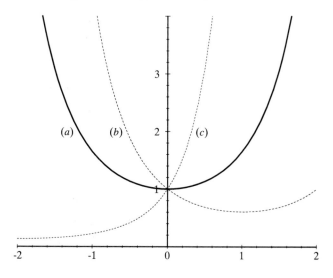

4.5 DISTRIBUTIONS OF FUNCTIONS OF A RANDOM VARIABLE

Let X be a random variable of the continuous type. If we consider a function of X, say $Y = u(X)$, then Y must also be a random variable that has its own distribution. If we can find its distribution function, say

$$G(y) = P(Y \leq y) = P[u(X) \leq y],$$

then its p.d.f. is given by $g(y) = G'(y)$. This **distribution function technique** has been illustrated by two results in Section 4.4:

1. If X is $N(\mu, \sigma^2)$, then $Z = (X - \mu)/\sigma$ is $N(0, 1)$.
2. If X is $N(\mu, \sigma^2)$, then $V = (X - \mu)^2/\sigma^2$ is $\chi^2(1)$.

Another example of the distribution function technique provides another important distribution called the **lognormal**.

Example 4.5.1

Say that X is $N(\mu, \sigma^2)$. If we let $W = e^X$, then the distribution function of W is

$$G(w) = P(W \leq w) = P(e^X \leq w) = P(X \leq \ln w), \qquad 0 < w.$$

That is,

$$G(w) = \int_{-\infty}^{\ln w} \frac{1}{\sigma\sqrt{2\pi}} \exp\left[-\frac{(x - \mu)^2}{2\sigma^2} \right] dx, \qquad 0 < w,$$

and thus the p.d.f. of W is

$$g(w) = G'(w) = \frac{1}{\sigma w \sqrt{2\pi}} \exp\left[-\frac{(\ln w - \mu)^2}{2\sigma^2} \right], \qquad 0 < w.$$

It is most important to observe that μ and σ^2 are not the mean and the variance of W, but those of $X = \ln W$, which has a normal distribution. Incidentally, it is easy to see why this distribution has been called the lognormal. See Figure 4.14 for some examples of lognormal p.d.f.'s.   ◇

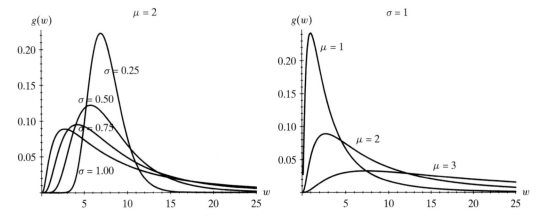

Figure 4.14   Lognormal distributions

By making a simple observation, we can sometimes "shortcut" the distribution function technique by using what is frequently called the **change-of-variable technique.** Let X be a continuous-type random variable with p.d.f. $f(x)$ with support $c_1 < x < c_2$. We begin this discussion by taking $Y = u(X)$ as a continuous increasing function of X with the inverse function $X = v(Y)$. Say the support of X, namely $c_1 < x < c_2$, maps onto $d_1 = u(c_1) < y < d_2 = u(c_2)$, the support of Y. Thus the distribution function of Y is

$$G(y) = P(Y \le y) = P[u(X) \le y] = P[X \le v(y)], \qquad d_1 < y < d_2,$$

since u and v are continuous increasing functions. Of course, $G(y) = 0$, $y \le d_1$, and $G(y) = 1$, $y \ge d_2$. Thus,

$$G(y) = \int_{c_1}^{v(y)} f(x)\, dx, \qquad d_1 < y < d_2.$$

Recall from calculus that the derivative, $G'(y) = g(y)$, of such an expression is given by

$$G'(y) = g(y) = f[v(y)][v'(y)], \qquad d_1 < y < d_2.$$

Of course, $G'(y) = g(y) = 0$ if $y < d_1$ or $y > d_2$. We may let $g(d_1) = g(d_2) = 0$.

For illustration of this change-of-variable technique, let us reconsider Example 4.5.1 with $W = e^X$, where X has p.d.f.

$$f(x) = \frac{1}{\sigma\sqrt{2\pi}} \exp\left[-\frac{(x-\mu)^2}{2\sigma^2}\right], \qquad -\infty < x < \infty.$$

Here $c_1 = -\infty$ and $c_2 = \infty$ and thus $d_1 = 0$ and $d_2 = \infty$; also $X = \ln W = v(W)$. Since $v'(w) = 1/w$, the p.d.f. of W is

$$g(w) = \frac{1}{\sigma\sqrt{2\pi}} \exp\left[-\frac{(\ln w - \mu)^2}{2\sigma^2}\right]\left(\frac{1}{w}\right), \qquad 0 < w < \infty,$$

which is the same result as obtained in Example 4.5.1.

Suppose now the function $Y = u(X)$ and its inverse $X = v(Y)$ were continuous decreasing functions. Thus the mapping of $c_1 < x < c_2$ would be $d_1 = u(c_1) > y > d_2 = u(c_2)$.

Since u and v are decreasing functions, we have

$$G(y) = P(Y \leq y) = P[u(X) \leq y] = P[X \geq v(y)] = \int_{v(y)}^{c_2} f(x)dx, \quad d_2 < y < d_1.$$

Accordingly, from calculus, we have

$$G'(y) = g(y) = f[v(y)][-v'(y)], \qquad d_2 < y < d_1,$$

and $G'(y) = g(y) = 0$, elsewhere. Note in both the increasing and decreasing cases, we could write

$$g(y) = f[v(y)]\,|v'(y)|, \qquad y \in R_y,$$

where R_y is the support of Y found by mapping the support of X, say R_x, onto it. The absolute value $|v'(y)|$ assures that $g(y)$ is nonnegative.

Example 4.5.2

Let X have the p.d.f.

$$f(x) = 3(1 - x)^2, \qquad 0 < x < 1.$$

Say $Y = (1 - X)^3 = u(X)$, a decreasing function of X. Thus $X = 1 - Y^{1/3} = v(Y)$ and $0 < x < 1$ is mapped onto $0 < y < 1$. Since

$$v'(y) = -\frac{1}{3y^{2/3}},$$

we have

$$g(y) = 3[1 - (1 - y^{1/3})]^2 \left| \frac{-1}{3y^{2/3}} \right|, \qquad 0 < y < 1.$$

That is,

$$g(y) = 3y^{2/3} \left(\frac{1}{3y^{2/3}} \right) = 1, \qquad 0 < y < 1;$$

so $Y = (1 - X)^3$ has the uniform distribution $U(0, 1)$.   ◇

As we have seen, it is sometimes easier to use the change-of-variable technique than the distribution function technique. However, there are many occasions that the latter is more convenient to use. As a matter of fact, we had to use the distribution function in finding the gamma distribution from the Poisson process (see Section 3.7). We again use the distribution function technique to prove two theorems involving the uniform distribution.

Theorem 4.5.1. Let Y have a distribution that is $U(0, 1)$. Let $F(x)$ have the properties of a distribution function of the continuous type with $F(a) = 0$, $F(b) = 1$, and suppose that $F(x)$ is strictly increasing on the support $a < x < b$, where a and b could be $-\infty$ and ∞, respectively. Then the random variable X defined by $X = F^{-1}(Y)$ is a continuous-type random variable with distribution function $F(x)$.

Proof.   The distribution function of X is

$$P(X \le x) = P[F^{-1}(Y) \le x], \qquad a < x < b.$$

However, since the function $F(x)$ is strictly increasing, $\{F^{-1}(Y) \le x\}$ is equivalent to $\{Y \le F(x)\}$ and hence

$$P(X \le x) = P[Y \le F(x)], \qquad a < x < b.$$

But Y is $U(0, 1)$; so $P(Y \le y) = y$ for $0 < y < 1$, and accordingly,

$$P(X \le x) = P[Y \le F(x)] = F(x), \qquad 0 < F(x) < 1.$$

That is, the distribution function of X is $F(x)$. $\square$

We now illustrate how Theorem 4.5.1 can be used to simulate observations from a given distribution.

Example 4.5.3

To see how we can simulate observations from an exponential distribution with a mean of $\theta = 10$, note that the distribution function of X is $F(x) = 1 - e^{-x/10}$ when $0 \le x < \infty$. Solving $y = F(x)$ for x yields $x = F^{-1}(y) = -10 \ln(1 - y)$. So given independent random numbers $y_1, y_2, \ldots, y_n$ from $U(0, 1)$, we would expect $x_i = -10 \ln(1 - y_i)$, $i = 1, 2, \ldots, n$, to represent n observations of an exponential random variable X with mean $\theta = 10$. Table 4.1 gives the values of 15 random numbers, y_i, along with the values of $x_i = -10 \ln(1 - y_i)$.

As a rough check on whether these 15 observations seem to come from an exponential distribution with mean 10, let us construct a q-q plot. We must order these 15 observations and plot them against the corresponding quantiles q_i of the exponential distribution with mean $\theta = 10$. These are the solution of

$$F(q_i) = 1 - e^{-q_i/10} = \frac{i}{16}$$

or, equivalently,

$$q_i = -10 \ln\left(1 - \frac{i}{16}\right).$$

The ordered x_i's and the respective quantiles, q_i, are given in Table 4.2.

TABLE 4.1 Random exponential observations

y	$x = -10\ln(1 - y)$
0.1514	1.6417
0.6697	11.0775
0.0527	0.5414
0.4749	6.4417
0.2900	3.4249
0.2354	2.6840
0.9662	33.8729
0.0043	0.0431
0.1003	1.0569
0.9192	25.1578
0.4971	6.8736
0.7293	13.0674
0.9118	24.2815
0.8225	17.2878
0.5915	8.9526

TABLE 4.2 Exponential quantiles

Ordered x	Quantile
0.0431	0.6454
0.5414	1.3353
1.0569	2.0764
1.6417	2.8768
2.6840	3.7469
3.4249	4.7000
6.4417	5.7536
6.8736	6.9315
8.9526	8.2668
11.0775	9.8083
13.0674	11.6315
17.2878	13.8629
24.2815	16.7398
25.1578	20.7944
33.8729	27.7259

Figure 4.15 shows the q-q plot for these data. The linearity of this q-q plot confirms the possibility of the exponential distribution being the correct model. The box plot (Figure 4.15) shows the skewness of the data as it should be for an exponential distribution. In addition, the sample mean is $\bar{x} = 10.4270$ and the sample standard

deviation is $s = 10.4311$, both of which are close to $\theta = 10$. We again caution the reader that in similar simulations, the fit is sometimes better and sometimes not as good. ◇

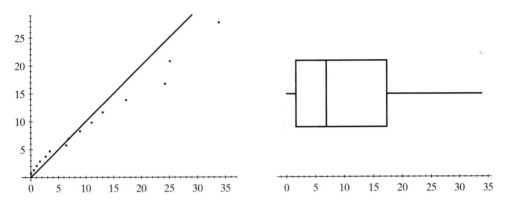

Figure 4.15   Exponential q-q plot and box plot

The following probability integral transformation theorem is the converse of Theorem 4.5.1.

Theorem 4.5.2.   Let X have the distribution function $F(x)$ of the continuous type that is strictly increasing on the support $a < x < b$. Then the random variable Y, defined by $Y = F(X)$, has a distribution that is $U(0, 1)$.

Proof.   Since $F(a) = 0$ and $F(b) = 1$, the distribution function of Y is

$$P(Y \leq y) = P[F(X) \leq y], \qquad 0 < y < 1.$$

However, $\{F(X) \leq y\}$ is equivalent to $\{X \leq F^{-1}(y)\}$; thus

$$P(Y \leq y) = P[X \leq F^{-1}(y)], \qquad 0 < y < 1.$$

Since $P(X \leq x) = F(x)$, we have that

$$P(Y \leq y) = P[X \leq F^{-1}(y)] = F[F^{-1}(y)] = y, \qquad 0 < y < 1,$$

which is the distribution function of a $U(0, 1)$ random variable.   □

Although in our statements and proofs of Theorems 4.5.1 and 4.5.2 we required $F(x)$ to be strictly increasing, this restriction can be dropped, and both theorems are still true. In our exposition we did not want to trouble students with certain difficulties that are experienced if $F(x)$ is not strictly increasing.

The change-of-variable technique can be used for a variable X of the discrete type, but there is one major difference since the p.d.f. $f(x) = P(X = x)$, $x \in R_x$, represents probability. Note that the support R_x consists of a countable number of points, say $c_1, c_2, c_3, \ldots$ Let $Y = u(X)$ be a one-to-one transformation with inverse

$X = v(Y)$. The function $y = u(x)$ maps R_x onto $d_1 = u(c_1), d_2 = u(c_2), d_3 = u(c_3), \ldots$, which we denote by R_y. Hence the p.d.f. of Y is

$$g(y) = P(Y = y) = P[u(X) = y] = P[X = v(y)], \qquad y \in R_y.$$

Since $P(X = x) = f(x)$, we have $g(y) = f[v(y)], y \in R_y$. Note that, in this discrete case, the value of the derivative, namely $|v'(y)|$, is not needed.

Example 4.5.4

Let X have a Poisson distribution with $\lambda = 4$ and thus the p.d.f. is

$$f(x) = \frac{4^x e^{-4}}{x!}, \qquad x = 0, 1, 2, \ldots.$$

If $Y = \sqrt{X}$, then, since $X = Y^2$, we have

$$g(y) = \frac{4^{y^2} e^{-4}}{(y^2)!}, \qquad y = 0, 1, \sqrt{2}, \sqrt{3}, \ldots. \qquad \diamond$$

One final observation concerns the situation in which the transformation $Y = u(X)$ is not one-to-one, as it has been up to this point in this section. For illustration in the last section, $V = Z^2$, where Z is $N(0, 1)$. Here $-\infty < z < \infty$ maps onto $0 \le v < \infty$, so

$$G(v) = P(Z^2 \le v) = P(-\sqrt{v} \le Z \le \sqrt{v})$$

$$= \int_{-\sqrt{v}}^{\sqrt{v}} f(z)dz, \qquad 0 \le v < \infty,$$

where

$$f(z) = \frac{1}{\sqrt{2\pi}} \exp\left(-\frac{z^2}{2}\right).$$

Thus

$$G'(v) = g(v) = f(\sqrt{v}) \left| \frac{1}{2\sqrt{v}} \right| + f(-\sqrt{v}) \left| \frac{-1}{2\sqrt{v}} \right|$$

$$= \frac{v^{1/2-1} e^{-v/2}}{\sqrt{\pi}\sqrt{2}}, \qquad 0 \le v < \infty,$$

which is the p.d.f. of the $\chi^2(1)$ distribution. That is, in this case of a two-to-one transformation, there is a need to sum two terms, each of which is similar to those of the one-to-one case; but here $x_1 = \sqrt{y}$ and $x_2 = -\sqrt{y}, 0 < y < \infty$, give the two inverse functions, respectively.

With careful thought, we can handle many situations which generalize this particular illustration.

Example 4.5.5

Let X have the p.d.f.

$$f(x) = \frac{x^2}{3}, \qquad -1 < x < 2.$$

The random variable $Y = X^2$ will have the support $0 \le y < 4$. However, for $0 < y < 1$, we obtain the two-to-one transformation represented by $x_1 = -\sqrt{y}$ for $-1 < x_1 < 0$, and $x_2 = \sqrt{y}$ for $0 < x_2 < 1$. On the other hand, if $1 < y < 4$, the one-to-one transformation is represented by $x_2 = \sqrt{y}, 1 < x_2 < 2$. Since

$$\frac{dx_1}{dy} = \frac{-1}{2\sqrt{y}} \qquad \text{and} \qquad \frac{dx_2}{dy} = \frac{1}{2\sqrt{y}},$$

we have that the p.d.f. of $Y = X^2$ is

$$g(y) = \begin{cases} \dfrac{(-\sqrt{y})^2}{3} \left| \dfrac{-1}{2\sqrt{y}} \right| + \dfrac{(\sqrt{y})^2}{3} \left| \dfrac{1}{2\sqrt{y}} \right| = \dfrac{\sqrt{y}}{3}, & 0 < y < 1, \\[4mm] \dfrac{(\sqrt{y})^2}{3} \left| \dfrac{1}{2\sqrt{y}} \right| = \dfrac{\sqrt{y}}{6}, & 1 < y < 4. \end{cases}$$

Note that when $0 < y < 1$, the p.d.f. is the sum of two terms but only one in case $1 < y < 4$; these different expressions in $g(y)$ correspond to the two types of transformations, namely the 2-to-1 and the 1-to-1 transformations, respectively. ◇

EXERCISES

4.5.1 Let X have the p.d.f. $f(x) = 4x^3, 0 < x < 1$. Find the p.d.f. of $Y = X^2$.

4.5.2 Let X have the p.d.f. $f(x) = xe^{-x^2/2}, 0 < x < \infty$. Find the p.d.f. of $Y = X^2$.

4.5.3 Let X have a gamma distribution with $\alpha = 3$ and $\theta = 2$. Determine the p.d.f. of $Y = \sqrt{X}$.

4.5.4 The p.d.f. of X is $f(x) = 2x, 0 < x < 1$.

(a) Find the distribution function of X.

(b) Describe how an observation of X can be simulated.

(c) Simulate 10 observations of X.

(d) Construct a q-q plot using the ordered observations and the corresponding quantiles for the given distribution.

4.5.5 Simulate a random sample of size $n = 10$ from an exponential distribution with mean $\theta = 3$. Construct a q-q plot of your data using the ordered observations and the corresponding quantiles for the exponential distribution with $\theta = 3$.

4.5.6 Let X have a **logistic distribution** with p.d.f.

$$f(x) = \frac{e^{-x}}{(1 + e^{-x})^2}, \qquad -\infty < x < \infty.$$

Show that

$$Y = \frac{1}{1 + e^{-X}}$$

has a $U(0, 1)$ distribution.

4.5.7 Let X be $N(0, 1)$. Find the p.d.f. of $Y = |X|$, a distribution that is often called the **half normal**. HINT: Here $y \in R_y = \{y : 0 < y < \infty\}$. Consider the two transformations $x_1 = -y, -\infty < x_1 < 0$, and $x_2 = y, 0 < y < \infty$.

4.5.8 If X is $N(\mu, \sigma^2)$, then $M(t) = E(e^{tX}) = \exp(\mu t + \sigma^2 t^2 / 2)$. Of course, $Y = e^X$ has a lognormal distribution. Using $M(t)$, find

(a) $E(Y) = E(e^X) = M(1)$.

(b) $E(Y^2) = E(e^{2X}) = M(2)$.

(c) $\text{Var}(Y)$.

4.5.9 Let X have the p.d.f. $f(x) = [\pi(1 + x^2)]^{-1}, -\infty < x < \infty$. Find the p.d.f. of $Y = X^2$.

4.5.10 Let X have the uniform distribution $U(-1, 3)$. Find the p.d.f. of $Y = X^2$.

4.5.11 The p.d.f. of X is $f(x) = \theta x^{\theta - 1}, 0 < x < 1, 0 < \theta < \infty$. Let $Y = -2\theta \ln X$. Find the moment-generating function of Y. How is Y distributed?

*4.6 MIXED DISTRIBUTIONS AND CENSORING

Thus far we have considered random variables that are either discrete or continuous. In most applications these are the types that are encountered. However, on some occasions, combinations of the two types of random variables are found. That is, in some experiments, positive probability is assigned to each of certain points and also is spread over an interval of outcomes, each point of which has zero probability. An illustration will help clarify these remarks.

Example 4.6.1

A bulb for a slide projector is tested by turning it on, letting it burn for 1 hour, and then turning it off. Let X equal the length of time that the bulb performs satisfactorily during this test. There is positive probability that the bulb will burn out when it is turned on; hence

$$0 < P(X = 0) < 1.$$

It could also burn out during the 1-hour time period; thus

$$P(0 < X < 1) > 0$$

with $P(X = x) = 0$ when $x \in (0, 1)$. In addition, $P(X = 1) > 0$. The act of turning it off after 1 hour so that the actual failure time beyond 1 hour is not observed is called censoring, which is considered later in this section.   ◇

The distribution function for a distribution of the mixed type will be a combination of those for the discrete and continuous types. That is, at each point of positive probability the distribution function will be discontinuous so that the height of the step there equals the corresponding probability; at all other points the distribution function will be continuous.

Example 4.6.2

Let X have a distribution function $F(x)$ defined by

$$F(x) = \begin{cases} 0, & x < 0, \\[2mm] \dfrac{x^2}{4}, & 0 \le x < 1, \\[2mm] \dfrac{1}{2}, & 1 \le x < 2, \\[2mm] \dfrac{x}{3}, & 2 \le x < 3, \\[2mm] 1, & 3 \le x. \end{cases}$$

This distribution function is depicted in Figure 4.16. Probabilities can be computed using $F(x)$; for illustration consider

$$P(0 < X < 1) = \frac{1}{4},$$

$$P(0 < X \le 1) = \frac{1}{2},$$

$$P(X = 1) = \frac{1}{4}. \qquad \diamond$$

Example 4.6.3

Consider the following game. An unbiased coin is tossed. If the outcome is heads, the player receives \$2. If the outcome is tails, the player spins a balanced spinner that has a scale from 0 to 1 and receives that fraction of a dollar associated with the point selected by the spinner. If X denotes the amount received, the space of X is $R = [0, 1) \cup \{2\}$. The distribution function of X is defined by

$$F(x) = \begin{cases} 0, & x < 0, \\[2mm] \dfrac{x}{2}, & 0 \le x < 1, \\[2mm] \dfrac{1}{2}, & 1 \le x < 2, \\[2mm] 1, & 2 \le x. \end{cases}$$

The graph of the distribution function $F(x)$ is given in Figure 4.17. $\qquad \diamond$

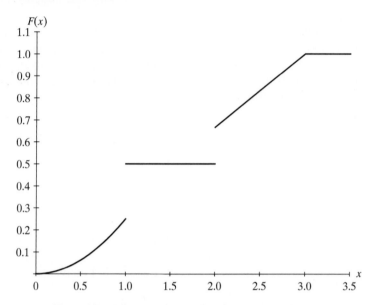

Figure 4.16   Mixed distribution function for Example 4.6.2

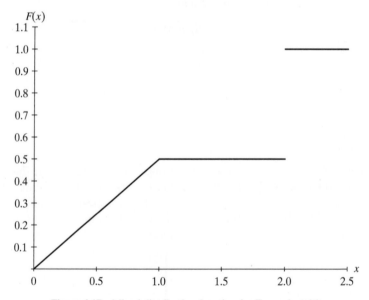

Figure 4.17   Mixed distribution function for Example 4.6.3

Suppose that the random variable X has a distribution of the mixed type. To find the expectation of the function $u(X)$ of X, a combination of a sum and a Riemann integral is used, as shown in Example 4.6.4.

Example 4.6.4

We shall find the mean and variance for the random variable given in Example 4.6.2. Note that there $F'(x) = x/2$ when $0 < x < 1$, and $F'(x) = 1/3$ when $2 < x < 3$; also $P(X = 1) = 1/4$ and $P(X = 2) = 1/6$. Accordingly, we have

$$\mu = E(X) = \int_0^1 x\left(\frac{x}{2}\right) dx + 1\left(\frac{1}{4}\right) + 2\left(\frac{1}{6}\right) + \int_2^3 x\left(\frac{1}{3}\right) dx$$

$$= \left[\frac{x^3}{6}\right]_0^1 + \frac{1}{4} + \frac{1}{3} + \left[\frac{x^2}{6}\right]_2^3 = \frac{19}{12};$$

$$\sigma^2 = E(X^2) - [E(X)]^2$$

$$= \int_0^1 x^2\left(\frac{x}{2}\right) dx + 1^2\left(\frac{1}{4}\right) + 2^2\left(\frac{1}{6}\right) + \int_2^3 x^2\left(\frac{1}{3}\right) dx - \left(\frac{19}{12}\right)^2$$

$$= \frac{31}{48}. \qquad \diamond$$

Frequently, in life testing, we know that the length of life, say X, exceeds the number b, but the exact value is unknown. This is called **censoring**. For instance, this can happen when a subject in a cancer study simply disappears; the investigator knows that the subject has lived a certain number of months, but the exact length of life is unknown. Or it might happen when an investigator does not have enough time in an investigation to observe the moments of deaths of all the animals, say rats, in some study. Censoring can also occur in the insurance industry; in particular, consider a loss with a limited-pay policy in which the top amount is exceeded but it is not known by how much.

Example 4.6.5

Reinsurance companies are concerned with large losses because they might agree, for illustration, to cover losses due to wind damages that are between $2 million and $10 million. Say that X equals the size of a wind loss in millions of dollars, and suppose that it has the distribution function

$$F(x) = 1 - \left(\frac{10}{10 + x}\right)^3, \qquad 0 \le x < \infty.$$

If losses beyond $10 million are reported only as 10, then the distribution function of this censored distribution is

$$F(x) = \begin{cases} 1 - \left(\dfrac{10}{10 + x}\right)^3, & 0 \le x < 10, \\ 1, & 10 \le x < \infty, \end{cases}$$

which has a jump of $[10/(10 + 10)]^3 = 1/8$ at $x = 10$. $\qquad \diamond$

EXERCISES

4.6.1 From the graph of the distribution function of X determine the following probabilities:

  (a) $P(X < 0)$.

  (b) $P(X < -1)$.

  (c) $P(X \leq -1)$.

  (d) $P(X < 1)$.

  (e) $P(-1 \leq X < 1/2)$.

  (f) $P(-1 < X \leq 1)$.

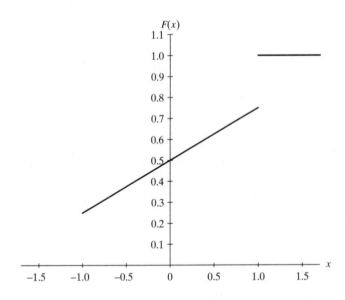

4.6.2 From the graph of the distribution function determine the following probabilities:

  (a) $P(-1/2 \leq X \leq 1/2)$.

  (b) $P(1/2 < X < 1)$.

  (c) $P(3/4 < X < 2)$.

  (d) $P(X > 1)$.

  (e) $P(2 < X < 3)$.

  (f) $P(2 < X \leq 3)$.

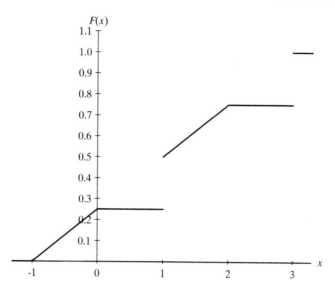

4.6.3 Let X be a random variable of the mixed type having the distribution function

$$F(x) = \begin{cases} 0, & x < 0, \\ \dfrac{x^2}{4}, & 0 \le x < 1, \\ \dfrac{x+1}{4}, & 1 \le x < 2, \\ 1, & 2 \le x. \end{cases}$$

(a) Carefully sketch the graph of $F(x)$.

(b) Find the mean and the variance of X.

(c) Find $P(1/4 < X < 1)$, $P(X = 1)$, $P(X = 1/2)$, and $P(1/2 \le X < 2)$.

4.6.4 Find the mean and variance of X if the distribution function of X is

$$F(x) = \begin{cases} 0, & x < 0, \\ 1 - \left(\dfrac{2}{3}\right)e^{-x}, & 0 \le x, \end{cases}$$

4.6.5 Consider the following game. A fair die is rolled. If the outcome is even, the player receives a number of dollars equal to the outcome on the die. If the outcome is odd, a number is selected at random from the interval $[0, 1)$ with a balanced spinner, and the player receives that fraction of a dollar associated with the point selected.

(a) Define and sketch the distribution function of X, the amount received.

(b) Find the expected value of X.

(c) Simulate 10 plays of this game.

(d) Is the average of your 10 plays close to the expected value calculated in part (b)?

4.6.6 Compute the means of both the uncensored and the censored distributions in Example 4.6.5.
HINT: Use integration by parts.

4.6.7 The distribution function of X is defined by

$$
F(x) = \begin{cases}
0, & -\infty < x < 0, \\[2mm]
\dfrac{1}{4}(x^2 + 1), & 0 \le x < 1, \\[2mm]
\dfrac{1}{4}(x + 2), & 1 \le x < 2, \\[2mm]
1, & 2 \le x < \infty.
\end{cases}
$$

(a) Sketch a graph of $y = F(x)$.
(b) Find $P(0 < X < 1)$.
(c) Find $P(1 \le X < 1.5)$.
(d) Find $\mu = E(X)$.
(e) Find $\sigma^2 = \text{Var}(X)$.

5

SAMPLING DISTRIBUTION THEORY

5.1 INDEPENDENT RANDOM VARIABLES

In Chapters 3 and 4 we introduced several distributions concerning one random variable. Each of these random variables could be thought of as a measurement on some random experiment. In this chapter, we consider the possibility of performing several random experiments or one random experiment several times in which each trial results in one measurement which can be considered a random variable. That is, we obtain one random variable from each experiment, and thus we obtain a collection of several random variables from the several experiments. Further suppose that these experiments are performed in such a way that the events associated with any one of them are independent of the events associated with others, and hence the corresponding random variables are, in some sense, independent.

To make this clear, say we have only two experiments, and the resulting random variables are denoted by X_1 and X_2, respectively. Suppose that the two spaces of X_1 and X_2 are R_1 and R_2, respectively. If $A_1 \subset R_1$ and $A_2 \subset R_2$, then the independence of events would imply that

$$P(X_1 \in A_1 \text{ and } X_2 \in A_2) = P(X_1 \in A_1)P(X_2 \in A_2).$$

For the moment say that X_1 and X_2 are random variables of the discrete type with probability density functions $f_1(x_1)$ and $f_2(x_2)$, respectively. If A_1 consists of only the point x_1 and A_2 consists of only the point x_2, then

$$P(X_1 = x_1 \text{ and } X_2 = x_2) = P(X_1 = x_1)P(X_2 = x_2)$$
$$= f_1(x_1)f_2(x_2), \qquad x_1 \in R_1, \qquad x_2 \in R_2.$$

If this is true for all $x_1 \in R_1$ and $x_2 \in R_2$, we say that X_1 and X_2 are **independent random variables** and $f_1(x_1)f_2(x_2)$ is the **joint probability density function of X_1 and X_2.**

Example 5.1.1

Let X_1 be the number of spots up on the cast of an unbiased die so that

$$f_1(x_1) = \frac{1}{6}, \qquad x_1 = 1, 2, 3, 4, 5, 6.$$

Let X_2 be the number of heads on 4 independent tosses of a fair coin so that

$$f_2(x_2) = \binom{4}{x_2}\left(\frac{1}{2}\right)^{x_2}\left(\frac{1}{2}\right)^{4-x_2}, \qquad x_2 = 0, 1, 2, 3, 4.$$

It is reasonable to assume that these two experiments are performed in such a way that the resulting events can be considered independent and hence X_1 and X_2 are independent random variables. Thus

$$P(X_1 = 4 \text{ and } X_2 = 3) = f_1(4)f_2(3) = \left(\frac{1}{6}\right)\binom{4}{3}\left(\frac{1}{2}\right)^4 = \frac{1}{24}.$$

In general, with $x_1 = 1, 2, 3, 4, 5, 6$, and $x_2 = 0, 1, 2, 3, 4$, we have that

$$P(X_1 = x_1 \text{ and } X_2 = x_2) = f_1(x_1)f_2(x_2) = \left(\frac{1}{6}\right)\binom{4}{x_2}\left(\frac{1}{2}\right)^4$$

is the joint p.d.f. of X_1 and X_2. Other probabilities can also be computed from the independence; for illustration,

$$P(X_1 = 1, 2 \text{ and } X_2 = 3, 4) = P(X_1 = 1, 2)P(X_2 = 3, 4)$$

$$= \left[\frac{2}{6}\right]\left[\sum_{x_2=3}^{4}\binom{4}{x_2}\left(\frac{1}{2}\right)^4\right]$$

$$= \left(\frac{2}{6}\right)\left(\frac{5}{16}\right) = \frac{5}{48},$$

because the events $\{X_1 = 1, 2\}$ and $\{X_2 = 3, 4\}$ are independent.   ◇

We often want to form functions of random variables, say $Y = u(X_1, X_2)$, which would itself be a random variable. In particular, linear functions of independent random variables are important in statistics. For illustration, in the preceding example, let $Y = X_1 + X_2$. Since the spaces of X_1 and X_2 are $R_1 = \{1, 2, 3, 4, 5, 6\}$ and $R_2 = \{0, 1, 2, 3, 4\}$, respectively, $Y = X_1 + X_2$ has the space $R = \{1, 2, 3, \cdots, 10\}$. The corresponding p.d.f., say $g(y)$, of $Y = X_1 + X_2$ can be computed as follows:

$$g(1) = P(X_1 = 1, X_2 = 0)$$

$$= \left(\frac{1}{6}\right)\left(\frac{1}{16}\right) = \frac{1}{96},$$

$$g(2) = P(X_1 = 2, X_2 = 0 \text{ or } X_1 = 1, X_2 = 1)$$

$$= \left(\frac{1}{6}\right)\left(\frac{1}{16}\right) + \left(\frac{1}{6}\right)\left(\frac{4}{16}\right) = \frac{5}{96},$$

$$g(3) = P(X_1 = 3, X_2 = 0 \text{ or } X_1 = 2, X_2 = 1 \text{ or } X_1 = 1, X_2 = 2)$$

$$= \left(\frac{1}{6}\right)\left(\frac{1}{16}\right) + \left(\frac{1}{6}\right)\left(\frac{4}{16}\right) + \left(\frac{1}{6}\right)\left(\frac{6}{16}\right) = \frac{11}{96},$$

and so on, until we obtain the p.d.f. of Y given by

y	1	2	3	4	5	6	7	8	9	10
$g(y)$	$\dfrac{1}{96}$	$\dfrac{5}{96}$	$\dfrac{11}{96}$	$\dfrac{15}{96}$	$\dfrac{16}{96}$	$\dfrac{16}{96}$	$\dfrac{15}{96}$	$\dfrac{11}{96}$	$\dfrac{5}{96}$	$\dfrac{1}{96}$

Of course, we note that

$$\sum_{y \in R} g(y) = \sum_{y=1}^{10} g(y) = 1.$$

The **mathematical expectation** (or **expected value**) of a function $u(X_1, X_2)$ of two independent random variables of the discrete type can be computed by

$$E[u(X_1, X_2)] = \sum_{x_2 \in R_2} \sum_{x_1 \in R_1} u(x_1, x_2) f_1(x_1) f_2(x_2).$$

The observant student notes that $Y = u(X_1, X_2)$ is a random variable with p.d.f., say $g(y)$, and space R. The expectation $E(Y)$ has been defined; and as stated formally in the next section (Theorem 5.2.1), we have

$$\sum_{x_2 \in R_2} \sum_{x_1 \in R_1} u(x_1, x_2) f_1(x_1) f_2(x_2) = \sum_{y \in R} y \, g(y).$$

While we do not prove this, we note in our illustration that

$$E(X_1 + X_2) = E(X_1) + E(X_2) = 3.5 + 2 = 5.5$$

and

$$E(Y) = \sum_{y=1}^{10} y \, g(y) = (1)\left(\frac{1}{96}\right) + (2)\left(\frac{5}{96}\right) + \cdots + (10)\left(\frac{1}{96}\right) = 5.5.$$

As in the case of one random variable, mathematical expectation E is a linear or distributive operator; that is, the expected value of a sum is the sum of the expected values. Also, the expected value of a constant c is again c because

$$\sum_{x_2 \in R_2} \sum_{x_1 \in R_1} c\, f_1(x_1) f_2(x_2) = c \sum_{x_2 \in R_2} f_2(x_2) \sum_{x_1 \in R_1} f_1(x_1) = (c)(1)(1) = c.$$

It is important to note that if $u(X_1, X_2)$ is the product $u_1(X_1)u_2(X_2)$, then

$$E[u_1(X_1)u_2(X_2)] = \sum_{x_2 \in R_2} \sum_{x_1 \in R_1} u_1(x_1)u_2(x_2) f_1(x_1) f_2(x_2)$$

$$= \sum_{x_2 \in R_2} u_2(x_2) f_2(x_2) \sum_{x_1 \in R_1} u_1(x_1) f_1(x_1)$$

$$= E[u_1(X_1)]E[u_2(X_2)].$$

That is, if X_1 and X_2 are independent, then the expected value of the product $u_1(X_1)u_2(X_2)$ is the product of the expected values of $u_1(X_1)$ and $u_2(X_2)$.

We also note that if $u(X_1, X_2) = X_1$, then

$$\sum_{x_2 \in R_2} \sum_{x_1 \in R_1} x_1 f_1(x_1) f_2(x_2) = \sum_{x_2 \in R_2} f_2(x_2) \sum_{x_1 \in R_1} x_1 f_1(x_1)$$

$$= (1) \sum_{x_1 \in R_1} x_1 f_1(x_1) = \mu_1,$$

the mean of X_1. Similarly, $E(X_2) = \mu_2$, the mean of X_2. In addition, if $u(X_1, X_2)$ equals $(X_1 - \mu_1)^2$ or $(X_2 - \mu_2)^2$, we obtain the variance of X_1 or X_2, respectively; that is

$$E[(X_1 - \mu_1)^2] = \sigma_1^2 \quad \text{and} \quad E[(X_2 - \mu_2)^2] = \sigma_2^2.$$

Sometimes the two random experiments are exactly the same. For illustration, we could cast the die twice resulting in X_1 and X_2, respectively. It is reasonable to say that the p.d.f. of X_1 is $f(x_1) = 1/6, x_1 = 1, 2, \ldots, 6$ and the p.d.f. of X_2 is $f(x_2) = 1/6, x_2 = 1, 2, \ldots, 6$. Assuming independence, which would be a fair way to perform the experiments, the joint p.d.f. is then

$$f(x_1)f(x_2) = \left(\frac{1}{6}\right)\left(\frac{1}{6}\right) = \frac{1}{36}, \quad x_1 = 1, 2, \ldots, 6; \quad x_2 = 1, 2, \ldots, 6.$$

In general, if the p.d.f. $f(x)$ of the independent random variables X_1 and X_2 is the same, then the joint p.d.f. is $f(x_1)f(x_2)$. Moreover, in this case, the collection of the two random variables X_1, X_2 is called a **random sample of size** $n = 2$ from a distribution with p.d.f. $f(x)$. Hence, in the two casts of the die, we say that we have a random sample of size $n = 2$ from the uniform distribution on the space $\{1, 2, 3, 4, 5, 6\}$.

Example 5.1.2

Let X_1 and X_2 be two independent random variables resulting from two casts of an unbiased die. We have that

$$E(X_1) = E(X_2) = \sum_{x=1}^{6} xf(x) = 3.5,$$

where $f(x) = 1/6, x = 1, 2, \ldots, 6$. Moreover,

$$\text{Var}(X_1) = \text{Var}(X_2) = \sum_{x=1}^{6} (x - 3.5)^2 f(x) = \frac{35}{12}.$$

In addition, from independence,

$$E(X_1 X_2) = E(X_1)E(X_2) = (3.5)(3.5) = 12.25$$

and

$$E[(X_1 - 3.5)(X_2 - 3.5)] = E(X_1 - 3.5)E(X_2 - 3.5) = 0.$$

If $Y = X_1 + X_2$, then

$$E(Y) = E(X_1) + E(X_2) = 3.5 + 3.5 = 7$$

and

$$\begin{aligned}
\text{Var}(Y) &= E[(X_1 + X_2 - 7)^2] = E\{[(X_1 - 3.5) + (X_2 - 3.5)]^2\} \\
&= E[(X_1 - 3.5)^2] + E[2(X_1 - 3.5)(X_2 - 3.5)] + E[(X_2 - 3.5)^2] \\
&= \text{Var}(X_1) + (2)(0) + \text{Var}(X_2) \\
&= (2)\left(\frac{35}{12}\right) = \frac{35}{6}.
\end{aligned}$$

$\diamond$

In Example 5.1.2 we can find the p.d.f. $g(y)$ of $Y = X_1 + X_2$. Since the space of Y is $\{2, 3, 4, \ldots, 12\}$, we have, by a rather straightforward calculation, that

$$g(2) = P(X_1 = 1, X_2 = 1) = f(1)f(1) = \left(\frac{1}{6}\right)\left(\frac{1}{6}\right) = \frac{1}{36},$$

$$g(3) = P(X_1 = 1, X_2 = 2 \text{ or } X_1 = 2, X_2 = 1)$$
$$= \left(\frac{1}{6}\right)\left(\frac{1}{6}\right) + \left(\frac{1}{6}\right)\left(\frac{1}{6}\right) = \frac{2}{36},$$

$$g(4) = P(X_1 = 1, X_2 = 3 \text{ or } X_1 = 2, X_2 = 2 \text{ or } X_1 = 3, X_2 = 1) = \frac{3}{36},$$

and so on. This results in the p.d.f. given by

y	2	3	4	5	6	7	8	9	10	11	12
$g(y)$	$\dfrac{1}{36}$	$\dfrac{2}{36}$	$\dfrac{3}{36}$	$\dfrac{4}{36}$	$\dfrac{5}{36}$	$\dfrac{6}{36}$	$\dfrac{5}{36}$	$\dfrac{4}{36}$	$\dfrac{3}{36}$	$\dfrac{2}{36}$	$\dfrac{1}{36}$

Using this p.d.f., it is simple to calculate

$$E(Y) = \sum_{y=2}^{12} y\, g(y) = 7$$

and

$$\mathrm{Var}(Y) = \sum_{y=2}^{12} (y - 7)^2\, g(y) = \frac{35}{6},$$

which agrees with the results of that example.

All of the definitions and results about two random variables of the discrete type can be carried over to two random variables of the continuous type. That is, in both cases, two random variables with respective probability density functions, $f_1(x)$ and $f_2(x)$, are **independent** if and only if the joint p.d.f. is $f_1(x_1) f_2(x_2)$. Again, with independence,

$$P(X_1 \in A_1, X_2 \in A_2) = P(X_1 \in A_1) P(X_2 \in A_2).$$

The definitions and results concerning expectations are similar, with integrals replacing summations. In particular, with independence,

$$E[(X_1 - \mu_1)(X_2 - \mu_2)] = E(X_1 - \mu_1) E(X_2 - \mu_2) = 0.$$

The continuous-type independent random variables can be thought of as a random sample of size $n = 2$ if each has the same p.d.f.

The notions about two independent random variables can be extended to n independent random variables, which can be thought of as measurements on the outcomes of n random experiments. That is, if $X_1, X_2, \ldots, X_n$ are **independent**, then the joint p.d.f. is the product of the respective probability density functions, namely

$$f_1(x_1) f_2(x_2) \cdots f_n(x_n),$$

and the expected value of the product $u_1(X_1) u_2(X_2) \cdots u_n(X_n)$ is the product of the expected values of $u_1(X_1), u_2(X_2), \ldots, u_n(X_n)$.

If all n of the distributions are the same, then the collection of n independent and identically distributed random variables, $X_1, X_2, \ldots, X_n$, is said to be a **random sample of size n from that common distribution**.

Example 5.1.3

Let X_1, X_2, X_3 be a random sample from a distribution with p.d.f.

$$f(x) = e^{-x}, \qquad 0 < x < \infty.$$

The joint p.d.f. of these three random variables is

$$(e^{-x_1})(e^{-x_2})(e^{-x_3}) = e^{-x_1-x_2-x_3}, \qquad 0 < x_i < \infty, i = 1, 2, 3.$$

The probability

$$P(0 < X_1 < 1, 2 < X_2 < 4, 3 < X_3 < 7)$$

$$= \left(\int_0^1 e^{-x_1} dx_1 \right) \left(\int_2^4 e^{-x_2} dx_2 \right) \left(\int_3^7 e^{-x_3} dx_3 \right)$$

$$= (1 - e^{-1})(e^{-2} - e^{-4})(e^{-3} - e^{-7}),$$

because of the independence of X_1, X_2, X_3. The expectation

$$E(X_1 X_2 X_3) = E(X_1)E(X_2)E(X_3)$$

$$= \left(\int_0^\infty x e^{-x} dx \right)^3 = 1^3 = 1,$$

since $E(X_1) = E(X_2) = E(X_3) = 1$. Also,

$$E[(X_1 - 1)(X_3 - 1)] = E(X_1 - 1)E(X_3 - 1) = 0. \qquad \diamond$$

EXERCISES

5.1.1 Let X_1 and X_2 be independent Poisson random variables with respective means $\lambda_1 = 2$ and $\lambda_2 = 3$. Find

(a) $P(X_1 = 3, X_2 = 5)$.

(b) $P(X_1 + X_2 = 1)$. HINT: Note that this event can occur if and only if $\{X_1 = 1, X_2 = 0\}$ or $\{X_1 = 0, X_2 = 1\}$.

5.1.2 Let X_1 and X_2 be independent random variables with respective binomial distributions $b(3, 1/2)$ and $b(5, 1/2)$. Determine

(a) $P(X_1 = 2, X_2 = 4)$.

(b) $P(X_1 + X_2 = 7)$.

5.1.3 Let X_1 and X_2 be independent random variables with probability density functions $f_1(x_1) = 2x_1, 0 < x_1 < 1$, and $f_2(x_2) = 4x_2^3, 0 < x_2 < 1$, respectively. Compute

(a) $P(0.5 < X_1 < 1 \text{ and } 0.4 < X_2 < 0.8)$.

(b) $E(X_1^2 X_2^3)$.

5.1.4 Let X_1 and X_2 be a random sample of size $n = 2$ from the exponential distribution with p.d.f. $f(x) = 2e^{-2x}, 0 < x < \infty$. Find

(a) $P(0.5 < X_1 < 1.0, 0.7 < X_2 < 1.2)$.

(b) $E[X_1(X_2 - 0.5)^2]$.

5.1.5 Let X_1, X_2 be a random sample of size $n = 2$ from a distribution with p.d.f. $f(x) = 1/3, x = 1, 2, 3$. If $Y = X_1 + X_2$, find the p.d.f. of Y. Compute the mean and the variance of Y in two ways.

5.1.6 Let X_1, X_2 be a random sample from $b(2, 0.5)$. If $Y = X_1 + X_2$, find the p.d.f. of Y. Compute the mean and the variance of Y in two ways.

5.1.7 Let X_1, X_2 be a random sample of size $n = 2$ from a distribution with p.d.f. $f(x) = 3x^2, 0 < x < 1$. Determine

(a) $P(\max X_i < 3/4) = P(X_1 < 3/4, X_2 < 3/4)$.

(b) The mean and the variance of $Y = X_1 + X_2$.

5.1.8 Let X_1, X_2, X_3 be a random sample of size $n = 3$ from the exponential distribution with p.d.f. $f(x) = e^{-x}, 0 < x < \infty$. Find

$$P(1 < \min X_i) = P(1 < X_1, 1 < X_2, 1 < X_3).$$

5.1.9 Let X_1, X_2, X_3 be three independent random variables with binomial distributions $b(4, 1/2)$, $b(6, 1/3)$, and $b(12, 1/6)$, respectively. Find

(a) $P(X_1 = 2, X_2 = 2, X_3 = 5)$.

(b) $E(X_1 X_2 X_3)$.

(c) The mean and the variance of $Y = X_1 + X_2$.

5.1.10 Say that X_1, X_2, X_3, X_4 is a random sample of size $n = 4$ from the normal distribution $N(75, 100)$. Determine

(a) $P(\max X_i < 85)$.

(b) $P(60 < X_1 < 80$ and $75 < X_3 < 90)$.

(c) $E(X_1^2 X_2^2 X_3^2)$.

(d) $E(2X_1 + 3X_2 + X_3 + X_4)$.

5.1.11 Let X_1, X_2 be a random sample of size $n = 2$ from the chi-square distribution $\chi^2(5)$.

(a) Compute the mean and the variance of $Y = X_1 + X_2$.

(b) Evaluate $E[e^{t(X_1+X_2)}] = E(e^{tX_1}e^{tX_2}) = E(e^{tX_1})E(e^{tX_2})$, the latter resulting from the independence of X_1 and X_2.

(c) Note that the expression in part (b) is the m.g.f. of $\chi^2(10)$. What are its mean and variance, and how do these compare with the results in part (a)?

5.1.12 Let X_1, X_2, X_3 denote a random sample of size $n = 3$ from a distribution with the geometric p.d.f.

$$f(x) = \left(\frac{3}{4}\right)\left(\frac{1}{4}\right)^{x-1}, \qquad x = 1, 2, 3, \ldots.$$

That is, X_1, X_2, and X_3 are mutually independent and each has this geometric distribution.

(a) Compute $P(X_1 = 1, X_2 = 3, X_3 = 1)$.

(b) Determine $P(X_1 + X_2 + X_3 = 5)$.

(c) If Y equals the maximum of X_1, X_2, X_3, find

$$P(Y \leq 2) = P(X_1 \leq 2)P(X_2 \leq 2)P(X_3 \leq 2).$$

5.1.13 Flip $n = 8$ fair coins and remove all that came up heads. Flip the other (tails) coins and remove the heads. Continue flipping the remaining coins until each has come up heads. We shall find the p.d.f. of Y, the number of trials needed. Let X_i equal the number of flips required to observe heads on coin i, $i = 1, 2, \ldots, 8$. Then $Y = \max(X_1, X_2, \ldots, X_8)$.

(a) Show that $P(Y \leq y) = [1 - (1/2)^y]^8$.

(b) Show that $P(Y = y) = [1 - (1/2)^y]^8 - [1 - (1/2)^{y-1}]^8$, $\quad y = 1, 2, \ldots$.

(c) Find $E(Y)$.

5.1.14 Let X_1, X_2, X_3 be a random sample of size $n = 3$ from a Poisson distribution with $\lambda = 5$.

(a) Find the mean and the variance of $Y = X_1 + X_2 + X_3$.

(b) Evaluate $E[e^{t(X_1+X_2+X_3)}] = E[e^{tX_1}]E[e^{tX_2}]E[e^{tX_3}]$.

(c) Note that the expression in part (b) is the m.g.f. of a Poisson distribution with $\lambda = 15$. How do the mean and the variance of this distribution compare with the results of part (a)?

5.2 DISTRIBUTIONS OF SUMS OF INDEPENDENT RANDOM VARIABLES

Functions of random variables are usually of interest in statistical applications; and, of course, we have already considered some functions, called **statistics**, of the observations of a random sample. Two important statistics are the sample mean $\overline{X}$ and the sample variance S^2. Of course, in a particular sample, say $x_1, x_2, \ldots, x_n$, we observed definite values of these statistics, $\overline{x}$ and s^2; however, we should recognize that each value is only one observation of the respective random variables, $\overline{X}$ and S^2. That is, each $\overline{X}$ or S^2 (or, more generally, any statistic) is also a random variable with its own distribution. In this chapter we determine the distributions of some of the important statistics and, more generally, distributions of functions of random variables. The derivations of distributions of these functions fall in that area of statistics usually referred to as **sampling distribution theory**.

We begin with an example that illustrates a way of finding the distribution of a sum of random variables in a random sample.

Example 5.2.1

Let us cast an unbiased four-sided die two independent times and observe the outcomes on these trials, say X_1 and X_2. This is equivalent to saying that X_1 and X_2 are the observations of a random sample of size $n = 2$ from a distribution with p.d.f.

$$f(x) = \frac{1}{4}, \quad x = 1, 2, 3, 4.$$

We shall give another way to find the p.d.f. of $Y = X_1 + X_2$. Note that the space of Y is $R = \{2, 3, 4, 5, 6, 7, 8\}$. To determine the p.d.f. of Y, say $g(y) = P(Y = y)$, $y \in R$, we first look at some special cases. Let $y = 3$. We note that the event

$\{Y = 3\}$ can occur in two mutually exclusive ways: $\{X_1 = 1, X_2 = 2\}$ and $\{X_1 = 2, X_2 = 1\}$. Thus recalling that X_1 and X_2 are independent, we obtain

$$g(3) = P(Y = 3) = P(X_1 = 1, X_2 = 2) + P(X_1 = 2, X_2 = 1)$$

$$= \left(\frac{1}{4}\right)\left(\frac{1}{4}\right) + \left(\frac{1}{4}\right)\left(\frac{1}{4}\right) = \frac{2}{16}.$$

Similarly,

$$g(5) = P(Y = 5)$$
$$= P(X_1 = 1, X_2 = 4) + P(X_1 = 2, X_2 = 3)$$
$$+ P(X_1 = 3, X_2 = 2) + P(X_1 = 4, X_2 = 1)$$
$$= \left(\frac{1}{4}\right)\left(\frac{1}{4}\right) + \left(\frac{1}{4}\right)\left(\frac{1}{4}\right) + \left(\frac{1}{4}\right)\left(\frac{1}{4}\right) + \left(\frac{1}{4}\right)\left(\frac{1}{4}\right) = \frac{4}{16}.$$

If we continue this process, we see that the p.d.f. of Y is given by the table

y	2	3	4	5	6	7	8
$g(y)$	$\frac{1}{16}$	$\frac{2}{16}$	$\frac{3}{16}$	$\frac{4}{16}$	$\frac{3}{16}$	$\frac{2}{16}$	$\frac{1}{16}$

$\diamond$

In general, we see that the p.d.f. in Example 5.2.1 is given by the following formula, called a **convolution formula** (see Exercise 5.2.14),

$$g(y) = P(Y = y) = \sum_{k=1}^{y-1} f(k)f(y - k),$$

where some of the summands could equal zero. If we like, we can give the p.d.f. by the formula

$$g(y) = \frac{4 - |y - 5|}{16}, \qquad y = 2, 3, 4, 5, 6, 7, 8.$$

Using the p.d.f. of Y, its mean is

$$E(Y) = \sum_{y=2}^{8} y\, g(y)$$

$$= 2\left(\frac{1}{16}\right) + 3\left(\frac{2}{16}\right) + 4\left(\frac{3}{16}\right) + 5\left(\frac{4}{16}\right) + 6\left(\frac{3}{16}\right) + 7\left(\frac{2}{16}\right) + 8\left(\frac{1}{16}\right)$$

$$= 5.$$

In Example 5.2.1, since $Y = X_1 + X_2$, we anticipate that $E(Y) = 5$, which is computed using the p.d.f. of Y, would be equal to $E(X_1 + X_2)$, which is computed using the joint p.d.f. of X_1 and X_2, namely

$$E(X_1 + X_2) = \sum_{x_2=1}^{4} \sum_{x_1=1}^{4} (x_1 + x_2)\left(\frac{1}{4}\right)\left(\frac{1}{4}\right).$$

Of course, this is true because

$$E(X_1 + X_2) = E(X_1) + E(X_2) = \frac{5}{2} + \frac{5}{2} = 5.$$

This special result with a linear function of two random variables extends to more general functions of several random variables. We accept the following without proof.

Theorem 5.2.1.   Let $X_1, X_2, \ldots, X_n$ be n independent random variables with joint p.d.f. $f_1(x_1) f_2(x_2) \cdots f_n(x_n)$. Let the random variable $Y = u(X_1, X_2, \ldots, X_n)$ have the p.d.f. $g(y)$. Then, in the discrete case,

$$E(Y) = \sum_y y\, g(y) = \sum_{x_1} \sum_{x_2} \cdots \sum_{x_n} u(x_1, x_2, \ldots, x_n) f_1(x_1) f_2(x_2) \cdots f_n(x_n),$$

provided that these summations exist. For random variables of the continuous type, integrals replace the summations.

The next theorem proves that the expected value of the product of functions of n independent random variables is the product of their expected values.

Theorem 5.2.2.   If $X_1, X_2, \ldots, X_n$ are independent random variables having p.d.f.'s $f_1(x_1), f_2(x_2), \ldots, f_n(x_n)$ and $E[u_i(X_i)]$, $i = 1, 2, \ldots, n$, exist, then

$$E[u_1(X_1)\, u_2(X_2) \cdots u_n(X_n)] = E[u_1(X_1)]E[u_2(X_2)] \cdots E[u_n(X_n)].$$

Proof.   In the discrete case, we have that

$$E[u_1(X_1)u_2(X_2) \cdots u_n(X_n)]$$

$$= \sum_{x_1} \sum_{x_2} \cdots \sum_{x_n} u_1(x_1)\, u_2(x_2) \cdots u_n(x_n) f_1(x_1) f_2(x_2) \cdots f_n(x_n)$$

$$= \sum_{x_1} u_1(x_1) f_1(x_1) \sum_{x_2} u_2(x_2) f_2(x_2) \cdots \sum_{x_n} u_n(x_n) f_n(x_n)$$

$$= E[u_1(X_1)]E[u_2(x_2)] \cdots E[u_n(X_n)].$$

In the proof for the continuous case, integrals replace summations.   ☐

Remark   Sometimes students recognize that $X^2 = (X)(X)$ and thus believe that $E(X^2)$ is equal to $[E(X)][E(X)] = [E(X)]^2$ because the theorem above states that the expected value of the product is the product of the expected values. However, note the hypothesis of independence in the theorem, and certainly X is not independent of itself. Incidentally, if $E(X^2)$ did equal $[E(X)]^2$, then the variance of X,

$$\sigma^2 = E(X^2) - [E(X)]^2,$$

would always equal zero. This really happens only in the case of degenerate (one-point) distributions.

Example 5.2.2

It is interesting to note that these two theorems allow us to determine the mean, the variance, and the moment-generating function of a function such as $Y = X_1 + X_2$, where X_1 and X_2 have been defined in Example 5.2.1. We have already seen that $\mu_Y = 5$.

The variance of Y is

$$\sigma_Y^2 = E[(Y - \mu_Y)^2] = E[(X_1 + X_2 - \mu_1 - \mu_2)^2],$$

where $\mu_i = E(X_i) = 5/2, i = 1, 2$. Thus

$$\sigma_Y^2 = E[[(X_1 - \mu_1) + (X_2 - \mu_2)]^2]$$
$$= E[(X_1 - \mu_1)^2 + 2(X_1 - \mu_1)(X_2 - \mu_2) + (X_2 - \mu_2)^2].$$

But from an earlier result about the expected value being a linear operator, we have

$$\sigma_Y^2 = E[(X_1 - \mu_1)^2] + 2E[(X_1 - \mu_1)(X_2 - \mu_2)] + E[(X_2 - \mu_2)^2].$$

However, since X_1 and X_2 are independent, then

$$E[(X_1 - \mu_1)(X_2 - \mu_2)] = E(X_1 - \mu_1)E(X_2 - \mu_2)$$
$$= (\mu_1 - \mu_1)(\mu_2 - \mu_2) = 0.$$

Thus

$$\sigma_Y^2 = \sigma_1^2 + \sigma_2^2,$$

where

$$\sigma_i^2 = E[(X_i - \mu_i)^2], \qquad i = 1, 2.$$

In the case in which X_1 and X_2 are the outcomes on two independent casts of a four-sided die, we have $\sigma_i^2 = 5/4, i = 1, 2$, and hence

$$\sigma_Y^2 = \frac{5}{4} + \frac{5}{4} = \frac{5}{2}.$$

Finally, the moment-generating function is

$$M_Y(t) = E(e^{tY}) = E[e^{t(X_1 + X_2)}] = E(e^{tX_1}e^{tX_2}).$$

The independence of X_1 and X_2 implies that

$$M_Y(t) = E(e^{tX_1})E(e^{tX_2}).$$

For our example in which X_1 and X_2 have the same p.d.f.,

$$f(x) = \frac{1}{4}, \qquad x = 1, 2, 3, 4,$$

and thus the same moment-generating function,

$$M_X(t) = \frac{1}{4}e^t + \frac{1}{4}e^{2t} + \frac{1}{4}e^{3t} + \frac{1}{4}e^{4t},$$

we have that $M_Y(t) = [M_X(t)]^2$ equals

$$\frac{1}{16} e^{2t} + \frac{2}{16} e^{3t} + \frac{3}{16} e^{4t} + \frac{4}{16} e^{5t} + \frac{3}{16} e^{6t} + \frac{2}{16} e^{7t} + \frac{1}{16} e^{8t}.$$

Note that the coefficient of e^{bt} is equal to the probability $P(Y = b)$; for illustration, $4/16 = P(Y = 5)$. This agrees with the result found in Example 5.2.1, and thus we see that we could find the distribution of Y by determining its moment-generating function.   ◇

In this section we are restricting attention to those functions that are linear combinations of independent random variables. We shall first prove an important theorem about the mean and the variance of such a linear combination, extending the results in Example 5.2.2.

Theorem 5.2.3.   If $X_1, X_2, \ldots, X_n$ are n independent random variables with respective means $\mu_1, \mu_2, \ldots, \mu_n$ and variances $\sigma_1^2, \sigma_2^2, \ldots, \sigma_n^2$, then the mean and the variance of $Y = \sum_{i=1}^{n} a_i X_i$, where $a_1, a_2, \ldots, a_n$ are real constants, are

$$\mu_Y = \sum_{i=1}^{n} a_i \mu_i \qquad \text{and} \qquad \sigma_Y^2 = \sum_{i=1}^{n} a_i^2 \sigma_i^2,$$

respectively.

Proof.   We have that

$$\mu_Y = E(Y) = E\left(\sum_{i=1}^{n} a_i X_i\right) = \sum_{i=1}^{n} a_i E(X_i) = \sum_{i=1}^{n} a_i \mu_i,$$

because the expected value of the sum is the sum of the expected values (i.e., E is a linear operator). Also,

$$\sigma_Y^2 = E[(Y - \mu_Y)^2] = E\left[\left(\sum_{i=1}^{n} a_i X_i - \sum_{i=1}^{n} a_i \mu_i\right)^2\right]$$

$$= E\left\{\left[\sum_{i=1}^{n} a_i (X_i - \mu_i)\right]^2\right\} = E\left[\sum_{i=1}^{n}\sum_{j=1}^{n} a_i a_j (X_i - \mu_i)(X_j - \mu_j)\right].$$

Again using the fact that E is a linear operator, we obtain

$$\sigma_Y^2 = \sum_{i=1}^{n}\sum_{j=1}^{n} a_i a_j E[(X_i - \mu_i)(X_j - \mu_j)].$$

However, if $i \neq j$, then from the independence of X_i and X_j we have

$$E[(X_i - \mu_i)(X_j - \mu_j)] = E(X_i - \mu_i)E(X_j - \mu_j) = (\mu_i - \mu_i)(\mu_j - \mu_j) = 0.$$

Thus the variance can be written as

$$\sigma_Y^2 = \sum_{i=1}^{n} a_i^2 E[(X_i - \mu_i)^2] = \sum_{i=1}^{n} a_i^2 \sigma_i^2. \qquad \square$$

We give two illustrations of the theorem.

Example 5.2.3

Let the independent random variables X_1 and X_2 have, respectively, means $\mu_1 = -4$ and $\mu_2 = 3$ and variances $\sigma_1^2 = 4$ and $\sigma_2^2 = 9$. The mean and the variance of $Y = 3X_1 - 2X_2$ are, respectively,

$$\mu_Y = (3)(-4) + (-2)(3) = -18$$

and

$$\sigma_Y^2 = (3)^2(4) + (-2)^2(9) = 72. \qquad \diamond$$

Example 5.2.4

Let $X_1, X_2, \ldots, X_n$ be a random sample of size n from a distribution with mean μ and variance σ^2. First let $Y = X_1 - X_2$; then

$$\mu_Y = \mu - \mu = 0 \quad \text{and} \quad \sigma_Y^2 = (1)^2\sigma^2 + (-1)^2\sigma^2 = 2\sigma^2.$$

Now consider the sample mean

$$\overline{X} = \frac{X_1 + X_2 + \cdots + X_n}{n},$$

which is a linear function with each $a_i = 1/n$. Then

$$\mu_{\overline{X}} = \sum_{i=1}^{n} \left(\frac{1}{n}\right)\mu = \mu \quad \text{and} \quad \sigma_{\overline{X}}^2 = \sum_{i=1}^{n} \left(\frac{1}{n}\right)^2 \sigma^2 = \frac{\sigma^2}{n}.$$

That is, the mean of $\overline{X}$ is that of the distribution from which the sample arose, but the variance of $\overline{X}$ is that of the underlying distribution divided by n. $\diamond$

In some applications it is sufficient to know the mean and variance of a linear combination of random variables, say Y. However, it is often helpful to know exactly how Y is distributed. The next theorem can frequently be used to find the distribution of a linear combination of independent random variables.

Theorem 5.2.4. If $X_1, X_2, \ldots, X_n$ are independent random variables with respective moment-generating functions $M_{X_i}(t)$, $i = 1, 2, 3, \ldots, n$, then the moment-generating function of $Y = \sum_{i=1}^{n} a_i X_i$ is

$$M_Y(t) = \prod_{i=1}^{n} M_{X_i}(a_i t).$$

Proof. The moment-generating function of Y is given by

$$M_Y(t) = E[e^{tY}] = E[e^{t(a_1 X_1 + a_2 X_2 + \cdots + a_n X_n)}]$$

$$= E[e^{a_1 t X_1} e^{a_2 t X_2} \cdots e^{a_n t X_n}]$$

$$= E[e^{a_1 t X_1}] E[e^{a_2 t X_2}] \cdots E[e^{a_n t X_n}]$$

using Theorem 5.2.2. However, since

$$E(e^{t X_i}) = M_{X_i}(t),$$

then

$$E(e^{a_i t X_i}) = M_{X_i}(a_i t).$$

Thus we have that

$$M_Y(t) = M_{X_1}(a_1 t) M_{X_2}(a_2 t) \cdots M_{X_n}(a_n t) = \prod_{i=1}^{n} M_{X_i}(a_i t). \qquad \square$$

A corollary follows immediately, and it will be used in some important examples.

Corollary 5.2.1. If $X_1, X_2, \ldots, X_n$ are observations of a random sample from a distribution with moment-generating function $M(t)$, then

(a) The moment-generating function of $Y = \sum_{i=1}^{n} X_i$ is

$$M_Y(t) = \prod_{i=1}^{n} M(t) = [M(t)]^n.$$

(b) The moment-generating function of $\overline{X} = \sum_{i=1}^{n} (1/n) X_i$ is

$$M_{\overline{X}}(t) = \prod_{i=1}^{n} M\left(\frac{t}{n}\right) = \left[M\left(\frac{t}{n}\right)\right]^n.$$

Proof. For part (a), let $a_i = 1$, $i = 1, 2, \ldots, n$, in Theorem 5.2.4. For part (b), take $a_i = 1/n$, $i = 1, 2, \ldots, n$. $\qquad \square$

The following examples and the exercises give some important applications of Theorem 5.2.4 and its corollary.

Example 5.2.5

Let $X_1, X_2, \ldots, X_n$ denote the outcomes on n Bernoulli trials. The moment-generating function of X_i, $i = 1, 2, \ldots, n$, is

$$M(t) = q + pe^t.$$

If

$$Y = \sum_{i=1}^{n} X_i,$$

then

$$M_Y(t) = \prod_{i=1}^{n}(q + pe^t) = (q + pe^t)^n.$$

Thus we again see that Y is $b(n, p)$.   ◇

Example 5.2.6

Let X_1, X_2, X_3 be the observations of a random sample of size $n = 3$ from the exponential distribution having mean θ and, of course, moment-generating function $M(t) = 1/(1-\theta t), t < 1/\theta$. The moment-generating function of $Y = X_1 + X_2 + X_3$ is

$$M_Y(t) = [(1 - \theta t)^{-1}]^3 = (1 - \theta t)^{-3}, \qquad t < 1/\theta,$$

which is that of a gamma distribution with parameters $\alpha = 3$ and θ. Thus Y has this distribution. On the other hand, the moment-generating function of $\overline{X}$ is

$$M_{\overline{X}}(t) = \left[\left(1 - \frac{\theta t}{3}\right)^{-1}\right]^3 = \left(1 - \frac{\theta t}{3}\right)^{-3}, \qquad t < 3/\theta;$$

and hence the distribution of $\overline{X}$ is gamma with parameters $\alpha = 3$ and $\theta/3$, respectively.   ◇

EXERCISES

5.2.1 Let X_1 and X_2 be observations of a random sample of size $n = 2$ from a distribution with p.d.f. $f(x) = x/6, x = 1, 2, 3$. Then find the p.d.f. of $Y = X_1 + X_2$. Determine the mean and the variance of the sum in two ways.

5.2.2 Let X_1 and X_2 be a random sample of size $n = 2$ from a distribution with p.d.f. $f(x) = 6x(1 - x), 0 < x < 1$. Find the mean and the variance of $Y = X_1 + X_2$.

5.2.3 Let X_1, X_2, X_3 be a random sample of size 3 from the distribution with p.d.f. $f(x) = 1/4$, $x = 1, 2, 3, 4$. For example, observe three independent rolls of a fair four-sided die.

   (a) Find the p.d.f. of $Y = X_1 + X_2 + X_3$.

   (b) Sketch a bar graph of the p.d.f. of Y.

5.2.4 Let X_1 and X_2 be two independent random variables with respective means μ_1 and μ_2 and variances σ_1^2 and σ_2^2. Show that the mean and the variance of $Y = X_1 X_2$ are $\mu_1 \mu_2$ and $\sigma_1^2 \sigma_2^2 + \mu_1^2 \sigma_2^2 + \mu_2^2 \sigma_1^2$, respectively. HINT: Note that $E(Y) = E(X_1)E(X_2)$ and $E(Y^2) = E(X_1^2)E(X_2^2)$.

5.2.5 Let X_1 and X_2 be two independent random variables with respective means 3 and 7 and variances 9 and 25. Compute the mean and the variance of $Y = -2X_1 + X_2$.

5.2.6 Let X_1 and X_2 have independent distributions $b(n_1, p)$ and $b(n_2, p)$. Find the moment-generating function of $Y = X_1 + X_2$. How is Y distributed?

5.2.7 Let X_1, X_2, X_3 be mutually independent random variables with Poisson distributions having means 2, 1, 4, respectively.

(a) Find the moment-generating function of the sum $Y = X_1 + X_2 + X_3$.

(b) How is Y distributed?

(c) Compute $P(3 \leq Y \leq 9)$.

5.2.8 Generalize Exercise 5.2.7 by showing that the sum of n independent Poisson random variables with respective means $\mu_1, \mu_2, \ldots, \mu_n$ is Poisson with mean

$$\mu_1 + \mu_2 + \cdots + \mu_n.$$

5.2.9 Let X_1, X_2, X_3, X_4, X_5 be a random sample of size 5 from a geometric distribution with $p = 1/3$.

(a) Find the moment-generating function of $Y = X_1 + X_2 + X_3 + X_4 + X_5$.

(b) How is Y distributed?

5.2.10 Let $W = X_1 + X_2 + \cdots + X_h$, a sum of h mutually independent and identically distributed exponential random variables with mean θ. Show that W has a gamma distribution with mean $h\theta$.

5.2.11 Let X_1, X_2, X_3 denote a random sample of size 3 from a gamma distribution with $\alpha = 7$ and $\theta = 5$.

(a) Find the moment-generating function of $Y = X_1 + X_2 + X_3$.

(b) How is Y distributed?

5.2.12 The moment-generating function of X is

$$M_X(t) = \left(\frac{1}{4}\right)(e^t + e^{2t} + e^{3t} + e^{4t});$$

the moment-generating function of Y is

$$M_Y(t) = \left(\frac{1}{3}\right)(e^t + e^{2t} + e^{3t});$$

X and Y are independent random variables. Let $W = X + Y$.

(a) Find the moment-generating function of W.

(b) Give the p.d.f. of W; that is, determine $P(W = w)$, $w = 2, 3, \ldots, 7$, from the moment-generating function of W.

5.2.13 Let X equal the outcome when a four-sided die is rolled. Let Y equal the outcome when a six-sided die is rolled. Let $W = X + Y$.

(a) Find the moment-generating function of W.

(b) Give the p.d.f. of W.

5.2.14 Let X and Y, with respective p.d.f.'s $f(x)$ and $g(y)$, be independent discrete random variables, each of whose support is a subset of the nonnegative integers $0, 1, 2, \ldots$. Show that the p.d.f. of $W = X + Y$ is given by the convolution formula

$$h(w) = \sum_{x=0}^{w} f(x)g(w - x), \qquad w = 0, 1, 2, \ldots.$$

HINT: Argue that $h(w) = P(W = w)$ is the probability of the $w + 1$ mutually exclusive events $(x, y = w - x), x = 0, 1, \ldots, w$.

5.2.15 Let X_1, X_2, X_3, X_4 be a random sample from a distribution having p.d.f. $f(x) = (x + 1)/6$, $x = 0, 1, 2$.

(a) Use Exercise 5.2.14 to find the p.d.f. of $W_1 = X_1 + X_2$.

(b) What is the p.d.f. of $W_2 = X_3 + X_4$?

(c) Now find the p.d.f. of $W = W_1 + W_2 = X_1 + X_2 + X_3 + X_4$.

5.2.16 Roll a fair four-sided die eight times and denote the outcomes by $X_1, X_2, \ldots, X_8$. Use Exercise 5.2.14 to find the p.d.f.'s of

(a) $X_1 + X_2$.

(b) $\displaystyle\sum_{i=1}^{4} X_i$.

(c) $\displaystyle\sum_{i=1}^{8} X_i$.

5.2.17 Given a fair four-sided die, let Y equal the number of rolls needed to observe each face at least once.

(a) Argue that $Y = X_1 + X_2 + X_3 + X_4$, where X_i has a geometric distribution with $p_i = (5 - i)/4, i = 1, 2, 3, 4$, and X_1, X_2, X_3, X_4 are independent.

(b) Find the mean and variance of Y.

(c) Find $P(Y = y)$, $y = 4, 5, 6$.

5.3 RANDOM FUNCTIONS ASSOCIATED WITH NORMAL DISTRIBUTIONS

In statistical applications, it is often assumed that the population from which a sample is taken is normally distributed, $N(\mu, \sigma^2)$. There is then interest in estimating the parameters μ and σ^2 or in testing conjectures about these parameters. The usual statistics that are used in these activities are the sample mean $\overline{X}$ and the sample variance S^2; thus we need to know something about the distribution of these statistics or functions of these statistics.

Theorem 5.3.1. If $X_1, X_2, \ldots, X_n$ are observations of a random sample of size n from the normal distribution $N(\mu, \sigma^2)$, then the distribution of the sample mean $\overline{X} = (1/n) \sum_{i=1}^{n} X_i$ is $N(\mu, \sigma^2/n)$.

Proof. Since the moment-generating function of each X is

$$M_X(t) = \exp\left(\mu t + \frac{\sigma^2 t^2}{2} \right),$$

the moment-generating function of

$$\overline{X} = \frac{1}{n} \sum_{i=1}^{n} X_i$$

is, from the corollary of Theorem 5.2.4, equal to

$$M_{\overline{X}}(t) = \left\{ \exp\left[\mu\left(\frac{t}{n}\right) + \frac{\sigma^2(t/n)^2}{2} \right] \right\}^n$$
$$= \exp\left[\mu t + \frac{(\sigma^2/n)t^2}{2} \right].$$

However, the moment-generating function uniquely determines the distribution of the random variable. Since this one is that associated with the normal distribution $N(\mu, \sigma^2/n)$, the sample mean $\overline{X}$ is $N(\mu, \sigma^2/n)$.   □

Theorem 5.3.1 shows that if $X_1, X_2, \ldots, X_n$ is a random sample from the normal distribution, $N(\mu, \sigma^2)$, then the probability distribution of $\overline{X}$ is also normal with the same mean μ but a variance σ^2/n. This means that $\overline{X}$ has a greater probability of falling in an interval containing μ than does a single observation, say X_1. For example, if $\mu = 50$, $\sigma^2 = 16$, $n = 64$, then $P(49 < \overline{X} < 51) = 0.9544$, whereas $P(49 < X_1 < 51) = 0.1974$. This is illustrated again in the next example.

Example 5.3.1

Let $X_1, X_2, \ldots, X_n$ be a random sample from the $N(50, 16)$ distribution. We know that the distribution of $\overline{X}$ is $N(50, 16/n)$. To illustrate the effect of n, the graph of the p.d.f. of $\overline{X}$ is given in Figure 5.1 for $n = 1, 4, 16$, and 64. When $n = 64$, compare the areas that represent $P(49 < \overline{X} < 51)$ and $P(49 < X_1 < 51)$.   ◇

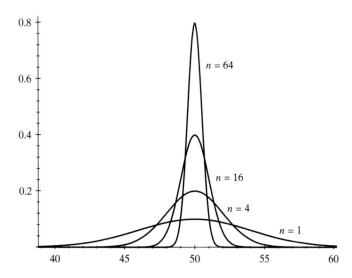

Figure 5.1   Normal sample mean p.d.f.'s

Before we can find the distribution of S^2, or $(n - 1)S^2/\sigma^2$, we give two preliminary theorems that are important in their own right. The first is a direct result of the material in Section 5.2 and gives an additive property for independent chi-square random variables.

Theorem 5.3.2. Let the distributions of $X_1, X_2, \ldots, X_k$ be $\chi^2(r_1), \chi^2(r_2), \ldots, \chi^2(r_k)$, respectively. If $X_1, X_2, \ldots, X_k$ are independent, then $Y = X_1 + X_2 + \cdots + X_k$ is $\chi^2(r_1 + r_2 + \cdots + r_k)$.

Proof. We give the proof for $k = 2$ but it is similar for a general k. The moment-generating function of Y is given by

$$M_Y(t) = E[e^{tY}] = E[e^{t(X_1+X_2)}] = E[e^{tX_1}e^{tX_2}].$$

Because X_1 and X_2 are independent, this last expectation can be factored so that

$$M_Y(t) = E[e^{tX_1}]E[e^{tX_2}] = (1 - 2t)^{-r_1/2}(1 - 2t)^{-r_2/2}, \qquad t < \frac{1}{2},$$

since X_1, and X_2 have chi-square distributions. Thus

$$M_Y(t) = (1 - 2t)^{-(r_1+r_2)/2}, \qquad t < \frac{1}{2},$$

the moment-generating function for a chi-square distribution with $r = r_1+r_2$ degrees of freedom. The uniqueness of the moment-generating function implies that Y is $\chi^2(r_1 + r_2)$. For the general k, Y is $\chi^2(r_1 + r_2 + \cdots + r_k)$. $\quad\square$

The next theorem combines and extends the results of Theorems 4.4.2 and 5.3.2 and gives one interpretation of degrees of freedom.

Theorem 5.3.3. Let $Z_1, Z_2, \ldots, Z_n$ have standard normal distributions, $N(0, 1)$. If these random variables are independent, then $W = Z_1^2 + Z_2^2 + \cdots + Z_n^2$ has a distribution that is $\chi^2(n)$.

Proof. By Theorem 4.4.2, Z_i^2 is $\chi^2(1)$ for $i = 1, 2, \ldots, n$. From Theorem 5.3.2, with $k = n$, $Y = W$, and $r_i = 1$, we see that W is $\chi^2(n)$. $\quad\square$

Corollary 5.3.1. If $X_1, X_2, \ldots, X_n$ are independent and have normal distributions $N(\mu_i, \sigma_i^2)$, $i = 1, 2, \ldots, n$, respectively, then the distribution of

$$W = \sum_{i=1}^{n} \frac{(X_i - \mu_i)^2}{\sigma_i^2}$$

is $\chi^2(n)$.

Proof. This follows since $Z_i = (X_i - \mu_i)/\sigma_i$ is $N(0, 1)$, $i = 1, 2, \ldots, n$. $\quad\square$

Note that the number of terms in the summation and the number of degrees of freedom are equal in Theorem 5.3.3 and the corollary.

The following theorem gives an important result that will be used in statistical applications. In connection with those, we will use the sample variance s^2 to estimate the variance, σ^2, when sampling from the normal distribution, $N(\mu, \sigma^2)$. More will be said about s^2 at the time of its use.

Theorem 5.3.4.   If $X_1, X_2, \ldots, X_n$ are observations of a random sample of size n from the normal distribution $N(\mu, \sigma^2)$,

$$\overline{X} = \frac{1}{n}\sum_{i=1}^{n} X_i$$

and

$$S^2 = \frac{1}{n-1}\sum_{i=1}^{n}(X_i - \overline{X})^2,$$

then

(a) $\overline{X}$ and S^2 are independent.

(b) $\dfrac{(n-1)S^2}{\sigma^2} = \dfrac{\displaystyle\sum_{i=1}^{n}(X_i - \overline{X})^2}{\sigma^2}$ is $\chi^2(n-1)$.

Proof.   We are not prepared to prove part (a) in this book; so we accept it without proof here. To prove part (b), note that

$$\sum_{i=1}^{n}\left(\frac{X_i - \mu}{\sigma}\right)^2 = \sum_{i=1}^{n}\left[\frac{(X_i - \overline{X}) + (\overline{X} - \mu)}{\sigma}\right]^2$$

$$= \sum_{i=1}^{n}\left(\frac{X_i - \overline{X}}{\sigma}\right)^2 + \frac{n(\overline{X} - \mu)^2}{\sigma^2} \qquad (5.1)$$

because the cross-product term is equal to

$$2\sum_{i=1}^{n}\frac{(\overline{X} - \mu)(X_i - \overline{X})}{\sigma^2} = \frac{2(\overline{X} - \mu)}{\sigma^2}\sum_{i=1}^{n}(X_i - \overline{X}) = 0.$$

But $Y_i = (X_i - \mu)/\sigma$, $i = 1, 2, \ldots n$, are standardized normal variables that are independent. Hence $W = \sum_{i=1}^{n} Y_i^2$ is $\chi^2(n)$ by Corollary 5.3.1. Moreover, since $\overline{X}$ is $N(\mu, \sigma^2/n)$, then

$$Z^2 = \left(\frac{\overline{X} - \mu}{\sigma/\sqrt{n}}\right)^2 = \frac{n(\overline{X} - \mu)^2}{\sigma^2}$$

is $\chi^2(1)$ by Theorem 4.4.2. In this notation, Equation 5.1 becomes

$$W = \frac{(n-1)S^2}{\sigma^2} + Z^2.$$

However, from part (a), $\overline{X}$ and S^2 are independent; thus Z^2 and S^2 are also independent. In the moment-generating function of W, this independence permits us to write

$$E[e^{tW}] = E[e^{t\{(n-1)S^2/\sigma^2+Z^2\}}] = E[e^{t(n-1)S^2/\sigma^2}e^{tZ^2}]$$
$$= E[e^{t(n-1)S^2/\sigma^2}]E[e^{tZ^2}].$$

Since W and Z^2 have chi-square distributions, we can substitute their moment-generating functions to obtain

$$(1 - 2t)^{-n/2} = E[e^{t(n-1)S^2/\sigma^2}](1 - 2t)^{-1/2}.$$

Equivalently, we have

$$E[e^{t(n-1)S^2/\sigma^2}] = (1 - 2t)^{-(n-1)/2}, \qquad t < \frac{1}{2}.$$

This, of course, is the moment-generating function of a $\chi^2(n-1)$ variable, and accordingly, $(n-1)S^2/\sigma^2$ has this distribution. $\square$

Combining the results of Corollary 5.3.1 and Theorem 5.3.4, we see that when sampling from a normal distribution,

$$U = \sum_{i=1}^{n} \frac{(X_i - \mu)^2}{\sigma^2}$$

is $\chi^2(n)$, and

$$W = \sum_{i=1}^{n} \frac{(X_i - \overline{X})^2}{\sigma^2}$$

is $\chi^2(n-1)$. That is, when the population mean, μ, in $\sum_{i=1}^{n}(X_i-\mu)^2$ is replaced by the sample mean, $\overline{X}$, one degree of freedom is lost. There are more general situations in which a degree of freedom is lost for each parameter estimated in certain chi-square random variables (see, e.g., Section 7.5).

Example 5.3.2

Let X_1, X_2, X_3, X_4 be a random sample of size 4 from the normal distribution, $N(76.4, 383)$. Then

$$U = \sum_{i=1}^{4} \frac{(X_i - 76.4)^2}{383}$$

is $\chi^2(4)$,

$$W = \sum_{i=1}^{4} \frac{(X_i - \overline{X})^2}{383}$$

is $\chi^2(3)$, and, for example,

$$P(0.711 \leq U \leq 7.779) = 0.90 - 0.05 = 0.85,$$
$$P(0.352 \leq W \leq 6.251) = 0.90 - 0.05 = 0.85. \qquad \diamond$$

In later sections we illustrate the importance of the chi-square distribution in applications.

We now prove another theorem which deals with linear functions of independent normally distributed random variables.

Theorem 5.3.5.   If $X_1, X_2, \ldots, X_n$ are n mutually independent normal variables with means $\mu_1, \mu_2, \ldots, \mu_n$ and variances $\sigma_1^2, \sigma_2^2, \ldots, \sigma_n^2$, respectively, then the linear function

$$Y = \sum_{i=1}^{n} c_i X_i$$

has the normal distribution

$$N\left(\sum_{i=1}^{n} c_i \mu_i, \sum_{i=1}^{n} c_i^2 \sigma_i^2\right).$$

Proof.   By Theorem 5.2.4, we have

$$M_Y(t) = \prod_{i=1}^{n} M_{X_i}(c_i t) = \prod_{i=1}^{n} \exp(\mu_i c_i t + \sigma_i^2 c_i^2 t^2 / 2)$$

because $M_{X_i}(t) = \exp(\mu_i t + \sigma_i^2 t^2 / 2)$, $i = 1, 2, \ldots, n$. Thus

$$M_Y(t) = \exp\left[\left(\sum_{i=1}^{n} c_i \mu_i\right) t + \left(\sum_{i=1}^{n} c_i^2 \sigma_i^2\right)\left(\frac{t^2}{2}\right)\right].$$

This is the moment-generating function of a distribution which is

$$N\left(\sum_{i=1}^{n} c_i \mu_i, \sum_{i=1}^{n} c_i^2 \sigma_i^2\right)$$

and thus Y has this normal distribution. $\qquad \square$

From Theorem 5.3.5 we can make the observation that the difference of two independent normally distributed random variables, say $Y = X_1 - X_2$, has the normal distribution $N(\mu_1 - \mu_2, \sigma_1^2 + \sigma_2^2)$.

Example 5.3.3

Let X_1 and X_2 equal the number of pounds of butterfat produced by two Holstein cows (one selected at random from those on the Koopman farm and one selected at random from those on the Vliestra farm, respectively) during the 305-day lactation period following the births of calves. Assume that the distribution of X_1 is $N(693.2, 22820)$ and the distribution of X_2 is $N(631.7, 19205)$. Moreover, if X_1 and X_2 are independent, we shall find $P(X_1 > X_2)$. That is, we shall find the probability that the butterfat produced by the Koopman farm cow exceeds that produced by the Vliestra farm cow. (Sketch p.d.f.'s on the same graph for these two normal distributions.) If we let $Y = X_1 - X_2$, then the distribution of Y is $N(693.2 - 631.7, 22820 + 19205)$. Thus,

$$P(X_1 > X_2) = P(Y > 0) = P\left(\frac{Y - 61.5}{\sqrt{42025}} > \frac{0 - 61.5}{205}\right)$$

$$= P(Z > -0.30) = 0.6179. \qquad \diamond$$

EXERCISES

5.3.1 Let $X_1, X_2, \ldots, X_{16}$ be a random sample from a normal distribution $N(77, 25)$. Compute

(a) $P(77 < \overline{X} < 79.5)$.

(b) $P(74.2 < \overline{X} < 78.4)$.

5.3.2 Let X be $N(50, 36)$. Using the same set of axes, sketch the graphs of the probability density functions of

(a) X.

(b) $\overline{X}$, the mean of a random sample of size 9 from this distribution.

(c) $\overline{X}$, the mean of a random sample of size 36 from this distribution.

5.3.3 Let X equal the widest diameter (in millimeters) of the fetal head measured between the 16th and 25th weeks of pregnancy. Assume that the distribution of X is $N(46.58, 40.96)$. Let $\overline{X}$ be the sample mean of a random sample of $n = 16$ observations of X.

(a) Give the values of $E(\overline{X})$ and $\text{Var}(\overline{X})$.

(b) Find $P(44.42 \leq \overline{X} \leq 48.98)$.

5.3.4 Let X equal the weight of the soap in a "6-pound" box. Assume that the distribution of X is $N(6.05, 0.0004)$.

(a) Find $P(X < 6.0171)$.

(b) If nine boxes of soap are selected at random from the production line, find the probability that at most two boxes weigh less than 6.0171 each.

(c) Let $\overline{X}$ be the sample mean of the nine boxes. Find $P(\overline{X} \leq 6.035)$.

5.3.5 Let $Z_1, Z_2, \ldots, Z_7$ be a random sample from the standard normal distribution $N(0, 1)$. Let $W = Z_1^2 + Z_2^2 + \cdots + Z_7^2$. Find $P(1.69 < W < 14.07)$.

5.3.6 If $X_1, X_2, \ldots, X_{16}$ is a random sample of size $n = 16$ from the normal distribution $N(50, 100)$, determine

(a) $P\left(796.2 \leq \sum_{i=1}^{16} (X_i - 50)^2 \leq 2630\right).$

(b) $P\left(726.1 \leq \sum_{i=1}^{16} (X_i - \overline{X})^2 \leq 2500\right).$

5.3.7 Let X equal the weight (in grams) of a nail of the type that is used for making decks. Assume that the distribution of X is $N(8.78, 0.16)$. Let $\overline{X}$ be the mean of a random sample of the weights of $n = 9$ nails.

(a) Sketch, on the same set of axes, the graphs of the p.d.f.'s of X and of $\overline{X}$.

(b) Let S^2 be the sample variance of the 9 weights. Find constants a and b so that $P(a \leq S^2 \leq b) = 0.90$. HINT: That probability is equivalent to $P(8a/0.16 \leq 8S^2/0.16 \leq 8b/0.16)$ and $8S^2/0.16$ is $\chi^2(8)$. Find $8a/0.16$ and $8b/0.16$ from the tables.

5.3.8 At a heat-treating company, iron castings and steel forgings are heat-treated to achieve desired mechanical properties and machinability. One steel forging is annealed to soften the part for each machining. Two lots of this part, made of 1020 steel, are heat-treated in two different furnaces. The specification for this part is 36–66 on the Rockwell G scale. Let X_1 and X_2 equal the respective hardness measurements for parts selected randomly from furnaces 1 and 2. Assume that the distributions of X_1 and X_2 are $N(47.88, 2.19)$ and $N(43.04, 14.89)$, respectively.

(a) Sketch the p.d.f.'s of X_1 and X_2 on the same graph.

(b) Compute $P(X_1 > X_2)$, assuming independence of X_1 and X_2.

5.3.9 Let X equal the force required to pull a stud out of a window that is to be inserted into an automobile. Assume that the distribution of X is $N(147.8, 12.3^2)$.

(a) Find $P(X < 163.3)$.

(b) If $\overline{X}$ is the mean and S^2 is the variance of a random sample of size $n = 25$ from this distribution of X, determine $P(\overline{X} \leq 150.9)$.

(c) Find constants a and b so that $P(a \leq S^2 \leq b) = 0.90$. (See the hint in Exercise 5.3.7.)

5.3.10 Suppose that the distribution of the weight of a prepackaged "1-pound" bag of carrots is $N(1.18, 0.07^2)$ and the distribution of the weight of a prepackaged "3-pound" bag of carrots is $N(3.22, 0.09^2)$. Selecting bags at random, find the probability that the sum of three 1-pound bags exceeds the weight of one 3-pound bag. HINT: First determine the distribution of Y, the sum of the three, and then compute $P(Y > W)$, where W is the weight of the 3-pound bag.

5.3.11 Let X denote the wing length in millimeters of a male gallinule and Y the wing length in millimeters of a female gallinule. Assume that X is $N(184.09, 39.37)$ and Y is $N(171.93, 50.88)$, and that X and Y are independent. If a male and a female gallinule are captured, what is the probability that X is greater than Y?

5.3.12 Suppose that for a particular population of students SAT mathematics scores are $N(529, 5732)$ and SAT verbal scores are $N(474, 6368)$. Select two students at random, and let X equal the first student's math score and Y the second student's verbal score. Find $P(X > Y)$.

5.3.13 Suppose that the length of life in hours, say X, of a light bulb manufactured by company A is $N(800, 14,400)$ and the length of life in hours, say Y, of a light bulb manufactured by company B is $N(850, 2500)$. One bulb is selected from each company and is burned until "death."

(a) Find the probability that the length of life of the bulb from company A exceeds the length of life of the bulb from company B by at least 15 hours.

(b) Find the probability that at least one of the bulbs "lives" for at least 920 hours.

5.3.14 Let X and Y equal the number of miles per gallon for compact cars and midsized cars, respectively, as reported in fuel economy ratings. Assume that $\mu_X = 24.5$, $\sigma_X = 3.8$, $\mu_Y = 21.3$, and $\sigma_Y = 2.7$. Let $\overline{X}$ and $\overline{Y}$ be the sample means of independent random samples of eight observations of X and Y, respectively.

(a) What are the values of the means and variances of $\overline{X}$ and $\overline{Y}$?

(b) Assuming that $\overline{X}$ and $\overline{Y}$ are each (approximately) normally distributed, how is $\overline{X} - \overline{Y}$ distributed?

(c) Find the (approximate) probability, $P(\overline{X} > \overline{Y})$.

5.3.15 Let X equal the weight of a fat-free Fig Newton cookie. Assume that the distribution of X is $N(14.22, 0.0854)$. These cookies are sold in packages that have a label weight of 340 grams. The number of cookies in a package is usually 24, 25, or 26. Assuming that a package is filled with a random sample of cookies, how many cookies should be put into a package to be quite certain (say with a probability of at least 0.95) that the total weight of the cookies exceeds 340 grams? Also keep in mind that having extra cookies in a package decreases profit.

5.4 THE CENTRAL LIMIT THEOREM

In Section 5.2 we found that the mean $\overline{X}$ of a random sample of size n from a distribution with mean μ and variance $\sigma^2 > 0$ is a random variable with the properties that

$$E(\overline{X}) = \mu \quad \text{and} \quad \text{Var}(\overline{X}) = \frac{\sigma^2}{n}.$$

Thus, as n increases, the variance of $\overline{X}$ decreases. Consequently, the distribution of $\overline{X}$ clearly depends on n, and we see that we are dealing with sequences of distributions.

In Theorem 5.3.1 we considered the p.d.f. of $\overline{X}$ when sampling from the normal distribution $N(\mu, \sigma^2)$. We showed that the distribution of $\overline{X}$ is $N(\mu, \sigma^2/n)$, and in Figure 5.1, by graphing the p.d.f.'s for several values of n, we illustrated that as n increases, the probability becomes concentrated in a small interval centered at μ. That is, as n increases, $\overline{X}$ tends to converge to μ, or $(\overline{X} - \mu)$ tends to converge to 0 in some probability sense.

In general, if we let

$$W = \frac{\sqrt{n}}{\sigma}(\overline{X} - \mu) = \frac{\overline{X} - \mu}{\sigma/\sqrt{n}},$$

where $\overline{X}$ is the mean of a random sample of size n from some distribution with mean μ and variance σ^2, then for each positive integer n,

$$E(W) = E\left[\frac{\overline{X} - \mu}{\sigma/\sqrt{n}}\right] = \frac{E(\overline{X}) - \mu}{\sigma/\sqrt{n}} = \frac{\mu - \mu}{\sigma/\sqrt{n}} = 0$$

and

$$\text{Var}(W) = E(W^2) = E\left[\frac{(\overline{X} - \mu)^2}{\sigma^2/n}\right] = \frac{E[(\overline{X} - \mu)^2]}{\sigma^2/n} = \frac{\sigma^2/n}{\sigma^2/n} = 1.$$

Thus, while $\overline{X} - \mu$ "degenerates" to zero, the factor $\sqrt{n}/\sigma$ in $\sqrt{n}(\overline{X} - \mu)/\sigma$ "spreads out" the probability enough to prevent this degeneration. What, then, is the distribution of W as n increases? One observation that might shed some light on the answer to this question can be made immediately. If the sample arises from a normal distribution, then, from Theorem 5.3.1, we know that $\overline{X}$ is $N(\mu, \sigma^2/n)$ and hence W is $N(0, 1)$ for each positive n. Thus, in the limit, the distribution of W must be $N(0, 1)$. So if the solution of the question does not depend on the underlying distribution (i.e., it is unique), the answer must be $N(0, 1)$. As we will see, this is exactly the case, and this result is so important it is called the central limit theorem, the proof of which is given in Section 5.6.

Theorem 5.4.1.   (Central Limit Theorem) If $\overline{X}$ is the mean of a random sample $X_1, X_2, \ldots, X_n$ of size n from a distribution with a finite mean μ and a finite positive variance σ^2, then the distribution of

$$W = \frac{\overline{X} - \mu}{\sigma/\sqrt{n}} = \frac{\sum\limits_{i=1}^{n} X_i - n\mu}{\sqrt{n}\,\sigma}$$

is $N(0, 1)$ in the limit as $n \to \infty$.

A practical use of the central limit theorem is approximating, when n is "sufficiently large," the distribution function of W, namely

$$P(W \le w) \approx \int_{-\infty}^{w} \frac{1}{\sqrt{2\pi}} e^{-z^2/2}\, dz = \Phi(w).$$

We present some illustrations of this application, discuss "sufficiently large," and try to give an intuitive feeling for the central limit theorem.

Example 5.4.1

Let $\overline{X}$ denote the mean of a random sample of size $n = 15$ from the distribution whose p.d.f. is $f(x) = (3/2)x^2$, $-1 < x < 1$. Here $\mu = 0$ and $\sigma^2 = 3/5$. Thus

$$P(0.03 \leq \overline{X} \leq 0.15) = P\left(\frac{0.03 - 0}{\sqrt{3/5}/\sqrt{15}} \leq \frac{\overline{X} - 0}{\sqrt{3/5}/\sqrt{15}} \leq \frac{0.15 - 0}{\sqrt{3/5}/\sqrt{15}}\right)$$

$$= P(0.15 \leq W \leq 0.75)$$

$$\approx \Phi(0.75) - \Phi(0.15)$$

$$= 0.7734 - 0.5596$$

$$= 0.2138. \qquad \diamond$$

Example 5.4.2

Let $X_1, X_2, \ldots, X_{20}$ denote a random sample of size 20 from the uniform distribution $U(0, 1)$. Here $E(X_i) = 1/2$ and $\text{Var}(X_i) = 1/12$, for $i = 1, 2, \ldots, 20$. If $Y = X_1 + X_2 + \cdots + X_{20}$, then

$$P(Y \leq 9.1) = P\left(\frac{Y - 20(1/2)}{\sqrt{20/12}} \leq \frac{9.1 - 10}{\sqrt{20/12}}\right) = P(W \leq -0.697)$$

$$\approx \Phi(-0.697)$$

$$= 0.2423.$$

Also,

$$P(8.5 \leq Y \leq 11.7) = P\left(\frac{8.5 - 10}{\sqrt{5/3}} \leq \frac{Y - 10}{\sqrt{5/3}} \leq \frac{11.7 - 10}{\sqrt{5/3}}\right)$$

$$= P(-1.162 \leq W \leq 1.317)$$

$$\approx \Phi(1.317) - \Phi(-1.162)$$

$$= 0.9061 - 0.1226$$

$$= 0.7835. \qquad \diamond$$

Example 5.4.3

Let $\overline{X}$ denote the mean of a random sample of size 25 from the distribution whose p.d.f. is $f(x) = x^3/4$, $0 < x < 2$. It is easy to show that $\mu = 8/5 = 1.6$ and $\sigma^2 = 8/75$. Thus

$$P(1.5 \leq \overline{X} \leq 1.65) = P\left(\frac{1.5 - 1.6}{\sqrt{8/75}/\sqrt{25}} \leq \frac{\overline{X} - 1.6}{\sqrt{8/75}/\sqrt{25}} \leq \frac{1.65 - 1.6}{\sqrt{8/75}/\sqrt{25}}\right)$$

$$= P(-1.531 \leq W \leq 0.765)$$

$$\approx \Phi(0.765) - \Phi(-1.531)$$

$$= 0.7779 - 0.0629$$

$$= 0.7150. \qquad \diamond$$

These examples have shown how the central limit theorem can be used for approximating certain probabilities concerning the mean $\overline{X}$ or the sum $Y = \sum_{i=1}^{n} X_i$ of a random sample. That is, $\overline{X}$ is approximately $N(\mu, \sigma^2/n)$, and Y is approximately $N(n\mu, n\sigma^2)$ when n is "sufficiently large," where μ and σ^2 are the mean and the variance of the underlying distribution from which the sample arose. Generally, if n is greater than 25 or 30, these approximations will be good. However, if the underlying distribution is symmetric, unimodal, and of the continuous type, a value of n as small as 4 or 5 can yield a very adequate approximation. Moreover, if the original distribution is approximately normal, $\overline{X}$ would have a distribution very close to normal when n equals 2 or 3. In fact, we know that if the sample is taken from $N(\mu, \sigma^2)$, $\overline{X}$ is exactly $N(\mu, \sigma^2/n)$ for every $n = 1, 2, 3, \dots$.

The following examples will help to illustrate the previous remarks and will give the reader a better intuitive feeling about the central limit theorem. In particular, we shall see how the size of n affects the distribution of $\overline{X}$ and $Y = \sum_{i=1}^{n} X_i$ for samples from several underlying distributions.

Example 5.4.4

Let X_1, X_2, X_3, X_4 be a random sample of size 4 from the uniform distribution $U(0, 1)$ with p.d.f. $f(x) = 1, 0 < x < 1$. Then $\mu = 1/2$ and $\sigma^2 = 1/12$. We shall compare the graph of the p.d.f. of

$$Y = \sum_{i=1}^{n} X_i$$

with the graph of the $N[n(1/2), n(1/12)]$ p.d.f. for $n = 2$ and 4, respectively. By methods given later (Section 11.6), we can determine that the p.d.f. of $Y = X_1 + X_2$ is

$$g(y) = \begin{cases} y, & 0 < y \le 1, \\ 2 - y, & 1 < y < 2. \end{cases}$$

This is the triangular p.d.f. that is graphed in Figure 5.2(a). In this figure the $N[2(1/2), 2(1/12)]$ p.d.f. is also graphed.

Moreover, the p.d.f. of $Y = X_1 + X_2 + X_3 + X_4$ is

$$g(y) = \begin{cases} \dfrac{y^3}{6}, & 0 \le y < 1, \\[2mm] \dfrac{-3y^3 + 12y^2 - 12y + 4}{6}, & 1 \le y < 2, \\[2mm] \dfrac{3y^3 - 24y^2 + 60y - 44}{6}, & 2 \le y < 3, \\[2mm] \dfrac{-y^3 + 12y^2 - 48y + 64}{6}, & 3 \le y \le 4. \end{cases}$$

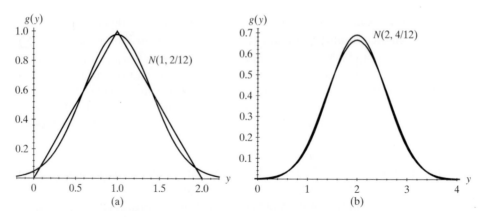

Figure 5.2 P.d.f.'s of sums of uniform random variables

This p.d.f. is graphed in Figure 5.2(b) along with the $N[4(1/2), 4(1/12)]$ p.d.f. If we are interested in finding $P(1.7 \le Y \le 3.2)$, this could be done by evaluating

$$\int_{1.7}^{3.2} g(y)\,dy,$$

which is tedious (see Exercise 5.4.10). It is much easier to use a normal approximation, which results in a number very close to the exact value. ◇

In Example 5.4.4 and Exercise 5.4.10 we show that even for a small value of n, such as $n = 4$, the sum of the sample items has an approximate normal distribution. The following example illustrates that for some underlying distributions (particularly skewed ones) n must be quite large to obtain a satisfactory approximation. In order to keep the scale on the horizontal axis the same for each value of n, we will use the following result.

Let $f(x)$ and $F(x)$ be the p.d.f. and distribution function of a random variable, X, of the continuous type having mean μ and variance σ^2. Let $W = (X - \mu)/\sigma$. The distribution function of W is given by

$$G(w) = P(W \le w) = P\left(\frac{X - \mu}{\sigma} \le w\right)$$

$$= P(X \le \sigma w + \mu) = F(\sigma w + \mu).$$

Thus the p.d.f. of W is given by

$$g(w) = F'(\sigma w + \mu) = \sigma f(\sigma w + \mu).$$

Example 5.4.5

Let $X_1, X_2, \ldots, X_n$ be a random sample of size n from a chi-square distribution with one degree of freedom. If

$$Y = \sum_{i=1}^{n} X_i,$$

then Y is $\chi^2(n)$, and $E(Y) = n$, $\mathrm{Var}(Y) = 2n$. Let

$$W = \frac{Y - n}{\sqrt{2n}}.$$

The p.d.f. of W is given by

$$g(w) = \sqrt{2n}\,\frac{(\sqrt{2n}\,w + n)^{n/2-1}}{\Gamma\!\left(\dfrac{n}{2}\right)2^{n/2}}\,e^{-(\sqrt{2n}\,w+n)/2}, \qquad -n/\sqrt{2n} < w < \infty.$$

Note that $w > -n/\sqrt{2n}$ corresponds to $y > 0$. In Figure 5.3(a) and (b), the graph of W is given along with the $N(0, 1)$ p.d.f. for $n = 20$ and 100, respectively.   ◇

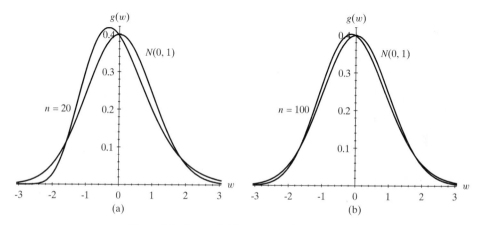

Figure 5.3   Sums of chi-square random variables

In order to have an intuitive feeling about how the sample size n affects the distribution of $W = (\overline{X} - \mu)/(\sigma/\sqrt{n})$, it is helpful to simulate values of W on a computer using different values of n and different underlying distributions. The next example illustrates this.

Remark   We simulate observations from a distribution of X having a continuous-type distribution function $F(x)$ as follows. Suppose that $F(a) = 0$, $F(b) = 1$, and $F(x)$ is strictly increasing for $a < x < b$. Let $Y = F(x)$ and let the distribution of Y be $U(0, 1)$. If y is an observed value of Y, then $x = F^{-1}(y)$ is an observed value of X (see Section 4.5). Thus, if y is the value of a computer-generated random number, then $x = F^{-1}(y)$ is the simulated value of X.

Example 5.4.6

Finding the exact distribution of the random variable $W = (\overline{X} - \mu)/(\sigma/\sqrt{n})$ is often difficult. However, let us consider some empirical evidence about the distribution of W by simulating random samples on the computer. Let $X_1, X_2, \ldots, X_n$ denote a random sample of size n from the distribution with p.d.f. $f(x)$, distribution function

$F(x)$, mean μ, and variance σ^2. We shall generate 1000 random samples of size n from this distribution and compute a value of W for each sample, thus obtaining 1000 observed values of W. A histogram of these 1000 values is constructed using 21 intervals of equal length. We depict the results of this experiment as follows:

(a) In Figure 5.4, $f(x) = (x + 1)/2$ and $F(x) = (x + 1)^2/4$ for $-1 < x < 1$; $\mu = 1/3$, $\sigma^2 = 2/9$; and $n = 2$ and 12.

(b) In Figure 5.5, $f(x) = (3/2)x^2$ and $F(x) = (x^3 + 1)/2$ for $-1 < x < 1$; $\mu = 0$, $\sigma^2 = 3/5$; and $n = 2$ and 12. [Sketch the graph of $y = f(x)$ and give an intuitive argument as to why the histogram for $n = 2$ looks the way it does.]

The $N(0, 1)$ p.d.f. has been superimposed on each histogram.   ◇

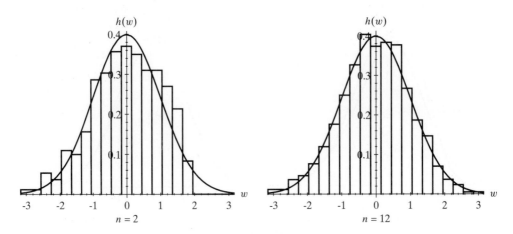

Figure 5.4   Sum of triangular random variables

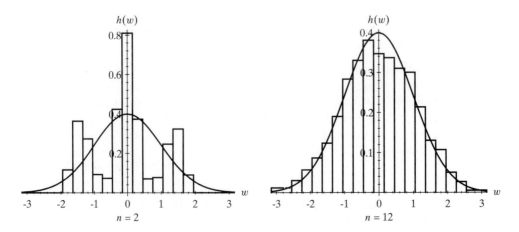

Figure 5.5   Sum of U-shaped random variables

Note very clearly that these examples have not proved anything. They are presented to give evidence of the truth of the central limit theorem.

So far all the illustrations have concerned distributions of the continuous type. However, the hypotheses for the central limit theorem do not require the distribution to be continuous. We consider applications of the central limit theorem for discrete-type distributions in the next section.

EXERCISES

5.4.1 Let $\overline{X}$ be the mean of a random sample of size 12 from the uniform distribution on the interval $(0, 1)$. Approximate $P(1/2 \leq \overline{X} \leq 2/3)$.

5.4.2 Let $Y = X_1 + X_2 + \cdots + X_{15}$ be the sum of a random sample of size 15 from the distribution whose p.d.f. is $f(x) = (3/2)x^2$, $-1 < x < 1$. Approximate $P(-0.3 \leq Y \leq 1.5)$.

5.4.3 Let $\overline{X}$ be the mean of a random sample of size 36 from an exponential distribution with mean 3. Approximate $P(2.5 \leq \overline{X} \leq 4)$.

5.4.4 Approximate $P(39.75 \leq \overline{X} \leq 41.25)$, where $\overline{X}$ is the mean of a random sample of size 32 from a distribution with mean $\mu = 40$ and variance $\sigma^2 = 8$.

5.4.5 Let $X_1, X_2, \ldots, X_{18}$ be a random sample of size 18 from a chi-square distribution with $r = 1$. Recall that $\mu = 1$, $\sigma^2 = 2$.

(a) How is $Y = \sum_{i=1}^{18} X_i$ distributed?

(b) Using the result of part (a), we see from Table IV that

$$P(Y \leq 9.390) = 0.05 \quad \text{and} \quad P(Y \leq 34.80) = 0.99.$$

Compare these two probabilities with the approximations found using the central limit theorem.

5.4.6 A random sample of size $n = 18$ is taken from the distribution with p.d.f $f(x) = 1 - x/2$, $0 \leq x \leq 2$.

(a) Find μ and σ^2.

(b) Find, approximately, $P(2/3 \leq \overline{X} \leq 5/6)$.

5.4.7 Let X equal the maximal oxygen intake of a human on a treadmill, where the measurements are in milliliters of oxygen per minute per kilogram of weight. Assume that for a particular population the mean of X is $\mu = 54.030$ and the standard deviation is $\sigma = 5.8$. Let $\overline{X}$ be the sample mean of a random sample of size $n = 47$. Find $P(52.761 \leq \overline{X} \leq 54.453)$, approximately.

5.4.8 Let X equal the weight in grams of a miniature candy bar. Assume that $\mu = E(X) = 24.43$ and $\sigma^2 = \text{Var}(X) = 2.20$. Let $\overline{X}$ be the sample mean of a random sample of $n = 30$ candy bars. Find

(a) $E(\overline{X})$.

(b) $\text{Var}(\overline{X})$.

(c) $P(24.17 \leq \overline{X} \leq 24.82)$, approximately.

5.4.9 Let X equal the birth weight in grams of a baby born in the Sudan. Assume that $E(X) = 3320$ and $\mathrm{Var}(X) = 660^2$. Let $\overline{X}$ be the sample mean of a random sample of size $n = 225$. Find $P(3233.76 \le \overline{X} \le 3406.24)$, approximately.

5.4.10 In Example 5.4.4, with $n = 4$, compute $P(1.7 \le Y \le 3.2)$ and compare this answer with the normal approximation of this probability.

5.4.11 Five measurements in "ohms per square" of the electronic conductive coating (that allows light to pass through it) on a thin, clear piece of glass were made and the average was calculated. This was repeated 150 times, yielding the following averages:

83.86	75.86	79.65	90.57	95.37	97.97	77.00	80.80	83.53	83.17
80.20	81.42	89.14	88.68	85.15	90.11	89.03	85.00	82.57	79.46
81.20	82.80	81.28	74.64	69.85	75.60	76.60	78.30	88.51	86.32
84.03	95.27	90.05	77.50	75.14	81.33	86.12	78.30	80.30	84.96
79.30	88.96	82.76	83.13	79.60	86.20	85.16	87.86	91.00	91.10
84.04	92.10	79.20	83.76	87.87	86.71	81.89	85.72	75.84	74.27
93.08	81.75	75.66	75.35	76.55	84.86	90.68	91.02	90.97	98.30
91.84	97.41	73.60	90.65	80.20	74.75	90.35	79.66	86.88	83.00
86.24	80.50	74.25	91.20	70.16	78.40	85.60	80.82	75.95	80.75
81.86	82.18	82.98	84.00	76.85	85.00	79.50	86.56	83.30	72.40
79.20	86.20	82.36	84.08	86.11	88.25	88.93	93.12	78.30	77.24
82.52	81.37	83.72	86.90	84.37	92.60	95.01	78.95	81.40	88.40
76.10	85.33	82.95	80.20	88.21	83.49	81.00	82.72	81.12	83.62
91.18	85.90	79.01	77.56	81.13	80.60	81.65	70.70	69.36	79.09
71.35	67.20	67.43	69.95	66.76	76.35	69.45	80.13	84.26	88.13

(a) Construct a frequency table for these 150 observations using 10 intervals of equal length and using $(66.495, 69.695)$ for the first class interval. Which is the modal class?

(b) Construct a relative frequency histogram for the grouped data.

(c) Superimpose a normal p.d.f. using the mean 82.661 and variance 42.2279 of these 150 values as the true mean and variance. Do these 150 means of five measurements seem to be normally distributed?

5.5 APPROXIMATIONS FOR DISCRETE DISTRIBUTIONS

In this section we illustrate how the normal distribution can be used to approximate probabilities for certain discrete-type distributions. One of the most important discrete distributions is the binomial distribution. To see how the central limit theorem can be applied, recall that a binomial random variable can be described as the sum of Bernoulli random variables. That is, let $X_1, X_2, \ldots, X_n$ be a random sample from a Bernoulli distribution with a mean $\mu = p$ and a variance $\sigma^2 = p(1 - p)$, where $0 < p < 1$. Then $Y = \sum_{i=1}^{n} X_i$ is $b(n, p)$. The central limit theorem states that the distribution of

$$W = \frac{Y - np}{\sqrt{np(1 - p)}} = \frac{\overline{X} - p}{\sqrt{p(1 - p)/n}}$$

is $N(0, 1)$ in the limit as $n \to \infty$. Thus, if n is sufficiently large, the distribution of Y is approximately $N[np, np(1 - p)]$, and probabilities for the binomial distribution $b(n, p)$ can be approximated using this normal distribution. A rule often stated is that n is "sufficiently large" if $np \geq 5$ and $n(1 - p) \geq 5$; and it can be used as a rough guide, although as p deviates more and more from 0.5, we need larger and larger sample sizes.

Note that we shall be approximating probabilities for a discrete distribution with probabilities for a continuous distribution. Let us discuss a reasonable proce-dure in this situation. If V is $N(\mu, \sigma^2)$, $P(a < V < b)$ is equivalent to the area bounded by the p.d.f. of V, the v-axis, $v = a$, and $v = b$. If Y is $b(n, p)$, recall that the probability histogram for Y was defined as follows. For each y such that $k - 1/2 < y < k + 1/2$, let

$$f(y) = \frac{n!}{k!(n - k)!} p(1 - p), \qquad k = 0, 1, 2, \ldots, n.$$

Then $P(Y = k)$ can be represented by the area of the rectangle with a height of $P(Y = k)$ and a base of length 1 centered at k. Figure 5.6 shows the graph of the probability histogram for the binomial distribution $b(4, 1/4)$. When using the normal distribution to approximate probabilities for the binomial distribution, areas under the p.d.f. for the normal distribution will be used to approximate areas of rectangles in the probability histogram for the binomial distribution.

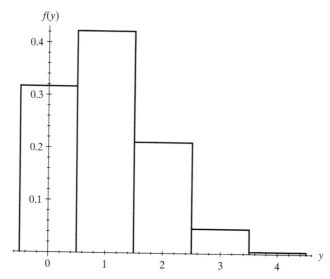

Figure 5.6   Probability histogram for $b(4, 1/4)$

Example 5.5.1

Let Y be $b(10, 1/2)$. Then, using the central limit theorem, $P(a < Y < b)$ can be approximated using the normal distribution with mean $10(1/2) = 5$ and variance $10(1/2)(1/2) = 5/2$. Figure 5.7 shows the graph of the probability histogram for

$b(10, 1/2)$ and the graph of the p.d.f. of the normal distribution $N(5, 5/2)$. Note that the area of the rectangle whose base is

$$\left(k - \frac{1}{2}, k + \frac{1}{2}\right)$$

and the area under the normal curve between $k - 1/2$ and $k + 1/2$ are approximately equal for each integer k.   ◇

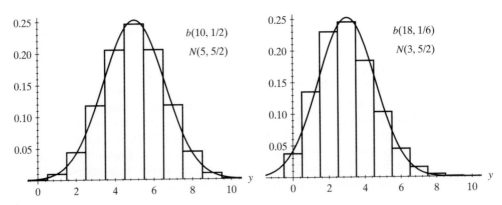

Figure 5.7   Normal approximation for the binomial distribution

Example 5.5.2

Let Y be $b(18, 1/6)$. Because $np = 18(1/6) = 3 < 5$, the normal approximation is not as good here. Figure 5.7 illustrates this by depicting the skewed probability histogram for the $b(18, 1/6)$ distribution and the symmetric p.d.f. of the normal distribution $N(3, 5/2)$.   ◇

Example 5.5.3

Let Y have the binomial distribution of Example 5.5.1 and Figure 5.7, namely $b(10, 1/2)$. Then

$$P(3 \leq Y < 6) = P(2.5 \leq Y \leq 5.5)$$

because $P(Y = 6)$ is not in the desired answer. But the latter equals

$$P\left(\frac{2.5 - 5}{\sqrt{10/4}} \leq \frac{Y - 5}{\sqrt{10/4}} \leq \frac{5.5 - 5}{\sqrt{10/4}}\right) \approx \Phi(0.316) - \Phi(-1.581)$$

$$= 0.6240 - 0.0570 = 0.5670.$$

Using Table II, we find that $P(3 \leq Y < 6) = 0.5683$.   ◇

Example 5.5.4

Let Y be $b(36, 1/2)$. Then, since $(36)(1/2) = 18$ and $(36)(1/2)(1/2) = 9$,

$$P(12 < Y \leq 18) = P(12.5 \leq Y \leq 18.5)$$

$$= P\left(\frac{12.5 - 18}{\sqrt{9}} \leq \frac{Y - 18}{\sqrt{9}} \leq \frac{18.5 - 18}{\sqrt{9}}\right)$$

$$\approx \Phi(0.167) - \Phi(-1.833)$$

$$= 0.5329.$$

Note that 12 was increased to 12.5 because $P(Y = 12)$ is not included in the desired probability. Using the binomial formula, we find that (you may verify this answer using your calculator)

$$P(12 < Y \leq 18) = P(13 \leq Y \leq 18) = 0.5334.$$

Also,

$$P(Y = 20) = P(19.5 \leq Y \leq 20.5)$$

$$= P\left(\frac{19.5 - 18}{\sqrt{9}} \leq \frac{Y - 18}{\sqrt{9}} \leq \frac{20.5 - 18}{\sqrt{9}}\right)$$

$$\approx \Phi(0.833) - \Phi(0.5)$$

$$= 0.1060.$$

Using the binomial formula, we have $P(Y = 20) = 0.1063$. So, in this situation, the approximation is extremely good.   ◇

Note that, in general, if Y is $b(n, p)$, n is large, and $k = 0, 1, \ldots, n$,

$$P(Y \leq k) \approx \Phi\left(\frac{k + 1/2 - np}{\sqrt{npq}}\right)$$

and

$$P(Y < k) \approx \Phi\left(\frac{k - 1/2 - np}{\sqrt{npq}}\right),$$

because in the first case k is included and in the second it is not.

We now show how the Poisson distribution with large enough mean can be approximated using a normal distribution.

Example 5.5.5

A random variable having a Poisson distribution with mean 20 can be thought of as the sum Y of the observations of a random sample of size 20 from a Poisson distribution with mean 1. Thus

$$W = \frac{Y - 20}{\sqrt{20}}$$

has a distribution that is approximately $N(0, 1)$, and the distribution of Y is approximately $N(20, 20)$ (see Figure 5.8). So, for illustration,

$$P(16 < Y \le 21) = P(16.5 \le Y \le 21.5)$$

$$= P\left(\frac{16.5 - 20}{\sqrt{20}} \le \frac{Y - 20}{\sqrt{20}} \le \frac{21.5 - 20}{\sqrt{20}}\right)$$

$$\approx \Phi(0.335) - \Phi(-0.783)$$

$$= 0.4142.$$

Note that 16 is increased to 16.5 because $Y = 16$ is not included in the event $\{16 < Y \le 21\}$. The answer using the Poisson formula is 0.4226.   ◇

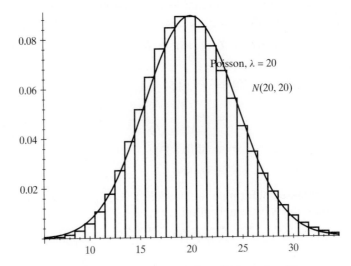

Figure 5.8   Normal approximation of Poisson, $\lambda = 20$

In general, if Y has a Poisson distribution with mean λ, then the distribution of

$$W = \frac{Y - \lambda}{\sqrt{\lambda}}$$

is approximately $N(0, 1)$ when λ is sufficiently large.

EXERCISES

5.5.1 Let the distribution of Y be $b(25, 1/2)$. Find the following probabilities in two ways: exactly using Table II and approximately using the central limit theorem. Compare the two results in each of the three cases.

(a) $P(10 < Y \le 12)$.

(b) $P(12 \le Y < 15)$.

(c) $P(Y = 12)$.

5.5.2 Among the gifted seventh-graders who score very high on a mathematics exam, approximately 20% are left-handed or ambidextrous. Let X equal the number of left-handed or ambidextrous students among a random sample of $n = 25$ gifted seventh-graders. Find $P(2 < X < 9)$

(a) Using Table II.

(b) Approximately, using the central limit theorem. NOTE: Since X has a skewed distribution, the approximation is not as good as for the symmetrical distribution where $p = 0.50$ even though $np = 5$.

5.5.3 A public opinion poll in southern California was conducted to determine whether southern Californians are prepared for the big earthquake that experts predict will devastate the region sometime in the next 50 years. It was learned that "60% have not secured objects in their homes that might fall and cause injury and damage during a temblor." In a random sample of $n = 864$ southern Californians, let X equal the number who "have not secured objects in their homes." Find $P(496 \le X \le 548)$, approximately.

5.5.4 Let X equal the number out of $n = 48$ mature aster seeds that will germinate when $p = 0.75$ is the probability that a particular seed germinates. Determine $P(35 \le X \le 40)$, approximately.

5.5.5 Let p equal the proportion of all college students who would say yes to the question, "Would you drink from the same glass as your friend if you suspected that this friend were an AIDS virus carrier?" Assume that $p = 0.10$. Let X equal the number of students out of a random sample of size $n = 100$ who would say yes to this question. Compute $P(X \le 11)$, approximately. NOTE: The value used for p was based on a poll conducted in a class at San Diego State University. It would be interesting for you to conduct a survey at your college to estimate the value of p.

5.5.6 In adults the pneumococcus bacterium causes 70% of pneumonia cases. In a random sample of $n = 84$ adults who have pneumonia, let X equal the number whose pneumonia was caused by the pneumococcus bacterium. Use the normal distribution to find, approximately, $P(X \le 52)$.

5.5.7 Let $X_1, X_2, \ldots, X_{48}$ be a random sample of size 48 from the distribution with p.d.f. $f(x) = 1/x^2, 1 < x < \infty$. Approximate the probability that at most 10 of these random variables have values greater than 4. HINT: Let the ith trial be a success if $X_i > 4, i = 1, 2, \ldots, 48$.

5.5.8 A candymaker produces mints that have a label weight of 20.4 grams. Assume that the distribution of the weights of these mints is $N(21.37, 0.16)$.

(a) Let X denote the weight of a single mint selected at random from the production line. Find $P(X < 20.857)$.

(b) During a particular shift 100 mints are selected at random and weighed. Let Y equal the number of these mints that weigh less than 20.857 grams. Find approximately $P(Y \le 5)$.

(c) Let $\overline{X}$ equal the sample mean of the 100 mints selected and weighed on a particular shift. Find $P(21.31 \le \overline{X} \le 21.39)$.

5.5.9 Let X equal the number of alpha particles emitted by barium–133 per second and counted by a Geiger counter. Assume that X has a Poisson distribution with $\lambda = 49$. Approximate $P(45 < X < 60)$.

5.5.10 Let X equal the number of alpha particles counted by a Geiger counter during 30 seconds. Assume that the distribution of X is Poisson with a mean of 4829. Determine approximately $P(4776 \le X \le 4857)$.

5.5.11 Let $X_1, X_2, \ldots, X_{30}$ be a random sample of size 30 from a Poisson distribution with a mean of 2/3. Approximate

(a) $P\left(15 < \sum_{i=1}^{30} X_i \le 22\right).$

(b) $P\left(21 \le \sum_{i=1}^{30} X_i < 27\right).$

5.5.12 In roulette, the probability of winning with a bet on red is $p = 18/38$. Let Y equal the number of winning bets out of 1000 independent bets that are placed. Find $P(Y > 500)$ approximately.

5.5.13 Let X denote the payoff for \$1 bets in the game Chuck-a-Luck. Then X can equal -1, 1, 2, or 3, such that $E(X) = -17/216$ and $\text{Var}(X) = 57,815/46,656 = 1.2392$. Given 300 independent observations of X, say $X_1, X_2, \ldots, X_{300}$, let $Y = \sum_{i=1}^{300} X_i$. Then Y represents the amount won after 300 bets. Approximate

(a) $P(Y \le -21).$

(b) $P(Y \ge 21).$

5.5.14 If X is $b(100, 0.1)$, find the approximate value of $P(12 \le X \le 14)$ using

(a) The normal approximation.

(b) The Poisson approximation.

(c) The binomial p.d.f.

5.5.15 Let $X_1, X_2, \ldots, X_{36}$ be a random sample of size 36 from the geometric distribution with p.d.f. $f(x) = (1/4)^{x-1}(3/4)$, $x = 1, 2, 3, \ldots.$ Approximate

(a) $P\left(46 \le \sum_{i=1}^{36} X_i \le 49\right).$

(b) $P(1.25 \le \overline{X} \le 1.50).$

HINT: Observe that the distribution of the sum is of the discrete type.

5.5.16 A die is rolled 24 independent times. Let Y be the sum of the 24 resulting values. Recalling that Y is a random variable of the discrete type, approximate

(a) $P(Y \ge 86).$

(b) $P(Y < 86).$

(c) $P(70 < Y \le 86).$

5.5.17 A student played the board game Pinocchio with his friend's nephews. In this game, to try to escape the great white whale, a player rolls a fair six-sided die 20 times, adding together these 20 outcomes. To escape the whale, the player must receive a score of at least 75. For a score of less than 75 the player is "eaten by the whale." Find, approximately, the probability of escaping the whale.

5.5.18 Let Y equal the sum of $n = 100$ Bernoulli trials. That is, Y is $b(100, p)$. For each of (i) $p = 0.1$, (ii) $p = 0.5$, and (iii) $p = 0.8$,

(a) Draw on the same graph the approximating normal p.d.f.'s.

(b) Find, approximately, $P(\,|\,Y/100 - p\,| \le 0.015).$

*5.6 LIMITING MOMENT-GENERATING FUNCTIONS

We would like to begin this section by showing that the binomial distribution can be approximated by the Poisson distribution when n is sufficiently large and p fairly small. Of course, this was proved in Section 3.7 by showing that under these conditions, the binomial p.d.f. is close to that of the Poisson. Here, however, we show that the binomial moment-generating function is close to that of the Poisson distribution. We do this by taking the limit of a moment-generating function.

Consider the moment-generating function of Y, which is $b(n, p)$. We shall take the limit of this as $n \to \infty$ such that $np = \lambda$ is a constant; thus $p \to 0$. The moment-generating function of Y is

$$M(t) = (1 - p + pe^t)^n.$$

Because $p = \lambda/n$, we have that

$$M(t) = \left[1 - \frac{\lambda}{n} + \frac{\lambda}{n}e^t \right]^n$$

$$= \left[1 + \frac{\lambda(e^t - 1)}{n} \right]^n.$$

Since

$$\lim_{n \to \infty} \left(1 + \frac{b}{n} \right)^n = e^b,$$

we have

$$\lim_{n \to \infty} = M(t) = e^{\lambda(e^t - 1)},$$

which exists for all real t. But this is the moment-generating function of a Poisson random variable with mean λ. Hence this Poisson distribution seems like a reasonable approximation to the binomial one when n is large and p is small. This approximation is usually found to be fairly successful if $n \geq 20$ and $p \leq 0.05$ and very successful if $n \geq 100$ and $p \leq 0.10$, but it is not bad if these bounds are violated somewhat. That is, it could be used in other situations too; we only want to stress that the approximation becomes better with larger n and smaller p.

Example 5.6.1

Let Y be $b(50, 1/25)$. Then

$$P(Y \leq 1) = \left(\frac{24}{25} \right)^{50} + 50 \left(\frac{1}{25} \right) \left(\frac{24}{25} \right)^{49} = 0.400.$$

Since $\lambda = np = 2$, the Poisson approximation is

$$P(Y \leq 1) \approx 0.406,$$

from Table III.   ◇

The preceding result illustrates the theorem we now state without proof.

Theorem 5.6.1.   If a sequence of moment-generating functions approaches a certain one, say $M(t)$, then the limit of the corresponding distributions must be the distribution corresponding to $M(t)$.

Remark   This theorem certainly appeals to one's intuition! In a more advanced course, the proof of this theorem is given and there the existence of the moment-generating function is not even needed, for we would use the characteristic function $\phi(t) = E(e^{itX})$ instead.

Theorem 5.6.1 is used to prove the central limit theorem. To help in the understanding of this proof, let us first consider a different problem, that of the limiting distribution of the mean $\overline{X}$ of a random sample $X_1, X_2, \ldots, X_n$, from a distribution with mean μ. If the distribution has moment-generating function $M(t)$, the moment-generating function of $\overline{X}$ is $[M(t/n)]^n$. But, by Taylor's expansion, there exists a number t_1 between 0 and t/n such that

$$M\left(\frac{t}{n}\right) = M(0) + M'(t_1)\frac{t}{n}$$

$$= 1 + \frac{\mu t}{n} + \frac{[M'(t_1) - M'(0)]t}{n}$$

because $M(0) = 1$ and $M'(0) = \mu$. Since $M'(t)$ is continuous at $t = 0$ and since $t_1 \to 0$ as $n \to \infty$, we know that

$$\lim_{n\to\infty} [M'(t_1) - M'(0)] = 0.$$

Thus, using a result from advanced calculus, we obtain

$$\lim_{n\to\infty}\left[M\left(\frac{t}{n}\right)\right]^n = \lim_{n\to\infty}\left\{1 + \frac{\mu t}{n} + \frac{[M'(t_1) - M'(0)]t}{n}\right\}^n$$

$$= e^{\mu t}$$

for all real t. But this limit is the moment-generating function of a degenerate distribution with all of the probability on μ. Accordingly, $\overline{X}$ has this limiting distribution, indicating that $\overline{X}$ converges to μ in a certain sense. This is one form of the law of large numbers.

We have seen that in some probability sense, $\overline{X}$ converges to μ in the limit, or, equivalently, $\overline{X} - \mu$ converges to zero. Let us multiply the difference $\overline{X} - \mu$ by some function of n so that the result will not converge to zero. In our search for such a function, it is natural to consider

$$W = \frac{\overline{X} - \mu}{\sigma/\sqrt{n}} = \frac{\sqrt{n}(\overline{X} - \mu)}{\sigma} = \frac{Y - n\mu}{\sqrt{n}\,\sigma},$$

where Y is the sum of the observations of the random sample. The reason for this is that W is a standardized random variable and has mean 0 and variance 1 for each positive integer n. We are now ready to prove the central limit theorem.

Proof   (of the central limit theorem): We first consider

$$E[\exp(tW)] = E\left\{\exp\left[\left(\frac{t}{\sqrt{n}\sigma}\right)\left(\sum_{i=1}^{n} X_i - n\mu\right)\right]\right\}$$

$$= E\left\{\exp\left[\left(\frac{t}{\sqrt{n}}\right)\left(\frac{X_1 - \mu}{\sigma}\right)\right]\cdots\exp\left[\left(\frac{t}{\sqrt{n}}\right)\left(\frac{X_n - \mu}{\sigma}\right)\right]\right\}$$

$$= E\left\{\exp\left[\left(\frac{t}{\sqrt{n}}\right)\left(\frac{X_1 - \mu}{\sigma}\right)\right]\right\}\cdots E\left\{\exp\left[\left(\frac{t}{\sqrt{n}}\right)\left(\frac{X_n - \mu}{\sigma}\right)\right]\right\},$$

which follows from the independence of $X_1, X_2, \ldots, X_n$. Then

$$E[\exp(tW)] = \left[m\left(\frac{t}{\sqrt{n}}\right)\right]^n, \qquad -h < \frac{t}{\sqrt{n}} < h,$$

where

$$m(t) = E\left\{\exp\left[t\left(\frac{X_i - \mu}{\sigma}\right)\right]\right\}, \qquad -h < t < h,$$

is the common moment-generating function of each

$$Y_i = \frac{X_i - \mu}{\sigma}, \qquad i = 1, 2, \ldots, n.$$

Since $E(Y_i) = 0$ and $E(Y_i^2) = 1$, it must be that

$$m(0) = 1, \qquad m'(0) = E\left(\frac{X_i - \mu}{\sigma}\right) = 0, \qquad m''(0) = E\left[\left(\frac{X_i - \mu}{\sigma}\right)^2\right] = 1.$$

Hence, using Taylor's formula with a remainder, we can find a number t_1 between 0 and t such that

$$m(t) = m(0) + m'(0)t + \frac{m''(t_1)t^2}{2} = 1 + \frac{m''(t_1)t^2}{2}.$$

By adding and subtracting $t^2/2$, we have that

$$m(t) = 1 + \frac{t^2}{2} + \frac{[m''(t_1) - 1]t^2}{2}.$$

Using this expression of $m(t)$ in $E[\exp(tW)]$, we can represent the moment-generating function of W by

$$
E[\exp(tW)] = \left\{ 1 + \frac{1}{2}\left(\frac{t}{\sqrt{n}}\right)^2 + \frac{1}{2}[m''(t_1) - 1]\left(\frac{t}{\sqrt{n}}\right)^2 \right\}^n
$$

$$
= \left\{ 1 + \frac{t^2}{2n} + \frac{[m''(t_1) - 1]t^2}{2n} \right\}^n, \qquad -\sqrt{n}\,h < t < \sqrt{n}\,h,
$$

where now t_1 is between 0 and $t/\sqrt{n}$. Since $m''(t)$ is continuous at $t = 0$ and $t_1 \to 0$ as $n \to \infty$, we have that

$$
\lim_{n \to \infty} [m''(t_1) - 1] = 1 - 1 = 0.
$$

Thus, using a result from advanced calculus, we have that

$$
\lim_{n \to \infty} E[\exp(tW)] = \lim_{n \to \infty} \left\{ 1 + \frac{t^2}{2n} + \frac{[m''(t_1) - 1]t^2}{2n} \right\}^n
$$

$$
= \lim_{n \to \infty} \left\{ 1 + \frac{t^2}{2n} \right\}^n = e^{t^2/2}
$$

for all real t. This means that the limiting distribution of

$$
W = \frac{\overline{X} - \mu}{\sigma/\sqrt{n}} = \frac{\sum\limits_{i=1}^{n} X_i - n\mu}{\sqrt{n}\sigma}
$$

is $N(0, 1)$. This completes the proof of the central limit theorem. □

Examples of the use of the central limit theorem as an approximating distribution have been given in Sections 5.4 and 5.5.

EXERCISES

5.6.1 Let Y be the number of defectives in a box of 50 articles taken from the output of a machine. Each article is defective with probability 0.01. What is the probability that $Y = 0, 1, 2,$ or 3 using

(a) The binomial distribution?

(b) The Poisson approximation?

5.6.2 The probability that a certain type of inoculation takes effect is 0.995. Use the Poisson distribution to approximate the probability that at most 2 out of 400 people given the inoculation find that it has not taken effect. HINT: Let $p = 1 - 0.995 = 0.005$.

5.6.3 Let S^2 be the sample variance of a random sample of size n from $N(\mu, \sigma^2)$. Show that the limit, as $n \to \infty$, of the moment-generating function of S^2 is $e^{\sigma^2 t}$. Thus, in the limit, the distribution of S^2 is degenerate with probability 1 at σ^2.

5.6.4 Let Y be $\chi^2(n)$. Use the central limit theorem to demonstrate that $W = (Y - n)/\sqrt{2n}$ has a limiting distribution function that is $N(0, 1)$. HINT: Think of Y as being the sum of a random sample from a certain distribution.

5.6.5 Let Y have a Poisson distribution with mean $3n$. Use the central limit theorem to show that $W = (Y - 3n)/\sqrt{3n}$ has a limiting distribution that is $N(0, 1)$.

6

ESTIMATION

6.1 POINT ESTIMATION

In earlier chapters we have alluded to estimating characteristics of the distribution from the corresponding ones of the sample, hoping that the latter would be reasonably close to the former. For example, the sample mean $\bar{x}$ can be thought of as an estimate of the distribution mean μ, and the sample variance s^2 can be used as an estimate of the distribution variance σ^2. Even the relative frequency histogram associated with a sample can be taken as an estimate of the p.d.f. of the underlying distribution. But how good are these estimates? What makes an estimate good? Can we say anything about the closeness of an estimate to an unknown parameter?

In this chapter we consider random variables for which the functional form of the p.d.f. is known, but the distribution depends on an unknown parameter, say θ, that may have any value in a set, say Ω, which is called the **parameter space**. For example, perhaps it is known that $f(x; \theta) = (1/\theta)e^{-x/\theta}$, $0 < x < \infty$ and that $\theta \in \Omega = \{\theta : 0 < \theta < \infty\}$. In certain instances, it might be necessary for the experimenter to select precisely one member of the family $\{f(x, \theta), \theta \in \Omega\}$ as the most likely p.d.f. of the random variable. That is, the experimenter needs a point estimate of the parameter θ, namely the value of the parameter that corresponds to the p.d.f. selected.

In estimation we take a random sample from the distribution to elicit some information about the unknown parameter θ. That is, we repeat the experiment n independent times, observe the sample, $X_1, X_2, \ldots, X_n$, and try to estimate the value of θ using the observations $x_1, x_2, \ldots, x_n$. The function of $X_1, X_2, \ldots, X_n$ used to estimate θ, say the statistic $u(X_1, X_2, \ldots, X_n)$, is called an **estimator** of θ. We want it to be such that the computed **estimate** $u(x_1, x_2, \ldots, x_n)$ is usually close to θ. Since we are estimating one member of $\theta \in \Omega$, such an estimator is often called a **point estimator**.

The following illustration should help motivate one principle that is often used in finding point estimates. Suppose that X is $b(1, p)$ so that the p.d.f. of X is

$$f(x; p) = p^x(1 - p)^{1-x}, \qquad x = 0, 1, \qquad 0 \le p \le 1.$$

We note that $p \in \Omega = \{p : 0 \le p \le 1\}$ where Ω represents the parameter space, that is, the space of all possible values of the parameter, p. Given a random sample $X_1, X_2, \ldots, X_n$, the problem is to find an estimator $u(X_1, X_2, \ldots, X_n)$ such that $u(x_1, x_2, \ldots, x_n)$ is a good point estimate of p, where $x_1, x_2, \ldots, x_n$ are the observed values of the random sample. Now the probability that $X_1, X_2, \ldots, X_n$ takes these particular values is (with Σx_i denoting $\sum_{i=1}^{n} x_i$)

$$P(X_1 = x_1, \ldots, X_n = x_n) = \prod_{i=1}^{n} p^{x_i}(1 - p)^{1-x_i} = p^{\Sigma x_i}(1 - p)^{n - \Sigma x_i},$$

which is the joint p.d.f. of $X_1, X_2, \ldots, X_n$ evaluated at the values observed. One reasonable way to proceed toward finding a good estimate of p is to regard this probability (or joint p.d.f.) as a function of p and find the value of p that maximizes it. That is, find the p value most likely to have produced these sample values. The joint p.d.f., when regarded as a function of p, is frequently called the **likelihood function**. Thus here the likelihood function is

$$\begin{aligned} L(p) &= L(p; x_1, x_2, \ldots, x_n) \\ &= f(x_1; p)f(x_2; p) \cdots f(x_n; p) \\ &= p^{\Sigma x_i}(1 - p)^{n - \Sigma x_i}, \qquad 0 \le p \le 1. \end{aligned}$$

To find the value of p that maximizes $L(p)$, we first take its derivative when $0 < p < 1$:

$$\frac{dL(p)}{dp} = (\Sigma x_i)p^{\Sigma x_i - 1}(1 - p)^{n - \Sigma x_i} - (n - \Sigma x_i)p^{\Sigma x_i}(1 - p)^{n - \Sigma x_i - 1}.$$

Setting this first derivative equal to zero gives us

$$p^{\Sigma x_i}(1 - p)^{n - \Sigma x_i}\left[\frac{\Sigma x_i}{p} - \frac{n - \Sigma x_i}{1 - p}\right] = 0.$$

Since $0 < p < 1$, this equals zero when

$$\frac{\Sigma x_i}{p} - \frac{n - \Sigma x_i}{1 - p} = 0. \tag{6.1}$$

Multiplying each member of Equation 6.1 by $p(1 - p)$ and simplifying, we obtain

$$\sum_{i=1}^{n} x_i - np = 0$$

or, equivalently,

$$p = \frac{\sum_{i=1}^{n} x_i}{n} = \bar{x}.$$

The corresponding statistic, namely $(\sum_{i=1}^{n} X_i)/n = \overline{X}$, is called the **maximum likelihood estimator** and is denoted by $\widehat{p}$; that is,

$$\widehat{p} = \frac{1}{n} \sum_{i=1}^{n} X_i = \overline{X}.$$

When finding a maximum likelihood estimator, it is often easier to find the value of the parameter that maximizes the natural logarithm of the likelihood function rather than the value of the parameter that maximizes the likelihood function itself. Because the natural logarithm function is an increasing function, the solutions will be the same. To see this, the example we have been considering gives us, for $0 < p < 1$,

$$\ln L(p) = \left(\sum_{i=1}^{n} x_i\right) \ln p + \left(n - \sum_{i=1}^{n} x_i\right) \ln(1 - p).$$

To find the maximum, we set the first derivative equal to zero to obtain

$$\frac{d\,[\ln L(p)]}{dp} = \left(\sum_{i=1}^{n} x_i\right)\left(\frac{1}{p}\right) + \left(n - \sum_{i=1}^{n} x_i\right)\left(\frac{-1}{1-p}\right) = 0,$$

which is the same as Equation 6.1. Thus the solution is $p = \overline{x}$ and the maximum likelihood estimator for p is $\widehat{p} = \overline{X}$.

Motivated by the preceding illustration, we present the formal definition of maximum likelihood estimators. This definition is used in both the discrete and continuous cases.

Let $X_1, X_2, \ldots, X_n$ be a random sample from a distribution that depends on one or more unknown parameters $\theta_1, \theta_2, \ldots, \theta_m$ with p.d.f. denoted by $f(x; \theta_1, \theta_2, \ldots, \theta_m)$. Suppose that $(\theta_1, \theta_2, \ldots, \theta_m)$ is restricted to a given parameter space Ω. Then the joint p.d.f. of $X_1, X_2, \ldots, X_n$, namely

$$L(\theta_1, \theta_2, \ldots, \theta_m) = f(x_1; \theta_1, \ldots, \theta_m) f(x_2; \theta_1, \ldots, \theta_m)$$
$$\cdots f(x_n; \theta_1, \ldots, \theta_m), \quad (\theta_1, \theta_2, \ldots, \theta_m) \in \Omega,$$

when regarded as a function of $\theta_1, \theta_2, \ldots, \theta_m$, is called the **likelihood function**. Say

$$[u_1(x_1, \ldots, x_n), u_2(x_1, \ldots, x_n), \ldots, u_m(x_1, \ldots, x_n)]$$

is that m-tuple in Ω that maximizes $L(\theta_1, \theta_2, \ldots, \theta_m)$. Then

$$\widehat{\theta_1} = u_1(X_1, \ldots, X_n),$$
$$\widehat{\theta_2} = u_2(X_1, \ldots, X_n),$$
$$\vdots$$
$$\widehat{\theta_m} = u_m(X_1, \ldots, X_n)$$

are maximum likelihood estimators of $\theta_1, \theta_2, \ldots, \theta_m$, respectively; and the corresponding observed values of these statistics, namely

$$u_1(x_1, \ldots, x_n), u_2(x_1, \ldots, x_n), \ldots, u_m(x_1, \ldots, x_n),$$

are called **maximum likelihood estimates**. In many practical cases, these estimators (and estimates) are unique. For many applications there is just one unknown parameter. In these cases the likelihood function is given by

$$L(\theta) = \prod_{i=1}^{n} f(x_i; \theta).$$

Some additional examples will help clarify these definitions.

Example 6.1.1

Let $X_1, X_2, \ldots, X_n$ be a random sample from the exponential distribution with p.d.f.

$$f(x; \theta) = \frac{1}{\theta} e^{-x/\theta}, \qquad 0 < x < \infty, \qquad \theta \in \Omega = \{\theta : 0 < \theta < \infty\}.$$

The likelihood function is given by

$$\begin{aligned}
L(\theta) &= L(\theta; x_1, x_2, \ldots, x_n) \\
&= \left(\frac{1}{\theta} e^{-x_1/\theta} \right) \left(\frac{1}{\theta} e^{-x_2/\theta} \right) \cdots \left(\frac{1}{\theta} e^{-x_n/\theta} \right) \\
&= \frac{1}{\theta^n} \exp\left(\frac{-\sum_{i=1}^{n} x_i}{\theta} \right), \qquad 0 < \theta < \infty.
\end{aligned}$$

The natural logarithm of $L(\theta)$ is

$$\ln L(\theta) = -(n) \ln(\theta) - \frac{1}{\theta} \sum_{i=1}^{n} x_i, \qquad 0 < \theta < \infty.$$

Thus

$$\frac{d\,[\ln L(\theta)]}{d\theta} = \frac{-n}{\theta} + \frac{\sum_{i=1}^{n} x_i}{\theta^2} = 0.$$

The solution of this equation for θ is

$$\theta = \frac{1}{n} \sum_{i=1}^{n} x_i = \bar{x}.$$

Note that

$$\frac{d\,[\ln L(\theta)]}{d\theta} = \frac{1}{\theta} \left(-n + \frac{n\bar{x}}{\theta} \right) \begin{array}{ll} > 0, & \theta < \bar{x}, \\ = 0, & \theta = \bar{x}, \\ < 0, & \theta > \bar{x}. \end{array}$$

Hence $\ln L(\theta)$ does have a maximum at $\bar{x}$, and thus the maximum likelihood estimator for θ is

$$\hat{\theta} = \bar{X} = \frac{1}{n} \sum_{i=1}^{n} X_i. \qquad\qquad \diamond$$

Example 6.1.2

Let $X_1, X_2, \ldots, X_n$ be a random sample from the geometric distribution with p.d.f. $f(x; p) = (1-p)^{x-1}p$, $x = 1, 2, 3, \ldots$ The likelihood function is given by

$$L(p) = (1-p)^{x_1-1}p(1-p)^{x_2-1}p \cdots (1-p)^{x_n-1}p$$
$$= p^n(1-p)^{\sum x_i - n}, \qquad 0 \le p \le 1.$$

The natural logarithm of $L(p)$ is

$$\ln L(p) = n \ln p + \left(\sum_{i=1}^{n} x_i - n\right)\ln(1-p), \qquad 0 < p < 1.$$

Thus, restricting p to $0 < p < 1$ so as to be able to take the derivative, we have

$$\frac{d \ln L(p)}{dp} = \frac{n}{p} - \frac{\displaystyle\sum_{i=1}^{n} x_i - n}{1-p} = 0.$$

Solving for p, we obtain

$$p = \frac{n}{\displaystyle\sum_{i=1}^{n} x_i} = \frac{1}{\bar{x}}.$$

So the maximum likelihood estimator of p is

$$\hat{p} = \frac{n}{\displaystyle\sum_{i=1}^{n} X_i} = \frac{1}{\bar{X}}.$$

This estimator agrees with our intuition because, in n observations of a geometric random variable, there are n successes in the $\sum_{i=1}^{n} x_i$ trials. Thus the estimate of p is the number of successes divided by the total number of trials. ◇

In the following important example we find the maximum likelihood estimators of the parameters associated with the normal distribution.

Example 6.1.3

Let $X_1, X_2, \ldots, X_n$ be a random sample from $N(\theta_1, \theta_2)$, where

$$\Omega = \{(\theta_1, \theta_2) : -\infty < \theta_1 < \infty, 0 < \theta_2 < \infty\}.$$

That is, here we let $\theta_1 = \mu$ and $\theta_2 = \sigma^2$. Then

$$L(\theta_1, \theta_2) = \prod_{i=1}^{n} \frac{1}{\sqrt{2\pi\theta_2}} \exp\left[-\frac{(x_i - \theta_1)^2}{2\theta_2}\right]$$

or, equivalently,

$$L(\theta_1, \theta_2) = \left(\frac{1}{\sqrt{2\pi\theta_2}}\right)^n \exp\left[\frac{-\sum_{i=1}^{n}(x_i - \theta_1)^2}{2\theta_2}\right], \qquad (\theta_1, \theta_2) \in \Omega.$$

The natural logarithm of the likelihood function is

$$\ln L(\theta_1, \theta_2) = -\frac{n}{2}\ln(2\pi\theta_2) - \frac{\sum_{i=1}^{n}(x_i - \theta_1)^2}{2\theta_2}.$$

The partial derivatives with respect to θ_1 and θ_2 are

$$\frac{\partial(\ln L)}{\partial\theta_1} = \frac{1}{\theta_2}\sum_{i=1}^{n}(x_i - \theta_1)$$

and

$$\frac{\partial(\ln L)}{\partial\theta_2} = \frac{-n}{2\theta_2} + \frac{1}{2\theta_2^2}\sum_{i=1}^{n}(x_i - \theta_1)^2.$$

The equation $\partial(\ln L)/\partial\theta_1 = 0$ has the solution $\theta_1 = \bar{x}$. Setting $\partial(\ln L)/\partial\theta_2 = 0$ and replacing θ_1 by $\bar{x}$ yields

$$\theta_2 = \frac{1}{n}\sum_{i=1}^{n}(x_i - \bar{x})^2.$$

By considering the usual condition on the second partial derivatives, these solutions do provide a maximum. Thus the maximum likelihood estimators of $\mu = \theta_1$ and $\sigma^2 = \theta_2$ are

$$\widehat{\theta_1} = \overline{X} \qquad \text{and} \qquad \widehat{\theta_2} = \frac{1}{n}\sum_{i=1}^{n}(X_i - \overline{X})^2 = V. \qquad \diamond$$

It is interesting to note that in each of our first illustration where $\widehat{p} = \overline{X}$ and Example 6.1.1, where $\widehat{\theta} = \overline{X}$, the expected value of the estimator is equal to the corresponding parameter. This observation leads to the following definition.

Definition 6.1.1. If $E[u(X_1, X_2, \ldots, X_n)] = \theta$, the statistic $u(X_1, X_2, \ldots X_n)$ is called an **unbiased estimator** of θ. Otherwise, it is said to be **biased**.

Example 6.1.4

We have shown that when sampling from $N(\theta_1 = \mu, \theta_2 = \sigma^2)$, the maximum likelihood estimators of μ and σ^2 are

$$\widehat{\theta_1} = \widehat{\mu} = \overline{X} \qquad \text{and} \qquad \widehat{\theta_2} = \widehat{\sigma^2} = \frac{(n-1)S^2}{n}.$$

Recalling that the distribution of $\overline{X}$ is $N(\mu, \sigma^2/n)$, we see that $E(\overline{X}) = \mu$ and thus $\overline{X}$ is an unbiased estimator of μ.

In Theorem 5.3.4 we showed that the distribution of $(n - 1)S^2/\sigma^2$ is $\chi^2(n - 1)$. Thus

$$E(S^2) = E\left[\frac{\sigma^2}{n - 1} \frac{(n - 1)S^2}{\sigma^2}\right] = \frac{\sigma^2}{n - 1}(n - 1) = \sigma^2.$$

That is, the sample variance

$$S^2 = \frac{1}{n - 1} \sum_{i=1}^{n} (X_i - \overline{X})^2$$

is an unbiased estimator of σ^2. Thus since

$$E(\widehat{\theta}_2) = \frac{n - 1}{n} E(S^2) = \frac{n - 1}{n} \sigma^2,$$

$\widehat{\theta}_2$ is a biased estimator of $\theta_2 = \sigma^2$.   ◇

Sometimes it is impossible to find maximum likelihood estimators in a convenient closed form and numerical methods must be used to maximize the likelihood function. For illustration, suppose that $X_1, X_2, \ldots, X_n$ is a random sample from a gamma distribution with parameters $\alpha = \theta_1$ and $\beta = \theta_2$, where $\theta_1 > 0, \theta_2 > 0$. It is difficult to maximize

$$L(\theta_1, \theta_2; x_1, \ldots, x_n) = \left[\frac{1}{\Gamma(\theta_1)\theta_2^{\theta_1}}\right]^n (x_1 x_2 \cdots x_n)^{\theta_1 - 1} \exp\left(-\sum_{i=1}^{n} x_i/\theta_2\right)$$

with respect to θ_1 and θ_2, owing to the presence of the gamma function $\Gamma(\theta_1)$. Thus numerical methods must be used to maximize L once $x_1, x_2, \ldots, x_n$ are observed.

There are other ways, however, to obtain easily point estimates of θ_1 and θ_2. For illustration, in the gamma distribution situation, let us simply equate the first two moments of the distribution to the corresponding moments of the sample. This seems like a reasonable way in which to find estimators, since the empirical distribution function $F_n(x)$ converges in some sense to $F(x)$, and hence corresponding moments should be about equal. Here in this illustration we have

$$\theta_1 \theta_2 = \overline{X}, \qquad \theta_1 \theta_2^2 = V,$$

the solutions of which are

$$\widetilde{\theta}_1 = \frac{\overline{X}^2}{V} \qquad \text{and} \qquad \widetilde{\theta}_2 = \frac{V}{\overline{X}}.$$

We say that these latter two statistics, $\widetilde{\theta}_1$ and $\widetilde{\theta}_2$, are respective estimators of θ_1 and θ_2 found by the **method of moments**.

To generalize this discussion, let $X_1, X_2, \ldots, X_n$ be a random sample of size n from a distribution with p.d.f. $f(x; \theta_1, \theta_2, \ldots, \theta_r)$, $(\theta_1, \ldots, \theta_r) \in \Omega$. The expectation $E(X^k)$ is frequently called the kth moment of the distribution, $k = 1, 2, 3, \ldots$. The sum $M_k = \sum_{i=1}^{n} X_i^k/n$ is the kth moment of the sample, $k = 1, 2, 3, \ldots$. The method of moments can be described as follows. Equate $E(X^k)$ to M_k, beginning with $k = 1$ and continuing until there are enough equations to provide unique solutions for $\theta_1, \theta_2, \ldots, \theta_r$, say $h_i(M_1, M_2, \ldots)$, $i = 1, 2, \ldots, r$, respectively. It should be noted that this could be done in an equivalent manner by equating $\mu = E(X)$ to $\overline{X}$ and $E[(X - \mu)^k]$ to $\sum_{i=1}^{n}(X_i - \overline{X})^k / n$, $k = 2, 3$, and so on until unique solutions for $\theta_1, \theta_2, \ldots, \theta_r$ are obtained. This alternative procedure was used in the preceding illustration. In most practical cases, the estimator $\tilde{\theta}_i = h_i(M_1, M_2, \ldots)$ of θ_i, found by the method of moments, is an estimator of θ_i that in some sense gets close to that parameter when n is large, $i = 1, 2, \ldots, r$.

The following two examples illustrate the method of moments technique for finding estimators, the first for a one-parameter family and the second for a two-parameter family.

Example 6.1.5

Let $X_1, X_2, \ldots, X_n$, be a random sample of size n from the distribution with p.d.f. $f(x; \theta) = \theta x^{\theta-1}$, $0 < x < 1$, $0 < \theta < \infty$. Sketch the graphs of this p.d.f. for $\theta = 1/4, 1$, and 4. Note that sets of observations for these three values of θ would look very different. How do we estimate the value of θ? The mean of this distribution is given by

$$E(X) = \int_0^1 x \theta x^{\theta-1} \, dx = \frac{\theta}{\theta + 1}.$$

We shall set the population mean equal to the sample mean and solve for θ. We have

$$\overline{x} = \frac{\theta}{\theta + 1}$$

or, solving for θ, we obtain the method of moments estimator,

$$\tilde{\theta} = \frac{\overline{X}}{1 - \overline{X}}.$$

Thus an estimate of θ by the method of moments is $\overline{x}/(1 - \overline{x})$.   ◇

Recall that in the method of moments, if there are two parameters that have to be estimated, the first two sample moments are set equal to the first two population moments that are given in terms of the unknown parameters. These two equations are then solved simultaneously for the unknown parameters.

Example 6.1.6

Let the distribution of X be $N(\mu, \sigma^2)$. Then

$$E(X) = \mu \quad \text{and} \quad E(X^2) = \sigma^2 + \mu^2.$$

Given a random sample of size n, the first two moments are given by

$$m_1 = \frac{1}{n}\sum_{i=1}^{n} x_i \quad \text{and} \quad m_2 = \frac{1}{n}\sum_{i=1}^{n} x_i^2.$$

We set $m_1 = E(X)$ and $m_2 = E(X^2)$ and solve for μ and σ^2. That is,

$$\frac{1}{n}\sum_{i=1}^n x_i = \mu \quad \text{and} \quad \frac{1}{n}\sum_{i=1}^n x_i^2 = \sigma^2 + \mu^2.$$

The first equation yields $\bar{x}$ as the estimate of μ. Replacing μ^2 with $\bar{x}^2$ in the second equation and solving for σ^2, we obtain

$$\frac{1}{n}\sum_{i=1}^n x_i^2 - \bar{x}^2 = \sum_{i=1}^n \frac{(x_i - \bar{x})^2}{n} = v$$

for the solution of σ^2. Thus the method of moment estimators for μ and σ^2 are $\tilde{\mu} = \bar{X}$ and $\tilde{\sigma}^2 = V$. Of course, $\tilde{\mu} = \bar{X}$ is unbiased, whereas $\tilde{\sigma}^2 = V$ is biased.   ◇

In Example 6.1.4 we showed that $\bar{X}$ and S^2 are unbiased estimators of μ and σ^2, respectively, when sampling from a normal distribution. This is also true when sampling from any population with finite variance σ^2. That is, $E(\bar{X}) = \mu$ and $E(S^2) = \sigma^2$, provided that the sample arises from a distribution with variance $\sigma^2 < \infty$ (see Exercise 6.1.13). Although S^2 is an unbiased estimator of σ^2, S is a biased estimator of σ. In Exercise 6.1.14, you are asked to show that when sampling from a normal distribution, cS is an unbiased estimator for σ where

$$c = \frac{\sqrt{n-1}\,\Gamma\left(\frac{n-1}{2}\right)}{\sqrt{2}\,\Gamma\left(\frac{n}{2}\right)}.$$

EXERCISES

6.1.1 Let $X_1, X_2, \ldots, X_n$ be a random sample from $N(\mu, \sigma^2)$ where the mean $\theta = \mu$ is such that $-\infty < \theta < \infty$ and σ^2 is a known positive number. Show that the maximum likelihood estimator for θ is $\hat{\theta} = \bar{X}$.

6.1.2 A random sample $X_1, X_2, \ldots, X_n$ of size n is taken from $N(\mu, \sigma^2)$, where the variance $\theta = \sigma^2$ is such that $0 < \theta < \infty$ and μ is a known real number. Show that the maximum likelihood estimator for θ is $\hat{\theta} = (1/n)\sum_{i=1}^n (X_i - \mu)^2$.

6.1.3 A random sample $X_1, X_2, \ldots, X_n$ of size n is taken from a Poisson distribution with a mean of λ, $0 < \lambda < \infty$.

(a) Show that the maximum likelihood estimator for λ is $\hat{\lambda} = \bar{X}$.

(b) Let X equal the number of flaws per 100 feet of a used computer tape. Assume that X has a Poisson distribution with a mean of λ. If 40 observations of X yielded 5 zeros, 7 ones, 12 twos, 9 threes, 5 fours, 1 five, and 1 six, find the maximum likelihood estimate of λ.

6.1.4 For determining half-lives of radioactive isotopes, it is important to know what the background radiation is in a given detector over a period of time. The following data were taken in a γ-ray detection experiment over 98 ten-second intervals:

58	50	57	58	64	63	54	64	59	41	43	56	60	50
46	59	54	60	59	60	67	52	65	63	55	61	68	58
63	36	42	54	58	54	40	60	64	56	61	51	48	50
60	42	62	67	58	49	66	58	57	59	52	54	53	53
57	43	73	65	45	43	57	55	73	62	68	55	51	55
53	68	58	53	51	73	44	50	53	62	58	47	63	59
59	56	60	59	50	52	62	51	66	51	56	53	59	57

Assume that these are observations of a Poisson random variable with mean λ.

(a) Find the values of $\overline{x}$ and s^2.

(b) What is the value of the maximum likelihood estimator of λ?

(c) Is S^2 an unbiased estimator of λ?

(d) Which of $\overline{x}$ and s^2 would you recommend for estimating λ? Why?

6.1.5 Let $X_1, X_2, \ldots, X_n$ be a random sample from distributions with the following probability density functions. In each case find the maximum likelihood estimator $\widehat{\theta}$.

(a) $f(x; \theta) = (1/\theta^2) x e^{-x/\theta}, \qquad 0 < x < \infty, 0 < \theta < \infty.$

(b) $f(x; \theta) = (1/2\theta^3) x^2 e^{-x/\theta}, \qquad 0 < x < \infty, 0 < \theta < \infty.$

(c) $f(x; \theta) = (1/2) e^{-|x-\theta|}, \qquad -\infty < x < \infty, -\infty < \theta < \infty.$

6.1.6 Find the maximum likelihood estimates for $\theta_1 = \mu$ and $\theta_2 = \sigma^2$ if a random sample of size 15 from $N(\mu, \sigma^2)$ yielded the following values:

31.5	36.9	33.8	30.1	33.9
35.2	29.6	34.4	30.5	34.2
31.6	36.7	35.8	34.5	32.7

6.1.7 Let $f(x; \theta) = \theta x^{\theta-1}, 0 < x < 1, \theta \in \Omega = \{\theta : 0 < \theta < \infty\}$. Let $X_1, X_2, \ldots, X_n$ denote a random sample of size n from this distribution.

(a) Sketch the p.d.f. of X for (i) $\theta = 1/2$, (ii) $\theta = 1$, and (iii) $\theta = 2$.

(b) Show that $\widehat{\theta} = -n/\ln \left(\prod_{i=1}^{n} X_i \right)$ is the maximum likelihood estimator of θ.

(c) For each of the following three sets of 10 observations from this distribution, calculate the values of the maximum likelihood estimates and the method of moments estimates for θ.

(d) For each set of data, sketch the empirical and theoretical distribution functions (using one of your estimates as the value of θ) on the same graph.

(i) 0.0256   0.3051   0.0278   0.8971   0.0739
     0.3191   0.7379   0.3671   0.9763   0.0102
(ii) 0.9960   0.3125   0.4374   0.7464   0.8278
     0.9518   0.9924   0.7112   0.2228   0.8609
(iii) 0.4698 0.3675   0.5991   0.9513   0.6049
     0.9917   0.1551   0.0710   0.2110   0.2154

6.1.8 Let $f(x; \theta) = (1/\theta)x^{(1-\theta)/\theta}, 0 < x < 1, 0 < \theta < \infty$.
   (a) Show that the maximum likelihood estimator of θ is $\widehat{\theta} = -(1/n) \sum_{i=1}^{n} \ln X_i$.
   (b) Show that $E(\widehat{\theta}) = \theta$ and thus $\widehat{\theta}$ is an unbiased estimator of θ.

6.1.9 Out of 50,000,000 instant winner lottery tickets, the proportion of winning tickets is p. Each day, for 20 consecutive days, a bettor purchased tickets, one at a time, until a winning ticket was purchased. The numbers of tickets that were purchased each day to obtain the winning ticket were

$$1 \quad 26 \quad 19 \quad 6 \quad 6 \quad 1 \quad 2 \quad 3 \quad 1 \quad 23$$
$$19 \quad 3 \quad 6 \quad 8 \quad 4 \quad 1 \quad 18 \quad 34 \quad 1 \quad 8$$

By making reasonable assumptions, find the maximum likelihood estimate of p based on these data.

6.1.10 Let $X_1, X_2, \ldots, X_n$ be a random sample of size n from the exponential distribution whose p.d.f. is $f(x; \theta) = (1/\theta)e^{-x/\theta}, 0 < x < \infty, 0 < \theta < \infty$.
   (a) Show that $\overline{X}$ is an unbiased estimator of θ.
   (b) Show that the variance of $\overline{X}$ is θ^2/n.
   (c) What is a good estimate of θ if a random sample of size 5 yielded the sample values 3.5. 8.1. 0.9, 4.4, and 0.5?

6.1.11 Let $X_1, X_2, \ldots, X_n$ be a random sample of size n from a geometric distribution for which p is the probability of success.
   (a) Use the method of moments to find a point estimate for p.
   (b) Explain intuitively why your estimate makes good sense.
   (c) Use the following data to give a point estimate of p:

$$3 \quad 34 \quad 7 \quad 4 \quad 19 \quad 2 \quad 1 \quad 19 \quad 43 \quad 2$$
$$22 \quad 4 \quad 19 \quad 11 \quad 7 \quad 1 \quad 2 \quad 21 \quad 15 \quad 16$$

6.1.12 Let $X_1, X_2, \ldots, X_n$ be a random sample from $b(1, p)$, i.e., n Bernoulli trials. Thus

$$Y = \sum_{i=1}^{n} X_i \text{ is } b(n, p).$$

   (a) Show that $\overline{X} = Y/n$ is an unbiased estimator of p.
   (b) Show that $\text{Var}(\overline{X}) = p(1 - p)/n$.
   (c) Show that $E[\overline{X}(1 - \overline{X})/n] = (n - 1)[p(1 - p)/n^2]$.
   (d) Find the value of c so that $c\overline{X}(1 - \overline{X})$ is an unbiased estimator of $\text{Var}(\overline{X}) = p(1 - p)/n$.

6.1.13 Let $X_1, X_2, \ldots, X_n$ be a random sample from a distribution having finite variance σ^2. Show that

$$S^2 = \sum_{i=1}^{n} \frac{(X_i - \overline{X})^2}{n-1}$$

is an unbiased estimator of σ^2. HINT: Write

$$S^2 = \frac{1}{n-1}\left(\sum_{i=1}^{n} X_i^2 - n\overline{X}^2\right)$$

and compute $E(S^2)$.

6.1.14 Let $X_1, X_2, \ldots, X_n$ be a random sample of size n from a normal distribution.

(a) Show that an unbiased estimator for σ is cS, where

$$c = \frac{\sqrt{n-1}\,\Gamma\left(\dfrac{n-1}{2}\right)}{\sqrt{2}\,\Gamma\left(\dfrac{n}{2}\right)}.$$

(b) Find the value of c when $n = 5$, $n = 6$.

6.1.15 Let $X_1, X_2, \ldots, X_n$ be a random sample from a continuous uniform distribution on the interval $(\theta - 1, \theta + 1)$.

(a) Find the method of moments estimator for θ.

(b) Is your estimator in part (a) an unbiased estimator for θ?

(c) Give a point estimate of θ using the following $n = 5$ observations of X:

$$6.61 \quad 7.70 \quad 6.98 \quad 8.36 \quad 7.26$$

(d) The method of moments estimator actually has greater variance than the estimator $[\min(X_i) + \max(X_i)]/2$, which is the maximum likelihood estimator of θ. Compute the value of this estimator for the $n = 5$ observations in part (c).

6.1.16 Given the following 25 observations from a gamma distribution with mean $\mu = \alpha\theta$ and variance $\sigma^2 = \alpha\theta^2$, use the method of moments estimators to find point estimates of α and θ.

$$6.9 \quad 7.3 \quad 6.7 \quad 6.4 \quad 6.3 \quad 5.9 \quad 7.0 \quad 7.1 \quad 6.5 \quad 7.6 \quad 7.2 \quad 7.1 \quad 6.1$$
$$7.3 \quad 7.6 \quad 7.6 \quad 6.7 \quad 6.3 \quad 5.7 \quad 6.7 \quad 7.5 \quad 5.3 \quad 5.4 \quad 7.4 \quad 6.9$$

6.1.17 Let $X_1, X_2, \ldots, X_n$ be a random sample of size n from the normal distribution $N(\mu, \sigma^2 = \theta)$, $0 < \theta < \infty$, where μ is known. Show that $(1/n)\sum_{i=1}^{n}(X_i - \mu)^2$ is an unbiased estimator of θ.

6.1.18 Let X equal the number of telephone calls per "five minutes" that are received at Great Lakes Pizza Co. in the evening. Assume that the distribution of X is Poisson with mean λ.

(a) Given n observations of X, find the method of moments estimate of λ.

(b) Give a point estimate of λ using the following 12 observations of X:

$$1 \quad 2 \quad 1 \quad 1 \quad 2 \quad 4 \quad 0 \quad 1 \quad 0 \quad 1 \quad 1 \quad 0$$

(c) Compare the values of $\overline{x}$ and s^2. Does this information support the assumption that X has a Poisson distribution?

6.1.19 A Geiger counter was set up in the physics laboratory to record the number of alpha-particle emissions of carbon–14 in half a second. Ten observations yielded the following data:

$$4 \quad 6 \quad 9 \quad 6 \quad 10 \quad 11 \quad 6 \quad 3 \quad 7 \quad 10$$

Give the value of an unbiased estimate for λ, the mean number of counts per half second, assuming that these are observations of a Poisson random variable.

6.1.20 An urn contains 64 balls of which N_1 are orange and N_2 are blue. A random sample of $n = 8$ balls is selected from the urn without replacement and X is equal to the number of orange balls in the sample. This experiment was repeated 30 times (the 8 balls being returned to the urn before each repetition), yielding the following data:

$$
\begin{array}{cccccccccccccccc}
3 & 0 & 0 & 1 & 1 & 1 & 1 & 3 & 1 & 1 & 2 & 0 & 1 & 3 & 1 \\
0 & 1 & 0 & 2 & 1 & 1 & 2 & 3 & 2 & 2 & 4 & 3 & 1 & 1 & 2
\end{array}
$$

Using these data, guess the value of N_1 and give a reason for your guess.

6.2 CONFIDENCE INTERVALS FOR MEANS

Given a random sample $X_1, X_2, \ldots, X_n$ from a normal distribution $N(\mu, \sigma^2)$, we shall now consider the closeness of $\overline{X}$, the unbiased estimator of μ, to the unknown mean μ. To do this, we use the error structure (distribution) of $\overline{X}$, namely that $\overline{X}$ is $N(\mu, \sigma^2/n)$, to construct what is called a confidence interval for the unknown parameter μ, when the variance σ^2 is known. For the probability $1 - \alpha$, we can find a number $z_{\alpha/2}$ from Table V in the Appendix such that

$$P\left(-z_{\alpha/2} \leq \frac{\overline{X} - \mu}{\sigma/\sqrt{n}} \leq z_{\alpha/2}\right) = 1 - \alpha.$$

For example, if $1 - \alpha = 0.95$, then $z_{\alpha/2} = z_{0.025} = 1.96$ and if $1 - \alpha = 0.90$, then $z_{\alpha/2} = z_{0.05} = 1.645$. Now recalling that $\sigma > 0$, we see that the following inequalities are equivalent:

$$-z_{\alpha/2} \leq \frac{\overline{X} - \mu}{\sigma/\sqrt{n}} \leq z_{\alpha/2},$$

$$-z_{\alpha/2}\left(\frac{\sigma}{\sqrt{n}}\right) \leq \overline{X} - \mu \leq z_{\alpha/2}\left(\frac{\sigma}{\sqrt{n}}\right),$$

$$-\overline{X} - z_{\alpha/2}\left(\frac{\sigma}{\sqrt{n}}\right) \leq -\mu \leq -\overline{X} + z_{\alpha/2}\left(\frac{\sigma}{\sqrt{n}}\right),$$

$$\overline{X} + z_{\alpha/2}\left(\frac{\sigma}{\sqrt{n}}\right) \geq \mu \geq \overline{X} - z_{\alpha/2}\left(\frac{\sigma}{\sqrt{n}}\right).$$

Thus, since the probability of the first of these is $1 - \alpha$, the probability of the last must also be $1 - \alpha$ because the latter is true if and only if the former is true. That is, we have

$$P\left[\overline{X} - z_{\alpha/2}\left(\frac{\sigma}{\sqrt{n}}\right) \le \mu \le \overline{X} + z_{\alpha/2}\left(\frac{\sigma}{\sqrt{n}}\right)\right] = 1 - \alpha.$$

So, the probability that the random interval

$$\left[\overline{X} - z_{\alpha/2}\left(\frac{\sigma}{\sqrt{n}}\right), \overline{X} + z_{\alpha/2}\left(\frac{\sigma}{\sqrt{n}}\right)\right]$$

includes the unknown mean μ is $1 - \alpha$.

Once the sample is observed and the sample mean computed to equal $\overline{x}$, the interval $[\overline{x} - z_{\alpha/2}(\sigma/\sqrt{n}), \overline{x} + z_{\alpha/2}(\sigma/\sqrt{n})]$ is a known interval. Since the probability that the random interval covers μ before the sample is drawn is equal to $1 - \alpha$, we now call the computed interval, $\overline{x} \pm z_{\alpha/2}(\sigma/\sqrt{n})$ (for brevity), a $100(1 - \alpha)\%$ **confidence interval** for the unknown mean μ. For illustration, $\overline{x} \pm 1.96(\sigma/\sqrt{n})$ is a 95% confidence interval for μ. The number $100(1 - \alpha)\%$, or equivalently, $1 - \alpha$, is called the **confidence coefficient**.

We see that the confidence interval for μ is centered at the point estimate $\overline{x}$ and is completed by subtracting and adding the quantity $z_{\alpha/2}(\sigma/\sqrt{n})$.

Note that as n increases, $z_{\alpha/2}(\sigma/\sqrt{n})$ decreases, resulting in a shorter confidence interval with the same confidence coefficient $1 - \alpha$. A shorter confidence interval indicates that we have more reliance in $\overline{x}$ as an estimate of μ. Statisticians who are not restricted by time, money, effort, or availability of observations can obviously make the confidence interval as short as they like by increasing the sample size n. For a fixed sample size n, the length of the confidence interval can also be shortened by decreasing the confidence coefficient $1 - \alpha$. But if this is done, we achieve a shorter confidence interval by losing some confidence.

Example 6.2.1

Let X equal the length of life of a 60-watt light bulb marketed by a certain manufacturer of light bulbs. Assume that the distribution of X is $N(\mu, 1296)$. If a random sample of $n = 27$ bulbs were tested until they burned out, yielding a sample mean of $\overline{x} = 1478$ hours, then a 95% confidence interval for μ is

$$\left[\overline{x} - z_{0.025}\left(\frac{\sigma}{\sqrt{n}}\right), \overline{x} + z_{0.025}\left(\frac{\sigma}{\sqrt{n}}\right)\right]$$

$$= \left[1478 - 1.96\left(\frac{36}{\sqrt{27}}\right), 1478 + 1.96\left(\frac{36}{\sqrt{27}}\right)\right]$$

$$= [1478 - 13.58, 1478 + 13.58]$$

$$= [1464.42, 1491.58]. \qquad \Diamond$$

The next example will help to give a better intuitive feeling for the interpretation of a confidence interval.

Example 6.2.2

Let $\bar{x}$ be the observed sample mean of five items of a random sample from the normal distribution $N(\mu, 16)$. A 90% confidence interval for the unknown mean μ is

$$\left[\bar{x} - 1.645\sqrt{\frac{16}{5}}, \bar{x} + 1.645\sqrt{\frac{16}{5}} \right].$$

For a particular sample this interval either does or does not contain the mean μ. However, if many such intervals were calculated, it should be true that about 90% of them contain the mean μ. Fifty random samples of size five from the normal distribution $N(50, 16)$ were simulated on a computer. A 90% confidence interval was calculated for each random sample, as if the mean were unknown. Figure 6.1(a) depicts each of these 50 intervals as a line segment. For these 50 intervals, 45 (or 90%) of them contain the mean, $\mu = 50$. In other simulations of 50 confidence intervals, the number of 90% confidence intervals containing the mean could be larger or smaller. [In fact, if W is the random variable that counts the number of 90% confidence intervals that contain the mean, then the distribution of W is $b(50, 0.90)$.] [See Example 6.2.8 for a discussion of Figure 6.1(b).]   ◇

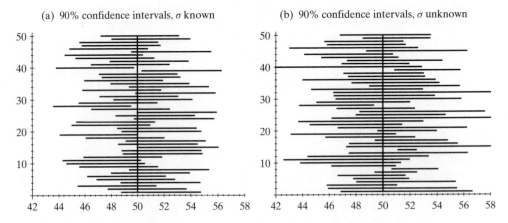

(a) 90% confidence intervals, σ known   (b) 90% confidence intervals, σ unknown

Figure 6.1   Confidence intervals using z and t

If we cannot assume that the distribution from which the sample arose is normal, we can still obtain an approximate confidence interval for μ. By the central limit theorem the ratio $(\bar{X} - \mu)/(\sigma/\sqrt{n})$ has, provided that n is large enough, the approximate normal distribution $N(0, 1)$ when the underlying distribution is not normal. In this case

$$P\left(-z_{\alpha/2} \le \frac{\bar{X} - \mu}{\sigma/\sqrt{n}} \le z_{\alpha/2} \right) \approx 1 - \alpha,$$

and

$$\left[\bar{x} - z_{\alpha/2}\left(\frac{\sigma}{\sqrt{n}}\right), \bar{x} + z_{\alpha/2}\left(\frac{\sigma}{\sqrt{n}}\right) \right]$$

is an approximate $100(1 - \alpha)\%$ confidence interval for μ.

The closeness of the approximate probability $1 - \alpha$ to the exact probability depends on both the underlying distribution and the sample size. When the underlying distribution is unimodal (has only one mode) and continuous, the approximation is usually quite good for even small n, such as $n = 5$. As the underlying distribution becomes "less normal" (i.e., badly skewed or discrete), a larger sample size might be required to keep a reasonably accurate approximation. But, in all cases, an n of at least 30 is usually quite adequate.

Example 6.2.3

Let X equal the amount of orange juice (in grams per day) consumed by an American. Suppose it is known that the standard deviation of X is $\sigma = 96$. To estimate the mean μ of X, an orange growers' association took a random sample of $n = 576$ Americans and found that they consumed, on the average, $x = 133$ grams of orange juice per day. Thus an approximate 90% confidence interval for μ is

$$133 \pm 1.645\left(\frac{96}{\sqrt{576}}\right) \quad \text{or} \quad [133 - 6.58, 133 + 6.58] = [126.42, 139.58]. \quad \diamond$$

If σ^2 is unknown and the sample size n is 30 or greater, we shall use the fact that the ratio $(\overline{X} - \mu)/(S/\sqrt{n})$ has an approximate normal distribution $N(0, 1)$. This statement is true whether or not the underlying distribution is normal. However, if the underlying distribution is badly skewed or contaminated with occasional outliers, most statisticians would prefer to have a larger sample size, say 50 or more, and even that might not produce good results. After this next example, we consider what to do when n is small.

Example 6.2.4

Lake Macatawa, an inlet lake on the east side of Lake Michigan, is divided into an east basin and a west basin. To measure the effect on the lake of salting city streets in the winter, students took 32 samples of water from the west basin and measured the amount of sodium in parts per million in order to make a statistical inference about the unknown mean μ. They obtained the following data:

13.0	18.5	16.4	14.8	19.4	17.3	23.2	24.9
20.8	19.3	18.8	23.1	15.2	19.9	19.1	18.1
25.1	16.8	20.4	17.4	25.2	23.1	15.3	19.4
16.0	21.7	15.2	21.3	21.5	16.8	15.6	17.6

For these data $\bar{x} = 19.07$ and $s^2 = 10.60$. Thus an approximate 95% confidence interval for μ is

$$\bar{x} \pm 1.96\left(\frac{s}{\sqrt{n}}\right) \qquad \text{or} \qquad 19.07 \pm 1.96\sqrt{\frac{10.60}{32}} \qquad \text{or} \qquad [17.94, 20.20]. \quad \diamond$$

So we have found a confidence interval for the mean μ of a normal distribution, assuming that the value of the standard deviation σ is known and when σ is unknown but the sample size is large. However, in many applications, the sample sizes are small and we do not know the value of the standard deviation, although in some cases we might have a very good idea about its value. For illustration, a manufacturer of light bulbs probably has a good notion from past experience of the value of the standard deviation of the length of life of different types of light bulbs. But certainly, most of the time, the investigator will not have any more idea about the standard deviation than about the mean—and frequently less. Let us consider how to proceed under these circumstances.

Suppose that the underlying distribution is normal and that σ^2 is unknown. We again consider the ratio

$$T = \frac{\bar{X} - \mu}{S/\sqrt{n}}$$

in which S has replaced σ in $(\bar{X} - \mu)/(\sigma/\sqrt{n})$. Note that this ratio can be written as follows:

$$T = \frac{\bar{X} - \mu}{S/\sqrt{n}} = \frac{\dfrac{\bar{X} - \mu}{\sigma/\sqrt{n}}}{\sqrt{\dfrac{(n-1)S^2}{\sigma^2}\Big/(n-1)}}.$$

We know from Theorems 5.3.1 and 5.3.4 that

$$Z = \frac{\bar{X} - \mu}{\sigma/\sqrt{n}} \quad \text{is} \quad N(0,1),$$

$$U = \frac{(n-1)S^2}{\sigma^2} \quad \text{is} \quad \chi^2(n-1),$$

and these two are independent. That is,

$$T = \frac{Z}{\sqrt{U/r}}, \qquad \text{where } r = n-1.$$

Any such T random variable is said to have a t distribution with r degrees of freedom. Here, of course, $r = n - 1$.

In Section 11.6 we derive the p.d.f. of $T = Z/\sqrt{U/r}$, where Z is $N(0, 1)$, U is $\chi^2(r)$, and Z and U are independent. However, so that we can use here some of the characteristics of this t distribution, we give some of its properties. First note that the distribution of T is determined completely by the number r. Since it is, in general, difficult to evaluate the distribution function of T, some values of $P(T \leq t)$ are found in Table VI for $r = 1, 2, 3, \ldots, 30$. Also, the graph of the p.d.f. of T is symmetrical with respect to the vertical axis $t = 0$ and is very similar to the graph of the p.d.f. of the standard normal distribution $N(0, 1)$. Figure 6.2 (a) shows the graphs of the probability density functions of T when $r = 1, 3,$ and 7 and of the standard normal distribution $N(0, 1)$. In this figure we see that the tails of the t distribution are heavier than those of a normal one; that is, there is more extreme probability in the t distribution than in the standardized normal one.

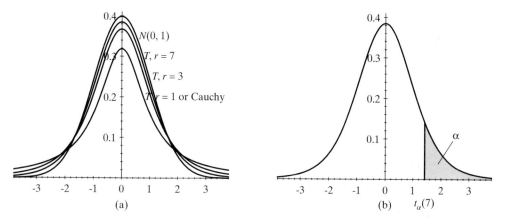

Figure 6.2 t distribution p.d.f.'s and right-tail probability

For notational purposes we shall let $t_\alpha(r)$ denote the constant for which

$$P(T \geq t_\alpha(r)) = \alpha,$$

when T has a t distribution with r degrees of freedom; that is, $t_\alpha(r)$ is the $100(1-\alpha)$ percentile (sometimes called the upper 100α percent point) of the t distribution with r degrees of freedom (see Figure 6.2 (b)). In this figure, $r = 7$ and $\alpha = 0.10$.

Let us consider some illustrations of the use of the t-table and this notation for right-tail probabilities.

Example 6.2.5

Let T have a t distribution with seven degrees of freedom. Then, from Table VI, we have

$$P(T \leq 1.415) = 0.90,$$

$$P(T \leq -1.415) = 1 - P(T < 1.415) = 0.10,$$

and

$$P(-1.895 < T < 1.415) = 0.90 - 0.05 = 0.85.$$

We also have, for example, $t_{0.10}(7) = 1.415$, $t_{0.90}(7) = -t_{0.10}(7) = -1.415$, and $t_{0.025}(7) = 2.365$.   ◇

Example 6.2.6

Let T have a t distribution with 14 degrees of freedom. Find a constant c such that $P(|T| < c) = 0.90$. From Table VI we see that $P(T \leq 1.761) = 0.95$ and therefore $c = 1.761 = t_{0.05}(14)$. ◇

We now use the fact that

$$T = \frac{\overline{X} - \mu}{S/\sqrt{n}}$$

has a t distribution with $r = n-1$ degrees of freedom, where S^2 is the usual unbiased estimator of σ^2. Select $t_{\alpha/2}(n-1)$ so that $P[T \geq t_{\alpha/2}(n-1)] = \alpha/2$. Then

$$1 - \alpha = P\left[-t_{\alpha/2}(n-1) \leq \frac{\overline{X} - \mu}{S/\sqrt{n}} \leq t_{\alpha/2}(n-1)\right]$$

$$= P\left[-t_{\alpha/2}(n-1)\left(\frac{S}{\sqrt{n}}\right) \leq \overline{X} - \mu \leq t_{\alpha/2}(n-1)\left(\frac{S}{\sqrt{n}}\right)\right]$$

$$= P\left[-\overline{X} - t_{\alpha/2}(n-1)\left(\frac{S}{\sqrt{n}}\right) \leq -\mu \leq -\overline{X} + t_{\alpha/2}(n-1)\left(\frac{S}{\sqrt{n}}\right)\right]$$

$$= P\left[\overline{X} - t_{\alpha/2}(n-1)\left(\frac{S}{\sqrt{n}}\right) \leq \mu \leq \overline{X} + t_{\alpha/2}(n-1)\left(\frac{S}{\sqrt{n}}\right)\right].$$

Thus the observations of a random sample provide $\overline{x}$ and s^2 and

$$\left[\overline{x} - t_{\alpha/2}(n-1)\left(\frac{S}{\sqrt{n}}\right), \overline{x} + t_{\alpha/2}(n-1)\left(\frac{S}{\sqrt{n}}\right)\right]$$

is a $100(1 - \alpha)\%$ confidence interval for μ.

Example 6.2.7

Let X equal the amount of butterfat in pounds produced by a typical cow during a 305-day milk production period between her first and second calves. Assume that the distribution of X is $N(\mu, \sigma^2)$. To estimate μ a farmer measured the butterfat production for $n = 20$ cows, yielding the following data:

$$
\begin{array}{cccccccccc}
481 & 537 & 513 & 583 & 453 & 510 & 570 & 500 & 457 & 555 \\
618 & 327 & 350 & 643 & 499 & 421 & 505 & 637 & 599 & 392
\end{array}
$$

For these data, $\overline{x} = 507.50$ and $s = 89.75$. Thus a point estimate of μ is $\overline{x} = 507.50$. Since $t_{0.05}(19) = 1.729$, a 90% confidence interval for μ is

$$507.50 \pm 1.729\left(\frac{89.75}{\sqrt{20}}\right)$$

$$507.50 \pm 34.70, \qquad \text{or equivalently,} \qquad [472.80, 542.20]. \qquad ◇$$

Let T have a t distribution with $n-1$ degrees of freedom. Then $t_{\alpha/2}(n-1) > z_{\alpha/2}$. Consequently, we would expect the interval $\bar{x} \pm z_{\alpha/2}(\sigma/\sqrt{n})$ to be shorter than the interval $\bar{x} \pm t_{\alpha/2}(n-1)(s/\sqrt{n})$. After all, we have more information, namely the value of σ, in constructing the first interval. However, the length of the second interval is very much dependent on the value of s. If the observed s is smaller than σ, a shorter confidence interval could result by the second scheme. But on the average, $\bar{x} \pm z_{\alpha/2}(\sigma/\sqrt{n})$ is the shorter of the two confidence intervals (Exercise 6.2.12).

Example 6.2.8

In Example 6.2.2, 50 confidence intervals were simulated for the mean of a normal distribution assuming that the variance was known. For those same data a 90% confidence interval was calculated for μ using $\bar{x} \pm 2.132(s/\sqrt{5})$. For those particular 50 intervals, 46 contained the mean $\mu = 50$. These are depicted in Figure 6.1(b). Note the different lengths of these intervals. Some are longer and some are shorter than the corresponding z intervals. The average length of these intervals is 7.137 which is quite close to the expected length of such an interval which is 7.169 (see Exercise 6.2.12). The length of the intervals that use z and $\sigma = 4$ is 5.885. ◇

If we are not able to assume that the underlying distribution is normal but μ and σ are both unknown, approximate confidence intervals for μ can still be constructed using

$$T = \frac{\bar{X} - \mu}{S/\sqrt{n}},$$

which now only has an approximate t distribution. Generally, this approximation is quite good for many nonnormal distributions (i.e., it is robust), in particular, if the underlying distribution is symmetric, unimodal, and of the continuous type. However, if the distribution is highly skewed, there is great danger using this approximation. In such a situation it would be safer to use certain nonparametric methods for finding a confidence interval for the median of the distribution, one of which is given in Section 10.2.

There is one other aspect of confidence intervals that should be mentioned. So far we have created only what are called **two-sided confidence intervals** for the mean μ. Sometimes it happens that you might want only a lower (or upper) bound on μ. We proceed as follows.

Say that $\bar{X}$ is the mean of a random sample of size n from the normal distribution $N(\mu, \sigma^2)$, where say for the moment that σ^2 is known. Then

$$P\left(\frac{\bar{X} - \mu}{\sigma/\sqrt{n}} \leq z_\alpha\right) = 1 - \alpha,$$

or equivalently,

$$P\left[\bar{X} - z_\alpha\left(\frac{\sigma}{\sqrt{n}}\right) \leq \mu\right] = 1 - \alpha.$$

Once $\overline{X}$ is observed to be equal to $\overline{x}$, then $[\overline{x} - z_\alpha(\sigma/\sqrt{n}), \infty)$ is a $100(1 - \alpha)\%$ **one-sided confidence interval** for μ. That is, with the confidence coefficient of $1 - \alpha$, $\overline{x} - z_\alpha(\sigma/\sqrt{n})$ is a lower bound for μ. Similarly, $(-\infty, \overline{x} + z_\alpha(\sigma/\sqrt{n})$ is a one-sided confidence interval for μ and $\overline{x} + z_\alpha(\sigma/\sqrt{n})$ provides an upper bound for μ with confidence coefficient $1 - \alpha$.

When σ is unknown, we would use $T = (\overline{X} - \mu)/(S/\sqrt{n})$ to find the corresponding lower or upper bounds for μ, $\overline{x} - t_\alpha(n-1)(s/\sqrt{n})$ and $\overline{x} + t_\alpha(n-1)(s/\sqrt{n})$.

EXERCISES

6.2.1 A random sample of size 16 from the normal distribution $N(\mu, 25)$ yielded $\overline{x} = 73.8$. Find a 95% confidence interval for μ.

6.2.2 A random sample of size 8 from $N(\mu, 72)$ yielded $\overline{x} = 85$. Find the following confidence intervals for μ: (a) 99%, (b) 95%, (c) 90%, and (d) 80%.

6.2.3 A pet store sells gerbil food in "2-pound" bags that are weighed on the platform of an old 25-pound scale. Suppose it is known that the standard deviation of weights is $\sigma = 0.12$ pound. If a sample of $n = 16$ bags of gerbil food were weighed carefully in a laboratory and the average weight was $\overline{x} = 2.09$ pounds, find an approximate 95% confidence interval for μ, the mean weight of gerbil food in the "2-pound" bags sold by the pet store.

6.2.4 Let X equal the weight in grams of a "52-gram" snack pack of candies. Assume that the distribution of X is $N(\mu, 4)$. A random sample of $n = 10$ observations of X yielded the following data:

$$
\begin{array}{ccccc}
55.95 & 56.54 & 57.58 & 55.13 & 57.48 \\
56.06 & 59.93 & 58.30 & 52.57 & 58.46
\end{array}
$$

(a) Give a point estimate for μ.

(b) Find the endpoints for a 95% confidence interval for μ.

(c) Based on these very limited data, what is the probability that an individual snack pack selected at random is filled with less than 52 grams of candy?

6.2.5 As a clue to the amount of organic waste in Lake Macatawa (see Example 6.2.4), a count was made of the number of bacteria colonies in 100 milliliters of water. The number of colonies, in hundreds, for $n = 30$ samples of water from the east basin yielded

$$
\begin{array}{cccccccccc}
93 & 140 & 8 & 120 & 3 & 120 & 33 & 70 & 91 & 61 \\
7 & 100 & 19 & 98 & 110 & 23 & 14 & 94 & 57 & 9 \\
66 & 53 & 28 & 76 & 58 & 9 & 73 & 49 & 37 & 92
\end{array}
$$

Find an approximate 90% confidence interval for the mean number, say μ_E, of colonies in 100 milliliters of water in the east basin.

6.2.6 To determine whether bacteria count was lower in the west basin of Lake Macatawa (Exercise 6.2.5) than in the east basin, $n = 37$ samples of water were taken from the west basin, and the number of bacteria colonies in 100 milliliters of water was counted. The sample characteristics were $\overline{x} = 11.95$ and $s = 11.80$, measured in hundreds of colonies. Find the approximate 95% confidence interval for the mean number of colonies, say μ_W, in 100 milliliters of water in the west basin.

6.2.7 Thirteen tons of cheese is stored in some old gypsum mines, including "22-pound" wheels (label weight). A random sample of $n = 9$ of these wheels yielded the following weights in pounds:

$$
\begin{array}{ccccc}
21.50 & 18.95 & 18.55 & 19.40 & 19.15 \\
22.35 & 22.90 & 22.20 & 23.10 &
\end{array}
$$

Assuming that the distribution of the weights of the wheels of cheese is $N(\mu, \sigma^2)$, find a 95% confidence interval for μ.

6.2.8 Assume that the yield per acre for a particular variety of soybeans is $N(\mu, \sigma^2)$. For a random sample of $n = 5$ plots, the yields in bushels per acre were 37.4, 48.8, 46.9, 55.0, and 44.0.

(a) Give a point estimate for μ.

(b) Find a 90% confidence interval for μ.

6.2.9 In a study of maximal aerobic capacity [*Journal of Applied Physiology* 65(6):2696 (December 1988)], 12 women were used as subjects, and one measurement that was made was blood plasma volume. The following data give their blood plasma volumes in liters:

$$
\begin{array}{cccccc}
3.15 & 2.99 & 2.77 & 3.12 & 2.45 & 3.85 \\
2.99 & 3.87 & 4.06 & 2.94 & 3.53 & 3.20
\end{array}
$$

Assume that these are observations of a normally distributed random variable X that has mean μ and standard deviation σ.

(a) Give the value of a point estimate of μ.

(b) Determine point estimates of σ^2 and σ.

(c) Find a 90% confidence interval for μ.

6.2.10 During the Friday night shift, $n = 28$ mints were selected at random from a production line and weighed. They had an average weight of $\bar{x} = 21.45$ grams and $s = 0.31$ gram. Give the lower endpoint of a 90% one-sided confidence interval for μ, the mean weight of all the mints.

6.2.11 A farm grows grapes for jelly. The following data are measurements of sugar in the grapes of a sample taken from each of 30 truckloads:

$$
\begin{array}{cccccccccc}
16.0 & 15.2 & 12.0 & 16.9 & 14.4 & 16.3 & 15.6 & 12.9 & 15.3 & 15.1 \\
15.8 & 15.5 & 12.5 & 14.5 & 14.9 & 15.1 & 16.0 & 12.5 & 14.3 & 15.4 \\
15.4 & 13.0 & 12.6 & 14.9 & 15.1 & 15.3 & 12.4 & 17.2 & 14.7 & 14.8
\end{array}
$$

Assume that these are observations of a random variable X that has mean μ and standard deviation σ.

(a) Give point estimates of μ and σ.

(b) Find an approximate 90% confidence interval for μ.

6.2.12 Let $X_1, X_2, \ldots, X_n$ be a random sample of size n from the normal distribution $N(\mu, \sigma^2)$. Calculate the expected length of a 95% confidence interval for μ assuming that $n = 5$ and the variance is (a) known and (b) unknown. HINT: To find $E(S)$, determine $E[\sqrt{(n-1)S^2/\sigma^2}]$, recalling that $(n-1)S^2/\sigma^2$ is $\chi^2(n-1)$. See also Exercise 6.1.14.

6.2.13 Students took $n = 35$ samples of water from the east basin of Lake Macatawa (see Example 6.2.4) and measured the amount of sodium in parts per million. For their data they calculated $\bar{x} = 24.11$ and $s^2 = 24.44$. Find an approximate 90% confidence interval for μ, the mean of the amount of sodium in parts per million.

6.2.14 A manufacturer of soap powder packages the soap in "6-pound" boxes. To check the filling machine, they took a sample of $n = 1219$ boxes and weighed them. Given that $\bar{x} = 6.05$ pounds and $s = 0.02$ pound, give the endpoints for a 99% confidence interval for μ, the mean weight of the boxes of soap filled by this machine.

6.2.15 Let X equal the weight (in pounds) of a "12-ounce" can of buttermilk biscuits. Assume that the distribution of X is $N(\mu, \sigma^2)$. Use the following 18 net weights to find a 95% one-sided confidence interval for μ that provides a lower bound for μ:

0.75	0.78	0.77	0.75	0.77	0.76	0.78	0.76	0.78
0.78	0.75	0.74	0.78	0.75	0.74	0.79	0.75	0.77

6.2.16 Let X equal the weight of an unbreaded $2.00 fish fry at a local restaurant in East Greenbush, NY. Assume that the distribution of X is $N(\mu, \sigma^2)$. A random sample of $n = 24$ weights was:

4.4	3.8	5.1	4.6	4.5	4.5	4.8	4.1
3.9	4.2	4.4	4.9	5.0	4.3	4.4	3.6
5.2	4.8	4.4	4.6	4.6	5.0	4.0	4.5

(a) Find point estimates of μ, σ^2, and σ.

(b) Find a 95% one-sided confidence interval for μ that gives a lower bound for μ.

(c) Does the assumption that these weights are normal seem to be justified? Why?

6.2.17 A study was conducted to measure the amount of cervical spine movement induced by different methods of gaining access to the mouth and nose to begin resuscitation of a helmet wearing football player and the time it takes to complete each method. One method is using a manual screwdriver to remove the side clips holding the face mask in place and then flipping the mask up. Twelve measured times in seconds for the manual screwdriver are:

33.8   31.6   28.5   29.9   29.8   26.0   35.7   27.2   29.1   32.1   26.1   24.1

Assume that these are independent observations of a normally distributed random variable that is $N(\mu, \sigma^2)$.

(a) Find point estimates of μ and σ.

(b) Find a 95% one-sided confidence interval for μ that provides an upper bound for μ.

(c) Does the assumption of normality seem to be justified? Why?

6.2.18 An interior automotive supplier places several electrical wires in a harness. A pull test measures the force required to pull spliced wires apart. A customer requires that each wire that is spliced into the harness withstand a pull force of 20 pounds. Let X equal the pull force required to pull 20 gauge wires apart. Assume that the distribution of X is $N(\mu, \sigma^2)$. The following data give 20 observations of X.

$$\begin{array}{cccccccccc} 28.8 & 24.4 & 30.1 & 25.6 & 26.4 & 23.9 & 22.1 & 22.5 & 27.6 & 28.1 \\ 20.8 & 27.7 & 24.4 & 25.1 & 24.6 & 26.3 & 28.2 & 22.2 & 26.3 & 24.4 \end{array}$$

(a) Find point estimates for μ and σ.

(b) Find a 99% one-sided confidence interval for μ that provides a lower bound for μ.

6.3 CONFIDENCE INTERVALS FOR DIFFERENCE OF TWO MEANS

Suppose that we are interested in comparing the means of two normal distributions. Let $X_1, X_2, \ldots, X_n$ and $Y_1, Y_2, \ldots, Y_m$ be, respectively, two independent random samples of sizes n and m from the two normal distributions $N(\mu_X, \sigma_X^2)$ and $N(\mu_Y, \sigma_Y^2)$. Suppose, for now, that σ_X^2 and σ_Y^2 are known. The random samples are independent and thus the respective sample means $\overline{X}$ and $\overline{Y}$ are also with distributions $N(\mu_X, \sigma_X^2/n)$ and $N(\mu_Y, \sigma_Y^2/m)$. Thus we know that the distribution of $W = \overline{X} - \overline{Y}$ is $N(\mu_X - \mu_Y, \sigma_X^2/n + \sigma_Y^2/m)$ and

$$P\left(-z_{\alpha/2} \le \frac{(\overline{X} - \overline{Y}) - (\mu_X - \mu_Y)}{\sqrt{\sigma_X^2/n + \sigma_Y^2/m}} \le z_{\alpha/2}\right) = 1 - \alpha,$$

which can be rewritten as

$$P[(\overline{X} - \overline{Y}) - z_{\alpha/2}\sigma_w \le \mu_X - \mu_Y \le (\overline{X} - \overline{Y}) + z_{\alpha/2}\sigma_w] = 1 - \alpha,$$

where $\sigma_w = \sqrt{\sigma_X^2/n + \sigma_Y^2/m}$ is the standard deviation of $\overline{X} - \overline{Y}$. Once the experiments have been performed and the means $\overline{x}$ and $\overline{y}$ computed, then

$$[\overline{x} - \overline{y} - z_{\alpha/2}\sigma_w, \; \overline{x} - \overline{y} + z_{\alpha/2}\sigma_w]$$

or, equivalently, $\overline{x} - \overline{y} \pm z_{\alpha/2}\sigma_w$ provides a $100(1 - \alpha)\%$ confidence interval for $\mu_X - \mu_Y$. Note that this interval is centered at the point estimate $\overline{x} - \overline{y}$ of $\mu_X - \mu_Y$ and completed by subtracting and adding the product of $z_{\alpha/2}$ and the standard deviation of the point estimator.

Example 6.3.1

In the preceding discussion, let $n = 15$, $m = 8$, $\overline{x} = 70.1$, $\overline{y} = 75.3$, $\sigma_X = 60$, $\sigma_Y = 40$, and $1 - \alpha = 0.90$. Thus $1 - \alpha/2 = 0.95 = \Phi(1.645)$. Hence

$$1.645\sigma_w = 1.645\sqrt{\frac{60}{15} + \frac{40}{8}} = 4.935,$$

and, since $\overline{x} - \overline{y} = -5.2$, we have that

$$[-5.2 - 4.935, -5.2 + 4.935] = [-10.135, -0.265]$$

is a 90% confidence interval for $\mu_X - \mu_Y$. Since the confidence interval does not include zero, we suspect that μ_Y is greater than μ_X. ◇

If the sample sizes are large and σ_X and σ_Y are unknown, we can replace σ_X^2 and σ_Y^2 with s_x^2 and s_y^2, where s_x^2 and s_y^2 are the values of the respective unbiased estimates of the variances. This means that

$$\bar{x} - \bar{y} \pm z_{\alpha/2}\sqrt{\frac{s_x^2}{n} + \frac{s_y^2}{m}}$$

serves as an approximate $100(1-\alpha)\%$ confidence interval for $\mu_X - \mu_Y$.

Now consider the problem of constructing confidence intervals for the difference of the means of two normal distributions when the variances are unknown but the sample sizes are small. Let $X_1, X_2, \ldots, X_n$ and $Y_1, Y_2, \ldots, Y_m$ be two independent random samples from the distributions $N(\mu_X, \sigma_X^2)$ and $N(\mu_Y, \sigma_Y^2)$, respectively. If the sample sizes are not large (say considerably smaller than 30), this problem can be a difficult one. However, even in these cases, if we can assume common, but unknown, variances, say $\sigma_X^2 = \sigma_Y^2 = \sigma^2$, there is a way out of our difficulty.

We know that

$$Z = \frac{\bar{X} - \bar{Y} - (\mu_X - \mu_Y)}{\sqrt{\sigma^2/n + \sigma^2/m}}$$

is $N(0, 1)$. Moreover, since the random samples are independent,

$$U = \frac{(n-1)S_X^2}{\sigma^2} + \frac{(m-1)S_Y^2}{\sigma^2}$$

is the sum of two independent chi-square random variables; thus U is $\chi^2(n+m-2)$. In addition, the independence of the sample means and sample variances implies that Z and U are independent. According to the definition of a T random variable,

$$T = \frac{Z}{\sqrt{U/(n+m-2)}}$$

has a t distribution with $n + m - 2$ degrees of freedom. That is,

$$T = \frac{\dfrac{\bar{X} - \bar{Y} - (\mu_X - \mu_Y)}{\sqrt{\sigma^2/n + \sigma^2/m}}}{\sqrt{\left[\dfrac{(n-1)S_X^2}{\sigma^2} + \dfrac{(m-1)S_Y^2}{\sigma^2}\right]\Big/(n+m-2)}}$$

$$= \frac{\bar{X} - \bar{Y} - (\mu_X - \mu_Y)}{\sqrt{\left[\dfrac{(n-1)S_X^2 + (m-1)S_Y^2}{n+m-2}\right]\left[\dfrac{1}{n} + \dfrac{1}{m}\right]}}$$

has a t distribution with $r = n + m - 2$ degrees of freedom. Thus, with $t_0 = t_{\alpha/2}(n+m-2)$,

$$P(-t_0 \leq T \leq t_0) = 1 - \alpha.$$

Solving the inequality for $\mu_X - \mu_Y$ yields

$$P\left(\overline{X} - \overline{Y} - t_0 S_P\sqrt{\frac{1}{n} + \frac{1}{m}} \le \mu_X - \mu_Y \le \overline{X} - \overline{Y} + t_0 S_P\sqrt{\frac{1}{n} + \frac{1}{m}}\right)$$

where the pooled estimator of the common standard deviation is

$$S_P = \sqrt{\frac{(n-1)S_X^2 + (m-1)S_Y^2}{n+m-2}}.$$

If $\overline{x}, \overline{y}$, and s_p are the observed values of $\overline{X}, \overline{Y}$, and S_P, then

$$\left[\overline{x} - \overline{y} - t_0 s_p\sqrt{\frac{1}{n} + \frac{1}{m}}, \overline{x} - \overline{y} + t_0 s_p\sqrt{\frac{1}{n} + \frac{1}{m}}\right]$$

is a $100(1-\alpha)\%$ confidence interval for $\mu_X - \mu_Y$.

Example 6.3.2

Suppose that scores on a standardized test in mathematics taken by students from large and small high schools are $N(\mu_X, \sigma^2)$ and $N(\mu_Y, \sigma^2)$, respectively, where σ^2 is unknown. If a random sample of $n = 9$ students from large high schools yielded $\overline{x} = 81.31$, $s_x^2 = 60.76$ and a random sample of $m = 15$ students from small high schools yielded $\overline{y} = 78.61$, $s_y^2 = 48.24$, the endpoints for a 95% confidence interval for $\mu_X - \mu_Y$ are given by

$$81.31 - 78.61 \pm 2.074\sqrt{\frac{8(60.76) + 14(48.24)}{22}}\sqrt{\frac{1}{9} + \frac{1}{15}}$$

because $t_{0.025}(22) = 2.074$. The 95% confidence interval is $[-3.65, 9.05]$.   ◇

The assumption of equal variances, namely $\sigma_X^2 = \sigma_Y^2$, can be modified somewhat so that we are still able to find a confidence interval for $\mu_X - \mu_Y$. That is, if we know the ratio σ_X^2/σ_Y^2 of the variances, we can still make this type of statistical inference using a random variable with a t distribution (see Exercise 6.3.6). However, if we do not know the ratio of the variances and yet suspect that the unknown σ_X^2 and σ_Y^2 differ by a great deal, what do we do? It is safest to return to

$$\frac{\overline{X} - \overline{Y} - (\mu_X - \mu_Y)}{\sqrt{\sigma_X^2/n + \sigma_Y^2/m}}$$

for the inference about $\mu_X - \mu_Y$ but replacing σ_X^2 and σ_Y^2 by their respective estimators S_X^2 and S_Y^2. That is, consider

$$W = \frac{\overline{X} - \overline{Y} - (\mu_X - \mu_Y)}{\sqrt{S_X^2/n + S_Y^2/m}}.$$

What is the distribution of W? As before, we note that if n and m are large enough and the underlying distributions are close to normal (or at least not badly skewed), W has an approximate normal distribution and a confidence interval for $\mu_X - \mu_Y$ can be found by considering

$$P(-z_{\alpha/2} \le W \le z_{\alpha/2}) \approx 1 - \alpha.$$

However, if n and m are smaller, then Welch has proposed a Student-t distribution as the approximating one for W. Welch's proposal was later modified by Aspin. [See A. A. Aspin, "Tables for Use in Comparisons Whose Accuracy Involves Two Variances, Separately Estimated," *Biometrika* **36**:290–296 (1949), with an appendix by B. L. Welch in which he makes the suggestion used here.] The approximating Student-t distribution has r degrees of freedom where

$$\frac{1}{r} = \frac{c^2}{n-1} + \frac{(1-c)^2}{m-1} \quad \text{and} \quad c = \frac{s_x^2/n}{s_x^2/n + s_y^2/m}.$$

An equivalent formula for r is

$$r = \frac{\left(\dfrac{s_x^2}{n} + \dfrac{s_y^2}{m} \right)^2}{\dfrac{1}{n-1}\left(\dfrac{s_x^2}{n} \right)^2 + \dfrac{1}{m-1}\left(\dfrac{s_y^2}{m} \right)^2}.$$

In particular, the assignment of r by this rule provides protection in the case in which the smaller sample size is associated with the larger variance by greatly reducing the number of degrees of freedom from the usual $n + m - 2$. Of course, this increases the value of $t_{\alpha/2}$. If r is not an integer, then use the greatest integer in r; that is, $[r]$ is the number of degrees of freedom associated with the approximating Student-t distribution. An approximate $100(1 - \alpha)$ percent confidence interval for $\mu_X - \mu_Y$ is given by

$$\bar{x} - \bar{y} \pm t_{\alpha/2}(r)\sqrt{\frac{s_x^2}{n} + \frac{s_y^2}{m}}.$$

In some applications, two measurements, say X and Y, are taken on the same subject. In these cases, X and Y are dependent random variables. Many times these are "before" and "after" measurements, such as weight before and after participating in a diet and exercise program. To compare the means of X and Y, it is not permissible to use the t statistics and confidence intervals that we just developed, because there X and Y are independent. Instead, we proceed as follows.

Let $(X_1, Y_1), (X_2, Y_2), \ldots, (X_n, Y_n)$ be n pairs of dependent measurements. Let $D_i = X_i - Y_i$, $i = 1, 2, \ldots, n$. Suppose that $D_1, D_2, \ldots, D_n$ could be thought of as a random sample from $N(\mu_D, \sigma_D^2)$, where μ_D and σ_D are the mean and standard deviation of each difference.

To form a confidence interval for $\mu_X - \mu_Y$, use

$$T = \frac{\overline{D} - \mu_D}{S_D/\sqrt{n}},$$

where $\overline{D}$ and S_D are the sample mean and sample standard deviation of the n differences. Thus T is a t statistic with $n - 1$ degrees of freedom. The endpoints for a $100(1 - \alpha)\%$ confidence interval for $\mu_D = \mu_X - \mu_Y$ are then

$$\overline{d} \pm t_{\alpha/2}(n-1) \frac{s_d}{\sqrt{n}},$$

where $\overline{d}$ and s_d are the observed mean and standard deviation of the sample. Of course, this is the same as the first confidence interval for a single mean in Section 6.2.

Example 6.3.3

An experiment was conducted to compare people's reaction times to a red light versus a green light. When signaled with either the red or the green light, the subject was asked to hit a switch to turn off the light. When the switch was hit, a clock was turned off and the reaction time in seconds was recorded. The following results give the reaction times for eight subjects:

Subject	Red (X)	Green (Y)	$D = X - Y$
1	0.30	0.43	-0.13
2	0.23	0.32	-0.09
3	0.41	0.58	-0.17
4	0.53	0.46	0.07
5	0.24	0.27	-0.03
6	0.36	0.41	-0.05
7	0.38	0.38	0.00
8	0.51	0.61	-0.10

For these data, $\overline{d} = -0.0625$ and $s_d = 0.0765$. To form a 95% confidence interval for $\mu_D = \mu_X - \mu_Y$, we find from Table VI that $t_{0.025}(7) = 2.365$. Thus the endpoints for the confidence interval are

$$-0.0625 \pm 2.365 \frac{0.07675}{\sqrt{8}} \qquad \text{or} \qquad [-0.1265, 0.0015].$$

In this very limited data set, zero is included in the confidence interval but is close to the endpoint 0.0015. We suspect that if more data were taken, zero might not be included in the confidence interval. Accordingly, if this actually happens, it would seem that people react faster to a red light. ◇

EXERCISES

6.3.1 The length of life of brand X light bulbs is assumed to be $N(\mu_X, 784)$. The length of life of brand Y light bulbs is assumed to be $N(\mu_Y, 627)$ and independent of that of X. If a random sample of $n = 56$ brand X light bulbs yielded a mean of $\overline{x} = 937.4$ hours and a random sample of size $m = 57$ brand Y light bulbs yielded a mean of $\overline{y} = 988.9$ hours, find a 90% confidence interval for $\mu_X - \mu_Y$.

6.3.2 Let $X_1, X_2, .., X_5$ be a random sample of SAT mathematics scores, assumed to be $N(\mu_X, \sigma^2)$, and let $Y_1, Y_2, \ldots, Y_8$ be an independent random sample of SAT verbal scores, assumed to be $N(\mu_Y, \sigma^2)$ If the following data are observed, find a 90% confidence interval for $\mu_X - \mu_Y$:

$$x_1 = 644 \quad x_2 = 493 \quad x_3 = 532 \quad x_4 = 462 \quad x_5 = 565$$

$$y_1 = 623 \quad y_2 = 472 \quad y_3 = 492 \quad y_4 = 661 \quad y_5 = 540$$
$$y_6 = 502 \quad y_7 = 549 \quad y_8 = 518$$

6.3.3 Independent random samples of the heights of adult males living in two countries yielded the following results: $n = 12$, $\overline{x} = 65.7$ inches, $s_x = 4$ inches and $m = 15$, $\overline{y} = 68.2$ inches, $s_y = 3$ inches. Find an approximate 98% confidence interval for the difference $\mu_X - \mu_Y$ of the means of the populations of heights. Assume that $\sigma_X^2 = \sigma_Y^2$.

6.3.4 [*Medicine and Science in Sports and Exercise* (January 1990).] Let X and Y equal, respectively, the blood volumes in milliliters for a male who is a paraplegic and participates in vigorous physical activities and a male who is able-bodied and participates in normal activities. Assume that X is $N(\mu_X, \sigma_X^2)$ and Y is $N(\mu_Y, \sigma_Y^2)$, respectively. Using the following $n = 7$ observations of X:

$$1612 \quad 1352 \quad 1456 \quad 1222 \quad 1560 \quad 1456 \quad 1924$$

and $m = 10$ observations of Y:

$$1082 \quad 1300 \quad 1092 \quad 1040 \quad 910$$
$$1248 \quad 1092 \quad 1040 \quad 1092 \quad 1288$$

(a) Give a point estimate for $\mu_X - \mu_Y$.

(b) Find a 95% confidence interval for $\mu_X - \mu_Y$. Since the variances σ_X^2 and σ_Y^2 might not be equal, use Welch's T.

6.3.5 A biologist who studies spiders was interested in comparing the lengths of female and male green lynx spiders. Assume that the length X of the male spider is approximately $N(\mu_X, \sigma_X^2)$ and the length Y of the female spider is approximately $N(\mu_Y, \sigma_Y^2)$. Find an approximate 95% confidence interval for $\mu_X - \mu_Y$ using $n = 30$ observations of X:

$$
\begin{array}{cccccc}
5.20 & 4.70 & 5.75 & 7.50 & 6.45 & 6.55 \\
4.70 & 4.80 & 5.95 & 5.20 & 6.35 & 6.95 \\
5.70 & 6.20 & 5.40 & 6.20 & 5.85 & 6.80 \\
5.65 & 5.50 & 5.65 & 5.85 & 5.75 & 6.35 \\
5.75 & 5.95 & 5.90 & 7.00 & 6.10 & 5.80 \\
\end{array}
$$

and $m = 30$ observations of Y:

8.25	9.95	5.90	7.05	8.45	7.55
9.80	10.80	6.60	7.55	8.10	9.10
6.10	9.30	8.75	7.00	7.80	8.00
9.00	6.30	8.35	8.70	8.00	7.50
9.50	8.30	7.05	8.30	7.95	9.60

where the measurements are in millimeters.

6.3.6 Let $\overline{X}, \overline{Y}, S_X^2$, and S_Y^2 be the respective sample means and unbiased estimates of the variances using samples of sizes n and m from the independent normal distributions $N(\mu_X, \sigma_X^2)$ and $N(\mu_Y, \sigma_Y^2)$, where μ_X, μ_Y, σ_X^2, and σ_Y^2 are unknown. If, however, $\sigma_X^2/\sigma_Y^2 = d$, a known constant, argue that

(a) $\dfrac{(\overline{X} - \overline{Y}) - (\mu_X - \mu_Y)}{\sqrt{d\sigma_Y^2/n + \sigma_Y^2/m}}$ is $N(0, 1)$.

(b) $\dfrac{(n-1)S_X^2}{d\sigma_Y^2} + \dfrac{(m-1)S_Y^2}{\sigma_Y^2}$ is $\chi^2(n+m-2)$.

(c) The two random variables in parts (a) and (b) are independent.

(d) With these results, construct a random variable (not depending upon σ_Y^2) that has a t distribution and can be used to construct a confidence interval for $\mu_X - \mu_Y$.

6.3.7 Students are weighed (in kilograms) at the beginning and the end of a semester-long health-fitness program. Let the random variable D equal the weight change for a student, postweight minus preweight. Assume that the distribution of D is $N(\mu_D, \sigma_D^2)$. A random sample of $n = 12$ female students yielded the following observations of D:

2.0	−0.5	1.4	−2.2	0.3	−0.8
3.7	−0.1	0.6	0.2	0.9	−0.1

(a) Give a point estimate of μ_D.

(b) Find a 95% confidence interval for μ_D.

6.3.8 Twenty-four 9th- and 10th-grade high school girls were put on an ultraheavy rope-jumping program. The following data give the time difference for each girl—"before program time" minus "after program time"— for the 40-yard dash:

0.28	0.01	0.13	0.33	−0.03	0.07	−0.18	−0.14
−0.33	0.01	0.22	0.29	−0.08	0.23	0.08	0.04
−0.30	−0.08	0.09	0.70	0.33	−0.34	0.50	0.06

(a) Give a point estimate of μ_D, the mean of the difference in race times.

(b) Find a 95% confidence interval for μ_D.

(c) Does it look like the rope-jumping program was effective?

312 Chap. 6 Estimation

6.3.9 The biomechanics lab at Hope College tested healthy old women and healthy young women to discover whether lower-extremity response time to a stimulus is a function of age. Let X and Y, respectively, equal the response times for these two groups when taking steps in the anterior direction. Find a 95% confidence interval for $\mu_X - \mu_Y$ given that $n = 60$ observations of X yielded $\bar{x} = 671$ and $s_x = 129$ while $m = 60$ observations of Y yielded $\bar{y} = 480$ and $s_y = 93$.

6.4 CONFIDENCE INTERVALS FOR VARIANCES

In this section we find confidence intervals for the variance of a normal distribution and for the ratio of the variances of two normal distributions. The confidence interval for the variance σ^2 is based on the sample variance

$$S^2 = \frac{1}{n-1} \sum_{i=1}^{n} (X_i - \bar{X})^2.$$

We use the fact that the distribution of $(n-1)S^2/\sigma^2$ is $\chi^2(n-1)$ to find a confidence interval for σ^2. Select constants a and b from Table IV with $n-1$ degrees of freedom such that

$$P\left(a \leq \frac{(n-1)S^2}{\sigma^2} \leq b\right) = 1 - \alpha.$$

One way to do this is by selecting a and b so that $a = \chi^2_{1-\alpha/2}(n-1)$ and $b = \chi^2_{\alpha/2}(n-1)$. That is, select a and b so that the probabilities in the two tails are equal. Then, solving the inequalities, we have

$$1 - \alpha = P\left(\frac{a}{(n-1)S^2} \leq \frac{1}{\sigma^2} \leq \frac{b}{(n-1)S^2}\right)$$

$$= P\left(\frac{(n-1)S^2}{b} \leq \sigma^2 \leq \frac{(n-1)S^2}{a}\right).$$

Thus the probability that the random interval $[(n-1)S^2/b, (n-1)S^2/a]$ contains the unknown σ^2 is $1 - \alpha$. Once the values of $X_1, X_2, \ldots, X_n$ are observed to be $x_1, x_2, \ldots, x_n$ and s^2 computed, then the interval $[(n-1)s^2/b, (n-1)s^2/a]$ is a $100(1-\alpha)\%$ confidence interval for σ^2. It follows that a $100(1-\alpha)\%$ confidence interval for σ, the standard deviation, is given by

$$\left[\sqrt{\frac{(n-1)s^2}{b}}, \sqrt{\frac{(n-1)s^2}{a}}\right] = \left[\sqrt{\frac{n-1}{b}}\,s, \sqrt{\frac{n-1}{a}}\,s\right].$$

Example 6.4.1

Assume that the time in days required for maturation of seeds of a species of guardi-ola, a flowering plant found in Mexico, is $N(\mu, \sigma^2)$. A random sample of $n = 13$ seeds, both parents having narrow leaves, yielded $\bar{x} = 18.97$ days and

$$12s^2 = \sum_{i=1}^{13} (x_i - \bar{x})^2 = 128.41.$$

A 90% confidence interval for σ^2 is

$$\left[\frac{128.41}{21.03}, \frac{128.41}{5.226} \right] = [6.11, 24.57]$$

because $5.226 = \chi^2_{0.95}(12)$ and $21.03 = \chi^2_{0.05}(12)$ from Table IV. The corresponding 90% confidence interval for σ is

$$[\sqrt{6.11}, \sqrt{24.57}] = [2.47, 4.96]. \qquad \diamond$$

Although a and b are generally selected so that the probabilities in the two tails are equal, the resulting $100(1 - \alpha)\%$ confidence interval is not the shortest that can be formed using the available data. Table X gives solutions for a and b that yield confidence intervals of minimum length for the standard deviation (see Exercise 6.4.14).

Example 6.4.2

Using the data in Example 6.4.1, a 90% confidence interval for σ of minimum length is (since $a = 5.940$ and $b = 24.202$ from Table X)

$$\left[\sqrt{\frac{128.41}{24.202}}, \sqrt{\frac{128.41}{5.940}} \right] = [2.30, 4.65].$$

The length of this interval is 2.35, whereas the length of the interval given in Example 6.4.1 is 2.49. To see why this new interval is shorter, carefully sketch a graph of the $\chi^2(12)$ p.d.f. and compare $\chi^2_{0.95}(12)$ to a and $\chi^2_{0.05}(12)$ to b on the graph. $\qquad \diamond$

There are occasions when it is of interest to compare the variances of two normal distributions. We do this by finding a confidence interval for σ_X^2/σ_Y^2 using the ratio of S_X^2/σ_X^2 and S_Y^2/σ_Y^2, where S_X^2 and S_Y^2 are the two sample variances based on two independent samples of sizes n and m from $N(\mu_X, \sigma_X^2)$ and $N(\mu_Y, \sigma_Y^2)$, respectively. However the reciprocal of that ratio can be rewritten as follows:

$$\frac{\dfrac{S_Y^2}{\sigma_Y^2}}{\dfrac{S_X^2}{\sigma_X^2}} = \frac{\left[\dfrac{(m-1)S_Y^2}{\sigma_Y^2} \right] \Big/ (m-1)}{\left[\dfrac{(n-1)S_X^2}{\sigma_X^2} \right] \Big/ (n-1)}.$$

Since $(m-1)S_Y^2/\sigma_Y^2$ and $(n-1)S_X^2/\sigma_X^2$ are independent chi-square variables with $(m-1)$ and $(n-1)$ degrees of freedom, respectively, this suggests that we define a new random variable, which is usually denoted by F.

If U and V are independent chi-square variables with r_1 and r_2 degrees of freedom, respectively, then

$$F = \frac{U/r_1}{V/r_2}$$

has an F distribution with r_1 and r_2 degrees of freedom. The distribution of F is derived in Section 11.6, and here we only mention how to find certain probabilities associated with this distribution. See Figure 11.12, Section 11.6, for an F p.d.f. graph.

Some values of the distribution function $P(F \le f)$ of the F distribution are given in Table VII. For notational purposes, if F has an F distribution with r_1 and r_2 degrees of freedom, we say that the distribution of F is $F(r_1, r_2)$. Furthermore, we will let $F_\alpha(r_1, r_2)$ denote the constant [the upper 100α percent point of $F(r_1, r_2)$] for which

$$P[F \ge F_\alpha(r_1, r_2)] = \alpha.$$

Example 6.4.3

If the distribution of F is $F(r_1, r_2)$, then from Table VII we see that

When $r_1 = 7, r_2 = 8$, $P(F \ge 3.50) = 0.95$, so $F_{0.05}(7, 8) = 3.50$.

When $r_1 = 9, r_2 = 4$, $P(F \ge 14.66) = 0.99$, so $F_{0.01}(9, 4) = 14.66$.   ◇

To use Table VII to find the F values corresponding to the 0.01, 0.025, and 0.05 cumulative probabilities we need to note the following: Since $F = (U/r_1)/(V/r_2)$, where U and V are independent and $\chi^2(r_1)$ and $\chi^2(r_2)$, respectively, then $1/F = (V/r_2)/(U/r_1)$ must have a distribution that is $F(r_2, r_1)$. Note the change in the order of the parameters in the distribution of $1/F$. Now if the distribution of F is $F(r_1, r_2)$, then

$$P[F \le F_{1-\alpha}(r_1, r_2)] = \alpha$$

and

$$P\left[\frac{1}{F} \ge \frac{1}{F_{1-\alpha}(r_1, r_2)}\right] = \alpha.$$

The complement of $\{1/F \ge 1/F_{1-\alpha}(r_1, r_2)\}$ is $\{1/F < 1/F_{1-\alpha}(r_1, r_2)\}$. Thus

$$P\left[\frac{1}{F} < \frac{1}{F_{1-\alpha}(r_1, r_2)}\right] = 1 - \alpha. \tag{6.2}$$

Since the distribution of $1/F$ is $F(r_2, r_1)$, by definition of $F_\alpha(r_2, r_1)$, we have

$$P\left[\frac{1}{F} < F_\alpha(r_2, r_1)\right] = 1 - \alpha. \tag{6.3}$$

From Equations 6.2 and 6.3 we see that

$$\frac{1}{F_{1-\alpha}(r_1, r_2)} = F_\alpha(r_2, r_1),$$

or

$$F_{1-\alpha}(r_1, r_2) = \frac{1}{F_\alpha(r_2, r_1)}. \tag{6.4}$$

The use of Equation 6.4 is illustrated in the next example.

Example 6.4.4

If the distribution of F is $F(4, 9)$, constants c and d such that

$$P(F \le c) = 0.01 \quad \text{and} \quad P(F \le d) = 0.05$$

are given by

$$c = F_{0.99}(4, 9) = \frac{1}{F_{0.01}(9, 4)} = \frac{1}{14.66} = 0.0682;$$

$$d = F_{0.95}(4, 9) = \frac{1}{F_{0.05}(9, 4)} = \frac{1}{6.00} = 0.1667.$$

Furthermore, if F is $F(6, 9)$, then

$$P(F \le 0.2439) = P\left(\frac{1}{F} \ge \frac{1}{0.2439}\right) = P\left(\frac{1}{F} \ge 4.100\right) = 0.05$$

because the distribution of $1/F$ is $F(9, 6)$.   ◇

We now show how the F distribution can be used to find a confidence interval for σ_X^2/σ_Y^2. Let $X_1, X_2, \ldots, X_n$ and $Y_1, Y_2, \ldots, Y_m$ be independent random samples of sizes n and m from two normal distributions $N(\mu_X, \sigma_X^2)$ and $N(\mu_Y, \sigma_Y^2)$, respectively. We know that $(n-1)S_X^2/\sigma_X^2$ is $\chi^2(n-1)$ and $(m-1)S_Y^2/\sigma_Y^2$ is $\chi^2(m-1)$. The independence of the samples implies that S_X and S_Y are independent so that

$$F = \frac{\dfrac{(m-1)S_Y^2}{\sigma_Y^2(m-1)}}{\dfrac{(n-1)S_X^2}{\sigma_X^2(n-1)}} = \frac{\dfrac{S_Y^2}{\sigma_Y^2}}{\dfrac{S_X^2}{\sigma_X^2}}$$

has an F distribution with $r_1 = m - 1$ and $r_2 = n - 1$ degrees of freedom. This is the ratio that we want to use to find a confidence interval for σ_X^2/σ_Y^2.

To form the confidence interval, select constants c and d from Table VII so that

$$1 - \alpha = P\left(c \le \frac{S_Y^2/\sigma_Y^2}{S_X^2/\sigma_X^2} \le d\right)$$

$$= P\left(c\frac{S_X^2}{S_Y^2} \le \frac{\sigma_X^2}{\sigma_Y^2} \le d\frac{S_X^2}{S_Y^2}\right).$$

Because of the limitations of Table VII, we generally let $c = F_{1-\alpha/2}(m-1, n-1) = 1/F_{\alpha/2}(n-1, m-1)$ and $d = F_{\alpha/2}(m-1, n-1)$. If s_x^2 and s_y^2 are the observed values of S_X^2 and S_Y^2, respectively, then

$$\left[\frac{1}{F_{\alpha/2}(n-1, m-1)} \frac{s_x^2}{s_y^2}, \; F_{\alpha/2}(m-1, n-1) \frac{s_x^2}{s_y^2} \right]$$

is a $100(1 - \alpha)\%$ confidence interval for σ_X^2/σ_Y^2. By taking square roots of both endpoints, we would obtain a $100(1 - \alpha)\%$ confidence interval for σ_X/σ_Y.

Example 6.4.5

In Example 6.4.1, denote σ^2 by σ_X^2. There $(n - 1)s_x^2 = 12s^2 = 128.41$. Assume that the time in days required for maturation of seeds of a species of guardiola, both parents having broad leaves, is $N(\mu_Y, \sigma_Y^2)$. A random sample of size $m = 9$ seeds yielded $\overline{y} = 23.20$ and

$$8s_y^2 = \sum_{i=1}^{8} (y_i - \overline{y})^2 = 36.72.$$

A 98% confidence interval for σ_X^2/σ_Y^2 is given by

$$\left[\left(\frac{1}{5.67} \right) \frac{(128.41)/12}{(36.72)/8}, \; (4.50) \frac{(128.41)/12}{(36.72)/8} \right] = [0.41, 10.49]$$

because $F_{0.01}(12, 8) = 5.67$ and $F_{0.01}(8, 12) = 4.50$. It follows that a 98% confidence interval for σ_X/σ_Y is

$$[\sqrt{0.41}, \sqrt{10.49}\,] = [0.64, 3.24]. \qquad \diamond$$

Although we are able to formally find a confidence interval for the ratio of two distribution variances and/or standard deviations, we should point out that these intervals are generally not too useful because they are often very wide.

EXERCISES

6.4.1 Let X equal the length (in centimeters) of a certain species of fish when caught in the spring. A random sample of $n = 13$ observations of X is

$$\begin{array}{ccccccc}
13.1 & 5.1 & 18.0 & 8.7 & 16.5 & 9.8 & 6.8 \\
12.0 & 17.8 & 25.4 & 19.2 & 15.8 & 23.0 &
\end{array}$$

(a) Give a point estimate of the standard deviation σ of this species of fish.

(b) Find a 95% confidence interval for σ.

6.4.2 A random sample of $n = 9$ wheels of cheese yielded the following weights in pounds, assumed to be $N(\mu, \sigma^2)$:

$$
\begin{array}{ccccc}
21.50 & 18.95 & 18.55 & 19.40 & 19.15 \\
22.35 & 22.90 & 22.20 & 23.10 &
\end{array}
$$

(a) Give a point estimate of σ.

(b) Find a 95% confidence interval for σ.

(c) Find a 90% confidence interval for σ.

6.4.3 A student who works in a blood lab tested 25 men for cholesterol levels and found the following values:

$$
\begin{array}{ccccccccc}
164 & 272 & 261 & 248 & 235 & 192 & 203 & 278 & 268 \\
230 & 242 & 305 & 286 & 310 & 345 & 289 & 326 & \\
335 & 297 & 328 & 400 & 228 & 194 & 338 & 252 &
\end{array}
$$

Assume that these values represent observations of a random sample taken from $N(\mu, \sigma^2)$.

(a) Calculate the sample mean and sample variance for these data.

(b) Find a 90% confidence interval for σ^2.

(c) Find a 90% confidence interval for σ.

(d) Find a 90% confidence interval for σ that has minimum length.

(e) Does the assumption of normality seem to be valid?

6.4.4 For the data in Exercise 6.2.9, find a 95% confidence interval for (a) σ^2, (b) σ, and (c) σ having minimum length.

6.4.5 Let $X_1, X_2, X_3, \ldots, X_n$ be a random sample from $N(\mu, \sigma^2)$, with known mean μ. Describe how you would construct a confidence interval for the unknown variance σ^2. HINT: Use the fact that $\sum_{i=1}^{n}(X_i - \mu)^2/\sigma^2$ is $\chi^2(n)$.

6.4.6 Let $X_1, X_2, \ldots, X_n$ be a random sample of size n from an exponential distribution with unknown mean of $\mu = \theta$.

(a) Show that the distribution of $W = (2/\theta)\sum_{i=1}^{n} X_i$ is $\chi^2(2n)$. HINT: Find the moment-generating function of W.

(b) Use W to construct a $100(1 - \alpha)\%$ confidence interval for θ.

(c) If $n = 7$ and $\bar{x} = 93.6$, give the endpoints for a 90% confidence interval for the mean θ.

6.4.7 Let X equal the time in seconds between phone calls at Great Lakes Pizza Company. Fifteen observations of X are:

$$
\begin{array}{cccccccc}
82 & 42 & 185 & 66 & 384 & 27 & 334 & 545 \\
650 & 127 & 35 & 45 & 285 & 133 & 120 &
\end{array}
$$

(a) Assuming that X has an exponential distribution with mean θ, use Exercise 6.4.6 to construct a 95% confidence interval for θ.

(b) Calculate the values of $\bar{x}$ and s. Do they support the assumption that X has an exponential distribution?

(c) Construct a q-q plot or a box-and-whiskers display to confirm or deny the assumption that X has an exponential distribution.

6.4.8 Let X and Y equal the weights of a phosphorus-free laundry detergent in a "6-pound" box and a "12-pound" box, respectively. Assume that the distributions of X and Y are $N(\mu_X, \sigma_X^2)$ and $N(\mu_Y, \sigma_Y^2)$, respectively. A random sample of $n = 10$ observations of X yielded a sample mean of $\bar{x} = 6.10$ pounds with a sample variance of $s_x^2 = 0.0040$, while an independent random sample of $m = 9$ observations of Y yielded a sample mean of $\bar{y} = 12.10$ pounds with a sample variance of $s_y^2 = 0.0076$.

    (a) Give a point estimate of σ_X^2/σ_Y^2.

    (b) Find a 95% confidence interval for σ_X^2/σ_Y^2.

6.4.9 Let X and Y equal the number of milligrams of tar in filtered and nonfiltered cigarettes, respectively. Assume that the distributions of X and Y are $N(\mu_X, \sigma_X^2)$ and $N(\mu_Y, \sigma_Y^2)$, respectively. A random sample of $n = 9$ observations of X were

$$0.9 \quad 1.1 \quad 0.1 \quad 0.7 \quad 0.3 \quad 0.9 \quad 0.8 \quad 1.0 \quad 0.4$$

and an independent random sample of $m = 11$ observations of Y were

$$1.5 \quad 0.9 \quad 1.6 \quad 0.5 \quad 1.4 \quad 1.9 \quad 1.0 \quad 1.2 \quad 1.3 \quad 1.6 \quad 2.1$$

    (a) Give a point estimate of σ_X^2/σ_Y^2.

    (b) Find a 98% confidence interval for σ_X^2/σ_Y^2.

6.4.10 A candymaker produces mints that have a label weight of 20.4 grams. For quality assurance, $n = 16$ mints were selected at random from the Wednesday morning shift, resulting in the statistics $\bar{x} = 21.95$ grams and $s_x = 0.197$. On Wednesday afternoon $m = 13$ mints were selected at random, giving $\bar{y} = 21.88$ grams and $s_y = 0.318$. Find a 90% confidence interval for σ_X/σ_Y, the ratio of the standard deviations of the mints produced by the morning and by the afternoon shifts, respectively.

6.4.11 Let X and Y equal the concentration in parts per billion of chromium in the blood for healthy persons and for persons with a suspected disease, respectively. Assume that the distributions of X and Y are $N(\mu_X, \sigma_X^2)$ and $N(\mu_Y, \sigma_Y^2)$, respectively. Using $n = 8$ observations of X:

$$15 \quad 23 \quad 12 \quad 18 \quad 9 \quad 28 \quad 11 \cdot 10$$

and $m = 10$ observations of Y:

$$25 \quad 20 \quad 35 \quad 15 \quad 40 \quad 16 \quad 10 \quad 22 \quad 18 \quad 32$$

    (a) Give a point estimate of σ_X^2/σ_Y^2.

    (b) Find a 95% confidence interval for σ_X^2/σ_Y^2.

6.4.12 Some nurses were interested in the effect of prenatal care on the birth weight of babies. Mothers were divided into two groups, and their babies' weights were compared. The birth weights in ounces of babies of mothers who had received 5 or fewer prenatal visits were

$$
\begin{array}{ccccccc}
49 & 108 & 110 & 82 & 93 & 114 & 134 \\
114 & 96 & 52 & 101 & 114 & 120 & 116
\end{array}
$$

and the birth weights of babies of mothers who had received 6 or more prenatal visits were

$$\begin{array}{ccccccc} 133 & 108 & 93 & 119 & 119 & 98 & 106 \\ 87 & 153 & 116 & 129 & 97 & 110 & 131 \end{array}$$

Assuming that these are, respectively, independent observations of X and Y, which are $N(\mu_X, \sigma_X^2)$ and $N(\mu_Y, \sigma_Y^2)$, find a 95% confidence interval for (a) σ_X^2/σ_Y^2 and (b) σ_X/σ_Y.

6.4.13 Using a random sample of size n from the normal distribution $N(\mu, \sigma^2)$, find the $100(1-\alpha)\%$ confidence interval for μ of minimum length based on the statistic $T = (\overline{X} - \mu)/(S/\sqrt{n})$ where let $\overline{X}$ and S^2 are the unbiased estimators of μ and σ^2, respectively. HINT: Show that (i) $1 - \alpha = P(a \leq T \leq b) = P[\overline{X} - b(S/\sqrt{n}) \leq \mu \leq \overline{X} - a(S/\sqrt{n})]$ and (ii) the length of the resulting confidence interval is $L = (s/\sqrt{n})(b - a)$. (iii) Minimize L subject to the condition $\int_a^b g(t)\, dt = 1 - \alpha$, where $g(t)$ is the p.d.f. of a t random variable with $r = n - 1$ degrees of freedom.

6.4.14 Let $X_1, X_2, \ldots, X_n$ be a random sample of size n from a normal distribution, $N(\mu, \sigma^2)$. Select a and b so that

$$P\left(a \leq \frac{(n-1)S^2}{\sigma^2} \leq b\right) = 1 - \alpha.$$

So a $100(1-\alpha)\%$ confidence interval for σ is $[\sqrt{(n-1)/b}\, s, \sqrt{(n-1)/a}\, s]$. Find values of a and b that minimize the length of this confidence interval. That is, minimize

$$k = s\sqrt{n-1}\left(\frac{1}{\sqrt{a}} - \frac{1}{\sqrt{b}}\right)$$

under the restriction

$$G(b) - G(a) = \int_a^b g(u)\, du = 1 - \alpha,$$

where $G(u)$ and $g(u)$ are the distribution function and p.d.f. of a $\chi^2(n{-}1)$ distribution, respectively. HINT: Due to the restriction, b is a function of a. In particular, by taking derivatives of the restricting equation with respect to a, show that $\dfrac{db}{da} = \dfrac{g(a)}{g(b)}$. Determine $\dfrac{dk}{da}$. By setting $\dfrac{dk}{da} = 0$, show that a and b must satisfy

$$a^{n/2}e^{-a/2} - b^{n/2}e^{-b/2} = 0.$$

This condition, along with the restriction, was used to calculate the values in Table X.

6.4.15 Let X and Y equal the playing times for a $33\frac{1}{3}$ rpm long-playing record album and a compact disk, respectively. Observe that $n = 9$ observations of X were

$$\begin{array}{ccccccccc} 40.83 & 43.18 & 35.72 & 38.68 & 37.17 & 39.75 & 24.76 & 34.58 & 33.98 \end{array}$$

and $m = 13$ independent observations of Y were

$$\begin{array}{ccccccc} 42.82 & 64.42 & 56.92 & 39.92 & 72.38 & 47.26 & 64.58 \\ 38.20 & 72.75 & 39.09 & 39.07 & 33.70 & 62.02 \end{array}$$

(a) Determine a 95% confidence interval for σ_X^2/σ_Y^2, assuming that the distributions of X and Y are $N(\mu_X, \sigma_X^2)$ and $N(\mu_Y, \sigma_Y^2)$, respectively.

(b) Does the assumption of normality seem to be valid for these data? Why?

6.5 CONFIDENCE INTERVALS FOR PROPORTIONS

We have suggested that the histogram is a good description of how the observations of a random sample are distributed. We might naturally inquire about the accuracy of those relative frequencies (or percentages) associated with the various classes. For illustration, in Example 1.1.10 concerning $n = 40$ candy bar weights, we found that the relative frequency of the class interval (22.25, 23.15) was $8/40 = 0.20$, that is, 20%. If we think of this collection of 40 weights as a random sample observed from a larger population of candy bar weights, how close is 20% to the true percentage (or 0.20 to the true proportion) of weights in that class interval for the entire population of weights for this type of candy bar?

In considering this problem, we generalize it somewhat by treating the class interval (22.25, 23.15) as "success." That is, there is some true probability of success p, namely the proportion of the population in that interval. Let Y equal the frequency of measurements in the interval out of the n observations so that under the assumptions of independence and constant probability p, Y has the binomial distribution $b(n, p)$. Thus the problem is to determine the accuracy of the relative frequency Y/n as an estimator of p. We solve this by finding for the unknown p a confidence interval based on Y/n. In general, when observing n Bernoulli trials with probability p of success on each trial, we shall find a confidence interval for p based on Y/n, where Y is the number of successes and Y/n is an unbiased point estimator for p.

In Section 5.5 we noted that

$$\frac{Y - np}{\sqrt{np(1 - p)}} = \frac{(Y/n) - p}{\sqrt{p(1 - p)/n}}$$

has an approximate normal distribution $N(0, 1)$, provided that n is large enough. This means that, for a given probability $1 - \alpha$, we can find a $z_{\alpha/2}$ in Table V such that

$$P\left[-z_{\alpha/2} \leq \frac{(Y/n) - p}{\sqrt{p(1 - p)/n}} \leq z_{\alpha/2}\right] \approx 1 - \alpha. \qquad (6.5)$$

If we proceed as we did when we found a confidence interval for μ in Section 6.2, we would obtain

$$P\left[\frac{Y}{n} - z_{\alpha/2}\sqrt{\frac{p(1 - p)}{n}} \leq p \leq \frac{Y}{n} + z_{\alpha/2}\sqrt{\frac{p(1 - p)}{n}}\right] \approx 1 - \alpha.$$

Unfortunately, the unknown parameter p appears in the endpoints of this inequality. There are two ways out of this dilemma. We could make an additional approximation, namely replacing p with Y/n in $p(1 - p)/n$ in the endpoints. That is, it is still true, if n is large enough, that

$$P\left[\frac{Y}{n} - z_{\alpha/2}\sqrt{\frac{(Y/n)(1 - Y/n)}{n}} \le p \le \frac{Y}{n} + z_{\alpha/2}\sqrt{\frac{(Y/n)(1 - Y/n)}{n}}\right] \approx 1 - \alpha.$$

Thus, for large n, if the observed Y equals y, the interval

$$\left[\frac{y}{n} - z_{\alpha/2}\sqrt{\frac{(y/n)(1 - y/n)}{n}} , \frac{y}{n} + z_{\alpha/2}\sqrt{\frac{(y/n)(1 - y/n)}{n}}\right]$$

serves as an approximate $100(1 - \alpha)\%$ confidence interval for p. Frequently, this is written as

$$\frac{y}{n} \pm z_{\alpha/2}\sqrt{\frac{(y/n)(1 - y/n)}{n}} \tag{6.6}$$

for brevity. This clearly notes, as does $\bar{x} \pm z_{\alpha/2}(\sigma/\sqrt{n})$ in Section 6.2, the reliability of the estimate y/n; namely, we are $100(1 - \alpha)\%$ confident that p is within $z_{\alpha/2}\sqrt{(y/n)(1 - y/n)/n}$ of $\hat{p} = y/n$.

A second way to solve for p, in the inequality in Equation 6.5, is to note that

$$\frac{|Y/n - p|}{\sqrt{p(1 - p)/n}} \le z_{\alpha/2}$$

is equivalent to

$$H(p) = \left(\frac{Y}{n} - p\right)^2 - \frac{z_{\alpha/2}^2\, p(1 - p)}{n} \le 0. \tag{6.7}$$

But $H(p)$ is a quadratic expression in p. Thus we can find those values of p for which $H(p) \le 0$ by finding the two zeros of $H(p)$. Letting $\hat{p} = Y/n$ and $z_0 = z_{\alpha/2}$ in Equation 6.7, we have

$$H(p) = \left(1 + \frac{z_0^2}{n}\right)p^2 - \left(2\hat{p} + \frac{z_0^2}{n}\right)p + \hat{p}^2.$$

By the quadratic formula, the zeros of $H(p)$ are, after simplifying,

$$\frac{\hat{p} + z_0^2/(2n) \pm z_0\sqrt{\hat{p}(1 - \hat{p})/n + z_0^2/(4n^2)}}{1 + z_0^2/n}, \tag{6.8}$$

and these zeros give the endpoints for an approximate $100(1 - \alpha)\%$ confidence interval for p. If n is large, $z_0^2/(2n)$, $z_0^2/(4n^2)$, and z_0^2/n are small. Thus the confidence intervals given by Equations 6.6 and 6.8 are approximately equal when n is large.

Example 6.5.1

Let us return to the example of the histogram of the candy bar weights, Example 1.1.10 with $n = 36$ and $y/n = 0.222$. If $1 - \alpha = 0.90$ so that $z_{\alpha/2} = 1.645$, then, using Equation 6.6,

$$0.222 \pm 1.645\sqrt{\frac{(0.222)(0.778)}{36}}$$

serves as an approximate 90% confidence interval for the true fraction p. That is, $[0.108, 0.336]$, which is the same as $[10.8\%, 33.6\%]$, is an approximate 90% confidence interval for the percentage of weights of the entire population in the interval $(22.25, 23.15)$. If we had used the endpoints given by Equation 6.8, the confidence interval would be $[0.130, 0.353]$. Because of the small sample size, there is a difference in the lengths of these intervals. If the sample size had been $n = 360$, the two 90% confidence intervals would have been $(0.186, 0.258)$ and $(0.188, 0.260)$, respectively, which differ very little.   ◇

Example 6.5.2

In a certain political campaign, one candidate has a poll taken at random among the voting population. The results are $y = 185$ out of $n = 351$ voters favor this candidate. Even though $y/n = 185/351 = 0.527$, should the candidate feel very confident of winning? An approximate 95% confidence interval for the fraction p of the voting population who favor this candidate is, using Equation 6.6,

$$0.527 \pm 1.96\sqrt{\frac{(0.527)(0.473)}{351}}$$

or, equivalently, $[0.475, 0.579]$. Thus there is a good possibility that p is less than 50%, and the candidate should certainly take this into account in campaigning.   ◇

One-sided confidence intervals are sometimes appropriate for p. For example, we may be interested in an upper bound on the proportion of defectives when manufacturing some item. Or we may be interested in a lower bound on the proportion of voters who favor a particular candidate. The one-sided confidence interval for p given by

$$\left[0, \frac{y}{n} + z_\alpha\sqrt{\frac{(y/n)(1 - y/n)}{n}} \right]$$

provides an upper bound for p, while

$$\left[\frac{y}{n} - z_\alpha\sqrt{\frac{(y/n)(1 - y/n)}{n}}, 1 \right]$$

provides a lower bound for p.

Frequently, there are two (or more) possible independent ways of performing an experiment; suppose that these have probabilities of success p_1 and p_2, respectively. Let n_1 and n_2 be the number of independent trials associated with these two methods, and let us say they result in Y_1 and Y_2 successes, respectively. In order to make a statistical inference about the difference $p_1 - p_2$, we proceed as follows.

Since the independent random variables Y_1/n_1 and Y_2/n_2 have respective means p_1 and p_2 and variances $p_1(1 - p_1)/n_1$ and $p_2(1 - p_2)/n_2$, we know from Section 5.2 that the difference $Y_1/n_1 - Y_2/n_2$ must have mean $p_1 - p_2$ and variance

$$\frac{p_1(1 - p_1)}{n_1} + \frac{p_2(1 - p_2)}{n_2}.$$

(Recall that the variances are added to get the variance of a difference of two independent random variables.) Moreover, the fact that Y_1/n_1 and Y_2/n_2 have approximate normal distributions would suggest that the difference

$$\frac{Y_1}{n_1} - \frac{Y_2}{n_2}$$

would have an approximate normal distribution with the foregoing mean and variance (see Theorem 5.3.5). That is,

$$\frac{(Y_1/n_1) - (Y_2/n_2) - (p_1 - p_2)}{\sqrt{p_1(1 - p_1)/n_1 + p_2(1 - p_2)/n_2}}$$

has an approximate normal distribution $N(0, 1)$. If we now replace p_1 and p_2 in the denominator of this ratio by Y_1/n_1 and Y_2/n_2, respectively, it is still true for large enough n_1 and n_2, that the new ratio will be approximately $N(0, 1)$. Thus, for a given $1 - \alpha$, we can find $z_{\alpha/2}$ from Table V so that

$$P\left[-z_{\alpha/2} \leq \frac{(Y_1/n_1) - (Y_2/n_2) - (p_1 - p_2)}{\sqrt{(Y_1/n_1)(1 - Y_1/n_1)/n_1 + (Y_2/n_2)(1 - Y_2/n_2)/n_2}} \leq z_{\alpha/2}\right] \approx 1 - \alpha.$$

This can be solved to obtain, once Y_1 and Y_2 are observed to be y_1 and y_2, an approximate $100(1 - \alpha)\%$ confidence interval

$$\frac{y_1}{n_1} - \frac{y_2}{n_2} \pm z_{\alpha/2}\sqrt{\frac{(y_1/n_1)(1 - y_1/n_1)}{n_1} + \frac{(y_2/n_2)(1 - y_2/n_2)}{n_2}}$$

for the unknown difference $p_1 - p_2$. Note again how this form indicates the reliability of the estimate $y_1/n_1 - y_2/n_2$ of the difference $p_1 - p_2$.

Example 6.5.3

Two detergents were tested for their ability to remove stains of a certain type. An inspector judged the first one to be successful on 63 out of 91 independent trials and the second one to be successful on 42 out of 79 independent trials. The respective relative frequencies of success are 0.692 and 0.532. An approximate 90% confidence interval for the difference $p_1 - p_2$ of the two detergents is

$$0.692 - 0.532 \pm 1.645\sqrt{\frac{(0.692)(0.308)}{91} + \frac{(0.532)(0.468)}{79}},$$

or, equivalently, [0.038, 0.282]. Accordingly, it seems that the first detergent is definitely better than the second one for removing these stains. ◇

EXERCISES

6.5.1 A machine shop manufactures toggle levers. A lever is flawed if a standard nut cannot be screwed onto the threads. Let p equal the proportion of flawed toggle levers that they manufacture. If there were 24 flawed levers out of a sample of 642 that were selected randomly from the production line,

    (a) Give a point estimate of p.

    (b) Find an approximate 95% confidence interval for p using Equation 6.6.

    (c) Find an approximate 95% confidence interval for p using Equation 6.8.

    (d) Find a one-sided 95% confidence interval for p that provides an upper bound for p.

6.5.2 Let p equal the proportion of letters mailed in the Netherlands that are delivered the next day. If $y = 142$ out of a random sample of $n = 200$ letters were delivered the day after they were mailed, find an approximate 90% confidence interval for p.

6.5.3 Let p equal the proportion of adult Americans who favor a law requiring a teenager to have her parents' consent before having an abortion. In a survey of 1000 adult Americans (conducted by *Time*/CNN and reported in *Time* on July 9, 1990), 690 said they favored such a law.

    (a) Give a point estimate of p.

    (b) Find an approximate 95% confidence interval for p.

6.5.4 Let p equal the proportion of Americans who favor the death penalty. If a random sample of $n = 1234$ Americans yielded $y = 864$ who favored the death penalty, find an approximate 95% confidence interval for p.

6.5.5 Let p equal the proportion of triathletes who suffered a training-related overuse injury during the past year. Out of 330 triathletes who responded to a survey, 167 indicated that they had suffered a training-related injury during the past year. Using these data,

    (a) Give a point estimate of p.

    (b) Find an approximate 90% confidence interval for p.

    (c) Do you think that the 330 triathletes who responded to the survey may be considered as a random sample from the population of triathletes?

6.5.6 Let p equal the proportion of Americans who select jogging as one of their recreational activities. If 1497 out of a random sample of 5757 selected jogging, find an approximate 98% confidence interval for p.

6.5.7 In order to estimate the percentage of a large class of college freshmen that had high school GPAs from 3.2 to 3.6 inclusive, a sample of $n = 50$ students was taken, and $y = 9$ students fell in this class. Give a 95% confidence interval for the percentage of this freshman class having a high school GPA of 3.2 to 3.6.

6.5.8 A proportion, p, that many public opinion polls estimate is the number of Americans who would say yes to the question, "If something were to happen to the president of the United States, do you think that the vice president would be qualified to take over as president?" In one such random sample of 1022 adults, 388 said yes.

    (a) Based on the data given, find a point estimate of p.

    (b) Find an approximate 90% confidence interval for p.

    (c) Give updated answers to this question if new poll results are available.

6.5.9 To obtain an estimate of the proportion, p, of New York City residents who feel that the quality of life in New York City has become worse in the past few years, a telephone poll by *Time*/CNN on August 2–5, 1990, revealed that 686 out of 1009 residents said that life has become worse.

   (a) Give a point estimate of p.

   (b) Find an approximate 98% confidence interval for p.

6.5.10 *Time* (January 17, 1994) reported that 58% of adult Americans said "yes" to the question "If you or your spouse were pregnant, would you want the unborn child tested for genetic defects?" This was based on a telephone poll of 500 adult Americans. Let p equal the proportion of all adult Americans who would say "yes" to this question.

   (a) Give the endpoints for a 90% confidence interval for p.

   (b) They claimed that their sampling error was 4.5%. What is their confidence level?

6.5.11 *Time* (January 29, 1996) reported that 48% of adult Americans "like the principle of a flat tax." They further claimed that this estimate had a sampling error of $\pm 3\%$. Let p equal the proportion of all adult Americans who like the principle of a flat tax.

   (a) Give the endpoints for a 95% confidence interval for p given that the sample size was $n = 800$.

   (b) Given that their reported sampling error was $\pm 3\%$ with $n = 800$, what is their confidence level?

6.5.12 *Time* (March 29, 1993) reported the proportions of adult Americans who favor "stricter gun-control laws." A telephone poll of 800 adult Americans, of whom 374 were gun owners and 426 did not own guns, showed that 206 gun owners and 338 non-gun owners favor stricter gun-control laws. Let p_1 and p_2 be the respective proportions of gun owners and non-gun owners who favor stricter gun-control laws.

   (a) Give point estimates of p_1 and p_2.

   (b) Find a 95% confidence interval for $p_1 - p_2$.

6.5.13 In developing countries in Africa and the Americas, let p_1 and p_2 be the respective proportions of women with nutritional anemia. Find an approximate 90% confidence interval for $p_1 - p_2$ given that a random sample of $n_1 = 2100$ African women yielded $y_1 = 840$ with nutritional anemia and a random sample of $n_2 = 1900$ women from the Americas yielded $y_2 = 323$ women with nutritional anemia.

6.5.14 A candy manufacturer selects mints at random from the production line and weighs them. For one week, the day shift weighed $n_1 = 194$ mints, and the night shift weighed $n_2 = 162$ mints. The numbers of these mints that weighed at most 21 grams was $y_1 = 28$ for the day shift and $y_2 = 11$ for the night shift. Let p_1 and p_2 denote the proportions of mints that weigh at most 21 grams for the day and night shifts, respectively.

   (a) Give a point estimate of p_1.

   (b) Give the endpoints for a 95% confidence interval for p_1.

   (c) Give a point estimate of $p_1 - p_2$.

   (d) Give the endpoints for a 95% confidence interval for $p_1 - p_2$.

6.5.15 Consider the following two groups of women: group 1, women who spend less than $500 annually on clothes; group 2, women who spend over $1000 annually on clothes. Let p_1 and p_2 equal the proportions of women in these two groups, respectively, who believe that clothes are too expensive. If 1009 out of a random sample of 1230 women from group 1 and 207 out of a random sample 340 from group 2 believe that clothes are too expensive,

  (a) Give a point estimate of $p_1 - p_2$.

  (b) Find an approximate 95% confidence interval for $p_1 - p_2$.

6.5.16 For developing countries in Asia (excluding China) and Africa, let p_1 and p_2 be the respective proportions of preschool children with chronic malnutrition (stunting). If respective random samples of $n_1 = 1300$ and $n_2 = 1100$ yielded $y_1 = 520$ and $y_2 = 385$ children with chronic malnutrition, find an approximate 95% confidence interval for $p_1 - p_2$.

6.5.17 The following question was asked in a *Newsweek* poll: "Would you prefer to live in a neighborhood with mostly whites, with mostly blacks, or in a neighborhood mixed half and half?" Let p_1 and p_2 equal the proportion of black and white adult respondents, respectively, who prefer "half and half." If 207 out of 305 black adults and 291 out of 632 white adults prefer "half and half":

  (a) Give a point estimate of $p_1 - p_2$.

  (b) Find an approximate 90% confidence interval for $p_1 - p_2$.

6.5.18 An environmental survey contained a question asking what the respondent thought was the major cause of air pollution in this country, giving the choices "automobiles," "factories," and "incinerators." Two versions of the test, A and B, were used. Let p_A and p_B be the respective proportions of people using forms A and B who select "factories." If 170 out of 460 people who used version A chose "factories" and 141 out of 440 people who used version B chose "factories,"

  (a) Find a 95% confidence interval for $p_A - p_B$.

  (b) Do the forms seem to be consistent concerning this answer? Why?

6.6 SAMPLE SIZE

In statistical consulting, the first question frequently asked is, "How large should the sample size be to estimate a mean?" In order to convince the inquirer that the answer will depend on the variation associated with the random variable under observation, the statistician could correctly respond, "Only one observation is needed, provided that the standard deviation of the distribution is zero." That is, if σ equals zero, then the value of that one observation would necessarily equal the unknown mean of the distribution. This, of course, is an extreme case and one that is not met in practice; however, it should help convince persons that the smaller the variance, the smaller the sample size needed to achieve a given degree of accuracy. This will become clearer as we consider several examples. Let us begin with a problem that involves a statistical inference about the unknown mean of a distribution.

Example 6.6.1

A mathematics department wishes to evaluate a new method of teaching calculus that does mathematics using a computer. At the end of the course, the evaluation will be made on the basis of scores of the participating students on a standard test. There is particular interest in estimating the mean score, μ, for students taking calculus using the computer. Thus there is a desire to determine the number of students, n, who are to be selected at random from a larger group of students to take the course taught using the computer. Since new computing equipment must be purchased, they cannot afford to let all of the students take calculus the new way. In addition, some of the staff question the value of this approach and hence do not want to expose every student to this new procedure. So, let us find the sample size n such that we are fairly confident that $\overline{x} \pm 1$ contains the unknown test mean μ. From past experience it is believed that the standard deviation associated with this type of test is about 15. (The mean is also known when students take the standard calculus course.) Accordingly, using the fact that the sample mean of the test scores, $\overline{X}$, is approximately $N(\mu, \sigma^2/n)$, we see that the interval given by $\overline{x} \pm 1.96(15/\sqrt{n})$ will serve as an approximate 95% confidence interval for μ. That is, we want

$$1.96\left(\frac{15}{\sqrt{n}}\right) = 1$$

or, equivalently,

$$\sqrt{n} = 29.4 \qquad \text{and thus} \qquad n \approx 864.36$$

or $n = 865$ because n must be an integer. $\diamond$

It is quite likely that it had not been anticipated that as many as 865 students would be needed in this study. If that is the case, the statistician must discuss with those involved in the experiment whether or not the accuracy and the confidence level could be relaxed some. For illustration, rather than requiring $\overline{x} \pm 1$ to be a 95% confidence interval for μ, possibly $\overline{x} \pm 2$ would be a satisfactory 80% one. If this modification is acceptable, we now have

$$1.282\left(\frac{15}{\sqrt{n}}\right) = 2$$

or, equivalently,

$$\sqrt{n} = 9.615 \qquad \text{and} \qquad n \approx 92.4.$$

Since n must be an integer, we would probably use 93 in practice. Most likely, the persons involved in this project would find this a more reasonable sample size. Of course, any sample size greater than 93 could be used. Then either the length of the confidence interval could be decreased from that of $\overline{x} \pm 2$ or the confidence coefficient could be increased from 80% or a combination of both. Also, since there might be some question of whether the standard deviation σ actually equals 15, the sample standard deviation s would no doubt be used in the construction of the interval. For example, suppose that the sample characteristics observed are

$$n = 145, \qquad \overline{x} = 77.2, \qquad s = 13.2;$$

then

$$\overline{x} \pm \frac{1.282s}{\sqrt{n}} \qquad \text{or} \qquad 77.2 \pm 1.41$$

provides an approximate 80% confidence interval for μ.

In general, if we want the $100(1-\alpha)\%$ confidence interval for μ, $\overline{x} \pm z_{\alpha/2}(\sigma \sqrt{n})$, to be no longer than that given by $\overline{x} \pm \varepsilon$, the sample size n is the solution of

$$\varepsilon = \frac{z_{\alpha/2}\sigma}{\sqrt{n}}, \qquad \text{where} \quad \Phi(z_{\alpha/2}) = 1 - \frac{\alpha}{2}.$$

That is,

$$n = \frac{z_{\alpha/2}^2 \sigma^2}{\varepsilon^2}, \tag{6.9}$$

where it is assumed that σ^2 is known. We sometimes call $\varepsilon = z_{\alpha/2}(\sigma/\sqrt{n})$ the **maximum error of the estimate**. If the experimenter has no idea about the value of σ^2, it may be necessary to first take a preliminary sample to estimate σ^2.

The type of statistic we see most often in newspapers and magazines is an estimate of a proportion p. We might, for example, want to know the percentage of the labor force that is unemployed or the percentage of voters favoring a certain candidate. Sometimes extremely important decisions are made on the basis of these estimates. If this is the case, we would most certainly desire short confidence intervals for p with large confidence coefficients. We recognize that these conditions will require a large sample size. On the other hand, if the fraction p being estimated is not too important, an estimate associated with a longer confidence interval with a smaller confidence coefficient is satisfactory; and thus a smaller sample size can be used.

Example 6.6.2

Suppose we know that the unemployment rate has been about 8% (0.08). However, we wish to update our estimate in order to make an important decision about the national economic policy. Accordingly, let us say we wish to be 99% confident that the new estimate of p is within 0.001 of the true p. If we assume Bernoulli trials (an assumption that might be questioned), the relative frequency y/n, based upon a large sample size n, provides the approximate 99% confidence interval

$$\frac{y}{n} \pm 2.576\sqrt{\frac{(y/n)(1 - y/n)}{n}}.$$

Although we do not know y/n exactly before sampling, we do know, since y/n will be near 0.08, that

$$2.576\sqrt{\frac{(y/n)(1 - y/n)}{n}} \approx 2.576\sqrt{\frac{(0.08)(0.92)}{n}}$$

and we want this number to equal 0.001. That is,

$$2.576\sqrt{\frac{(0.08)(0.92)}{n}} = 0.001$$

or, equivalently,

$$\sqrt{n} = 2576\sqrt{0.0736} \qquad \text{and} \qquad n \approx 488,394.$$

That is, under our assumptions, such a sample size is needed in order to achieve the reliability and the accuracy desired. Because n is so large, we would probably be willing to increase the error, say to 0.01 and perhaps reduce the confidence level to 98%. The required sample size required would then be

$$\sqrt{n} = (2.326/0.01)\sqrt{0.0736} \qquad \text{and} \qquad n \approx 3,982,$$

which is a more reasonable sample size.   ◇

From Example 6.6.2 we hope that the student will recognize how important it is to know the sample size (or length of the confidence interval and confidence coefficient) before he or she can place much weight on a statement such as 51% of the voters seem to favor candidate A, 46% favor candidate B, and 3% are undecided. Is this statement based on a sample of 100 or 2000 or 10,000 voters? If we assume Bernoulli trials, the approximate 95% confidence intervals for the fraction of voters favoring candidate A in these cases are, respectively, [0.41, 0.61], [0.49, 0.53], and [0.50, 0.52]. Quite obviously, the first interval, with $n = 100$, does not assure candidate A of the support of at least half the voters, whereas the interval with $n = 10,000$ is more convincing.

In general, to find the required sample size to estimate p, recall that the point estimate of p is $\widehat{p} = y/n$ and an approximate $1 - \alpha$ confidence interval for p is

$$\widehat{p} \pm z_{\alpha/2}\sqrt{\frac{\widehat{p}\,(1 - \widehat{p}\,)}{n}}.$$

Suppose we want an estimate of p that is within ε of the unknown p with $100(1-\alpha)\%$ confidence where $\varepsilon = z_{\alpha/2}\sqrt{\widehat{p}\,(1 - \widehat{p}\,)/n}$ is the **maximum error of the point estimate** $\widehat{p} = y/n$. Since $\widehat{p}$ is unknown before the experiment is run, we cannot use the value of $\widehat{p}$ in our determination of n. However, if it is known that p is about equal to p^*, the necessary sample size n is the solution of

$$\varepsilon = \frac{z_{\alpha/2}\sqrt{p^*(1 - p^*)}}{\sqrt{n}}.$$

That is,

$$n = \frac{z_{\alpha/2}^2 p^*(1 - p^*)}{\varepsilon^2}. \qquad (6.10)$$

It is often true, however, that we do not have a strong prior idea about p, as we did in Example 6.6.2 about the rate of unemployment. It is interesting to observe that no matter what value p takes between zero and one, it is always true that $p^*(1 - p^*) \le 1/4$. Hence,

$$n = \frac{z_{\alpha/2}^2 p^*(1 - p^*)}{\varepsilon^2} \le \frac{z_{\alpha/2}^2}{4\varepsilon^2}.$$

Thus, if we want the $100(1-\alpha)\%$ confidence interval for p to be no longer than that given by $y/n \pm \varepsilon$, a solution for n that provides this protection is

$$n = \frac{z_{\alpha/2}^2}{4\varepsilon^2}. \tag{6.11}$$

Example 6.6.3

A possible gubernatorial candidate wants to assess initial support among the voters before making a candidacy announcement. If the fraction p of voters who are favorable, without any advance publicity, is around 0.15, the candidate will enter the race. From a poll of n voters selected at random, the candidate would like the estimate y/n to be within 0.03 of p. That is, the decision will be based on a 95% confidence interval of the form $y/n \pm 0.03$. Since the candidate has no idea about the magnitude of p, a consulting statistician formulates the equation

$$n = \frac{(1.96)^2}{4(0.03)^2} = 1067.11.$$

Thus the sample size should be around 1068 to achieve the desired reliability and accuracy. Suppose that 1068 voters around the state were selected at random and interviewed and $y = 214$ express support for this candidate. Then $\widehat{p} = 214/1068 = 0.20$ is a point estimate of p and an approximate 95% confidence interval for p is

$$0.20 \pm 1.96\sqrt{(0.20)(0.80)n} \qquad \text{or} \qquad 0.20 \pm 0.024.$$

That is, we are 95% confident that p belongs to the interval $[0.176, 0.224]$. On the basis of this sample, the candidate decided to run for office. Note that for a confidence coefficient of 95%, we found a sample size so that the maximum error of the estimate would be 0.03. From the data that were collected, the maximum error of the estimate is only 0.024. We ended up with a smaller error because we found the sample size assuming that $p = 0.50$ while, in fact, p is closer to 0.20. ◇

Suppose that you want to estimate the proportion p of a student body that favors a new policy. How large should the sample be? If p is close to 1/2 and you want to be 95% confident that the maximum error of the estimate is $\varepsilon = 0.02$, then

$$n = \frac{(1.96)^2}{4(0.02)^2} = 2401.$$

Such a sample size makes sense at a large university. However, if you are a student at a small college, the entire enrollment could be less than 2401. Thus we now give a procedure that can be used to determine the sample size when the population is small relative to the desired sample size.

Let N equal the size of a population and assume that N_1 of them have a certain characteristic C (e.g., favor a new policy). Let $p = N_1/N$. Then $1 - p = 1 - N_1/N$. If we take a sample of size n without replacement, then X, the number of observations with the characteristic C, has a hypergeometric distribution. The mean and variance of X are

$$\mu = n\left(\frac{N_1}{N}\right) = np$$

and

$$\sigma^2 = n\left(\frac{N_1}{N}\right)\left(1 - \frac{N_1}{N}\right)\left(\frac{N - n}{N - 1}\right) = np(1 - p)\left(\frac{N - n}{N - 1}\right).$$

The mean and variance of X/n are

$$E\left(\frac{X}{n}\right) = \frac{\mu}{n} = p$$

and

$$\text{Var}\left(\frac{X}{n}\right) = \frac{\sigma^2}{n^2} = \frac{p(1 - p)}{n}\left(\frac{N - n}{N - 1}\right).$$

To find an approximate confidence interval for p, we can use the normal approximation:

$$P\left[-z_{\alpha/2} \leq \frac{(X/n) - p}{\sqrt{\dfrac{p(1 - p)}{n}\left(\dfrac{N - n}{N - 1}\right)}} \leq z_{\alpha/2}\right] \approx 1 - \alpha.$$

Thus

$$1 - \alpha \approx P\left[\frac{X}{n} - z_{\alpha/2}\sqrt{\frac{p(1 - p)}{n}\left(\frac{N - n}{N - 1}\right)} \leq p \leq \frac{X}{n} + z_{\alpha/2}\sqrt{\frac{p(1 - p)}{n}\left(\frac{N - n}{N - 1}\right)}\right].$$

An approximate $1 - \alpha$ confidence interval for p is, replacing p under the radical with $\widehat{p} = x/n$,

$$\widehat{p} \pm z_{\alpha/2}\sqrt{\frac{\widehat{p}(1 - \widehat{p})}{n}\left(\frac{N - n}{N - 1}\right)}.$$

This is similar to the confidence interval for p when the distribution of X is $b(n, p)$. If N is large relative to n, then

$$\frac{N - n}{N - 1} = \frac{1 - n/N}{1 - 1/N} \approx 1,$$

so in this case the two intervals are essentially equal.

Suppose now that we are interested in determining the sample size n that is required to have $1 - \alpha$ confidence that the maximum error of the estimate of p is ε. We let

$$\varepsilon = z_{\alpha/2} \sqrt{\frac{p(1-p)}{n}\left(\frac{N-n}{N-1}\right)}$$

and solve for n. After some simplification we obtain

$$n = \frac{N z_{\alpha/2}^2 \, p(1-p)}{(N-1)\varepsilon^2 + z_{\alpha/2}^2 \, p(1-p)}$$

$$= \frac{z_{\alpha/2}^2 p(1-p)/\varepsilon^2}{\dfrac{N-1}{N} + \dfrac{z_{\alpha/2}^2 \, p(1-p)/\varepsilon^2}{N}}.$$

If we let

$$m = \frac{z_{\alpha/2}^2 p^*(1-p^*)}{\varepsilon^2}$$

as given by Equation 6.10, we then choose for our sample size n

$$n = \frac{m}{1 + \dfrac{m-1}{N}}.$$

If we know nothing about p, set $p^* = 1/2$ to determine m. For example, if the size of the student body is $N = 4000$ and $1 - \alpha = 0.95$, $\varepsilon = 0.02$, and we let $p^* = 1/2$, then $m = 2401$ and

$$n = \frac{2401}{1 + 2400/4000} = 1501$$

rounded up to the nearest integer. Thus we would sample approximately 37.5% of the student body.

Example 6.6.4

Suppose that a college of $N = 3000$ students is interested in assessing student support for a new form for teacher evaluation. To estimate the proportion p in favor of the new form, how large a sample is required so that with 95% confidence the maximum error of the estimate of p is $\varepsilon = 0.03$? If we assume that p is completely unknown, we use $p^* = 1/2$ to obtain

$$m = \frac{(1.96)^2}{4(0.03)^2} = 1068,$$

rounding up to the nearest integer. Thus the desired sample size is

$$n = \frac{1068}{1 + 1067/3000} = 788,$$

rounding up to the nearest integer.   ◇

EXERCISES

6.6.1 Let X equal the tarsus length for a male grackle. Assume that the distribution of X is $N(\mu, 4.84)$. Find the sample size n that is needed so that we are 95% confident that the maximum error of the estimate of μ is 0.4.

6.6.2 Let X equal the excess weight of soap in a "1000-gram" bottle. Assume that the distribution of X is $N(\mu, 169)$. What sample size is required so that we have 95% confidence that the maximum error of the estimate of μ is 1.5?

6.6.3 A company packages powdered soap in "6-pound" boxes. The sample mean and standard deviation of the soap in these boxes are currently 6.09 and 0.02 pounds. If the mean fill can be lowered by 0.01 pound, $14,000 would be saved per year. Adjustments were made in the filling equipment, but it can be assumed that the standard deviation remains unchanged.

  (a) How large a sample is needed so that the maximum error of the estimate of the new μ is $\varepsilon = 0.001$ with 90% confidence?

  (b) A random sample of size $n = 1219$ yielded $\bar{x} = 6.048$ and $s = 0.022$. Calculate a 90% confidence interval for μ.

  (c) Estimate the savings per year with these new adjustments.

  (d) Estimate the proportion of boxes that will now weigh less than 6 pounds.

6.6.4 The length in centimeters of $n = 29$ fish (species nezumia) yielded an average length of $\bar{x} = 16.82$ and $s^2 = 34.9$. Determine the size of a new sample so that $\bar{x} \pm 0.5$ is an approximate 95% confidence interval for μ.

6.6.5 A quality engineer wanted to be 98% confident that the maximum error of the estimate of the mean strength, μ, of the left hinge on a vanity cover molded by a machine is 0.25. A preliminary sample of size $n = 32$ parts yielded a sample mean of $\bar{x} = 35.68$ and a standard deviation of $s = 1.723$.

  (a) How large a sample is required?

  (b) Does this seem to be a reasonable sample size, noting that destructive testing is needed to obtain the data?

6.6.6 A light bulb manufacturer sells a light bulb that has a mean life of 1450 hours with a standard deviation of 33.7 hours. A new manufacturing process is being tested and there is interest in knowing the mean life μ of the new bulbs. How large a sample is required so that $\bar{x} \pm 5$ is a 95% confidence interval for μ? You may assume that the change in the standard deviation is minimal.

6.6.7 For a public opinion poll for a close presidential election, let p denote the proportion of voters who favor candidate A. How large a sample should be taken if we want the maximum error of the estimate of p to be equal to

  (a) 0.03 with 95% confidence?

  (b) 0.02 with 95% confidence?

  (c) 0.03 with 90% confidence?

6.6.8 Let p equal the proportion of all college and university students who would say "yes" to the question, "Would you drink from the same glass as your friend if you suspected that this friend were an AIDS virus carrier?" Find the sample size required to be 95% confident that the maximum error of the estimate of p is 0.025. [For your information, it was reported in *Sociology and Social Research* **72**(2) (January 1988) that 30 out of 375 San Diego State University students answered "yes" to this question.]

6.6.9 A die has been loaded to change the probability of rolling a 6. In order to estimate p, the new probability of rolling a 6, how many times must the die be rolled so that we are 99% confident that the maximum error of the estimate of p is $\varepsilon = 0.02$?

6.6.10 Some college professors and students examined 137 Canadian geese for patent schistosome in the year they hatched. Of these 137 birds, 54 were infected. They were interested in estimating p, the proportion of infected birds of this type. For future studies determine the sample size n so that the estimate of p is within $\varepsilon = 0.04$ of the unknown p with 90% confidence.

6.6.11 According to the Department of Health Care of the Elderly, Sherwood Hospital, Nottingham, England, only 25% of the patients who were using canes had canes of the correct length. Suppose that you were interested in estimating the proportion p of Americans who used canes of the correct length. How large a sample is required so that, with 95% confidence, the maximum error of the estimate of p is 0.04?

6.6.12 A seed distributor claims that 80% of its beet seeds will germinate. How many seeds must be tested for germination in order to estimate p, the true proportion that will germinate, so that the maximum error of the estimate is $\varepsilon = 0.03$ with 90% confidence?

6.6.13 Some dentists were interested in studying the fusion of embryonic rat palates by using a standard transplantation technique. When no treatment is used, the probability of fusion approximately equals 0.89. They would like to estimate p, the probability of fusion, when vitamin A is lacking.

(a) How large a sample n of rat embryos is needed for $y/n \pm 0.10$ to be a 95% confidence interval for p?

(b) If $y = 44$ out of $n = 60$ palates showed fusion, give a 95% confidence interval for p.

6.6.14 Let p equal the proportion of New York City residents who feel that the quality of life in New York City has become worse in the past few years. To update the estimate given in Exercise 6.5.9, how large a sample is required to be 98% confident that the maximum error of the estimate of p is 0.025?

6.6.15 Let p equal the proportion of triathletes who suffered a training-related overuse injury during the past year (see Exercise 6.5.5). How large a sample would be required to estimate p so that with 95% confidence the maximum error of the estimate of p is 0.04?

6.6.16 Let p equal the proportion of college students who favor a new policy for alcohol consumption on campus. How large a sample is required to estimate p so that with 95% confidence the maximum error of the estimate of p is 0.04 when the size of the student body is

(a) $N = 1500$?

(b) $N = 15,000$?

(c) $N = 25,000$?

6.6.17 Out of 1000 welds that have been made on a tower, it is suspected that 15% of the welds are defective. To estimate p, the proportion of defective welds, how many welds must be inspected to have 95% confidence, approximately, that the maximum error of the estimate of p is 0.04?

6.6.18 If Y_1/n and Y_2/n are the respective independent relative frequencies of successes associated with the two binomial distributions $b(n, p_1)$ and $b(n, p_2)$, compute n such that the approximate probability that the random interval $Y_1/n - Y_2/n \pm 0.05$ covers $p_1 - p_2$ is at least 0.80. HINT: Take $p_1^* = p_2^* = 1/2$ to provide an upper bound for n.

7

TESTS OF STATISTICAL HYPOTHESES

7.1 TESTS ABOUT PROPORTIONS

A first major area of statistical inference, namely estimation of parameters, was introduced in Chapter 6. In this chapter we consider a second major one, **tests of statistical hypotheses**. This very important topic is introduced through an illustration.

Suppose a manufacturer of a certain printed circuit observes that about $p = 0.05$ of the circuits fail. An engineer and statistician working together suggest some changes that might improve the design of the product. To test this new procedure, it was agreed that $n = 200$ circuits would be produced using the proposed method and then checked. Let Y equal the number of these 200 circuits that fail. Clearly, if the number of failures, Y, is such that $Y/200$ is about equal to 0.05, then it seems that the new procedure has not resulted in an improvement. On the other hand, if Y is small so that $Y/200$ is about 0.01 or 0.02, we might believe that the new method is better than the old. On the other hand, if $Y/200$ is 0.08 or 0.09, the proposed method has perhaps caused a greater proportion of failures.

What we need to establish is a formal rule that tells us when to accept the new procedure as an improvement. In addition, we must know the consequences of this rule. For an example of a rule, we could accept the new procedure as an improvement if $Y \leq 5$ or $Y/n \leq 0.025$. We do note, however, that the probability of failure could still be about $p = 0.05$ even with the new procedure, and yet we could observe 5 or fewer failures in $n = 200$ trials. That is, we would accept the new method as being an improvement when, in fact, it was not. This decision is a mistake which we call a **type I error**. On the other hand, the new procedure might actually improve the product so that p is much smaller, say $p = 0.02$, and yet we could observe $y = 7$ failures so that $y/200 = 0.035$. Thus we would not accept the new method as resulting in an improvement when in fact it had. This decision would also be a mistake which we call a **type II error**. We must study the probabilities of these two types of errors to understand fully the consequences of our rule.

Let us begin by modeling the situation. If we believe these trials, using the new procedure, are independent and have about the same probability of failure on each trial, then Y is binomial $b(200, p)$. We wish to make a statistical inference about p using the unbiased estimator $\widehat{p} = Y/200$. Of course, we could construct a one-sided confidence interval, say one that has 95% confidence of providing an upper bound for p, obtaining

$$\left[0, \widehat{p} + 1.645\sqrt{\frac{\widehat{p}(1 - \widehat{p})}{200}} \right].$$

This inference is very appropriate and many statisticians simply do this. If the limits of this confidence interval contain 0.05, they would not say the new procedure is necessarily better, at least until more data are taken. If, on the other hand, the upper limit of this confidence interval is less than 0.05, then they feel 95% confident that the true p is now less than 0.05. Hence they would support the fact that the new procedure has improved the manufacturing of these printed circuits.

Although this use of confidence intervals is highly appropriate and later we indicate the relationship of confidence intervals and tests of hypotheses, every student of statistics should also have some understanding of the basic concepts in the latter area. Here, in our illustration, we are testing whether or not the probability of failure has or has not decreased from 0.05 when the new manufacturing procedure is used. The *no change* hypothesis, H_0: $p = 0.05$, is called the **null hypothesis**. Since H_0: $p = 0.05$ specifies the distribution completely it is called a **simple hypothesis**; thus H_0: $p = 0.05$ is a **simple null hypothesis**. The *research worker's* (here the engineer and/or the statistician) hypothesis H_1: $p < 0.05$ is called the **alternative hypothesis**. Since H_1: $p < 0.05$ does not completely specify the distribution, it is a **composite hypothesis** because it is composed of many simple hypotheses. Our rule of rejecting H_0 and accepting H_1 if $Y \leq 5$, and otherwise accepting H_0 is called a **test of a statistical hypothesis**. We now see that the two types of errors can be recorded as follows.

Type I error: rejecting H_0 and accepting H_1 when H_0 is true
Type II error: accepting H_0 when H_1 is true, that is, when H_0 is false

Since, in our illustration, we make a type I error if $Y \leq 5$ when in fact $p = 0.05$, we can calculate the probability of this error, which we denote by α and call the **significance level of the test**. Under our assumptions, it is

$$\alpha = P(Y \leq 5; \ p = 0.05) = \sum_{y=0}^{5} \binom{200}{y} (0.05)^y (0.95)^{200-y}.$$

Since n is rather large and p is small, these binomial probabilities can be approximated very well by Poisson probabilities with $\lambda = 200(0.05) = 10$. That is, from the

Poisson table, the probability of the type I error is

$$\alpha \approx \sum_{y=0}^{5} \frac{10^y e^{-10}}{y!} = 0.067;$$

thus, the approximate significance level of this test is $\alpha = 0.067$. (The exact value of α using the binomial distribution is 0.062.)

   This value of α is reasonably small. However, what about the probability of type II error in case p has been improved to 0.02, say? This error occurs if $Y > 5$ when, in fact, $p = 0.02$; hence its probability, denoted by β, is

$$\beta = P(Y > 5; \; p = 0.02) = \sum_{y=6}^{200} \binom{200}{y}(0.02)^y (0.98)^{200-y}.$$

Again we use the Poisson approximation, here with $\lambda = 200(0.02) = 4$, to obtain

$$\beta \approx 1 - \sum_{y=0}^{5} \frac{4^y e^{-4}}{y!} = 1 - 0.785 = 0.215.$$

(The exact probability using the binomial distribution is 0.213, so the approximation is very good.)   The engineer and the statistician who created this new procedure probably are not too pleased with this answer. That is, they note that if their new procedure of manufacturing circuits has actually decreased the probability of failure to 0.02 from 0.05 (*a big improvement*), there is still a good chance, 0.215, that H_0: $p = 0.05$ is accepted and their improvement rejected. Thus, in their eyes, this test of H_0: $p = 0.05$ against H_1: $p = 0.02$ is unsatisfactory. More will be said about modifying tests in Section 7.2 so that satisfactory values of the probabilities of the two types of errors, namely α and β, can be obtained.

   Without worrying more about the probability of the type II error here, we present a frequently used procedure for testing H_0: $p = p_0$, where p_0 is some specified probability of success. This test is based upon the fact that the number of successes, Y, in n independent Bernoulli trials is such that Y/n has an approximate normal distribution, $N[p_0, p_0(1 - p_0)/n]$, provided that H_0: $p = p_0$ is true and n is large. Suppose that the alternative hypothesis is H_1: $p > p_0$; that is, it has been hypothesized by a research worker that something has been done to increase the probability of success. Consider the test of H_0: $p = p_0$ against H_1: $p > p_0$ that rejects H_0 and accepts H_1 if and only if

$$Z = \frac{Y/n - p_0}{\sqrt{p_0(1 - p_0)/n}} \geq z_\alpha.$$

That is, if Y/n exceeds p_0 by z_α standard deviations of Y/n, we reject H_0 and accept the hypothesis H_1: $p > p_0$. Since, under H_0, Z is approximately $N(0, 1)$, the approximate probability of this occurring when H_0: $p = p_0$ is true is α. That is, the significance level of this test is approximately α.

If the alternative is H_1: $p < p_0$ instead of H_1: $p > p_0$, then the appropriate α-level test is given by $Z \leq -z_\alpha$. That is, if Y/n is smaller than p_0 by z_α standard deviations of Y/n, we accept H_1: $p < p_0$.

Example 7.1.1

It was claimed that many commercially manufactured dice are not fair because the "spots" are really indentations, so that, for example, the 6-side is lighter than the 1-side. Let p equal the probability of rolling a 6 with one of these dice. To test H_0: $p = 1/6$ against the alternative hypothesis H_1: $p > 1/6$, several of these dice will be rolled to yield a total of $n = 8000$ observations. Let Y equal the number of times that six resulted in the 8000 trials. The test statistic is

$$Z = \frac{Y/n - 1/6}{\sqrt{(1/6)(5/6)/n}} = \frac{Y/8000 - 1/6}{\sqrt{(1/6)(5/6)/8000}}.$$

If we use a significance level of $\alpha = 0.05$, the critical region (see below) is

$$z \geq z_{0.05} = 1.645.$$

The results of the experiment yielded $y = 1389$, so that the calculated value of the test statistic is

$$z = \frac{1389/8000 - 1/6}{\sqrt{(1/6)(5/6)/8000}} = 1.670.$$

Since

$$z = 1.670 > 1.645,$$

the null hypothesis is rejected and these experimental results indicate that these dice favor a 6 more than a fair die would. (You could perform your own experiment to check out other dice; see also Exercise 7.1.6.)   ◇

Tests of H_0: $p = p_0$ against H_1: $p < p_0$ or H_0: $p = p_0$ against H_1: $p > p_0$ are called *one-sided* tests because the alternative hypotheses are *one-sided*. There are times when *two-sided* alternatives and tests are appropriate as in H_1: $p \neq p_0$. For illustration, suppose that the pass rate in the usual beginning statistics course is p_0. There has been an intervention (say some new teaching method) and it is not known whether the pass rate will increase, decrease, or stay about the same. Thus we test the null (no change) hypothesis H_0: $p = p_0$ against the two-sided alternative H_1: $p \neq p_0$. A test of the approximate significance level α for doing this is to reject H_0: $p = p_0$ if

$$|Z| = \frac{|Y/n - p_0|}{\sqrt{p_0(1 - p_0)/n}} \geq z_{\alpha/2},$$

since, under H_0, $P(|Z| \geq z_{\alpha/2}) \approx \alpha$. These tests of approximate significance level α are summarized in Table 7.1. The rejection region for H_0 is often called the **critical region** of the test, and we use that terminology in the table.

Often there is interest in tests about p_1 and p_2, the probabilities of success for two different distributions or the proportions of two different populations having a certain characteristic. For example, if p_1 and p_2 denote the respective proportions

TABLE 7.1 Tests of hypotheses for one proportion

H_0	H_1	Critical Region
$p = p_0$	$p > p_0$	$z = \dfrac{y/n - p_0}{\sqrt{p_0(1-p_0)/n}} \geq z_\alpha$
$p = p_0$	$p < p_0$	$z = \dfrac{y/n - p_0}{\sqrt{p_0(1-p_0)/n}} \leq -z_\alpha$
$p = p_0$	$p \neq p_0$	$|z| = \dfrac{|y/n - p_0|}{\sqrt{p_0(1-p_0)/n}} \geq z_{\alpha/2}$

of homeowners and renters who vote in favor of a proposal to reduce property tax, a politician might be interested in testing $H_0: p_1 = p_2$ against the one-sided alternative hypothesis $H_1: p_1 > p_2$.

Let Y_1 and Y_2 represent, respectively, the numbers of observed successes in n_1 and n_2 independent trials with probabilities of success p_1 and p_2. Recall that the distribution of $\widehat{p}_1 = Y_1/n_1$ is approximately $N[p_1, p_1(1-p_1)/n_1]$ and the distribution of $\widehat{p}_2 = Y_2/n_2$ is approximately $N[p_2, p_2(1-p_2)/n_2]$. Thus the distribution of $\widehat{p}_1 - \widehat{p}_2 = Y_1/n_1 - Y_2/n_2$ is $N[p_1-p_2, p_1(1-p_1)/n_1 + p_2(1-p_2)/n_2]$ approximately. It follows that the distribution of

$$Z = \frac{Y_1/n_1 - Y_2/n_2 - (p_1 - p_2)}{\sqrt{p_1(1-p_1)/n_1 + p_2(1-p_2)/n_2}} \tag{7.1}$$

is approximately $N(0,1)$. To test $H_0: p_1 - p_2 = 0$ or, equivalently, $H_0: p_1 = p_2$, let $p = p_1 = p_2$ be the common value under H_0. We shall estimate p with $\widehat{p} = (Y_1 + Y_2)/(n_1 + n_2)$. Replacing p_1 and p_2 in the denominator of Equation 7.1 with this estimate, we obtain the test statistic

$$Z = \frac{\widehat{p}_1 - \widehat{p}_2 - 0}{\sqrt{\widehat{p}(1-\widehat{p})(1/n_1 + 1/n_2)}}$$

which has an approximate $N(0,1)$ distribution when the null hypothesis is true.

The three possible alternative hypotheses and their critical regions are summarized in Table 7.2.

Remark In testing both $H_0: p = p_0$ and $H_0: p_1 = p_2$, statisticians sometimes use different denominators for z. In the first, $\sqrt{p_0(1-p_0)/n}$ can be replaced by $\sqrt{(y/n)(1-y/n)/n}$; and, in the second, the following denominator can be used:

$$\sqrt{\frac{\widehat{p}_1(1-\widehat{p}_1)}{n_1} + \frac{\widehat{p}_2(1-\widehat{p}_2)}{n_2}}.$$

We do not have a strong preference one way or the other since the two methods provide about the same numerical result. The substitutions do provide better estimates of the standard deviations of the numerators when the null hypotheses are clearly

TABLE 7.2   Tests of hypotheses for two proportions

H_0	H_1	Critical Region
$p_1 = p_2$	$p_1 > p_2$	$z = \dfrac{\widehat{p}_1 - \widehat{p}_2}{\sqrt{\widehat{p}\,(1-\widehat{p})(1/n_1 + 1/n_2)}} \geq z_\alpha$
$p_1 = p_2$	$p_1 < p_2$	$z = \dfrac{\widehat{p}_1 - \widehat{p}_2}{\sqrt{\widehat{p}\,(1-\widehat{p})(1/n_1 + 1/n_2)}} \leq -z_\alpha$
$p_1 = p_2$	$p_1 \neq p_2$	$|z| = \dfrac{|\widehat{p}_1 - \widehat{p}_2|}{\sqrt{\widehat{p}\,(1-\widehat{p})(1/n_1 + 1/n_2)}} \geq z_{\alpha/2}$

false. There is some advantage to this if the null hypothesis is likely to be false. In addition, the substitutions also tie together the use of confidence intervals and tests of hypotheses. For example, if the null hypothesis is H_0: $p = p_0$, then the alternative hypothesis H_1: $p < p_0$ is rejected if

$$z = \frac{\widehat{p} - p_0}{\sqrt{\dfrac{\widehat{p}\,(1-\widehat{p})}{n}}} \leq -z_\alpha.$$

This is equivalent to the statement that

$$p_0 \notin \left[0, \widehat{p} + z_\alpha \sqrt{\frac{\widehat{p}\,(1-\widehat{p})}{n}}\,\right],$$

where the latter is a one-sided confidence interval providing an upper bound for p. Or if the alternative hypothesis is H_1: $p \neq p_0$, then H_0 is rejected if

$$\frac{|\widehat{p} - p_0|}{\sqrt{\dfrac{\widehat{p}\,(1-\widehat{p})}{n}}} \geq z_{\alpha/2}.$$

This is equivalent to

$$p_0 \notin \left[\widehat{p} - z_{\alpha/2}\sqrt{\frac{\widehat{p}\,(1-\widehat{p})}{n}}, \widehat{p} + z_{\alpha/2}\sqrt{\frac{\widehat{p}\,(1-\widehat{p})}{n}}\,\right],$$

where the latter is a confidence interval for p. However, using the forms given in Tables 7.1 and 7.2, we do get better approximations to α-level significance tests. Thus there are trade-offs and it is difficult to say that one is better than the other. Fortunately, the numerical answers are about the same.

EXERCISES

7.1.1 Bowl A contains 100 red balls and 200 white balls; bowl B contains 200 red balls and 100 white balls. Let p denote the probability of drawing a red ball from a bowl, but say p is unknown, since it is unknown whether bowl A or bowl B is being used. We shall test the simple null hypothesis H_0: $p = 1/3$ against the simple alternative hypothesis H_1: $p = 2/3$. Draw three balls at random, one at a time and with replacement from the selected bowl. Let X equal the number of red balls drawn. Then let the critical region be $C = \{x : x = 2, 3\}$. What are the values of α and β, the probabilities of type I and type II errors, respectively?

7.1.2 A bowl contains two red balls, two white balls, and a fifth ball that is either red or white. Let p denote the probability of drawing a red ball from the bowl. We shall test the simple null hypothesis H_0: $p = 3/5$ against the simple alternative hypothesis H_1: $p = 2/5$. Draw four balls at random from the bowl, one at a time and with replacement. Let X equal the number of red balls drawn.

(a) Define a critical region C for this test in terms of X.

(b) For the critical region C defined in part (a), find the values of α and β.

7.1.3 Let Y be $b(100, p)$. To test H_0: $p = 0.08$ against H_1: $p < 0.08$, we reject H_0 and accept H_1 if and only if $Y \leq 6$.

(a) Determine the significance level α of the test.

(b) Find the probability of the type II error if in fact $p = 0.04$.

7.1.4 Let p denote the probability that for a particular tennis player, the first serve is good. Since $p = 0.40$, this player decided to take lessons in order to increase p. When the lessons are completed, the hypothesis H_0: $p = 0.40$ will be tested against H_1: $p > 0.40$ based on $n = 25$ trials. Let y equal the number of first serves that are good, and let the critical region be defined by $C = \{y : y \geq 13\}$.

(a) Determine $\alpha = P(Y \geq 13; p = 0.40)$. Use Table II in the Appendix.

(b) Find $\beta = P(Y < 13)$ when $p = 0.60$; that is, $\beta = P(Y \leq 12; p = 0.60)$. Use Table II.

7.1.5 Let Y be $b(192, p)$. We reject H_0: $p = 0.75$ and accept H_1: $p > 0.75$ if and only if $Y \geq 152$. Use the normal approximation to determine

(a) $\alpha = P(Y \geq 152; p = 0.75)$.

(b) $\beta = P(Y < 152)$ when $p = 0.80$.

7.1.6 To determine whether the 1-side on a commercially manufactured die is heavy and the 6-side is light (see Example 7.1.1), some students kept track of the number of observed 1s when observing $n = 8000$ rolls of the dice. Let p equal the probability of rolling a 1 with such a die. We shall test the null hypothesis H_0: $p = 1/6$ against the alternative hypothesis H_1: $p < 1/6$.

(a) Define the test statistic and an $\alpha = 0.05$ critical region.

(b) If $y = 1265$ ones were observed in 8000 rolls, calculate the value of the test statistic and state your conclusion.

(c) Is 1/6 in the 95% one-sided confidence interval providing an upper bound for p?

7.1.7 If a newborn baby has a birth weight that is less than 2500 grams (5.5 pounds), we say that the baby has a low birth weight. The proportion of babies with a low birth weight is an indicator of nutrition (or lack of nutrition) for the mothers. For the United States, approximately 7% of babies have a low birth weight. Let p equal the proportion of babies born in the Sudan who weigh less than 2500 grams. We shall test the null hypothesis H_0: $p = 0.07$ against the alternative hypothesis H_1: $p > 0.07$. If $y = 23$ babies out of a random sample of $n = 209$ babies weighed less than 2500 grams, what is your conclusion at a significance level of

(a) $\alpha = 0.05$?

(b) $\alpha = 0.01$?

7.1.8 It was claimed that 75% of all dentists recommend a certain brand of gum for their gum-chewing patients. A consumer group doubted this claim and decided to test H_0: $p = 0.75$ against the alternative hypothesis H_1: $p < 0.75$, where p is the proportion of dentists who recommend this brand of gum. A survey of 390 dentists found that 273 recommended this brand of gum. Which hypothesis would you accept if the significance level is

(a) $\alpha = 0.05$?

(b) $\alpha = 0.01$?

7.1.9 It was claimed that the proportion of Americans who select jogging as one of their recreational activities is $p = 0.25$. A shoe manufacturer thought that p was larger than 0.25. They decided to test the null hypothesis H_0: $p = 0.25$ against the alternative hypothesis H_1: $p > 0.25$. If $y = 1497$ out of a random sample of $n = 5757$ selected jogging, what is your conclusion at a significance level of

(a) $\alpha = 0.05$?

(b) $\alpha = 0.025$?

7.1.10 Let p equal the proportion of drivers who use a seat belt in a state that does not have a mandatory seat belt law. It was claimed that $p = 0.14$. An advertising campaign was conducted to increase this proportion. Two months after the campaign, $y = 104$ out of a random sample of $n = 590$ drivers were wearing their seat belts. Was the campaign successful?

(a) Define the null and alternative hypotheses.

(b) Define a critical region with an $\alpha = 0.01$ significance level.

(c) What is your conclusion?

7.1.11 The management of the Tiger baseball team decided to sell only low-alcohol beer in their ballpark to help combat rowdy fan conduct. They claimed that more than 40% of the fans would approve of this decision. Let p equal the proportion of Tiger fans on opening day who approved of this decision. We shall test the null hypothesis H_0: $p = 0.40$ against the alternative hypothesis H_1: $p > 0.40$.

(a) Define a critical region that has an $\alpha = 0.05$ significance level.

(b) If out of a random sample of $n = 1278$ fans, $y = 550$ said that they approved of this new policy, what is your conclusion?

7.1.12 An official with the NCAA claimed that only $p = 0.01 = 1\%$ of college football players used anabolic steroids. Others believed that $p > 0.01$.

   (a) Define the test statistic and critical region that has an approximate significance level of $\alpha = 0.005$.

   (b) If $y = 17$ out of a random sample of $n = 546$ football players tested positive in an offseason testing program, what is your conclusion?

7.1.13 Let p equal the proportion of women who agree that "men are basically selfish and self-centered." Suppose that in the past, it was believed that $p = 0.40$. It is now claimed that p has increased.

   (a) Define the null and alternative hypotheses, a test statistic and critical region that has an approximate $\alpha = 0.01$ significance level. Sketch a standard normal p.d.f. and illustrate this critical region.

   (b) The *Detroit Free Press* (April 26, 1990) reported that $y = 1260$ out of a random sample of $n = 3000$ women agree with the statement. What is the conclusion of your test? Locate the calculated value of your test statistic on the p.d.f. in part (a).

7.1.14 A physician at a local hospital wrote that "In medicine we are now asked to perform quality assurance studies by various regulatory governmental or quasi-governmental agencies." He wrote: "Dr. X was notified that for a particular procedure, his complication rate of 20% last year ($N = 20$) was unacceptable for continued privileges. So he took a one-month remedial course for improving his technique. His complication rate is now 15% ($N = 15$). Has he really improved?" (You may assume that some rounding off was done in the letter and that the 15% is really 2/15.)

7.1.15 According to a population census in 1986, the percentage of males who are 18 or 19 years old that are married was 3.7%. We shall test whether this percentage increased from 1986 to 1988.

   (a) Define the null and alternative hypotheses.

   (b) Define a critical region that has an approximate significance level of $\alpha = 0.01$. Sketch a standard normal p.d.f. to illustrate this critical region.

   (c) If $y = 20$ out of a random sample of $n = 300$ males, each 18 or 19 years old, were married (U.S. Bureau of the Census, *Statistical Abstract of the United States: 1988*), what is your conclusion? Show the calculated value of the test statistic on your figure in part (b).

7.1.16 Let p equal the proportion of yellow candies in a package of mixed colors. It is claimed that $p = 0.20$.

   (a) Define the test statistic and critical region with a significance level of $\alpha = 0.05$ for testing H_0: $p = 0.20$ against a two-sided alternative hypothesis.

   (b) To perform the test, each of 20 students counted the number of yellow candies, y, and the total number, n, in a 48.1-gram package, yielding the following ratios, y/n: 8/56, 13/55, 12/58, 13/56, 14/57, 5/54, 14/56, 15/57, 11/54, 13/55, 10/57, 8/59, 10/54, 11/55, 12/56, 11/57, 6/54, 7/58, 12/58, 14/58. If each student makes a test of H_0: $p = 0.20$, what proportion of the students rejected the null hypothesis?

   (c) If we may assume that the null hypothesis is true, what proportion of the students would you have expected to reject the null hypothesis?

(d) For each of the 20 ratios in part (b), a 95% confidence interval for p can be calculated. What proportion of these 95% confidence intervals contain $p = 0.20$?

(e) If the 20 results are pooled so that $\sum_{i=1}^{20} y_i$ equals the number of yellow candies and $\sum_{i=1}^{20} n_i$ equals the total sample size, do we reject H_0: $p = 0.20$?

7.1.17 A machine shop that manufactures toggle levers has both a day and a night shift. A toggle lever is defective if a standard nut cannot be screwed onto the threads. Let p_1 and p_2 be the proportion of defective levers among those manufactured by the day and night shifts, respectively. We shall test the null hypothesis, H_0: $p_1 = p_2$, against a two-sided alternative hypothesis based on two random samples, each of 1000 levers taken from the production of the respective shifts.

(a) Define the test statistic and a critical region that has an $\alpha = 0.05$ significance level. Sketch a standard normal p.d.f. illustrating this critical region.

(b) If $y_1 = 37$ and $y_2 = 53$ defectives were observed for the day and night shifts, respectively, calculate the value of the test statistic. Locate the calculated test statistic on your figure in part (a) and state your conclusion.

7.1.18 *Time* (April 18, 1994) reported the results of a telephone poll of 800 adult Americans, 605 of them nonsmokers, who were asked the following question: "Should the federal tax on cigarettes be raised by \$1.25 to pay for health care reform?" Let p_1 and p_2 equal the proportions of nonsmokers and smokers, respectively, who would say "yes" to this question. Given that $y_1 = 351$ nonsmokers and $y_2 = 41$ smokers said "yes",

(a) With $\alpha = 0.05$, test H_0: $p_1 = p_2$ against H_1: $p_1 \neq p_2$.

(b) Find a 95% confidence interval for $p_1 - p_2$. Is this in agreement with the conclusion of part (a)?

(c) Find a 95% confidence interval for p, the proportion of adult Americans who would say "yes."

7.1.19 Let p_m and p_f be the respective proportions of male and female white-crowned sparrows that return to their hatching site. Give the endpoints for a 95% confidence interval for $p_m - p_f$, given that 124 out of 894 males and 70 out of 700 females returned. (*The Condor*, 1992, pp. 117–133.) Does this agree with the conclusion of a test of H_0: $p_m = p_f$ against H_1: $p_m \neq p_f$ with $\alpha = 0.05$?

7.1.20 For developing countries in Africa and the Americas, let p_1 and p_2 be the respective proportions of babies with a low birth weight (below 2500 grams). We shall test H_0: $p_1 = p_2$ against the alternative hypothesis H_1: $p_1 > p_2$.

(a) Define a critical region that has an $\alpha = 0.05$ significance level.

(b) If respective random samples of sizes $n_1 = 900$ and $n_2 = 700$ yielded $y_1 = 135$ and $y_2 = 77$ babies with a low birth weight, what is your conclusion?

(c) What would your decision be with a significance level of $\alpha = 0.01$?

7.2 POWER AND SAMPLE SIZE

In Section 7.1 we introduced some terminology connected with tests of statistical hypotheses. With two more examples we continue to examine a number of concepts associated with tests of statistical hypotheses.

Example 7.2.1

Let X equal the breaking strength of a steel bar. If the steel bar is manufactured by process I, X is $N(50, 36)$. It is hoped that if process II (a new process) is used, X will be $N(55, 36)$. Given a large number of steel bars manufactured by process II, how could we test whether the increase in the mean breaking strength was realized?

That is, we are assuming X is $N(\mu, 36)$ and μ is equal to 50 or 55. We want to test the null hypothesis $H_0: \mu = 50$ against alternative hypothesis $H_1: \mu = 55$. Note that each of these hypotheses completely specifies the distribution of X. That is, H_0 states that X is $N(50, 36)$, and H_1 states that X is $N(55, 36)$. A hypothesis that completely specifies the distribution of X is called a simple hypothesis; otherwise, it is called a composite hypothesis (composed of at least two simple hypotheses). For example, $H_1: \mu > 50$ would be a composite hypothesis because it is composed of all normal distributions with $\sigma^2 = 36$ and means greater than 50. In order to test which of the two hypotheses, H_0 or H_1, is true, we shall set up a rule based on the breaking strengths $x_1, x_2, \ldots, x_n$ of n bars (the observed values of a random sample of size n from this new normal distribution). The rule leads to a decision to accept or reject H_0; so it is necessary to partition the sample space into two parts, say C and C', so that if $(x_1, x_2, \ldots, x_n) \in C$, H_0 is rejected, and if $(x_1, x_2, \ldots, x_n) \in C'$, H_0 is accepted (not rejected). The rejection region C for H_0 is called the critical region for the test. Often, the partitioning of the sample space is specified in terms of the values of a statistic called the test statistic. In this illustration we could let $\bar{X}$ be the test statistic and, for example, take $C = \{(x_1, x_2, \ldots, x_n): \bar{x} \geq 53\}$. We could then define the critical region as those values of the test statistic for which H_0 is rejected. That is, the given critical region is equivalent to defining $C = \{\bar{x}: \bar{x} \geq 53\}$ in the $\bar{x}$ space. If $(x_1, x_2, \ldots, x_n) \in C$ when H_0 is true, H_0 would be rejected when it is true. Such an error is a type I error. If $(x_1, x_2, \ldots, x_n) \in C'$ when H_1 is true, H_0 would be accepted when in fact H_1 is true. Such an error is a type II error. The probability of a type I error is called the significance level of the test and is denoted by α. That is,

$$\alpha = P[(X_1, X_2, \ldots, X_n) \in C; H_0]$$

is the probability that $(X_1, X_2, \ldots, X_n)$ falls in C when H_0 is true. The probability of a type II error is denoted by β; that is,

$$\beta = P[(X_1, X_2, \ldots, X_n) \in C'; H_1]$$

is the probability of accepting (failing to reject) H_0 when it is false. For illustration, suppose that $n = 16$ bars were tested and $C = \{\bar{x}: \bar{x} \geq 53\}$. Then $\overline{X}$ is $N(50, 36/16)$ when H_0 is true and is $N(55, 36/16)$ when H_1 is true. Thus

$$\alpha = P(\overline{X} \geq 53; H_0) = P\left(\frac{\overline{X} - 50}{6/4} \geq \frac{53 - 50}{6/4}; H_0\right)$$

$$= 1 - \Phi(2) = 0.0228$$

and

$$\beta = P(\overline{X} < 53; H_1) = P\left(\frac{\overline{X} - 55}{6/4} < \frac{53 - 55}{6/4}; H_1\right)$$

$$= \Phi\left(-\frac{4}{3}\right) = 1 - 0.9087 = 0.0913.$$

See Figure 7.1 for the graphs of the probability density functions of $\overline{X}$ when H_0 and H_1 are true, respectively. Note that a decrease in the size of α leads to an increase in the size of β, and vice versa. Both α and β can be decreased if the sample size n is increased. $\diamond$

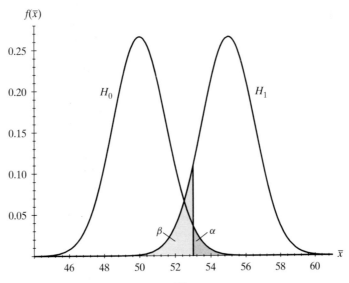

Figure 7.1   P.d.f. of $\overline{X}$ under H_0 and H_1

The next example introduces a new concept using a test about p, the probability of success. The sample size is kept small so that Table II can be used for finding probabilities. The application is one that you can actually perform.

Example 7.2.2

Assume that when given a name tag, a person puts it on either the right or left side. Let p equal the probability that the name tag is placed on the right side. We shall test the null hypothesis H_0: $p = 1/2$ against the composite alternative hypothesis H_1: $p < 1/2$. (Included with the null hypothesis are those values of p that are greater than 1/2. That is, we could think of H_0 as H_0: $p \geq 1/2$.) We shall give name tags to a random sample of $n = 20$ people, denoting the placements of their name tags with Bernoulli random variables, $X_1, X_2, \ldots, X_{20}$, where $X_i = 1$ if a person places the name tag on the right and $X_i = 0$ if a person places the name tag on the left. We can then use for our test statistic $Y = \sum_{i=1}^{20} X_i$ which has a binomial distribution, $b(20, p)$. Say that the critical region is defined by $C = \{y: y \leq 6\}$ or, equivalently, by $\{(x_1, x_2, \ldots, x_{20}): \sum_{i=1}^{20} x_i \leq 6\}$. Since Y is $b(20, 1/2)$ if $p = 1/2$, the significance level of the corresponding test is

$$\alpha = P\left(Y \leq 6; \ p = \frac{1}{2}\right) = \sum_{y=0}^{6} \binom{20}{y}\left(\frac{1}{2}\right)^{20} = 0.0577,$$

using Table II. Of course, the probability β of a type II error has different values with different values of p selected from the composite alternative hypothesis H_1: $p < 1/2$. For illustration, with $p = 1/4$,

$$\beta = P\left(7 \leq Y \leq 20; \ p = \frac{1}{4}\right) = \sum_{y=7}^{20} \binom{20}{y}\left(\frac{1}{4}\right)^{y}\left(\frac{3}{4}\right)^{20-y} = 0.2142,$$

whereas with $p = 1/10$,

$$\beta = P\left(7 \leq Y \leq 20; \ p = \frac{1}{10}\right) = \sum_{y=7}^{20} \binom{20}{y}\left(\frac{1}{10}\right)^{y}\left(\frac{9}{10}\right)^{20-y} = 0.0024.$$

Instead of considering the probability β of accepting H_0 when H_1 is true, we could compute the probability K of rejecting H_0 when H_1 is true. After all, β and $K = 1 - \beta$ provide the same information. Since K is a function of p, we denote this explicitly by writing $K(p)$. The probability

$$K(p) = \sum_{y=0}^{6} \binom{20}{y} p^y (1-p)^{20-y}, \qquad 0 < p \leq \frac{1}{2},$$

is called the **power function of the test**. Of course, $\alpha = K(1/2) = 0.0577$, $1 - K(1/4) = 0.2142$ and $1 - K(1/10) = 0.0024$. The value of the power function at a specified p is called the **power** of the test at that point. For illustration, $K(1/4) = 0.7858$ and $K(1/10) = 0.9976$ are the powers at $p = 1/4$ and $p = 1/10$, respectively. An acceptable power function is one that assumes small values when H_0 is true and larger values when p differs much from $p = 1/2$. See Figure 7.2 for a graph of this power function. ◇

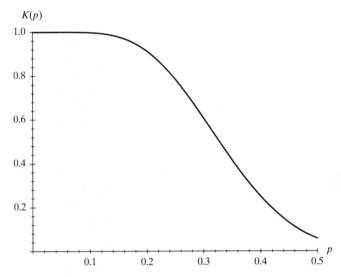

Figure 7.2   Power function: $K(p) = P(Y \le 6; p)$, where Y is $b(20, p)$

In Example 7.2.2 we introduced the new concept of the power function of a test. We now show how the sample size can be selected so as to create a test with appropriate power.

Let $X_1, X_2, \ldots, X_n$ be a random sample of size n from the normal distribution $N(\mu, 100)$, which we can suppose is a possible distribution of scores of students in a statistics course that uses a new method of teaching (e.g., computer-related materials). We wish to decide between H_0: $\mu = 60$ (*no change* hypothesis because, let us say, this was the mean by the previous method of teaching) and the research worker's hypothesis H_1: $\mu > 60$. Let us consider a sample of size $n = 25$. Of course, the sample mean $\overline{X}$ is the maximum likelihood estimator of μ, and thus it seems reasonable to base our decision on this statistic. Initially, we use the rule to reject H_0 and accept H_1 if and only if $\overline{x} \ge 62$. What are the consequences of this test? These are summarized in the power function of the test.

We first find the probability of rejecting H_0: $\mu = 60$ for various values of $\mu \ge 60$. The probability of rejecting H_0 is given by

$$K(\mu) = P(\overline{X} \ge 62; \mu)$$

because this test calls for the rejection of H_0: $\mu = 60$ when $\overline{x} \ge 62$. When the new process has the general mean μ, we know that $\overline{X}$ has the normal distribution $N(\mu, 100/25 = 4)$. Accordingly,

$$K(\mu) = P\left(\frac{\overline{X} - \mu}{2} \ge \frac{62 - \mu}{2}; \mu\right)$$

$$= 1 - \Phi\left(\frac{62 - \mu}{2}\right), \qquad 60 \le \mu,$$

is the probability of rejecting $H_0: \mu = 60$ using this particular test. Several values of $K(\mu)$ are given in Table 7.3. Figure 7.3 depicts the graph of the function $K(\mu)$.

TABLE 7.3   Values of the power function

μ	$K(\mu)$
60	0.1587
61	0.3085
62	0.5000
63	0.6915
64	0.8413
65	0.9332
66	0.9772

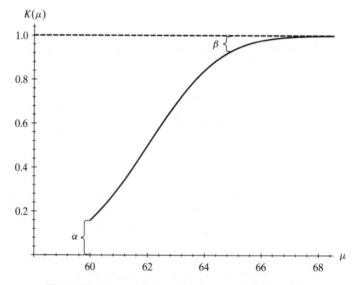

Figure 7.3   Power function: $K(\mu) = 1 - \Phi([62 - \mu]/2)$

The probability $K(\mu)$ of rejecting $H_0: \mu = 60$ is called the power function of the test. At the value μ_1 of the parameter, $K(\mu_1)$ is the power at μ_1. The power at $\mu = 60$ is $K(60) = 0.1587$, and this is the probability of rejecting $H_0: \mu = 60$ when H_0 is true. That is, $K(60) = 0.1587 = \alpha$ is the probability of a type I error and is called the significance level of the test.

The power at $\mu = 65$ is $K(65) = 0.9332$, and this is the probability of making the correct decision (namely, rejecting $H_0: \mu = 60$ when $\mu = 65$). Hence we are pleased that here it is large. When $\mu = 65, 1 - K(65) = 0.0668$ is the probability of not rejecting $H_0: \mu = 60$ when $\mu = 65$; that is, it is the probability of a type

II error and is denoted by $\beta = 0.0668$. These α and β values are displayed on Figure 7.3. Clearly, the probability $\beta = 1 - K(\mu_1)$ of a type II error depends on which value, say μ_1, is taken in the alternative hypothesis $H_1\colon \mu > 60$. Thus while $\beta = 0.0668$ when $\mu = 65$, β is equal to $1 - K(63) = 0.3085$ when $\mu = 63$.

Frequently, statisticians like to have the significance level α smaller than 0.1587, say around 0.05 or less, because it is a probability of an error, namely a type I error. Hence if we would like $\alpha = 0.05$, then we can no longer use, with $n = 25$, the critical region $\overline{x} \geq 62$, but $\overline{x} \geq c$, where c is selected so that

$$K(60) = P(\overline{X} \geq c; \mu = 60) = 0.05.$$

However, when $\mu = 60$, $\overline{X}$ is $N(60, 4)$ and

$$K(60) = P\left(\frac{\overline{X} - 60}{2} \geq \frac{c - 60}{2}; \mu = 60\right)$$

$$= 1 - \Phi\left(\frac{c - 60}{2}\right) = 0.05.$$

From Table Va in the Appendix we have that

$$\frac{c - 60}{2} = 1.645 = z_{0.05} \qquad \text{and} \qquad c = 60 + 3.29 = 63.29.$$

Although this change reduces α from 0.1587 to 0.05, it increases β from 0.0668 at $\mu = 65$ to

$$\beta = 1 - P(\overline{X} \geq 63.29; \mu = 65)$$

$$= 1 - P\left(\frac{\overline{X} - 65}{2} \geq \frac{63.29 - 65}{2}; \mu = 65\right)$$

$$= \Phi(-0.855) = 0.1963.$$

In general, without changing the sample size or the type of test of the hypothesis, a decrease in α causes an increase in β, and a decrease in β causes an increase in α. Both probabilities α and β of the two types of errors can be decreased only by increasing the sample size or, in some way, constructing a better test of the hypothesis.

For example, if $n = 100$ and we desire a test with significance level $\alpha = 0.05$, then

$$\alpha = P(\overline{X} \geq c; \mu = 60) = 0.05$$

means, since $\overline{X}$ is $N(\mu, 100/100 = 1)$,

$$P\left(\frac{\overline{X} - 60}{1} \geq \frac{c - 60}{1}; \mu = 60\right) = 0.05$$

and $c - 60 = 1.645$. Thus $c = 61.645$. The power function is

$$K(\mu) = P(\overline{X} \geq 61.645; \mu)$$

$$= P\left(\frac{\overline{X} - \mu}{1} \geq \frac{61.645 - \mu}{1}; \mu\right) = 1 - \Phi(61.645 - \mu).$$

In particular, this means that β at $\mu = 65$ is

$$\beta = 1 - K(\mu) = \Phi(61.645 - 65) = \Phi(-3.355) \approx 0;$$

so, with $n = 100$, both α and β have decreased from their respective original values of 0.1587 and 0.0668 when $n = 25$.

Rather than guess at the value of n, an ideal power function determines the sample size. Let us use a critical region of the form $\overline{x} \geq c$. Further, suppose that we want $\alpha = 0.025$ and, when $\mu = 65$, $\beta = 0.05$. Thus, since $\overline{X}$ is $N(\mu, 100/n)$,

$$0.025 = P(\overline{X} \geq c; \mu = 60) = 1 - \Phi\left(\frac{c - 60}{10/\sqrt{n}}\right)$$

and

$$0.05 = 1 - P(\overline{X} \geq c; \mu = 65) = \Phi\left(\frac{c - 65}{10/\sqrt{n}}\right).$$

That is,

$$\frac{c - 60}{10/\sqrt{n}} = 1.96 \quad \text{and} \quad \frac{c - 65}{10/\sqrt{n}} = -1.645.$$

Solving these equations simultaneously for c and $10/\sqrt{n}$, we obtain

$$\frac{10}{\sqrt{n}} = \frac{5}{3.605};$$

$$c = 60 + 1.96\frac{5}{3.605} = 62.718.$$

Thus

$$\sqrt{n} = 7.21 \quad \text{and} \quad n = 51.98.$$

Since n must be an integer, we would use $n = 52$ and obtain $\alpha = 0.025$ and $\beta = 0.05$, approximately.

For a number of years there has been another value associated with a statistical test, and most statistical computer programs print this out automatically; it is called the **probability value** or, for brevity, p-value. The p-**value** associated with a test is the probability that we obtain the observed value of the test statistic or a value that is more extreme in the direction of the alternative hypothesis, calculated when H_0 is true. Rather than select the critical region ahead of time, the p-value of a test can be reported and the reader then makes a decision.

Say that we are testing H_0: $\mu = 60$ against H_1: $\mu > 60$ with a sample mean $\overline{X}$ based on $n = 52$ observations. Suppose that we obtain the observed sample mean of $\overline{x} = 62.75$. If we compute the probability of obtaining an $\overline{x}$ of that value of 62.75 or greater when $\mu = 60$, then we obtain the p-value associated with $\overline{x} = 62.75$. That is,

$$p\text{-value} = P(\overline{X} \geq 62.75; \mu = 60)$$

$$= P\left(\frac{\overline{X} - 60}{10/\sqrt{52}} \geq \frac{62.75 - 60}{10/\sqrt{52}}; \mu = 60\right)$$

$$= 1 - \Phi\left(\frac{62.75 - 60}{10/\sqrt{52}}\right) = 1 - \Phi(1.983) = 0.0237.$$

If this p-value is small, we tend to reject the hypothesis H_0: $\mu = 60$. For example, rejection of H_0: $\mu = 60$ if the p-value is less than or equal to 0.025 is exactly the same as rejection if $\overline{x} \geq 62.718$. That is, $\overline{x} = 62.718$ has a p-value of 0.025. To help the reader keep the definition of p-value in mind, we note that it can be thought of as that **tail-end probability**, under H_0, of the distribution of the statistic, here $\overline{X}$, beyond the observed value of the statistic. See Figure 7.4 for the p-value associated with $\overline{x} = 62.75$.

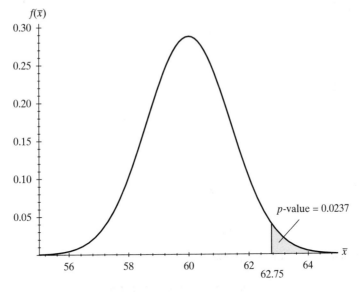

Figure 7.4 p-value

Example 7.2.3

Suppose that in the past, a golfer's scores have been (approximately) normally distributed with mean $\mu = 90$ and $\sigma^2 = 9$. After taking some lessons, the golfer has reason to believe that the mean μ has decreased. (We assume that σ^2 is still about 9.) To test the null hypothesis H_0: $\mu = 90$ against the alternative hypothesis H_1: $\mu < 90$, the golfer plays 16 games, computing the sample mean $\overline{x}$. If $\overline{x}$ is small, say $\overline{x} \leq c$, then H_0 is rejected and H_1 accepted; that is, it seems as if the mean μ has actually decreased after the lessons. If $c = 88.5$, then the power function of the test is

$$K(\mu) = P(\overline{X} \leq 88.5; \mu)$$

$$= P\left(\frac{\overline{X} - \mu}{3/4} \leq \frac{88.5 - \mu}{3/4}; \mu\right) = \Phi\left(\frac{88.5 - \mu}{3/4}\right)$$

because 9/16 is the variance of $\overline{X}$. In particular,

$$\alpha = K(90) = \Phi(-2) = 1 - 0.9772 = 0.0228.$$

If, in fact, the true mean is equal to $\mu = 88$ after the lessons, the power is $K(88) = \Phi(2/3) = 0.7475$. If $\mu = 87$, then $K(87) = \Phi(2) = 0.9772$. An observed sample mean of $\overline{x} = 88.25$ has a

$$p\text{-value} = P(\overline{X} \leq 88.25; \mu = 90)$$

$$= \Phi\left(\frac{88.25 - 90}{3/4}\right) = \Phi\left(-\frac{7}{3}\right) = 0.0098,$$

and this would lead to rejection at $\alpha = 0.0228$ (or even $\alpha = 0.01$).   ◇

The next example is an extension of Example 7.2.2.

Example 7.2.4

To test H_0: $p = 1/2$ against H_1: $p < 1/2$, we take a random sample of Bernoulli trials, $X_1, X_2, \ldots, X_n$, and use for our test statistic $Y = \sum_{i=1}^{n} X_i$ which has a binomial distribution $b(n, p)$. Let the critical region be defined by $C = \{y: y \leq c\}$. The power function for this test is defined by $K(p) = P(Y \leq c; p)$. We shall find the values of n and c so that $K(1/2) = 0.05$ and $K(1/4) = 0.90$, approximately. That is, we would like the significance level to be $\alpha = K(1/2) = 0.05$ and the power at $p = 1/4$ to equal 0.90. We proceed as follows:

$$0.05 = P\left(Y \leq c; p = \frac{1}{2}\right) = P\left(\frac{Y - n/2}{\sqrt{n(1/2)(1/2)}} \leq \frac{c - n/2}{\sqrt{n(1/2)(1/2)}}\right)$$

implies that

$$(c - n/2)/\sqrt{n/4} \approx -1.645;$$

and

$$0.90 = P\left(Y \leq c; p = \frac{1}{4}\right) = P\left(\frac{Y - n/4}{\sqrt{n(1/4)(3/4)}} \leq \frac{c - n/4}{\sqrt{n(1/4)(3/4)}}\right)$$

implies that

$$(c - n/4)/\sqrt{3n/16} \approx 1.282.$$

Therefore,

$$\frac{n}{4} \approx 1.645\sqrt{\frac{n}{4}} + 1.282\sqrt{\frac{3n}{16}} \quad \text{and} \quad \sqrt{n} \approx 4(1.378) = 5.512.$$

Thus n is about equal to 31, and from either of the first two approximate equalities we find that c is about equal to 10.9. Using $n = 31$ and $c = 10.9$ means that $K(1/2) = 0.05$ and $K(1/4) = 0.90$ are only approximate. In fact, since Y must be an integer, we could let $c = 10.5$. Then, with $n = 31$,

$$\alpha = K\left(\frac{1}{2}\right) = P\left(Y \leq 10.5; \ p = \frac{1}{2}\right) \approx 0.0362;$$

$$K\left(\frac{1}{4}\right) = P\left(Y \leq 10.5; \ p = \frac{1}{4}\right) \approx 0.8729.$$

Or, we could let $c = 11.5$ and $n = 31$, in which case

$$\alpha = K\left(\frac{1}{2}\right) = P\left(Y \leq 11.5; \ p = \frac{1}{2}\right) \approx 0.0558;$$

$$K\left(\frac{1}{4}\right) = P\left(Y \leq 11.5; \ p = \frac{1}{4}\right) \approx 0.9235. \qquad \diamond$$

EXERCISES

7.2.1 A certain-size bag is designed to hold 25 pounds of potatoes. A farmer fills such bags in the field. Assume that the weight X of potatoes in a bag is $N(\mu, 9)$. We shall test the null hypothesis $H_0: \mu = 25$ against the alternative hypothesis $H_1: \mu < 25$. Let X_1, X_2, X_3, X_4 be a random sample of size 4 from this distribution, and let the critical region C for this test be defined by $\overline{x} \leq 22.5$, where $\overline{x}$ is the observed value of $\overline{X}$.

(a) What is the power function $K(\mu)$ of this test? In particular, what is the significance level $\alpha = K(25)$ for your test?

(b) If the random sample of four bags of potatoes yielded the values $x_1 = 21.24$, $x_2 = 24.81$, $x_3 = 23.62$, $x_4 = 26.82$, would you accept or reject H_0 using your test?

(c) What is the p-value associated with the $\overline{x}$ in part (b)?

7.2.2 Let X equal the number of milliliters in a bottle that has a label volume of 350 milliliters. Assume that the distribution of X is $N(\mu, 4)$. To test the null hypothesis $H_0: \mu = 355$ against the alternative hypothesis $H_1: \mu < 355$, let the critical region be defined by $C = \{\overline{x} : \overline{x} \leq 354.05\}$, where $\overline{x}$ is the sample mean of the contents of a random sample of $n = 12$ bottles.

(a) Find the power function $K(\mu)$ for this test.

(b) What is the (approximate) significance level of this test?

(c) Find the values of $K(354.05)$ and $K(353.1)$ and sketch the graph of the power function.

(d) Use the following 12 observations to state your conclusion of this test:

$$350 \quad 353 \quad 354 \quad 356 \quad 353 \quad 352$$
$$354 \quad 355 \quad 357 \quad 353 \quad 354 \quad 355$$

(e) What is the approximate p-value of this test?

7.2.3 Assume that SAT mathematics scores of students who attend small liberal arts colleges are $N(\mu, 8100)$. We shall test $H_0: \mu = 530$ against the alternative hypothesis $H_1: \mu < 530$. Given a random sample of size $n = 36$ SAT mathematics scores, let the critical region be defined by $C = \{\overline{x}: \overline{x} \le 510.77\}$, where $\overline{x}$ is the observed mean of the sample.

(a) Find the power function, $K(\mu)$, for this test.

(b) What is the value of the significance level of this test?

(c) What is the value of $K(510.77)$?

(d) Sketch the graph of the power function.

(e) What is the p-value associated (i) with $\overline{x} = 507.35$ and (ii) with $\overline{x} = 497.45$?

7.2.4 Let X be $N(\mu, 100)$. To test $H_0: \mu = 80$ against $H_1: \mu > 80$, let the critical region be defined by $C = \{(x_1, x_2, \ldots, x_{25}): \overline{x} \ge 83\}$, where $\overline{x}$ is the sample mean of a random sample of size $n = 25$ from this distribution.

(a) What is the power function $K(\mu)$ for this test?

(b) What is the significance level of this test?

(c) What are the values of $K(80)$, $K(83)$, and $K(86)$?

(d) Sketch the graph of the power function.

(e) What is the p-value corresponding to $\overline{x} = 83.41$?

7.2.5 Let X equal the yield of alfalfa in tons per acre per year. Assume that X is $N(1.5, 0.09)$. It is hoped that new fertilizer will increase the average yield. We shall test the null hypothesis $H_0: \mu = 1.5$ against the alternative hypothesis $H_1: \mu > 1.5$. Assume that the variance continues to equal $\sigma^2 = 0,09$ with the new fertilizer. Using $\overline{X}$, the mean of a random sample of size n, as the test statistic, reject H_0 if $\overline{x} \ge c$. Find n and c so that the power function $K(\mu) = P(\overline{X} \ge c : \mu)$ is such that $\alpha = K(1.5) = 0.05$ and $K(1.7) = 0.95$.

7.2.6 Let X equal the number of pounds of butterfat produced by a Holstein cow during the 305-day milking period following the birth of a calf. Assume that the distribution of X is $N(\mu, 140^2)$. To test the null hypothesis $H_0: \mu = 715$ against the alternative hypothesis $H_1: \mu < 715$, let the critical region be defined by $C = \{\overline{x}: \overline{x} \le 668.94\}$, where $\overline{x}$ is the sample mean of $n = 25$ butterfat weights from 25 cows selected at random.

(a) Find the power function $K(\mu)$ for this test.

(b) What is the significance level of this test?

(c) What are the values of $K(668.94)$ and $K(622.88)$?

(d) Sketch a graph of the power function.

(e) State your conclusion using the following 25 observations of X:

425	710	661	664	732	714	934	761	744
653	725	657	421	573	535	602	537	405
874	791	721	849	567	468	975		

(f) What is the approximate p-value of this test?

7.2.7 In Exercise 7.2.6, let $C = \{\overline{x}: \overline{x} \le c\}$ be the critical region. Find values for n and c so that the significance level of this test is $\alpha = 0.05$ and the power at $\mu = 650$ is 0.90.

7.2.8 Let X have a Bernoulli distribution with p.d.f.

$$f(x; p) = p^x (1 - p)^{1-x}, \qquad x = 0, 1, \qquad 0 \le p \le 1.$$

We would like to test the null hypothesis $H_0: p \le 0.4$ against the alternative hypothesis H_1: $p > 0.4$. For the test statistic use $Y = \sum_{i=1}^{n} X_i$, where $X_1, X_2, \ldots, X_n$ is a random sample of size n from this Bernoulli distribution. Let the critical region be of the form $C = \{y : y \ge c\}$.

(a) Let $n = 100$. On the same set of axes, sketch the graphs of the power functions corresponding to the three critical regions, $C_1 = \{y : y \ge 40\}$, $C_2 = \{y : y \ge 50\}$, and $C_3 = \{y : y \ge 60\}$. Use the normal approximation to compute the probabilities.

(b) Let $C = \{y : y \ge 0.45n\}$. On the same set of axes, sketch the graphs of the power functions corresponding to the three samples of sizes 10, 100, and 1000.

7.2.9 Let p denote the probability that for a particular tennis player, the first serve is good. Since $p = 0.40$, this player decided to take lessons in order to increase p. When the lessons are completed, the hypothesis $H_0: p = 0.40$ will be tested against $H_1: p > 0.40$ based on $n = 25$ trials. Let y equal the number of first serves that are good, and let the critical region be defined by $C = \{y : y \ge 14\}$.

(a) Find the power function $K(p)$ for this test.

(b) What is the value of the significance level, $\alpha = K(0.40)$? Use Table II.

(c) Evaluate $K(p)$ at $p = 0.45, 0.50, 0.60, 0.70, 0.80$, and 0.90. Use Table II.

(d) Sketch the graph of the power function.

(e) If $y = 15$ following the lessons, would H_0 be rejected?

(f) What is the p-value associated with $y = 15$?

7.2.10 Let $X_1, X_2, \ldots, X_8$ be a random sample of size $n = 8$ from a Poisson distribution with mean λ. Reject the simple null hypothesis $H_0: \lambda = 0.5$ and accept $H_1: \lambda > 0.5$ if the observed sum $\sum_{i=1}^{8} x_i \ge 8$.

(a) Compute the significance level α of the test.

(b) Find the power function $K(\lambda)$ of the test as a sum of Poisson probabilities.

(c) Using Table III, determine $K(0.75)$, $K(1)$, and $K(1.25)$.

7.2.11 Let p equal the fraction defective of a certain manufactured item. To test $H_0: p = 1/26$ against $H_1: p > 1/26$, we inspect n items selected at random and let Y be the number of defective items in this sample. We reject H_0 if the observed $y \ge c$. Find n and c so that $\alpha = K(1/26) = 0.05$ and $K(1/10) = 0.90$ approximately, where $K(p) = P(Y \ge c; p)$. HINT: Use either the normal or Poisson approximation to help solve this exercise.

7.2.12 Let X_1, X_2, X_3 be a random sample of size $n = 3$ from an exponential distribution with mean $\theta > 0$. Reject the simple null hypothesis $H_0: \theta = 2$ and accept the composite alternative hypothesis $H_1: \theta < 2$ if the observed sum $\sum_{i=1}^{3} x_i \le 2$.

(a) What is the power function $K(\theta)$ written as an integral?

(b) Using integration by parts, define the power function as a summation.

(c) With the help of Table III, determine $\alpha = K(2)$, $K(1)$, $K(1/2)$, and $K(1/4)$.

7.3 TESTS ABOUT ONE MEAN AND ONE VARIANCE

In several of the examples and exercises in Section 7.2 we assumed that we were sampling from a normal distribution and that the variance was known. The null hypothesis was generally of the form $H_0: \mu = \mu_0$. There are essentially three possibilities for the alternative hypothesis, namely that μ has increased: (i) $H_1: \mu > \mu_0$; μ has decreased; (ii) $H_1: \mu < \mu_0$; or μ has changed, but it is not known if it has increased or decreased, which leads to a two-sided alternative hypothesis (iii): $H_1: \mu \neq \mu_0$.

To test $H_0: \mu = \mu_0$ against one of these three alternative hypotheses, a random sample is taken from the distribution, and an observed sample mean, $\overline{x}$, that is close to μ_0 supports H_0. The closeness of $\overline{x}$ to μ_0 is measured in terms of standard deviations of $\overline{X}$, $\sigma/\sqrt{n}$, which is sometimes called the **standard error of the mean**. Thus the test statistic could be defined by

$$Z = \frac{\overline{X} - \mu_0}{\sqrt{\sigma^2/n}} = \frac{\overline{X} - \mu_0}{\sigma/\sqrt{n}}, \tag{7.2}$$

and the critical regions, at a significance level α, for the three respective alternative hypotheses would be (i) $z \geq z_\alpha$, (ii) $z \leq -z_\alpha$, and (iii) $|z| \geq z_{\alpha/2}$. In terms of $\overline{x}$ these three critical regions become (i) $\overline{x} \geq \mu_0 + z_\alpha(\sigma/\sqrt{n})$, (ii) $\overline{x} \leq \mu_0 - z_\alpha(\sigma/\sqrt{n})$, and (iii) $|\overline{x} - \mu_0| \geq z_{\alpha/2}(\sigma/\sqrt{n})$.

These tests and critical regions are summarized in Table 7.4. The underlying assumption is that the distribution is $N(\mu, \sigma^2)$ and σ^2 is known.

TABLE 7.4   Tests of hypotheses about one mean, variance known

H_0	H_1	Critical Region
$\mu = \mu_0$	$\mu > \mu_0$	$z \geq z_\alpha$ or $\overline{x} \geq \mu_0 + z_\alpha \sigma/\sqrt{n}$
$\mu = \mu_0$	$\mu < \mu_0$	$z \leq -z_\alpha$ or $\overline{x} \leq \mu_0 - z_\alpha \sigma/\sqrt{n}$
$\mu = \mu_0$	$\mu \neq \mu_0$	$|z| \geq z_{\alpha/2}$ or $|\overline{x} - \mu_0| \geq z_{\alpha/2}\sigma/\sqrt{n}$

Thus far we have assumed that the variance σ^2 was known. We now take a more realistic position and assume that the variance is unknown. Suppose that our null hypothesis is $H_0: \mu = \mu_0$ and the two-sided alternative hypothesis is $H_1: \mu \neq \mu_0$. If a random sample $X_1, X_2, \ldots, X_n$ is taken from a normal distribution $N(\mu, \sigma^2)$, we recall from Section 6.2 that a confidence interval for μ was based on

$$T = \frac{\overline{X} - \mu}{\sqrt{S^2/n}} = \frac{\overline{X} - \mu}{S/\sqrt{n}}.$$

This suggests that T might be a good statistic to use for the test of $H_0: \mu = \mu_0$ with μ replaced by μ_0. In addition, it is the natural statistic to use if we replace σ^2/n by its unbiased estimator S^2/n in $(\overline{X} - \mu_0)/\sqrt{\sigma^2/n}$ in Equation 7.2. If $\mu = \mu_0$, we know that T has a t distribution with $n - 1$ degrees of freedom. Thus, with $\mu = \mu_0$,

$$P[|T| \geq t_{\alpha/2}(n - 1)] = P\left[\frac{|\overline{X} - \mu_0|}{S/\sqrt{n}} \geq t_{\alpha/2}(n - 1)\right] = \alpha.$$

Accordingly, if $\overline{x}$ and s are the sample mean and sample standard deviation, the rule that rejects $H_0: \mu = \mu_0$ and accepts $H_1: \mu \neq \mu_0$ if and only if

$$|t| = \frac{|\overline{x} - \mu_0|}{s/\sqrt{n}} \geq t_{\alpha/2}(n - 1)$$

provides a test of this hypothesis with significance level α. It should be noted that this rule is equivalent to rejecting $H_0: \mu = \mu_0$ if μ_0 is not in the open $100(1 - \alpha)\%$ confidence interval with end points $\overline{x} - t_{\alpha/2}(n-1)(s/\sqrt{n})$ and $\overline{x} + t_{\alpha/2}(n-1)(s/\sqrt{n})$.

Table 7.5 summarizes tests of hypotheses for a single mean, along with the three possible alternative hypotheses, when the underlying distribution is $N(\mu, \sigma^2)$, σ^2 is unknown, $t = (\overline{x} - \mu_0)/(s/\sqrt{n})$, and $n \leq 30$. If $n > 30$, use Table 7.4 for approximate tests with σ replaced by s.

TABLE 7.5   Tests of hypotheses for one mean, variance unknown

H_0	H_1	Critical Region
$\mu = \mu_0$	$\mu > \mu_0$	$t \geq t_\alpha(n-1)$ or $\overline{x} \geq \mu_0 + t_\alpha(n-1)s/\sqrt{n}$
$\mu = \mu_0$	$\mu < \mu_0$	$t \leq -t_\alpha(n-1)$ or $\overline{x} \leq \mu_0 - t_\alpha(n-1)s/\sqrt{n}$
$\mu = \mu_0$	$\mu \neq \mu_0$	$|t| \geq t_{\alpha/2}(n-1)$ or $|\overline{x} - \mu_0| \geq t_{\alpha/2}(n-1)s/\sqrt{n}$

Example 7.3.1

Let X (in millimeters) equal the growth in 15 days of a tumor induced in a mouse. Assume that the distribution of X is $N(\mu, \sigma^2)$. We shall test the null hypothesis $H_0: \mu = \mu_0 = 4.0$ millimeters against the two-sided alternative hypothesis $H_1: \mu \neq 4.0$. If we use $n = 9$ observations and a significance level of $\alpha = 0.10$, the critical region is

$$|t| = \frac{|\overline{x} - 4.0|}{s/\sqrt{9}} \geq t_{\alpha/2}(8) = 1.860.$$

If we are given that $n = 9, \overline{x} = 4.3$, and $s = 1.2$, we see that

$$t = \frac{4.3 - 4.0}{1.2/\sqrt{9}} = \frac{0.3}{0.4} = 0.75.$$

Thus

$$|t| = |0.75| < 1.860$$

and we accept (do not reject) $H_0: \mu = 4.0$ at the $\alpha = 10\%$ significance level (see Figure 7.5). ◇

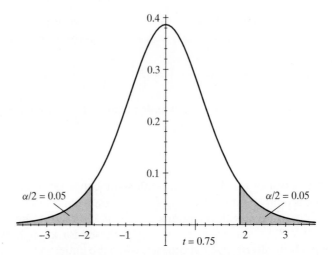

Figure 7.5   Test about mean of tumor growths

Remark   In discussing the test of a statistical hypothesis, the word *accept H_0* might better be replaced by *do not reject H_0*. That is, if, in Example 7.3.1, $\bar{x}$ is close enough to 4.0 so that we accept $\mu = 4.0$, we do not want that acceptance to imply that μ is actually equal to 4.0. We want to say that the data do not deviate enough from $\mu = 4.0$ for us to reject that hypothesis; that is, we do not reject $\mu = 4.0$ with these observed data. With this understanding, we sometimes accept, and sometimes fail to reject or do not reject, the null hypothesis.

The next example illustrates the use of the t-statistic with a one-sided alternative hypothesis.

Example 7.3.2

In attempting to control the strength of the wastes discharged into a nearby river, a paper firm has taken a number of measures. Members of the firm believe that they have reduced the oxygen-consuming power of their wastes from a previous mean μ of 500 (measured in permanganate in parts per million). They plan to test H_0: $\mu = 500$ against H_1: $\mu < 500$, using readings taken on $n = 25$ consecutive days. If these 25 values can be treated as a random sample, then the critical region, for a significance level of $\alpha = 0.01$, is

$$t = \frac{\bar{x} - 500}{s/\sqrt{25}} \leq -t_{0.01}(24) = -2.492.$$

The observed values of the sample mean and sample standard deviation were $\bar{x} = 308.8$ and $s = 115.15$.

Since

$$t = \frac{308.8 - 500}{115.15/\sqrt{25}} = -8.30 < -2.492,$$

we clearly reject the null hypothesis and accept H_1: $\mu < 500$. It should be noted, however, that although an improvement has been made, there still might exist the question of whether the improvement is adequate. The one-sided 99% confidence interval for μ,

$$[0, 308.8 + 2.492(115.25/\sqrt{25})] = [0, 366.191],$$

provides an upper bound for μ and may help the company answer this question.   ◇

Many times, there is interest in comparing the means of two different distributions or populations. We must consider two situations: that in which X and Y are dependent, and that in which X and Y are independent. We consider the independent case in Section 7.4.

If X and Y are dependent, let $W = X - Y$, and the hypothesis that $\mu_X = \mu_Y$ would be replaced with the hypothesis H_0: $\mu_W = 0$. For example, suppose that X and Y equal the resting pulse rate for a person before and after taking an 8-week program in aerobic dance. We would be interested in testing H_0: $\mu_W = 0$ (no change) against H_1: $\mu_W > 0$ (the aerobic dance program decreased the resting pulse rate). Because X and Y are measurements on the same person, X and Y are clearly dependent. If we can assume that the distribution of W is (approximately) $N(\mu_W, \sigma^2)$, then the appropriate t-test for a single mean could be used, selecting from Table 7.5. This is often called a **paired t-test**.

Example 7.3.3

Twenty-four girls in the 9th and 10th grades were put on an ultraheavy rope-jumping program. Someone thought that such a program would increase their speed when running the 40-yard dash. Let W equal the difference in time to run the 40-yard dash—the "before-program time" minus the "after-program time." Assume that the distribution of W is (approximately) $N(\mu_W, \sigma_W^2)$. We shall test the null hypothesis H_0: $\mu_W = 0$ against the alternative hypothesis H_1: $\mu_W > 0$. The test statistic and critical region that has an $\alpha = 0.05$ significance level are given by

$$t = \frac{\overline{w} - 0}{s_w/\sqrt{24}} \geq t_{0.05}(23) = 1.714.$$

The following data give the difference in time that it took each girl to run the 40-yard dash, with positive numbers indicating a faster time after the program:

0.28	0.01	0.13	0.33	−0.03	0.07	−0.18	−0.14
−0.33	0.01	0.22	0.29	−0.08	0.23	0.08	0.04
−0.30	−0.08	0.09	0.70	0.33	−0.34	0.50	0.06

For these data, $\overline{w} = 0.079$ and $s_w = 0.255$. Thus the observed value of the test statistic is

$$t = \frac{0.079 - 0}{0.255/\sqrt{24}} = 1.518.$$

Since $1.518 < 1.714$, the null hypothesis is not rejected. Note, however, that $t_{0.10}(23) = 1.319$ and $t = 1.518 > 1.319$. Thus, the null hypothesis would be rejected at an $\alpha = 0.10$ significance level. Another way of saying this is that

$$0.05 < p\text{-value} < 0.10.$$

It would be instructive for you to draw a figure illustrating this.   ◇

Since there has been no previous example of a test of a hypothesis concerning variances, an example follows a brief discussion and then rules are given formally.

A psychology professor claims that the variance of IQ scores for college students is equal to $\sigma^2 = 100$. To test this claim, it is decided to test the hypothesis $H_0\colon \sigma^2 = 100$ against a two-sided alternative hypothesis $H_1\colon \sigma^2 \neq 100$. A random sample of n students will be selected, and the test will be based on the observed unbiased estimate s^2 of the variance σ^2 of their IQ scores. The hypothesis H_0 will be rejected if s^2 differs "too much" from $\sigma^2 = 100$. That is, H_0 will be rejected if $s^2 \leq c_1$ or $s^2 \geq c_2$ for some constants $c_1 < 100$ and $c_2 > 100$. Generally, c_1 and c_2 are selected so that

$$P(S^2 \leq c_1) = \alpha/2 \qquad \text{and} \qquad P(S^2 \geq c_2) = \alpha/2.$$

If we may assume that IQ scores are normally distributed, then the distribution of $(n-1)S^2/100$ is $\chi^2(n-1)$ when H_0 is true. Thus the chi-square probability table can be used to give us

$$P\left[\frac{(n-1)S^2}{100} \leq \frac{(n-1)c_1}{100}\right] = \alpha/2 \quad \text{and} \quad P\left[\frac{(n-1)S^2}{100} \geq \frac{(n-1)c_2}{100}\right] = \alpha/2.$$

Accordingly,

$$\frac{(n-1)c_1}{100} = \chi^2_{1-\alpha/2}(n-1) \qquad \text{and} \qquad \frac{(n-1)c_2}{100} = \chi^2_{\alpha/2}(n-1).$$

Hence

$$c_1 = [100/(n-1)]\chi^2_{1-\alpha/2}(n-1) \qquad \text{and} \qquad c_2 = [100/(n-1)]\chi^2_{\alpha/2}(n-1).$$

This test can also be based on the value of the chi-square statistic

$$\chi^2 = \frac{(n-1)S^2}{100}.$$

We would reject the null hypothesis if

$$\chi^2 \leq \chi^2_{1-\alpha/2}(n-1) \qquad \text{or} \qquad \chi^2 \geq \chi^2_{\alpha/2}(n-1).$$

Example 7.3.4

Suppose that in the previous discussion $n = 23$ and $\alpha = 0.05$. To test $H_0: \sigma^2 = 100$ against $H_1: \sigma^2 \neq 100$, we let $\chi^2 = (n-1)S^2/\sigma_0^2 = 22S^2/100$. Because

$$\chi^2_{0.975}(22) = 10.98 \qquad \text{and} \qquad \chi^2_{0.025}(22) = 36.78,$$

H_0 will be rejected if

$$\chi^2 = \frac{22s^2}{100} \leq 10.98 \qquad \text{or} \qquad \chi^2 = \frac{22s^2}{100} \geq 36.78$$

or, equivalently, if

$$s^2 \leq c_1 = \frac{100(10.98)}{22} = 49.91 \qquad \text{or} \qquad s^2 \geq c_2 = \frac{100(36.78)}{22} = 167.18.$$

(Exercise 7.3.14 gives some insight as to why these critical values are so far apart.) Given that the observed value of the sample variance was $s^2 = 147.82$, the hypothesis $H_0: \sigma^2 = 100$ was not rejected. Note that the 95% confidence interval for σ^2,

$$\left[\frac{(22)(147.82)}{36.78}, \frac{(22)(147.82)}{10.98} \right] = [88.42, 296.18],$$

contains $\sigma^2 = 100$. Also, the observed value of the chi-square test statistic is

$$\chi^2 = \frac{22(147.82)}{100} = 32.52.$$

Because

$$10.98 < 32.52 < 36.78,$$

we would again accept H_0 as expected (see Figure 7.6).

If $H_1: \sigma^2 > 100$ had been the alternative hypothesis, $H_0: \sigma^2 = 100$ would have been rejected if

$$\chi^2 = \frac{22s^2}{100} \geq \chi^2_{0.05}(22) = 33.92$$

or, equivalently,

$$s^2 \geq \frac{100\chi^2_{0.05}(22)}{22} = \frac{(100)(33.92)}{22} = 154.18.$$

Because

$$\chi^2 = 32.52 < 33.92 \qquad \text{and} \qquad s^2 = 147.82 < 154.18,$$

H_0 would not be rejected in favor of this one-sided alternative hypothesis for $\alpha = 0.05$, although we suspect a slightly larger data set might lead to rejection.   ◇

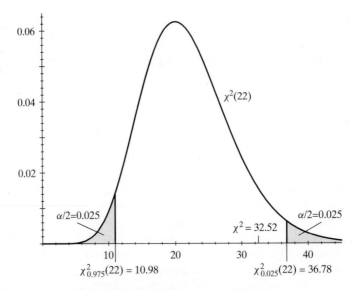

Figure 7.6   Test about variance of IQ scores

Table 7.6 summarizes tests of hypotheses for a single variance. The critical region is given in terms of the sample variance. The critical region is also given in terms of the chi-square test statistic

$$\chi^2 = \frac{(n-1)s^2}{\sigma_0^2}.$$

TABLE 7.6   Tests about the variance

H_0	H_1	Critical Region
$\sigma^2 = \sigma_0^2$	$\sigma^2 > \sigma_0^2$	$s^2 \geq \dfrac{\sigma_0^2\, \chi_\alpha^2(n-1)}{n-1}$ or $\chi^2 \geq \chi_\alpha^2(n-1)$
$\sigma^2 = \sigma_0^2$	$\sigma^2 < \sigma_0^2$	$s^2 \leq \dfrac{\sigma_0^2\, \chi_{1-\alpha}^2(n-1)}{n-1}$ or $\chi^2 \leq \chi_{1-\alpha}^2(n-1)$
$\sigma^2 = \sigma_0^2$	$\sigma^2 \neq \sigma_0^2$	$s^2 \leq \dfrac{\sigma_0^2\, \chi_{1-\alpha/2}^2(n-1)}{n-1}$ or $s^2 \geq \dfrac{\sigma_0^2\, \chi_{\alpha/2}^2(n-1)}{n-1}$ or $\chi^2 \leq \chi_{1-\alpha/2}^2(n-1)$ or $\chi^2 \geq \chi_{\alpha/2}^2(n-1)$

There are two ways of viewing a statistical test. One of these is through the p-value of the test; this is becoming more popular and is included in most computer printouts. (This was discussed in Section 7.2.) After observing the test statistic, the p-value is the probability, under H_0, of the test statistic being as extreme (in the direction of rejection of H_0) as the observed one. That is, the p-value is the tail-end probability. For illustration, in Example 7.2.3, if the golfer averaged $\overline{x} = 87.94$, then

$$p\text{-value} = P(\overline{X} \le 87.94) = P\left(\frac{\overline{X} - 90}{3/4} \le \frac{87.94 - 90}{3/4}\right) = 0.0027.$$

The fact that the p-value < 0.05 is equivalent to the fact that $\overline{x} < 88.77$ because $P(\overline{X} \le 88.77; \mu = 90) = 0.05$. Since $\overline{x} = 87.94$ is an observed value of a random variable, namely $\overline{X}$, then the p-value, a function of $\overline{x}$, is also an observed value of a random variable. That is, before the random experiment is performed, the probability that the p-value is less than α is equal to α when the null hypothesis is true. Many statisticians believe that the observed p-value provides an understandable measure of the truth of H_0: The smaller the p-value, the less they believe in H_0.

Three additional examples of the p-value are given by referring to Examples 7.3.1, 7.3.2, and 7.3.4. In two-sided tests for means and proportions, the p-value is the probability of the extreme values in both directions. With the mouse data (Example 7.3.1), the p-value is

$$p\text{-value} = P(|T| \ge 0.75).$$

In Table VI we see that if T has a t-distribution with 8 degrees of freedom, then $P(T \ge 0.706) = 0.25$. Thus $P(|T| \ge 0.706) = 0.50$ and the p-value will be a little smaller than 0.50. In fact, $P(|T| \ge 0.75) = 0.47$ (a probability that can be found with certain computer programs), which is not less than $\alpha = 0.10$, and thus we do not reject H_0 at that significance level. In the example concerned with waste (Example 7.3.2), the p-value is essentially zero, since $P(Z \le -8.30) \approx 0$, where Z is a standardized normal variable, and we reject H_0. In the example about the variance of IQ scores (Example 7.3.4), with the one-sided alternative, $H_1: \sigma^2 > 100$, the p-value is

$$p\text{-value} = P(W \ge 32.52),$$

where $W = 22S^2/100$ has a chi-square distribution with 22 degrees of freedom when H_0 is true. From Table IV we see that

$$P(W \ge 30.81) = 0.10 \qquad \text{and} \qquad P(W \ge 33.92) = 0.05.$$

Thus

$$0.05 < p\text{-value} = P(W \ge 32.52) < 0.10$$

and the null hypothesis would not be rejected at an $\alpha = 0.05$ significance level. To find the p-value for a two-sided alternative for tests about the variance, double the tail probability beyond the chi-square test statistic if the sample variance is larger than σ_0^2, or double the tail probability below the chi-square test statistic if the sample variance is below σ_0^2. So in Example 7.3.4, since $s^2 = 147.82 > 100$, the p-value equals $2[P(W \ge 32.52)]$ and thus

$$0.10 < p\text{-value} < 0.20.$$

The other way of looking at tests of hypotheses is through the consideration of confidence intervals, particularly for two-sided alternatives and the corresponding tests. For example, with the mouse data (Example 7.3.1), a 90% confidence interval for the unknown mean is

$$4.3 \pm (1.86)(1.2)/\sqrt{9} \quad \text{or} \quad [3.56, 5.04],$$

since $t_{0.05}(8) = 1.86$. Note that this confidence interval covers the hypothesized value $\mu = 4.0$ and we do not reject H_0: $\mu = 4.0$. If the confidence interval did not cover $\mu = 4.0$, then we would have rejected H_0: $\mu = 4.0$. Many statisticians believe that estimation is much more important than tests of hypotheses and accordingly approach statistical tests through confidence intervals. For one-sided tests, we use one-sided confidence intervals (see Example 7.3.2).

EXERCISES

7.3.1 Assume that IQ scores for a certain population are approximately $N(\mu, 100)$. To test H_0: $\mu = 110$ against the one-sided alternative hypothesis H_1: $\mu > 110$, we take a random sample of size $n = 16$ from this population and observe $\bar{x} = 113.5$. Do we accept or reject H_0 at

   (a) The 5% significance level?

   (b) The 10% significance level?

   (c) What is the p-value of this test?

7.3.2 Assume that the weight of cereal in a "10-ounce box" is $N(\mu, \sigma^2)$, To test H_0: $\mu = 10.1$ against H_1: $\mu > 10.1$, we take a random sample of size $n = 16$ and observe that $\bar{x} = 10.4$ and $s = 0.4$.

   (a) Do we accept or reject H_0 at the 5% significance level?

   (b) What is the approximate p-value of this test?

7.3.3 Let X equal the Brinell hardness measurement of ductile iron subcritically annealed. Assume that the distribution of X is $N(\mu, 100)$. We shall test the null hypothesis H_0: $\mu = 170$ against the alternative hypothesis H_1: $\mu > 170$ using $n = 25$ observations of X.

   (a) Define the test statistic and a critical region that has a significance level of $\alpha = 0.05$. Sketch a figure showing this critical region.

   (b) A random sample of $n = 25$ observations of X yielded the following measurements:

170	167	174	179	179	156	163	156	187
156	183	179	174	179	170	156	187	
179	183	174	187	167	159	170	179	

   Calculate the value of the test statistic and clearly give your conclusion.

   (c) Give the approximate p-value of this test.

7.3.4 Let X equal the thickness of spearmint gum manufactured for vending machines. Assume that the distribution of X is $N(\mu, \sigma^2)$. The target thickness is 7.5 hundredths of an inch. We shall test the null hypothesis H_0: $\mu = 7.5$ against a two-sided alternative hypothesis using 10 observations.

   (a) Define the test statistic and critical region for an $\alpha = 0.05$ significance level. Sketch a figure illustrating this critical region.

(b) Calculate the value of the test statistic and clearly give your decision using the following $n = 10$ thicknesses in hundredths of an inch for pieces of gum that were selected randomly from the production line:

$$7.65 \quad 7.60 \quad 7.65 \quad 7.70 \quad 7.55$$
$$7.55 \quad 7.40 \quad 7.40 \quad 7.50 \quad 7.50$$

(c) Is $\mu = 7.50$ contained in a 95% confidence interval for μ?

7.3.5 The mean birth weight in the United States is $\mu = 3315$ grams with a standard deviation of $\sigma = 575$. Let X equal the birth weight in grams in Jerusalem. Assume that the distribution of X is $N(\mu, \sigma^2)$. We shall test the null hypothesis $H_0: \mu = 3315$ against the alternative hypothesis $H_1: \mu < 3315$ using a random sample of size $n = 30$.

(a) Define a critical region that has a significance level of $\alpha = 0.05$.

(b) If the random sample of $n = 30$ yielded $\bar{x} = 3189$ and $s = 488$, what is your conclusion?

(c) What is the approximate p-value of your test?

7.3.6 Let X equal the forced vital capacity (FVC) in liters for a female college student. (This is the amount of air that a student can force out of her lungs.) Assume that the distribution of X is (approximately) $N(\mu, \sigma^2)$. Suppose it is known that $\mu = 3.4$ liters. A volleyball coach claims that the FVC of volleyball players is greater than 3.4. She plans to test her claim using a random sample of size $n = 9$.

(a) Define the null hypothesis.

(b) Define the alternative (coach's) hypothesis.

(c) Define the test statistic.

(d) Define a critical region for which $\alpha = 0.05$. Draw a figure illustrating your critical region.

(e) Calculate the value of the test statistic given that the random sample yielded the following forced vital capacities:

$$3.4 \quad 3.6 \quad 3.8 \quad 3.3 \quad 3.4 \quad 3.5 \quad 3.7 \quad 3.6 \quad 3.7$$

(f) What is your conclusion?

(g) What is the approximate p-value of this test?

7.3.7 Vitamin B_6 is one of the vitamins in a multiple-vitamin pill manufactured by a pharmaceutical company. The pills are produced with a mean of 50 milligrams of vitamin B_6 per pill. The company believes that there is a deterioration of 1 milligram per month, so that after 3 months they expect that $\mu = 47$. A consumer group suspects that $\mu < 47$ after 3 months.

(a) Define a critical region to test $H_0: \mu = 47$ against $H_1: \mu < 47$ at an $\alpha = 0.05$ significance level based on a random sample of size $n = 20$.

(b) If the 20 pills yielded a mean of $\bar{x} = 46.94$ with a standard deviation of $s = 0.15$, what is your conclusion?

(c) What is the approximate p-value of this test?

7.3.8 Assume that the birth weight in grams of a baby born in the United States is $N(3315, 525^2)$, boys and girls combined. Let X equal the weight of a baby girl who is born at home in Ottawa County and assume that the distribution of X is $N(\mu_X, \sigma_X^2)$.

(a) Using 11 observations of X, give the test statistic and critical region for testing $H_0: \mu_X = 3315$ against the alternative hypothesis $H_1: \mu_X > 3315$ (home-born babies are heavier) if $\alpha = 0.01$.

(b) Calculate the value of the test statistic and give your conclusion using the following weights:

$$3119 \quad 2657 \quad 3459 \quad 3629 \quad 3345 \quad 3629$$
$$3515 \quad 3856 \quad 3629 \quad 3345 \quad 3062$$

(c) What is the approximate p-value of the test?

(d) Give the test statistic and critical region for testing $H_0: \sigma_X^2 = 525^2$ against the alternative hypothesis $H_1: \sigma_X^2 < 525^2$ (less variation of weights of home-born babies) if $\alpha = 0.05$.

(e) Calculate the value of your test statistic and state your conclusion.

(f) What is the approximate p-value of this second test?

7.3.9 Let Y equal the weight in grams of a baby boy who is born at home in Ottawa County and assume that the distribution of Y is $N(\mu_Y, \sigma_Y^2)$. Using the following weights:

$$4082 \quad 3686 \quad 4111 \quad 3686 \quad 3175 \quad 4139$$
$$3686 \quad 3430 \quad 3289 \quad 3657 \quad 4082$$

answer the questions in Exercise 7.3.8. What other hypothesis is suggested by these two exercises?

7.3.10 A company that manufactures brackets for an automaker regularly selects brackets from the production line and performs a torque test. The goal is for mean torque to equal 125. Let X equal the torque and assume that X is $N(\mu, \sigma^2)$. We shall use a sample of size $n = 15$ to test $H_0: \mu = 125$ against a two-sided alternative hypothesis.

(a) Give the test statistic and a critical region with significance level $\alpha = 0.05$. Sketch a figure illustrating the critical region.

(b) Use the following observations to calculate the value of the test statistic and state your conclusion:

$$128 \quad 149 \quad 136 \quad 114 \quad 126 \quad 142 \quad 124 \quad 136$$
$$122 \quad 118 \quad 122 \quad 129 \quad 118 \quad 122 \quad 129$$

7.3.11 Let X equal the number of pounds of butterfat produced by a Holstein cow during the 305-day milking period following the birth of a calf. In Exercise 7.2.6 we assumed that the distribution of X is $N(\mu, 140^2)$. We shall now test the null hypothesis $H_0: \sigma^2 = 140^2$ against the alternative hypothesis $H_1: \sigma^2 > 140^2$

(a) Give the test statistic and a critical region that has a significance level of $\alpha = 0.05$, assuming that there are $n = 25$ observations.

(b) Using the data in Exercise 7.2.6, calculate the value of the test statistic and provide your conclusion.

7.3.12 In May the fill weights of "6-pound" boxes of laundry soap had a mean of 6.13 pounds with a standard deviation of 0.095. The goal was to decrease the standard deviation. The company decided to adjust the filling machines and then test $H_0: \sigma = 0.095$ against $H_1: \sigma < 0.095$. In June a random sample of size $n = 20$ yielded $\overline{x} = 6.10$ and $s = 0.065$.

(a) At an $\alpha = 0.05$ significance level, was the company successful?

(b) What is the approximate p-value of your test?

7.3.13 The mean birth weight in the United States is $\mu = 3315$ grams with a standard deviation of $\sigma = 575$. Let X equal the birth weight in Rwanda. Assume that the distribution of X is $N(\mu, \sigma^2)$. We shall test the hypothesis $H_0: \sigma = 575$ against the alternative hypothesis $H_1: \sigma < 575$ at an $\alpha = 0.10$ significance level.

(a) What is your decision if a random sample of size $n = 81$ yielded $\overline{x} = 2819$ and $s = 496$?

(b) What is the approximate p-value of this test?

7.3.14 Let $X_1, X_2, \ldots, X_{23}$ be a random sample from a normal distribution that has variance $\sigma^2 = 100$. Let $S^2 = (1/22) \sum_{i=1}^{23} (X_i - \overline{X})^2$ be the sample variance. Of course, $22S^2/100$ is $\chi^2(22)$. Show that $\text{Var}(S^2) = 10{,}000/11$. Thus the standard deviation of S^2 is 30.15, and this helps explain the critical region $s^2 \leq 49.91$ or $s^2 \geq 167.18$ in Example 7.3.4.

7.3.15 Let X be $N(\mu, \sigma^2)$.

(a) Is the hypothesis $H_0: \sigma^2 = 0.04$ rejected in favor of the two-sided alternative hypothesis $H_1: \sigma^2 \neq 0.04$ if a random sample of size $n = 13$ yielded $s^2 = 0.058$? Let $\alpha = 0.05$.

(b) Is $\sigma^2 = 0.04$ contained in a 95% confidence interval for σ^2?

7.3.16 Let $X_1, X_2, \ldots, X_{19}$ be a random sample of size $n = 19$ from the normal distribution $N(\mu, \sigma^2)$.

(a) Find a critical region, C, of size $\alpha = 0.05$ for testing $H_0: \sigma^2 = 30$ against $H_1: \sigma^2 = 80$.

(b) Find the approximate value of β, the probability of type II error, for the critical region C of part (a).

7.3.17 Each of 51 golfers hit three golf balls of brand X and three golf balls of brand Y in a random order. Let X_i and Y_i equal the averages of the distances traveled by the brand X and brand Y golf balls hit by the ith golfer, $i = 1, 2, \ldots, 51$. Let $W_i = X_i - Y_i$, $i = 1, 2, \ldots, 51$. Test $H_0: \mu_W = 0$ against $H_1: \mu_W > 0$, where μ_W is the mean of the differences. If $\overline{w} = 2.07$ and $s_w^2 = 84.63$, would H_0 be accepted or rejected at an $\alpha = 0.05$ significance level?

7.3.18 To test whether a golf ball of brand A can be hit a greater distance off the tee than a golf ball of brand B, each of 17 golfers hit a ball of each brand, eight hitting ball A before ball B and nine hitting ball B before ball A. Assume that the differences of the paired A distance and B distance are approximately normally distributed and test the null hypothesis $H_0: \mu_D = 0$

against the alternative hypothesis H_1: $\mu_D > 0$ using a paired t-test with the 17 differences. Let $\alpha = 0.05$.

Golfer	Distance for Ball A	Distance for Ball B
1	265	252
2	272	276
3	246	243
4	260	246
5	274	275
6	263	246
7	255	244
8	258	245
9	276	259
10	274	260
11	274	267
12	269	267
13	244	251
14	212	222
15	235	235
16	254	255
17	224	231

7.3.19 A vendor of milk products produces and sells low-fat dry milk to a company that uses it to produce baby formula. In order to determine the fat content of the milk, both the company and the vendor take a sample from each lot and test it for fat content in percent. Ten sets of paired test results are as follows

Lot Number	Company Test Results (X)	Vendor Test Results (Y)
1	0.50	0.79
2	0.58	0.71
3	0.90	0.82
4	1.17	0.82
5	1.14	0.73
6	1.25	0.77
7	0.75	0.72
8	1.22	0.79
9	0.74	0.72
10	0.80	0.91

Let μ_D denote the mean of the differences. Test H_0: $\mu_D = 0$ against H_1: $\mu_D > 0$ using a paired t-test with the differences. Let $\alpha = 0.05$.

7.3.20 A company that manufactures motors receives reels of 10,000 terminals per reel. Before using a reel of terminals, 20 terminals are randomly selected to be tested. The test is the amount of pressure needed to pull the terminal apart from its mate. This amount of pressure should continue to increase from test to test as the terminal is "roughed up." (Since this is destructive testing, a terminal that is tested cannot be used in a motor.) Let W equal the difference of the pressures: "test 1 pressure" minus "test 2 pressure." Assume that the distribution of W is $N(\mu_W, \sigma_W^2)$. We shall test the null hypothesis H_0: $\mu_W = 0$ against the alternative hypothesis H_1: $\mu_W < 0$ using 20 pairs of observations.

(a) Give the test statistic and a critical region that has a significance level of $\alpha = 0.05$. Sketch a figure illustrating this critical region.

(b) Use the following data to calculate the value of the test statistic and clearly state your conclusion.

Terminal	Test 1	Test 2	Terminal	Test 1	Test 2
1	2.5	3.8	11	7.3	8.2
2	4.0	3.9	12	7.2	6.6
3	5.2	4.7	13	5.9	6.8
4	4.9	6.0	14	7.5	6.6
5	5.2	5.7	15	7.1	7.5
6	6.0	5.7	16	7.2	7.5
7	5.2	5.0	17	6.1	7.3
8	6.6	6.2	18	6.3	7.1
9	6.7	7.3	19	6.5	7.2
10	6.6	6.5	20	6.5	6.7

(c) What would the conclusion be if $\alpha = 0.01$?

(d) What is the approximate p-value of this test?

7.3.21 The following data give the times of 22 male swimmers who entered the 50 yard freestyle race in the conference championship meet. For these data, x is the swimmer's best time for the season and y is the swimmer's time in the meet.

$$(24.97, 23.98) \quad (24.76, 23.63) \quad (23.80, 23.61) \quad (22.84, 23.58) \quad (24.07, 23.57)$$
$$(22.93, 22.83) \quad (23.41, 22.75) \quad (22.10, 22.74) \quad (23.08, 22.70) \quad (23.59, 22.62)$$
$$(22.38, 22.39) \quad (22.91, 22.34) \quad (22.08, 21.95) \quad (22.46, 21.73) \quad (21.34, 21.68)$$
$$(22.76, 21.66) \quad (21.82, 21.68) \quad (21.80, 21.58) \quad (22.26, 21.57) \quad (21.36, 21.35)$$
$$(22.98, 23.17) \quad (23.80, 22.53)$$

Let $d = x - y$, a swimmer's best time minus the swimmer's meet time. Assume that the distribution of D is $N(\mu_D, \sigma_D^2)$. Test the null hypothesis H_0: $\mu_D = 0$ against the alternative hypothesis H_1: $\mu_D > 0$. You may select the significance level. Interpret your conclusion.

7.4 TESTS OF THE EQUALITY OF TWO NORMAL DISTRIBUTIONS

Let independent random variables X and Y have normal distributions $N(\mu_X, \sigma_X^2)$ and $N(\mu_Y, \sigma_Y^2)$, respectively. There are times when we are interested in testing whether the distributions of X and Y are the same. So if the assumption of normality is valid, we would be interested in testing whether the two variances are equal and whether the two means are equal.

We first consider a test of the equality of the two means. When X and Y are independent and normally distributed, we can test hypotheses about their means using the same t-statistic that was used for constructing a confidence interval for $\mu_X - \mu_Y$ in Section 6.3. Recall that the t-statistic used for constructing the confidence interval assumed that the variances of X and Y are equal. That is why we shall later consider a test for the equality of two variances.

We begin with an example and then give a table that lists some hypotheses and critical regions. A botanist is interested in comparing the growth response of dwarf pea stems to two different levels of the hormone indoleacetic acid (IAA). Using 16-day-old pea plants, the botanist obtains 5-millimeter sections and floats these sections on solutions with different hormone concentrations to observe the effect of the hormone on the growth of the pea stem. Let X and Y denote, respectively, the independent growths that can be attributed to the hormone during the first 26 hours after sectioning for $(0.5)(10)^{-4}$ and 10^{-4} levels of concentration of IAA. The botanist would like to test the null hypothesis $H_0: \mu_X - \mu_Y = 0$ against the alternative hypothesis $H_1: \mu_X - \mu_Y < 0$. If we can assume X and Y are independent and normally distributed with common variance, respective random samples of sizes n and m give a test based on the statistic

$$T = \frac{\overline{X} - \overline{Y}}{\sqrt{\{[(n-1)S_X^2 + (m-1)S_Y^2]/(n+m-2)\}(1/n + 1/m)}} \quad (7.3)$$
$$= \frac{\overline{X} - \overline{Y}}{S_P\sqrt{1/n + 1/m}}$$

where

$$S_P = \sqrt{\frac{(n-1)S_X^2 + (m-1)S_Y^2}{n+m-2}}. \quad (7.4)$$

T has a t distribution with $r = n+m-2$ degrees of freedom when H_0 is true and the variances are (approximately) equal. The hypothesis H_0 will be rejected in favor of H_1 if the observed value of T is less than $-t_\alpha(n+m-2)$.

Example 7.4.1

In the preceding discussion, the botanist measured the growths of pea stem segments, in millimeters, for $n = 11$ observations of X:

$$0.8 \quad 1.8 \quad 1.0 \quad 0.1 \quad 0.9 \quad 1.7 \quad 1.0 \quad 1.4 \quad 0.9 \quad 1.2 \quad 0.5$$

and $m = 13$ observations of Y:

$$1.0 \quad 0.8 \quad 1.6 \quad 2.6 \quad 1.3 \quad 1.1 \quad 2.4$$
$$1.8 \quad 2.5 \quad 1.4 \quad 1.9 \quad 2.0 \quad 1.2$$

For these data, $\bar{x} = 1.03$, $s_x^2 = 0.24$, $\bar{y} = 1.66$, and $s_y^2 = 0.35$. The critical region for testing $H_0: \mu_X - \mu_Y = 0$ against $H_1: \mu_X - \mu_Y < 0$ is $t \leq -t_{0.05}(22) = -1.717$, where t is given in Equation 7.3. Since

$$t = \frac{1.03 - 1.66}{\sqrt{\{[10(0.24) + 12(0.35)]/(11 + 13 - 2)\}(1/11 + 1/13)}}$$
$$= -2.81 < -1.717,$$

H_0 is clearly rejected at an $\alpha = 0.05$ significance level. Notice that the approximate p-value of this test is 0.005 because $-t_{0.005}(22) = -2.819$ (see Figure 7.7). Also, the sample variances do not differ too much; thus most statisticians would use this two-sample t-test.

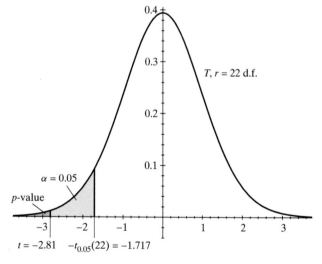

Figure 7.7   Critical region and p-value for pea stem growths

It is also instructive to construct box-and-whisker diagrams to gain a visual comparison of the two samples. For these two sets of data, the five-number summaries (minimum, three quartiles, maximum) are

$$0.1 \quad 0.8 \quad 1.0 \quad 1.4 \quad 1.8$$

for the X sample and

$$0.8 \quad 1.15 \quad 1.6 \quad 2.2 \quad 2.6$$

for the Y sample. The two box plots are shown in Figure 7.8. ◇

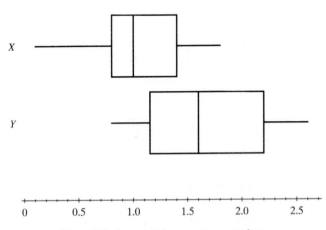

Figure 7.8 Box plots for pea stem growths

Based on independent random samples of sizes n and m, let $\bar{x}$, $\bar{y}$, and s_p^2 represent the observed unbiased estimates of the respective parameters μ_X, μ_Y, and $\sigma_X^2 = \sigma_Y^2$ of two normal distributions with common variance. Then α-level tests of certain hypotheses are given in Table 7.7, when $\sigma_X^2 = \sigma_Y^2$. If the common variance assumption is violated, but not too badly, the test is satisfactory but the significance levels are only approximate. The t-statistic along with s_p are given in Equations 7.3 and 7.4, respectively.

TABLE 7.7 Tests of hypotheses for equality of two means

H_0	H_1	Critical Region
$\mu_X = \mu_Y$	$\mu_X > \mu_Y$	$t \geq t_\alpha(n+m-2)$ or $\bar{x} - \bar{y} \geq t_\alpha(n+m-2)s_p\sqrt{1/n + 1/m}$
$\mu_X = \mu_Y$	$\mu_X < \mu_Y$	$t \leq -t_\alpha(n+m-2)$ or $\bar{x} - \bar{y} \leq -t_\alpha(n+m-2)s_p\sqrt{1/n + 1/m}$
$\mu_X = \mu_Y$	$\mu_X \neq \mu_Y$	$|t| \geq t_{\alpha/2}(n+m-2)$ or $|\bar{x} - \bar{y}| \geq t_{\alpha/2}(n+m-2)s_p\sqrt{1/n + 1/m}$

Example 7.4.2

A product is packaged using a machine with 24 filler heads numbered 1 to 24, with the odd-numbered heads on one side of the machine and the even on the other side. Let X and Y equal the fill weights in grams when a package is filled by an odd-numbered head and an even-numbered head, respectively. Assume that the distributions of X and Y are $N(\mu_X, \sigma^2)$ and $N(\mu_Y, \sigma^2)$, respectively, and that X and Y are independent. We would like to test the null hypothesis $H_0: \mu_X - \mu_Y = 0$ against the alternative hypothesis $H_1: \mu_X - \mu_Y \neq 0$. To perform the test, after the machine has been set up and is running, we shall select one package at random from each filler head and weigh it. The test statistic is that given by Equation 7.3 with $n = m = 12$. At an $\alpha = 0.10$ significance level, the critical region is $|t| \geq t_{0.05}(22) = 1.717$.

For the $n = 12$ observations of X,

$$
\begin{array}{cccccc}
1071 & 1076 & 1070 & 1083 & 1082 & 1067 \\
1078 & 1080 & 1075 & 1084 & 1075 & 1080
\end{array}
$$

$\bar{x} = 1076.75$ and $s_x^2 = 29.30$. For the $m = 12$ observations of Y,

$$
\begin{array}{cccccc}
1074 & 1069 & 1075 & 1067 & 1068 & 1079 \\
1082 & 1064 & 1070 & 1073 & 1072 & 1075
\end{array}
$$

$\bar{y} = 1072.33$ and $s_y^2 = 26.24$. The calculated value of the test statistic is

$$
t = \frac{1076.75 - 1072.33}{\sqrt{\dfrac{11(29.30) + 11(26.24)}{22} \left(\dfrac{1}{12} + \dfrac{1}{12} \right)}} = 2.05.
$$

Since

$$
|t| = |2.05| = 2.05 > 1.717,
$$

the null hypothesis is rejected at an $\alpha = 0.10$ significance level. Note, however, that

$$
|t| = 2.05 < 2.074 = t_{0.025}(22)
$$

so that the null hypothesis would not be rejected at an $\alpha = 0.05$ significance level. That is, the p-value is between 0.05 and 0.10.

Again it is instructive to construct box plots on the same graph for these two sets of data. The box plots in Figure 7.9 were constructed using the five-number summary for the observations of X—1067, 1072, 1077, 1081.5, 1084— and the five-number summary for the observations of Y— 1064, 1068.25, 1072.5, 1075, 1082. It looks like additional sampling would be advisable to test that the filler heads on the two sides of the machine are filling in a similar manner. If not, some corrective action needs to be taken.   ◇

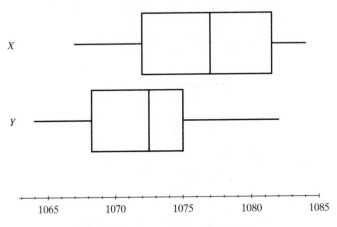

Figure 7.9   Box plots for fill weights

We would like to give two modifications of tests about two means. If we are able to assume that we know the variances of X and Y, then the appropriate test statistic to use for testing $H_0: \mu_X = \mu_Y$ is

$$Z = \frac{\overline{X} - \overline{Y}}{\sqrt{\dfrac{\sigma_X^2}{n} + \dfrac{\sigma_Y^2}{m}}}, \tag{7.5}$$

which has a standard normal distribution when the null hypothesis is true and, of course, the populations are normally distributed. If the variances are unknown and the sample sizes are large, replace σ_X^2 with S_X^2 and σ_Y^2 with S_Y^2 in Equation 7.5. The resulting statistic will have an approximate $N(0, 1)$ distribution.

Example 7.4.3

The target thickness for fruit-flavored gum and for fruit-flavored bubble gum is 6.7 hundredths of an inch. Let the independent random variables X and Y equal the respective thicknesses of these gums in hundredths of an inch and assume that their distributions are $N(\mu_X, \sigma_X^2)$ and $N(\mu_Y, \sigma_Y^2)$, respectively. Because bubble gum has more elasticity than regular gum, it seems as if it would be harder to roll it out to the correct thickness. Thus we shall test the null hypothesis $H_0: \mu_X = \mu_Y$ against the alternative hypothesis $H_1: \mu_X < \mu_Y$ using samples of sizes $n = 50$ and $m = 40$.

Because the variances are unknown and the sample sizes are large, the test statistic that is used is

$$Z = \frac{\overline{X} - \overline{Y}}{\sqrt{\dfrac{S_X^2}{50} + \dfrac{S_Y^2}{40}}}.$$

At an approximate significance level of $\alpha = 0.01$, the critical region is

$$z \le -z_{0.01} = -2.326.$$

The observed values of X were

6.85	6.60	6.70	6.75	6.75	6.90	6.85	6.90	6.70	6.85
6.60	6.70	6.75	6.70	6.70	6.70	6.55	6.60	6.95	6.95
6.80	6.80	6.70	6.75	6.60	6.70	6.65	6.55	6.55	6.60
6.60	6.70	6.80	6.75	6.60	6.75	6.50	6.75	6.70	6.65
6.70	6.70	6.55	6.65	6.60	6.65	6.60	6.65	6.80	6.60

for which $\overline{x} = 6.701$ and $s_x = 0.108$. The observed values of Y were

7.10	7.05	6.70	6.75	6.90	6.90	6.65	6.60	6.55	6.55
6.85	6.90	6.60	6.85	6.95	7.10	6.95	6.90	7.15	7.05
6.70	6.90	6.85	6.95	7.05	6.75	6.90	6.80	6.70	6.75
6.90	6.90	6.70	6.70	6.90	6.90	6.70	6.70	6.90	6.95

for which $\overline{y} = 6.841$ and $s_y = 0.155$. Since the calculated value of the test statistic is

$$z = \frac{6.701 - 6.841}{\sqrt{0.108^2/50 + 0.155^2/40}} = -4.848 < -2.326,$$

the null hypothesis is clearly rejected.

The box-and-whisker diagrams in Figure 7.10 were constructed using the five-number summary of the observations of X— 6.50, 6.60, 6.70, 6.75, 6.95— and the five-number summary of the observations of Y— 6.55, 6.70, 6.90, 6.94, 7.15. This graphical display also confirms our conclusion.   ◇

Remark   To have satisfactory tests, our assumptions must be satisfied reasonably well. As long as the underlying distributions are not highly skewed, the normal assumptions are not too critical as $\overline{X}$ and $\overline{Y}$ have approximate normal distributions by the central limit theorem. As distributions become nonnormal and highly skewed, the sample mean and sample variance become more dependent and that causes problems using the Student-t as an approximating distribution for T. In these cases, nonparametric methods described in Chapter 10 could be used.

When the distributions are close to normal but the variances seem to differ by a great deal, the t-statistic should again be avoided, particularly if the sample sizes are also different. In that case use Z or the modification by substituting the sample variances for the distribution variances. In the latter situation, if n and m are large

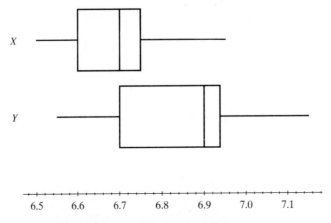

Figure 7.10   Box plots for gum thicknesses

enough, there is no problem. With small n and m, most statisticians would use Welch's suggestion (or other modifications of it); that is, use an approximating Student-t distribution with $[r]$ degrees of freedom, where

$$\frac{1}{r} = \frac{c^2}{n-1} + \frac{(1-c)^2}{m-1} \quad \text{and} \quad c = \frac{s_x^2}{s_x^2 + s_y^2}.$$

An equivalent formula for r is

$$r = \frac{\left(\dfrac{s_x^2}{n} + \dfrac{s_y^2}{m}\right)^2}{\dfrac{1}{n-1}\left(\dfrac{s_x^2}{n}\right)^2 + \dfrac{1}{m-1}\left(\dfrac{s_y^2}{m}\right)^2}.$$

We actually give a test for the equality of variances. This could be used to decide whether to use T or a modification of Z. However, most statisticians do not place much confidence in this test of $\sigma_X^2 = \sigma_Y^2$ and would use a modification of Z (possibly Welch's) if they suspected that the variances differ greatly.

We now give a test for the equality of two variances when sampling from normal populations. Let the independent random variables X and Y have respective distributions that are $N(\mu_X, \sigma_X^2)$ and $N(\mu_Y, \sigma_Y^2)$. To test the null hypothesis H_0: $\sigma_X^2/\sigma_Y^2 = 1$ (or, equivalently, $\sigma_X^2 = \sigma_Y^2$), take random samples of n observations of X and m observations of Y. Recall that $(n-1)S_X^2/\sigma_X^2$ and $(m-1)S_Y^2/\sigma_Y^2$ have independent chi-square distributions $\chi^2(n-1)$ and $\chi^2(m-1)$, respectively. Thus, when H_0 is true,

$$F = \frac{\dfrac{(n-1)S_X^2}{\sigma_X^2(n-1)}}{\dfrac{(m-1)S_Y^2}{\sigma_Y^2(m-1)}} = \frac{S_X^2}{S_Y^2}$$

has an F distribution with $r_1 = n - 1$ and $r_2 = m - 1$ degrees of freedom. This F statistic is our test statistic. When H_0 is true, we would expect the observed value of F to be close to 1.

Three possible alternative hypotheses along with critical regions of size α are summarized in Table 7.8. Recall that $1/F$, the reciprocal of F, has an F distribution with $m - 1$ and $n - 1$ degrees of freedom so all critical regions may be written in terms of right-tail rejection regions so that the critical values can be selected easily from Table VII.

TABLE 7.8   Tests of hypotheses of the equality of variances

H_0	H_1	Critical Region
$\sigma_X^2 = \sigma_Y^2$	$\sigma_X^2 > \sigma_Y^2$	$\dfrac{s_x^2}{s_y^2} \geq F_\alpha(n-1, m-1)$
$\sigma_X^2 = \sigma_Y^2$	$\sigma_X^2 < \sigma_Y^2$	$\dfrac{s_y^2}{s_x^2} \geq F_\alpha(m-1, n-1)$
$\sigma_X^2 = \sigma_Y^2$	$\sigma_X^2 \neq \sigma_Y^2$	$\dfrac{s_x^2}{s_y^2} \geq F_{\alpha/2}(n-1, m-1)$ or
		$\dfrac{s_y^2}{s_x^2} \geq F_{\alpha/2}(m-1, n-1)$

Example 7.4.4

A biologist who studies spiders believes that not only do female green lynx spiders tend to be longer than their male counterparts but also that the lengths of the female spiders seem to vary more than those of the male spiders. We shall test whether the latter belief is true. Suppose that the distribution of the length X of male spiders is $N(\mu_X, \sigma_X^2)$ and the length Y of female spiders is $N(\mu_Y, \sigma_Y^2)$, and X and Y are independent. We shall test $H_0: \sigma_X^2/\sigma_Y^2 = 1$ (i.e., $\sigma_X^2 = \sigma_Y^2$) against the alternative hypothesis $H_1: \sigma_X^2/\sigma_Y^2 < 1$ (i.e., $\sigma_X^2 < \sigma_Y^2$.) If we use $n = 30$ and $m = 30$ observations of X and Y, respectively, a critical region that has a significance level of $\alpha = 0.01$ is

$$\frac{s_y^2}{s_x^2} \geq F_{0.01}(29, 29) = 2.42,$$

approximately, using interpolation in Table VII. In Exercise 6.3.5, $n = 30$ observations of X yielded $\bar{x} = 5.917$ and $s_x^2 = 0.4399$, while $m = 30$ observations of Y

Figure 7.11   Test of variances for pea stem growths

yielded $\overline{y} = 8.153$ and $s_y^2 = 1.4100$. Since

$$\frac{s_y^2}{s_x^2} = \frac{1.4100}{0.4399} = 3.2053 > 2.42,$$

the null hypothesis is rejected in favor of the biologist's belief (see Exercise 7.4.12 for a continuation of this example).   ◇

In Examples 7.4.1 and 7.4.2, we used a t-statistic for testing the equality of means that assumed the variances were equal. In the next two examples we shall test whether that assumption is valid.

Example 7.4.5

For Example 7.4.1, given $n = 11$ observations of X and $m = 13$ observations of Y, where X is $N(\mu_X, \sigma_X^2)$ and Y is $N(\mu_Y, \sigma_Y^2)$, we shall test the null hypothesis H_0: $\sigma_X^2/\sigma_Y^2 = 1$ against a two-sided alternative hypothesis. At an $\alpha = 0.05$ significance level, H_0 is rejected if

$$s_x^2/s_y^2 \geq F_{0.025}(10, 12) = 3.37$$

or

$$s_y^2/s_x^2 \geq F_{0.025}(12, 10) = 3.62.$$

Using the data in Example 7.4.1, we obtain

$$s_x^2/s_y^2 = 0.24/0.35 = 0.686 \qquad \text{and} \qquad s_y^2/s_x^2 = 1.458,$$

so we do not reject H_0. Thus the assumption of equal variances for the t-statistic that was used in Example 7.4.1 seems to be valid. See Figure 7.11, noting that $F_{0.975}(10, 12) = 1/F_{0.025}(12, 10) = 1/3.62 = 0.276$.   ◇

Example 7.4.6

For Example 7.4.2, we shall test the null hypothesis $H_0: \sigma_X^2/\sigma_Y^2 = 1$ against a two-sided alternative hypothesis. Since

$$\frac{s_x^2}{s_y^2} = \frac{29.30}{26.24} = 1.12 < F_{0.025}(11, 11)$$

because $F_{0.025}(11, 11)$ must be between $F_{0.025}(12, 12) = 3.28$ and $F_{0.025}(10, 10) = 3.72$ from Table VII, the assumption of equal variances is confirmed.   ◇

EXERCISES

7.4.1 The botanist in Example 7.4.1 is really interested in testing for synergistic interaction. That is, given two hormones, gibberellin (GA_3) and indoleacetic acid (IAA), let X_1 and X_2 equal the growth responses (in millimeters) of dwarf pea stem segments to GA3 and IAA, respectively and separately. Let $X = X_1 + X_2$. Let Y equal the growth response when both hormones are present. Assuming that X is $N(\mu_X, \sigma^2)$ and Y is $N(\mu_Y, \sigma^2)$, the botanist is interested in testing the hypothesis $H_0: \mu_X = \mu_Y$ against the alternative hypothesis of synergistic interaction $H_1: \mu_X < \mu_Y$.

  (a) Using $n = m = 10$ observations of X and Y, define the test statistic and critical region. Sketch a figure of the t p.d.f. and show the critical region on your figure. Let $\alpha = 0.05$.

  (b) Given $n = 10$ observations of X:

$$2.1 \quad 2.6 \quad 2.6 \quad 3.4 \quad 2.1 \quad 1.7 \quad 2.6 \quad 2.6 \quad 2.2 \quad 1.2$$

  and $m = 10$ observations of Y:

$$3.5 \quad 3.9 \quad 3.0 \quad 2.3 \quad 2.1 \quad 3.1 \quad 3.6 \quad 1.8 \quad 2.9 \quad 3.3$$

  calculate the value of the test statistic and state your conclusion. Locate the test statistic on your figure.

  (c) Construct two box plots on the same figure. Does this confirm your conclusion?

7.4.2 Let X and Y denote the weights in grams of male and female gallinules, respectively. Assume that X is $N(\mu_X, \sigma_X^2)$ and Y is $N(\mu_Y, \sigma_Y^2)$.

  (a) Given $n = 16$ observations of X and $m = 13$ observations of Y, define a test statistic and critical region for testing the null hypothesis $H_0: \mu_X = \mu_Y$ against the one-sided alternative hypothesis $H_1: \mu_X > \mu_Y$. Let $\alpha = 0.01$. (Assume that the variances are equal.)

  (b) Given that $\bar{x} = 415.16$, $s_x^2 = 1356.75$, $\bar{y} = 347.40$, and $s_y^2 = 692.21$, calculate the value of the test statistic and state your conclusion.

  (c) Test whether the assumption of equal variances is valid. Let $\alpha = 0.05$.

  (d) Despite the fact that $\sigma_X^2 = \sigma_Y^2$ is accepted in part (c), let us say we suspect the equality is not valid. Thus use the test proposed by Welch.

7.4.3 Let X equal the weight in grams of a low-fat strawberry kudo and Y the weight of a low-fat blueberry kudo. Assume that the distributions of X and Y are $N(\mu_X, \sigma_X^2)$ and $N(\mu_Y, \sigma_Y^2)$, respectively. Use the following $n = 9$ observations of X:

$$21.7 \quad 21.0 \quad 21.2 \quad 20.7 \quad 20.4 \quad 21.9 \quad 20.2 \quad 21.6 \quad 20.6$$

and $m = 13$ observations of Y:

$$21.5 \quad 20.5 \quad 20.3 \quad 21.6 \quad 21.7 \quad 21.3 \quad 23.0 \quad 21.3 \quad 18.9 \quad 20.0 \quad 20.4 \quad 20.8 \quad 20.3$$

to answer the following questions:

(a) Test the null hypothesis H_0: $\sigma_X^2/\sigma_Y^2 = 1$ against a two-sided alternative hypothesis. You may select the significance level.

(b) Test the null hypothesis H_0: $\mu_X = \mu_Y$ against a two-sided alternative hypothesis. You may select the significance level. Also select the appropriate test statistic, depending on your answer to part (a).

(c) Construct and interpret box-and-whisker diagrams to support your conclusions.

7.4.4 Among the data collected for the World Health Organization air quality monitoring project is a measure of suspended particles in $\mu g/m^3$. Let X and Y equal the concentration of suspended particles in $\mu g/m^3$ in the city center (commercial district), for Melbourne and Houston, respectively. Using $n = 13$ observations of X and $m = 16$ observations of Y, we shall test H_0: $\mu_X = \mu_Y$ against H_1: $\mu_X < \mu_Y$.

(a) Define the test statistic and critical region, assuming that the variances are equal. Let $\alpha = 0.05$.

(b) If $\bar{x} = 72.9$, $s_x = 25.6$, $\bar{y} = 81.7$, and $s_y = 28.3$, calculate the value of the test statistic and state your conclusion.

(c) Give limits for the p-value of this test.

(d) Test whether the assumption of equal variances is valid. Let $\alpha = 0.05$.

7.4.5 Some nurses in a county public health organization conducted a survey of women who had received inadequate prenatal care. They used information from birth certificates to select mothers for the survey. The mothers that were selected were divided into two groups: 14 mothers who said they had 5 or fewer prenatal visits and 14 mothers who said they had 6 or more prenatal visits. Let X and Y equal the respective birth weights of the babies from these two sets of mothers and assume that the distribution of X is $N(\mu_X, \sigma^2)$ and the distribution of Y is $N(\mu_Y, \sigma^2)$,

(a) Define the test statistic and critical region for testing H_0: $\mu_X - \mu_Y = 0$ against H_1: $\mu_X - \mu_Y < 0$. Let $\alpha = 0.05$.

(b) Given that the observations of X were

$$
\begin{array}{ccccccc}
49 & 108 & 110 & 82 & 93 & 114 & 134 \\
114 & 96 & 52 & 101 & 114 & 120 & 116
\end{array}
$$

and the observations of Y were

$$
\begin{array}{ccccccc}
133 & 108 & 93 & 119 & 119 & 98 & 106 \\
131 & 87 & 153 & 116 & 129 & 97 & 110
\end{array}
$$

calculate the value of the test statistic and state your conclusion.

(c) Approximate the p-value.

(d) Construct box plots on the same figure for these two sets of data. Do the box plots support your conclusion?

(e) Test whether the assumption of equal variances is valid. Let $\alpha = 0.05$.

7.4.6 Let X and Y equal the forces required to pull stud 3 and stud 4 out of a window that has been manufactured for an automobile. Assume that the distributions of X and Y are $N(\mu_X, \sigma_X^2)$ and $N(\mu_Y, \sigma_Y^2)$, respectively.

(a) If $m = n = 10$ observations are selected randomly, define a test statistic and critical region for testing $H_0: \mu_X - \mu_Y = 0$ against a two-sided alternative hypothesis. Let $\alpha = 0.05$. Assume that the variances are equal.

(b) Given $n = 10$ observations of X:

$$111 \quad 120 \quad 139 \quad 136 \quad 138 \quad 149 \quad 143 \quad 145 \quad 111 \quad 123$$

and $m = 10$ observations of Y:

$$152 \quad 155 \quad 133 \quad 134 \quad 119 \quad 155 \quad 142 \quad 146 \quad 157 \quad 149$$

calculate the value of the test statistic and clearly state your conclusion.

(c) What is the approximate p-value of this test?

(d) Construct box plots on the same figure for these two sets of data. Do the box plots confirm your decision in part (b)?

(e) Test whether the assumption of equal variances is valid.

7.4.7 Let X and Y equal the number of milligrams of tar in filtered and nonfiltered cigarettes, respectively. Assume that the distributions of X and Y are $N(\mu_X, \sigma_X^2)$ and $N(\mu_Y, \sigma_Y^2)$, respectively. We shall test the null hypothesis $H_0: \mu_X - \mu_Y = 0$ against the alternative hypothesis H_1: $\mu_X - \mu_Y < 0$ using random samples of sizes $n = 9$ and $m = 11$ observations of X and Y, respectively.

(a) Define the test statistic and a critical region that has an $\alpha = 0.01$ significance level. Sketch a figure illustrating this critical region.

(b) Given the $n = 9$ observations of X:

$$0.9 \quad 1.1 \quad 0.1 \quad 0.7 \quad 0.4 \quad 0.9 \quad 0.8 \quad 1.0 \quad 0.4$$

and the $m = 11$ observations of Y:

$$1.5 \quad 0.9 \quad 1.6 \quad 0.5 \quad 1.4 \quad 1.9 \quad 1.0 \quad 1.2 \quad 1.3 \quad 1.6 \quad 2.1$$

calculate the value of the test statistic and clearly state your conclusion. Locate the value of the test statistic on your figure.

7.4.8 In Exercise 6.3.4, we let X and Y equal the blood volumes in milliliters for males who are paraplegics participating in vigorous physical activities and males who are able bodied participating in normal activities. We assumed that X was $N(\mu_X, \sigma_X^2)$ and Y was $N(\mu_Y, \sigma_Y^2)$. Using $n = 7$ observations of X:

$$1612 \quad 1352 \quad 1456 \quad 1222 \quad 1560 \quad 1456 \quad 1924$$

and $m = 10$ observations of Y:

$$
\begin{array}{ccccc}
1082 & 1300 & 1092 & 1040 & 910 \\
1248 & 1092 & 1040 & 1092 & 1288
\end{array}
$$

you were asked to find a 95% confidence interval for $\mu_X - \mu_Y$.

(a) To construct the confidence interval, you would like to assume that the variances are equal. Is this a valid assumption? Test the hypothesis that the variances are equal against a two-sided alternative hypothesis. Let $\alpha = 0.05$.

(b) Test the hypothesis that the means are equal against a two-sided alternative hypothesis. Let $\alpha = 0.05$.

(c) Did 0 belong to the 95% confidence interval in Exercise 6.3.4? Did you reject the null hypothesis in this exercise? Are your answers to these questions compatible?

7.4.9 An office furniture manufacturer installed a new adhesive application process. To compare the new process to the old process, random samples were selected from the two processes, and "pull tests" were performed to determine the number of pounds of pressure that were required to pull apart the glued parts (destructive testing). Let X and Y denote the pounds of pressure needed for the new and old processes, respectively.

(a) Based on $n = m = 24$ observations, define the test statistic and critical region for testing $H_0: \mu_X - \mu_Y = 0$ against $H_1: \mu_X - \mu_Y > 0$. Let $\alpha = 0.05$. State assumptions.

(b) Use the following $n = 24$ observations of X:

$$
\begin{array}{cccccccc}
1250 & 1210 & 990 & 1310 & 1320 & 1200 & 1290 & 1360 \\
1120 & 1360 & 1310 & 1110 & 1320 & 980 & 950 & 1430 \\
960 & 1050 & 1310 & 1240 & 1420 & 1170 & 1470 & 1060
\end{array}
$$

and the following $m = 24$ observations of Y:

$$
\begin{array}{cccccccc}
1180 & 1360 & 1310 & 1190 & 920 & 1060 & 1440 & 1010 \\
1310 & 980 & 1310 & 1030 & 960 & 800 & 1280 & 1080 \\
930 & 1050 & 1010 & 1310 & 940 & 860 & 1450 & 1070
\end{array}
$$

to calculate the value of the test statistic and clearly state your conclusion.

(c) What is the approximate p-value of this test?

(d) Construct two box plots on the same graph. Do the box plots support your conclusion?

7.4.10 Let X and Y denote the tarsus lengths of male and female grackles, respectively. Assume that X is $N(\mu_X, \sigma_X^2)$ and Y is $N(\mu_Y, \sigma_Y^2)$. Given that $n = 25$, $\bar{x} = 33.80$, $s_x^2 = 4.88$, $m = 29$, $\bar{y} = 31.66$, and $s_y^2 = 5.81$, test

(a) $H_0: \sigma_X^2/\sigma_Y^2 = 1$ against a two-sided alternative with $\alpha = 0.02$.

(b) $H_0: \mu_X = \mu_Y$ against $H_1: \mu_X > \mu_Y$ with $\alpha = 0.01$.

7.4.11 Let X equal the fill weight in April and Y the fill weight in June for an "8-pound" box of bleach. We shall test the null hypothesis $H_0: \mu_X - \mu_Y = 0$ against the alternative hypothesis $H_1: \mu_X - \mu_Y > 0$ given that $n = 90$ observations of X yielded $\bar{x} = 8.10$, $s_x = 0.117$ and $m = 110$ observations of Y yielded $\bar{x} = 8.07$ and $s_y = 0.054$.

(a) What is your conclusion if $\alpha = 0.05$? HINT: Do the variances seem to be equal?

(b) What is the approximate p-value of this test?

7.4.12 Let X and Y denote the lengths of male and female green lynx spiders, respectively. Assume that the distributions of X and Y are $N(\mu_X, \sigma_X^2)$ and $N(\mu_Y, \sigma_Y^2)$. In Example 7.4.4, we concluded that $\sigma_Y^2 > \sigma_X^2$. Thus, use the modification of Z to test the hypothesis $H_0: \mu_X - \mu_Y = 0$ against the alternative hypothesis $H_1: \mu_X - \mu_Y < 0$.

(a) Define the test statistic and a critical region that has a significance level of $\alpha = 0.025$.

(b) Using the sample characteristics given in Example 7.4.4, calculate the value of the test statistic and state your conclusion. (The data are listed in Exercise 6.3.5.)

(c) Draw two box-and-whisker diagrams on the same figure. Does this confirm both the conclusion of this exercise and that of Example 7.4.4?

7.4.13 To measure air pollution in a home, let X and Y equal the amount of suspended particulate matter (in $\mu g/m^3$) measured during a 24-hour period in a home in which there is no smoker and a home in which there is a smoker, respectively. [Assume that the distributions of the independent random variables X and Y are $N(\mu_X, \sigma_X^2)$ and $N(\mu_Y, \sigma_Y^2)$, respectively.] We shall test the null hypothesis $H_0: \sigma_X^2/\sigma_Y^2 = 1$ against the one-sided alternative hypothesis $H_1: \sigma_X^2/\sigma_Y^2 > 1$. If a random sample of size $n = 9$ yielded $\bar{x} = 93$ and $s_x = 12.9$ while a random sample of size $m = 11$ yielded $\bar{y} = 132$ and $s_y = 7.1$, define a critical region and give your conclusion if $\alpha = 0.05$. Now test $H_0: \mu_X = \mu_Y$ against $H_1: \mu_X < \mu_Y$ if $\alpha = 0.05$.

7.4.14 Let X and Y equal the times in days required for maturation of guardiola seeds from narrow-leaved and broad-leaved parents, respectively. Assume that X is $N(\mu_X, \sigma_X^2)$ and Y is $N(\mu_Y, \sigma_Y^2)$, and X and Y are independent. Test the hypothesis $H_0: \sigma_X^2/\sigma_Y^2 = 1$ against the alternative hypothesis $H_1: \sigma_X^2/\sigma_Y^2 > 1$ if a sample size $n = 13$ yielded $\bar{x} = 18.97$, $s_x^2 = 9.88$ and a sample of size $m = 9$ yielded $\bar{y} = 23.20$, $s_y^2 = 4.08$. Let $\alpha = 0.05$.

7.4.15 A random sample of $n = 7$ brand X light bulbs yielded $\bar{x} = 891$ hours and $s_x^2 = 9201$. A random sample of $m = 10$ brand Y light bulbs yielded $\bar{y} = 592$ hours and $s_y^2 = 4856$. Use these data to test $H_0: \sigma_X^2/\sigma_Y^2 = 1$ against $H_1: \sigma_X^2/\sigma_Y^2 > 1$. Let $\alpha = 0.05$ and state assumptions.

7.5 CHI-SQUARE GOODNESS-OF-FIT TESTS

We now consider applications of the very important chi-square statistic first proposed by Karl Pearson in 1900. As the reader will see, it is a very adaptable test statistic and can be used for many different types of tests. In particular, one application of it allows us to test the appropriateness of different probabilistic models and, in this sense, is a competitor of the Kolmogorov-Smirnov test (see Section 10.7).

So that the reader can get some idea why Pearson first proposed his chi-square statistic, we begin with the binomial case. That is, let Y_1 be $b(n, p_1)$, where $0 < p_1 < 1$. According to the central limit theorem,

$$Z = \frac{Y_1 - np_1}{\sqrt{np_1(1 - p_1)}}$$

has a distribution that is approximately $N(0, 1)$ when n is large, particularly when $np_1 \geq 5$ and $n(1 - p_1) \geq 5$. Thus it is not surprising that $Q_1 = Z^2$ is approximately $\chi^2(1)$. If we let $Y_2 = n - Y_1$ and $p_2 = 1 - p_1$, we see that Q_1 may be written as

$$Q_1 = \frac{(Y_1 - np_1)^2}{np_1(1 - p_1)} = \frac{(Y_1 - np_1)^2}{np_1} + \frac{(Y_1 - np_1)^2}{n(1 - p_1)}.$$

Since

$$(Y_1 - np_1)^2 = (n - Y_1 - n[1 - p_1])^2 = (Y_2 - np_2)^2,$$

we have

$$Q_1 = \frac{(Y_1 - np_1)^2}{np_1} + \frac{(Y_2 - np_2)^2}{np_2}.$$

Let us now carefully consider each term in this last expression for Q_1. Of course, Y_1 is the number of "successes," and np_1 is the expected number of "successes"; that is, $E(Y_1) = np_1$. Similarly, Y_2 and np_2 are, respectively, the number and the expected number of "failures." So each numerator consists of the square of a difference of the observed number and expected number. Note that Q_1 can be written as

$$Q_1 = \sum_{i=1}^{2} \frac{(Y_i - np_i)^2}{np_i}, \tag{7.6}$$

and we have seen intuitively that it has an approximate chi-square distribution with one degree of freedom. In a sense, Q_1 measure the "closeness" of the observed numbers to the corresponding expected numbers. For example, if the observed values of Y_1 and Y_2 equal their expected values, then the computed Q_1 is equal to $q_1 = 0$; but if they differ much from them, then the computed $Q_1 = q_1$ is relatively large.

To generalize, we let an experiment have k (instead of only two) mutually exclusive and exhaustive outcomes, say $A_1, A_2, \ldots, A_k$. Let $p_i = P(A_i)$ and thus $\sum_{i=1}^{k} p_i = 1$. The experiment is repeated n independent times, and we let Y_i represent the number of times the experiment results in A_i, $i = 1, 2, \ldots, k$. This is the first occasion in this book that we have considered the joint distribution of two or more random variables that are not independent (that is, they are dependent). However, this joint distribution is a straightforward generalization of the binomial distribution; this generalization is as follows.

In considering the joint p.d.f.,

$$f(y_1, y_2, \ldots, y_{k-1}) = P(Y_1 = y_1, Y_2 = y_2, \ldots, Y_{k-1} = y_{k-1}),$$

where $y_1, y_2, \ldots, y_{k-1}$ are nonnegative integers such that $y_1 + y_2 + \cdots + y_{k-1} \le n$. Note that we do not need to consider Y_k since once the other $k - 1$ random variables are observed to equal $y_1, y_2, \ldots, y_{k-1}$, respectively, we know that

$$Y_k = n - y_1 - y_2 - \cdots - y_{k-1} = y_k, \quad \text{say.}$$

The probability of each particular arrangement of y_1 A_1's, y_2 A_2's, $\ldots$, y_k A_k's is, from the independence of the trials,

$$p_1^{y_1} p_2^{y_2} \cdots p_k^{y_k}.$$

The number of such arrangements is the multinomial coefficient

$$\binom{n}{y_1, y_2, \ldots, y_k} = \frac{n!}{y_1! \, y_2! \cdots y_k!}.$$

Hence the product of these two expressions gives the joint p.d.f. of $Y_1, Y_2, \ldots, Y_{k-1}$, namely,

$$f(y_1, y_2, \ldots, y_{k-1}) = \frac{n!}{y_1! \, y_2! \cdots y_k!} \, p_1^{y_1} \, p_2^{y_2} \cdots p_k^{y_k},$$

recalling that $y_k = n - y_1 - y_2 - \cdots - y_{k-1}$.

Pearson then constructed an expression similar to Q_1 (Equation 7.6), which involves Y_1 and $Y_2 = n - Y_1$, that we denote by Q_{k-1}, which involves $Y_1, Y_2, \ldots, Y_{k-1}$, and $Y_k = n - Y_1 - Y_2 - \cdots - Y_{k-1}$, namely,

$$Q_{k-1} = \sum_{i=1}^{k} \frac{(Y_i - np_i)^2}{np_i}.$$

He argued that Q_{k-1} has an approximate chi-square distribution with $k-1$ degrees of freedom in much the same way we argued that Q_1 is approximately $\chi^2(1)$. We accept this fact as the proof is beyond the level of this book.

Some writers suggest that n should be large enough so that $np_i \geq 5$, $i = 1, 2, \ldots, k$, to be certain that the approximating distribution is adequate. This is probably good advice for the beginner to follow, although we have seen the approximation work very well when $np_i \geq 1$, $i = 1, 2, \ldots, k$. The important thing to guard against is allowing some particular np_i to become so small that the corresponding term in q_{k-1}, namely $(y_i - np_i)^2/np_i$, tends to dominate the others because of its small denominator. In any case it is important to realize that Q_{k-1} has only an approximate chi-square distribution.

We shall now show how we can use the fact that Q_{k-1} is approximately $\chi^2(k-1)$ to test hypotheses about probabilities of various outcomes. Let an experiment have k mutually exclusive and exhaustive outcomes, $A_1, A_2, \ldots, A_k$. We would like to test whether $p_i = P(A_i)$ is equal to a known number p_{i0}, $i = 1, 2, \ldots, k$. That is, we shall test the hypothesis

$$H_0: \; p_i = p_{i0}, \quad i = 1, 2, \ldots, k.$$

In order to test such a hypothesis, we shall take a sample of size n, that is repeat the experiment n independent times. We tend to favor H_0 if the observed number of times that A_i occurred, say y_i, and the number of times A_i was expected to occur if H_0 were true, namely np_{i0}, are approximately equal. That is, if

$$q_{k-1} = \sum_{i=1}^{k} \frac{(y_i - np_{i0})^2}{np_{i0}}$$

is "small," we tend to favor H_0. Since the distribution of Q_{k-1} is approximately $\chi^2(k-1)$, we shall reject H_0 if $q_{k-1} \geq \chi_\alpha^2(k-1)$, where α is the desired significance level of the test.

Example 7.5.1

If persons are asked to record a string of random digits, such as

$$3 \quad 7 \quad 2 \quad 4 \quad 1 \quad 9 \quad 7 \quad 2 \quad 1 \quad 5 \quad 0 \quad 8 \dots,$$

we usually find that they are reluctant to record the same or even the two closest numbers in adjacent positions. And yet, in true random digit generation, the probability of the next digit being the same as the preceding one is $p_{10} = 1/10$, the probability of the next being only one away from the preceding (assuming that 0 is one away from 9) is $p_{20} = 2/10$, and the probability of all other possibilities is $p_{30} = 7/10$. We shall test one person's concept of a random sequence by asking him to record a string of 51 digits that seems to represent a random generation. Thus we shall test

$$H_0: \quad p_1 = p_{10} = \frac{1}{10}, \quad p_2 = p_{20} = \frac{2}{10}, \quad p_3 = p_{30} = \frac{7}{10}.$$

The critical region for an $\alpha = 0.05$ significance level is $q_2 \geq \chi^2_{0.05}(2) = 5.991$. The sequence of digits were as follows:

$$
\begin{array}{ccccccccccccc}
5 & 8 & 3 & 1 & 9 & 4 & 6 & 7 & 9 & 2 & 6 & 3 & 0 \\
8 & 7 & 5 & 1 & 3 & 6 & 2 & 1 & 9 & 5 & 4 & 8 & 0 \\
3 & 7 & 1 & 4 & 6 & 0 & 4 & 3 & 8 & 2 & 7 & 3 & 9 \\
8 & 5 & 6 & 1 & 8 & 7 & 0 & 3 & 5 & 2 & 5 & 2 &
\end{array}
$$

We went through this listing and observed how many times the next digit was the same as or was one away from the preceding one.

	Frequency	Expected Number
Same	0	$50(1/10) = 5$
One away	8	$50(2/10) = 10$
Other	42	$50(7/10) = 35$
	50	50

The computed chi-square statistic is

$$\frac{(0-5)^2}{5} + \frac{(8-10)^2}{10} + \frac{(42-35)^2}{35} = 6.8 > 5.991 = \chi^2_{0.05}(2).$$

Thus we would say that this string of 51 digits does not seem to be random.   ◇

One major disadvantage in the use of the chi-square test is that it is a many-sided test. That is, the alternative hypothesis is very general, and it would be difficult to restrict alternatives to situations such as, with $k = 3$, $H_1: p_1 > p_{10}$, $p_2 > p_{20}$, $p_3 < p_{30}$. As a matter of fact, some statisticians would probably test H_0 against this particular alternative H_1 by using a linear function of Y_1, Y_2, and Y_3. However, this sort of discussion is beyond the scope of this book because it involves knowing more about the distributions of linear functions of the dependent random variables Y_1, Y_2, and Y_3. In any case, the student who truly recognizes that this chi-square statistic tests H_0: $p_i = p_{i0}$, $i = 1, 2, \ldots, k$, against all alternatives can usually appreciate the fact that it is more difficult to reject H_0 at a given significance level α using the chi-square statistic than it would be if some appropriate "one-sided" test statistic were available.

Many experiments yield a set of data, say $x_1, x_2, \ldots, x_n$, and the experimenter is often interested in determining whether these data can be treated as the observed values of a random sample $X_1, X_2, \ldots, X_n$ from a given distribution. That is, would this proposed distribution be a reasonable probabilistic model for these sample items? To see how the chi-square test can help us answer questions of this sort, consider a very simple example.

Example 7.5.2

Let X denote the number of heads that occur when four coins are tossed at random. Under the assumptions that the four coins are independent and the probability of heads on each coin is 1/2, X is $b(4, 1/2)$. One hundred repetitions of this experiment resulted in 0, 1, 2, 3, and 4 heads being observed on 7, 18, 40, 31, and 4 trials, respectively. Do these results support the assumptions? That is, is $b(4, 1/2)$ a reasonable model for the distribution of X? To answer this, we begin by letting $A_1 = \{0\}$, $A_2 = \{1\}$, $A_3 = \{2\}$, $A_4 = \{3\}$, $A_5 = \{4\}$. If $p_{i0} = P(X \in A_i)$ when X is $b(4, 1/2)$, then

$$p_{10} = p_{50} = \binom{4}{0}\left(\frac{1}{2}\right)^4 = \frac{1}{16} = 0.0625,$$

$$p_{20} = p_{40} = \binom{4}{1}\left(\frac{1}{2}\right)^4 = \frac{4}{16} = 0.25,$$

$$p_{30} = \binom{4}{2}\left(\frac{1}{2}\right)^4 = \frac{6}{16} = 0.375.$$

At an approximate $\alpha = 0.05$ significance level, the null hypothesis

$$H_0: \quad p_i = p_{i0}, \qquad i = 1, 2, \ldots, 5,$$

is rejected if the observed value of Q_4 is greater than $\chi^2_{0.05}(4) = 9.488$. If we use the 100 repetitions of this experiment that resulted in the observed values of $Y_1, Y_2, \ldots, Y_5$ of $y_1 = 7$, $y_2 = 18$, $y_3 = 40$, $y_4 = 31$, and $y_5 = 4$, the computed value of Q_4 is

$$q_4 = \frac{(7 - 6.25)^2}{6.25} + \frac{(18 - 25)^2}{25} + \frac{(40 - 37.5)^2}{37.5} + \frac{(31 - 25)^2}{25} + \frac{(4 - 6.25)^2}{6.25}$$

$$= 4.47.$$

Since $4.47 < 9.488$, the hypothesis is not rejected. That is, the data support the hypothesis that $b(4, 1/2)$ is a reasonable probabilistic model for X.   ◇

Thus far all the hypotheses H_0 tested with the chi-square statistic Q_{k-1} have been simple ones (i.e., completely specified, namely, in H_0: $p_i = p_{i0}, i = 1, 2, \ldots, k$, each p_{i0} has been known). This is not always the case, and it frequently happens that $p_{10}, p_{20}, \ldots, p_{k0}$ are functions of one or more unknown parameters. For example, suppose that the hypothesized model for X in Example 7.5.2 was H_0: X is $b(4, p)$, $0 < p < 1$. Then

$$p_{i0} = P(X \in A_i) = \frac{4!}{(i-1)!(5-i)!} p^{i-1}(1-p)^{5-i}, \qquad i = 1, 2, \ldots, 5,$$

which is a function of the unknown parameter p. Of course, if H_0: $p_i = p_{i0}$, $i = 1, 2, \ldots, 5$, is true, for large n,

$$Q_4 = \sum_{i=1}^{5} \frac{(Y_i - np_{i0})^2}{np_{i0}}$$

still has an approximate chi-square distribution with four degrees of freedom. The difficulty is that when $Y_1, Y_2, \ldots, Y_5$ are observed to be equal to $y_1, y_2, \ldots, y_5$, Q_4 cannot be computed, since $p_{10}, p_{20}, \ldots, p_{50}$ (and hence Q_4) are functions of the unknown parameter p.

One way out of the difficulty would be to estimate p from the data and then carry out the computations using this estimate. It is interesting to note the following. Say the estimation of p is carried out by minimizing Q_4 with respect to p yielding $\widetilde{p}$. This $\widetilde{p}$ is sometimes called a **minimum chi-square estimator** of p. If then this $\widetilde{p}$ is used in Q_4, the statistic Q_4 still has an approximate chi-square distribution but with only $4 - 1 = 3$ degrees of freedom. That is, the number of degrees of freedom of the approximating chi-square distribution is reduced by one for each parameter estimated by the minimum chi-square technique. We accept this result without proof (as it is a rather difficult one). Although we have considered this when p_{i0}, $i = 1, 2, \ldots, k$, is a function of only one parameter, it holds when there is more than one unknown parameter, say r. Hence, in a more general situation, the test would be completed by computing Q_{k-1} using Y_i and the estimated p_{i0}, $i = 1, 2, \ldots, k$, to obtain q_{k-1} (i.e., q_{k-1} is the minimized chi-square). This value q_{k-1} would then be compared to a critical value $\chi_\alpha^2(k-1-r)$. In our special case, the computed (minimized) chi-square q_4 would be compared to $\chi_\alpha^2(3)$.

There is still one trouble with all of this: It is usually very difficult to find minimum chi-square estimators. Hence most statisticians usually use some reasonable method (maximum likelihood is satisfactory) of estimating the parameters. They then compute q_{k-1}, recognizing that it is somewhat larger than the minimized chi-square, and compare it with $\chi_\alpha^2(k-1-r)$. Note that this provides a slightly larger probability of rejecting H_0 than would the scheme in which the minimized chi-square were used because this computed q_{k-1} is larger than the minimum q_{k-1}.

Example 7.5.3

Let X denote the number of alpha particles emitted by barium–133 in 1/10 of a second. The following 50 observations of X were taken with a Geiger counter in a fixed position:

$$
\begin{array}{cccccccccc}
7 & 4 & 3 & 6 & 4 & 4 & 5 & 3 & 5 & 3 \\
5 & 5 & 3 & 2 & 5 & 4 & 3 & 3 & 7 & 6 \\
6 & 4 & 3 & 11 & 9 & 6 & 7 & 4 & 5 & 4 \\
7 & 3 & 2 & 8 & 6 & 7 & 4 & 1 & 9 & 8 \\
4 & 8 & 9 & 3 & 9 & 7 & 7 & 9 & 3 & 10 \\
\end{array}
$$

The experimenter is interested in determining whether X has a Poisson distribution. To test H_0: X is Poisson, we first estimate the mean of X, say λ, with the sample mean, $\bar{x} = 5.4$, of these 50 observations. We then partition the set of outcomes for this experiment into the sets $A_1 = \{0, 1, 2, 3\}$, $A_2 = \{4\}$, $A_3 = \{5\}$, $A_4 = \{6\}$, $A_5 = \{7\}$, and $A_6 = \{8, 9, 10, \ldots\}$. We combined $\{0, 1, 2, 3\}$ into one set A_1 and $\{8, 9, 10, \ldots\}$ into another A_6 so that the expected number of outcomes for each set is at least five when H_0 is true. In Table 7.9 the data are grouped, and the estimated probabilities specified by the hypothesis that X has a Poisson distribution with an estimated $\hat{\lambda} = \bar{x} = 5.4$ are given. Since one parameter was estimated, Q_{6-1} has an approximate chi-square distribution with $r = 5 - 1 = 4$ degrees of freedom. Since

$$
q_5 = \frac{[13 - 50(0.213)]^2}{50(0.213)} + \cdots + \frac{[10 - 50(0.178)]^2}{50(0.178)}
$$

$$
= 2.763 < 9.488 = \chi^2_{0.05}(4),
$$

H_0 is not rejected at the 5% significance level. That is, with only these data, we are quite willing to accept the model that X has a Poisson distribution.   ◇

TABLE 7.9   Grouped Geiger counter data

	A_1	A_2	Outcome A_3	A_4	A_5	A_6
Frequency	13	9	6	5	7	10
Probability	0.213	0.160	0.173	0.156	0.120	0.178

Let us now consider the problem of testing a model for the distribution of a random variable W of the continuous type. That is, if $F(w)$ is the distribution function of W, we wish to test

$$
H_0 : \quad F(w) = F_0(w),
$$

where $F_0(w)$ is some known distribution function of the continuous type.

In Section 10.7 we will use a Kolmogorov-Smirnov test for H_0. Also recall that we have considered problems of this type using q-q plots. In order to use the chi-square statistic, we must partition the set of possible values of W into k sets. One way this can be done is as follows. Partition the interval $[0, 1]$ into k sets with the points $b_0, b_1, b_2, \ldots, b_k$, where

$$0 = b_0 < b_1 < b_2 < \cdots < b_k = 1.$$

Let $a_i = F_0^{-1}(b_i)$, $i = 1, 2, \ldots, k - 1$; $A_1 = (-\infty, a_1]$, $A_i = (a_{i-1}, a_i]$ for $i = 2, 3, \ldots, k-1$, and $A_k = (a_{k-1}, \infty)$; $p_i = P(W \in A_i)$, $i = 1, 2, \ldots, k$. Let Y_i denote the number of times the observed value of W belongs to A_i, $i = 1, 2, \ldots, k$, in n independent repetitions of the experiment. Then $Y_1, Y_2, \ldots, Y_k$ have a multinomial distribution with parameters n, $p_1, p_2, \ldots, p_{k-1}$. Also let $p_{i0} = P(W \in A_i)$ when the distribution function of W is $F_0(w)$. The hypothesis that we actually test is a modification of H_0, namely

$$H_0': \; p_i = p_{i0}, \qquad i = 1, 2, \ldots, k.$$

This hypothesis is rejected if the observed value of the chi-square statistic

$$Q_{k-1} = \sum_{i=1}^{k} \frac{(Y_i - np_{i0})^2}{np_{i0}}$$

is at least as great as $\chi_\alpha^2(k - 1)$. If the hypothesis H_0': $p_i = p_{i0}$, $i = 1, 2, \ldots, k$, is not rejected, we do not reject the hypothesis H_0: $F(w) = F_0(w)$.

Example 7.5.4

Let W denote the outcome of a random experiment. Let $F(w)$ denote the distribution function of W, and let

$$F_0(w) = \begin{cases} 0, & w < -1, \\ \dfrac{1}{2}(w^3 + 1), & -1 \le w < 1, \\ 1, & 1 \le w. \end{cases}$$

Note that $f_0(w) = F_0'(w) = (3/2)w^2$, $-1 < w < 1$, is the hypothesized p.d.f. of W, and it is U-shaped. We shall test the hypothesis H_0: $F(w) = F_0(w)$ against all alternatives. The interval $[0, 1]$ can be partitioned into 10 sets of equal probability with the points $b_i = i/10$, $i = 0, 1, \ldots, 10$. If $a_i = F_0^{-1}(b_i) = (2b_i - 1)^{1/3}$, $i = 1, 2, \ldots, 9$, then the sets $A_1 = [-1, a_1]$, $A_2 = (a_1, a_2]$, $\ldots$, $A_{10} = (a_9, 1]$ will each have probability $1/10$ when H_0 is true. If a random sample of size $n = 50$ is observed, then $50(1/10) = 5$ is the expected number of outcomes for each set A_i, $i = 1, 2, \ldots, 10$. Letting Y_i denote the number of outcomes in A_i, we see that the hypothesis is rejected if the calculated value of

$$Q_9 = \sum_{i=1}^{10} \frac{(Y_i - 5)^2}{5}$$

is greater than 16.92 at an approximate 5% significance level. The outcomes of 50 independent repetitions of this experiment simulated on the computer were

$$
\begin{array}{rrrrrrrrrr}
-0.75 & -0.28 & -0.59 & 0.26 & -0.89 & -0.60 & 0.63 & -0.68 & -0.62 & 0.75 \\
-0.81 & 0.99 & 0.76 & -0.88 & 0.35 & -0.87 & 0.68 & 0.81 & -0.82 & -0.90 \\
 0.65 & 0.43 & 0.56 & 0.62 & 0.96 & 0.72 & -0.67 & -0.40 & -0.97 & -0.96 \\
 0.83 & 0.90 & 0.83 & -0.55 & 0.69 & 0.75 & 0.91 & 0.88 & -0.70 & 0.96 \\
-0.98 & -0.93 & -0.77 & -0.96 & -0.29 & 0.83 & -0.93 & 0.96 & 0.79 & -0.76 \\
\end{array}
$$

A summary of these data is given in Table 7.10. The values of $a_0, a_1, \ldots, a_{10}$ are $-1.00, -0.928, -0.843, -0.737, -0.585, 0.00, 0.585, 0.737, 0.843, 0.928,$ and 1.00, respectively. The calculated value of Q_9 is

$$
q_9 = \frac{(6-5)^2}{5} + \frac{(4-5)^2}{5} + \cdots + \frac{(4-5)^2}{5} = 4.0 \leq 16.92 = \chi^2_{0.05}(9),
$$

and thus the hypothesis is not rejected.

TABLE 7.10 Summary of U-shaped data

Outcome	Frequency	Outcome	Frequency
$A_1 = [-1.000, -0.928]$	6	$A_6 = (0.000, 0.585]$	4
$A_2 = (-0.928, -0.843]$	4	$A_7 = (0.585, 0.737]$	6
$A_3 = (-0.843, -0.737]$	5	$A_8 = (0.737, 0.843]$	8
$A_4 = (-0.737, -0.585]$	6	$A_9 = (0.843, 0.928]$	3
$A_5 = (-0.585, 0.000]$	4	$A_{10} = (0.928, 1.000]$	4

Figure 7.12 shows the relative frequency histogram of the data, with unequal class intervals, along with the hypothesized p.d.f. superimposed. ◇

Let $X_1, X_2, \ldots, X_n$ denote a random sample of size n from a distribution with mean μ and positive variance σ^2. Let $W = (\overline{X} - \mu)/(\sigma/\sqrt{n})$, where $\overline{X}$ is the sample mean. From the central limit theorem, we claim that W has a distribution that is approximately $N(0, 1)$ when n is sufficiently large. For what values of n can we say that W is approximately $N(0, 1)$? The answer will, of course, depend on the distribution from which the sample is taken. For example, if the underlying distribution is normal, $n = 1$ is sufficient. In the next example, we shall give a partial answer for the U-shaped distribution used in Example 7.5.4.

Example 7.5.5

Let $X_1, X_2, \ldots, X_n$ denote a random sample of size n from the distribution with p.d.f. $f(x) = (3/2)x^2, -1 < x < 1$. For this distribution $\mu = 0$ and $\sigma^2 = 3/5$. We shall test the hypothesis that $W = (\overline{X} - 0)/\sqrt{3/(5n)}$ is approximately $N(0, 1)$ when $n = 7$. One thousand observations of W were generated by the computer.

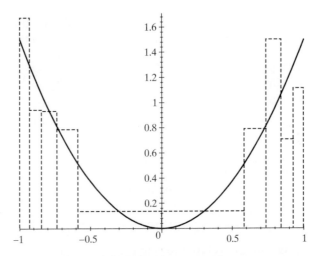

Figure 7.12   U-shaped p.d.f.

Table 7.11 gives 19 sets, their frequencies, and their expected values when sampling from $N(0, 1)$. For these data,

$$q_{18} = \frac{(3 - 5.4)^2}{5.4} + \frac{(4 - 6.8)^2}{6.8} + \cdots + \frac{(5 - 5.4)^2}{5.4} = 39.988.$$

Since $39.988 > 28.87 = \chi^2_{0.05}(18)$, the hypothesis is rejected when $n = 7$. In Exercise 7.5.17 you will be asked to test this hypothesis for other values of n. It would be instructive for you to draw a relative frequency histogram for the data in Table 7.11 with the $N(0, 1)$ p.d.f. superimposed on the same graph (see Figure 5.5).   ◇

TABLE 7.11   Grouped values of w_n

A_i	Y_i	$1000 p_{i0}$	A_i	Y_i	$1000 p_{i0}$
$(-\infty, -2.55]$	3	5.4	$(0.15, 0.45]$	115	114.0
$(-2.55, -2.25]$	4	6.8	$(0.45, 0.75]$	94	99.8
$(-2.25, -1.95]$	21	13.4	$(0.75, 1.05]$	83	79.7
$(-1.95, -1.65]$	29	23.9	$(1.05, 1.35]$	77	58.4
$(-1.65, -1.35]$	43	39.0	$(1.35, 1.65]$	29	39.0
$(-135, -1.05]$	79	58.4	$(1.65, 1.95]$	18	23.9
$(-1.05, -0.75]$	90	79.7	$(1.95, 2.25]$	9	13.4
$(-0.75, -0.45]$	85	99.8	$(2.25, 2.55]$	7	6.8
$(-0.45, -0.15]$	122	114.0	$(2.55, +\infty)$	5	5.4
$(-0.15, 0.15]$	87	119.2			

It is also true in dealing with models of random variables of the continuous type that we must frequently estimate unknown parameters. For example, let H_0 be that W is $N(\mu, \sigma^2)$, where μ and σ^2 are unknown. With a random sample $W_1, W_2, \ldots, W_n$, we first can estimate μ and σ^2, possibly with $\overline{w}$ and s_w^2. We partition the space $\{w : -\infty < w < \infty\}$ into k mutually disjoint sets $A_1, A_2, \ldots, A_k$. We then use the estimates of μ and σ^2, say $\overline{w}$ and s^2, to estimate

$$p_{i0} = \int_{A_i} \frac{1}{s\sqrt{2\pi}} \exp\left[-\frac{(w - \overline{w})^2}{2s^2}\right] dw,$$

$i = 1, 2, \ldots, k$. Using the observed frequencies $y_1, y_2, \ldots, y_k$ of $A_1, A_2, \ldots, A_k$, respectively, from the observed random sample $w_1, w_2, \ldots, w_n$, and $\widehat{p}_{10}, \widehat{p}_{20}, \ldots, \widehat{p}_{k0}$ estimated with $\overline{w}$ and $s^2 = s_w^2$, we compare the computed

$$q_{k-1} = \sum_{i=1}^{k} \frac{(y_i - n\widehat{p}_{i0})^2}{n\widehat{p}_{i0}}$$

to $\chi_\alpha^2(k - 1 - 2)$. This value q_{k-1} will again be somewhat larger than that which would be found using minimum chi-square estimation, and certain caution should be observed. Several exercises illustrate the procedure in which one or more parameters must be estimated. Finally, it should be noted that the methods given in this section frequently are classified under the more general title of goodness-of-fit tests. In particular, then, the tests in this section would be **chi-square goodness-of-fit tests**.

EXERCISES

7.5.1 A 1-pound bag of candy-coated chocolate-covered peanuts contained 224 pieces of candy colored brown, orange, green, and yellow. Test the null hypothesis that the machine filling these bags treats the four colors of candy equally likely; that is, test

$$H_0: \; p_B = p_O = p_G = p_Y = \frac{1}{4}.$$

The observed values were 42 brown, 64 orange, 53 green, and 65 yellow. You may select the significance level or give an approximate p-value.

7.5.2 A particular brand of candy-coated chocolate comes in five different colors that we shall denote as $A_1 = \{\text{brown}\}$, $A_2 = \{\text{yellow}\}$, $A_3 = \{\text{orange}\}$, $A_4 = \{\text{green}\}$, and $A_5 = \{\text{coffee}\}$. Let p_i equal the probability that the color of a piece of candy selected at random belongs to A_i, $i = 1, 2, \ldots, 5$. Test the null hypothesis

$$H_0: \; p_1 = 0.4, \, p_2 = 0.2, \, p_3 = 0.2, \, p_4 = 0.1, \, p_5 = 0.1$$

using a random sample of $n = 580$ pieces of candy whose colors yielded the respective frequencies $224, 119, 130, 48$, and 59. You may select the significance level or give an approximate p-value.

7.5.3 In the Michigan daily lottery, each weekday a three-digit integer is generated one digit at a time. Let p_i denote the probability of generating digit i, $i = 0, 1, \ldots, 9$. Use the following 50 digits to test $H_0: p_0 = p_1 = \cdots = p_9 = 1/10$. Let $\alpha = 0.05$.

1	6	9	9	3	8	5	0	6	7
4	7	5	9	4	6	5	6	4	4
4	8	0	9	3	2	1	5	4	5
7	3	2	1	4	6	7	1	3	4
4	8	8	6	1	6	1	2	8	8

7.5.4 In a large bin of crocus bulbs it is claimed that 1/4 will produce yellow crocuses, 1/4 will produce white crocuses, and 1/2 will produce purple crocuses. If 40 bulbs produced 6 yellow, 7 white, and 27 purple crocuses, would the claim be rejected at the $\alpha = 0.05$ significance level?

7.5.5 In a biology laboratory the mating of two red-eyed fruit flies yielded $n = 432$ offspring, among which 254 were red-eyed, 69 were brown-eyed, 87 were scarlet-eyed, and 22 were white-eyed. Use these data to test, with $\alpha = 0.05$, the hypothesis that the ratio among the offspring would be 9:3:3:1, respectively.

7.5.6 In a biology laboratory students test the Mendelian theory of inheritance using corn. The Mendelian theory of inheritance claims that frequencies of the four categories smooth and yellow, wrinkled and yellow, smooth and purple, and wrinkled and purple will occur in the ratio 9:3:3:1. If a student counted 124, 30, 43, and 11, respectively, for these four categories, would these data support the Mendelian theory? Let $\alpha = 0.05$.

7.5.7 Let X equal the number of female children in a three-child family. We shall use a chi-square goodness-of-fit statistic to test the null hypothesis that the distribution of X is $b(3, 0.5)$.

(a) Define the test statistic and critical region using an $\alpha = 0.05$ significance level.

(b) Among students who were taking statistics, 52 came from families with 3 children. For these families, $x = 0, 1, 2,$ and 3 for 5, 17, 24, and 6 families, respectively. Calculate the value of the test statistic and state your conclusion, considering how the sample was selected.

7.5.8 It has been claimed that when spinning a penny, the probability of observing heads in $p = 0.30$. Three students got together and they would each spin a penny and record the number of heads out of the three spins, say X. They repeated this $n = 200$ times, observing 0, 1, 2, and 3 heads 57, 95, 38, and 10 times, respectively. Use these data to test the hypotheses that X is $b(3, 0.30)$. Give limits for the p-value of this test. In addition, out of the 600 spins, calculate the number of heads occurring and then a 95% confidence interval for p.

7.5.9 In Example 3.7.3, it is claimed that X has a Poisson distribution and 100 observations of X are given in Table 3.2. Use a chi-square goodness-of-fit statistic to test whether this is true. Since $\bar{x} = 5.59$, let the estimate of λ be 5.6 so that you can use Table III. Let $A_1 = \{x \leq 1\}$ and $A_{11} = \{x \geq 11\}$. Use an $\alpha = 0.05$ significance level.

7.5.10 In Exercise 3.7.12, we are asked whether the data look like observations of a Poisson random variable with mean $\lambda = 3$. Use a chi-square goodness-of-fit test to answer this question. Use $\alpha = 0.05$.

7.5.11 While testing a used tape for bad records, a computer operator counted the number of flaws per 100 feet of tape. Let X equal this random variable. Test the null hypothesis that X has a Poisson distribution with a mean of $\lambda = 2.4$ given that 40 observations of X yielded 5 zeros, 7 ones, 12 two, 9 threes, 5 fours, 1 five, and 1 six. Let $\alpha = 0.05$. HINT: Combine five and six into one set; that is, the last set would be all x values ≥ 5.

7.5.12 Let X equal the distance between bad records on a used computer tape. Test the hypothesis that the distribution of X is exponential using the following 90 observations of X and 10 classes of equal probability. Use $\bar{x} = 42.2$ as an estimate of θ. Let $\alpha = 0.05$.

30	79	38	47	22	52	36	36	7	57
3	22	30	14	8	32	15	21	12	12
6	67	6	7	35	78	28	74	5	9
37	1	3	3	44	160	50	27	61	15
39	44	130	18	6	1	32	116	23	12
58	101	68	53	58	21	21	7	79	41
80	33	71	81	17	10	13	49	21	56
107	21	17	64	14	36	26	1	54	207
64	238	25	51	82	8	2	3	43	87

7.5.13 In Exercise 4.2.16, we are asked whether the times that bees spend in low-density flower patches are observations of an exponential random variable. Test whether this is true using $k = 10$ classes of equal probability. Note that θ must be estimated using $\bar{x}$. Let $\alpha = 0.05$.

7.5.14 Let X equal the amount of butterfat (in pounds) produced by 90 cows during a 305-day milk production period following their first calf. Test the hypothesis that the distribution of X is $N(\mu, \sigma^2)$ using $k = 10$ classes of equal probability. You may use $\bar{x} = 511.633$ and $s_x = 87.576$ as estimates of μ and σ, respectively.

486	537	513	583	453	510	570	500	458	555
618	327	350	643	500	497	421	505	637	599
392	574	492	635	460	696	593	422	499	524
539	339	472	427	532	470	417	437	388	481
537	489	418	434	466	464	544	475	608	444
573	611	586	613	645	540	494	532	691	478
513	583	457	612	628	516	452	501	453	643
541	439	627	619	617	394	607	502	395	470
531	526	496	561	491	380	345	274	672	509

7.5.15 A sample of 100 2.2-kilohm resistors was tested, yielding the following measurements:

2.17	2.18	2.17	2.19	2.23	2.18	2.16	2.23	2.29	2.20
2.24	2.13	2.18	2.22	2.21	2.22	2.23	2.18	2.23	2.25
2.24	2.20	2.21	2.20	2.25	2.15	2.20	2.25	2.16	2.19
2.20	2.18	2.18	2.19	2.22	2.19	2.20	2.19	2.22	2.21
2.22	2.18	2.18	2.28	2.19	2.23	2.21	2.19	2.21	2.21
2.19	2.18	2.21	2.17	2.20	2.18	2.18	2.21	2.22	2.18
2.22	2.23	2.19	2.23	2.18	2.19	2.18	2.21	2.18	2.15
2.20	2.23	2.20	2.20	2.23	2.20	2.19	2.22	2.17	2.20
2.20	2.17	2.19	2.19	2.25	2.19	2.19	2.20	2.20	2.19
2.21	2.14	2.24	2.21	2.19	2.23	2.18	2.22	2.23	2.19

(a) Group these data into nine classes of equal length with the class boundaries for the first class being 2.125–2.145.

(b) Use a chi-square goodness-of-fit statistic to test whether these are observations of a normally distributed random variable, after grouping the first two classes and the last two classes to avoid small expected numbers. Estimate μ and σ with the mean and standard deviation of the grouped data. Let $\alpha = 0.05$.

7.5.16 Let $X_1, X_2, \ldots, X_n$ denote a random sample of size n from the distribution with p.d.f. $f(x) = (x + 1)/2$, $-1 < x < 1$. For this distribution $\mu = 1/3$, and $\sigma^2 = 2/9$. Let $W_n = (\overline{X} - 1/3)/\sqrt{2/(9n)}$. One thousand observations of W_n were simulated by the computer when $n = 2, 7$, and 12. These data were grouped and are listed in the following table. Test the hypothesis H_0: W_n is $N(0, 1)$, at $\alpha = 0.05$, for

(a) $n = 2$.

(b) $n = 7$.

(c) $n = 12$.

(d) Based on these limited data, for what values of n would you accept H_0, when the underlying distribution is this skewed one?

NOTE: Figure 5.4 uses the data listed for $n = 2$ and $n = 12$.

A_i	Y_i ($n = 2$)	Y_i ($n = 7$)	Y_i ($n = 12$)
$(-\infty, -2.55]$	7	6	7
$(-2.55, -2.25]$	16	10	11
$(-2.25, -1.95]$	11	18	14
$(-1.95, -1.65]$	33	29	23
$(-1.65, -1.35]$	30	42	36
$(-1.35, -1.05]$	47	60	53
$(-1.05, -0.75]$	86	80	75
$(-0.75, -0.45]$	91	87	98
$(-0.45, -0.15]$	107	112	121
$(-0.15, 0.15]$	111	108	112
$(0.15, 0.45]$	105	120	115
$(0.45, 0.75]$	93	101	113
$(0.75, 1.05]$	93	86	80
$(1.05, 1.35]$	81	64	56
$(1.35, 1.65]$	64	36	44
$(1.65, 1.95]$	25	22	21
$(1.95, 2.25]$	0	15	11
$(2.25, 2.55]$	0	3	7
$(2.55, +\infty)$	0	1	3

7.5.17 Let $X_1, X_2, \ldots, X_n$ denote a random sample of size n from the distribution with p.d.f. $f(x) = (3/2)x^2$, $-1 < x < 1$. Let $W_n = (\overline{X} - 0)/\sqrt{3/(5n)}$. One thousand observations of W_n were simulated using a computer when $n = 2, 8$, and 12. These data were grouped and are listed in the following table. Test the hypothesis H_0: W_n is $N(0, 1)$, at $\alpha = 0.05$, for

(a) $n = 2$.

(b) $n = 8$.

(c) $n = 12$.

(d) Draw the relative frequency histogram for each value of n.

(e) Based on these limited data, for what values of n would you accept the hypothesis that W_n is about $N(0, 1)$ when the underlying distribution has this U-shape?

NOTE: Figure 5.5 uses the data listed for $n = 2$ and $n = 12$.

A_i	Y_i $(n = 2)$	Y_i $(n = 8)$	Y_i $(n = 12)$
$(-\infty, -2.55]$	0	5	7
$(-2.55, -2.25]$	0	6	9
$(-2.25, -1.95]$	0	14	17
$(-1.95, -1.65]$	35	11	26
$(-1.65, -1.35]$	109	51	37
$(-1.35, -1.05]$	82	41	57
$(-1.05, -0.75]$	28	89	85
$(-0.75, -0.45]$	22	93	99
$(-0.45, -0.15]$	127	97	114
$(-0.15, 0.15]$	243	138	104
$(0.15, 0.45]$	112	117	101
$(0.45, 0.75]$	21	108	93
$(0.75, 1.05]$	23	78	90
$(1.05, 1.35]$	74	49	64
$(1.35, 1.65]$	97	51	39
$(1.65, 1.95]$	27	25	32
$(1.95, 2.25]$	0	16	15
$(2.25, 2.55]$	0	9	8
$(2.55, +\infty)$	0	2	3

7.6 CONTINGENCY TABLES

In this section we demonstrate the flexibility of the chi-square test. We first look at a method for testing whether two or more multinomial distributions are equal, sometimes called a test for homogeneity. Then we consider a test for independence of attributes of classification. Both of these lead to a similar test statistic.

Suppose that each of two independent experiments can end in one of the k mutually exclusive and exhaustive events $A_1, A_2, \ldots, A_k$. Let

$$p_{ij} = P(A_i), \qquad i = 1, 2, \ldots, k, \quad j = 1, 2.$$

That is, $p_{11}, p_{21}, \ldots, p_{k1}$ are the probabilities of the events in the first experiment and $p_{12}, p_{22}, \ldots, p_{k2}$ are those associated with the second experiment. Let the experiments be repeated n_1 and n_2 independent times, respectively. Also let

$Y_{11}, Y_{21}, \ldots, Y_{k1}$ be the frequencies of $A_1, A_2, \ldots, A_k$ associated with the n_1 independent trials of the first experiment. Similarly, let $Y_{12}, Y_{22}, \ldots, Y_{k2}$ be the respective frequencies associated with the n_2 trials of the second experiment. Of course, $\sum_{i=1}^{k} Y_{ij} = n_j$, $j = 1, 2$. From the sampling distribution theory corresponding to the basic chi-square test, we know that each of

$$\sum_{i=1}^{k} \frac{(Y_{ij} - n_j p_{ij})^2}{n_j p_{ij}}, \qquad j = 1, 2,$$

has an approximate chi-square distribution with $k - 1$ degrees of freedom. Since the two experiments are independent (and thus the two chi-square statistics are independent), the sum

$$\sum_{j=1}^{2} \sum_{i=1}^{k} \frac{(Y_{ij} - n_j p_{ij})^2}{n_j p_{ij}}$$

is approximately chi-square with $k - 1 + k - 1 = 2k - 2$ degrees of freedom.

Usually, p_{ij}, $i = 1, 2, \ldots, k$, $j = 1, 2$, are unknown, but frequently we wish to test the hypothesis

$$H_0: \quad p_{11} = p_{12}, \, p_{21} = p_{22}, \ldots, p_{k1} = p_{k2},$$

that is, the hypothesis that the corresponding probabilities associated with the two independent experiments are equal. Under H_0, we can estimate the unknown

$$p_{i1} = p_{i2}, \qquad i = 1, 2, \ldots, k,$$

by using the relative frequency $(Y_{i1} + Y_{i2})/(n_1 + n_2)$, $i = 1, 2, \ldots, k$. That is, if H_0 is true, we can say that the two experiments are actually parts of a large one in which $Y_{i1} + Y_{i2}$ is the frequency of the event A_i, $i = 1, 2, \ldots, k$. Note that we only have to estimate the $k - 1$ probabilities $p_{i1} = p_{i2}$ using

$$\frac{Y_{i1} + Y_{i2}}{n_1 + n_2}, \qquad i = 1, 2, \ldots, k - 1,$$

since the sum of the k probabilities must equal 1. That is, the estimator of $p_{k1} = p_{k2}$ is

$$1 - \frac{Y_{11} + Y_{12}}{n_1 + n_2} - \cdots - \frac{Y_{k-1,1} + Y_{k-1,2}}{n_1 + n_2} = \frac{Y_{k1} + Y_{k2}}{n_1 + n_2}.$$

Substituting these estimators, we have that

$$Q = \sum_{j=1}^{2} \sum_{i=1}^{k} \frac{[Y_{ij} - n_j(Y_{i1} + Y_{i2})/(n_1 + n_2)]^2}{n_j(Y_{i1} + Y_{i2})/(n_1 + n_2)}$$

has an approximate chi-square distribution with $2k - 2 - (k - 1) = k - 1$ degrees of freedom. Here $k - 1$ is subtracted from $2k - 2$ because that is the number of estimated parameters. The critical region for testing H_0 is of the form

$$q \geq \chi_\alpha^2(k - 1).$$

Example 7.6.1

To test two methods of instruction, 50 students are selected at random from each of two groups. At the end of the instruction period each student is assigned a grade (A, B, C, D, or F) by an evaluating team. The data are recorded as follows:

Group	Grade					Total
	A	B	C	D	F	
I	8	13	16	10	3	50
II	4	9	14	16	7	50

Accordingly, if the hypothesis H_0 that the corresponding probabilities are equal is true, the respective estimates of the probabilities are

$$\frac{8+4}{100} = 0.12, 0.22, 0.30, 0.26, \frac{3+7}{100} = 0.10.$$

Thus the estimates of $n_1 p_{i1} = n_2 p_{i2}$ are 6, 11, 15, 13, 5, respectively. Hence the computed value of Q is

$$q = \frac{(8-6)^2}{6} + \frac{(13-11)^2}{11} + \frac{(16-15)^2}{15} + \frac{(10-13)^2}{13} + \frac{(3-5)^2}{5}$$

$$+ \frac{(4-6)^2}{6} + \frac{(9-11)^2}{11} + \frac{(14-15)^2}{15} + \frac{(16-13)^2}{13} + \frac{(7-5)^2}{5}$$

$$= \frac{4}{6} + \frac{4}{11} + \frac{1}{15} + \frac{9}{13} + \frac{4}{5} + \frac{4}{6} + \frac{4}{11} + \frac{1}{15} + \frac{9}{13} + \frac{4}{5} = 5.18.$$

Now, under H_0, Q has an approximate chi-square distribution with $k-1 = 4$ degrees of freedom so the $\alpha = 0.05$ critical region is $q \geq 9.488 = \chi^2_{0.05}(4)$. Here $q = 5.18 < 9.488$, and hence H_0 is not rejected at the 5% significance level. Furthermore, the p-value for $q = 5.18$ is 0.268, which is greater than most significance levels. Thus with these data, we cannot say that there is a difference between the two methods of instruction. ◇

It is fairly obvious how this procedure can be extended to testing the equality of h independent multinomial distributions. That is, let

$$p_{ij} = P(A_i), \quad i = 1, 2, \ldots, k, \quad j = 1, 2, \ldots, h,$$

and test

$$H_0: \ p_{i1} = p_{i2} = \cdots = p_{ih} = p_i, \quad i = 1, 2, \ldots, k.$$

Repeat the jth experiment n_j independent times and let $Y_{1j}, Y_{2j}, \ldots, Y_{kj}$ denote the frequencies of the respective events $A_1, A_2, \ldots, A_k$. Now

$$Q = \sum_{j=1}^{h} \sum_{i=1}^{k} \frac{(Y_{ij} - n_j p_{ij})^2}{n_j p_{ij}}$$

has an approximate chi-square distribution with $h(k-1)$ degrees of freedom. Under H_0, we must estimate $k-1$ probabilities using

$$\widehat{p}_i = \frac{\sum_{j=1}^{h} Y_{ij}}{\sum_{j=1}^{h} n_j}, \qquad j = 1, 2, \ldots, k-1,$$

because the estimate of p_k follows from $\widehat{p}_k = 1 - \widehat{p}_1 - \widehat{p}_2 - \cdots - \widehat{p}_{k-1}$. We use these estimates to obtain

$$Q = \sum_{j=1}^{h} \sum_{i=1}^{k} \frac{(Y_{ij} - n_j \widehat{p}_i)^2}{n_j \widehat{p}_i},$$

which has an approximate chi-square distribution with its degrees of freedom given by $h(k-1) - (k-1) = (h-1)(k-1)$.

Let us see how we can use the procedures above to test the equality of two or more independent distributions that are not necessarily multinomial. Suppose first that we are given random variables U and V with distribution functions $F(u)$ and $G(v)$, respectively. It is sometimes of interest to test the hypothesis H_0: $F(x) = G(x)$ for all x. We have previously considered tests such as $\mu_U = \mu_V$, $\sigma_U^2 = \sigma_V^2$, and have used q-q plots for graphical comparisons of two random samples. Now we shall only assume that the distributions are independent and of the continuous types.

We are interested in testing the hypothesis H_0: $F(x) = G(x)$ for all x. This hypothesis will be replaced by another one. Partition the real line into k mutually disjoint sets $A_1, A_2, \ldots, A_k$. Let

$$p_{i1} = P(U \in A_i), \qquad i = 1, 2, \ldots, k$$

and

$$p_{i2} = P(V \in A_i), \qquad i = 1, 2, \ldots, k.$$

We observe that if $F(x) = G(x)$ for all x, then $p_{i1} = p_{i2}$, $i = 1, 2, \ldots, k$. We replace the hypothesis H_0: $F(x) = G(x)$ with the less restrictive hypothesis H_0': $p_{i1} = p_{i2}$, $i = 1, 2, \ldots, k$. That is, we are now essentially interested in testing the equality of two multinomial distributions.

Let n_1 and n_2 denote the number of independent observations of U and V, respectively. For $i = 1, 2, \ldots, k$, let Y_{ij} denote the number of these observations of U and V, $j = 1, 2$, respectively, that fall into a set A_i. At this point, we proceed to make the test of H_0' as described earlier. Of course, if H_0' is rejected at the (approximate) significance level α, then H_0 is rejected with the same probability. However, if H_0' is true, H_0 is not necessarily true. Thus if H_0' is accepted, it is probably better to say that we do not reject H_0 than to say H_0 is accepted.

In applications, the question of how to select $A_1, A_2, \ldots, A_k$ is frequently raised. Obviously, there is not a single choice for k nor the dividing marks of the partition. But it is interesting to observe that the combined sample can be used in this selection without upsetting the approximate distribution of Q. For example, suppose that $n_1 = n_2 = 20$. We could easily select the dividing marks of the partition so that $k = 4$ and one fourth of the combined sample falls in each of the four sets.

Example 7.6.2

Select, at random, 20 cars of each of two comparable major-brand models. All 40 cars are submitted to accelerated life testing; that is, they are driven many miles over very poor roads in a short time and their failure times are recorded (in weeks):

Brand U:	25	31	20	42	39	19	35	36	44	26
	38	31	29	41	43	36	28	31	25	38
Brand V:	28	17	33	25	31	21	16	19	31	27
	23	19	25	22	29	32	24	20	34	26

If we use 23.5, 28.5, and 34.5 as dividing marks, we note that exactly one fourth of the 40 cars fall in each of the resulting four sets. Thus the data can be summarized as follows:

Brand	A_1	A_2	A_3	A_4	Total
U	2	4	4	10	20
V	8	6	6	0	20

The estimate of each p_i is $10/40 = 1/4$, which multiplied by $n_j = 20$ gives 5. Hence the computed Q is

$$q = \frac{(2-5)^2}{5} + \frac{(4-5)^2}{5} + \frac{(4-5)^2}{5} + \frac{(10-5)^2}{5} + \frac{(8-5)^2}{5}$$

$$+ \frac{(6-5)^2}{5} + \frac{(6-5)^2}{5} + \frac{(0-5)^2}{5}$$

$$= \frac{72}{5} = 14.4 > 7.815 = \chi^2_{0.05}(3).$$

Also the p-value is 0.0028. Hence it seems that the two brands of cars have different distributions for the length of life under accelerated life testing.   ◇

Again it should be clear how this can be extended to more than two distributions, and this extension will be illustrated in the exercises.

Now let us suppose that a random experiment results in an outcome that can be classified by two different attributes, such as height and weight. Assume that the first attribute is assigned to one and only one of k mutually exclusive and exhaustive events, say $A_1, A_2, \ldots, A_k$, and the second attribute falls in one and only one of h mutually exclusive and exhaustive events, say $B_1, B_2, \ldots, B_h$. Let the probability of $A_i \cap B_j$ be defined by

$$p_{ij} = P(A_i \cap B_j), \qquad i = 1, 2, \ldots, k, \qquad j = 1, 2, \ldots, h.$$

The random experiment is to be repeated n independent times, and Y_{ij} will denote the frequency of the event $A_i \cap B_j$. Since there are kh such events as $A_i \cap B_j$, the random variable

$$Q_{kh-1} = \sum_{j=1}^{h}\sum_{i=1}^{k} \frac{(Y_{ij} - np_{ij})^2}{np_{ij}}$$

has an approximate chi-square distribution with $kh-1$ degrees of freedom, provided that n is large.

Suppose that we wish to test the hypothesis of the independence of the A and B attributes, namely,

$$H_0: \quad P(A_i \cap B_j) = P(A_i)P(B_j), \quad i = 1, 2, \ldots, k, \quad j = 1, 2, \ldots, h.$$

Let us denote $P(A_i)$ by $p_{i\cdot}$, and $P(B_j)$ by $p_{\cdot j}$; that is,

$$p_{i\cdot} = \sum_{j=1}^{h} p_{ij} = P(A_i), \quad \text{and} \quad p_{\cdot j} = \sum_{i=1}^{k} p_{ij} = P(B_j),$$

and, of course,

$$1 = \sum_{j=1}^{h}\sum_{i=1}^{k} p_{ij} = \sum_{j=1}^{h} p_{\cdot j} = \sum_{i=1}^{k} p_{i\cdot}.$$

Then the hypothesis can be formulated as

$$H_0: \quad p_{ij} = p_{i\cdot}p_{\cdot j}, \quad i = 1, 2, \ldots, k, \quad j = 1, 2, \ldots, h.$$

To test H_0, we can use Q_{kh-1} with p_{ij} replaced by $p_{i\cdot}p_{\cdot j}$. But if $p_{i\cdot}, i = 1, 2, \ldots, k$, and $p_{\cdot j}, j = 1, 2, \ldots, h$, are unknown, as they usually are in the applications, we cannot compute Q_{kh-1} once the frequencies are observed. In such a case we estimate these unknown parameters by

$$\widehat{p}_{i\cdot} = \frac{y_{i\cdot}}{n}, \quad \text{where } y_{i\cdot} = \sum_{j=1}^{h} y_{ij}$$

is the observed frequency of $A_i, i = 1, 2, \ldots, k$; and

$$\widehat{p}_{\cdot j} = \frac{y_{\cdot j}}{n}, \quad \text{where } y_{\cdot j} = \sum_{i=1}^{k} y_{ij}$$

is the observed frequency of $B_j, j = 1, 2, \ldots, h$. Since $\sum_{i=1}^{k} p_{i\cdot} = \sum_{j=1}^{h} p_{\cdot j} = 1$, we actually estimate only $k-1+h-1 = k+h-2$ parameters. So if these estimates are used in Q_{kh-1}, with $p_{ij} = p_{i\cdot}p_{\cdot j}$, then, according to the rule stated earlier, the random variable

$$Q = \sum_{j=1}^{h}\sum_{i=1}^{k} \frac{[Y_{ij} - n(Y_{i\cdot}/n)(Y_{\cdot j}/n)]^2}{n(Y_{i\cdot}/n)(Y_{\cdot j}/n)}$$

has an approximate chi-square distribution with $kh - 1 - (k + h - 2) = (k-1)(h-1)$ degrees of freedom, provided that H_0 is true. The hypothesis H_0 is rejected if the computed value of this statistic exceeds $\chi_\alpha^2[(k-1)(h-1)]$.

Example 7.6.3

A random sample of 400 undergraduate students at the University of Iowa was taken. The students in the sample were classified according to the college in which they were enrolled and according to their gender. These results are recorded in Table 7.12; this table is called a $k \times h$ **contingency table**, where here $k = 2$ and $h = 5$. (Do not be concerned about the numbers in parentheses at this point.) Incidentally, these numbers do actually reflect the composition of the undergraduate colleges at Iowa, but they were modified a little to make the computations easier in this first example.

TABLE 7.12   Undergraduates at the University of Iowa

	College					
Gender	Business	Engineering	Liberal Arts	Nursing	Pharmacy	Total
Male	21	16	145	2	6	190
	(16.625)	(9.5)	(152)	(7.125)	(4.75)	
Female	14	4	175	13	4	210
	(18.375)	(10.5)	(168)	(7.875)	(5.25)	
Total	35	20	320	15	10	400

We desire to test the null hypothesis H_0: $p_{ij} = p_{i\cdot} p_{\cdot j}$, $i = 1, 2$ and $j = 1, 2, 3, 4, 5$, that the college in which a student enrolls is independent of the gender of that student. Under H_0, estimates of the probabilities are

$$\widehat{p}_{1\cdot} = \frac{190}{400} = 0.475 \quad \text{and} \quad \widehat{p}_{2\cdot} = \frac{210}{400} = 0.525$$

and

$$\widehat{p}_{\cdot 1} = \frac{35}{400} = 0.0875, \widehat{p}_{\cdot 2} = 0.05, \widehat{p}_{\cdot 3} = 0.8, \widehat{p}_{\cdot 4} = 0.0375, \widehat{p}_{\cdot 5} = 0.025.$$

The expected numbers $n(y_{i\cdot}/n)(y_{\cdot j}/n)$ are computed as follows:

$$400(0.475)(0.0875) = 16.625,$$
$$400(0.525)(0.0875) = 18.375,$$
$$400(0.475)(0.05) = 9.5,$$

and so on. These are the values recorded in the parentheses in Table 7.12. The computed chi-square statistic is

$$q = \frac{(21 - 16.625)^2}{16.625} + \frac{(14 - 18.375)^2}{18.375} + \cdots + \frac{(4 - 5.25)^2}{5.25}$$

$$= 1.15 + 1.04 + 4.45 + 4.02 + 0.32 + 0.29 + 3.69$$

$$+ 3.34 + 0.33 + 0.30 = 18.93.$$

Since the number of degrees of freedom equals $(k-1)(h-1) = 4$, this $q = 18.93 > 13.28 = \chi_{0.01}^2(4)$, and we reject H_0 at the $\alpha = 0.01$ significance level. Moreover, since the first two terms of q come from the business college, the next two from engineering, and so on, it is clear that the enrollments in engineering and nursing are more highly dependent on gender than in the other colleges because they have contributed the most to the value of the chi-square statistic. It is also interesting to note that one expected number is less than five, namely 4.75. However, as the associated term in q does not contribute an unusual amount to the chi-square value, it does not concern us.   ◇

It is fairly obvious how to extend the testing procedure above to more than two attributes. For example, if the third attribute falls in one and only one of m mutually exclusive and exhaustive events, say $C_1, C_2, \ldots, C_m$, then we test the independence of the three attributes by using

$$Q = \sum_{r=1}^{m} \sum_{j=1}^{h} \sum_{i=1}^{k} \frac{[Y_{ijr} - n(Y_{i..}/n)(Y_{.j.}/n)(Y_{..r}/n)]^2}{n(Y_{i..}/n)(Y_{.j.}/n)(Y_{..r}/n)},$$

where Y_{ijr}, $Y_{i..}$, $Y_{.j.}$, and $Y_{..r}$ are the respective observed frequencies of the events $A_i \cap B_j \cap C_r$, A_i, B_j, and C_r in n independent trials of the experiment. If n is large and if the three attributes are independent, then Q has an approximate chi-square distribution with $khm - 1 - (k - 1) - (h - 1) - (m - 1) = khm - k - h - m + 2$ degrees of freedom.

Rather than explore this extension further, it is more instructive to note some interesting uses of contingency tables.

Example 7.6.4

Say we observed 30 values $x_1, x_2, \ldots, x_{30}$ that are claimed to be the values of a random sample. That is, the corresponding random variables $X_1, X_2, \ldots, X_{30}$ were supposed to be mutually independent and to have the same distribution. Say, however, by looking at the 30 values we detect an upward trend that indicates there might have been some dependence and/or the random variables did not actually have the same distribution. One simple way to test if they could be thought of as being observed values of a random sample is the following. Mark each x high (H) or low (L) depending on whether it is above or below the sample median. Then divide the x values into three groups: $x_1, \ldots, x_{10}$; $x_{11}, \ldots, x_{20}$; $x_{21}, \ldots, x_{30}$. Certainly, if the observations are those of a random sample we would expect five H's and five L's in each group. That is, the attribute classified as H or L should be independent of the group number. The summary of these data provides a 3×2 contingency table. For example, say that the 30 values are

5.6	8.2	7.8	4.8	5.5	8.1	6.7	7.7	9.3	6.9
8.2	10.1	7.5	6.9	11.1	9.2	8.7	10.3	10.7	10.0
9.2	11.6	10.3	11.7	9.9	10.6	10.0	11.4	10.9	11.1

The median can be taken to be the average of the two middle observations in magnitude, namely, 9.2 and 9.3. Marking each item H or L after comparing it with this median, we obtain the following 3×2 contingency table.

Group	L	H	Total
1	9	1	10
2	5	5	10
3	1	9	10
	15	15	30

Here each $n(y_{i.}/n)(y_{.j}/n) = 30(10/30)(15/30) = 5$ so that the computed value of Q is

$$q = \frac{(9-5)^2}{5} + \frac{(1-5)^2}{5} + \frac{(5-5)^2}{5} + \frac{(5-5)^2}{5} + \frac{(1-5)^2}{5} + \frac{(9-5)^2}{5}$$
$$= 12.8 > 5.991 = \chi^2_{0.05}(2),$$

since here $(k-1)(h-1) = 2$ degrees of freedom. (The p-value is 0.0017.) Hence we reject the conjecture that these 30 values could be the items of a random sample. Obviously, modifications could be made to this scheme: dividing the sample into more (or less) than three groups and rating items differently, such as low (L), middle (M), and high (H). ◇

It cannot be emphasized enough that the chi-square statistic can be used fairly effectively in almost any situation in which there should be independence. For illustration, suppose that we have a group of workers who have essentially the same qualifications (training, experience, etc.). Many believe that salary and gender of the workers should be independent attributes; yet there have been several claims in special cases that there is a dependence— or discrimination— in calculations associated with such a problem.

Example 7.6.5

Two groups of workers have the same qualifications for a particular type of work. Their experience in salaries is summarized by the following 2×5 contingency table, in which the upper bound of each salary range is not included in that listing.

Group	Salary (Thousands of Dollars)					Total
	12–14	14–16	16–18	18–20	20+	
1	6	11	16	14	13	60
2	5	9	8	6	2	30
	11	20	24	20	15	90

To test if the group assignment and the salaries seem to be independent with these data at the $\alpha = 0.05$ significance level, we compute

$$q = \frac{[6 - 90(60/90)(11/90)]^2}{90(60/90)(11/90)} + \cdots + \frac{[2 - 90(30/90)(15/90)]^2}{90(30/90)(15/90)}$$

$$= 4.752 < 9.488 = \chi^2_{0.05}(4).$$

Also, the p-value is 0.313. Hence with these limited data, group assignment and salaries seem to be independent.   ◇

Before turning to the exercises, note that we could have thought of the last two examples in this section as testing the equality of two or more multinomial distributions. In Example 7.6.4 the three groups define three binomial distributions; and in Example 7.6.5 the two groups define two multinomial distributions. What would have happened if we had used the computations outlined earlier in this section? It is interesting to note that we obtain exactly the same value of chi-square and in each case the number of degrees of freedom is equal to $(k - 1)(h - 1)$. Hence it makes no difference whether we think of it as a test of independence or a test of the equality of several multinomial distributions. Our advice is to use the terminology that seems most natural for the particular situations.

EXERCISES

7.6.1 We wish to test to see if two groups of nurses distribute their time in six different categories about the same way. That is, the hypothesis under consideration is H_0: $p_{i1} = p_{i2}$, $i = 1, 2, \ldots, 6$. To test this, nurses are observed at random throughout several days, each observation resulting in a mark in one of the six categories. The summary is given by the following frequency table:

Group	Category						Total
	1	2	3	4	5	6	
I	95	36	71	21	45	32	300
II	53	26	43	18	32	28	200

Use a chi-square test with $\alpha = 0.05$.

7.6.2 Suppose that a third group of nurses was observed along with groups I and II of Exercise 7.6.1, resulting in the respective frequencies 130, 75, 136, 33, 61, and 65. Test H_0: $p_{i1} = p_{i2} = p_{i3}$, $i = 1, 2, \ldots, 6$, at the $\alpha = 0.025$ significance level.

7.6.3 Each of two comparable classes of 15 students responded to two different methods of instructions with the following scores on a standardized test:

Class U:	91	42	39	62	55	82	67	44
	51	77	61	52	76	41	59	
Class V:	80	71	55	67	61	93	49	78
	57	88	79	81	63	51	75	

Use a chi-square test with $\alpha = 0.05$ to test the equality of the distributions of test scores by dividing the combined sample into three equal parts (low, middle, high).

7.6.4 Suppose that a third class (W) of 15 students was observed along with classes U and V of Exercise 7.6.3, resulting in scores of

91	73	67	83	59	98	87	69
78	80	65	94	82	74	85	

Again use a chi-square test with $\alpha = 0.05$ to test the equality of the three distributions by dividing the combined sample into three equal parts.

7.6.5 In a contingency table, 1015 individuals are classified by gender and by whether they favor, oppose, or have no opinion on a complete ban on smoking in public places. Test the null hypothesis that gender and opinion on smoking in public places are independent. Give the approximate p-value of this test.

	Ban Smoking in Public Places			
Gender	Favor	Oppose	No Opinion	Total
Male	262	231	10	503
Female	302	205	5	512
	564	436	15	1015

7.6.6 A random survey of 100 students asked each student to select the most preferred form of recreational activity from 5 choices. Test whether the choice is independent of the gender of the respondent. Approximate the p-value of the test. Would we reject at $\alpha = 0.05$?

	Recreational Choice					
Gender	Basketball	Baseball Softball	Swimming	Jogging Running	Tennis	Total
Male	21	5	9	12	13	60
Female	9	3	1	15	12	40
	30	8	10	27	25	100

7.6.7 A random sample of 100 students were classified by gender and by the "instrument" that they "played." Test whether the selection of instrument is independent of the gender of the respondent. Approximate the p-value of this test.

		Instrument				
Gender	Piano	Woodwind	Brass	String	Vocal	Total
Male	4	11	15	6	9	45
Female	7	18	6	6	18	55
	11	29	21	12	27	100

7.6.8 A student who uses the college recreational facilities was interested in whether there is a difference between the facilities used by men and women. Test the null hypothesis that facility and gender are independent attributes using the following data. Use $\alpha = 0.05$.

	Facility		
Gender	Racquetball Court	Track	Total
Male	51	30	81
Female	43	48	91
	94	78	172

7.6.9 A survey of high school girls classified them by two attributes of classification: whether or not they participated in sports and whether or not they had (an) older brother(s). Test the null hypothesis that these two attributes of classification are independent.

	Participated in Sports		
Older Brother(s)	Yes	No	Total
Yes	12	8	20
No	13	27	40
	25	35	60

Approximate the p-value of this test. Do we reject the null hypothesis if $\alpha = 0.05$?

7.6.10 A random sample of 50 women who were tested for cholesterol were classified according to age and cholesterol level and grouped in the following contingency table.

	Cholesterol Level			
Age	< 180	180–210	> 210	Total
< 50	5	11	9	25
≥ 50	4	3	18	25
	9	14	27	50

Test the null hypothesis H_0: age and cholesterol level are independent attributes of classification. What is your conclusion if $\alpha = 0.01$?

7.6.11 Although high school grades and testing scores, such as SAT or ACT, can be used to predict first-year college grade-point average (GPA), many educators claim that a more important factor influencing that GPA is the living conditions of students. In particular, it is claimed that the roommate of the student will have a great influence on his or her grades. To test this, suppose that we selected at random 200 students and classified each according to the following two attributes:

(a) Ranking of the student's roommate from 1 to 5, from a person who was difficult to live with and discouraged scholarship to one who was congenial but encouraged scholarship.

(b) The student's first-year GPA.

Say this gives the following 5×4 contingency table.

Rank of Roommate	Under 2.00	2.00–2.69	2.70–3.19	3.20–4.00	Total
1	8	9	10	4	31
2	5	11	15	11	42
3	6	7	20	14	47
4	3	5	22	23	53
5	1	3	11	12	27
	23	35	78	64	200

Grade-Point Average

Compute the chi-square statistic used to test the independence of the two attributes and compare it to the critical value associated with $\alpha = 0.05$.

7.6.12 A random sample of 395 people were classified by their age and by whether or not they change channels during programs when watching television. Use the following data to test the null hypothesis H_0: Changing channels and age are independent attributes.

(a) Give a significance level, α, for which H_0 is rejected. Why?

(b) Give a value of the significance level, α, for which H_0 is not rejected. Why?

Change?	18–24	25–34	35–49	50–64	Total
Yes	60	54	46	41	201
No	40	44	53	57	194
	100	98	99	98	395

Age

7.6.13 A study was conducted to determine the media credibility for reporting news. Those surveyed were asked to give their age, gender, education, and the most credible medium. Test whether

(a) Media credibility and age are independent.

(b) Media credibility and gender are independent.

(c) Media credibility and education are independent.

(d) Give the approximate p-value for each test.

	Most Credible Medium			
Age	Newspaper	Television	Radio	Total
Under 35	30	68	10	108
35–54	61	79	20	160
Over 54	98	43	21	162
	189	190	51	430

	Most Credible Medium			
Gender	Newspaper	Television	Radio	Total
Male	92	108	19	219
Female	97	81	32	210
	189	189	51	429

	Most Credible Medium			
Education	Newspaper	Television	Radio	Total
Grade school	45	22	6	73
High school	94	115	30	239
College	49	52	13	114
	188	189	49	426

7.6.14 In a psychology experiment 140 students were divided into majors emphasizing left-hemi-sphere brain skills (e.g., philosophy, physics, and mathematics), and into majors emphasizing right-hemisphere skills (e.g., art, music, theater, and dance). They were also classified into one of three groups on the basis of hand posture (right noninverted, left inverted, left noninverted).

	LH	RH
RN	89	29
LI	5	4
LN	5	8

Do these data show sufficient evidence to reject the claim that the choice of college major is independent of hand posture? Let $\alpha = 0.025$.

7.6.15 A random sample of $n = 1362$ persons were classified according to the respondent's education level and whether the respondent was Protestant, Catholic, or Jewish. Use these data to test at an $\alpha = 0.05$ significance level the hypothesis that these attributes of classification are independent.

Education Level	Protestant	Catholic	Jewish
Less than high school	359	140	5
High school or junior college	462	200	17
Bachelor's degree	88	39	2
Graduate degree	37	10	3

7.6.16 In the Michigan daily lottery one three-digit integer is generated 6 days a week. Twenty weeks of lottery numbers have been classified by the day of the week and the magnitude of the number. At the 10% significance level, test whether these attributes of classification are independent.

	Magnitude of Number		
Days	000–499	500–999	Total
Monday and Tuesday	22	18	40
Wednesday and Thursday	19	21	40
Friday and Saturday	13	27	40
	54	66	120

8

LINEAR MODELS

8.1 TESTS OF THE EQUALITY OF SEVERAL MEANS

Frequently, experimenters want to compare more than two treatments: yields of several different corn hybrids, results due to three or more teaching techniques, or miles per gallon obtained from many different types of compact cars. Sometimes the different treatment distributions of the resulting observations are due to changing the level of a certain factor, (e.g., different doses of a given drug). Thus the consideration of the equality of the different means of the various distributions comes under the analysis of a **one-factor experiment**.

In Section 7.4 we discussed how to compare two normal distributions. More generally, let us consider m normal distributions with unknown means $\mu_1, \mu_2, \ldots, \mu_m$, respectively, and an unknown but common variance σ^2. One inference that we wish to consider is a test of the equality of the m means, namely $H_0: \mu_1 = \mu_2 = \cdots = \mu_m = \mu$, μ unspecified, against all possible alternative hypotheses H_1. In order to test this hypothesis we shall take independent random samples from these distributions. Let $X_{i1}, X_{i2}, \ldots, X_{in_i}$ represent a random sample of size n_i from the normal distribution $N(\mu_i, \sigma^2)$, $i = 1, 2, \ldots, m$. In Table 8.1 we have indicated these random samples along with the row means (sample means) where, with $n = n_1 + n_2 + \cdots + n_m$,

$$\overline{X}_{..} = \frac{1}{n} \sum_{i=1}^{m} \sum_{j=1}^{n_i} X_{ij} \quad \text{and} \quad \overline{X}_{i.} = \frac{1}{n_i} \sum_{j=1}^{n_i} X_{ij}, \qquad i = 1, 2, \ldots, m.$$

The dot in the notation for the means, $\overline{X}_{..}$ and $\overline{X}_{i.}$, indicates the index over which the average is taken. Here $\overline{X}_{..}$ is an average taken over both indices while $\overline{X}_{i.}$ is just taken over the index j.

TABLE 8.1   One-factor random samples

	Observations				Mean
X_1:	X_{11}	X_{12}	$\cdots$	X_{1n_1}	$\overline{X}_{1\cdot}$
X_2:	X_{21}	X_{22}	$\cdots$	X_{2n_2}	$\overline{X}_{2\cdot}$
$\vdots$	$\vdots$	$\vdots$	$\vdots$	$\vdots$	$\vdots$
X_m:	X_{m1}	X_{m2}	$\cdots$	X_{mn_m}	$\overline{X}_{m\cdot}$
			Grand mean:		$\overline{X}_{\cdot\cdot}$

To determine a critical region for a test of H_0, we shall first partition the sum of squares associated with the variance of the combined samples into two parts. This sum of squares is given by

$$SS(TO) = \sum_{i=1}^{m} \sum_{j=1}^{n_i} (X_{ij} - \overline{X}_{\cdot\cdot})^2$$

$$= \sum_{i=1}^{m} \sum_{j=1}^{n_i} (X_{ij} - \overline{X}_{i\cdot} + \overline{X}_{i\cdot} - \overline{X}_{\cdot\cdot})^2$$

$$= \sum_{i=1}^{m} \sum_{j=1}^{n_i} (X_{ij} - \overline{X}_{i\cdot})^2 + \sum_{i=1}^{m} \sum_{j=1}^{n_i} (\overline{X}_{i\cdot} - \overline{X}_{\cdot\cdot})^2$$

$$+ 2 \sum_{i=1}^{m} \sum_{j=1}^{n_i} (X_{ij} - \overline{X}_{i\cdot})(\overline{X}_{i\cdot} - \overline{X}_{\cdot\cdot}).$$

The last term of the right-hand member of this identity may be written as

$$2 \sum_{i=1}^{m} \left[(\overline{X}_{i\cdot} - \overline{X}_{\cdot\cdot}) \sum_{j=1}^{n_i} (X_{ij} - \overline{X}_{i\cdot}) \right] = 2 \sum_{i=1}^{m} (\overline{X}_{i\cdot} - \overline{X}_{\cdot\cdot})(n_i \overline{X}_{i\cdot} - n_i \overline{X}_{i\cdot}) = 0,$$

and the preceding term may be written as

$$\sum_{i=1}^{m} \sum_{j=1}^{n_i} (\overline{X}_{i\cdot} - \overline{X}_{\cdot\cdot})^2 = \sum_{i=1}^{m} n_i (\overline{X}_{i\cdot} - \overline{X}_{\cdot\cdot})^2.$$

Thus

$$SS(TO) = \sum_{i=1}^{m} \sum_{j=1}^{n_i} (X_{ij} - \overline{X}_{i\cdot})^2 + \sum_{i=1}^{m} n_i (\overline{X}_{i\cdot} - \overline{X}_{\cdot\cdot})^2.$$

For notation let

$$SS(TO) = \sum_{i=1}^{m} \sum_{j=1}^{n_i} (X_{ij} - \overline{X}_{..})^2, \qquad \text{the total sum of squares.}$$

$$SS(E) = \sum_{i=1}^{m} \sum_{j=1}^{n_i} (X_{ij} - \overline{X}_{i.})^2, \qquad \begin{array}{l} \text{the sum of squares within treatments,} \\ \text{groups, or classes, often called the error} \\ \text{sum of squares.} \end{array}$$

$$SS(T) = \sum_{i=1}^{m} n_i (\overline{X}_{i.} - \overline{X}_{..})^2, \qquad \begin{array}{l} \text{the sum of squares among the differ-} \\ \text{ent treatments, groups, or classes, of-} \\ \text{ten called the between treatment sum of} \\ \text{squares.} \end{array}$$

Thus

$$SS(TO) = SS(E) + SS(T).$$

When H_0 is true, we may regard X_{ij}, $i = 1, 2, \ldots, m$, $j = 1, 2, \ldots, n_i$, as a random sample of size $n = n_1 + n_2 + \cdots + n_m$ from the normal distribution $N(\mu, \sigma^2)$. Then $SS(TO)/(n-1)$ is an unbiased estimator of σ^2 because $SS(TO)/\sigma^2$ is $\chi^2(n-1)$ so that $E[SS(TO)/\sigma^2] = n-1$ and $E[SS(TO)/(n-1)] = \sigma^2$. An unbiased estimator of σ^2 based only on the sample from the ith distribution is

$$W_i = \frac{\displaystyle\sum_{j=1}^{n_i} (X_{ij} - \overline{X}_{i.})^2}{n_i - 1} \qquad \text{for } i = 1, 2, \ldots, m,$$

because $(n_i - 1)W_i/\sigma^2$ is $\chi^2(n_i - 1)$. Thus

$$E\left[\frac{(n_i - 1)W_i}{\sigma^2}\right] = n_i - 1,$$

and so

$$E(W_i) = \sigma^2, \qquad i = 1, 2, \ldots, m.$$

It follows that the sum of m of these independent chi-square random variables, namely

$$\sum_{i=1}^{m} \frac{(n_i - 1)W_i}{\sigma^2} = \frac{SS(E)}{\sigma^2},$$

is also chi-square with $(n_1 - 1) + (n_2 - 1) + \cdots + (n_m - 1) = n - m$ degrees of freedom. Hence $SS(E)/(n - m)$ is an unbiased estimator of σ^2. We now have that

$$\frac{SS(TO)}{\sigma^2} = \frac{SS(E)}{\sigma^2} + \frac{SS(T)}{\sigma^2},$$

where

$$\frac{SS(TO)}{\sigma^2} \text{ is } \chi^2(n - 1) \qquad \text{and} \qquad \frac{SS(E)}{\sigma^2} \text{ is } \chi^2(n - m).$$

Because SS(T) $\geq$ 0, there is a theorem (see following the remark) that states that SS(E) and SS(T) are independent and the distribution of SS(T)/σ^2 is $\chi^2(m-1)$.

Remark The sums of squares, SS(T), SS(E), and SS(TO), are examples of **quadratic forms** in the variables X_{ij}, $i = 1, 2, \ldots, m$, $j = 1, 2, \ldots, n_i$. That is, each term in these sums of squares is of second degree in X_{ij}. Furthermore, the coefficients of the variables are real numbers, so these sums of squares are called **real quadratic forms**. The following theorem, stated without proof, is used in this chapter. [For a proof, see R. V. Hogg and A. T. Craig, *Introduction to Mathematical Statistics*, 5th ed. (Upper Saddle River, NJ: Prentice Hall, 1995).]

Theorem 8.1.1. Let $Q = Q_1 + Q_2 + \cdots + Q_k$, where $Q, Q_1, \ldots, Q_k$ are $k+1$ real quadratic forms in n mutually independent random variables normally distributed with the same variance σ^2. Let Q/σ^2, Q_1/σ^2, $\ldots$, Q_{k-1}/σ^2 have chi-square distributions with $r, r_1, \ldots, r_{k-1}$ degrees of freedom, respectively. If Q_k is nonnegative, then

(a) $Q_1, \ldots, Q_k$ are mutually independent, and hence,

(b) Q_k/σ^2 has a chi-square distribution with $r - (r_1 + \cdots + r_{k-1}) = r_k$ degrees of freedom.

Since under H_0, SS(T)/σ^2 is $\chi^2(m-1)$, we have $E[\text{SS(T)}/\sigma^2] = m - 1$ and hence $E[\text{SS(T)}/(m - 1)] = \sigma^2$. Now the estimator of σ^2, which is based on SS(E), namely SS(E)/$(n - m)$, is always unbiased whether H_0 is true or false. However, if the means $\mu_1, \mu_2, \ldots, \mu_m$ are not equal, the expected value of the estimator based on SS(T) will be greater than σ^2. To make this last statement clear, we have

$$E[\text{SS(T)}] = E\left[\sum_{i=1}^{m} n_i(\overline{X}_{i.} - \overline{X}_{..})^2\right] = E\left[\sum_{i=1}^{m} n_i\overline{X}_{i.}^2 - n\overline{X}_{..}^2\right]$$

$$= \sum_{i=1}^{m} n_i\{\text{Var}(\overline{X}_{i.}) + [E(\overline{X}_{i.})]^2\} - n\{\text{Var}(\overline{X}_{..}) + [E(\overline{X}_{..})]^2\}$$

$$= \sum_{i=1}^{m} n_i\left\{\frac{\sigma^2}{n_i} + \mu_i^2\right\} - n\left\{\frac{\sigma^2}{n} + \overline{\mu}^2\right\}$$

$$= (m - 1)\sigma^2 + \sum_{i=1}^{m} n_i(\mu_i - \overline{\mu})^2,$$

where $\overline{\mu} = (1/n)\sum_{i=1}^{m} n_i\mu_i$. If $\mu_1 = \mu_2 = \cdots = \mu_m = \mu$,

$$E\left(\frac{\text{SS(T)}}{m - 1}\right) = \sigma^2.$$

If the means are not all equal,

$$E\left[\frac{\text{SS(T)}}{m - 1}\right] = \sigma^2 + \sum_{i=1}^{m} n_i\frac{(\mu_i - \overline{\mu})^2}{m - 1} > \sigma^2.$$

Exercise 8.1.4 also illustrates the fact that the estimator using SS(T) is usually greater than that using SS(E) when H_0 is false.

We can base our test of H_0 on the ratio of $SS(T)/(m-1)$ and $SS(E)/(n-m)$, both of which are unbiased estimators of σ^2, provided that $H_0: \mu_1 = \mu_2 = \cdots = \mu_m$ is true so that, under H_0, the ratio would assume values near 1. However, as the means $\mu_1, \mu_2, \ldots, \mu_m$ begin to differ, this ratio tends to become large, since $E[SS(T)/(m-1)]$ gets larger. Under H_0, the ratio

$$\frac{SS(T)/(m-1)}{SS(E)/(n-m)} = \frac{[SS(T)/\sigma^2]/(m-1)}{[SS(E)/\sigma^2]/(n-m)} = F$$

has an F distribution with $m-1$ and $n-m$ degrees of freedom because $SS(T)/\sigma^2$ and $SS(E)/\sigma^2$ are independent chi-square variables. We would reject H_0 if the observed value of F is too large because this would indicate that we have a relatively large SS(T), which suggests that the means are unequal. Thus the critical region is of the form $F \geq F_\alpha(m-1, n-m)$.

The information for tests of the equality of several means is often summarized in an **analysis-of-variance table** or **ANOVA** table like that given in Table 8.2, where the mean square (MS) is the sum of squares (SS) divided by its degrees of freedom.

TABLE 8.2 Analysis-of-variance table

Source	Sum of Squares (SS)	Degrees of Freedom	Mean Square (MS)	F-Ratio
Treatment	SS(T)	$m-1$	$MS(T) = \dfrac{SS(T)}{m-1}$	$\dfrac{MS(T)}{MS(E)}$
Error	SS(E)	$n-m$	$MS(E) = \dfrac{SS(E)}{n-m}$	
Total	SS(TO)	$n-1$		

Example 8.1.1

Let X_1, X_2, X_3, X_4 be independent random variables that have normal distributions $N(\mu_i, \sigma^2)$, $i = 1, 2, 3, 4$. We shall test

$$H_0 : \mu_1 = \mu_2 = \mu_3 = \mu_4 = \mu$$

against all alternatives based on a random sample of size $n_i = 3$ from each of the four distributions. A critical region of size $\alpha = 0.05$ is given by

$$F = \frac{SS(T)/(4-1)}{SS(E)/(12-4)} \geq 4.07 = F_{0.05}(3, 8).$$

TABLE 8.3   Illustrative data

	Observations			$\overline{X}_{i\cdot}$
X_1:	13	8	9	10
X_2:	15	11	13	13
X_3:	8	12	7	9
X_4:	11	15	10	12
$\overline{X}_{\cdot\cdot}$				11

The observed data are given in Table 8.3. (Clearly, these data are not observations from normal distributions. They were selected to illustrate the calculations.)

For these data, the calculated SS(TO), SS(E), and SS(T) are

$$SS(TO) = (13 - 11)^2 + (8 - 11)^2 + \cdots + (15 - 11)^2 + (10 - 11)^2 = 80;$$

$$SS(E) = (13 - 10)^2 + (8 - 10)^2 + \cdots + (15 - 12)^2 + (10 - 12)^2 = 50;$$

$$SS(T) = 3[(10 - 11)^2 + (13 - 11)^2 + (9 - 11)^2 + (12 - 11)^2] = 30.$$

Note that since SS(TO) = SS(E) + SS(T), only two of the three values need to be calculated from the data directly. Here the computed value of F is

$$\frac{30/3}{50/8} = 1.6 < 4.07,$$

and H_0 is not rejected. The p-value is the probability, under H_0, of observing an F that is at least as large as this observed F. It is often given by computer programs.

The information for this example is summarized in the ANOVA table, Table 8.4. Again we note that (here and elsewhere) the F statistic is the ratio of two appropriate mean squares.   ◇

TABLE 8.4   ANOVA table for illustrative data

Source	Sum of Squares (SS)	Degrees of Freedom	Mean Square (MS)	F-Ratio	p-Value
Treatment	30	3	30/3	1.6	0.264
Error	50	8	50/8		
Total	80	11			

Formulas that sometimes simplify the calculations of SS(TO), SS(T), and SS(E) are

$$SS(TO) = \sum_{i=1}^{m} \sum_{j=1}^{n_i} X_{ij}^2 - \frac{1}{n} \left[\sum_{i=1}^{m} \sum_{j=1}^{n_i} X_{ij} \right]^2,$$

$$SS(T) = \sum_{i=1}^{m} \frac{1}{n_i} \left[\sum_{j=1}^{n_i} X_{ij} \right]^2 - \frac{1}{n} \left[\sum_{i=1}^{m} \sum_{j=1}^{n_i} X_{ij} \right]^2,$$

and

$$SS(E) = SS(TO) - SS(T).$$

It is interesting to note that in these formulas each square is divided by the number of observations in the sum being squared: X_{ij}^2 by 1, $(\sum_{j=1}^{n_i} X_{ij})^2$ by n_i, and $(\sum_{i=1}^{m} \sum_{j=1}^{n_i} X_{ij})^2$ by n. These formulas are used in Example 8.1.2. Although these formulas are useful, you are encouraged to use appropriate statistical packages on a computer to aid you with these calculations.

   If the sample sizes are all at least equal to 7, insight can be gained by plotting on the same figure box-and-whisker diagrams for each of the samples. This is also illustrated in Example 8.1.2.

Example 8.1.2

   A window that is manufactured for an automobile has five studs for attaching it. A company that manufactures these windows performs "pull-out tests" to determine the force needed to pull a stud out of the window. Let X_i, $i = 1, 2, 3, 4, 5$, equal the force required at position i and assume that the distribution of X_i is $N(\mu_i, \sigma^2)$. We shall test the null hypothesis H_0: $\mu_1 = \mu_2 = \mu_3 = \mu_4 = \mu_5$ using 7 independent observations at each position. At an $\alpha = 0.01$ significance level, H_0 is rejected if the computed

$$F = \frac{SS(T)/(5-1)}{SS(E)/(35-5)} \geq 4.02 = F_{0.01}(4, 30).$$

The observed data along with certain sums are given in Table 8.5. For these data,

$$SS(TO) = 556{,}174 - \frac{1}{35}(4334)^2 = 19{,}500.97,$$

$$SS(T) = \frac{1}{7}[645^2 + 721^2 + 970^2 + 1017^2 + 981^2]$$

$$- \frac{1}{35}(4334)^2 = 16{,}672.11,$$

$$SS(E) = 19{,}500.97 - 16{,}672.11 = 2828.86.$$

Since the computed F is

$$F = \frac{16{,}672.11/4}{2828.86/30} = 44.20,$$

TABLE 8.5 Pull-out test data

Observations							$\sum_{j=1}^{7} x_{ij}$	$\sum_{j=1}^{7} x_{ij}^2$
X_1:	92	90	87	105	86	83	102	645	59,847
X_2:	100	108	98	110	114	97	94	721	74,609
X_3:	143	149	138	136	139	120	145	970	134,936
X_4:	147	144	160	149	152	131	134	1017	148,367
X_5:	142	155	119	134	133	146	152	981	138,415
								4334	556,174

TABLE 8.6 ANOVA table for pull-out tests

Source	Sum of Squares (SS)	Degrees of Freedom	Mean Square (MS)	F
Treatment	16,672.11	4	4,168.03	44.20
Error	2,828.86	30	94.30	
Total	19,500.97	34		

the null hypothesis is clearly rejected. This information is summarized in Table 8.6.

But why is H_0 rejected? The box-and-whisker diagrams shown in Figure 8.1 help to answer this question. It looks like the forces required to pull out studs in positions 1 and 2 are similar and those in positions 3, 4, and 5 are quite similar but different from positions 1 and 2 (see Exercise 8.1.12). An examination of the window would confirm that this is the case. ◇

As with the two-sample t-test, the F-test works quite well even if the underlying distributions are nonnormal, unless they are highly skewed or the variances are quite different. In the latter cases, we might need to transform the observations to make the data more symmetric with about the same variances or to use certain nonparametric methods that are beyond the scope of this course.

EXERCISES

8.1.1 Let μ_1, μ_2, μ_3 be, respectively, the means of three normal distributions with a common but unknown variance σ^2. In order to test, at the $\alpha = 0.05$ significance level, the hypothesis H_0: $\mu_1 = \mu_2 = \mu_3$ against all possible alternative hypotheses, we take a random sample of size 4 from each of these distributions. Determine whether we accept or reject H_0 if the observed values from these three distributions are, respectively,

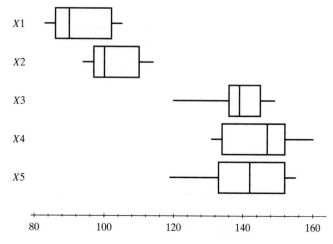

Figure 8.1   Box plots for pull-out tests

$$
\begin{array}{lcccc}
X_1: & 5 & 9 & 6 & 8 \\
X_2: & 11 & 13 & 10 & 12 \\
X_3: & 10 & 6 & 9 & 9
\end{array}
$$

8.1.2 Let μ_i be the average yield in bushels per acre of variety i of corn, $i = 1, 2, 3, 4$. In order to test, at the 5% significance level, the hypothesis $H_0: \mu_1 = \mu_2 = \mu_3 = \mu_4$, four test plots for each of the four varieties of corn are planted. Determine whether we accept or reject H_0 if the yield in bushels per acre of the four varieties of corn are, respectively,

$$
\begin{array}{lcccc}
X_1: & 68.82 & 76.99 & 74.30 & 78.73 \\
X_2: & 86.84 & 75.69 & 77.87 & 76.18 \\
X_3: & 90.16 & 78.84 & 80.65 & 83.58 \\
X_4: & 61.58 & 73.51 & 74.57 & 70.75
\end{array}
$$

8.1.3 Four groups of three pigs each were fed individually four different feeds for a specified length of time to test the hypothesis $H_0: \mu_1 = \mu_2 = \mu_3 = \mu_4$, where μ_i is the mean weight gain for each of the feeds, $i = 1, 2, 3, 4$. Determine whether the null hypothesis is accepted or rejected at a 5% significance level if the observed weight gains are, respectively,

$$
\begin{array}{lccc}
X_1: & 194.11 & 182.80 & 187.43 \\
X_2: & 216.06 & 203.50 & 216.88 \\
X_3: & 178.10 & 189.20 & 181.33 \\
X_4: & 197.11 & 202.68 & 209.18
\end{array}
$$

8.1.4 For the following set of data show that the computed $\mathrm{SS(E)}/(n-m) = 1$ and $\mathrm{SS(T)}/(m-1) = 75$. This suggests that the unbiased estimate of σ^2 based on $\mathrm{SS(T)}$ is usually greater than σ^2 when the true means are unequal.

$$X_1: \quad 4 \quad 5 \quad 6$$
$$X_2: \quad 9 \quad 10 \quad 11$$
$$X_3: \quad 14 \quad 15 \quad 16$$

8.1.5 In a tulip display area, several varieties of tulips have long stems. The lengths of the flower stems were measured for four varieties of tulips, the colors being white, pink, red trimmed with yellow, and dark red. We shall call them varieties 1, 2, 3, and 4. Let μ_i, $i = 1, 2, 3, 4$, denote the mean of the lengths of the stems for variety i. Test the hypothesis $H_0: \mu_1 = \mu_2 = \mu_3 = \mu_4$ against all alternatives, making reasonable assumptions. Use a 5% significance level. Clearly specify the critical region for this test.

X_1:	13.75	13.00	14.25	12.00	12.25
X_2:	13.75	14.50	12.75	14.50	13.25
X_3:	12.75	12.50	12.00	14.00	13.00
X_4:	16.75	14.25	14.50	15.50	13.75

8.1.6 Use the data of Example 1.4.3 concerning the strengths of three different types of beams to make a formal F-test of the equality of the three means. Use $\alpha = 0.05$.

8.1.7 Montgomery considers the strengths of a synthetic fiber that is possibly affected by the percentage of cotton in the fiber. Five levels of this percentage are considered with five observations taken at each level.

Percentage of Cotton	Tensile Strength (pounds per square inch)				
15	7	7	15	11	9
20	12	17	12	18	18
25	14	18	18	19	19
30	19	25	22	19	23
35	7	10	11	15	11

Use the F-test, with $\alpha = 0.05$, to see if there are differences in the breaking strengths due to the percentages of cotton used. [D. C. Montgomery, *Design and Analysis of Experiments*, 2nd ed. (New York: Wiley, 1984), p. 51.]

8.1.8 Let X_1, X_2, X_3, X_4 equal the cholesterol level of a woman under the age of 50, a man under 50, a woman 50 or older, and a man 50 or older, respectively. Assume that the distribution of X_i is $N(\mu_i, \sigma^2)$, $i = 1, 2, 3, 4$. We shall test the null hypothesis $H_0: \mu_1 = \mu_2 = \mu_3 = \mu_4$ using 7 observations of each X_i.

(a) Give a critical region for an $\alpha = 0.05$ significance level.

(b) Construct an ANOVA table and state your conclusion using the following data:

X_1:	221	213	202	183	185	197	162
X_2:	271	192	189	209	227	236	142
X_3:	262	193	224	201	161	178	265
X_4:	192	253	248	278	232	267	289

(c) Give bounds on the p-value for this test.

(d) Construct box-and-whisker diagrams for each set of data on the same figure and give an interpretation.

8.1.9 Let X_i, $i = 1, 2, 3, 4$ equal the distance that a golf ball travels when hit from a tee, where i denotes the index of the ith manufacturer. Assume that the distribution of X_i is $N(\mu_i, \sigma^2)$, $i = 1, 2, 3, 4$ when hit by a certain golfer. We shall test the null hypothesis $H_0: \mu_1 = \mu_2 = \mu_3 = \mu_4$ using 3 observations of each random variable.

(a) Give a critical region for an $\alpha = 0.05$ significance level.

(b) Construct an ANOVA table and state your conclusion using the following data:

$$
\begin{array}{llll}
X_1: & 240 & 221 & 265 \\
X_2: & 286 & 256 & 272 \\
X_3: & 259 & 245 & 232 \\
X_4: & 239 & 215 & 223
\end{array}
$$

(c) What would your conclusion be if $\alpha = 0.025$?

(d) What is the approximate p-value of this test?

8.1.10 Different sizes of nails are packaged in "1-pound" boxes. Let X_i equal the weight of a box with nail size $(4i)C$, $i = 1, 2, 3, 4, 5$, where $4C$, $8C$, $12C$, $16C$, and $20C$ are the sizes of the sinkers from smallest to largest. Assume that the distribution of X_i is $N(\mu_i, \sigma^2)$, To test the null hypothesis that the mean weights of "1-pound" boxes are all equal for different sizes of nails, we shall use random samples of size 7, weighing the nails to the nearest hundredth of a pound.

(a) Give a critical region for an $\alpha = 0.05$ significance level.

(b) Construct an ANOVA table and state your conclusion using the following data:

$$
\begin{array}{llllllll}
X_1: & 1.03 & 1.04 & 1.07 & 1.03 & 1.08 & 1.06 & 1.07 \\
X_2: & 1.03 & 1.10 & 1.08 & 1.05 & 1.06 & 1.06 & 1.05 \\
X_3: & 1.03 & 1.08 & 1.06 & 1.02 & 1.04 & 1.04 & 1.07 \\
X_4: & 1.10 & 1.10 & 1.09 & 1.09 & 1.06 & 1.05 & 1.08 \\
X_5: & 1.04 & 1.06 & 1.07 & 1.06 & 1.05 & 1.07 & 1.05
\end{array}
$$

(c) Construct box-and-whisker diagrams for each set of data on the same figure and give an interpretation.

8.1.11 The driver of a diesel-powered automobile decided to test the quality of three types of diesel fuel sold in the area based on mpg. Test the null hypothesis that the three means are equal using the following data. Make the usual assumptions and take $\alpha = 0.05$.

$$
\begin{array}{llllll}
\text{Brand A:} & 38.7 & 39.2 & 40.1 & 38.9 \\
\text{Brand B:} & 41.9 & 42.3 & 41.3 \\
\text{Brand C:} & 40.8 & 41.2 & 39.5 & 38.9 & 40.3
\end{array}
$$

8.1.12 Based on the box-and-whisker diagrams in Figure 8.1, it looks like the means of X_1 and X_2 could be equal and also that the means of X_3, X_4, and X_5 could be equal but different from the first two.

(a) Using the data in Example 8.1.2, test H_0: $\mu_1 = \mu_2$ against a two-sided alternative hypothesis using a t-test and an F-test. Let $\alpha = 0.05$. Do F and t tests give the same result?

(b) Using the data in Example 8.1.2, test H_0: $\mu_3 = \mu_4 = \mu_5$. Let $\alpha = 0.05$.

8.1.13 For an aerosol product, there are three weights: the tare weight (container weight), the concentrate weight, and the propellant weight. Let X_1, X_2, X_3 denote the propellant weights on three different days. Assume that each of these independent random variables has a normal distribution with common variance and respective means, μ_1, μ_2, μ_3. We shall test the null hypothesis H_0: $\mu_1 = \mu_2 = \mu_3$ using 9 observations of each of the random variables.

(a) Give a critical region for an $\alpha = 0.01$ significance level.

(b) Construct an ANOVA table and state your conclusion using the following data:

X_1:	43.06	43.32	42.63	42.86	43.05
	42.87	42.94	42.80	42.36	
X_2:	42.33	42.81	42.13	42.41	42.39
	42.10	42.42	41.42	42.52	
X_3:	42.83	42.57	42.96	43.16	42.25
	42.24	42.20	41.97	42.61	

(c) Construct box-and-whisker diagrams for each set of data on the same figure and give an interpretation.

8.1.14 A particular process puts a coating on a piece of glass so that it is sensitive to touch. Randomly throughout the day, pieces of glass are selected from the production line and the resistance is measured at 12 different locations on the glass. On each of three different days, December 6, December 7, and December 22, the following data give the means of the 12 measurements on each of 11 pieces of glass.

December 6:	175.05	177.44	181.94	176.51	182.12	164.34
	163.20	168.12	171.26	171.92	167.87	
December 7:	175.93	176.62	171.39	173.90	178.34	172.90
	174.67	174.27	177.16	184.13	167.21	
December 22:	167.27	161.48	161.86	173.83	170.75	172.90
	173.27	170.82	170.93	173.89	177.68	

(a) Use these data to test whether the means on all three days are equal.

(b) Use box-and-whisker diagrams to confirm your answer.

8.2 TWO-FACTOR ANALYSIS OF VARIANCE

The test of the equality of several means considered in Section 8.1 is an example of a statistical inference method called the analysis of variance (ANOVA). This method derives its name from the fact that the quadratic form $SS(TO) = (n-1)S^2$, the total sum of squares about the combined sample mean, is decomposed into component parts and analyzed. In this section, other problems in the analysis of variance will be investigated; here we restrict our considerations to the two-factor case, but the reader can see how it can be extended to three-factor and other cases.

Consider a situation in which it is desirable to investigate the effects of two factors that influence an outcome of an experiment. For example, a teaching method (lecture, discussion, computer-assisted, television, etc.) and the size of a class might influence a student's score on a standard test; or the type of car and the grade of gasoline used might change the number of miles per gallon. In the latter example, if the number of miles per gallon is not affected by the grade of gasoline, we would no doubt use the least expensive grade.

The first analysis of variance model that we discuss is referred to as a **two-way classification with one observation per cell**. In particular, assume that there are two factors (attributes), one of which has a levels and the other b levels. There are thus $n = ab$ possible combinations, each of which determines a cell. Let us think of these cells as being arranged in a rows and b columns. Here we take one observation per cell, and we denote the observation in the ith row and jth column by X_{ij}. Further assume that X_{ij} is $N(\mu_{ij}, \sigma^2)$, $i = 1, 2, \ldots, a$, and $j = 1, 2, \ldots, b$; and the $n = ab$ random variables are independent. [The assumptions of normality and homogeneous (same) variances can be somewhat relaxed in applications with little change in the significance levels of the resulting tests.] We shall assume that the means μ_{ij} are composed of a row effect, a column effect, and an overall effect in some additive way, namely $\mu_{ij} = \mu + \alpha_i + \beta_j$, where $\sum_{i=1}^{a} \alpha_i = 0$ and $\sum_{j=1}^{b} \beta_j = 0$. The parameter α_i represents the ith row effect, and the parameter β_j represents the jth column effect.

Remark   There is no loss in generality in assuming that

$$\sum_{i=1}^{a} \alpha_i = \sum_{j=1}^{b} \beta_j = 0.$$

To see this, let $\mu_{ij} = \mu' + \alpha_i' + \beta_j'$. Write

$$\overline{\alpha}' = \left(\frac{1}{a}\right) \sum_{i=1}^{a} \alpha_i' \quad \text{and} \quad \overline{\beta}' = \left(\frac{1}{b}\right) \sum_{j=1}^{b} \beta_j'.$$

We have

$$\mu_{ij} = (\mu' + \overline{\alpha}' + \overline{\beta}') + (\alpha_i' - \overline{\alpha}') + (\beta_j' - \overline{\beta}') = \mu + \alpha_i + \beta_j,$$

where $\sum_{i=1}^{a} \alpha_i = 0$ and $\sum_{j=1}^{b} \beta_j = 0$. The reader is asked to find μ, α_i, and β_j for one display of μ_{ij} in Exercise 8.2.2.

To test the hypothesis that there is no row effect, we would test $H_A: \alpha_1 = \alpha_2 = \cdots = \alpha_a = 0$, since $\sum_{i=1}^{a} \alpha_i = 0$. Similarly, to test that there is no column effect, we would test $H_B: \beta_1 = \beta_2 = \cdots = \beta_b = 0$, since $\sum_{j=1}^{b} \beta_j = 0$. To test these hypotheses, we shall again partition the total sum of squares into several component parts. Letting

$$\overline{X}_{i\cdot} = \frac{1}{b}\sum_{j=1}^{b}X_{ij}, \qquad \overline{X}_{\cdot j} = \frac{1}{a}\sum_{i=1}^{a}X_{ij}, \qquad \overline{X}_{\cdot\cdot} = \frac{1}{ab}\sum_{i=1}^{a}\sum_{j=1}^{b}X_{ij},$$

we have

$$SS(TO) = \sum_{i=1}^{a}\sum_{j=1}^{b}(X_{ij} - \overline{X}_{\cdot\cdot})^2$$

$$= \sum_{i=1}^{a}\sum_{j=1}^{b}[(\overline{X}_{i\cdot} - \overline{X}_{\cdot\cdot}) + (\overline{X}_{\cdot j} - \overline{X}_{\cdot\cdot}) + (X_{ij} - \overline{X}_{i\cdot} - \overline{X}_{\cdot j} + \overline{X}_{\cdot\cdot})]^2$$

$$= b\sum_{i=1}^{a}(\overline{X}_{i\cdot} - \overline{X}_{\cdot\cdot})^2 + a\sum_{j=1}^{b}(\overline{X}_{\cdot j} - \overline{X}_{\cdot\cdot})^2$$

$$+ \sum_{i=1}^{a}\sum_{j=1}^{b}(X_{ij} - \overline{X}_{i\cdot} - \overline{X}_{\cdot j} + \overline{X}_{\cdot\cdot})^2$$

$$= SS(A) + SS(B) + SS(E),$$

where SS(A) is the sum of squares among levels of factor A or among rows, SS(B) is the sum of squares among levels of factor B or among columns, and SS(E) is the error or residual sum of squares. In Exercise 8.2.4 the reader is asked to show that the three "cross-product" terms in the square of the trinomial sum to zero. The distribution of the error sum of squares does not depend on the mean μ_{ij}, provided that the additive model is correct, and hence its distribution is the same whether H_A or H_B is true or not; thus SS(E) acts as a "measuring stick" as did SS(E) in Section 8.1. This can be seen more clearly by writing

$$SS(E) = \sum_{i=1}^{a}\sum_{j=1}^{b}(X_{ij} - \overline{X}_{i\cdot} - \overline{X}_{\cdot j} + \overline{X}_{\cdot\cdot})^2$$

$$= \sum_{i=1}^{a}\sum_{j=1}^{b}[X_{ij} - (\overline{X}_{i\cdot} - \overline{X}_{\cdot\cdot}) - (\overline{X}_{\cdot j} - \overline{X}_{\cdot\cdot}) - \overline{X}_{\cdot\cdot}]^2$$

and noting the similarity of the summand in the right-hand member with

$$X_{ij} - \mu_{ij} = X_{ij} - \alpha_i - \beta_j - \mu.$$

We now show that $SS(A)/\sigma^2$, $SS(B)/\sigma^2$, and $SS(E)/\sigma^2$ are independent chi-square variables, provided that both H_A and H_B are true, that is, when all the means μ_{ij} have a common value μ. To do this, first note that $SS(TO)/\sigma^2$ is $\chi^2(ab-1)$. In addition, from Section 8.1, we see that expressions such as $SS(A)/\sigma^2$ and $SS(B)/\sigma^2$

are chi-square variables, namely $\chi^2(a-1)$ and $\chi^2(b-1)$, by replacing the n_i of Section 8.1 by a and b, respectively. Obviously, $SS(E) \geq 0$, and hence by Theorem 8.1.1 we have that $SS(A)/\sigma^2$, $SS(B)/\sigma^2$, and $SS(E)/\sigma^2$ are independent chi-square variables with $a-1$, $b-1$, and $ab-1-(a-1)-(b-1) = (a-1)(b-1)$ degrees of freedom, respectively.

To test the hypothesis $H_A: \alpha_1 = \alpha_2 = \cdots = \alpha_a = 0$, we shall use the row sum of squares $SS(A)$ and the residual sum of squares $SS(E)$. When H_A is true, $SS(A)/\sigma^2$ and $SS(E)/\sigma^2$ are independent chi-square variables with $a-1$ and $(a-1)(b-1)$ degrees of freedom, respectively. Thus $SS(A)/(a-1)$ and $SS(E)/[(a-1)(b-1)]$ are both unbiased estimators of σ^2 when H_A is true. However, $E[SS(A)/(a-1)] > \sigma^2$ when H_A is not true, and hence we would reject H_A when

$$F_A = \frac{SS(A)/[\sigma^2(a-1)]}{SS(E)/[\sigma^2(a-1)(b-1)]} = \frac{SS(A)/(a-1)}{SS(E)/[(a-1)(b-1)]}$$

is "too large." Since F_A has an F distribution with $a-1$ and $(a-1)(b-1)$ degrees of freedom when H_A is true, H_A is rejected if the observed value of $F_A \geq F_\alpha[a-1, (a-1)(b-1)]$.

Similarly, the test of the hypothesis $H_B: \beta_1 = \beta_2 = \cdots = \beta_b = 0$ against all alternatives can be based on

$$F_B = \frac{SS(B)/[\sigma^2(b-1)]}{SS(E)/[\sigma^2(a-1)(b-1)]} = \frac{SS(B)/(b-1)}{SS(E)/[(a-1)(b-1)]}$$

which has an F distribution with $b-1$ and $(a-1)(b-1)$ degrees of freedom, provided that H_B is true.

In Table 8.7 we give the ANOVA table that summarizes the information needed for these tests of hypotheses. The formulas for F_A and F_B show that each of them is a ratio of two mean squares.

TABLE 8.7 Two-way ANOVA table, one observation per cell

Source	Sum of Squares (SS)	Degrees of Freedom	Mean Square (MS)	F
Factor A (row)	$SS(A)$	$a-1$	$MS(A) = \dfrac{SS(A)}{a-1}$	$\dfrac{MS(A)}{MS(E)}$
Factor B (column)	$SS(B)$	$b-1$	$MS(B) = \dfrac{SS(B)}{b-1}$	$\dfrac{MS(B)}{MS(E)}$
Error	$SS(E)$	$(a-1)(b-1)$	$MS(E) = \dfrac{SS(E)}{(a-1)(b-1)}$	
Total	$SS(TO)$	$ab-1$		

Example 8.2.1

Each of three cars is driven with each of four different brands of gasoline. The number of miles per gallon driven for each of the $ab = (3)(4) = 12$ different combinations is recorded in Table 8.8.

TABLE 8.8   Gas mileage data

Car	1	2	3	4	$\overline{X}_{i\cdot}$
		Gasoline			
1	16	18	21	21	19
2	14	15	18	17	16
3	15	15	18	16	16
$\overline{X}_{\cdot j}$	15	16	19	18	17

We would like to test whether we can expect the same mileage for each of these four brands of gasoline. In our notation, we test the hypothesis

$$H_B : \beta_1 = \beta_2 = \beta_3 = \beta_4 = 0$$

against all alternatives. At a 1% significance level we shall reject H_B if the computed F, namely

$$\frac{\mathrm{SS(B)}/(4 - 1)}{\mathrm{SS(E)}/[(3 - 1)(4 - 1)]} \geq 9.78 = F_{0.01}(3, 6).$$

We have

$$\mathrm{SS(B)} = 3[(15 - 17)^2 + (16 - 17)^2 + (19 - 17)^2 + (18 - 17)^2] = 30;$$

$$\mathrm{SS(E)} = (16 - 19 - 15 + 17)^2 + (14 - 16 - 15 + 17)^2 + \cdots$$
$$+ (16 - 16 - 18 + 17)^2 = 4.$$

Hence the computed F is

$$\frac{30/3}{4/6} = 15 > 9.78,$$

and the hypothesis H_B is rejected. That is, the gasolines seem to give different performances (at least with these three cars). The information for this example is summarized in Table 8.9. $\diamond$

In a two-way classification problem, particular combinations of the two factors might interact differently from that expected from the additive model. For example, in Example 8.2.1 gasoline 3 seemed to be the best gasoline and car 1 the best car; however, it sometimes happens that the two best do not "mix" well and the joint performance is poor. That is, there might be a strange interaction between this combination of car and gasoline and accordingly the joint performance is not as good as expected. Sometimes it can happen that we get good results from a combination of

TABLE 8.9 ANOVA table for gas mileage data

Source	Sum of Squares (SS)	Degrees of Freedom	Mean Square (MS)	F	p-Value
Row (A)	24	2	12	18	0.0029
Column (B)	30	3	10	15	0.0034
Error	4	6	2/3		
Total	58	11			

some of the poorer levels of each factor. This is called interaction, and it frequently occurs in practice (e.g., in chemistry). In order to test for possible interaction, we shall consider a two-way classification problem in which $c > 1$ independent observations are taken per cell.

Assume that X_{ijk}, $i = 1, 2, \ldots, a$; $j = 1, 2, \ldots, b$; and $k = 1, 2, \ldots, c$, are $n = abc$ random variables that are mutually independent and have normal distributions with a common, but unknown, variance σ^2. The mean of each X_{ijk}, $k = 1, 2, \ldots, c$, is $\mu_{ij} = \mu + \alpha_i + \beta_j + \gamma_{ij}$, where $\sum_{i=1}^a \alpha_i = 0$, $\sum_{j=1}^b \beta_j = 0$, $\sum_{i=1}^a \gamma_{ij} = 0$, and $\sum_{j=1}^b \gamma_{ij} = 0$. The parameter γ_{ij} is called the **interaction** associated with cell (i, j). That is, the interaction between the ith level of one classification and the jth level of the other classification is γ_{ij}. The reader is asked to determine μ, α_i, β_j, and γ_{ij} for some given μ_{ij} in Exercise 8.2.6.

To test the hypotheses that (a) the row effects are equal to zero, (b) the column effects are equal to zero, and (c) there is no interaction, we shall again partition the total sum of squares into several component parts. Letting

$$\overline{X}_{ij.} = \frac{1}{c} \sum_{k=1}^c X_{ijk},$$

$$\overline{X}_{i..} = \frac{1}{bc} \sum_{j=1}^b \sum_{k=1}^c X_{ijk},$$

$$\overline{X}_{.j.} = \frac{1}{ac} \sum_{i=1}^a \sum_{k=1}^c X_{ijk},$$

$$\overline{X}_{...} = \frac{1}{abc} \sum_{i=1}^a \sum_{j=1}^b \sum_{k=1}^c X_{ijk},$$

we have

$$SS(TO) = \sum_{i=1}^{a}\sum_{j=1}^{b}\sum_{k=1}^{c}(X_{ijk} - \overline{X}_{...})^2$$

$$= bc\sum_{i=1}^{a}(\overline{X}_{i..} - \overline{X}_{...})^2 + ac\sum_{j=1}^{b}(\overline{X}_{.j.} - \overline{X}_{...})^2 +$$

$$c\sum_{i=1}^{a}\sum_{j=1}^{b}(\overline{X}_{ij.} - \overline{X}_{i..} - \overline{X}_{.j.} + \overline{X}_{...})^2 + \sum_{i=1}^{a}\sum_{j=1}^{b}\sum_{k=1}^{c}(X_{ijk} - \overline{X}_{ij.})^2$$

$$= SS(A) + SS(B) + SS(AB) + SS(E),$$

where SS(A) is the row sum of squares or the sum of squares among levels of factor A, SS(B) is the column sum of squares or the sum of squares among levels of factor B, SS(AB) is the interaction sum of squares, and SS(E) is the error sum of squares. Again we can show that the cross-product terms sum to zero.

To consider the joint distribution of SS(A), SS(B), SS(AB), and SS(E), let us assume that all the means equal the same value μ. Of course, we know that SS(TO)/σ^2 is $\chi^2(abc-1)$. Also, by letting the n_i of Section 8.1 equal bc and ac, respectively, we know that SS(A)/σ^2 and SS(B)/σ^2 are $\chi^2(a-1)$ and $\chi^2(b-1)$. Moreover,

$$\frac{\displaystyle\sum_{k=1}^{c}(X_{ijk} - \overline{X}_{ij.})^2}{\sigma^2}$$

is $\chi^2(c-1)$; hence SS(E)/σ^2 is the sum of ab independent chi-square variables such as this and thus is $\chi^2[ab(c-1)]$. Of course, SS(AB) $\geq$ 0; so, according to Theorem 8.1.1, we have that SS(A)/σ^2, SS(B)/σ^2, SS(AB)/σ^2, and SS(E)/σ^2 are mutually independent chi-square variables with $a-1$, $b-1$, $(a-1)(b-1)$, and $ab(c-1)$ degrees of freedom, respectively.

To test the hypotheses concerning row, column, and interaction effects, we form F statistics in which the numerators are affected by deviations from the respective hypotheses whereas the denominator is a function of SS(E), whose distribution depends only on the value of σ^2 and not on the values of the cell means. Hence SS(E) acts as our measuring stick here.

The statistic for testing the hypothesis

$$H_{AB}: \gamma_{ij} = 0, \qquad i = 1, 2, \ldots, a, \quad j = 1, 2, \ldots, b,$$

against all alternatives is

$$F_{AB} = \frac{c\sum_{i=1}^{a}\sum_{j=1}^{b}(\overline{X}_{ij\cdot} - \overline{X}_{i\cdot\cdot} - \overline{X}_{\cdot j\cdot} + \overline{X}_{\cdots})^2/[\sigma^2(a-1)(b-1)]}{\sum_{i=1}^{a}\sum_{j=1}^{b}\sum_{k=1}^{c}(X_{ijk} - \overline{X}_{ij\cdot})^2/[\sigma^2 ab(c-1)]}$$

$$= \frac{\text{SS}(AB)/[(a-1)(b-1)]}{\text{SS}(E)/[ab(c-1)]},$$

which has an F distribution with $(a-1)(b-1)$ and $ab(c-1)$ degrees of freedom when H_{AB} is true. If the computed $F_{AB} \geq F_{\alpha}[(a-1)(b-1), ab(c-1)]$, we reject H_{AB} and say there is a difference among the means, since there seems to be interaction. Most statisticians do *not* proceed to test row and column effects if H_{AB} is rejected.

The statistic for testing the hypothesis

$$H_A : \alpha_1 = \alpha_2 = \cdots = \alpha_a = 0$$

against all alternatives is

$$F_A = \frac{bc\sum_{i=1}^{a}(\overline{X}_{i\cdot\cdot} - \overline{X}_{\cdots})^2/[\sigma^2(a-1)]}{\sum_{i=1}^{a}\sum_{j=1}^{b}\sum_{k=1}^{c}(X_{ijk} - \overline{X}_{ij\cdot})^2/[\sigma^2 ab(c-1)]} = \frac{\text{SS}(A)/(a-1)}{\text{SS}(E)/[ab(c-1)]},$$

which has an F distribution with $a-1$ and $ab(c-1)$ degrees of freedom when H_A is true. The statistic for testing the hypothesis

$$H_B : \beta_1 = \beta_2 = \cdots = \beta_b = 0$$

against all alternatives is

$$F_B = \frac{ac\sum_{j=1}^{b}(\overline{X}_{\cdot j\cdot} - \overline{X}_{\cdots})^2/[\sigma^2(b-1)]}{\sum_{i=1}^{a}\sum_{j=1}^{b}\sum_{k=1}^{c}(X_{ijk} - \overline{X}_{ij\cdot})^2/[\sigma^2 ab(c-1)]} = \frac{\text{SS}(B)/(b-1)}{\text{SS}(E)/[ab(c-1)]},$$

which has an F distribution with $b-1$ and $ab(c-1)$ degrees of freedom when H_B is true. Each of these hypotheses is rejected if the observed value of F is greater than a given constant that is selected to yield the desired significance level.

In Table 8.10 we give the ANOVA table that summarizes the information needed for these tests of hypotheses.

TABLE 8.10   Two-way ANOVA table, c observation per cell

Source	Sum of Squares (SS)	Degrees of Freedom	Mean Square (MS)	F
Factor A (row)	SS(A)	$a-1$	$MS(A) = \dfrac{SS(A)}{a-1}$	$\dfrac{MS(A)}{MS(E)}$
Factor B (column)	SS(B)	$b-1$	$MS(B) = \dfrac{SS(B)}{b-1}$	$\dfrac{MS(B)}{MS(E)}$
Factor AB (interaction)	SS(AB)	$(a-1)(b-1)$	$MS(AB) = \dfrac{SS(AB)}{(a-1)(b-1)}$	$\dfrac{MS(AB)}{MS(E)}$
Error	SS(E)	$ab(c-1)$	$MS(E) = \dfrac{SS(E)}{ab(c-1)}$	
Total	SS(TO)	$abc-1$		

Remark   It should be noted that when the hypotheses are not true, each of F_{AB}, F_A, and F_B has a noncentral F distribution. Although they are not too common, tables of probabilities for the *noncentral F* distribution are available in extensive statistical libraries.

Example 8.2.2

Consider the following experiment. One hundred eight people were randomly divided into six groups with 18 people in each group. Each person was given sets of three numbers to add. The three numbers were either in a "down array" or an "across array," representing the two levels of factor A. The levels of factor B are determined by the number of digits in the numbers to be added: one-digit, two-digit, or three-digit numbers. Table 8.11 illustrates this with a sample problem for each

TABLE 8.11   Arrays for numbers of digits

Type of Array	Number of Digits		
	1	2	3
Down	5	25	259
	3	69	567
	8	37	130
Across	$5+3+8=$	$25+69+37=$	$259+567+130=$

cell; note, however, that an individual person only works problems of one of these types. Each person was placed in one of the six groups and was told to work as many problems as possible in 90 seconds. The measurement that was recorded was the average number of problems worked correctly in two trials.

Whenever this many subjects are used, a computer becomes an invaluable tool. A computer program provided the summary in Table 8.12 of the sample means of the rows, the columns, and the six cells. Each cell mean is the average for 18 people.

TABLE 8.12   Cell, row, and column means when adding numbers

Type of Array	Number of Digits			Row Mean
	1	2	3	
Down	23.806	10.694	6.278	13.593
Across	26.056	6.750	3.944	12.250
Column mean	24.931	8.722	5.111	

Simply considering these means, we can see clearly that there is a column effect. It is not surprising that it is easier to add one-digit than three-digit numbers.

The most interesting feature of these results is that they show the possibility of interaction. The largest cell mean occurs when adding one-digit numbers in an across array. Note, however, that for two- and three-digit numbers the down arrays have larger means than the across arrays.

The computer program provided the ANOVA table given in Table 8.13. The number of degrees of freedom for SS(E) is not in our F table. However, the right

TABLE 8.13   ANOVA table when adding numbers

Source	Sum of Squares	Degrees of Freedom	Mean Square	F	p-Value
Factor A (array)	48.67	1	48.669	2.885	0.0925
Factor B (number of digits)	8022.73	2	4011.363	237.778	< 0.0001
Interaction	185.92	2	92.961	5.510	0.0053
Error	1720.76	102	16.870		
Total	9978.08	107			

column, obtained from the computer printout, provides the p-value of each test, namely the probability of obtaining an F as large as or larger than the calculated F-ratio. Note, for example, that to test for interaction $F = 5.51$ and the p-value is 0.006. Thus the hypothesis of no interaction would be rejected at the $\alpha = 0.05$ or $\alpha = 0.01$ significance level but not with $\alpha = 0.001$.   ◇

EXERCISES

8.2.1 For the data given in Example 8.2.1, test the hypothesis $H_A: \alpha_1 = \alpha_2 = \alpha_3 = 0$ against all alternatives at the 5% significance level.

8.2.2 With $a = 3$ and $b = 4$, find μ, α_i, and β_j if μ_{ij}, $i = 1, 2, 3$ and $j = 1, 2, 3, 4$, are given by

$$
\begin{array}{cccc}
6 & 3 & 7 & 8 \\
10 & 7 & 11 & 12 \\
8 & 5 & 9 & 10
\end{array}
$$

Note that in an "additive" model such as this one, one row (column) can be determined by adding a constant value to each of the elements of another row (column).

8.2.3 We wish to compare compressive strengths of concrete corresponding to $a = 3$ different drying methods (treatments). Concrete is mixed in batches that are just large enough to produce three cylinders. Although care is taken to achieve uniformity, we expect some variability among the $b = 5$ batches used to obtain the following compressive strengths. (There is little reason to suspect interaction and hence only one observation is taken in each cell.)

Treatment	Batch				
	B_1	B_2	B_3	B_4	B_5
A_1	52	47	44	51	42
A_2	60	55	49	52	43
A_3	56	48	45	44	38

(a) Use the 5% significance level and test $H_A: \alpha_1 = \alpha_2 = \alpha_3 = 0$ against all alternatives.

(b) Use the 5% significance level and test $H_B: \beta_1 = \beta_2 = \beta_3 = \beta_4 = \beta_5 = 0$ against all alternatives. [See R. V. Hogg and J. Ledolter, *Applied Statistics for Engineers and Physical Scientists*, 2nd ed. (New York: Macmillan, 1992).]

8.2.4 Show that the cross-product terms formed from $(\overline{X}_{i\cdot} - \overline{X}_{\cdot\cdot})$, $(\overline{X}_{\cdot j} - \overline{X}_{\cdot\cdot})$, and $(X_{ij} - \overline{X}_{i\cdot} - \overline{X}_{\cdot j} + \overline{X}_{\cdot\cdot})$ sum to zero, $i = 1, 2, \ldots a$ and $j = 1, 2, \ldots, b$. HINT: For example, write

$$
\sum_{i=1}^{a} \sum_{j=1}^{b} (\overline{X}_{\cdot j} - \overline{X}_{\cdot\cdot})(X_{ij} - \overline{X}_{i\cdot} - \overline{X}_{\cdot j} + \overline{X}_{\cdot\cdot})
$$

$$
= \sum_{j=1}^{b} (\overline{X}_{\cdot j} - \overline{X}_{\cdot\cdot}) \sum_{i=1}^{a} [(X_{ij} - \overline{X}_{\cdot j}) - (\overline{X}_{i\cdot} - \overline{X}_{\cdot\cdot})]
$$

and sum each term in the inner summation as grouped here to get zero.

8.2.5 A psychology student was interested in testing how food consumption by rats would be affected by a particular drug. She used two levels of one attribute, namely drug and placebo, and four levels of a second attribute, namely male (M), castrated (C), female (F), and ovariectomized (O). For each cell she observed five rats. The amount of food consumed in grams per 24 hours is listed in Table 8.14. Test the hypotheses using a 5% significance level for each.

(a) H_{AB}: $\gamma_{ij} = 0, i = 1, 2, j = 1, 2, 3, 4$.

(b) H_A: $\alpha_1 = \alpha_2 = 0$.

(c) H_B: $\beta_1 = \beta_2 = \beta_3 = \beta_4 = 0$.

(d) How could you modify this model so that there are three attributes of classification, each with two levels?

TABLE 8.14   Food consumption by rats

	M	C	F	O
Drug:	22.56	16.54	18.58	18.20
	25.02	24.64	15.44	14.56
	23.66	24.62	16.12	15.54
	17.22	19.06	16.88	16.82
	22.58	20.12	17.58	14.56
Placebo:	25.64	22.50	17.82	19.74
	28.84	24.48	15.76	17.48
	26.00	25.52	12.96	16.46
	26.02	24.76	15.00	16.44
	23.24	20.62	19.54	15.70

8.2.6 With $a = 3$ and $b = 4$, find μ, α_i, β_j, and γ_{ij}, if μ_{ij}, $i = 1, 2, 3$ and $j = 1, 2, 3, 4$, are given by

$$
\begin{array}{cccc}
6 & 7 & 7 & 12 \\
10 & 3 & 11 & 8 \\
8 & 5 & 9 & 10
\end{array}
$$

Note the difference between the layout here and that in Exercise 8.2.2. Does the interaction help explain these differences?

8.2.7 In order to test whether four brands of gasoline give equal performance in terms of mileage, each of three cars was driven with each of the four brands of gasoline. Then each of the $(3)(4) = 12$ possible combinations was repeated four times. The number of miles per gallon for each of the four repetitions in each cell is recorded in Table 8.15. Test each of the hypotheses H_{AB}, H_A, H_B at the 5% significance level.

8.2.8 The data in Table 8.16 could represent the outcomes of some experiment, but in actuality they were generated on a computer. Recall that $\mu_{ij} = \mu + \alpha_i + \beta_j + \gamma_{ij}$ is the mean of cell (i, j) and in the simulation, we took $\mu = 15$, $\alpha_1 = -3$, $\alpha_2 = 2$, $\alpha_3 = 1$, $\beta_1 = -1.5$, $\beta_2 = 1.5$, $\gamma_{11} = 4$, $\gamma_{12} = -4$, $\gamma_{21} = -2$, $\gamma_{22} = 2$, $\gamma_{31} = -2$, $\gamma_{32} = 2$, and $\sigma^2 = 9$. Use the data and the two-factor analysis-of-variance model with four repetitions per cell to test the hypotheses

TABLE 8.15   Mileage performance

Car	Brand of Gasoline			
	1	2	3	4
1	21.0 14.9	16.3 20.0	15.8 19.4	17.8 17.3
	16.2 18.8	15.2 21.6	14.5 14.8	18.2 20.4
2	20.6 19.5	15.5 16.8	16.6 13.7	18.1 17.1
	20.8 18.9	17.4 19.4	18.2 16.1	21.5 19.1
3	14.2 13.1	17.4 18.1	15.2 16.7	16.3 16.4
	16.8 17.4	16.4 16.9	17.7 18.1	17.9 18.8

H_{AB}: no interaction, H_A: no row effect, and H_B: no column effect, each at the $\alpha = 0.05$ significance level.

TABLE 8.16   Data

14.552	13.980
17.024	5.777
10.132	7.638
16.979	12.089
11.378	20.825
12.578	17.333
17.709	20.329
11.037	25.277
12.345	22.223
11.044	21.844
9.458	16.176
16.156	17.946

8.2.9 Hogg and Ledolter report on an engineer in a textile mill who studies the effects of temperature and time in a process involving dye on the brightness of a synthetic fabric (brightness is measured on a 50-point scale). Three observations were taken at each combination of temperature and time.

Time (cycles)	Temperature		
	350°F	375°F	400°F
40	38, 32, 30	37, 35, 40	36, 39, 43
50	40, 45, 36	39, 42, 46	39, 48, 47

Construct the ANOVA table and conduct tests for the interaction first and then, if appropriate, the mean effects. Use $\alpha = 0.05$ for each test. [R. V. Hogg and J. Ledolter, *Applied Statistics for Engineers and Physical Scientists*, 2nd ed. (New York: Macmillan, 1992).]

8.2.10 There is another way of looking at Exercise 8.1.8, namely as a two-factor analysis-of-variance problem with the levels of gender being female and male, the levels of age being less than 50 and at least 50, and the measurement for each subject being their cholesterol level. The data would then be set up as follows:

Gender	Age < 50	Age ≥ 50
Female	221	262
	213	193
	202	224
	183	201
	185	161
	197	178
	162	265
Male	271	192
	192	253
	189	248
	209	278
	227	232
	236	267
	142	289

(a) Test H_{AB}: $\gamma_{ij} = 0, i = 1, 2; j = 1, 2$ (no interaction).
(b) Test H_A: $\alpha_1 = \alpha_2 = 0$ (no row effect).
(c) Test H_B: $\beta_1 = \beta_2 = 0$ (no column effect).
Use a 5% significance level for each test.

8.3 SIMPLE REGRESSION PROBLEM

There is often interest in the relation between two variables, for example, a student's scholastic aptitude test score in mathematics and this same student's grade in calculus. Frequently, one of these variables, say x, is known in advance of the other, and hence there is interest in predicting a future random variable Y. Since Y is a random variable, we cannot predict its future observed value $Y = y$ with certainty. Thus let us first concentrate on the problem of estimating the mean of Y, that is, $E(Y)$. Now $E(Y)$ is usually a function of x; for example, in our illustration with the calculus grade, say Y, we would expect $E(Y)$ to increase with increasing mathematics aptitude score x. Sometimes $E(Y) = \mu(x)$ is assumed to be of a given form, such as linear or quadratic or exponential; that is, $\mu(x)$ could be assumed to be equal to $\alpha + \beta x$ or $\alpha + \beta x + \gamma x^2$ or $\alpha e^{\beta x}$. To estimate $E(Y) = \mu(x)$, or equivalently the parameters α, β, and γ, we observe the random variable Y for each of n different values of x, say $x_1, x_2, \ldots, x_n$. Once the n independent experiments have been performed, we have n pairs of known numbers $(x_1, y_1), (x_2, y_2), \ldots, (x_n, y_n)$. These pairs are then used to estimate the mean $E(Y)$. Problems like this are often classified under **regression** because $E(Y) = \mu(x)$ is frequently called a regression curve.

Remark   A model for the mean like $\alpha + \beta x + \gamma x^2$, is called a **linear model** because it is linear in the parameters, α, β, and γ. Thus $\alpha e^{\beta x}$ is not a linear model because it is not linear in α and β. Note that in Sections 8.1 and 8.2, all the means were linear in the parameters and hence linear models. That is the reason for the title of this chapter.

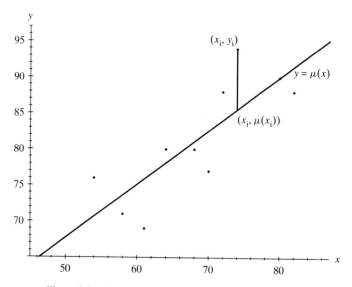

Figure 8.2   Scatter plot and least squares regression line

Let us begin with the case in which $E(Y) = \mu(x)$ is a linear function. The data points are $(x_1, y_1), (x_2, y_2), \ldots, (x_n, y_n)$; so the first problem is that of fitting a straight line to the set of data (see Figure 8.2). Recall that we considered a similar problem in Section 1.6. There we used the method of least squares to find the best-fitting line. In this chapter we will not only find the line of best fit, but will also make some statistical inferences about the unknown parameters.

In addition to assuming that the mean of Y is a linear function, we assume that for a particular value of x, the value of Y will differ from its mean by a random amount ε. We further assume that the distribution of ε is $N(0, \sigma^2)$. So we have for our linear model

$$Y_i = \alpha_1 + \beta x_i + \varepsilon_i,$$

where ε_i, for $i = 1, 2, \ldots, n$, are independent and $N(0, \sigma^2)$.

We shall now find point estimates for α_1, β, and σ^2. For convenience we let $\alpha_1 = \alpha - \beta \bar{x}$ so that

$$Y_i = \alpha + \beta(x_i - \bar{x}) + \varepsilon_i, \qquad \text{where} \quad \bar{x} = \frac{1}{n} \sum_{i=1}^{n} x_i.$$

Then Y_i is equal to a constant, $\alpha + \beta(x_i - \bar{x})$, plus a normal random variable ε_i. Hence $Y_1, Y_2, \ldots, Y_n$ are mutually independent normal variables with respective means $\alpha + \beta(x_i - \bar{x})$, $i = 1, 2, \ldots, n$, and unknown variance σ^2. Their joint p.d.f. is therefore the product of the individual probability density functions; that is, the likelihood function equals

$$L(\alpha, \beta, \sigma^2) = \prod_{i=1}^{n} \frac{1}{\sqrt{2\pi\sigma^2}} \exp\left\{ -\frac{[y_i - \alpha - \beta(x_i - \bar{x})]^2}{2\sigma^2} \right\}$$

$$= \left(\frac{1}{2\pi\sigma^2} \right)^{n/2} \exp\left\{ -\frac{\sum_{i=1}^{n} [y_i - \alpha - \beta(x_i - \bar{x})]^2}{2\sigma^2} \right\}.$$

To maximize $L(\alpha, \beta, \sigma^2)$, or, equivalently, to minimize

$$-\ln L(\alpha, \beta, \sigma^2) = \frac{n}{2} \ln(2\pi\sigma^2) + \frac{\sum_{i=1}^{n} [y_i - \alpha - \beta(x_i - \bar{x})]^2}{2\sigma^2},$$

we must select α and β to minimize

$$H(\alpha, \beta) = \sum_{i=1}^{n} [y_i - \alpha - \beta(x_i - \bar{x})]^2.$$

Since $|y_i - \alpha - \beta(x_i - \bar{x})| = |y_i - \mu(x_i)|$ is the vertical distance from the point (x_i, y_i) to the line $y = \mu(x)$, we note that $H(\alpha, \beta)$ represents the sum of the squares of those distances. Thus selecting α and β so that the sum of the squares is minimized means that we are fitting the straight line to the data by the **method of least squares**.

To minimize $H(\alpha, \beta)$, we find the two first partial derivatives

$$\frac{\partial H(\alpha, \beta)}{\partial \alpha} = 2 \sum_{i=1}^{n} [y_i - \alpha - \beta(x_i - \bar{x})](-1),$$

$$\frac{\partial H(\alpha, \beta)}{\partial \beta} = 2 \sum_{i=1}^{n} [y_i - \alpha - \beta(x_i - \bar{x})][-(x_i - \bar{x})].$$

Setting $\partial H(\alpha, \beta)/\partial \alpha = 0$, we obtain

$$\sum_{i=1}^{n} y_i - n\alpha - \beta \sum_{i=1}^{n} (x_i - \bar{x}) = 0.$$

Since

$$\sum_{i=1}^{n} (x_i - \bar{x}) = 0,$$

we have that

$$\sum_{i=1}^{n} y_i - n\alpha = 0$$

and thus

$$\hat{\alpha} = \bar{Y}.$$

The equation $\partial H(\alpha, \beta)/\partial \beta = 0$ yields, with α replaced by $\bar{y}$,

$$\sum_{i=1}^{n} (y_i - \bar{y})(x_i - \bar{x}) - \beta \sum_{i=1}^{n} (x_i - \bar{x})^2 = 0$$

or, equivalently,

$$\hat{\beta} = \frac{\displaystyle\sum_{i=1}^{n} (Y_i - \bar{Y})(x_i - \bar{x})}{\displaystyle\sum_{i=1}^{n} (x_i - \bar{x})^2} = \frac{\displaystyle\sum_{i=1}^{n} Y_i(x_i - \bar{x})}{\displaystyle\sum_{i=1}^{n} (x_i - \bar{x})^2}.$$

Thus, to find the mean line of best fit, $\mu(x) = \alpha + \beta(x_i - \bar{x})$, we use

$$\hat{\alpha} = \bar{y} \tag{8.1}$$

and

$$\hat{\beta} = \frac{\displaystyle\sum_{i=1}^{n} y_i(x_i - \bar{x})}{\displaystyle\sum_{i=1}^{n} (x_i - \bar{x})^2} = \frac{\displaystyle\sum_{i=1}^{n} x_i y_i - \left(\frac{1}{n}\right)\left(\displaystyle\sum_{i=1}^{n} x_i\right)\left(\displaystyle\sum_{i=1}^{n} y_i\right)}{\displaystyle\sum_{i=1}^{n} x_i^2 - \left(\frac{1}{n}\right)\left(\displaystyle\sum_{i=1}^{n} x_i\right)^2}. \tag{8.2}$$

It is also true that $\beta = r s_y / s_x$ (see Section 1.6).

To find the maximum likelihood estimator of σ^2, consider the partial derivative

$$\frac{\partial[-\ln L(\alpha, \beta, \sigma^2)]}{\partial(\sigma^2)} = \frac{n}{2\sigma^2} - \frac{\displaystyle\sum_{i=1}^{n} [y_i - \alpha - \beta(x_i - \bar{x})]^2}{2(\sigma^2)^2}.$$

Setting this equal to zero and replacing α and β by their solutions $\widehat{\alpha}$ and $\widehat{\beta}$, we obtain

$$\widehat{\sigma}^2 = \frac{1}{n}\sum_{i=1}^{n}[Y_i - \widehat{\alpha} - \widehat{\beta}(x_i - \overline{x})]^2. \tag{8.3}$$

A formula for $n\widehat{\sigma}^2$ useful in its calculation is

$$n\widehat{\sigma}^2 = \sum_{i=1}^{n} y_i^2 - \left(\frac{1}{n}\right)\left(\sum_{i=1}^{n} y_i\right)^2 - \widehat{\beta}\left(\sum_{i=1}^{n} x_i y_i\right) + \widehat{\beta}\left(\frac{1}{n}\right)\left(\sum_{i=1}^{n} x_i\right)\left(\sum_{i=1}^{n} y_i\right). \tag{8.4}$$

Note that the summand in Equation 8.3 for $\widehat{\sigma}^2$ is the square of the difference between the value of Y_i and the predicted mean of Y_i. Let $\widehat{Y}_i = \widehat{\alpha} + \widehat{\beta}(x_i - \overline{x})$, the predicted mean value of Y_i. The difference

$$Y_i - \widehat{Y}_i = Y_i - \widehat{\alpha} - \widehat{\beta}(x_i - \overline{x})$$

is called the ith **residual**, $i = 1, 2, \ldots, n$. The maximum likelihood estimate of σ^2 is then the sum of the squares of the residuals divided by n. It should always be true that the sum of the residuals is equal to zero. However, in practice, due to round off, the sum of the observed residuals, $y_i - \widehat{y}_i$, sometimes differs slightly from zero.

Example 8.3.1

The data plotted in Figure 8.2 are 10 pairs of test scores of 10 students in a psychology class, x being the score on a preliminary test and y the score on the final examination. The values of x and y are shown in Table 8.17. The sums that are needed to calculate estimates of the parameters are also given. Of course, the estimates of α and β have to be found before the residuals can be calculated. (The sum of the squares of the y values is included for use in a later section.)

TABLE 8.17 Calculations for test score data

x	y	x^2	xy	y^2	$\widehat{y}$	$y - \widehat{y}$	$(y - \widehat{y})^2$
70	77	4,900	5,390	5,929	82.561566	−5.561566	30.931016
74	94	5,476	6,956	8,836	85.529956	8.470044	71.741645
72	88	5,184	6,336	7,744	84.045761	3.954239	15.636006
68	80	4,624	5,440	6,400	81.077371	−1.077371	1.160728
58	71	3,364	4,118	5,041	73.656395	−2.656395	7.056434
54	76	2,916	4,104	5,776	70.688004	5.311996	28.217302
82	88	6,724	7,216	7,744	91.466737	−3.466737	12.018265
64	80	4,096	5,120	6,400	78.108980	1.891020	3.575957
80	90	6,400	7,200	8,100	89.982542	0.017458	0.000305
61	69	3,721	4,209	4,761	75.882687	−6.882687	47.371380
683	813	47,405	56,089	66,731		0.000001	217.709038

Thus $\widehat{\alpha} = 813/10 = 81.3$,

$$\widehat{\beta} = \frac{56{,}089 - (683)(813)/10}{47{,}405 - (683)(683)/10} = \frac{561.1}{756.1} = 0.742.$$

Since $\bar{x} = 683/10 = 68.3$, the least squares regression line is

$$\widehat{y} = 81.3 + (0.742)(x - 68.3).$$

The maximum likelihood estimate of σ^2 is

$$\widehat{\sigma^2} = \frac{217.709038}{10} = 21.7709. \qquad \diamond$$

We shall now consider the problem of finding the distributions of $\widehat{\alpha}$, $\widehat{\beta}$, and $\widehat{\sigma^2}$ (or distributions of functions of these estimators). We would like to be able to say something about the error of the estimates, find confidence intervals for the parameters, and in Section 8.4, test some hypotheses and predict future values of Y.

During the preceding discussion we have treated $x_1, x_2, \ldots, x_n$ as constants. Of course, many times they can be set by the experimenter; for example, a chemist in experimentation might produce a compound at many different temperatures. But it is also true that these numbers might be observations on an earlier random variable, such as an SAT score or preliminary test grade (as in Example 8.3.1), but we consider the problem on the condition that these x values are given in either case. Thus, in finding the distributions of $\widehat{\alpha}$, $\widehat{\beta}$, and $\widehat{\sigma^2}$, the only random variables are $Y_1, Y_2, \ldots, Y_n$. More will be said about the other situation in Section 11.5.

Since $\widehat{\alpha}$ is a linear function of independent and normally distributed random variables, $\widehat{\alpha}$ has a normal distribution with mean

$$E(\widehat{\alpha}) = E\left(\frac{1}{n}\sum_{i=1}^{n} Y_i\right) = \frac{1}{n}\sum_{i=1}^{n} E(Y_i)$$

$$= \frac{1}{n}\sum_{i=1}^{n} [\alpha + \beta(x_i - \bar{x})] = \alpha,$$

and variance

$$\text{Var}(\widehat{\alpha}) = \sum_{i=1}^{n} \left(\frac{1}{n}\right)^2 \text{Var}(Y_i) = \frac{\sigma^2}{n}.$$

The estimator $\widehat{\beta}$ is also a linear function of $Y_1, Y_2, \ldots, Y_n$ and hence has a normal distribution with mean

$$E(\widehat{\beta}) = \frac{\displaystyle\sum_{i=1}^{n} (x_i - \bar{x}) E(Y_i)}{\sum_{i=1}^{n} (x_i - \bar{x})^2}$$

$$= \frac{\displaystyle\sum_{i=1}^{n} (x_i - \bar{x})[\alpha + \beta(x_i - \bar{x})]}{\displaystyle\sum_{i=1}^{n} (x_i - \bar{x})^2}$$

$$= \frac{\alpha \displaystyle\sum_{i=1}^{n} (x_i - \bar{x}) + \beta \displaystyle\sum_{i=1}^{n} (x_i - \bar{x})^2}{\displaystyle\sum_{i=1}^{n} (x_i - \bar{x})^2} = \beta$$

and variance

$$\mathrm{Var}(\widehat{\beta}) = \sum_{i=1}^{n} \left[\frac{x_i - \bar{x}}{\sum_{i=1}^{n} (x_i - \bar{x})^2} \right]^2 \mathrm{Var}(Y_i)$$

$$= \frac{\displaystyle\sum_{i=1}^{n} (x_i - \bar{x})^2}{\left[\displaystyle\sum_{i=1}^{n} (x_i - \bar{x})^2 \right]^2} \sigma^2 = \frac{\sigma^2}{\displaystyle\sum_{i=1}^{n} (x_i - \bar{x})^2}.$$

It can be shown (Exercise 8.3.4) that

$$\sum_{i=1}^{n} [Y_i - \alpha - \beta(x_i - \bar{x})]^2 = \sum_{i=1}^{n} \{(\widehat{\alpha} - \alpha) + (\widehat{\beta} - \beta)(x_i - \bar{x})$$

$$+ [Y_i - \widehat{\alpha} - \widehat{\beta}(x_i - \bar{x})]\}^2$$

$$= n(\widehat{\alpha} - \alpha)^2 + (\widehat{\beta} - \beta)^2 \sum_{i=1}^{n} (x_i - \bar{x})^2$$

$$+ \sum_{i=1}^{n} [Y_i - \widehat{\alpha} - \widehat{\beta}(x_i - \bar{x})]^2. \qquad (8.5)$$

From the fact that Y_i, $\widehat{\alpha}$, and $\widehat{\beta}$ have normal distributions, we know that each of

$$\frac{[Y_i - \alpha - \beta(x_i - \bar{x})]^2}{\sigma^2}, \quad \frac{(\widehat{\alpha} - \alpha)^2}{\dfrac{\sigma^2}{n}}, \quad \frac{(\widehat{\beta} - \beta)^2}{\dfrac{\sigma^2}{\sum_{i=1}^{n}(x_i - \bar{x})^2}}$$

has a chi-square distribution with one degree of freedom. Since $Y_1, Y_2, \ldots, Y_n$ are mutually independent, then

$$\frac{\sum_{i=1}^{n}[Y_i - \alpha - \beta(x_i - \bar{x})]^2}{\sigma^2}$$

is $\chi^2(n)$. That is, the left-hand member of Equation 8.5 divided by σ^2 is $\chi^2(n)$ and is equal to the sum of two $\chi^2(1)$ variables and

$$\frac{\sum_{i=1}^{n}[Y_i - \widehat{\alpha} - \widehat{\beta}(x_i - \bar{x})]^2}{\sigma^2} = \frac{n\widehat{\sigma^2}}{\sigma^2} \geq 0.$$

Thus we might then guess that $n\widehat{\sigma^2}/\sigma^2$ is $\chi^2(n-2)$. This is true, and, moreover, $\widehat{\alpha}$, $\widehat{\beta}$, and $\widehat{\sigma^2}$ are mutually independent (see Theorem 8.1.1).

Suppose now that we are interested in forming a confidence interval for β, the slope of the line. We can use the fact that

$$T_1 = \frac{\sqrt{\sum_{i=1}^{n}(x_i - \bar{x})^2}\left(\dfrac{\widehat{\beta} - \beta}{\sigma}\right)}{\sqrt{\dfrac{n\widehat{\sigma^2}}{\sigma^2(n-2)}}} = \frac{\widehat{\beta} - \beta}{\sqrt{\dfrac{n\widehat{\sigma^2}}{(n-2)\sum_{i=1}^{n}(x_i - \bar{x})^2}}}$$

has a t distribution with $n - 2$ degrees of freedom. Thus

$$P\left[-t_{\alpha/2}(n-2) \leq \frac{\widehat{\beta} - \beta}{\sqrt{\dfrac{n\widehat{\sigma^2}}{(n-2)\sum_{i=1}^{n}(x_i - \bar{x})^2}}} \leq t_{\alpha/2}(n-2)\right] = 1 - \alpha.$$

It follows that

$$\left[\widehat{\beta} - t_{\alpha/2}(n-2)\sqrt{\frac{n\widehat{\sigma}^2}{(n-2)\sum_{i=1}^{n}(x_i - \bar{x})^2}} \, , \right.$$

$$\left. \widehat{\beta} + t_{\alpha/2}(n-2)\sqrt{\frac{n\widehat{\sigma}^2}{(n-2)\sum_{i=1}^{n}(x_i - \bar{x})^2}} \, \right]$$

is a $100(1-\alpha)\%$ confidence interval for β.

Similarly,

$$T_2 = \frac{\dfrac{\sqrt{n}(\widehat{\alpha} - \alpha)}{\sigma}}{\sqrt{\dfrac{n\widehat{\sigma}^2}{\sigma^2(n-2)}}}$$

$$= \frac{\widehat{\alpha} - \alpha}{\sqrt{\dfrac{\widehat{\sigma}^2}{n-2}}}$$

has a t distribution with $n-2$ degrees of freedom. Thus T_2 can be used to make inferences about α (see Exercise 8.3.5). The fact that $n\widehat{\sigma}^2/\sigma^2$ has a chi-square distribution with $n-2$ degrees of freedom can be used to make inferences about the variance σ^2 (see Exercise 8.3.6).

EXERCISES

8.3.1 The midterm and final exam scores of 10 students in a statistics course are tabulated as shown.

(a) Calculate the least squares regression line for these data.

(b) Plot the points and the least squares regression line on the same graph.

(c) Find the value of $\widehat{\sigma}^2$.

Midterm	Final	Midterm	Final
70	87	67	73
74	79	70	83
80	88	64	79
84	98	74	91
80	96	82	94

8.3.2 The final course grade in calculus was predicted on the basis of the student's high school grade-point average in mathematics, SAT score in mathematics, and score on a mathematics entrance examination. The predicted grades X and the earned grades Y for 10 students are given (2.0 represents a C, 2.3 a C+, 2.7 a B-, etc.).

(a) Calculate the least squares regression line for these data.

(b) Plot the points and the least squares regression line on the same graph.

(c) Find the value of $\widehat{\sigma}^2$.

x	y	x	y
2.0	1.3	2.7	3.0
3.3	3.3	4.0	4.0
3.7	3.3	3.7	3.0
2.0	2.0	3.0	2.7
2.3	1.7	2.3	3.0

8.3.3 Students' scores on the mathematics portion of the ACT examination, X, and on the final examination in first semester calculus (200 points possible), Y, are given.

(a) Calculate the least squares regression line for these data.

(b) Plot the points and the least squares regression line on the same graph.

x	y	x	y
25	138	20	100
20	84	25	143
26	104	26	141
26	112	28	161
28	88	25	124
28	132	31	118
29	90	30	168
32	183		

8.3.4 Show that

$$\sum_{i=1}^{n} [Y_i - \alpha - \beta(x_i - \bar{x})]^2 = n(\widehat{\alpha} - \alpha)^2 + (\widehat{\beta} - \beta)^2 \sum_{i=1}^{n} (x_i - \bar{x})^2$$

$$+ \sum_{i=1}^{n} [Y_i - \widehat{\alpha} - \widehat{\beta}(x_i - \bar{x})]^2.$$

Remark   In Exercises 8.3.5 to 8.3.12, to make the required statistical inferences, we need the distributional assumptions made in Section 8.3.

8.3.5 Show that the endpoints for a $100(1 - \alpha)\%$ confidence interval for α are

$$\widehat{\alpha} \pm t_{\alpha/2}(n - 2)\sqrt{\frac{\widehat{\sigma^2}}{n - 2}}.$$

8.3.6 Show that a $100(1 - \alpha)\%$ confidence interval for σ^2 is

$$\left[\frac{n\widehat{\sigma^2}}{\chi^2_{\alpha/2}(n - 2)}, \frac{n\widehat{\sigma^2}}{\chi^2_{1-\alpha/2}(n - 2)} \right].$$

8.3.7 Find 95% confidence intervals for α, β, and σ^2 for the data in Exercise 8.3.1.

8.3.8 Find 95% confidence intervals for α, β, and σ^2 for the data in Exercise 8.3.2.

8.3.9 In a "48.1-gram" package of candies, let X equal the number of pieces of candy and let Y equal the total weight of the candies. For each of 5 different values of X, 4 observations of Y are given.

x	y	x	y
54	48.8	56	50.8
54	49.4	56	50.9
54	49.2	57	50.7
54	50.4	57	51.6
55	49.5	57	51.3
55	49.0	57	50.8
55	50.2	58	52.1
55	48.9	58	51.3
56	49.9	58	51.4
56	50.1	58	52.0

(a) Calculate the least squares regression line for these data.

(b) Plot the points and the least squares regression line on the same graph.

(c) Find the value of $\widehat{\sigma^2}$,

(d) Find a 95% confidence interval for β under the usual assumptions.

8.3.10 The Federal Trade Commission measured the number of milligrams of tar and carbon monoxide (CO) per cigarette for all domestic cigarettes. Let x and y equal the measurements of tar and CO, respectively, for 100-millimeter filtered and mentholated cigarettes. A sample of 12 brands yielded the following data:

Brand	x	y
Capri	9	6
Carlton	4	6
Kent	14	14
Kool Milds	12	12
Marlboro Lights	10	12
Merit Ultras	5	7
Now	3	4
Salem	17	18
Triumph	6	8
True	7	8
Vantage	8	13
Virginia Slims	15	13

(a) Calculate the least squares regression line for these data.

(b) Plot the points and the least squares regression line on the same graph.

(c) Find point estimates for α, β, and σ^2.

(d) Find 95% confidence intervals for α, β, and σ^2 under the usual assumptions.

8.3.11 Let X and Y equal the lengths in inches of a foot and a hand, respectively. The following measurements were made on 15 women.

x	y	x	y	x	y
9.00	6.50	10.00	7.00	9.25	7.00
8.50	6.25	9.50	6.50	10.00	7.50
9.25	7.25	9.00	7.00	10.00	7.25
9.75	7.00	9.25	7.00	9.75	7.25
9.00	6.75	9.50	7.00	9.50	7.25

(a) Calculate the least squares regression line for these data.

(b) Plot the points and the least squares regression line on the same graph.

(c) Find point estimates for α, β, and σ^2.

(d) Find 95% confidence intervals for α, β, and σ^2 under the usual assumptions.

8.3.12 Let x and y equal the ACT scores in social science and natural science for a student who is applying for admission to a small liberal arts college. A sample of $n = 15$ such students yielded the following data:

x	y	x	y	x	y
32	28	30	27	26	32
23	25	17	23	16	22
23	24	20	30	21	28
23	32	17	18	24	31
26	31	18	18	30	26

(a) Calculate the least squares regression line for these data.

(b) Plot the points and the least squares regression line on the same graph.

(c) Find point estimates for α, β, and σ^2.

(d) Find 95% confidence intervals for α, β, and σ^2 under the usual assumptions.

8.4 MORE REGRESSION

In Section 8.3 we considered the estimation of the parameters of a very simple regression curve, namely a straight line. In addition to the point and interval estimates, we can perform tests of hypotheses about these parameters. For illustration, with the same model as that in Section 8.3, we could test the hypothesis $H_0: \beta = \beta_0$ by using a t random variable, namely

$$T_1 = \frac{\widehat{\beta} - \beta_0}{\sqrt{\dfrac{n\widehat{\sigma^2}}{(n-2)\displaystyle\sum_{i=1}^{n}(x_i - \overline{x})^2}}}.$$

The null hypothesis along with three possible alternative hypotheses are given in Table 8.18.

TABLE 8.18   Tests about the slope of the regression line

H_0	H_1	Critical Region
$\beta = \beta_0$	$\beta > \beta_0$	$t_1 \geq t_\alpha(n-2)$
$\beta = \beta_0$	$\beta < \beta_0$	$t_1 \leq -t_\alpha(n-2)$
$\beta = \beta_0$	$\beta \neq \beta_0$	$|t_1| \geq t_{\alpha/2}(n-2)$

Often we let $\beta_0 = 0$ and test the hypothesis $H_0: \beta = 0$. That is, we test the null hypothesis that the slope is equal to zero.

Example 8.4.1

Let x equal a student's preliminary test score in a psychology course and y the same student's score on the final examination. With $n = 10$ students, we shall test H_0: $\beta = 0$ against H_1: $\beta \neq 0$. At an $\alpha = 0.01$ significance level, the critical region is $|t_1| \geq t_{0.005}(8) = 3.355$. Using the data in Example 8.3.1, the observed value of T_1 is

$$t_1 = \frac{0.742 - 0}{\sqrt{10(21.7709)/8(756.1)}} = \frac{0.742}{0.1897} = 3.911.$$

Thus we reject H_0. $\diamond$

There is another way of looking at tests of H_0: $\beta = 0$ that gives additional insight into determining how much of the variation in the data is explained by the linear model. This approach is similar to that used in the study of analysis of variance.

We take the total sum of squares and partition it into the variation due to the linear model and an error sum of squares. Within the total sum of squares, SS(TO), we add and subtract the same quantity, recalling that $\widehat{\alpha} = \overline{Y}$. We have

$$\begin{aligned}
\text{SS(TO)} &= \sum_{i=1}^{n} (Y_i - \overline{Y})^2 \\
&= \sum_{i=1}^{n} [\widehat{\beta}(x_i - \overline{x}) + Y_i - \widehat{\alpha} - \widehat{\beta}(x_i - \overline{x})]^2 \\
&= \sum_{i=1}^{n} \widehat{\beta}^2(x_i - \overline{x})^2 + \sum_{i=1}^{n} [Y_i - \widehat{\alpha} - \widehat{\beta}(x_i - \overline{x})]^2 \\
&= \text{SS(R)} + \text{SS(E)}.
\end{aligned} \tag{8.6}$$

The cross-product term is equal to zero (see Exercise 8.4.6). When $\beta = 0$, $Y_i = \alpha + \varepsilon_i$; SS(TO)/$(n-1)$ is an unbiased estimator of σ^2; and SS(TO)/σ^2 is $\chi^2(n-1)$. Also,

$$\begin{aligned}
\frac{\text{SS(R)}}{\sigma^2} &= \frac{\widehat{\beta}^2 \sum_{i=1}^{n} (x_i - \overline{x})^2}{\sigma^2} \\
&= \left[\frac{\widehat{\beta} - 0}{\dfrac{\sigma}{\sqrt{\sum_{i=1}^{n} (x_i - \overline{x})^2}}} \right]^2.
\end{aligned}$$

We have noted near the end of Section 8.3 that $\widehat{\beta}$ is $N[\beta, \sigma^2/\sum_{i=1}^{n}(x_i - \overline{x})^2]$. Since $\beta = 0$, we see that SS(R)/σ^2 is the square of a standard normal random variable and thus is $\chi^2(1)$. Furthermore, SS(E)≥ 0, so by Theorem 8.1.1 we see that the distribution of $SS(E)/\sigma^2$ is $\chi^2(n-2)$ and that $SS(R)/\sigma^2$ and $SS(E)/\sigma^2$ are independent random variables. Also, if $\beta = 0$,

$$E[MS(R)] = E\left[\frac{SS(R)}{1}\right]$$

$$= E\left[\{\sigma^2\}\left\{\frac{SS(R)}{\sigma^2}\right\}\right] = \sigma^2;$$

$$E[MS(E)] = E\left[\frac{SS(E)}{n-2}\right]$$

$$= E\left[\left\{\frac{\sigma^2}{n-2}\right\}\left\{\frac{SS(E)}{\sigma^2}\right\}\right]$$

$$= \sigma^2.$$

Thus both $MS(R)$ and $MS(E)$ provide unbiased estimates of σ^2 when $\beta = 0$. $MS(R)$ and $MS(E)$ are called the **mean square due to regression** and the **mean square due to error**.

Because $MS(R)$ and $MS(E)$ are independent chi-square random variables,

$$F = \frac{\dfrac{SS(R)/\sigma^2}{1}}{\dfrac{SS(E)/\sigma^2}{n-2}} = \frac{MS(R)}{MS(E)}$$

has an F distribution with 1 and $n-2$ degrees of freedom under $H_0 : \beta = 0$.

If $\beta \neq 0$, it is still true that $MS(E) = SS(E)/(n-2)$ is an unbiased estimator of σ^2, However, $MS(R)$ will tend to overestimate σ^2. Thus to test $H_0 : \beta = 0$ against $H_1 : \beta \neq 0$, we can define the critical region by

$$F = \frac{MS(R)}{MS(E)} \geq F_\alpha(1, n-2).$$

(Note that $T_1^2 = F$ with $\beta_0 = 0$ and thus T_1^2 has an F distribution with 1 and $n-2$ degrees of freedom under H_0. See Exercise 11.6.5.)

We can summarize this test using an ANOVA table like that given in Table 8.19.

TABLE 8.19 ANOVA table for testing $H_0 : \beta = 0$

Source	SS	d.f.	MS	F
Regression	SS(R)	1	MS(R) = SS(R)/1	$\dfrac{MS(R)}{MS(E)}$
Error	SS(E)	$n-2$	MS(E) = SS(E)/(n-2)	
Total	SS(TO)	$n-1$		

Example 8.4.2

We repeat Example 8.3.1 by constructing an ANOVA table. From that example,

$$SS(R) = \widehat{\beta}^2 \sum_{i=1}^{n} (x_i - \overline{x})^2 = \left(\frac{561.1}{756.1}\right)^2 (756.1) = 416.391,$$

$$SS(E) = \sum_{i=1}^{n} (y_i - \widehat{y}_i)^2 = 217.709,$$

$$SS(TO) = \sum_{i=1}^{n} y_i^2 - \left(\frac{1}{10}\right)\left(\sum_{i=1}^{n} y_i\right)^2 = 66731 - 66096.9 = 634.100.$$

These are summarized in an ANOVA table:

Source	SS	d.f.	MS	F
Regression	416.391	1	416.3910	15.3008
Error	217.709	8	27.2136	
Total	634.100	9		

Since $F = 15.3008 > 11.26 = F_{0.01}(1, 8)$, the null hypothesis H_0: $\beta = 0$ is rejected at the $\alpha = 0.01$ significance level. Note that $t_1^2 = (3.9111)^2 = 15.2967 \approx 15.3006 = F$, the difference being due to round-off error.   ◇

We have noted that $\widehat{Y} = \widehat{\alpha} + \widehat{\beta}(x - \overline{x})$ is a point estimate for the mean of Y for some given x, or we could think of this as a prediction of the value of Y for this given x. But how close is $\widehat{Y}$ to the mean of Y or to Y itself? We shall now find a confidence interval for $\alpha + \beta(x - \overline{x})$ and a prediction interval for Y, given a particular value of x.

To find a confidence interval for

$$E(Y) = \mu(x) = \alpha + \beta(x - \overline{x}),$$

let

$$\widehat{Y} = \widehat{\alpha} + \widehat{\beta}(x - \overline{x}).$$

Recall that $\widehat{Y}$ is a linear combination of normally and independently distributed random variables so that $\widehat{Y}$ has a normal distribution. Furthermore

$$E(\widehat{Y}) = E[\widehat{\alpha} + \widehat{\beta}(x - \overline{x})]$$
$$= \alpha + \beta(x - \overline{x})$$

and

$$\text{Var}(\widehat{Y}) = \text{Var}[\widehat{\alpha} + \widehat{\beta}(x - \overline{x})]$$

$$= \frac{\sigma^2}{n} + \frac{\sigma^2}{\displaystyle\sum_{i=1}^{n}(x_i - \overline{x})^2}(x - \overline{x})^2$$

$$= \sigma^2 \left[\frac{1}{n} + \frac{(x - \overline{x})^2}{\displaystyle\sum_{i=1}^{n}(x_i - \overline{x})^2} \right].$$

Recall that the distribution of $n\widehat{\sigma}^2/\sigma^2$ is $\chi^2(n-2)$. Since $\widehat{\alpha}$ and $\widehat{\beta}$ are independent of $\widehat{\sigma^2}$, we can form the t-statistic

$$T = \frac{\dfrac{\widehat{\alpha} + \widehat{\beta}(x - \overline{x}) - [\alpha + \beta(x - \overline{x})]}{\sigma\sqrt{\dfrac{1}{n} + \dfrac{(x - \overline{x})^2}{\sum_{i=1}^{n}(x_i - \overline{x})^2}}}}{\sqrt{\dfrac{n\widehat{\sigma}^2}{(n-2)\sigma^2}}}$$

which has a t-distribution with $r = n - 2$ degrees of freedom. Select $t_{\alpha/2}(n-2)$ from Table VI so that

$$P[-t_{\alpha/2}(n-2) \leq T \leq t_{\alpha/2}(n-2)] = 1 - \alpha.$$

This becomes

$$P[\widehat{\alpha} + \widehat{\beta}(x - \overline{x}) - ct_{\alpha/2}(n-2) \leq \alpha + \beta(x - \overline{x}) \leq$$
$$\widehat{\alpha} + \widehat{\beta}(x - \overline{x}) + ct_{\alpha/2}(n-2)] = 1 - \alpha,$$

where

$$c = \sqrt{\frac{n\widehat{\sigma}^2}{n-2}}\sqrt{\frac{1}{n} + \frac{(x - \overline{x})^2}{\sum_{i=1}^{n}(x_i - \overline{x})^2}}.$$

Thus the endpoints for a $100(1 - \alpha)\%$ confidence interval for $\mu(x) = \alpha + \beta(x - \overline{x})$ are

$$\widehat{\alpha} + \widehat{\beta}(x - \overline{x}) \pm ct_{\alpha/2}(n-2).$$

Note that the width of this interval depends on the particular value of x because c depends on x (see Example 8.4.3).

We have used $(x_1, y_1), (x_2, y_2), \ldots, (x_n, y_n)$ to estimate α and β. Suppose that we are given a value of x, say x_{n+1}. A point estimate of the corresponding value of Y is

$$\widehat{y}_{n+1} = \widehat{\alpha} + \widehat{\beta}(x_{n+1} - \overline{x}).$$

However, $\widehat{y}_{n+1}$ is just one possible value of the random variable

$$Y_{n+1} = \alpha + \beta(x_{n+1} - \overline{x}) + \varepsilon_{n+1},$$

What can we say about possible values for Y_{n+1}? We shall now obtain a prediction interval for Y_{n+1} when $x = x_{n+1}$ that is similar to the confidence interval for the mean of Y when $x = x_{n+1}$.

We have that

$$Y_{n+1} = \alpha + \beta(x_{n+1} - \overline{x}) + \varepsilon_{n+1},$$

where ε_{n+1} is $N(0, \sigma^2)$ and $\overline{x} = (1/n)\sum_{i=1}^{n} x_i$. Now

$$W = Y_{n+1} - \widehat{\alpha} - \widehat{\beta}(x_{n+1} - \overline{x})$$

is a linear combination of normally and independently distributed random variables, so W has a normal distribution. The mean of W is

$$E(W) = E[Y_{n+1} - \widehat{\alpha} - \widehat{\beta}(x_{n+1} - \overline{x})]$$
$$= \alpha + \beta(x_{n+1} - \overline{x}) - \alpha - \beta(x_{n+1} - \overline{x}) = 0.$$

Since $Y_{n+1}, \widehat{\alpha}$ and $\widehat{\beta}$ are independent, the variance of W is

$$\text{Var}(W) = \sigma^2 + \frac{\sigma^2}{n} + \frac{\sigma^2}{\sum_{i=1}^{n}(x_i - \overline{x})^2}(x_{n+1} - \overline{x})^2$$

$$= \sigma^2\left[1 + \frac{1}{n} + \frac{(x_{n+1} - \overline{x})^2}{\sum_{i=1}^{n}(x_i - \overline{x})^2}\right].$$

Recall that $n\widehat{\sigma^2}/[(n-2)\sigma^2]$ is $\chi^2(n-2)$. Since $Y_{n+1}, \widehat{\alpha}$, and $\widehat{\beta}$ are independent of $\widehat{\sigma^2}$, we can form the t-statistic

$$T = \frac{Y_{n+1} - \widehat{\alpha} - \widehat{\beta}(x_{n+1} - \overline{x})}{\sigma\sqrt{1 + \frac{1}{n} + \frac{(x_{n+1} - \overline{x})^2}{\sum_{i=1}^{n}(x_i - \overline{x})^2}} \Big/ \sqrt{\frac{n\widehat{\sigma^2}}{(n-2)\sigma^2}}}$$

which has a t-distribution with $r = n - 2$ degrees of freedom. Select a constant $t_{\alpha/2}(n-2)$ from Table VI so that

$$P[-t_{\alpha/2}(n-2) \leq T \leq t_{\alpha/2}(n-2)] = 1 - \alpha.$$

Solving this inequality for Y_{n+1}, we have

$$P[\widehat{\alpha} + \widehat{\beta}(x_{n+1} - \overline{x}) - d\, t_{\alpha/2}(n-2) \leq Y_{n+1} \leq$$
$$\widehat{\alpha} + \widehat{\beta}(x_{n+1} - \overline{x}) + dt_{\alpha/2}(n-2)] = 1 - \alpha,$$

where

$$d = \sqrt{\frac{n\widehat{\sigma^2}}{n-2}} \sqrt{1 + \frac{1}{n} + \frac{(x_{n+1} - \overline{x})^2}{\sum_{i=1}^{n}(x_i - \overline{x})^2}}.$$

Thus the endpoints for a $100(1-\alpha)\%$ prediction interval for Y_{n+1} are

$$\widehat{\alpha} + \widehat{\beta}(x_{n+1} - \overline{x}) \pm dt_{\alpha/2}(n-2).$$

We shall now illustrate a 95% confidence interval for $\mu(x)$ and a 95% prediction interval for Y for a given value of x using the data in Example 8.3.1. To find such intervals, the calculating formulas that are given in Equations 8.1, 8.2, and 8.4 can be used.

Example 8.4.3

To find a 95% confidence interval for $\mu(x)$ using the data in Example 8.3.1, note that we have already found that $\overline{x} = 68.3$, $\widehat{\alpha} = 81.3$, $\widehat{\beta} = 561.1/756.1 = 0.7421$, and $\widehat{\sigma^2} = 21.7709$. We also need

$$\sum_{i=1}^{n}(x_i - \overline{x})^2 = \sum_{i=1}^{n} x_i^2 - \left(\frac{1}{n}\right)\left(\sum_{i=1}^{n} x_i\right)^2$$

$$= 47,405 - \frac{683^2}{10} = 756.1.$$

For 95% confidence, $t_{0.025}(8) = 2.306$. When $x = 60$, the endpoints for a 95% confidence interval for $\mu(60)$ are

$$81.3 + 0.7421(60 - 68.3) \pm \left[\sqrt{\frac{10(21.7709)}{8}}\sqrt{\frac{1}{10} + \frac{(60 - 68.3)^2}{756.1}}\right](2.306)$$

or

$$75.1406 \pm 5.2589.$$

Similarly, when $x = 70$ the endpoints for a 95% confidence interval for $\mu(70)$ are

$$82.5616 \pm 3.8761.$$

Note that the lengths of these intervals depend on the particular value of x. A 95% confidence band for $\mu(x)$ is graphed in Figure 8.3 along with the scatter diagram and $\widehat{y} = \widehat{\alpha} + \widehat{\beta}(x - \overline{x})$.

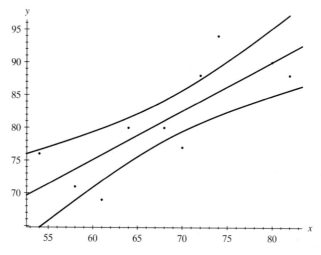

Figure 8.3   A 95% confidence interval for $\mu(x)$

The endpoints for a 95% prediction interval for Y when $x = 60$ are

$$81.3 + 0.7421(60 - 68.3) \pm \left[\sqrt{\frac{10(21.7709)}{8}} \sqrt{1.1 + \frac{(60 - 68.3)^2}{756.1}} \right](2.306)$$

or

$$75.1406 \pm 13.1289.$$

Note that this interval is much wider than that for $\mu(60)$. In Figure 8.4 the 95% prediction band for Y is graphed along with the scatter diagram and the least squares regression line. ◇

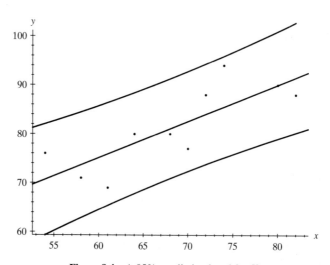

Figure 8.4   A 95% prediction band for Y

We now generalize the simple regression model to the **multiple regression** case. Say we observe several x-values, say $x_1, x_2, \ldots, x_k$, along with the y-value. For example, suppose that x_1 equals the student's ACT composite score, x_2 equals the student's high school class rank, and y equals the student's first-year GPA in college. We want to estimate a regression function $E(Y) = \mu(x_1, x_2, \ldots, x_k)$ from some observed data. If

$$\mu(x_1, x_2, \ldots, x_k) = \beta_1 x_1 + \beta_2 x_2 + \cdots + \beta_k x_k,$$

then we say that we have a **linear model** because this expression is linear in the coefficients $\beta_1, \beta_2, \ldots, \beta_k$.

Note, for illustration, that the model in Section 8.3 is linear in $\alpha = \beta_1$ and $\beta = \beta_2$, with $x_1 = 1$ and $x_2 = x$, giving the mean $\alpha + \beta x$. (For convenience, there the mean of the x-values was subtracted from x.) Suppose, however, that we had wished to use the cubic function $\beta_1 + \beta_2 x + \beta_3 x^2 + \beta_4 x^3$ as the mean. This cubic expression still provides a linear model (i.e., linear in the β-values), and we would take $x_1 = 1$, $x_2 = x$, $x_3 = x^2$, and $x_4 = x^3$.

Say our n observation points are

$$(x_{1j}, x_{2j}, \ldots, x_{kj}, y_j), \qquad j = 1, 2, \ldots, n.$$

To fit the linear model $\beta_1 x_1 + \beta_2 x_2 + \cdots + \beta_k x_k$ by the method of least squares, we minimize

$$G = \sum_{j=1}^{n} (y_j - \beta_1 x_{1j} - \beta_2 x_{2j} - \cdots - \beta_k x_{kj})^2.$$

If we equate to zero the k first partial derivatives

$$\frac{\partial G}{\partial \beta_i} = \sum_{j=1}^{n} (-2)(y_j - \beta_1 x_{1j} - \beta_2 x_{2j} - \cdots - \beta_k x_{kj})(x_{ij}), \qquad i = 1, 2, \ldots, k,$$

we obtain the k **normal equations**

$$\beta_1 \sum_{j=1}^{n} x_{1j}^2 + \beta_2 \sum_{j=1}^{n} x_{1j} x_{2j} + \cdots + \beta_k \sum_{j=1}^{n} x_{1j} x_{kj} = \sum_{j=1}^{n} x_{1j} y_j,$$

$$\beta_1 \sum_{j=1}^{n} x_{2j} x_{1j} + \beta_2 \sum_{j=1}^{n} x_{2j}^2 + \cdots + \beta_k \sum_{j=1}^{n} x_{2j} x_{kj} = \sum_{j=1}^{n} x_{2j} y_j,$$

$$\vdots \qquad\qquad \vdots \qquad\qquad \ddots \qquad\qquad \vdots \qquad\qquad \vdots$$

$$\beta_1 \sum_{j=1}^{n} x_{kj} x_{1j} + \beta_2 \sum_{j=1}^{n} x_{kj} x_{2j} + \cdots + \beta_k \sum_{j=1}^{n} x_{kj}^2 = \sum_{j=1}^{n} x_{kj} y_j.$$

The solution of these k equations provides the least squares estimates of $\beta_1, \beta_2, \ldots, \beta_k$. These estimates are also maximum likelihood estimates of $\beta_1, \beta_2, \ldots, \beta_k$, provided that the random variables $Y_1, Y_2, \ldots, Y_n$ are mutually independent and Y_j is $N(\beta_1 x_{1j} + \beta_2 x_{2j} + \cdots + \beta_k x_{kj}, \sigma^2)$, $j = 1, 2, \ldots, k$.

By the method of least squares, fit $y = \beta_1 x_1 + \beta_2 x_2 + \beta_3 x_3$ to the five observed points (x_1, x_2, x_3, y):

$$(1, 1, 0, 4), (1, 0, 1, 3), (1, 2, 3, 2), (1, 3, 0, 6), (1, 0, 0, 1).$$

Note that $x_1 = 1$ in each point; so we are really fitting $y = \beta_1 + \beta_2 x_2 + \beta_3 x_3$. Since

$$\sum_{j=1}^{5} x_{1j}^2 = 5, \quad \sum_{j=1}^{5} x_{1j} x_{2j} = 6, \quad \sum_{j=1}^{5} x_{1j} x_{3j} = 4, \quad \sum_{j=1}^{5} x_{1j} y_j = 16,$$

$$\sum_{j=1}^{5} x_{2j} x_{1j} = 6, \quad \sum_{j=1}^{5} x_{2j}^2 = 14, \quad \sum_{j=1}^{5} x_{2j} x_{3j} = 6, \quad \sum_{j=1}^{5} x_{2j} y_j = 26,$$

$$\sum_{j=1}^{5} x_{3j} x_{1j} = 4, \quad \sum_{j=1}^{5} x_{3j} x_{2j} = 6, \quad \sum_{j=1}^{5} x_{3j}^2 = 10, \quad \sum_{j=1}^{5} x_{3j} y_j = 9,$$

the normal equations are

$$5\beta_1 + 6\beta_2 + 4\beta_3 = 16,$$
$$6\beta_1 + 14\beta_2 + 6\beta_3 = 26,$$
$$4\beta_1 + 6\beta_2 + 10\beta_3 = 9.$$

Solving these three linear equations in three unknowns, we obtain

$$\widehat{\beta}_1 = \frac{274}{112}, \quad \widehat{\beta}_2 = \frac{127}{112}, \quad \widehat{\beta}_3 = -\frac{85}{112}.$$

Thus the least squares fit is

$$y = \frac{274x_1 + 127x_2 - 85x_3}{112}.$$

If x_1 always equals 1, then the equation reads

$$y = \frac{274 + 127x_2 - 85x_3}{112}.$$

It is interesting to observe that the usual two-sample problem is actually a linear model. Let $\beta_1 = \mu_1$ and $\beta_2 = \mu_2$ and consider n pairs of (x_1, x_2) that equal $(1, 0)$ and m pairs that equal $(0, 1)$. This would require each of the first n Y-variables, namely $Y_1, Y_2, \ldots, Y_n$, to have the mean

$$\beta_1 \cdot 1 + \beta_2 \cdot 0 = \beta_1 = \mu_1$$

and the next m Y variables, namely $Y_{n+1}, Y_{n+2}, \ldots, Y_{n+m}$, to have the mean

$$\beta_1 \cdot 0 + \beta_2 \cdot 1 = \beta_2 = \mu_2.$$

This is the background of the two-sample problem with the usual $X_1, X_2, \ldots, X_n$ and $Y_1, Y_2, \ldots, Y_m$ replaced by $Y_1, Y_2, \ldots, Y_n$ and $Y_{n+1}, Y_{n+2}, \ldots, Y_{n+m}$.

Clearly, this two-sample problem can be extended to three or more samples. For illustration, let $\beta_1 = \mu_1, \beta_2 = \mu_2$, and $\beta_3 = \mu_3$ and have n_1 triples of (x_1, x_2, x_3) equal to $(1, 0, 0)$, n_2 triples equal to $(0, 1, 0)$, and n_3 triples equal to $(0, 0, 1)$.

EXERCISES

8.4.1 For the data given in Exercise 8.3.1,

  (a) Test $H_0: \beta = 0$ against $H_1: \beta > 0$ at the $\alpha = 0.025$ significance level using a t-test.

  (b) Test $H_0: \beta = 0$ against $H_1: \beta \neq 0$ at the $\alpha = 0.05$ significance level by setting up the ANOVA table and using an F statistic.

  (c) Find a 95% confidence interval for $\mu(x)$ when $x = 68, 75$, and 82.

  (d) Find a 95% prediction interval for Y when $x = 68, 75$, and 82.

8.4.2 For the data given in Exercise 8.3.2,

  (a) Test $H_0: \beta = 0$ against $H_1: \beta > 0$ at the $\alpha = 0.025$ significance level using a t-test.

  (b) Test $H_0: \beta = 0$ against $H_1: \beta \neq 0$ at the $\alpha = 0.05$ significance level by setting up the ANOVA table and using an F statistic.

  (c) Find a 95% confidence interval for $\mu(x)$ when $x = 2, 3$, and 4.

  (d) Find a 95% prediction interval for Y when $x = 2, 3$, and 4.

8.4.3 For the data given in Exercise 8.3.3,

  (a) Test $H_0: \beta = 0$ against $H_1: \beta > 0$ at the $\alpha = 0.025$ significance level using a t-test.

  (b) Test $H_0: \beta = 0$ against $H_1: \beta \neq 0$ at the $\alpha = 0.05$ significance level by setting up the ANOVA table and using an F statistic.

  (c) Find a 95% confidence interval for $\mu(x)$ when $x = 20, 25$, and 30.

  (d) Find a 95% prediction interval for Y when $x = 20, 25$, and 30.

8.4.4 For the candy data given in Exercise 8.3.9, with the usual assumptions,

  (a) Test $H_0: \beta = 0$ against $H_1: \beta > 0$ at an $\alpha = 0.01$ significance level.

  (b) Find a 90% confidence interval for $\mu(x)$ when $x = 54, 56$, and 58.

  (c) Determine a 90% prediction interval for Y when $x = 54, 56$, and 58.

8.4.5 For the cigarette data in Exercise 8.3.10, with the usual assumptions,

  (a) Find a 95% confidence interval for $\mu(x)$ when $x = 5, 10$, and 15.

  (b) Determine a 95% prediction interval for Y when $x = 5, 10$, and 15.

8.4.6 Show that the cross-product term in Equation 8.6 is equal to zero, i.e.,

$$\sum_{i=1}^{n} [\widehat{\beta}(x_i - \bar{x})][Y_i - \widehat{\alpha} - \widehat{\beta}(x_i - \bar{x})] = 0.$$

8.4.7 For the ACT scores in Exercise 8.3.12, with the usual assumptions,

  (a) Test $H_0: \beta = 0$ against $H_1: \beta > 0$ at an $\alpha = 0.05$ significance level.

  (b) Find a 95% confidence interval for $\mu(x)$ when $x = 17, 20, 23, 26$, and 29.

  (c) Determine a 90% prediction interval for Y when $x = 17, 20, 23, 26$, and 29.

8.4.8 A computer center recorded the number of programs it maintained during each of 10 consecutive years.

  (a) Calculate the least squares regression line for these data.

  (b) Plot the points and the line on the same graph.

  (c) Find a 95% prediction interval for the number of programs in year 11 under the usual assumptions.

Year	Number of Programs
1	430
2	480
3	565
4	790
5	885
6	960
7	1200
8	1380
9	1530
10	1591

8.4.9 By the method of least squares, fit the cubic $y = \beta_1 + \beta_2 x + \beta_3 x^2 + \beta_4 x^3$ to the 10 observed data points (x, y): $(0, 1), (-1, -3), (0, 3), (1, 3), (-1, -1), (2, 10), (0, 0), (-2, -9), (-1, -2),$ $(2, 8)$.

8.4.10 By the method of least squares, fit the regression plane $y = \beta_1 x_1 + \beta_2 x_2$ to the 12 observations of (x_1, x_2, y): $(1, 1, 6), (0, 2, 3), (3, 0, 10), (-2, 0, -4),(-1, 2, 0),(0, 0, 1),$ $(2, 1, 8),(-1, -1, -2),(0, -3, -3), (2, 1, 5), (1, 1, 1), (-1, 0, -2)$.

8.4.11 Explain why the model $\mu(x) = \beta_1 e^{\beta_2 x}$ is not a linear model. Would the logarithm, $\ln \mu(x)$, be a linear model?

8.4.12 The following data give the times of 22 male swimmers who entered the 50 yard freestyle race in the conference championship meet. For these data, x is the swimmer's best time for the season and y is the swimmer's time in the meet.

x	y	x	y	x	y	x	y
24.97	23.98	24.76	23.63	23.80	23.61	22.84	23.58
24.07	23.57	22.98	23.17	22.93	22.83	23.41	22.75
22.10	22.74	23.08	22.70	23.59	22.62	23.80	22.53
22.38	22.39	22.91	22.34	22.08	21.95	22.46	21.73
21.34	21.68	22.76	21.66	21.82	21.68	21.80	21.58
22.26	21.57	21.36	21.35				

(a) Graph the points and the regression line for these data.

(b) Calculate and plot the residuals. Does linear regression seem to be appropriate?

(c) Find a 90% confidence band for the regression line. Interpret.

(d) Find a 90% prediction band for these data. Interpret.

(e) Find 95% confidence intervals for α, β, and σ^2.

9

QUALITY IMPROVEMENT THROUGH STATISTICAL METHODS

9.1 SHEWHART CONTROL CHARTS

Statistical methods can be used in many scientific fields, such as medical research, engineering, chemistry, and psychology. Often, it is necessary to compare two ways of doing something, say the old way and a possible new way. We collect data on each, quite possibly in a laboratory situation, and try to decide if the new is actually better than the old. Needless to say, it would be terrible to change to a new way at great expense only to find out that it is really not any better than the old. That is, suppose the lab results indicate, by some statistical method, that the new is seemingly better than the old. Can we actually extrapolate those outcomes in the lab to the situations in the real world? Clearly, statisticians cannot make these decisions, but they should be made by some professional who knows that specialty very well. The statistical analysis might provide helpful guidelines, but we still need the expert to make the final decision.

However, even before investigating possible changes in any process, it is first extremely important to determine exactly what the process in question is doing at the present time. Often, people in charge of an organization do not understand the capabilities of many of its processes. Simply measuring what is going on often leads to improvements. In many cases measurement is easy, such as determining the diameter of a bolt, but sometimes it is extremely difficult, as in evaluating good teaching or many other service activities. But if at all possible, we encourage those involved to begin to "listen" to their processes; that is, they should measure what is going on in your organization. These measurements alone often are the beginning of desirable improvements. While most of our remarks in this chapter concern measurements made in manufacturing, service industries frequently find them just as useful.

At one time, some manufacturing plants would make parts to be used in the construction of some piece of equipment. Say a particular line in the plant making a certain part might produce several hundreds of them each day. These items would then be sent on to an inspection cage where they would be checked for goodness, often several days or even weeks later. Occasionally, the inspectors would discover many defectives among the items made two weeks ago, say. There was little that could be done at that point except scrap or rework the defective parts, both expensive outcomes.

In the 1920s, W. A. Shewhart, who was working for AT&T Bell Labs, recognized that this was an undesirable situation and suggested that with some suitable frequency, a sample of these parts should be taken as they were being made. If their sample indicated the items were satisfactory, the manufacturing process would continue. But if these sampled parts were not satisfactory, corrections should be made then so that things were satisfactory. This idea led to what are commonly called *Shewhart control charts*, which are the basis of what was called *statistical quality control* in those early days; today, it is often referred to as *statistical process control*.

These control charts consist of calculated values of a statistic, say $\bar{x}$, plotted in time sequence. That is, in making products, every so often (each hour, each day, each week depending upon how many items are being produced) a sample of size n of them is taken, and they are measured, resulting in the observations $x_1, x_2, \ldots, x_n$. The average $\bar{x}$ and the standard deviation s are computed. This is done k times, and the k values of $\bar{x}$ and s are averaged resulting in $\bar{\bar{x}}$ and $\bar{s}$, respectively; usually, k is equal to some number between 10 and 30.

If the true mean μ and standard deviation σ of the process were known, then the central limit theorem states that almost all of the $\bar{x}$ values would plot between $\mu - 3\sigma/\sqrt{n}$ and $\mu + 3\sigma/\sqrt{n}$, unless the system has actually changed. However, we know neither μ nor σ, and thus μ is estimated by $\bar{\bar{x}}$ and $3\sigma/\sqrt{n}$ by $A_3\bar{s}$, where A_3 is a factor depending upon n that can be found in books on statistical quality control. For example, a few values of A_3 (and some other constants that will be used later) are given in Table 9.1 for typical values of n.

TABLE 9.1 Some constants used with control charts

n	A_3	B_3	B_4	A_2	D_3	D_4
4	1.63	0	2.27	0.73	0	2.28
5	1.43	0	2.09	0.58	0	2.11
6	1.29	0.03	1.97	0.48	0	2.00
8	1.10	0.185	1.815	0.37	0.14	1.86
10	0.98	0.28	1.72	0.31	0.22	1.78
20	0.68	0.51	1.49	0.18	0.41	1.59

The estimates of $\mu \pm 3\sigma/\sqrt{n}$ are called the upper control limit (UCL), $\bar{\bar{x}}+A_3\bar{s}$, and the lower control limit (LCL), $\bar{\bar{x}} - A_3\bar{s}$, and $\bar{\bar{x}}$ provides the estimate of the centerline. A typical plot is given in Figure 9.1. Here, in the 13th sampling period, $\bar{x}$ is outside the control limits, indicating the process has changed and some investigation and action are needed to correct this change, which seems like a shift upward in the process.

Figure 9.1   Typical control chart

It should be noted that there is a control chart for the s values, too. From sampling distribution theory, values of B_3 and B_4 have been determined and are given in Table 9.1 so that we know that almost all s values should be between $B_3\bar{s}$ and $B_4\bar{s}$ if there is no change in the underlying distribution. So again, if an individual s value is outside these control limits, some action should be taken as it seems as if there has been a change in the variation of the underlying distribution.

Often, when these charts are first constructed after $k = 10$ to 30 sampling periods, many points fall outside the control limits. A team consisting of workers, the foreman, the supervisor, an engineer, and even a statistician should try to find the reasons why this has occurred, and the situation should be corrected. After this is done and the points plot within the control limits, the process is "in statistical control." However, being in statistical control is not a guarantee of satisfaction with the products. Since $A_3\bar{s}$ is an estimate of $3\sigma/\sqrt{n}$, then $\sqrt{n}A_3\bar{s}$ is an estimate of 3σ, and with an underlying distribution close to a normal one, almost all items would be between $\bar{\bar{x}} \pm \sqrt{n}A_3\bar{s}$. If these limits are too wide, then corrections must be made again.

If the variation is under control (i.e., $\bar{x}$ and s within their control limits), we say the variations seen in $\bar{x}$ and s are due to common causes. If products made under such a system with these existing common causes are satisfactory, then the production continues. If either $\bar{x}$ or s, however, is outside the control limits, that is an indication that some special causes are at work, which must be corrected. That is, a team should investigate why and some action should be taken.

Example 9.1.1

A company produces a storage console. Twice a day nine critical characteristics are tested on five consoles that are selected randomly from the production line. One of these characteristics is the time it takes the lower storage component door to open completely. Table 9.2 lists the opening times in seconds for the consoles that were tested during one week. Also included in the table are the sample means, sample standard deviations, and the ranges.

TABLE 9.2   Console opening times

Group	x_1	x_2	x_3	x_4	x_5	$\bar{x}$	s	R
1	1.2	1.8	1.7	1.3	1.4	1.48	0.26	0.6
2	1.5	1.2	1.0	1.0	1.8	1.30	0.35	0.8
3	0.9	1.6	1.0	1.0	1.0	1.10	0.28	0.7
4	1.3	0.9	0.9	1.2	1.2	1.06	0.18	0.4
5	0.7	0.8	0.9	0.6	0.8	0.76	0.11	0.3
6	1.2	0.9	1.1	1.0	1.0	1.04	0.10	0.3
7	1.1	0.9	1.1	1.0	1.4	1.10	0.19	0.5
8	1.4	0.9	0.9	1.1	1.0	1.06	0.19	0.5
9	1.3	1.4	1.1	1.5	1.6	1.38	0.19	0.5
10	1.6	1.5	1.4	1.3	1.5	1.46	0.11	0.3
						$\bar{\bar{x}} = 1.17$	$\bar{s} = 0.20$	$\bar{R} = 0.49$

The upper control limit (UCL) and the lower control limit (LCL) for $\bar{x}$ are found using A_3 from Table 9.1 with $n = 5$ as follows:

$$UCL = \bar{\bar{x}} + A_3\bar{s} = 1.17 + 1.43(0.20) = 1.46$$

and

$$LCL = \bar{\bar{x}} - A_3\bar{s} = 1.17 - 1.43(0.20) = 0.88.$$

These control limits and the sample means are plotted on the $\bar{x}$-chart in Figure 9.2. There should be some concern about the fifth sampling period; thus there should be an investigation to determine why that $\bar{x}$ is below the LCL.

Figure 9.2 $\bar{x}$-chart

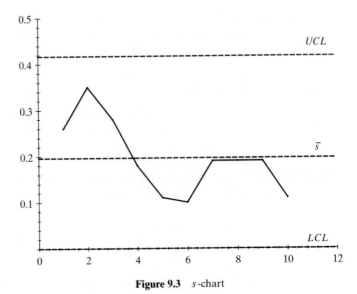

Figure 9.3 s-chart

The UCL and LCL for s are found using B_3 and B_4 in Table 9.1 and $n = 5$ as follows:

$$UCL = B_4\bar{s} = 2.09(0.20) = 0.42$$

and

$$LCL = B_3\bar{s} = 0(0.20) = 0.$$

These control limits and the sample standard deviations are plotted on the s-chart in Figure 9.3.

Almost all of the observations should lie between $\bar{\bar{x}} \pm \sqrt{n}A_3\bar{s}$, namely,

$$1.17 + \sqrt{5}\,(1.43)(0.20) = 1.81$$

and

$$1.17 - \sqrt{5}\,(1.43)(0.20) = 0.53.$$

This is illustrated in Figure 9.4, in which all 50 observations do fall within these control limits.   ◇

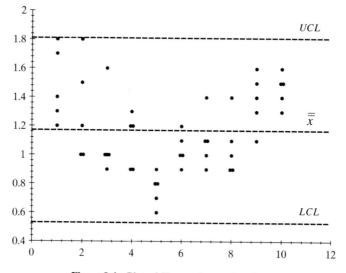

Figure 9.4   Plot of 50 console opening times

In most books on statistical quality control, there is an alternative way of constructing the limits on an $\bar{x}$-chart. For each sample, compute the range, R, which is defined in Section 1.3 as the absolute value of the difference of the extremes of the sample. This computation is much easier than that for calculating s. After k samples are taken, compute the average of these R-values, obtaining $\bar{R}$ as well as $\bar{\bar{x}}$. The statistic $A_2\bar{R}$ serves as an estimate of $3\sigma/\sqrt{n}$, where A_2 is found in Table 9.1. Thus the estimates of $\mu \pm 3\sigma/\sqrt{n}$, namely $\bar{\bar{x}} \pm A_2\bar{R}$, can be used as the UCL and LCL of an $\bar{x}$-chart.

In addition, $\sqrt{n}A_2\overline{R}$ is an estimate of 3σ; so with an underlying distribution that is close to a normal one, we find that almost all observations are within the limits $\overline{\overline{x}} \pm \sqrt{n}A_2\overline{R}$. Moreover, an R-chart can be constructed with centerline $\overline{R}$ and control limits equal to $D_3\overline{R}$ and $D_4\overline{R}$, where D_3 and D_4 are given in Table 9.1 and were determined so that almost all R-values should be between the control limits if there is no change in the underlying distribution. Thus an R falling outside those limits would indicate a change in the spread of the underlying distribution, and some corrective action should be considered. The use of R, rather than s, is illustrated in the next example.

Example 9.1.2

Using the data in Example 9.1.1, we compute UCL and LCL for an x-bar chart using $\overline{\overline{x}} \pm A_2\overline{R}$ as follows:

$$UCL = \overline{\overline{x}} + A_2\overline{R} = 1.17 + 0.58(0.49) = 1.45$$

and

$$LCL = \overline{\overline{x}} - A_2\overline{R} = 1.17 - 0.58(0.49) = 0.89.$$

Note that these are very close to the limits that were found for the $\overline{x}$-chart in Figure 9.2 using $\overline{\overline{x}} \pm A_3\overline{s}$. In addition, almost all of the observations should lie within the limits $\overline{\overline{x}} \pm \sqrt{n}A_2\overline{R}$. These limits are

$$UCL = 1.17 + \sqrt{5}(0.58)(0.49) = 1.81$$

and

$$LCL = 1.17 - \sqrt{5}(0.58)(0.49) = 0.53.$$

Note that these are the same as the limits found in Example 9.1.1 and plotted in Figure 9.4.

An R-chart can be constructed with centerline $\overline{R} = 0.49$ and control limits given by

$$UCL = D_4\overline{R} = 2.11(0.49) = 1.03$$

and

$$LCL = D_3\overline{R} = 0(0.49) = 0.$$

Figure 9.5 illustrates this control chart for the range, and we see its pattern is similar to that of the s-chart in Figure 9.3.   ◇

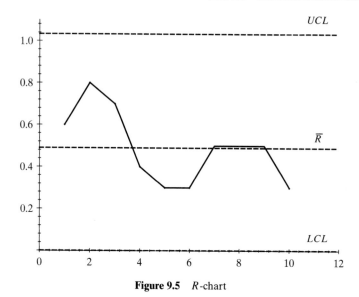

Figure 9.5 R-chart

There are two other Shewhart control charts, the p- and c- charts. The central limit theorem, which provided a justification for the three sigma limits in the $\bar{x}$-chart, also justifies the control limits in the p-chart. Suppose the number of defectives among n items that are selected randomly, say D, has a binomial distribution $b(n, p)$. The limits $p \pm 3\sqrt{p(1-p)/n}$ should include almost all of the D/n values. However, p must be approximated by observing k values of D, say $D_1, D_2, \ldots . D_k$, and computing what is called $\bar{p}$ in the statistical quality control literature, namely,

$$\bar{p} = \frac{D_1 + D_2 + \cdots + D_k}{kn}.$$

Thus the LCL and UCL for the fraction defective, D/n, are given by

$$LCL = \bar{p} - 3\sqrt{\bar{p}(1-\bar{p})/n}$$

and

$$UCL = \bar{p} + 3\sqrt{\bar{p}(1-\bar{p})/n}.$$

If the process is in control, almost all D/n values are between LCL and UCL. This may not be satisfactory and improvements might be needed to decrease $\bar{p}$. If satisfactory, however, let the process continue under these common causes of variation until a point, D/n, outside the control limits would indicate that some special cause has changed the variation. Incidentally, if D/n is below the LCL, this might very well indicate that some type of change for the better has been made, and we want to find out why. In general, outlying statistics can often suggest that good (as well as bad) breakthroughs have been made.

The following example gives the results of a simple experiment that you can easily duplicate.

Example 9.1.3

Let D_i equal the number of yellow candies in a 1.69-ounce bag. Because the number of pieces of candy varies slightly from bag to bag, we shall use an average value for n when we construct the control limits. Table 9.3 lists, for 20 packages, the number of pieces of candy in the package, the number of yellow ones, and the proportion of yellow ones.

TABLE 9.3   Data on yellow candies

Package	n_i	D_i	$\bar{p}_i$
1	56	8	0.14
2	55	13	0.24
3	58	12	0.21
4	56	13	0.23
5	57	14	0.25
6	54	5	0.09
7	56	14	0.25
8	57	15	0.26
9	54	11	0.20
10	55	13	0.24
11	57	10	0.18
12	59	8	0.14
13	54	10	0.19
14	55	11	0.20
15	56	12	0.21
16	57	11	0.19
17	54	6	0.11
18	58	7	0.12
19	58	12	0.21
20	58	14	0.24
	1124	219	

For these data

$$\bar{p} = \frac{219}{1124} = 0.195 \quad \text{and} \quad \bar{n} = \frac{1124}{20} \approx 56.$$

Thus the UCL and LCL are given by

$$LCL = \bar{p} - 3\sqrt{\bar{p}(1-\bar{p})/56} = 0.195 - 3\sqrt{0.195(0.805)/56} = 0.036$$

and

$$UCL = \bar{p} + 3\sqrt{\bar{p}(1-\bar{p})/56} = 0.195 + 3\sqrt{0.195(0.805)/56} = 0.354.$$

The control chart for p is depicted in Figure 9.6. (For your information the "true" value for p is 0.20.)   ◇

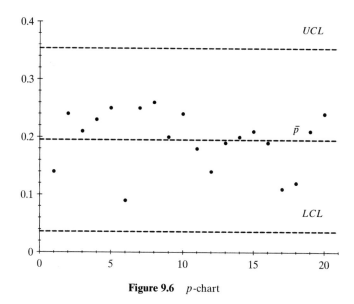

Figure 9.6 p-chart

Consider the following explanation of the c-chart. Suppose the number of flaws, say C, on some product has a Poisson distribution with parameter λ. If λ is sufficiently large, as in Example 5.5.5, we considered approximating the discrete Poisson distribution with the continuous $N(\lambda, \lambda)$ distribution. Thus the interval from $\lambda - 3\sqrt{\lambda}$ to $\lambda + 3\sqrt{\lambda}$ contains virtually all of the C values. Since λ is unknown, however, it must be approximated by $\bar{c}$, the average of the k values, $c_1, c_2, \ldots . c_k$. Hence the two control limits for C are computed as

$$LCL = \bar{c} - 3\sqrt{\bar{c}} \quad \text{and} \quad UCL = \bar{c} + 3\sqrt{\bar{c}}.$$

The remarks made about the $\bar{x}$- and $\bar{p}$-charts apply to the c-chart as well, but we must remember that each c value is the number of flaws on one manufactured item, not an average $\bar{x}$ or a fraction defective D/n.

Remark   In observing time-sequence plots, as with control charts, do not read too much into a short sequence of points. For example, after a pep talk by the coach of a women's golf team, three successive decreasing scores might indicate that the coach had some influence. But even if the process has not changed, the probability of three decreasing scores (ruling out ties) is 1/6, since there are 3! = 6 equally likely ways of arranging three points. Often businesspeople are worse offenders because they might think two increasing sales points indicate improvement. We know that with no change in the system, the probability of that is 1/2. Now maybe 4 points that successively increase after an intervention would be cause for claiming improvement because the probability of such an event is 1/4! = 1/24 if no change has actually been made. This probability is small enough to justify a celebration.

EXERCISES

9.1.1 It is important to control the viscosity of liquid dishwasher soap so that it flows out of the container but does not run out too rapidly. Thus samples are taken randomly throughout the day and the viscosity is measured. Use the following 20 sets of 5 observations for this exercise:

Observations					$\bar{x}$	s	R
158	147	158	159	169	158.20	7.79	22
151	166	151	143	169	156.00	11.05	26
153	174	151	164	185	165.40	14.33	34
168	140	180	176	154	163.60	16.52	40
160	187	145	164	158	162.80	15.29	42
169	153	149	144	157	154.40	9.48	25
156	183	157	140	162	159.60	15.47	43
158	160	180	154	160	162.40	10.14	26
164	168	154	158	164	161.60	5.55	14
159	153	170	158	170	162.00	7.65	17
150	161	169	166	154	160.00	7.97	19
157	138	155	134	165	149.80	13.22	31
161	172	156	145	153	157.40	10.01	27
143	152	152	156	163	153.20	7.26	20
179	157	135	172	143	157.20	18.63	44
154	165	145	152	145	152.20	8.23	20
171	189	144	154	147	161.00	18.83	45
187	147	159	167	151	162.20	15.85	40
153	168	148	188	152	161.80	16.50	40
165	155	140	157	176	158.60	13.28	36

(a) Calculate the values of $\bar{\bar{x}}$, $\bar{s}$, and $\bar{R}$.

(b) Construct an $\bar{x}$-chart using the value of A_3 and $\bar{s}$.

(c) Construct an s-chart.

(d) Construct an $\bar{x}$-chart using A_2 and $\bar{R}$.

(e) Construct an R-chart.

(f) Do the charts indicate that viscosity is in statistical control?

9.1.2 It is necessary to control the percentage of solids in a product, so samples are taken randomly throughout the day and the percentage of solids is measured. Use the following 20 sets of 5 observations for this exercise:

Observations					$\overline{x}$	s	R
69.8	71.3	65.6	66.3	70.1	68.62	2.51	5.7
71.9	69.6	71.9	71.1	71.7	71.24	0.97	2.3
71.9	69.8	66.8	68.3	64.4	68.24	2.86	7.5
64.2	65.1	63.7	66.2	61.9	64.22	1.61	4.3
66.1	62.9	66.9	67.3	63.3	65.30	2.06	4.4
63.4	67.2	67.4	65.5	66.2	65.94	1.61	4.0
67.5	67.3	66.9	66.5	65.5	66.74	0.79	2.0
63.9	64.6	62.3	66.2	67.2	64.84	1.92	4.9
66.0	69.8	69.7	71.0	69.8	69.26	1.90	5.0
66.0	70.3	65.5	67.0	66.8	67.12	1.88	4.8
67.6	68.6	66.5	66.2	70.4	67.86	1.71	4.2
68.1	64.3	65.2	68.0	65.1	66.14	1.78	3.8
64.5	66.6	65.2	69.3	62.0	65.52	2.69	7.3
67.1	68.3	64.0	64.9	68.2	66.50	1.96	4.3
67.1	63.8	71.4	67.5	63.7	66.70	3.17	7.7
60.7	63.5	62.9	67.0	69.6	64.74	3.53	8.9
71.0	68.6	68.1	67.4	71.7	69.36	1.88	4.3
69.5	61.5	63.7	66.3	68.6	65.92	3.34	8.0
66.7	75.2	79.0	75.3	79.2	75.08	5.07	12.5
77.3	67.2	69.3	67.9	65.6	69.46	4.58	11.7

(a) Calculate the values of $\overline{\overline{x}}$, $\overline{s}$, and $\overline{R}$.

(b) Construct an $\overline{x}$-chart using the value of A_3 and $\overline{s}$.

(c) Construct an s-chart.

(d) Construct an $\overline{x}$-chart using A_2 and $\overline{R}$.

(e) Construct an R-chart.

( ) Do the charts indicate that the percentage of solids in this product is in statistical control?

9.1.3 It is important to control the net weight of a packaged item; thus items are selected randomly throughout the day from the production line and their weights are recorded. Use the following 20 sets of 5 weights (in grams) for this exercise. (Note that a weight recorded here is the actual weight minus 330.)

Observations					$\bar{x}$	s	R
7.97	8.10	7.73	8.26	7.30	7.866	0.3723	0.96
8.11	7.26	7.99	7.88	8.88	8.024	0.5800	1.62
7.60	8.23	8.07	8.51	8.05	8.092	0.3309	0.91
8.44	4.35	4.33	4.48	3.89	5.098	1.8815	4.55
5.11	4.05	5.62	4.13	5.01	4.784	0.6750	1.57
4.79	5.25	5.19	5.23	3.97	4.886	0.5458	1.28
4.47	4.58	5.35	5.86	5.61	5.174	0.6205	1.39
5.82	4.51	5.38	5.01	5.54	5.252	0.5077	1.31
5.06	4.98	4.13	4.58	4.35	4.620	0.3993	0.93
4.74	3.77	5.05	4.03	4.29	4.376	0.5199	1.28
4.05	3.71	4.73	3.51	4.76	4.152	0.5748	1.25
3.94	5.72	5.07	5.09	4.61	4.886	0.6599	1.78
4.63	3.79	4.69	5.13	4.66	4.580	0.4867	1.34
4.30	4.07	4.39	4.63	4.47	4.372	0.2079	0.56
4.05	4.14	4.01	3.95	4.05	4.040	0.0693	0.19
4.20	4.50	5.32	4.42	5.24	4.736	0.5094	1.12
4.54	5.23	4.32	4.66	3.86	4.522	0.4999	1.37
5.02	4.10	5.08	4.94	5.18	4.864	0.4360	1.08
4.80	4.73	4.82	4.69	4.27	4.662	0.2253	0.55
4.55	4.76	4.45	4.85	4.02	4.526	0.3249	0.83

(a) Calculate the values of $\bar{\bar{x}}$, $\bar{s}$, and $\bar{R}$.

(b) Construct an $\bar{x}$-chart using the value of A_3 and $\bar{s}$.

(c) Construct an s-chart.

(d) Construct an $\bar{x}$-chart using A_2 and $\bar{R}$.

(e) Construct an R-chart.

(f) Do the charts indicate that these fill weights are in statistical control?

9.1.4 To give some indication of how the values in Table 9.1 are calculated, values of A_3 are found in this exercise. Let $X_1, X_2, \ldots, X_n$ be a random sample of size n from the normal distribution $N(\mu, \sigma^2)$. Let S^2 equal the sample variance of this random sample.

(a) Show that $E[S^2] = \sigma^2$ using the fact that $Y = (n-1)S^2/\sigma^2$ has a distribution that is $\chi^2(n-1)$.

(b) Using the $\chi^2(n-1)$ p.d.f., find the value of $E(\sqrt{Y})$.

(c) Show that

$$E\left[\frac{\sqrt{n-1}\,\Gamma\left(\dfrac{n-1}{2}\right)}{\sqrt{2}\,\Gamma\left(\dfrac{n}{2}\right)}\,S\right] = \sigma.$$

(d) Verify that

$$\frac{3}{\sqrt{n}}\left[\frac{\sqrt{n-1}\,\Gamma\left(\dfrac{n-1}{2}\right)}{\sqrt{2}\,\Gamma\left(\dfrac{n}{2}\right)}\right] = A_3,$$

found in Table 9.1 for $n = 5$ and $n = 6$. Thus $A_3\bar{s}$ approximates $3\sigma/\sqrt{n}$.

9.1.5 A company has been producing bolts that are about $\overline{p} = 0.02$ defective, and this is satisfactory. To monitor the quality of this process, 100 bolts are selected at random each hour and the number of defective bolts counted. With $\overline{p} = 0.02$, compute the UCL and LCL of the $\overline{p}$-chart. Then suppose that over the next 24 hours, the following numbers of defective bolts are observed:

    4 1 1 0 5 2 1 3 4 3 1 0 0 4 1 1 6 2 0 0 2 8 7 5

Would any action have been required during this time?

9.1.6 In the past, $n = 50$ fuses are tested each hour and $\overline{p} = 0.03$ have been defective. Calculate the UCL and LCL. After a production error, say the true p shifts to $p = 0.05$.

(a) What is the probability that the next observation exceeds the UCL?

(b) What is the probability that at least one of the next five observations exceeds the UCL? HINT: Assume independence and compute the probability that none of the next five exceeds the UCL.

9.1.7 In a woolen mill, 100-yard pieces are inspected. In the last 20 observations the following numbers of flaws were found:

    2 4 0 1 0 3 4 1 1 2 4 0 0 1 0 3 2 3 5 0

(a) Compute the control limits of the c-chart.

(b) Is the process in statistical control?

9.1.8 In 50-foot tin strips, we find that the number of blemishes average about $\bar{c} = 1.4$. Calculate the control limits. Say that the process has gone out of control and this average has increased to 3.

(a) What is the probability that the next observation will exceed the UCL?

(b) What is the probability that at least 1 of the next 10 observations will exceed the UCL?

9.2 GENERAL FACTORIAL AND 2^k FACTORIAL DESIGNS

In Section 8.2 we studied two-factor experiments in which the A factor is performed at a levels and the B factor has b levels. Without replications, we need ab level combinations and, with c replications with each of these combinations, we need a total of abc experiments.

Let us consider a situation with three factors, say A, B, and C with a, b, and c levels, respectively. Here there are a total of abc level combinations; and if at each of these combinations we have d replications, there is need for $abcd$ experiments. Once these experiments are run, in some random order, and the data collected, there are computer programs available to calculate the entries in the ANOVA table, as in Table 9.4.

TABLE 9.4   ANOVA table

Source	SS	d.f.	MS	F
A	SS(A)	$a-1$	MS(A)	MS(A)/MS(E)
B	SS(B)	$b-1$	MS(B)	MS(B)/MS(E)
C	SS(C)	$c-1$	MS(C)	MS(C)/MS(E)
AB	SS(AB)	$(a-1)(b-1)$	MS(AB)	MS(AB)/MS(E)
AC	SS(AC)	$(a-1)(c-1)$	MS(AC)	MS(AC)/MS(E)
BC	SS(BC)	$(b-1)(c-1)$	MS(BC)	MS(BC)/MS(E)
ABC	SS(ABC)	$(a-1)(b-1)(c-1)$	MS(ABC)	MS(ABC)/MS(E)
Error	SS(E)	$abc(d-1)$	MS(E)	
Total	SS(TO)	$abcd-1$		

The main effects (A, B, and C) and the two-factor interactions (AB, AC, and BC) have the same interpretations as in the two-factor ANOVA. The three-factor interaction represents that part of the model for the means μ_{ijh}; $i = 1, 2, \ldots, a$; $j = 1, 2, \ldots, b$; $h = 1, 2, \ldots, c$ that cannot be explained by a model including only the main effects and two-factor interactions. In particular, if for each fixed h, the "plane" created by μ_{ijh} is "parallel" to the "plane" created by every other fixed h, then the three-factor interaction is equal to zero. Usually, higher-order interactions tend to be small.

In the testing sequence, we test the three-factor interaction first by checking to see whether or not

$$\text{MS(ABC)/MS(E)} \geq F_\alpha[(a-1)(b-1)(c-1), abc(d-1)].$$

If this inequality holds, the ABC interaction is significant at the α level. We would then not continue testing the two-factor interactions and the main effects with those F values, but analyze the data otherwise. For example, we could, for each fixed h, look at a two-factor ANOVA for factors A and B. Of course, if the inequality does not hold, we next check the two-factor interactions with the appropriate F values. If these are not significant, we check the main effects, A, B, and C.

Factorial experiments with three or more factors require many experiments, particularly if each factor has several levels. Often, in the health, social, and physical sciences, experimenters want to consider several factors (maybe as many as 10 or 20 or hundreds), and they cannot afford to run that many experiments. This is particularly true with preliminary or screening investigations in which they want to detect the factors that seem most important. In these cases they often consider factorial experiments such that each of k factors is run only at two levels, frequently without replication. We consider only this situation, although the reader should recognize that there are many variations of this. In particular, there are methods for investigating only *fractions of these* 2^k *designs.* The reader interested in more information should refer to a good book on design of experiments, such as that by Box, Hunter, and Hunter.[1] Many statisticians in industry believe these statistical methods are the most useful in improving product and process designs. Hence it is clearly an extremely important topic as many industries are greatly concerned about the quality of their products.

In factorial experiments in which each of the k factors is considered at only two levels, those levels are selected at some reasonable low and high values. That is, with the help of someone in the field, the typical range of each factor is considered. For illustration, if we are considering baking temperatures in the range of 300 to $375°F$, a representative low is selected, say $320°F$, and a representative high is selected, say $355°F$. There is no formula for these selections, and someone familiar with the experiment would help make these selections. Often, it happens that only two different types of material are considered, say fabric, and one is called low and the other high.

Thus we select a low and high for each factor; these will be coded as -1 and $+1$ or, more simply, as $-$ and $+$, respectively. We give three 2^k designs, for $k = 2, 3,$ and 4, in standard order in Tables 9.5, 9.6, and 9.7, respectively. From these three tables, we can easily note what is meant by standard order. The A column starts with a

TABLE 9.5 2^2 design

	2^2 Design		
Run	A	B	Observation
1	$-$	$-$	X_1
2	$+$	$-$	X_2
3	$-$	$+$	X_3
4	$+$	$+$	X_4

[1]   Box, Hunter, and Hunter, *Statistics for Experimenters*, John Wiley & Sons, Inc., NY, 1978.

TABLE 9.6 2^3 design

Run	2³ Design			Observation
	A	B	C	
1	−	−	−	X_1
2	+	−	−	X_2
3	−	+	−	X_3
4	+	+	−	X_4
5	−	−	+	X_5
6	+	−	+	X_6
7	−	+	+	X_7
8	+	+	+	X_8

TABLE 9.7 2^4 design

Run	A	B	C	D	Observation
	2⁴ Design				
1	−	−	−	−	X_1
2	+	−	−	−	X_2
3	−	+	−	−	X_3
4	+	+	−	−	X_4
5	−	−	+	−	X_5
6	+	−	+	−	X_6
7	−	+	+	−	X_7
8	+	+	+	−	X_8
9	−	−	−	+	X_9
10	+	−	−	+	X_{10}
11	−	+	−	+	X_{11}
12	+	+	−	+	X_{12}
13	−	−	+	+	X_{13}
14	+	−	+	+	X_{14}
15	−	+	+	+	X_{15}
16	+	+	+	+	X_{16}

minus sign and then the sign alternates. The B column begins with two minus signs, then the signs alternate in blocks of two. The C column has 4 minus signs and then 4 plus signs and so on. The D column starts with 8 minus signs and then 8 plus signs. It is easy to extend this idea to 2^k designs, where $k \geq 5$. For illustration, under the E column in a 2^5 design, we have 16 minus signs followed by 16 plus signs, which account for the 32 experiments.

To be absolutely certain what these runs mean, consider run 12 in Table 9.7: A is set at its high level, B at its high, C at its low, and D at its high level. The value X_{12} is the random observation resulting from this one combination of these four settings.

It must be emphasized, however, that the runs are not necessarily performed in the order $1, 2, 3, \ldots, 2^k$; but they should be performed in a random order, if at all possible. That is, in a 2^3 design, we might perform the experiment in the order: 3, 2, 8, 6, 5, 1, 4, 7 if this, in fact, was a random selection of a permutation of the first eight positive integers.

Once all 2^k experiments have been run, it is possible to consider the total sum of squares

$$\sum_{i=1}^{2^k} (X_i - \overline{X})^2$$

and decompose it very easily into $2^k - 1$ parts, which represent the respective meas/-ure/-ments (estimators) of the k main effects, $\binom{k}{2}$ two-factor interactions; $\binom{k}{3}$ three-factor interactions, and so on, until we have the one k-factor interaction. We

TABLE 9.8 2^3 design decomposition

Run	\multicolumn{3}{c}{2^3 Design}			AB	AC	BC	ABC	Observation
	A	B	C	AB	AC	BC	ABC	Observation
1	−	−	−	+	+	+	−	X_1
2	+	−	−	−	−	+	+	X_2
3	−	+	−	−	+	−	+	X_3
4	+	+	−	+	−	−	−	X_4
5	−	−	+	+	−	−	+	X_5
6	+	−	+	−	+	−	−	X_6
7	−	+	+	−	−	+	−	X_7
8	+	+	+	+	+	+	+	X_8

illustrate this decomposition with the 2^3 design in Table 9.8. Note the column AB is found by formally multiplying the elements of column A by the corresponding ones in B. Similarly, AC is found by multiplying column A and column C, and so on, until column ABC is the product of the corresponding elements of columns A, B, C. We then construct seven linear forms using these seven columns of signs with the corresponding observations. The resulting measures (estimates) of the main effects (A, B, C), the two-factor interactions (AB, AC, BC), and the three-factor interaction (ABC) are then found by dividing these linear forms by $2^k = 2^3 = 8$

(some statisticians divide by $2^{k-1} = 2^{3-1} = 4$). These are denoted by

$$[A] = (-X_1 + X_2 - X_3 + X_4 - X_5 + X_6 - X_7 + X_8)/8,$$
$$[B] = (-X_1 - X_2 + X_3 + X_4 - X_5 - X_6 + X_7 + X_8)/8,$$
$$[C] = (-X_1 - X_2 - X_3 - X_4 + X_5 + X_6 + X_7 + X_8)/8,$$
$$[AB] = (+X_1 - X_2 - X_3 + X_4 + X_5 - X_6 - X_7 + X_8)/8,$$
$$[AC] = (+X_1 - X_2 + X_3 - X_4 - X_5 + X_6 - X_7 + X_8)/8,$$
$$[BC] = (+X_1 + X_2 - X_3 - X_4 - X_5 - X_6 + X_7 + X_8)/8,$$
$$[ABC] = (-X_1 + X_2 + X_3 - X_4 + X_5 - X_6 - X_7 + X_8)/8.$$

With assumptions of normality, mutual independence, and common variance σ^2, under the overall null hypothesis of the equality of all the means, each of these measures has a normal distribution with mean zero and variance $\sigma^2/8$ (in general $\sigma^2/2^k$). This implies that the square of each measure divided by $\sigma^2/8$ is $\chi^2(1)$. Moreover, it can be shown (see Exercise 9.2.2) that

$$\sum_{i=1}^{8} (X_i - \overline{X})^2$$

$$= 8([A]^2 + [B]^2 + [C]^2 + [AB]^2 + [AC]^2 + [BC]^2 + [ABC]^2).$$

So by Theorem 8.1.1, the terms on the right-hand side, divided by σ^2, are mutually independent random variables, each being $\chi^2(1)$. While it requires a little more theory, it follows that the linear forms [A], [B], [C], [AB], [AC], [BC], and [ABC] are mutually independent $N(0, \sigma^2/8)$ random variables.

Since we have assumed that we have not run any replications, how can we obtain an estimate of σ^2 to see if any of the main effects or interactions are significant? To help us, we fall back on the use of a q-q plot because, under the overall null hypothesis, those seven measures are mutually independently normally distributed with the same mean and variance. Thus a q-q plot of the normal percentiles against the corresponding ordered values of the measures should be about on a straight line if, in fact, the null hypothesis is true. If one of these points is "out of line," we might believe that the overall null hypothesis is not true and that the effect associated with it is significant. It is possible that two or three points might be out of line, then all corresponding effects (main or interaction) should be investigated. Clearly, this is not a formal test, but it has been extremely successful in practice.

As an illustration, we use the data from an experiment designed to evaluate the effects of laundering on a certain fire-retardant treatment for fabrics. These data, somewhat modified, were taken from *Experimental Statistics, National Bureau of Standards Handbook 91* by Mary G. Natrella (Washington, DC: U.S. Government Printing Office, 1963). Factor A is the type of fabric (sateen or monk's cloth), factor

B corresponds to two different fire-retardant treatments, and factor C describes the laundering conditions (no laundering, after one laundering). The observations are inches burned, measured on a standard size fabric after a flame test. They are, in standard order

$$x_1 = 41.0, \quad x_2 = 30.5, \quad x_3 = 47.5, \quad x_4 = 27.0,$$
$$x_5 = 39.5, \quad x_6 = 26.5, \quad x_7 = 48.0, \quad x_8 = 27.5.$$

Thus the measures of the effects are

$$[A] = (-41.0 + 30.5 - 47.5 + 27.0 - 39.5 + 26.5 - 48.0 + 27.5)/8 = -8.06,$$
$$[B] = (-41.0 - 30.5 + 47.5 + 27.0 - 39.5 - 26.5 + 48.0 + 27.5)/8 = 1.56,$$
$$[C] = (-41.0 - 30.5 - 47.5 - 27.0 + 39.5 + 26.5 + 48.0 + 27.5)/8 = 0.56,$$
$$[AB] = (+41.0 - 30.5 - 47.5 + 27.0 + 39.5 - 26.5 - 48.0 + 27.5)/8 = -2.19,$$
$$[AC] = (+41.0 - 30.5 + 47.5 - 27.0 - 39.5 + 26.5 - 48.0 + 27.5)/8 = -0.31,$$
$$[BC] = (+41.0 + 30.5 - 47.5 - 27.0 - 39.5 - 26.5 + 48.0 + 27.5)/8 = 0.81,$$
$$[ABC] = (-41.0 + 30.5 + 47.5 - 27.0 + 39.5 - 26.5 - 48.0 + 27.5)/8 = 0.31.$$

In Table 9.9 we order these seven measures, determine their percentiles, and find the corresponding percentiles of the standard normal distribution.

TABLE 9.9   Seven measures ordered

Identity of Effect	Ordered Effect	Percentile	Percentile from $N(0, 1)$
[A]	−8.06	12.5	−1.15
[AB]	−2.19	25.0	−0.67
[AC]	−0.31	37.5	−0.32
[ABC]	0.31	50.0	0.00
[C]	0.56	62.5	0.32
[BC]	0.81	75.0	0.67
[B]	1.56	87.5	1.15

The q-q plot is given in Figure 9.7. Each point has been identified with its effect. A straight line fits six of those points reasonably well, but the point associated with $[A] = -8.06$ is far from this straight line. Hence the main effect of factor A (the type of fabric) seems to be significant. It is interesting to note that the laundering factor, C, does not seem to be a significant factor.

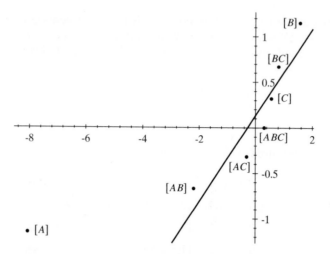

Figure 9.7 q-q plot of normal percentiles vs. estimated effects

EXERCISES

9.2.1 Write out a 2^2 design, displaying the A, B, and AB columns for the four runs.

(a) If X_1, X_2, X_3, X_4 are the four observations for the respective runs in standard order, write out the three linear forms, [A], [B], and [AB], that measure the two main effects and the interaction. These linear forms should include the divisor $2^2 = 4$.

(b) Show that $\sum_{i=1}^{4} (X_i - \overline{X})^2 = 4([A]^2 + [B]^2 + [AB]^2)$.

(c) Under the null hypothesis that all the means are equal and with the usual assumptions (normality, mutually independent, and common variance), what can you say about the distributions of the expressions in part (b) after each is divided by σ^2?

9.2.2 Show that in a 2^3 design, it is true that

$$\sum_{i=1}^{8} (X_i - \overline{X})^2 = 8([A]^2 + [B]^2 + [C]^2 + [AB]^2 + [AC]^2 + [BC]^2 + [ABC]^2).$$

HINT: Since both the right and left members of this equation are symmetric in the variables $X_1, X_2, \ldots, X_8$, it is only necessary to show that the corresponding coefficients of $X_1 X_i$, $i = 1, 2, \ldots, 8$, are the same in each member of the equation. Of course, recall $\overline{X} = (X_1 + X_2 + \cdots + X_8)/8$.

9.2.3 Show that the unbiased estimator of the variance σ^2 from a sample of size $n = 2$ is one-half of the square of the difference of the two observations. Thus, if a 2^k design is replicated, say with X_{i1} and X_{i2}, $i = 1, 2, \ldots, 2^k$, show that the estimate of the common σ^2 is

$$\frac{1}{2^{k+1}} \sum_{i=1}^{2^k} (X_{i1} - X_{i2})^2 = MS(E).$$

Under the usual assumptions, this implies that each of $2^k[A]^2/MS(E)$, $2^k[B]^2/MS(E)$, $2^k[AB]^2/MS(E)$, and so on, has an $F(1, 2^k)$ distribution under the null hypothesis. This, of course, would provide tests for the significance of the various effects, including interactions.

9.2.4 Hogg and Ledolter note that percent yields from a certain chemical reaction for changing temperature (factor A), reaction time (factor B), and concentration (factor C) are $x_1 = 79.7$, $x_2 = 74.3$, $x_3 = 76.7$, $x_4 = 70.0$, $x_5 = 84.0$, $x_6 = 81.3$, $x_7 = 87.3$, $x_8 = 73.7$ in standard order with a 2^3 design.

(a) Estimate the main effects, the three two-factor interactions, and the three-factor interaction.

(b) Construct an appropriate q-q plot to see if any of these effects seem to be significantly larger than the others. [R. V. Hogg and J. Ledolter, *Applied Statistics for Engineers and Physical Scientists*, 2nd ed. (New York: Macmillan, 1992).]

9.2.5 Box, Hunter, and Hunter[2] studied the effects of catalyst charge (10 pounds $= -1$, 20 pounds $= +1$), temperature ($220°C = -1$, $240°C = +1$), pressure (50 psi $= -1$, 80 psi $= +1$), and concentration (10% $= -1$, 12% $= +1$) on percent conversion (X) of a certain chemical. The results of a 2^4 design, in standard order, are

$$x_1 = 71, \quad x_2 = 61, \quad x_3 = 90, \quad x_4 = 82, \quad x_5 = 68, \quad x_6 = 61,$$
$$x_7 = 87, \quad x_8 = 80, \quad x_9 = 61, \quad x_{10} = 50, \quad x_{11} = 89,$$
$$x_{12} = 83, \quad x_{13} = 59, \quad x_{14} = 51, \quad x_{15} = 85, \quad x_{16} = 78.$$

(a) Estimate the main effects and the two-, three-, and four-factor interactions.

(b) Construct an appropriate q-q plot and assess the significance of the various effects.

9.3 MORE ON DESIGN OF EXPERIMENTS

Many observations that are taken occur without active participation of an experimenter. The analysis of such data represents a passive use of statistics that is often extremely helpful in making future decisions. However, in trying to create better products or services, there are frequently a number of factors (independent variables or "parameters") that we can control, and we want to set these at levels which will produce the best products.

For example, in making a cake mix, a company must decide how much sugar, flour, vanilla, salt, butter, egg mix, and milk to use, and then also recommend what time and temperature should be used in baking the mixture. Each of these variables (from amount of sugar to temperature) is called a factor, and the experimenters must decide which level is best for each factor. If we can list nine or so factors for making a cake, think of how many factors would be associated with a more complicated product.

Much can be learned about a process by considering the most influential factors and changing them according to a specified plan to try to find the best level of each. How does a cake mix company know that a certain mixture should be baked at 350°F for 35 minutes? Why not use 325°F for 40 minutes? Or 375°F for 30 minutes? The company must experiment with these levels of time and temperature, as well as with

---
[2] Ibid.

those of the other factors, to determine the optimal. Only by experimenting can the company determine the best level for each factor.

Such investigations involve collecting and analyzing data from carefully designed experiments. Since a poorly designed experiment may not shed light on a particular question that we want answered, it is important to plan an experiment before starting to perform it. Trying to rescue a poorly designed experiment after it has been run and the data have been collected usually leads to less-than-satisfactory results. Thus it is important to design the experiment ahead of time so that the validity of the experimental results is assured. The *design of experiments* is a very important area of statistics that teaches us how to perform experiments to maximize our knowledge of the situation under consideration with a relatively few number of experimental runs.

As an example of such experimentation, let us return to the time and temperature needed to bake the best cake. Suppose that this is a completely new cake mix and the company has no idea about the right time and temperature, except that they will probably be in the ranges of 25–50 minutes and 250–400°F, respectively. Suppose that we have some measure of the "goodness" of a cake, possibly scores of professional tasters who grade each cake on a 100-point scale.

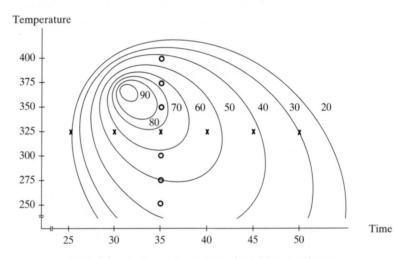

Figure 9.8   Goodness response to cake-baking experiment

Unknown to anyone, suppose that the real goodness response to the cakes made at different times and temperatures is given in Figure 9.8. That is, this response surface is depicted by contour curves which are like "isotherm curves." For example, a cake baked at 300°F for 35 minutes would get a score of 59 from the tasters. (Although there would be variation among tasters, let us say they average 59 for such a cake.) Clearly, if we knew this response surface ahead of time, we should bake this cake mix at about 363°F for about 31 minutes to obtain a score above 90. However, we do *not* know this surface. It is our job to find out something about it so that we can determine the optimal time and temperature at which to bake this cake. We do this by baking the cake mixture at different times and temperatures.

One procedure, called "changing one factor at a time," is sometimes used; but as we see in Figure 9.8, that scheme often misses the optimal levels of the factors. In it, we set each of the factors except the first at some middle value. In our case, we only have two factors; so we set the second at 325°F, say. Now let us run the experiment 6 times: at 25, 30, 35, 40, 45, and 50 minutes. The respective scores are *about* 20, 58, 68, 53, 41, and 30 by interpolating between contour curves at the x points. That is, with the temperature set at 325°F, the maximum score occurs at 35 minutes. Thus we now set the time at 35 minutes and perform six more experiments with respective temperatures of 250, 275, 300, 350, 375, and 400°F, recalling that 35 minutes and 325°F has already produced a score of about 68. The seven scores, with 68 inserted in the correct position, are about 43, 50, 59, 68, 76, 63, and 42. Thus it looks as if the maximum occurs at 350°F for 35 minutes; also, the maximum value is 76. Clearly, we have missed the true optimum levels of these two factors by using the "changing one factor at a time" scheme.

Wouldn't it have been much better to run at first four experiments changing the levels of each factor? That is, take two middle values of each factor: say 35 and 40 minutes, 300°F and 325°F. Then run four experiments at (35, 300°F), (35, 325°F), (40, 300°F), and (40, 325°F), obtaining scores of about 59, 68, 53, and 53, respectively. These values are plotted in Figure 9.9.

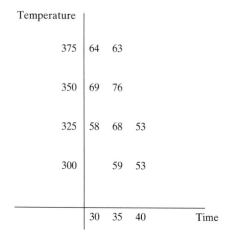

Figure 9.9   Results of cake-baking experiments

It seems fairly clear that the surface is increasing as we move to the left and upward. Thus we might run three more experiments at (30, 325°F), (30, 350°F), and (35, 350°F), obtaining scores of about 58, 69, and 76, respectively; these are also noted in Figure 9.9. Clearly, we must continue upward and we perform two more experiments at (30, 375°F) and (35, 375°F) obtaining 64 and 63. At this point, it seems as if we should investigate points around times of 30 to 35 minutes and temperatures in the neighborhood of $350°$ F, which will just about give us an optimal solution. But note how changing both factors in experimenting is better than changing one at a time. This is even more true as the number of factors increases.

In the preceding discussion, in which we first ran four experiments, with two factors at two levels each, is often called a 2^2 factorial experiment. If we had three factors involved and ran each at two levels, that would give us $2 \times 2 \times 2 = 8$ experimental runs, which is called a 2^3 factorial experiment. In general, with k factors at 2 levels each, we have a 2^k factorial experiment. Note that with $k = 10$ factors and only two levels each, we would need $2^{10} = 1024$ experimental runs. Often this is too many, and in advanced courses on design of experiments we learn how to run only a fraction of this number without losing too much information (see also Section 9.2).

Let us say that we now wish to compare two different methods of doing something (say teaching statistics). In essence we have only one factor, teaching methods, and we have only two levels: method I (M_1) and method 2 (M_2). If we had 50 students who were participating in this study, it would be desirable to place 25 in each group. The first group uses M_1 and the second M_2. Clearly, if the investigators favor M_2 over M_1, it would not be fair to let them place the better students in the second group. The assignments should be made at random so that each combination of 50 students taken 25 at a time has the same chance of being the first group. One way of doing this is placing the 50 names on equal size slips of paper and then drawing 25 from a bowl at random.

Suppose that we knew a little more about these students and were able to pair them up by mathematical ability, say. Student S_1 is about the same as S_2, S_3 about the same as S_4, and so on. Then let us select one from each pair at random (like with the flip of a coin) for the first group. We would say that the pair S_1, S_2 is a block; pair S_3, S_4 is a block; and so on. This procedure of assigning students to the respective groups is called **blocking**.

To illustrate blocking, suppose that we wish to test the wearing ability of two different types of soles for tennis shoes. Say we have 30 grade-school children who would participate. One way to select 15 for each group would be at random, and the first group would wear shoes with the first sole and the second group the second sole. After one month, say, we would compare the amount of wear for each group. A better way would be to block as follows: Let each student use a pair of shoes that has both types of soles, one on the right foot and the other on the left, as we recognize that each foot must take essentially the same number of steps (to be safe, we might let 15 students, selected at random, wear one type of sole on the right foot and the other 15 wear the other type on the right foot). In this experiment each student is a block. Sometimes in experiments, siblings or twins or two rats from the same litter could be used as a block.

Actually, a lot of good design of experiments is just plain common sense, but sometimes it is helpful for statisticians to note a few things about careful designs for those interested in running important experiments and studies. Our simple recommendation is to block what you can and randomize what you cannot block. However, if a person is involved in a complicated experiment or study, he or she should study advanced methods in the design of experiments or see a friendly statistician.

A block design can be used with more than two levels of one factor. Two illustrations of a *randomized block design* are given.

1. Suppose that we wish to test three methods of teaching statistics, say M_1, M_2, and M_3; and we have selected 60 students with whom to experiment. We could divide these 60 into three groups at random to be used with the respective methods, M_1, M_2, and M_3. This would be good randomized design, but it might happen "by the luck of the draw" that one of the methods had better math students than the other two and hence would seemingly have an advantage. Possibly a better way would be to put these students into 4 blocks, consisting of 15 students each. The first block would be the best math students as measured by the highest 15 ACT math scores. The second block would consist of the 15 students with the next highest ACT math scores. This is continued until we have 4 blocks, each with 15 students. After doing this, we would select 5 students at random from each block to use M_1, 5 students to use M_2, and 5 students to use M_3. This seems fairer for now each method has the same number of students with essentially the same math ability as measured by the ACT math exam. We now observe the scores on some evaluating test at the end of the semester and compare the differences.

2. Suppose that we wish to test four different feeds, say F_1, F_2, F_3, F_4, to be used on newly weaned pigs. It is recognized that there might be difference in pigs from different litters, some sows producing better pigs than others. Hence we block as follows. Select 4 pigs (after ruling out the runts) from each of several litters. The four from one litter are assigned at random to each of the four feeds. This is done for each litter; that is, we have blocked on sows' litters. So the experiment seems perfectly fair as each feed has one pig from each litter. At the end of the experiment we compare the differences of the weights gained with each feed.

EXERCISES

These and similar problems could be assigned as projects if time permits.

9.3.1 Design an experiment to find out how the distance to the basket and the angle from the perpendicular to the backboard affects your shooting percentage in basketball. Make certain you randomize the order of the positions from which you shoot to avoid "warm up" and "fatigue" effects. Display your results. (Essentially, you are estimating a response surface and you might draw contours.)

9.3.2 Suppose that we want to test the effects of two factors in making popcorn. One factor is the type of popper with two levels: oil-based or air-based. The other factor, type of popcorn, has three levels: gourmet, national brand, generic brand. Clearly, one response measurement would be the amount of popcorn that is obtained from $\frac{1}{2}$ cup of popcorn. But you could also try some sort of taste measurement, the amount of popcorn obtained from 10 cents' worth of each of popcorn, or some combination of taste and cost. What are your conclusions?

9.3.3 Suppose that we plan to compare two detergents, say A and B. We will wash equally dirty white clothes with each detergent, also changing two other factors: hot or cold water, soft or hard water. Thus, we have a three-factor experiment in which each factor has two levels; a 2^3 factorial experiment is under consideration and, without replication, we needed 8 runs. Replications might be desirable using 16 or 24 runs. What is the response variable?

9.3.4 We want to investigate the effects of five factors on the growth of pinto beans. They are

- Soaking fluid: water or regular beer
- Salinity: no salt or salt
- Acidity: no vinegar or vinegar
- Temperature: refrigerator or room
- Soaking time: 2 hours or 4 hours

Use 3 tablespoons of soaking fluid. For salt add $\frac{1}{4}$ teaspoon to soaking liquid. For vinegar, add 1 teaspoon to the soaking fluid. Note that this is a 2^5 factorial experiment and, without replications, 32 runs are needed to cover all possible combinations.

9.3.5 Thirty-six overweight undergraduate college males agree to participate in a study of the effectiveness of four reducing plans, say R_1, R_2, R_3, R_4. Before starting these plans, the ideal weight of each participant is determined and from that how much he is overweight. These amounts of overweight ranged roughly from 15 to 80 pounds. Explain how you might run a randomized block design with three blocks, in which you block on pounds overweight.

9.4 QUALITY IMPROVEMENT BY UNDERSTANDING VARIABILITY

As frequently happens in studying a textbook like this, we are so involved in the mathematics of probability, distributions, sampling distributions, and even descriptive statistics that we miss the real meanings of some of these concepts, particularly as far as the role they play in day-to-day activities. Even the definitions and distributions of the t and F random variables say something about understanding variability. Both t and F are defined as the ratio of two independent random variables. The resulting distributions have heavier tails than the respective random variables in the numerators. For example, the t random variable has a standardized normal random variable in the numerator and the p.d.f. of t is much heavier in the tails than that of $N(0, 1)$. That is, there is a tendency to get more outliers in sampling from the t distribution than from $N(0, 1)$.

We often observe the situation in which the denominator is a random variable, creating a ratio that has more variability than we think it should have. For illustration, suppose that we observe successive sample averages, say $\bar{x}_1, \bar{x}_2, \ldots, \bar{x}_k$. However, the sample sizes change from one sample to the next, say $n_1, n_2, \ldots, n_k$. If these sample sizes vary enough, the $\bar{x}$-values will look as if they arose from an approximate t-distribution. That is why in practice we often see more variation in a plot of sample means than we might expect; there are different sample sizes.

In Chapter 1, we noted a few important things about variability. For instance, it explains what is commonly called the "sophomore jinx": Some freshmen (rookies) have great first years (above their averages) and then most of them do worse the second year. The same explanation can be applied to movies that have sequels; often the original was outstanding, but even with the same cast and same director, the sequel is usually worse than the original.

We had also noted that workers who do very well (above their averages) and who thus might be given some reward, usually do worse during the next work period. On the other hand, those who have bad periods (below average) and are reprimanded frequently do better the next. Yet it is wrong to think that a policy of reprimand is better than that of a pat on the back. Both situations can be explained by understanding the worker's pattern of variation. If the employer really wants to improve the outcome of the process, he or she must consider ways of improving the level (average). The workers can sometimes make small adjustments, but they need road maps to make the major ones. Thus it is really the responsibility of management to improve the process substantially. That is, management must realize that working harder with techniques that have failed will not improve the situation much. Changes must be made, and often data and the resulting statistical analysis can suggest changes for the better.

It is also disturbing to see the following situation: Suppose that the manager has created a team (say of 10) of outstanding workers, and it is time to give raises. The workers are ranked from one to ten. (It is always true that 10 workers, no matter how good or how bad, will always get the ranks 1, 2, 3, . . . , 10.) Then say they get raises according to their ranks, which might have been determined simply by a random process for this particular period. That is, in the next period, their ranks might be entirely different. How does the small raise make the one with lowest rank feel if, in fact, all 10 are members of an outstanding team and there might be little or no difference among the persons ranked 1 and 10? Not good, and he or she is less likely to help the ones with higher ranks in a future period. That is, this does not promote teamwork. As a matter of fact, there is always this danger in any kind of reward, such as the "worker of the month." Others are not likely to help that person the next month, because they want to be the worker of the month and receive the corresponding bonus.

A better way of rewarding a good team would be to give them essentially the same raises. Clearly, a good manager will continuously monitor performance and supply appropriate feedback to the workers. Of course, if one worker is consistently on the high side, then he or she should be considered for promotion or a substantial raise. On the other hand, if some worker is on the low end most of the time, some help for this worker is in order: maybe additional training or even a transfer to a different department for which the worker's talents are more suitable. It is important not to demean this worker because each of us needs to take pride in our accomplishments, even though they may be little. If at all possible, firing the worker should be avoided, possibly by finding some other suitable job in the company.

Some teachers will announce to a class of 30 students that there will be only so many A's, so many B's, and so on. This is wrong! If those students are competing for those given number of A's, why would anyone want to help anyone else? It completely destroys any sort of teamwork, and yet that is what students must learn to do when they are on the job later in life. Clearly, it is better to say that all of them can earn an A (they probably won't) and encourage them to work together. In that way, the teacher has a better opportunity to improve the level of the entire class which, after all, is the real objective. Too frequently, mathematics and science courses discourage interaction among students, and teachers must do everything possible to break down those barriers.

Many of these preceding observations are those of W. Edwards Deming, an esteemed statistician who went to Japan after World War II and taught the Japanese how to make quality products. For his work there, he was awarded the Emperor's Medal and the Japanese established the Deming Prize to be awarded each year to the company or individual contributing the most to quality improvement. One of the things Deming stressed is the need for "profound knowledge," and a major item in that is understanding variation and statistical theory.

In making quality products, you want to reduce the variation as much as possible and move the level closer to the target. Deming believed that barriers between departments, between management and workers, and among workers must be broken down to improve communications and the ability to work as a team. The lines of communication, all the way from suppliers to customers, must be open to help reduce variation and improve products and services. For example, he argued that a company should not buy only on price tag but have a few reliable suppliers (possibly one) for a single part because that will reduce variation. That is, many different suppliers would obviously increase variation. Moreover, he argued that you should become partners, friends if you like, with your suppliers. You learn to trust the other; in that way, you can use methods like "just in time," in which your inventory can be kept reasonably small, to keep costs down.

If each of us thinks about these ideas, we might become obsessed by understanding variation, and that might make a big difference in our everyday lives. For illustration, once we have selected good suppliers, we continue to go to the same barber, the same service station, the same clothier, the same bank, and on and on. We like to buy items, even if a little more expensive, from places that will give us good service if something goes wrong. If this does not happen, then of course we must consider changing suppliers.

Clearly, listening to the customer can help improve the quality of our products and services. We can then meet—or even exceed—the expectations of our customers. We should continually try to improve by reducing the variation and moving the process to a better level (high is not always good, as in golf). Although the customer—as well as the supplier—must be part of the total team, more often management must continue to look for better ways to do things, ways that the customer never would have imagined. For example, in the early days of automobiles not many owners would have thought of driving on pneumatic tires. Harvey Firestone did, and in this way exceeded the expectations of those early customers.

Deming also preached constancy of purpose. If management ideas tend to change too much, employees really do not know what to do and cannot do their best.

That is, they are mixed up, increasing variation. It is better for them to receive a constant signal from their employer, a signal that changes only if research dictates ways of improving.

More training and education for everyone associated with a company also decreases variation by teaching how to make the product more uniform. Workers must know that they do not have to be afraid to make suggestions to improve the process. Often, being team members will make it easier for workers to speak up without fear of reprisal.

Many of us remember playing a game called "telephone": One whispers a message to the next person, who whispers the message to the next, and so on. The message at the end is compared to the original message, and it is usually much different from the original. This is like trying to hit a target (say at zero) with a random variable X_1. Then starting with X_1 as the center and adding on another error X_2, and so on, creating the sum of the errors, $X_1 + X_2 + \cdots + X_n$, which has an ever-increasing variance with independent errors. Deming would say, "We are off to the Milky Way." Yet we actually do this in business and industry by letting worker train worker. Once errors are introduced, they will stay there, and others will add on and on. (Incidentally, we might say the same thing about too many layers of management.) To decrease variation, wouldn't it be better to have a master instructor train each worker (or have fewer layers of management)?

Deming also noted that requiring quotas does not help the quality. A foreman who has a quota of 100 per day will often ship out 90 good and 10 bad items just to make the quota. Clearly, it would reduce the variation and satisfy the customer better if only the 90 good ones were sent on.

This leads to the point that a final inspection of products often does not really improve the quality. With such a procedure, you can only eliminate the bad ones and send on the good ones. Improvements in the design of the products and manufacturing processes are needed. If these are done well, often with the help of statistical methods, that final inspection can be eliminated. That is, improvements should be made early in the manufacturing process rather than try to correct things at an end inspection by weeding out the bad items.

Once you recognize that there is variation in almost every process, you begin to think like a statistician. Often you want to reduce that variation and center what variation that remains at the right level, because then "the doors will fit better." But while we want to reduce undesirable variation, we do not want to reduce all variation: Most of us do not want to eat Chinese food every night even though we enjoy it from time to time. Certain variation is "the spice of life."

This desire to understand variation is the key to statistical thinking, and it can affect everything we do, including our relationships with others. All of us have our "ups and downs," and we must seriously think about how to deal with others, recognizing this variation.

At one state university, the legislators appropriated enough money for 2 percent raises one year. The administrators thought this was not adequate; so they had the "brilliant" idea to give one-third of the faculty 6 percent raises and zero to the other two-thirds. Thus, in a department of 18, six received raises and 12 did not. Who ranked those 18? Was there much difference between the sixth and seventh? Or between the fifth and tenth for that matter? So the person who received a raise might have people on either side of him or her that got nothing. It created a terrible morale problem. Deming certainly would have suggested to give everybody a 2 percent raise and blame it on the legislators. Now Deming recognized differences among people and accordingly they must be treated differently, but certainly not just due to chance variation.

It is the authors' opinion that this quality improvement must begin with the individual by creating a personal vision, goals, and principles by which to live. Hopefully, one of those individuals is the chief of the organization, for then things go much smoother. Bob Galvin, who was the CEO of Motorola during the 1980s, believes that "quality improvement is a daily, personal, priority obligation," and his company made great improvements in the quality of their products during that period. Much of this progress was due to understanding variation and statistical methods. Even Motorola's program was called Six-Sigma, in which *sigma* referred to a standard deviation, a name that is a delight to a statistician. But the description of that program is another story for another day.

Each of us should use as much statistical thinking as we know to help us build trust with the persons with whom we deal directly or indirectly. Let us not do what those administrators of that university did when they did not understand the variability associated with ranking people. That action destroyed any trust that was there and created a mean competition among the members of that faculty. Why should anyone help others when their gain only hurts that person? Although statistical thinking cannot solve all problems, it most certainly can help in many situations. We hope you use what statistics you know to benefit you the rest of your life.

10

NONPARAMETRIC METHODS

10.1 ORDER STATISTICS

The **order statistics** are the observations of the random sample arranged, or ordered, in magnitude from the smallest to the largest. In recent years, the importance of order statistics has increased owing to the more frequent use of nonparametric inferences and robust procedures. However, order statistics have always been prominent because, among other things, they are needed to determine rather simple statistics such as the sample median, the sample range, and the empirical distribution function. Recall that in Section 1.3 we discussed observed order statistics in connection with descriptive and exploratory statistical methods. We consider certain interesting aspects about their distributions in this section; and in Section 10.2 we demonstrate how they can be used to construct confidence intervals for percentiles.

In most of our discussions about order statistics, we will assume that the random sample arises from a continuous-type distribution. This means, among other things, that the probability of any two sample observations being equal is zero. That is, the probability is *one* that the observations can be ordered from smallest to largest without having two equal values. Of course, in practice, we do frequently observe *ties*; but if the probability of this is small, the following distribution theory will hold approximately. Thus, in the discussion here, we are assuming that the probability of ties is zero.

Example 10.1.1

The values $x_1 = 0.62$, $x_2 = 0.98$, $x_3 = 0.31$, $x_4 = 0.81$, and $x_5 = 0.53$ are the $n = 5$ observed values of five independent trials of an experiment with p.d.f. $f(x) = 2x, 0 < x < 1$. The observed order statistics are

$$y_1 = 0.31 < y_2 = 0.53 < y_3 = 0.62 < y_4 = 0.81 < y_5 = 0.98.$$

Recall from Section 1.3 that the middle observation in the ordered arrangement, here $y_3 = 0.62$, is called the sample median, and the difference of the largest and the smallest, here

$$y_5 - y_1 = 0.98 - 0.31 = 0.67,$$

is called the sample range.   ◇

If $X_1, X_2, \ldots, X_n$ are observations of a random sample of size n from a continuous-type distribution, we let the random variables

$$Y_1 < Y_2 < \cdots < Y_n$$

denote the order statistics of that sample. That is,

$$Y_1 = \text{smallest of } X_1, X_2, \ldots, X_n,$$
$$Y_2 = \text{second smallest of } X_1, X_2, \ldots, X_n,$$

$$\vdots$$

$$Y_n = \text{largest of } X_1, X_2, \ldots, X_n.$$

There is a very simple method for determining the distribution function of the rth order statistic, Y_r. This procedure depends on the binomial distribution and is illustrated in Example 10.1.2.

Example 10.1.2

Let $Y_1 < Y_2 < Y_3 < Y_4 < Y_5$ be the order statistics of a random sample X_1, X_2, X_3, X_4, X_5 of size $n = 5$ from the distribution with p.d.f. $f(x) = 2x$, $0 < x < 1$. Consider $P(Y_4 < 1/2)$. For the event $\{Y_4 < 1/2\}$ to occur, at least four of the random variables X_1, X_2, X_3, X_4, X_5 must be less than $1/2$ because Y_4 is the fourth smallest among the five observations. Thus if the event $\{X_i < 1/2\}$, $i = 1, 2, \ldots, 5$, is called "success," we must have at least four successes in the five mutually independent trials, each of which has probability of success

$$P\left(X_i \le \frac{1}{2}\right) = \int_0^{1/2} 2x \, dx = \left(\frac{1}{2}\right)^2 = \frac{1}{4}.$$

Thus

$$P\left(Y_4 \le \frac{1}{2}\right) = \binom{5}{4}\left(\frac{1}{4}\right)^4\left(\frac{3}{4}\right) + \left(\frac{1}{4}\right)^5 = 0.0156.$$

In general, if $0 < y < 1$, then the distribution function of Y_4 is

$$G(y) = P(Y_4 < y) = \binom{5}{4}(y^2)^4(1 - y^2) + (y^2)^5,$$

since this represents the probability of at least four "successes" in five independent trials, each of which has probability of success

$$P(X_i < y) = \int_0^y 2x \, dx = y^2.$$

The p.d.f. of Y_4 is therefore, for $0 < y < 1$,

$$g(y) = G'(y) = \binom{5}{4} 4(y^2)^3 (2y)(1 - y^2) + \binom{5}{4}(y^2)^4(-2y) + 5(y^2)^4(2y)$$

$$= \frac{5!}{3! \, 1!} (y^2)^3 (1 - y^2)(2y), \qquad 0 < y < 1.$$

Note that in this example, the distribution function of each X is $F(x) = x^2$ when $0 < x < 1$. Thus

$$g(y) = \frac{5!}{3! \, 1!} [F(y)]^3 [1 - F(y)] f(y), \qquad 0 < y < 1. \qquad \diamondsuit$$

The preceding example should make the following generalization easier to read. Let $Y_1 < Y_2 < \cdots < Y_n$ be the order statistics of a random sample of size n from a distribution of the continuous type with distribution function $F(x)$ and p.d.f. $F'(x) = f(x)$, where $0 < F(x) < 1$ for $a < x < b$ and $F(a) = 0$, $F(b) = 1$. (It is possible that $a = -\infty$ and/or $b = +\infty$.) The event that the rth order statistic Y_r is at most y, $\{Y_r \leq y\}$, can occur if and only if at least r of the n observations are less than or equal to y. That is, here the probability of "success" on each trial is $F(y)$ and we must have at least r successes. Thus

$$G_r(y) = P(Y_r \leq y) = \sum_{k=r}^{n} \binom{n}{k} [F(y)]^k [1 - F(y)]^{n-k}.$$

That is, rewriting this slightly, we have

$$G_r(y) = \sum_{k=r}^{n-1} \binom{n}{k} [F(y)]^k [1 - F(y)]^{n-k} + [F(y)]^n.$$

Thus the p.d.f. of Y_r is

$$g_r(y) = G_r'(y) = \sum_{k=r}^{n-1} \binom{n}{k} (k)[F(y)]^{k-1} f(y)[1 - F(y)]^{n-k}$$

$$+ \sum_{k=r}^{n-1} \binom{n}{k} [F(y)]^k (n-k)[1 - F(y)]^{n-k-1}[-f(y)]$$

$$+ n[F(y)]^{n-1} f(y). \tag{10.1}$$

But since

$$\binom{n}{k}k = \frac{n!}{(k-1)!\,(n-k)!} \quad \text{and} \quad \binom{n}{k}(n-k) = \frac{n!}{k!\,(n-k-1)!},$$

we have that the p.d.f. of Y_r is

$$g_r(y) = \frac{n!}{(r-1)!\,(n-r)!}\,[F(y)]^{r-1}[1-F(y)]^{n-r}f(y), \qquad a < y < b,$$

which is the first term of the first summation in $g_r(y) = G_r'(y)$, Equation 10.1. The remaining terms in $g_r(y) = G_r'(y)$ sum to zero because the second term of the first summation (when $k = r+1$) equals the negative of the first term in the second summation (when $k = r$), and so on. Finally, the last term of the second summation equals the negative of $n[F(y)]^{n-1}f(y)$. To see this clearly, the student is urged to write out a number of terms in these summations (see Exercise 10.1.4).

It is worth noting that the p.d.f. of the smallest order statistic is

$$g_1(y) = n[1-F(y)]^{n-1}f(y), \qquad a < y < b,$$

and the p.d.f. of the largest order statistic is

$$g_n(y) = n[F(y)]^{n-1}f(y), \qquad a < y < b.$$

Remark   There is one very satisfactory way to construct heuristically the expression for the p.d.f. of Y_r. To do this, we must recall the multinomial probability and then consider the probability element $g_r(y)(\Delta y)$ of Y_r. If the length Δy is *very* small, $g_r(y)(\Delta y)$ represents approximately the probability

$$P(y < Y_r \le y + \Delta y).$$

Thus we want the probability $g_r(y)(\Delta y)$ that $(r-1)$ items fall less than y, $(n-r)$ items are greater than $y + \Delta y$, and one item falls between y and $y + \Delta y$. Recall that the probabilities on a single trial are

$$P(X \le y) = F(y)$$
$$P(X > y + \Delta y) = 1 - F(y + \Delta y) \approx 1 - F(y)$$
$$P(y < X \le y + \Delta y) \approx f(y)(\Delta y).$$

Thus the multinomial probability is approximately

$$g_r(y)(\Delta y) = \frac{n!}{(r-1)!\,1!\,(n-r)!}\,[F(y)]^{r-1}[1-F(y)]^{n-r}[f(y)(\Delta y)].$$

If we divide both sides by the length Δy, the formula for $g_r(y)$ results.

Example 10.1.3

Let $Y_1 < Y_2 < Y_3 < Y_4$ be the order statistics of a random sample of size $n = 4$ from a distribution with the uniform of p.d.f. $f(x) = 1, 0 < x < 1$. Since $F(x) = x$, when $0 \le x < 1$, the p.d.f. of Y_3 is

$$g_3(y) = \frac{4!}{2! \, 1!} y^2 (1 - y), \qquad 0 < y < 1.$$

Thus, for illustration,

$$P\left(\frac{1}{3} < Y_3 < \frac{2}{3}\right) = \int_{1/3}^{2/3} 12 y^2 (1 - y) \, dy$$

$$= \left[4y^3 - 3y^4\right]_{1/3}^{2/3} = \frac{13}{27}. \qquad \diamond$$

Example 10.1.4

Let $Y_1 < Y_2 < \cdots < Y_7$ be the order statistics of a random sample of size $n = 7$ from a distribution with p.d.f. $f(x) = 3(1 - x)^2, 0 < x < 1$. Compute the probability that the sample median is less than $1 - \sqrt[3]{0.6}$; that is, find $P(Y_4 < 1 - \sqrt[3]{0.6})$. We could find the p.d.f. of Y_4. However, note that the probability of a single observation being less than $1 - \sqrt[3]{0.6}$ is

$$\int_0^{1 - \sqrt[3]{0.6}} 3(1 - x)^2 \, dx = \left[-(1 - x)^3\right]_0^{1 - \sqrt[3]{0.6}}$$

$$= 1 - \left(\sqrt[3]{.6}\right)^3 = 0.4.$$

Thus, from Table II,

$$P(Y_4 < 1 - \sqrt[3]{0.6}) = \sum_{k=4}^{7} \binom{7}{k} (0.4)^k (0.6)^{7-k}$$

$$= 1.0000 - 0.7102 = 0.2898. \qquad \diamond$$

Now, in Example 10.1.4, it is easy enough to look up the resulting probability in Table II. Suppose that the sample size is much larger and we wish to know certain probabilities associated with a given order statistic, such as $P(Y_r < y)$. Of course, these probabilities can be expressed in terms of binomial probabilities, which may be difficult to evaluate but in turn can be approximated by normal probabilities, provided that n is large. This is illustrated in Example 10.1.5.

Example 10.1.5

Let $Y_1 < Y_2 < \cdots < Y_{100}$ be the order statistics of a random sample of size $n = 100$ from a continuous distribution having median $m = 68.1$ (i.e., the probability that a single observation X is less than 68.1 is $1/2$). What then is the probability that the 55th order statistic, Y_{55}, is less than $m = 68.1$? Since a binomial distribution with $n = 100$ and $p = 1/2$ has mean 50 and standard deviation 5, we have that

$$P(Y_{55} < 68.1) = \sum_{k=55}^{100} \binom{100}{k} \left(\frac{1}{2}\right)^k \left(\frac{1}{2}\right)^{100-k}$$

$$= 1 - \sum_{k=0}^{54} \binom{100}{k} \left(\frac{1}{2}\right)^k \left(\frac{1}{2}\right)^{100-k}$$

$$\approx 1 - \Phi\left(\frac{54.5 - 50}{5}\right) = 1 - \Phi(0.9)$$

$$= 1 - 0.8159 = 0.1841. \qquad \diamond$$

Recall that in Theorem 4.5.2, we proved that if X has a distribution function $F(x)$ of the continuous type, then $F(X)$ has a uniform distribution on the interval zero to one. If $Y_1 < Y_2 < \cdots < Y_n$ are the order statistics of a random sample $X_1, X_2, \ldots, X_n$ of size n, then

$$F(Y_1) < F(Y_2) < \cdots < F(Y_n)$$

because F is a nondecreasing function and the probability of an equality is again zero. Note that this last display could be looked upon as an ordering of the mutually independent random variables $F(X_1), F(X_2), \ldots, F(X_n)$, each of which is $U(0, 1)$. That is,

$$W_1 = F(Y_1) < W_2 = F(Y_2) < \cdots < W_n = F(Y_n)$$

can be thought of as the order statistics of a random sample of size n from that uniform distribution. Since the distribution function of $U(0, 1)$ is $G(w) = w, 0 < w < 1$, the p.d.f. of the rth order statistic $W_r = F(Y_r)$ is

$$h_r(w) = \frac{n!}{(r-1)!\,(n-r)!}\, w^{r-1}(1-w)^{n-r}, \qquad 0 < w < 1.$$

Of course, the mean, $E(W_r) = E[F(Y_r)]$ of $W_r = F(Y_r)$, is given by the integral

$$E(W_r) = \int_0^1 w \, \frac{n!}{(r-1)!\,(n-r)!}\, w^{r-1}(1-w)^{n-r}\, dw.$$

This can be evaluated by integrating by parts several times, but it is easier to obtain the answer if we rewrite it as follows:

$$E(W_r) = \left(\frac{r}{n+1}\right) \int_0^1 \frac{(n+1)!}{r!\,(n-r)!}\, w^r (1-w)^{n-r}\, dw.$$

The integrand in this last expression can be thought of as the p.d.f. of the $(r+1)$st order statistic of a random sample of size $n+1$ from a $U(0,1)$ distribution. Hence the integral must equal 1 and

$$E(W_r) = \frac{r}{n+1}, \qquad r = 1, 2, \ldots, n.$$

There is an extremely interesting interpretation of $W_r = F(Y_r)$. Note that $F(Y_r)$ is the cumulated probability up to and including Y_r or, equivalently, the area under $f(x) = F'(x)$ but less than Y_r. Hence $F(Y_r)$ can be treated as a random area. Since $F(Y_{r-1})$ is also a random area, $F(Y_r) - F(Y_{r-1})$ is the random area under $f(x)$ between Y_{r-1} and Y_r. The expected value of the random area between any two adjacent order statistics is then

$$E[F(Y_r) - F(Y_{r-1})] = E[F(Y_r)] - E[F(Y_{r-1})]$$

$$= \frac{r}{n+1} - \frac{r-1}{n+1} = \frac{1}{n+1}.$$

Also, it is easy to show (see Exercise 10.1.6)

$$E[F(Y_1)] = \frac{1}{n+1} \qquad \text{and} \qquad E[1 - F(Y_n)] = \frac{1}{n+1}.$$

That is, the order statistics $Y_1 < Y_2 < \cdots < Y_n$ partition the support of X into $n+1$ parts and thus create $n+1$ areas under $f(x)$ and above the x axis. "On the average," each of the $n+1$ areas equals $1/(n+1)$.

If we recall that the $(100p)$th percentile π_p is such that the area under $f(x)$ to the left of π_p is p, the preceding discussion suggests that we let Y_r be an estimator of π_p, where $p = r/(n+1)$. For this reason, we define the $(100p)$**th percentile of the sample** as Y_r where $r = (n+1)p$. In case $(n+1)p$ is not an integer, we use a weighted average (or an average) of the two adjacent order statistics Y_r and Y_{r+1}, where r is the greatest integer $[(n+1)p]$ in $(n+1)p$. In particular, the sample median is

$$\tilde{m} = \begin{cases} Y_{(n+1)/2}, & \text{when } n \text{ is odd,} \\ \dfrac{Y_{n/2} + Y_{(n/2)+1}}{2}, & \text{when } n \text{ is even.} \end{cases}$$

Example 10.1.6

Let X equal the weight of soap in a "1000-gram" bottle. A random sample of $n = 12$ observations of X yielded the following weights that have been ordered:

$$\begin{array}{cccccc} 1013 & 1019 & 1021 & 1024 & 1026 & 1028 \\ 1033 & 1035 & 1039 & 1040 & 1043 & 1047 \end{array}$$

Since $n = 12$ is even, the sample median is

$$\tilde{m} = \frac{y_6 + y_7}{2} = \frac{1028 + 1033}{2} = 1030.5.$$

The location of the 25th percentile (or first quartile) is

$$(n + 1)(0.25) = (12 + 1)(0.25) = 3.25.$$

Thus the first quartile, using a weighted average, is

$$\tilde{q}_1 = y_3 + (0.25)(y_4 - y_3) = (0.75)y_3 + (0.25)y_4$$
$$= (0.75)(1021) + (0.25)(1024) = 1021.75.$$

Similarly, the 75th percentile (or third quartile) is

$$\tilde{q}_3 = y_9 + (0.75)(y_{10} - y_9) = (0.25)y_9 + (0.75)y_{10}$$
$$= (0.25)(1039) + (0.75)(1040) = 1039.75$$

because $(12 + 1)(0.75) = 9.75$. Since $(12 + 1)(0.60) = 7.8$, the 60th percentile is

$$\tilde{\pi}_{0.60} = (0.2)y_7 + (0.8)y_8 = (0.2)(1033) + (0.8)(1035) = 1034.6. \qquad \diamond$$

We close this section with examples, one of the type that will be used in Section 10.2 to find confidence intervals for percentiles and a second concerning a maximum likelihood estimator.

Example 10.1.7

Let $Y_1 < Y_2 < \cdots < Y_{13}$ be the order statistics of a random sample of size $n = 13$ from a continuous-type distribution with 35th percentile $\pi_{0.35} = 73.247$. To find $P(Y_3 < \pi_{0.35} < Y_7)$, note that the event $\{Y_3 < \pi_{0.35} < Y_7\}$ happens if and only if there are at least three but less than seven "successes," where the probability of success is $p = 0.35$. Thus

$$P(Y_3 < \pi_{0.35} < Y_7) = \sum_{k=3}^{6} \binom{13}{k}(0.35)^k (0.65)^{13-k}$$

$$= 0.8705 - 0.1132 = 0.7573,$$

from Table II. $\qquad \diamond$

Example 10.1.8

Let $X_1, X_2, \ldots, X_n$ denote a random sample from a distribution with p.d.f.

$$f(x; \theta) = e^{-(x-\theta)}, \qquad \theta \leq x < \infty,$$

where $0 < \theta < \infty$. The likelihood function is

$$L(\theta) = e^{-n\bar{x}+n\theta}, \qquad \theta \leq x_i, \quad i = 1, 2, \ldots, n,$$

and zero otherwise. Thus

$$\ln L(\theta) = -n\bar{x} + n\theta, \qquad \theta \leq x_i, \quad i = 1, 2, \ldots, n,$$

and the derivative

$$D_\theta[\ln L(\theta)] = n$$

is positive for all $\theta \leq x_i$, $i = 1, 2, \ldots, n$. That is, $\ln L(\theta)$ and hence $L(\theta)$ is an increasing function of θ. To achieve the maximum value of $L(\theta)$, we accordingly make θ as large as possible. But since $\theta \le x_i$, $i = 1, 2, \ldots, n$, the largest value of θ is the smallest x-value. That is, the maximum likelihood estimator is $\hat{\theta} = Y_1$, the first order statistic.

$\diamond$

EXERCISES

10.1.1 One of the tasks performed by a computer operator is that of testing tapes in order to detect bad records (i.e., records that could not store information in the form of positive or negative charges). Let X equal the distance in feet between bad records. Ten observations of X are

$$67 \quad 6 \quad 7 \quad 35 \quad 78 \quad 28 \quad 74 \quad 5 \quad 9 \quad 37$$

(a) Find the order statistics.
(b) Find the median and 80th percentile of the sample.
(c) Determine the first and third quartiles (i.e., 25th and 75th percentiles) of the sample.

10.1.2 Let X equal the forced vital capacity (the volume of air a person can expel from their lungs) for a male freshman. Seventeen observations of X, which have been ordered, are

$$3.7 \quad 3.8 \quad 4.0 \quad 4.3 \quad 4.7 \quad 4.8 \quad 4.9 \quad 5.0$$
$$5.2 \quad 5.4 \quad 5.6 \quad 5.6 \quad 5.6 \quad 5.7 \quad 6.2 \quad 6.8 \quad 7.6$$

(a) Find the median, the first quartile, and the third quartile.
(b) Find the 35th and 65th percentiles.

10.1.3 Let $Y_1 < Y_2 < Y_3 < Y_4 < Y_5$ be the order statistics of a random sample of size $n = 5$ from an exponential distribution that has a mean of $\theta = 3$.
(a) Find the p.d.f. of the sample median Y_3.
(b) Compute the probability that Y_4 is less than 5.
(c) Determine $P(1 < Y_1)$.

10.1.4 In the expression for $g_r(y) = G'_r(y)$ in Equation 10.1 let $n = 6$, $r = 3$, and write out the summations, showing that the "telescoping" suggested in the text is achieved.

10.1.5 Let $Y_1 < Y_2 < \cdots < Y_8$ be the order statistics of a random sample of size $n = 8$ from a continuous-type distribution with 70th percentile $\pi_{0.7} = 27.3$.
(a) Determine $P(Y_7 < 27.3)$.
(b) Find $P(Y_5 < 27.3 < Y_8)$.

10.1.6 Let $W_1 < W_2 < \cdots < W_n$ be the order statistics of a random sample of size n from a $U(0, 1)$ distribution.
(a) Find the p.d.f. of W_1 and that of W_n.
(b) Use the results of part (a) to verify that $E(W_1) = 1/(n+1)$ and $E(W_n) = n/(n+1)$.

10.1.7 Let $Y_1 < Y_2 < \cdots < Y_{72}$ be the order statistics of a random sample of size $n = 72$ from a distribution of the continuous type having the $100(1/3)$th percentile $\pi_{1/3} = 7.2$.
(a) Approximate the probability that $\{Y_{20} < \pi_{1/3} = 7.2\}$.
(b) Find approximately the probability $P(Y_{18} < \pi_{1/3} < Y_{30})$ by first noting that the event $\{Y_{18} < \pi_{1/3} < Y_{30}\}$ means that we must have at least 18 but less than 30 "successes."

10.1.8 Let $W_1 < W_2 < \cdots < W_n$ be the order statistics of a random sample of size n from a $U(0, 1)$ distribution.

(a) Show that $E(W_r^2) = r(r + 1)/(n + 1)(n + 2)$ using a technique similar to that used in determining that $E(W_r) = r/(n + 1)$.

(b) Find the variance of W_r.

10.1.9 Let $Y_1 < Y_2 < \cdots < Y_{19}$ be the order statistics of a random sample of size $n = 19$ from the exponential distribution with mean θ.

(a) What is the p.d.f. of Y_1?

(b) Find, using integration, the value of $E[F(Y_1)]$, where F is the distribution function of the exponential distribution.

10.1.10 Let $Y_1 < Y_2 < \cdots < Y_n$ be the order statistics of a random sample of size n from the uniform distribution $U(0, \theta)$.

(a) Show that the maximum likelihood estimator of θ is Y_n.

(b) Show that $E(Y_n) = n\theta/(n + 1)$ and $\text{Var}(Y_n) = n\theta^2/(n + 1)^2(n + 2)$.

(c) Find a constant c so that cY_n is an unbiased estimator of θ.

(d) Show that a $100(1 - \alpha)\%$ confidence interval for θ is $[y_n, y_n/\alpha^{1/n}]$.

10.1.11 Let $Y_1 < Y_2 < Y_3$ be the order statistics of a random sample X_1, X_2, X_3 of size 3 from the uniform distribution $U(\theta - 1/2, \theta + 1/2)$. Let $W_1 = \overline{X} = (1/3)\sum_{i=1}^{3} X_i$, the sample mean; $W_2 = Y_2$, the sample median; and $W_3 = (Y_1 + Y_3)/2$, the sample midrange.

(a) Give a reason why each of W_1, W_2, and W_3 is a possible estimator of θ.

(b) Given that $E(W_1) = E(W_2) = E(W_3) = \theta$ and $\text{Var}(W_1) = 1/36$, $\text{Var}(W_2) = 1/20$, $\text{Var}(W_3) = 1/40$, which of these three estimators is the best?

(c) The following data give 20 sets of observations of y_1, y_2, y_3 that were generated using $\theta = 1/2$. For each replication, calculate the values of $w_1 = \overline{y} = \overline{x}$, $w_2 = y_2$, and $w_3 = (y_1 + y_3)/2$.

(d) Find the sample mean and sample variance of the 20 observations of (i) W_1, (ii) W_2, and (iii) W_3.

(e) From this small set of empirical evidence, which estimator seems to be the best? Why?

Replication	y_1	y_2	y_3
1	0.3098	0.3146	0.6144
2	0.0887	0.2601	0.3247
3	0.3638	0.7087	0.7960
4	0.2159	0.7038	0.8607
5	0.6016	0.7653	0.7908
6	0.4995	0.6317	0.9305
7	0.0341	0.2097	0.9026
8	0.1400	0.6444	0.9620
9	0.1485	0.5929	0.6025
10	0.1668	0.3718	0.5973
11	0.0371	0.4175	0.9272
12	0.2395	0.3198	0.3275
13	0.3199	0.3566	0.7422
14	0.7278	0.8002	0.8480
15	0.0294	0.5282	0.9613
16	0.0108	0.2354	0.4948
17	0.0981	0.4908	0.8840
18	0.1937	0.4930	0.7527
19	0.0436	0.7345	0.9981
20	0.5599	0.9046	0.9591

10.1.12 Let $X_1, X_2, \ldots, X_n$ be a random sample of size n from a two-parameter (shifted) exponential distribution with p.d.f.

$$f(x; \alpha, \theta) = \frac{1}{\theta} e^{-(x-\alpha)/\theta}, \qquad \alpha \leq x < \infty$$

where $-\infty < \alpha < \infty, 0 < \theta < \infty$.

(a) Find the mean and variance of this distribution.

(b) Find the maximum likelihood estimators for α and θ.

(c) Use the following data to give point estimates of α and θ:

$$18.218 \quad 16.117 \quad 8.609 \quad 85.387 \quad 20.668 \quad 34.232 \quad 37.941 \quad 31.183$$

10.1.13 Let X equal the weight in grams of peanuts in a "14-gram" snack package. The order statistics of a random sample of size $n = 16$ are

$$13.9 \quad 14.4 \quad 14.6 \quad 14.7 \quad 14.7 \quad 15.2 \quad 15.2 \quad 15.2$$
$$15.3 \quad 15.4 \quad 15.4 \quad 15.5 \quad 15.6 \quad 15.6 \quad 15.9 \quad 16.4$$

(a) Find the 30th percentile of this sample.

(b) Find $P(Y_2 < \pi_{0.30} < Y_8)$, where Y_2 and Y_8 are the respective second and eighth order statistics in the sample before the sample is observed.

10.2 CONFIDENCE INTERVALS FOR PERCENTILES

In Section 10.1 we defined the sample percentiles in terms of the order statistics and noted that the sample percentiles can be used to estimate the corresponding distribution percentiles. In this section we use the order statistics to construct confidence intervals for the unknown distribution percentiles. Since little is assumed about the underlying distribution in the construction of these confidence intervals, they are often called **distribution-free confidence intervals**.

If $Y_1 < Y_2 < Y_3 < Y_4 < Y_5$ are the order statistics of a random sample of size $n = 5$ from a continuous-type distribution, then the sample median Y_3 could be thought of as an estimator of the distribution median $\pi_{0.5}$. We shall let $m = \pi_{0.5}$. We could simply use the sample median Y_3 as an estimator of the distribution median m. However, we are certain that all of us recognize that, with only a sample of size 5, we would be quite lucky if the observed $Y_3 = y_3$ were very close to m. Thus we now describe how a confidence interval can be constructed for m.

Instead of simply using Y_3 as an estimator of m, let us also compute the probability that the random interval (Y_1, Y_5) includes m. That is, let us determine $P(Y_1 < m < Y_5)$. This is easy if we follow the procedure given in Section 10.1. Again say that we have success if an individual item, say X, is less than m; thus the probability of success on one of the independent trials is $P(X < m) = 0.5$. In order for the first order statistic Y_1 to be less than m and the last order statistic Y_5 to be greater than m, we must have at least one success but not five successes. That is,

$$P(Y_1 < m < Y_5) = \sum_{k=1}^{4} \binom{5}{k} \left(\frac{1}{2}\right)^k \left(\frac{1}{2}\right)^{5-k}$$

$$= 1 - \left(\frac{1}{2}\right)^5 - \left(\frac{1}{2}\right)^5 = \frac{15}{16}.$$

So the probability that the random interval (Y_1, Y_5) includes m is $15/16 \approx 0.94$. Suppose that this random sample is actually taken and the order statistics are observed to equal $y_1 < y_2 < y_3 < y_4 < y_5$, respectively. Then (y_1, y_5) is a 94% confidence interval for m.

It is interesting to note what happens as the sample size increases. Let $Y_1 < Y_2 < \cdots < Y_n$ be the order statistics of a random sample of size n from a distribution of the continuous type. Thus $P(Y_1 < m < Y_n)$ is the probability that there is at least one "success" but not n successes, where the probability of success on each trial is $P(X < m) = 0.5$. Consequently,

$$P(Y_1 < m < Y_n) = \sum_{k=1}^{n-1} \binom{n}{k} \left(\frac{1}{2}\right)^k \left(\frac{1}{2}\right)^{n-k}$$

$$= 1 - \left(\frac{1}{2}\right)^n - \left(\frac{1}{2}\right)^n = 1 - \left(\frac{1}{2}\right)^{n-1}.$$

This probability increases as n increases so that the corresponding confidence interval (y_1, y_n) would have a very large confidence coefficient, $1 - (1/2)^{n-1}$. Unfortunately, the interval (y_1, y_n) tends to get wider as n increases, and thus we are not "pinning down" m very well. However, if we used the interval (y_2, y_{n-1}) or (y_3, y_{n-2}) we would obtain shorter intervals but also smaller confidence coefficients. Let us investigate this possibility further.

With the order statistics $Y_1 < Y_2 < \cdots < Y_n$ associated with a random sample of size n from a continuous-type distribution, consider $P(Y_i < m < Y_j)$, where $i < j$. For example, we might want

$$P(Y_2 < m < Y_{n-1}) \quad \text{or} \quad P(Y_3 < m < Y_{n-2}).$$

On each of the n independent trials we say that we have success if that X is less than m; thus the probability of success on each trial is $P(X < m) = 0.5$. Consequently, to have the ith order statistic Y_i less than m and the jth order statistic greater than m, we must have at least i successes but fewer than j successes (or else $Y_j < m$). That is,

$$P(Y_2 < m < Y_j) = \sum_{k=i}^{j-1} \binom{n}{k} \left(\frac{1}{2}\right)^k \left(\frac{1}{2}\right)^{n-k} = 1 - \alpha.$$

For particular values of n, i, and j, this probability, say $1 - \alpha$, which is the sum of probabilities from a binomial distribution, can be calculated directly or approximated by an area under the normal p.d.f. provided that n is large enough. The observed interval (y_i, y_j) could then serve as a $100(1 - \alpha)\%$ confidence interval for the unknown distribution median.

Example 10.2.1

The lengths in centimeters of $n = 9$ fish of a particular species (*nezumia*) captured off the New England coast were 32.5, 27.6, 29.3, 30.1, 15.5, 21.7, 22.8, 21.2, 19.0. Thus the observed order statistics are

$$15.5 < 19.0 < 21.2 < 21.7 < 22.8 < 27.6 < 29.3 < 30.1 < 32.5.$$

Before the sample is drawn, we know that

$$P(Y_2 < m < Y_8) = \sum_{k=2}^{7} \binom{9}{k} \left(\frac{1}{2}\right)^k \left(\frac{1}{2}\right)^{9-k} = 0.9805 - 0.0195 = 0.9610,$$

from Table II. Thus the confidence interval $(y_2 = 19.0, y_8 = 30.1)$ for m, the median of the lengths of all fish of this species, has a 96.1% confidence coefficient. ◇

So that the student need not compute many of these probabilities, we give in Table 10.1 the necessary information for constructing confidence intervals of the form (y_i, y_{n+1-i}) for the unknown m for sample sizes $n = 5, 6, \ldots, 20$. The subscript i is selected so that the confidence coefficient $P(Y_i < m < Y_{n+1-i})$ is greater than 90% and as close to 95% as possible.

For sample sizes larger than 20, we approximate those binomial probabilities with areas under the normal curve. To illustrate how good these approximations

TABLE 10.1 Information for confidence intervals for m

n	$(i, n+1-i)$	$P(Y_i < m < Y_{n+1-i})$
5	(1, 5)	0.9376
6	(1, 6)	0.9688
7	(1, 7)	0.9844
8	(2, 7)	0.9296
9	(2, 8)	0.9610
10	(2, 9)	0.9786
11	(3, 9)	0.9346
12	(3, 10)	0.9614
13	(3, 11)	0.9776
14	(4, 11)	0.9426
15	(4, 12)	0.9648
16	(5, 12)	0.9232
17	(5, 13)	0.9510
18	(5, 14)	0.9692
19	(6, 14)	0.9364
20	(6, 15)	0.9586

are, we compute the probability corresponding to $n = 16$ in Table 10.1. Here, using Table II, we have

$$1 - \alpha = P(Y_5 < m < Y_{12}) = \sum_{k=5}^{11} \binom{16}{k} \left(\frac{1}{2}\right)^k \left(\frac{1}{2}\right)^{16-k}$$

$$= P(W = 5, 6, \ldots, 11)$$

$$= 0.9616 - 0.0384$$

$$= 0.9232,$$

where W is $b(16, 1/2)$. The normal approximation gives

$$1 - \alpha = P(4.5 < W < 11.5) = P\left(\frac{4.5 - 8}{2} < \frac{W - 8}{2} < \frac{11.5 - 8}{2}\right)$$

because W has mean $np = 8$ and variance $np(1-p) = 4$. The standardized variable $Z = (W - 8)/2$ has an approximate normal distribution. Thus

$$1 - \alpha \approx \Phi\left(\frac{3.5}{2}\right) - \Phi\left(\frac{-3.5}{2}\right) = \Phi(1.75) - \Phi(-1.75)$$

$$= 0.9599 - 0.0401 = 0.9198.$$

This compares very favorably with the probability 0.9232 recorded in Table 10.1.

The argument used to find a confidence interval for the median m of a distribution of the continuous type can be applied to any percentile π_p. In this case we say that we have success on a single trial if that X is less than π_p. Thus the probability of success on each of the independent trials is $P(X < \pi_p) = p$. Accordingly, with $i < j, 1 - \alpha = P(Y_i < \pi_p < Y_j)$ is the probability that we have at least i successes but fewer than j successes. Thus

$$1 - \alpha = P(Y_i < \pi_p < Y_j) = \sum_{k=i}^{j-1} \binom{n}{k} p^k (1-p)^{n-k}.$$

Once the sample is observed and the order statistics determined, the known interval (y_i, y_j) could serve as a $100(1-\alpha)\%$ confidence interval for the unknown distribution percentile π_p.

Example 10.2.2

Let the following numbers represent the order statistics of the $n = 27$ observations obtained in a random sample from a certain population of incomes (measured in hundreds of dollars):

161	180	192	205	229	264
169	183	193	213	241	291
171	184	196	221	243	317
174	186	200	222	256	376
179	187	204			

Say we are interested in estimating the 25th percentile $\pi_{0.25}$ of the population. Since $(n+1)p = 28(1/4) = 7$, the 7th order statistic, namely $y_7 = 183$, would be a point estimate of $\pi_{0.25}$. To find a confidence interval for $\pi_{0.25}$, let us move down and up a few order statistics from y_7, say to y_4 and y_{10}. What is the confidence coefficient associated with the interval (y_4, y_{10})? Before the sample was drawn, we had

$$1 - \alpha = P(Y_4 < \pi_{0.25} < Y_{10}) = \sum_{k=4}^{9} \binom{27}{k} (0.25)^k (0.75)^{27-k}$$

$$= P(3.5 < W < 9.5),$$

where W is $b(27, 1/4)$ with mean $27/4 = 6.75$ and variance $81/16$. Hence

$$1 - \alpha \approx \Phi\left(\frac{9.5 - 6.75}{9/4}\right) - \Phi\left(\frac{3.5 - 6.75}{9/4}\right)$$

$$= \Phi\left(\frac{11}{9}\right) - \Phi\left(-\frac{13}{9}\right) = 0.8149.$$

Thus ($y_4 = 174$, $y_{10} = 187$) serves as an 81.49% confidence interval for $\pi_{0.25}$. It should be noted that we could choose other intervals, such as ($y_3 = 171$, $y_{11} = 192$), and these would have different confidence coefficients. The persons involved in the study must select the desired confidence coefficient, and then the appropriate order statistics are taken, usually quite symmetrically about the $(n+1)p$th order statistic.   ◇

When the number of observations is large, it is important to be able to determine rather easily the order statistics. A stem-and-leaf diagram, as introduced in Section 1.3, can be helpful in determining the needed order statistics. This is illustrated in the next example.

The measurements of butterfat produced by $n = 90$ cows during a 305-day milk production period following their first calf are summarized in the following ordered stem-and-leaf diagram in which each leaf consists of two digits:

Stem	Leaves									Frequency
2s	74									1
2•										0
3*										0
3t	27	39								2
3f	45	50								2
3s										0
3•	80	88	92	94	95					5
4*	17	18								2
4t	21	22	27	34	37	39				6
4f	44	52	53	53	57	58				6
4s	60	64	66	70	70	72	75	78		8
4•	81	86	89	91	92	94	96	97	99	9
5*	00	00	01	02	05	09	10	13	13 16	10
5t	24	26	31	32	32	37	37	39		8
5f	40	41	44	55						4
5s	61	70	73	74						4
5•	83	83	86	93	99					5
6*	07	08	11	12	13	17	18	19		8
6t	27	28	35	37						4
6f	43	43	45							3
6s	72									1
6•	91	96								2

From this display it is quite easy to see that $y_8 = 392$. It takes a little more work to show that $y_{38} = 494$ and $y_{53} = 526$. The interval (494, 526) serves as a confidence interval for the unknown median m of all butterfat production for the given breed of cows. Its confidence coefficient is

$$P(Y_{38} < m < Y_{53}) = \sum_{k=38}^{52} \binom{90}{k} \left(\frac{1}{2}\right)^k \left(\frac{1}{2}\right)^{90-k}$$

$$\approx \Phi\left(\frac{52.5 - 45}{\sqrt{22.5}}\right) - \Phi\left(\frac{37.5 - 45}{\sqrt{22.5}}\right)$$

$$= \Phi(1.58) - \Phi(-1.58) = 0.8858.$$

Similarly, $(y_{17} = 437, y_{29} = 470)$ is a confidence interval for the first quartile $\pi_{0.25}$ with confidence coefficient

$$P(Y_{17} < \pi_{0.25} < Y_{29}) \approx \Phi\left(\frac{28.5 - 22.5}{\sqrt{16.875}}\right) - \Phi\left(\frac{16.5 - 22.5}{\sqrt{16.875}}\right)$$

$$= \Phi(1.46) - \Phi(-1.46) = 0.8558.$$

EXERCISES

10.2.1 Let $Y_1 < Y_2 < Y_3 < Y_4 < Y_5 < Y_6$ be the order statistics of a random sample of size $n = 6$ from a distribution of the continuous type having $(100p)$th percentile π_p. Compute

(a) $P(Y_2 < \pi_{0.5} < Y_5)$.

(b) $P(Y_1 < \pi_{0.25} < Y_4)$.

(c) $P(Y_4 < \pi_{0.9} < Y_6)$.

10.2.2 For twelve 1995 model automobiles whose horsepower is between 290 and 390, the following measurements give the time in seconds for the car to go from 0 to 60 mph:

$$6.2 \quad 7.9 \quad 6.6 \quad 6.4 \quad 5.2 \quad 4.9 \quad 5.5 \quad 7.0 \quad 6.6 \quad 5.4 \quad 5.3 \quad 5.1$$

(a) Find a 96.14% confidence interval for the median, m.

(b) The interval (y_1, y_7) could serve as a confidence interval for $\pi_{0.3}$. Find it and give its confidence coefficient.

10.2.3 A sample of $n = 9$ electrochromic mirrors was used to measure the following low-end reflectivity percentages:

$$7.12 \quad 7.22 \quad 6.78 \quad 6.31 \quad 5.99 \quad 6.58 \quad 7.80 \quad 7.40 \quad 7.05$$

(a) Find the endpoints for an approximate 95% confidence interval for the median, m.

(b) The interval (y_3, y_7) could serve as a confidence interval for m. Find it and give its confidence coefficient.

10.2.4 Let m denote the median weight of "80-pound" bags of water softener pellets. Use the following random sample of $n = 14$ weights to find an approximate 95% confidence interval for m.

80.51	80.28	80.40	80.35	80.38	80.28	80.27
80.16	80.59	80.56	80.32	80.27	80.53	80.32

(a) Find a 94.26% confidence interval for m.

(b) The interval (y_6, y_{12}) could serve as a confidence interval for $\pi_{0.6}$. What is its confidence coefficient?

10.2.5 Use the following weights of an observed random sample of $n = 11$ pieces of candy to find an approximate 95% confidence interval for the median weight, m. Also give the exact confidence level.

$$2.76 \quad 2.96 \quad 2.67 \quad 2.97 \quad 2.77 \quad 2.72$$
$$2.74 \quad 2.84 \quad 2.61 \quad 2.82 \quad 2.72$$

10.2.6 Use the 81 weights of mints listed in Exercise 1.3.2 to find an approximate 95% confidence interval for (a) $\pi_{0.25}$, (b) $\pi_{0.5}$, and $\pi_{0.75}$.

10.2.7 A biologist who studies spiders selected a random sample of 20 male green lynx spiders (a spider that does not weave a web but chases and leaps on its prey) and measured in millimeters the lengths of one of the front legs of the 20 spiders. Use the following measurements to construct a confidence interval for m that has a confidence coefficient about equal to 0.95:

$$15.10 \quad 13.55 \quad 15.75 \quad 20.00 \quad 15.45$$
$$13.60 \quad 16.45 \quad 14.05 \quad 16.95 \quad 19.05$$
$$16.40 \quad 17.05 \quad 15.25 \quad 16.65 \quad 16.25$$
$$17.75 \quad 15.40 \quad 16.80 \quad 17.55 \quad 19.05$$

10.2.8 The biologist (Exercise 10.2.7) also selected a random sample of 20 female green lynx spiders and measured in millimeters the length of one of their front legs. Use the following data to construct a confidence interval for m that has a confidence coefficient about equal to 0.95:

$$15.85 \quad 18.00 \quad 11.45 \quad 15.60 \quad 16.10$$
$$18.80 \quad 12.85 \quad 15.15 \quad 13.30 \quad 16.65$$
$$16.25 \quad 16.15 \quad 15.25 \quad 12.10 \quad 16.20$$
$$14.80 \quad 14.60 \quad 17.05 \quad 14.15 \quad 15.85$$

10.2.9 The following 25 observations give the time in seconds between submissions of computer programs to a printer queue:

$$79 \quad 315 \quad 445 \quad 350 \quad 136 \quad 723 \quad 198 \quad 75 \quad 161$$
$$13 \quad 215 \quad 24 \quad 57 \quad 152 \quad 238 \quad 288 \quad 272$$
$$9 \quad 315 \quad 11 \quad 51 \quad 98 \quad 620 \quad 244 \quad 34$$

(a) Construct a stem-and-leaf diagram.

(b) Give the values of y_3, y_9, y_{10}, y_{13}, and y_{17}.

(c) Give point estimates of $\pi_{0.25}$, m, and $\pi_{0.75}$.

(d) Find the following confidence intervals and give the confidence level (using Table II) for

    (i) (y_3, y_{10}), a confidence interval for $\pi_{0.25}$.

    (ii) (y_9, y_{17}), a confidence interval for the median m.

    (iii) (y_{16}, y_{23}), a confidence interval for $\pi_{0.75}$.

10.2.10 Let X equal the weight in grams of a miniature candy bar. A random sample of candy bar weights yielded the following observations of X:

$$22.4 \quad 23.7 \quad 22.4 \quad 23.6 \quad 24.0 \quad 24.5 \quad 23.4 \quad 23.9$$
$$22.8 \quad 22.5 \quad 23.4 \quad 23.0 \quad 22.8 \quad 22.6 \quad 24.0 \quad 23.3$$

(a) Give a point estimate of the median, $m = \pi_{0.50}$.

(b) Find a confidence interval for m with confidence level between 90% and 95%. Give the value of the exact confidence coefficient.

10.2.11 Let X equal the amount of fluoride in a certain brand of toothpaste. The specifications are 0.85–1.10 mg/g. The following are 25 observations of X.

$$0.98 \quad 0.94 \quad 1.06 \quad 0.95 \quad 1.02 \quad 0.88 \quad 0.98 \quad 1.00 \quad 1.01$$
$$0.95 \quad 0.99 \quad 0.92 \quad 0.93 \quad 0.92 \quad 0.89 \quad 0.92 \quad 0.98$$
$$0.90 \quad 0.98 \quad 1.00 \quad 0.97 \quad 0.87 \quad 0.95 \quad 0.91 \quad 0.90$$

(a) Construct an ordered stem-and-leaf diagram.

(b) Give a point estimate of the median, $m = \pi_{0.50}$.

(c) Find an approximate 95% confidence interval for the median m. Give the exact confidence level.

(d) Give a point estimate for the first quartile.

(e) Find an approximate 90% confidence interval for the first quartile and give the exact confidence level.

(f) Give a point estimate for the third quartile.

(g) Find an approximate 95% confidence interval for the third quartile and give the exact confidence level.

10.2.12 When placed in solutions of varying ionic strength, paramecia grow blisters in order to counteract the flow of water. The following 60 measurements in microns are blister lengths:

$$
\begin{array}{cccccccc}
7.42 & 5.73 & 3.80 & 5.20 & 11.66 & 8.51 & 6.31 & 8.49 \\
10.31 & 6.92 & 7.36 & 5.92 & 6.74 & 8.93 & 9.61 & 11.38 \\
12.78 & 11.43 & 6.57 & 13.50 & 10.58 & 8.03 & 10.07 & 8.71 \\
10.09 & 11.16 & 7.22 & 10.10 & 6.32 & 10.30 & 10.75 & 11.51 \\
11.55 & 11.41 & 9.40 & 4.74 & 6.52 & 12.10 & 6.01 & 5.73 \\
7.57 & 7.80 & 6.84 & 6.95 & 8.93 & 8.92 & 5.51 & 6.71 \\
10.40 & 13.44 & 9.33 & 8.57 & 7.08 & 8.11 & 13.34 & 6.58 \\
8.82 & 7.70 & 12.22 & 7.46 & & & &
\end{array}
$$

(a) Construct an ordered stem-and-leaf diagram.

(b) Give a point estimate of the median, $m = \pi_{0.50}$.

(c) Find an approximate 95% confidence interval for m. Give the exact confidence level.

(d) Give a point estimate for the 40th percentile, $\pi_{0.40}$.

(e) Find an approximate 90% confidence interval for $\pi_{0.40}$ and give the exact confidence level.

10.3 BINOMIAL TESTS FOR PERCENTILES

In Section 10.2 we found ways of constructing interval estimates for certain characteristics of the distribution that did not depend too much on distributional assumptions. That is, we found distribution-free interval estimates for the $(100p)$th percentile of a distribution of the continuous type. We now take up some tests of statistical hypotheses that are **distribution free**, provided that the null hypotheses are true. In this section we find a distribution-free test of the hypothesis H_0: $\pi_p = \pi_0$, where π_0 is given. It should be noted here, however, that the powers of these tests are not distribution free when the null hypotheses are false. Moreover, the resulting nonnull distribution theory is frequently quite complicated; so little will be said here about the power functions.

We begin by considering a hypothesis about the median, $m = \pi_{0.5}$, of a distribution of the continuous type. Further, we assume that the median m is unique. Suppose it is known from past experience that the median length of sunfish in a particular polluted lake was $m = 3.7$ inches. During the past two years the lake was "cleaned up," and the conjecture is made that now $m > 3.7$ inches. We consider a procedure for testing this conjecture. In particular, we describe a procedure for testing the null hypothesis H_0: $m = 3.7$ against the alternative hypothesis H_1: $m > 3.7$.

Let X denote the length of a sunfish selected at random from the lake. If the null hypothesis H_0: $m = 3.7$ is true, $P(X \leq 3.7; H_0) = 0.5$. However, if the alternative hypothesis H_1: $m > 3.7$ is true, $P(X \leq 3.7; H_0) < 0.5$. If we take a random sample of $n = 10$ fish, we would expect about half the fish to be shorter than 3.7 inches if H_0 is true. However, if H_1 is true, we would expect less than half the fish to be shorter than 3.7 inches. We shall decide on the basis of the number, say Y, of fish shorter than 3.7 inches whether to accept H_0 or H_1. Thus Y can be thought of as the number of "successes" in 10 Bernoulli trials with probability of success given by $p = P(X \leq 3.7)$. If H_0 is true, $p = 1/2$ and Y is $b(10, 1/2)$; whereas if H_1 is true, $p < 1/2$ and Y is $b(10, p)$. We reject H_0 and accept H_1 if and only if the observed value y of Y is sufficiently small, say $y \leq c$. From Table II we find that $P(Y \leq 2; H_0) = 0.0547$. Thus, if we let the critical region be defined by $C = \{y : y \leq 2\}$, then $\alpha = 0.0547$.

Suppose that the lengths of 10 sunfish selected at random from this lake were 5.0, 3.9, 5.2, 5.5, 2.8, 6.1, 6.4, 2.6, 1.7, and 4.3 inches. Since $y = 3$ of these lengths are shorter than 3.7, we would not reject H_0: $m = 3.7$ at the $\alpha = 0.0547$ significance level by considering only these few data.

Example 10.3.1

Let X denote the length of time in seconds between two calls entering a college switchboard. Let m be the unique median of this continuous-type distribution. We test the null hypothesis H_0: $m = 6.2$ against the alternative hypothesis H_1: $m < 6.2$. If Y is the number of lengths of time in a random sample of size 20 that are less than 6.2, the critical region $C = \{y : y \geq 14\}$ has a significance level of $\alpha = 0.0577$ using Table II. A random sample of size 20 yielded the following data:

6.8	5.7	6.9	5.3	4.1	9.8	1.7	7.0
2.1	19.0	18.9	16.9	10.4	44.1	2.9	2.4
4.8	18.9	4.8	7.9				

Since $y = 9$, the null hypothesis is not rejected. ◇

In many places in the literature the test we have just described is called the **sign test**. The reason for this terminology is that the test is based on a statistic Y that is equal to the number of negative signs among (here m_0 is the hypothesized median)

$$X_1 - m_0, \quad X_2 - m_0, \quad \ldots, \quad X_n - m_0.$$

The sign test can also be used to test the hypothesis that two continuous-type random variables X and Y are such that $p = P(X > Y) = 1/2$. To test the hypothesis H_0: $p = 1/2$ against an appropriate alternative hypothesis, consider the independent pairs $(X_1, Y_1), (X_2, Y_2), \ldots, (X_n, Y_n)$. Let W denote the number of pairs for which $X_i - Y_i > 0$. When H_0 is true, W is $b(n, 1/2)$, and the test can be based upon the statistic W. It is important to recognize that X_i and Y_i do not need to be independent and are frequently not independent in practice. If they were independent, then there are many ways in which $X_1, X_2, \ldots, X_n$ and $Y_1, Y_2, \ldots, Y_n$ can be paired up. If, however, X is the length of the right foot of a person and Y the length of the corresponding left foot, there is a natural pairing. Here H_0: $p = P(X > Y) = 1/2$ suggests that either foot of a particular person is equally likely to be longer.

Example 10.3.2

Freshmen in a health dynamics course have their percentage of body fat measured at the beginning (x) and at the end (y) of the semester. These measurements are given for 26 students in Table 10.2. Also recorded is a plus sign when $x_i > y_i$ and a minus sign with $x_i < y_i$. Note that, in theory, the case $x_i = y_i$ should not happen. However, because measurements are rounded off, ties do occur in applications. When a tie does occur, the subject is dropped and the test is run with a reduced sample size. We test the hypothesis H_0: $p = P(X > Y) = 1/2$ against the one-sided alternative H_1: $p > 1/2$. That is, the alternative hypothesis is that the students' percentage of body fat decreases during the semester. Let the critical region be defined by $C = \{w : w \geq 17\}$, where w is the number of "plus signs." The distribution of W is $b(25, 1/2)$ when H_0 is true, so we see that $\alpha = P(W \geq 17; p = 1/2) = 0.0539$ from Table II. Since $w = 16$, the null hypothesis is not rejected with these data; although they suggest, with more data of the same type, we might be able to reject H_0. The p-value of this test is

$$p\text{-value} = P(W \geq 16; p = 1/2) = 0.1148. \qquad \diamond$$

TABLE 10.2 Body fat percentages, beginning (x) and end (y) of semester

x	y	Sign	x	y	Sign
35.4	33.6	+	22.4	21.0	+
28.8	31.9	−	23.5	24.5	−
10.6	10.5	+	24.1	21.9	+
16.7	15.6	+	22.5	21.7	+
14.6	14.0	+	17.5	17.9	−
8.8	13.9	−	16.9	14.9	+
17.9	8.7	+	11.7	17.5	−
17.8	17.6	+	8.3	11.7	−
9.3	8.9	+	7.9	10.2	−
23.6	23.6	0	20.7	17.7	+
15.6	13.7	+	26.8	24.1	+
24.3	24.7	−	20.6	20.4	+
23.8	25.3	−	25.1	21.9	+

The sign test can also be used in comparisons when the outcomes are not numerical. For example, each of a number of coffee drinkers selected at random is asked if some freshly perked coffee is preferred over some instant coffee.

There are some objections to the sign test, among them being that the data must be paired and that the test does not take into account the magnitude of differences. How do we decide when to use the sign test? In the situation in which we need a test about a location parameter of a distribution, would we use the sign test to test $m = m_0$ or a test based on $\overline{X}$ to test $\mu = \mu_0$? In particular, if the distribution is symmetric, then $m = \mu$, and hence each procedure would be testing the hypothesis that the common value of m and μ is equal to a known constant. In more advanced theory and practice, we discover that if there are outliers (i.e., extreme x values that deviate greatly from most of the other values), the sign test is usually better than that based upon $\overline{X}$. This certainly appeals to our intuition because a few extreme x values can influence the average $\overline{x}$ a great deal; but in the sign test, each extreme value is associated with only one sign no matter how far it is from m. Hence, in cases with highly skewed data or data with outliers, the distribution-free sign test would probably be preferred to that based on X. In Section 10.4 we consider another distribution-free test, first proposed by Wilcoxon, which does account somewhat for the magnitudes of the deviations $X_1 - m_0, X_2 - m_0, \ldots, X_n - m_0$ and is an excellent test in many situations.

Another advantage of the sign test is the fact that it can easily be generalized to percentiles other than the median. To test the hypothesis that the $(100p)$th percentile π_p of a continuous-type distribution is equal to a specified value π_0, let Y be the number of the items of the random sample of size n that are less than π_0. If H_0: $\pi_p = \pi_0$ is true, then Y is $b(n, p)$; where here, of course, p is a known probability. If, for example, the alternative hypothesis is H_1: $\pi_p > \pi_0$, then the critical region would be of the form $C = \{y : y \leq c\}$.

Example 10.3.3

Suppose, from past testing, we know that the 25th percentile of ninth-grade general mathematics students taking a standardized examination is 62.4. Statisticians believe that by introducing more statistics involving real problems into such a course, the students will see the usefulness of mathematics. It is hoped that this approach will motivate the poorer students in these classes and increase some of the lower percentiles. In particular, they believe that if these courses were changed as suggested, then H_1: $\pi_{0.25} > 62.4$. To test H_0: $\pi_{0.25} = 62.4$ against H_1, 192 general mathematics students were selected at random and given this new type of course using more statistics. Let Y be the number of students scoring less than 62.4. Of course, if H_0 is true, then Y is $b(192, 1/4)$. If H_1 is true, we would expect smaller values of Y, and thus the critical region is of the form $y \leq c$. To determine c, we can use the fact that Y, under H_0, has an approximate normal distribution with mean $192(1/4) = 48$ and variance $192(1/4)(3/4) = 36$. Thus, for the significance level to be about 0.05, we want

$$c \approx 48 - (1.645)(6) = 38.13.$$

Accordingly, if we take $c = 38$, then the significance level is

$$P(Y \leq 38) = P(Y < 38.5) = P\left(\frac{Y - 48}{6} < \frac{-9.5}{6}\right)$$

$$\approx \Phi(-1.583) = 0.0567.$$

Suppose that after the course is over, these 192 students take the standardized examination and only 31 have grades less than 62.4. We would reject H_0: $\pi_{0.25} = 62.4$ and accept H_1: $\pi_{0.25} > 62.4$; that is, it seems as if the 25th percentile is greater in the new type of general mathematics course than that associated with the old one.   ◇

EXERCISES

10.3.1 On any given day at a candy warehouse, 13 trucks are loaded with candy. It is claimed that the median weight m of loads of candy is 40,000 pounds.

(a) Test, at the $\alpha = 0.0461$ significance level, the null hypothesis H_0: $m = 40,000$ against the one-sided alternative hypothesis H_1: $m < 40,000$ using the following observations:

$$41,195 \quad 39,485 \quad 41,229 \quad 36,840 \quad 38,050 \quad 40,890 \quad 38,345$$
$$34,930 \quad 39,245 \quad 31,031 \quad 40,780 \quad 38,050 \quad 30,906$$

(b) What is the p-value of this test?

10.3.2 The forced vital capacity (FVC) gives a measurement of the amount of air that a person can exhale from his lungs. Let m equal the median FVC in liters for male college freshmen. Use the following 17 observations to test, at the $\alpha = 0.0768$ significance level, the null hypothesis H_0: $m = 4.7$ against the two-sided alternative hypothesis H_1: $m \neq 4.7$ (deleting any observation that is equal to 4.7):

$$4.3 \quad 5.0 \quad 5.7 \quad 6.2 \quad 4.8 \quad 4.7 \quad 5.6 \quad 5.2 \quad 3.7$$
$$4.0 \quad 5.6 \quad 6.8 \quad 4.9 \quad 7.6 \quad 5.4 \quad 3.8 \quad 5.6$$

10.3.3 A course in economics was taught to two groups of students, one in a classroom situation and the other by TV. There were 24 students in each group. These students were first paired according to cumulative grade-point averages and background in economics, and then assigned to the courses by a flip of a coin (this was repeated 24 times). At the end of the course each class was given the same final examination. Use the sign test to test ($\alpha = 0.0662$) that the two methods of teaching are equally effective against a two-sided alternative. The differences in final scores for each pair of students, the TV student's score having been subtracted from the corresponding classroom student's scores, were as follows:

$$14 \quad -4 \quad -6 \quad -2 \quad -1 \quad 18$$
$$6 \quad 12 \quad 8 \quad -4 \quad 13 \quad 7$$
$$2 \quad 6 \quad 21 \quad 7 \quad -2 \quad 11$$
$$-3 \quad -14 \quad -2 \quad 17 \quad -4 \quad -5$$

10.3.4 The administration claims that professor A is more popular than professor B. We shall test the hypothesis H_0 of no difference in popularity against the administration's claim H_1, at an approximate 0.05 significance level. If 14 out of 20 students prefer professor A to professor B, is H_0 rejected and the administration's claim supported?

10.3.5 In a certain region, it was known that about 20% of the drivers exceeded the speed limit of 55 by more than 5 miles per hour. The highway patrol then established a new enforcement procedure and claimed that this percentage had been reduced substantially. How would you test the null hypothesis H_0: $\pi_{0.8} = 60$ against the patrol's claim H_1: $\pi_{0.8} < 60$ at $\alpha = 0.05$ if you could observe the speeds of 100 drivers selected at random? If, in fact, out of the 100 drivers, you found that the speeds of 89 were under 60 miles per hour, what decision would you make?

10.3.6 To test whether a golf ball of brand A can be hit a greater distance off the tee than a golf ball of brand B, each of 17 golfers hit a ball of each brand, eight hitting ball A before ball B and nine hitting ball B before ball A. Let m_D denote the median of the differences of the distances of ball A and ball B. Use the sign test to test the null hypothesis H_0: $m_D = 0$ against H_1: $m_D > 0$. HINT: In case of ties, delete that pair of observations and use a reduced sample size.

Golfer	Distance for Ball A	Distance for Ball B
1	265	252
2	272	276
3	246	243
4	260	246
5	274	275
6	263	246
7	255	244
8	258	245
9	276	259
10	274	260
11	274	267
12	269	267
13	244	251
14	212	222
15	235	235
16	254	255
17	224	231

10.3.7 In Exercise 10.3.6 assume that the differences for each golfer, distance for ball A minus distance for ball B, are observations of a normal random variable with mean μ_D. Test the null hypothesis H_0: $\mu_D = 0$ against the alternative hypothesis H_1: $\mu_D > 0$ using a t-test with the 17 differences.

10.3.8 For each student who takes introductory statistics, let X and Y denote the students' ACT math and ACT verbal scores, respectively. Let $D = X - Y$ and m_D denote the median of the differences. Test H_0: $m_D = 0$ against the two-sided alternative hypothesis H_1: $m_D \neq 0$ at an approximate $\alpha = 0.10$ significance level.

Student	ACT Math	ACT Verbal
1	16	19
2	18	17
3	22	18
4	20	23
5	17	20
6	25	21
7	21	24
8	23	18
9	24	18
10	31	25
11	27	29
12	28	24
13	30	24
14	27	23
15	28	24

10.3.9 In Exercise 7.3.19, a company and a vendor measured the fat content of samples of milk from the same lot. Let $D = X - Y$ and let m_D denote the median of the differences.

(a) Test H_0: $m_D = 0$ against H_1: $m_D > 0$ using the sign test. Let $\alpha = 0.05$, approximately.

(b) How does this conclusion compare with that of Exercise 7.3.19?

10.3.10 Let m equal the median FVC in liters for college freshmen women (see Exercise 10.3.2). We shall use the sign test to test the null hypothesis H_0: $m = 3.05$ against the alternative hypothesis H_1: $m > 3.05$ using 25 observations.

(a) Define a critical region that has an approximate significance level of $\alpha = 0.05$.

(b) State your conclusion using the following data:

$$4.2 \quad 5.3 \quad 3.5 \quad 4.3 \quad 3.7 \quad 3.5 \quad 3.5 \quad 2.8 \quad 3.5$$
$$3.7 \quad 3.5 \quad 3.7 \quad 2.7 \quad 2.8 \quad 3.3 \quad 2.7 \quad 3.0$$
$$3.1 \quad 3.0 \quad 3.7 \quad 3.3 \quad 3.4 \quad 2.3 \quad 2.6 \quad 3.2$$

(c) What is the p-value of your test?

10.3.11 Let U and V equal a student's ACT score in social science and in natural science, respectively. Let $p = P(U > V)$. We shall use the sign test to test the null hypothesis H_0: $p = 0.5$ against a two-sided alternative hypothesis using the following 15 pairs, (u, v):

$$(32, 28) \quad (30, 27) \quad (26, 32) \quad (23, 25) \quad (17, 23)$$
$$(16, 22) \quad (23, 24) \quad (20, 30) \quad (21, 28) \quad (23, 32)$$
$$(17, 18) \quad (24, 31) \quad (26, 31) \quad (18, 18) \quad (30, 26)$$

(a) Define a critical region that has a significance level that is close to $\alpha = 0.05$. Give the exact significance level.

(b) What is your conclusion?

10.3.12 Thirteen tons of Wisconsin cheese were packaged in "22-pound" wheels and stored underground in old gypsum mines. The manufacturer claimed that the median weight of wheels is 22 pounds, while the Michigan Department of Agriculture suspects that the median is less than 22 pounds.

(a) Define the null (manufacturer's) and alternative (Department of Agriculture's) hypotheses.

(b) Based on $n = 9$ observations, define a critical region that has an approximate 10% significance level.

(c) Given the weights 21.50, 18.95, 18.55, 19.40, 19.15, 22.35, 22.90, 22.20, and 23.10, what is your conclusion?

(d) What is the p-value of this test?

10.4 THE WILCOXON TEST

Let X be a continuous-type random variable. Let m denote the median of X. To test the hypothesis H_0: $m = m_0$ against an appropriate alternative hypothesis, the sign test, discussed in the preceding section, can be used. That is, if $X_1, X_2, \ldots, X_n$ denote the items of a random sample from this distribution and if we let Y equal the number of negative differences among $X_1 - m_0, X_2 - m_0, \ldots, X_n - m_0$, then Y has the binomial distribution $b(n, 1/2)$ under H_0 and is the test statistic for the sign test. One major objection to this test is that it does not take into account the magnitude of the differences $X_1 - m_0, \ldots, X_n - m_0$.

In this section we discuss a test that does take into account the magnitude of the differences $|X_i - m_0|$, $i = 1, 2, \ldots, n$. However, in addition to the assumption that the random variable X is of the continuous type, we must also assume that the p.d.f. of X is symmetric about the median in order to find the distribution of this new statistic. Because of the continuity assumption, we assume, in the following discussion, that no two observations are equal and that no observation is equal to the median.

We are interested in testing the hypothesis H_0: $m = m_0$ where m_0 is some given constant. With our random sample $X_1, X_2, \ldots, X_n$, we rank the differences $X_1 - m_0, X_2 - m_0, \ldots, X_n - m_0$ in ascending order according to magnitude. That is, for $i = 1, 2, \ldots, n$, we let R_i denote the rank of $|X_i - m_0|$ among $|X_1 - m_0|$, $|X_2 - m_0|, \ldots, |X_n - m_0|$. Note that $R_1, R_2, \ldots, R_n$ is a permutation of the first n positive integers, $1, 2, \ldots, n$. Now with each R_i we associate the sign of the difference $X_i - m_0$; that is, if $X_i - m_0 > 0$, we use R_i, but if $X_i - m_0 < 0$, we use $-R_i$. The Wilcoxon statistic W is the sum of these n signed ranks.

Example 10.4.1

Consider the sunfish example in Section 10.3. There we considered testing H_0: $m = 3.7$ against the alternative hypothesis H_1: $m > 3.7$. The observed lengths of the $n = 10$ fish were

$$x_i: 5.0 \quad 3.9 \quad 5.2 \quad 5.5 \quad 2.8 \quad 6.1 \quad 6.4 \quad 2.6 \quad 1.7 \quad 4.3$$

Thus we have

$X_i - m_0$:	1.3	0.2	1.5	1.8	-0.9	2.4	2.7	-1.1	-2.0	0.6
$\lvert x_i - m_0 \rvert$:	1.3	0.2	1.5	1.8	0.9	2.4	2.7	1.1	2.0	0.6
Rank:	5	1	6	7	3	9	10	4	8	2
Signed rank:	5	1	6	7	-3	9	10	-4	-8	2

Therefore, the Wilcoxon statistic is equal to

$$W = 5 + 1 + 6 + 7 - 3 + 9 + 10 - 4 - 8 + 2 = 25.$$

Incidentally, the positive answer seems reasonable because the number of the 10 lengths that are less than 3.7 is 3, which is the statistic used in the sign test.   ◇

If the hypothesis H_0: $m = m_0$ is true, about one-half of the differences would be negative and thus about one-half of the signs would be negative. Thus it seems that the hypothesis H_0: $m = m_0$ is supported if the observed value of W is close to zero. If the alternative hypothesis is H_1: $m > m_0$, we would reject H_0 if the observed $W = w$ is too large, since, in this case, the larger deviations $\lvert X_i - m_0 \rvert$ would usually be associated with observations for which $x_i - m_0 > 0$. That is, the critical region would be of the form $\{w : w \geq c_1\}$. If the alternative hypothesis is H_1: $m < m_0$, the critical region would be of the form $\{w : w \leq c_2\}$. Of course, the critical region would be of the form $\{w : w \leq c_3 \text{ or } w \geq c_4\}$ for a two-sided alternative hypothesis H_1: $m \neq m_0$. In order to find the values of c_1, c_2, c_3, c_4 that yield desired significance levels, it is necessary to determine the distribution of W under H_0. We now consider certain characteristics of this distribution.

When H_0: $m = m_0$ is true,

$$P(X_i < m_0) = P(X_i > m_0) = \frac{1}{2}, \qquad i = 1, 2, \ldots, n.$$

Hence the probability is $1/2$ that a negative sign is associated with the rank R_i of $\lvert X_i - m_0 \rvert$. Moreover, the assignments of these n signs are independent because $X_1, X_2, \ldots, X_n$ are mutually independent. In addition, W is a sum that contains the integers $1, 2, \ldots, n$, each integer with a positive or negative sign. Since the underlying distribution is symmetric, it seems intuitively obvious that W has the same distribution as the random variable

$$V = \sum_{i=1}^{n} V_i,$$

where $V_1, V_2, \ldots, V_n$ are independent and

$$P(V_i = i) = P(V_i = -i) = \frac{1}{2}, \qquad i = 1, 2, \ldots, n.$$

That is, V is a sum that contains the integers $1, 2, \ldots, n$, and these integers receive their algebraic signs by independent assignments.

Since W and V have the same distribution, their means and variances are equal, and we can easily find those of V. Now the mean of V_i is

$$E(V_i) = -i\left(\frac{1}{2}\right) + i\left(\frac{1}{2}\right) = 0$$

and thus

$$E(W) = E(V) = \sum_{i=1}^{n} E(V_i) = 0.$$

The variance of V_i is

$$\mathrm{Var}(V_i) = E(V_i^2) = (-i)^2\left(\frac{1}{2}\right) + (i)^2\left(\frac{1}{2}\right) = i^2.$$

Thus

$$\mathrm{Var}(W) = \mathrm{Var}(V) = \sum_{i=1}^{n} \mathrm{Var}(V_i) = \sum_{i=1}^{n} i^2 = \frac{n(n+1)(2n+1)}{6}.$$

We shall not try to find the distribution of W in general, since that p.d.f. does not have a convenient expression. However, we demonstrate how we could find the distribution of W (or V) with enough patience and computer support. Recall that the moment-generating function of V_i is

$$M_i(t) = e^{t(-i)}\left(\frac{1}{2}\right) + e^{t(+i)}\left(\frac{1}{2}\right) = \frac{e^{-it} + e^{it}}{2}, \qquad i = 1, 2, \ldots, n.$$

Let $n = 2$; so the moment-generating function of $V_1 + V_2$ is

$$M(t) = E[e^{t(V_1 + V_2)}].$$

From the independence of V_1 and V_2, we obtain

$$
\begin{aligned}
M(t) &= E(e^{tV_1})E(e^{tV_2}) \\
&= \left(\frac{e^{-t} + e^{t}}{2}\right)\left(\frac{e^{-2t} + e^{2t}}{2}\right) \\
&= \frac{e^{-3t} + e^{-t} + e^{t} + e^{3t}}{4}.
\end{aligned}
$$

This means that each of the points $-3, -1, 1, 3$ in the support of $V_1 + V_2$ has probability $1/4$.

Next let $n = 3$, so the moment-generating function of $V_1 + V_2 + V_3$ is

$$
\begin{aligned}
M(t) &= E[e^{t(V_1 + V_2 + V_3)}] \\
&= E[e^{t(V_1 + V_2)}] E(e^{t V_3}) \\
&= \left(\frac{e^{-3t} + e^{-t} + e^{t} + e^{3t}}{4} \right) \left(\frac{e^{-3t} + e^{3t}}{2} \right) \\
&= \frac{e^{-6t} + e^{-4t} + e^{-2t} + 2e^{0} + e^{-2t} + e^{-4t} + e^{-6t}}{8}.
\end{aligned}
$$

Thus the points $-6, -4, -2, 0, 2, 4, 6$ in the support of $V_1 + V_2 + V_3$ have the respective probabilities $1/8, 1/8, 1/8, 2/8, 1/8, 1/8, 1/8$. Obviously, this procedure can be continued for $n = 4, 5, 6, \ldots$, but it is rather tedious. Fortunately, however, even though $V_1, V_2, \ldots, V_n$ are not identically distributed random variables, the sum V of them still has an approximate normal distribution. To obtain this normal approximation for V (or W), a more general form of the central limit theorem, due to Liapounov, can be used that allows us to say that the standardized random variable

$$
Z = \frac{W - 0}{\sqrt{n(n + 1)(2n + 1)/6}}
$$

is approximately $N(0, 1)$ when H_0 is true. We accept this without proof, and thus we can approximate probabilities such as $P(W \geq c; H_0) \approx P(Z \geq z_\alpha; H_0)$ when the sample size n is sufficiently large using this normal distribution. However, the next example illustrates this approximation.

Example 10.4.2

The moment-generating function of W or of V is given by

$$
M(t) = \prod_{i=1}^{n} \frac{e^{-it} + e^{it}}{2}.
$$

Using a computer algebra system (CAS) such as *Maple*, it is possible to expand $M(t)$ and find the coefficients of e^{kt} which equal $P(W = k)$. In Figure 10.1 we have drawn a probability histogram for the distribution of W along with the approximating $N[0, n(n + 1)(2n + 1)/6]$ p.d.f. for $n = 4$ (a poor approximation) and for $n = 10$. It is important to note that the widths of the rectangles in the probability histogram are equal to 2, so the "half-unit correction for continuity" now becomes a unit correction. ◇

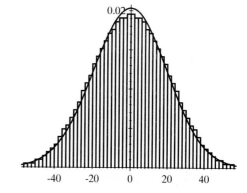

Figure 10.1   Wilcoxon distribution for $n = 4$ and $n = 10$

Example 10.4.3

Let m be the median of a symmetric distribution of the continuous type. To test the hypothesis H_0: $m = 160$ against the alternative hypothesis H_1: $m > 160$, we take a random sample of size $n = 16$. For an approximate significance level of $\alpha = 0.05$, H_0 is rejected if the computed $W = w$ is such that

$$z = \frac{w}{\sqrt{16(17)(33)/6}} \geq 1.645,$$

or

$$w \geq 1.645\sqrt{\frac{16(17)(33)}{6}} = 63.626.$$

Say that the observed values of a random sample are 176.9, 158.3, 152.1, 158.8, 172.4, 169.8, 159.7, 162.7, 156.6, 174.5, 184.4, 165.2, 147.8, 177.8, 160.1, 160.5. In Table 10.3 the magnitudes of the differences $|x_i - 160|$ have been ordered and ranked. Those differences $x_i - 160$ that were negative have been underlined and the ranks are under the ordered values. For this set of data

$$w = 1 - 2 + 3 - 4 - 5 + 6 + \cdots + 16 = 60.$$

TABLE 10.3   Ordered absolute differences from 160

0.1	0.3	0.5	1.2	1.7	2.7	3.4	5.2
1	2	3	4	5	6	7	8
7.9	9.8	12.2	12.4	14.5	16.9	17.8	24.4
9	10	11	12	13	14	15	16

Since $60 < 63.626$, H_0 is not rejected at the 0.05 significance level. It is interesting to note that H_0 would have been rejected at $\alpha = 0.10$, since the approximate

p-value is, making a unit correction for continuity,

$$p\text{-value} = P(W \geq 60)$$

$$= P\left(\frac{W - 0}{\sqrt{(16)(17)(33)/6}} \geq \frac{59 - 0}{\sqrt{(16)(17)(33)/6}} \right)$$

$$\approx P(Z \geq 1.525) = 0.0636.$$

(Using *Maple*, the p-value is equal to $4{,}251/65{,}536 = 0.0649$.) This would indicate to us that the data are too few to reject H_0, but if the pattern continues, we shall most certainly reject with a larger sample size.   ◇

Although theoretically we could ignore the possibilities that $x_i = m_0$, for some i, and that $|x_i - m_0| = |x_j - m_0|$ for some $i \neq j$, these situations do occur in applications. Usually, in practice, if $x_i = m_0$ for some i, that observation is deleted and the test is performed with a reduced sample size. If the absolute values of the differences from m_0 of two or more observations are equal, each observation is assigned the average of the corresponding ranks. The change this causes in the distribution of W is not very great, and thus we continue using the same normal approximation.

We now give an example that has some tied observations. This example also illustrates that the Wilcoxon rank sum test can be used to test hypotheses about the median of paired data.

Example 10.4.4

In Example 10.3.2 we gave some paired data for percentage of body fat measured at the beginning and the end of a semester. Let m equal the median of the differences, $x - y$. We shall use the Wilcoxon statistic to test the null hypothesis H_0: $m = 0$ against the alternative hypothesis H_1: $m > 0$. Since there are $n = 25$ nonzero differences, we reject H_0 if

$$z = \frac{w - 0}{\sqrt{(25)(26)(51)/6}} \geq 1.645$$

or, equivalently, if

$$w \geq 1.645 \sqrt{\frac{(25)(26)(51)}{6}} = 122.27$$

at an approximate $\alpha = 0.05$ significance level. From Table 10.2 we can calculate the differences, yielding

1.8	−3.1	0.1	1.1	0.6	−5.1	9.2	0.2	0.4
0.0	1.9	−0.4	−1.5	1.4	−1.0	2.2	0.8	−0.4
2.0	−5.8	−3.4	−2.3	3.0	2.7	0.2	3.2	

In Table 10.4 we list the ordered absolute values, underlining those that were originally negative. The rank is under each observation. Note that in the case of ties, the average of the ranks of the tied measurements is given.

TABLE 10.4   Ordered absolute values, changes in percentage of body fat

0.1	0.2	0.2	0.4	0.4	0.4	0.6	0.8	1.0	1.1	1.4	1.5	1.8
1	2.5	2.5	5	5	5	7	8	9	10	11	12	13

1.9	2.0	2.2	2.3	2.7	3.0	3.1	3.2	3.4	5.1	5.8	9.2
14	15	16	17	18	19	20	21	22	23	24	25

The value of the Wilcoxon statistic is

$$w = 1 + 2.5 + 2.5 + 5 - 5 - 5 + \cdots + 25 = 51.$$

Since $51 < 122.27$, we fail to reject the null hypothesis. The approximate p-value of this test is

$$p\text{-value} = P(W \geq 51) = P(W \geq 50)$$
$$\approx P\left(Z \geq \frac{50 - 0}{\sqrt{(25)(26)(51)/6}}\right)$$
$$= P(Z \geq 0.673) = 0.2505.$$ ◇

EXERCISES

10.4.1 In Exercise 10.3.1 it is claimed that the median weight m of loads of candy is 40,000 pounds.

(a) Use those 13 observations and the Wilcoxon statistic to test, at an approximate significance level of $\alpha = 0.05$, the null hypothesis H_0: $m = 40,000$ against the one-sided alternative hypothesis H_1: $m < 40,000$. For convenience, we list those 13 observations again.

$$41,195 \quad 39,485 \quad 41,229 \quad 36,840 \quad 38,050 \quad 40,890 \quad 38,345$$
$$34,930 \quad 39,245 \quad 31,031 \quad 40,780 \quad 38,050 \quad 30,906$$

(b) What is the approximate p-value of this test? (How does it compare with the p-value in Exercise 10.3.1?)

10.4.2 In Exercise 10.3.2 we tested the null hypothesis that the median, m, of the forced vital capacity for male college freshmen is equal to 4.7 using the sign test. Now use the Wilcoxon statistic to test H_0: $m = 4.7$ against H_1: $m \neq 4.7$ using the $n = 17$ observations in Exercise 10.3.2. Use an approximate significance level of $\alpha = 0.05$.

10.4.3 With the data in Exercise 10.3.3, use the Wilcoxon statistic to test the hypothesis that the median of the differences in final scores equals zero.

10.4.4 The outcomes on $n = 10$ simulations of a Cauchy random variable were -1.9415, 0.5901, -5.9848, -0.0790, -0.7757, -1.0962, 9.3820, -74.0216, -3.0678, 3.8545. For the Cauchy distribution, the mean does not exist; but for this one, the median is believed to equal zero. Use the Wilcoxon test and these data to test H_0: $m = 0$ against the alternative hypothesis H_1: $m \neq 0$. Let $\alpha = 0.05$, approximately.

10.4.5 For the golf ball example in Exercise 10.3.6, use the Wilcoxon statistic to test the hypothesis H_0: $m_D = 0$ against H_1: $m_D > 0$. Let $\alpha = 0.05$, approximately.

10.4.6 For the milk fat content example in Exercises 7.3.19 and 10.3.9, use the Wilcoxon statistic to test the hypothesis H_0: $m_D = 0$ against H_1: $m_D > 0$. Let $\alpha = 0.05$, approximately.

10.4.7 Let x equal a student's GPA in the fall semester and y the same student's GPA in the spring semester. Let m equal the median of the differences, $x - y$. We shall test the null hypothesis H_0: $m = 0$ against an appropriate alternative hypothesis that you select based on your past experience. Use a Wilcoxon test and the following 15 observations of paired data to test this hypothesis:

x	y	x	y
2.88	3.22	3.98	3.76
3.67	3.49	4.00	3.96
2.76	2.54	3.39	3.52
2.34	2.17	2.59	2.36
2.46	2.53	2.78	2.62
3.20	2.98	2.85	3.06
3.17	2.98	3.25	3.16
2.90	2.84		

10.4.8 Let m equal the median of the posttest grip strengths in the right arms of male freshmen in the health dynamics course. We shall use observations on $n = 15$ such students to test the null hypothesis H_0: $m = 50$ against the alternative hypothesis H_1: $m > 50$.

(a) Using the sign test, define a critical region that has an approximate significance level of $\alpha = 0.05$.

(b) Given the observed values

$$58 \quad 52.5 \quad 46 \quad 57.5 \quad 52 \quad\quad 45.5 \quad 65.5 \quad 71$$
$$57 \quad 54 \quad\quad 48 \quad 58 \quad\quad 35.5 \quad 44 \quad\quad 53$$

what is your conclusion?

(c) What is the p-value of this test?

(d) Perform the same test using the Wilcoxon statistic. Let $\alpha = 0.05$. What is your conclusion?

(e) What is the approximate p-value using the Wilcoxon statistic?

10.4.9 Let X equal the weight in pounds of a "1-pound" bag of carrots. Let m equal the median weight of a population of these bags. Test the null hypothesis H_0: $m = 1.14$ against the alternative hypothesis H_1: $m > 1.14$.

(a) With a sample of size $n = 14$, use the sign test to define a critical region. Use $\alpha = 0.10$, approximately.

(b) What is your conclusion if the observed weights were

$$1.12 \quad 1.13 \quad 1.19 \quad 1.25 \quad 1.06 \quad 1.31 \quad 1.12$$
$$1.23 \quad 1.29 \quad 1.17 \quad 1.20 \quad 1.11 \quad 1.18 \quad 1.23$$

(c) What is the p-value of your test?

(d) Use the Wilcoxon statistic to perform the same test. What is your conclusion?

(e) What is the approximate p-value using the Wilcoxon statistic?

10.4.10 In Exercise 10.3.10, we used the sign test to test $H_0: \ m = 3.05$ against the alternative hypothesis $H_1: m > 3.05$.

(a) For $n = 25$ observations, define a critical region using the Wilcoxon statistic with a significance level of $\alpha = 0.05$.

(b) What is your conclusion using the data in Exercise 10.3.10?

(c) What is the approximate p-value using the Wilcoxon statistic?

10.5 TWO-SAMPLE DISTRIBUTION-FREE TESTS

In Sections 10.3 and 10.4 we considered some distribution-free tests based on a sample from one distribution or a sample of paired dependent random variables. In this section we consider corresponding tests associated with the characteristics of two distributions.

The first test corresponds to the sign test and is called the **median test**. Let m_X and m_Y be the respective medians of two distributions of the continuous type. By taking independent random samples $X_1, X_2, \ldots, X_{n_1}$, and $Y_1, Y_2, \ldots, Y_{n_2}$ from these two distributions, respectively, we wish to test the hypothesis $H_0: \ m_X = m_Y$. To do this, combine the two samples and count the number, say V, of X values in the lower half of this combined sample. If $H_0: \ m_X = m_Y$ is true, then we would expect V to equal some number around $n_1/2$. If, as an alternative, $m_X < m_Y$, we would expect V to be somewhat larger; and, of course, the alternative $m_X > m_Y$ would suggest a smaller value of V.

Let us see what this means in terms of the distribution functions $F(x)$ and $G(y)$ of the respective distributions. Of course, if $F(z) = G(z)$, then $H_0: \ m_X = m_Y$ is true. Since we cannot find the distribution of V knowing only that $m_X = m_Y$, we shall find its distribution assuming that $F(z) = G(z)$. If $F(z) \geq G(z)$, $m_X \leq m_Y$, as depicted in Figure 10.2. If the observed value of V is quite large—that is, if the number of values of X falling below the median of the combined sample is large—we would suspect that $m_X < m_Y$. Thus the critical region for testing $H_0: \ m_X = m_Y$ against $H_1: \ m_X < m_Y$ is of the form $v \geq c$, where c is to be determined to yield the desired significance level [when $F(z) = G(z)$]. Similarly, the critical region for testing $H_0: \ m_X = m_Y$ against $H_1: \ m_X > m_Y$ is of the form $v \leq c$.

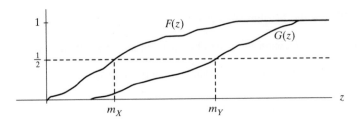

Figure 10.2   Two distribution functions with $F(z) \geq G(z), m_X < m_Y$

When $F(z) = G(z)$ is true and still assuming continuous-type distributions, we shall argue that V has a hypergeometric distribution. To simplify the discussion, say that $n_1 + n_2 = 2k$, where k is a positive integer. To compute $P(V = v)$, we need the probability that exactly v of $X_1, X_2, \ldots, X_{n_1}$ are in the lower half of the ordered combined sample. (Under our assumptions, the probability is zero that any two of the $2k$ random variables are equal.) The smallest k of the $n_1 + n_2 = 2k$ items can be selected in any one of $\binom{2k}{k}$ ways, each having the same probability, provided that $F(z) = G(z)$. Of these $\binom{2k}{k}$ ways, the number in which exactly v of the n_1 values of X and $k - v$ of the n_2 values of Y appear in the lower k items is $\binom{n_1}{v}\binom{n_2}{k - v}$. Hence

$$h(v) = P(V = v) = \frac{\binom{n_1}{v}\binom{n_2}{k - v}}{\binom{n_1 + n_2}{k}}, \qquad v = 0, 1, \ldots, n_1,$$

with the understanding that $\binom{j}{i} = 0$ if $i > j$.

Example 10.5.1

Let X and Y denote the weights of ground cinnamon in "115-gram" tins packaged by companies A and B, respectively. We shall test the hypothesis $H_0: m_X = m_Y$ against the one-sided alternative hypothesis $H_1: m_X < m_Y$. The weights of $n_1 = 8$ and $n_2 = 8$ tins of cinnamon packaged by companies A and B, respectively, selected at random, yielded the following observations of X:

$$117.1 \quad 121.3 \quad 127.8 \quad 121.9 \quad 117.4 \quad 124.5 \quad 119.5 \quad 115.1$$

and the following observations of Y:

$$123.5 \quad 125.3 \quad 126.5 \quad 127.9 \quad 122.1 \quad 125.6 \quad 129.8 \quad 117.2$$

The critical region is of the form $v \geq c$. To determine the value of c when $F(z) = G(z)$, we compute $P(V = v)$ for $v = 6, 7$, and 8. Using Table I,

$$\binom{n_1 + n_2}{k} = \binom{8 + 8}{8} = 12{,}870.$$

Thus

$$h(8) = \frac{\binom{8}{8}\binom{8}{0}}{12{,}870} = \frac{(1)(1)}{12{,}870} = \frac{1}{12{,}870};$$

$$h(7) = \frac{\binom{8}{7}\binom{8}{1}}{12{,}870} = \frac{(8)(8)}{12{,}870} = \frac{64}{12{,}870};$$

$$h(6) = \frac{\binom{8}{6}\binom{8}{2}}{12{,}870} = \frac{(28)(28)}{12{,}870} = \frac{784}{12{,}870}.$$

Since

$$h(8) + h(7) + h(6) = \frac{849}{12{,}870} = 0.066,$$

we shall reject H_0 if $v \geq 6$ at an $\alpha = 0.066$ significance level. The combined ordered sample of weights is listed below with the observations of X underlined.

115.1	117.1	117.2	117.4	119.5	121.3	121.9	122.1
123.5	124.5	125.3	125.6	126.5	127.8	127.9	129.8

We see that $v = 6$, and thus H_0 is rejected with $\alpha = 0.066$. Note that the p-value of this test is

$$p\text{-value} = P(V \geq 6) = 0.066. \qquad \diamond$$

Of course, the median test can be generalized as easily as the sign test. Instead of letting V be the number of X values in the lower half of the combined sample of size $n_1 + n_2$, let V be the number of X values in the lower i values of the combined sample. Thus, if $i/(n_1 + n_2) = p$, we would use V to test the equality of the $(100p)$th percentiles of the two distributions. If V is much larger than $n_1 p$, we would suspect that the $(100p)$th percentile of the X distribution is smaller than that of the Y distribution. If V is much smaller than $n_1 p$, we would guess that it would be the other way around. Since, under $F(z) = G(z)$, all orderings of the n_1 values of X and n_2 values of Y have the same probability, the p.d.f. of V is

$$h(v) = P(V = v) = \frac{\binom{n_1}{v}\binom{n_2}{i - v}}{\binom{n_1 + n_2}{i}}, \qquad v = 0, 1, 2, \ldots, n_1.$$

Example 10.5.2

Using the data in Example 10.5.1 we shall test the null hypothesis that the first quartiles of X and Y are equal against the alternative hypothesis that the first quartile of X is less than the first quartile of Y. Let V equal the number of X values in the lower four values of the combined sample because $i/(n_1 + n_2) = 4/(8 + 8) = 0.25$. The critical region is of the form $v \geq c$. To determine the value of c, we use the p.d.f. of V given by

$$h(v) = P(V = v) = \frac{\binom{8}{v}\binom{8}{4 - v}}{\binom{16}{4}}, \qquad v = 0, 1, 2, 3, 4.$$

Now

$$h(4) = \frac{70}{1820} = 0.038;$$

$$h(3) = \frac{448}{1820} = 0.246.$$

Thus we take for our critical region $C = \{v: v \geq 4\}$, which gives a significance level of $\alpha = 0.038$. From Example 10.5.1 we see that three of the first four observations in the ordered arrangement are values of X and thus $v = 3$, so we fail to reject the null hypothesis. Furthermore, the p-value of this test is

$$p\text{-value} = P(V \geq 3) = 0.246 + 0.038 = 0.284. \qquad \diamond$$

There is another method (due to Wilcoxon) for testing the equality of the medians of two distributions of the continuous type that uses the magnitudes of the observations. For this test it is assumed that the populations have similar shapes. Order the combined sample of $X_1, X_2, \ldots, X_{n_1}$, and $Y_1, Y_2, \ldots, Y_{n_2}$ in increasing order of magnitude. Assign to the ordered values the ranks $1, 2, 3, \ldots, n_1 + n_2$. In the case of ties, assign the average of the ranks associated with the tied values. Let W equal the sum of the ranks of $Y_1, Y_2, \ldots, Y_{n_2}$. If the distribution of Y is shifted to the right of that of X, the values of Y would tend to be larger than the values of X and W would usually be larger than expected when $F(z) = G(z)$. Thus the critical region for testing H_0: $m_X = m_Y$ against H_1: $m_X < m_Y$ would be of the form $w \geq c$. Similarly, if the alternative hypothesis is $m_X > m_Y$, the critical region would be of the form $w \leq c$.

We shall not derive the distribution of W. However, if n_1 and n_2 are both greater than 7, a normal approximation can be used. With $F(z) = G(z)$, the mean and variance of W are

$$\mu_W = \frac{n_2(n_1 + n_2 + 1)}{2}$$

and

$$\text{Var}(W) = \frac{n_1 n_2(n_1 + n_2 + 1)}{12};$$

thus the statistic

$$Z = \frac{W - n_2(n_1 + n_2 + 1)/2}{\sqrt{n_1 n_2 (n_1 + n_2 + 1)/12}}$$

is approximately $N(0, 1)$.

Example 10.5.3

We illustrate the two-sample Wilcoxon test using the data given in Example 10.5.1. The critical region for testing H_0: $m_X = m_Y$ against H_1: $m_X < m_Y$ is of the form $w \geq c$. Since $n_1 = n_2 = 8$, at an approximate $\alpha = 0.05$ significance level, H_0 is rejected if

$$z = \frac{w - 8(8 + 8 + 1)/2}{\sqrt{[(8)(8)(8 + 8 + 1)]/12}} > 1.645,$$

that is, if

$$w > 1.645\sqrt{\frac{(8)(8)(17)}{12}} + 4(17) = 83.66.$$

From the data we see that the computed W is

$$w = 3 + 8 + 9 + 11 + 12 + 13 + 15 + 16 = 87 > 83.66.$$

Thus H_0 is rejected, an action consistent with that of the median test. The p-value of this test is, making a half-unit correction for continuity,

$$p\text{-value} = P(W \geq 87)$$

$$= P\left(\frac{W - 68}{\sqrt{90.667}} \geq \frac{86.5 - 68}{\sqrt{90.667}}\right)$$

$$\approx P(Z \geq 1.943) = 0.0260. \qquad \diamond$$

Remark   It should be mentioned that, shortly after Wilcoxon proposed his test, Mann and Whitney suggested a test based on the estimate of the probability

$$P(X < Y).$$

Namely, they let U equal the number of times that $X_i < Y_j$, $i = 1, 2, \ldots, n_1$ and $j = 1, 2, \ldots, n_2$. Using the data in Example 10.5.1, we find that the computed U is $u = 51$ among all $n_1 n_2 = (8)(8) = 64$ pairs of (X, Y). Thus the estimate of $P(X < Y)$ is $51/64$ or, in general, $u/n_1 n_2$. At the time of the Mann-Whitney suggestion, it was noted that U was just a linear function of Wilcoxon's W and hence really provided the same test. That relationship is

$$U = W - \frac{n_2(n_2 + 1)}{2},$$

which in our special case is

$$51 = 87 - \frac{8(9)}{2} = 87 - 36.$$

Thus we often read about the test of Mann, Whitney, and Wilcoxon.

It should be noted here that the median and Wilcoxon tests are much less sensitive to extreme values than is Student's test based on $\overline{X} - \overline{Y}$. Therefore, if there is much skewness or contamination, these proposed distribution-free tests are much safer. In particular, that of Wilcoxon is quite good and does not lose too much in case the distributions are close to normal ones. It is important to note that the one-sample Wilcoxon requires symmetry of the underlying distribution, but the two-sample Wilcoxon does not and thus can be used for skewed distributions.

The next example illustrates that it is possible to accept the null hypothesis that the medians of two distributions are equal, when in fact the two distributions could be different. A quantile-quantile plot is used to help to detect this difference.

Example 10.5.4

Let X have a Cauchy distribution and let Y have a standard normal distribution $N(0, 1)$. Of course, the assumptions of the Wilcoxon test are not satisfied, but the distributions of X and Y have "similar shapes." We shall simulate on the computer random samples of sizes $n_1 = n_2 = 12$ from these distributions and use the Wilcoxon statistic to test the null hypothesis H_0: $m_X = m_Y$ against a two-sided alternative hypothesis. The $n_1 = 12$ observations of X, which have been ordered, are

$$
\begin{array}{cccccc}
-9.7465 & -1.9458 & -1.5203 & -1.3990 & -0.7215 & -0.1802 \\
0.1219 & 0.2104 & 0.9365 & 2.3121 & 7.5518 & 13.7523
\end{array}
$$

The $n_2 = 12$ observations of Y, which have been ordered, are

$$
\begin{array}{cccccc}
-1.5124 & -1.3661 & -1.0712 & -0.6178 & -0.3631 & -0.2465 \\
0.1484 & 0.2688 & 0.4528 & 1.0480 & 1.2914 & 1.7841
\end{array}
$$

The combined ordered samples, with the observations of X underlined and the ranks of the observations listed below each observation, are

-9.7465	-1.9458	-1.5203	-1.5124	-1.3990	-1.3661
1	2	3	4	5	6
-1.0712	-0.7215	-0.6178	-0.3631	-0.2465	-0.1802
7	8	9	10	11	12
0.1219	0.1484	0.2104	0.2688	0.4528	0.9365
13	14	15	16	17	18
1.0480	1.2914	1.7841	2.3121	7.5518	13.7523
19	20	21	22	23	24

The value of the Wilcoxon statistic, the sum of the Y ranks, is

$$
\begin{aligned}
w &= 4 + 6 + 7 + 9 + 10 + 11 + 14 + 16 + 17 + 19 + 20 + 21 \\
&= 154.
\end{aligned}
$$

Of course, the sum of the X ranks is $300 - 154 = 146$. Now

$$z = \frac{154 - [12(25)/2]}{\sqrt{[(12)(12)(25)]/12}}$$

$$= \frac{154 - 150}{\sqrt{300}} = 0.23,$$

so that H_0 is clearly accepted. That is, we accept the null hypothesis that the medians of X and Y are equal, as we expected. However, if we look at the data more closely by making a q-q plot (Figure 10.3) it is clear that the distributions of X and Y are not equal because the distribution of X has "thicker tails" than does the distribution of Y.   ◇

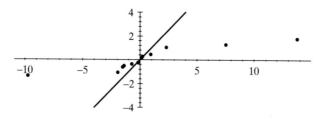

Figure 10.3 q-q plot of Cauchy vs. $N(0, 1)$

EXERCISES

10.5.1 Let us compare the failure times of a certain type of light bulb produced by two different manufacturers, X and Y, by testing 10 bulbs selected at random from each of the outputs. The data, in hundreds of hours used before failure, are

$$X: \quad 5.6 \quad 4.6 \quad 6.8 \quad 4.9 \quad 6.1 \quad 5.3 \quad 4.5 \quad 5.8 \quad 5.4 \quad 4.7$$
$$Y: \quad 7.2 \quad 8.1 \quad 5.1 \quad 7.3 \quad 6.9 \quad 7.8 \quad 5.9 \quad 6.7 \quad 6.5 \quad 7.1$$

(a) Use the one-sided median test at an approximate 9% significance level. What is the p-value?

(b) Use the Wilcoxon test to test the equality of medians of the two processes at the approximate 5% significance level. What is the p-value?

(c) Modify the median test to test the equality of the 25th percentiles of the two processes. HINT: Let V equal the number of X scores in the lower five values of the combined sample.

(d) Construct and interpret a q-q plot of these data.

10.5.2 A pharmaceutical company is interested in testing the effect of humidity on the weight of pills that are sold in aluminum packaging. Let X and Y denote the respective weights of pills and their packaging when the packaging is good and when it is defective after the pill has spent one week in a chamber containing 100% humidity and heated to $30°$C.

(a) Use the Wilcoxon test to test H_0: $m_X = m_Y$ against H_0: $m_X - m_Y < 0$ using the following random samples of $n_1 = 12$ observations of X and $n_2 = 12$ observations of Y. What is the p-value?

$$
\begin{array}{lllllll}
X: & 0.7565 & 0.7720 & 0.7776 & 0.7750 & 0.7494 & 0.7615 \\
 & 0.7741 & 0.7701 & 0.7712 & 0.7719 & 0.7546 & 0.7719 \\
Y: & 0.7870 & 0.7750 & 0.7720 & 0.7876 & 0.7795 & 0.7972 \\
 & 0.7815 & 0.7811 & 0.7731 & 0.7613 & 0.7816 & 0.7851
\end{array}
$$

(b) Use the median test. What is the p-value?

(c) Construct and interpret a q-q plot of these data.

10.5.3 Let X and Y equal the sizes of grocery orders from a southside and a northside foodstore of the same chain, respectively. We shall test the null hypothesis H_0: $m_X = m_Y$ against a two-sided alternative using the following observations that have been ordered:

$$
\begin{array}{llllll}
X: & 5.13 & 8.22 & 11.81 & 13.77 & 15.36 \\
 & 23.71 & 31.39 & 34.65 & 40.17 & 75.58 \\
Y: & 4.42 & 6.47 & 7.12 & 10.50 & 12.12 \\
 & 12.57 & 21.29 & 33.14 & 62.84 & 72.05
\end{array}
$$

(a) Use the median test.

(b) Use the Wilcoxon test.

(c) Construct a q-q plot and interpret it.

10.5.4 Let X and Y denote the heights of blue spruce trees, measured in centimeters, in two large fields. We shall compare these heights by measuring 12 trees selected at random from each of the fields. Use the statistic W, the sum of the ranks of the observations of Y in the combined sample, to test at $\alpha = 0.05$, approximately, the hypothesis H_0: $m_X = m_Y$, against the alternative hypothesis H_0: $m_X < m_Y$, based on $n_1 = 12$ observations of X:

$$
\begin{array}{llllll}
90.4 & 77.2 & 75.9 & 83.2 & 84.0 & 90.2 \\
87.6 & 67.4 & 77.6 & 69.3 & 83.3 & 72.7
\end{array}
$$

and $n_2 = 12$ observations of Y:

$$
\begin{array}{llllll}
92.7 & 78.9 & 82.5 & 88.6 & 95.0 & 94.4 \\
73.1 & 88.3 & 90.4 & 86.5 & 84.7 & 87.5
\end{array}
$$

10.5.5 A company manufactures and packages soap powder in "6-pound" boxes. The quality assurance department was interested in comparing the fill weights of packages from the east and west lines. Taking random samples from the two lines they obtained the following weights:

$$
\begin{array}{llllll}
\text{East line } (X): & 6.06 & 6.04 & 6.11 & 6.06 & 6.06 \\
 & 6.07 & 6.06 & 6.08 & 6.05 & 6.09 \\
\text{West line } (Y): & 6.08 & 6.03 & 6.04 & 6.07 & 6.11 \\
 & 6.08 & 6.08 & 6.10 & 6.06 & 6.04
\end{array}
$$

Let m_X and m_Y denote the median weights for the east and west lines, respectively. Test H_0: $m_X = m_Y$ against a two-sided alternative hypothesis using

(a) The median test with $\alpha = 0.025$, approximately.

(b) The Wilcoxon test with $\alpha = 0.05$, approximately.

(c) Construct and interpret a q-q plot of these data.

10.5.6 A charter bus line has 48-passenger and 38-passenger buses. Let m_{48} and m_{38} denote the median number of miles traveled per day by the respective buses. With $\alpha = 0.05$, use the Wilcoxon statistic to test H_0: $m_{48} = m_{38}$ against the one-sided alternative H_1: $m_{48} > m_{38}$ using the following data, which give the numbers of miles traveled per day for respective random samples of sizes 9 and 11:

48-passenger buses:	331	308	300	414	253
	323	452	396	104	
38-passenger buses:	248	393	260	355	279	184
	386	450	432	196	197

10.5.7 (a) Use the Wilcoxon statistic to test H_0: $m_X = m_Y$ against a two-sided alternative using the following observations of X and Y that have been ordered for your convenience:

X:	-2.3864	-2.2171	-1.9148	-1.9097	-1.4883
	-1.2007	-1.1077	-0.3601	0.4325	1.0598
	1.3035	1.5241	1.7133	1.7656	2.4912
Y:	-1.7613	-0.9391	-0.7437	-0.5530	-0.2469
	0.0647	0.2031	0.3219	0.3579	0.6431
	0.6557	0.6724	0.6762	0.9041	1.3571

(b) Construct a q-q plot and interpret it. NOTE: These data were simulated on a computer. The distribution of X is $U(-2.5, 2.5)$ and the distribution of Y is $N(0, 1)$.

10.5.8 Let X and Y equal the CPU time in days per month for two different computers on a college campus. Let m_X and m_Y equal the medians of X and Y. Using $n_1 = n_2 = 14$ observations of each of X and Y, test the null hypothesis H_0: $m_X = m_Y$ against the alternative hypothesis H_1: $m_X < m_Y$. Use a normal approximation and the two-sample Wilcoxon test. Let $\alpha = 0.05$, approximately.

X:	7.1	8.5	9.9	13.8	2.2	2.4	2.3
	1.0	3.9	4.6	5.3	5.0	3.9	6.6
Y:	7.9	15.4	16.2	9.6	4.9	8.9	9.4
	15.2	12.1	4.3	5.0	2.4	2.3	5.4

10.5.9 Let X equal the playing time in minutes for a $33\frac{1}{3}$ rpm long-playing album and let Y equal the playing time for a compact disk. We shall test the null hypothesis H_0: $m_X = m_Y$ against the alternative hypothesis H_1: $m_X < m_Y$. Give the p-value for each of the two tests. The observed data are

$$
\begin{array}{llllll}
X: & 40.83 & 43.18 & 35.72 & 38.68 & 37.17 \\
   & 39.75 & 24.76 & 34.58 & 33.98 \\
Y: & 42.82 & 64.42 & 56.92 & 39.92 & 72.38 & 47.26 & 64.58 \\
   & 38.20 & 72.75 & 39.09 & 39.07 & 33.70 & 62.02
\end{array}
$$

What is your conclusion using

(a) The median test?

(b) The Wilcoxon test?

10.5.10 In Exercise 7.4.9 a comparison was made between a new and an old process for applying adhesive. Let X and Y equal the number of pounds of pressure needed to pull the glued parts apart using the new and the old process, respectively. Let m_X and m_Y equal the respective medians. We shall test $H_0: m_X = m_Y$ against the alternative $H_1: m_X > m_Y$ using 24 observations of each of X and Y.

(a) At an $\alpha = 0.05$ significance level, what is your conclusion using the Wilcoxon statistic?

(b) What is the approximate p-value of this test?

(c) How does this result compare with that in Exercise 7.4.9?

10.6 RUN TEST AND TEST FOR RANDOMNESS

Under the assumption that the random variables X and Y are of the continuous type and have distribution functions $F(x)$ and $G(y)$, respectively, we describe another test of the hypothesis $H_0: F(z) = G(z)$. This new test can also be used to test for randomness. For these tests we need the concept of runs, which we now define.

Suppose that we have n_1 observations of the random variable X and n_2 observations of the random variable Y. The combination of two sets of independent observations into one collection of $n_1 + n_2$ observations, placed in ascending order of magnitude, might yield an arrangement

$$\underline{y\ y\ y}\ \underline{x\ x}\ \underline{y}\ \underline{x}\ \underline{y}\ \underline{x\ x}\ \underline{y\ y}$$

where x denotes an observation of X and y an observation of Y in the ordered arrangement. We have underlined groups of successive values of X and Y. Each underlined group is called a **run**. Thus we have a run of three values of Y, followed by a run of two values of X, followed by a run of one value of Y, and so on. In this example there are seven runs.

We give two more examples to show what might be indicated by the number of runs. If the five x's and seven y's had the orderings

$$\underline{x\ x\ x\ x}\ \underline{y}\ \underline{x}\ \underline{y\ y\ y\ y\ y\ y}$$

we might suspect that $F(z) \geq G(z)$ (see Figure 10.2). Note that there are four runs in this ordering. The ordered arrangement

$$\underline{y\ y\ y}\ \underline{x\ x}\ \underline{y}\ \underline{x\ x\ x}\ \underline{y\ y\ y}$$

might suggest that the medians of the two distributions are equal but that the spread of the Y distribution is greater than the spread of the X distribution, for example, that $\sigma_Y > \sigma_X$. These examples suggest that the hypothesis $F(z) = G(z)$ should be rejected if the number of runs is too small, where a small number of runs could be caused by differences in the location or in the spread of the two distributions.

Let the random variable R equal the number of runs in the combined ordered sample of n_1 observations of X and n_2 observations of Y. We shall find the distribution of R when $F(z) = G(z)$ and then describe a test of the hypothesis H_0: $F(z) = G(z)$.

Under H_0, all permutations of the n_1 observations of X and n_2 observations of Y have equal probabilities. We can select the n_1 positions for the n_1 values of X in

$$\binom{n_1 + n_2}{n_1}$$

ways, the probability of each arrangement being

$$\frac{1}{\binom{n_1 + n_2}{n_1}}.$$

To find $P(R = r)$, we must determine the number of permutations that yield r runs.

First suppose that $r = 2k$, where k is a positive integer. In this case the n_1 ordered values of X and the n_2 ordered values of Y must each be separated into k runs. We can form k runs of the n_1 values of X by inserting $k - 1$ dividers into the $n_1 - 1$ spaces between the values of X, with no more than one divider per space. This can be done in

$$\binom{n_1 - 1}{k - 1}$$

ways (see Exercise 10.6.4). Similarly, k runs of the n_2 values of Y can be formed in

$$\binom{n_2 - 1}{k - 1}$$

ways. These two sets of runs can be placed together to form $r = 2k$ runs, of which

$$\binom{n_1 - 1}{k - 1}\binom{n_2 - 1}{k - 1}$$

begin with a run of x's and

$$\binom{n_2 - 1}{k - 1}\binom{n_1 - 1}{k - 1}$$

begin with a run of y's. Thus

$$P(R = 2k) = \frac{2\binom{n_1 - 1}{k - 1}\binom{n_2 - 1}{k - 1}}{\binom{n_1 + n_2}{n_2}}, \tag{10.2}$$

where $2k$ is an element of the space of R.

When $r = 2k + 1$, it is possible to have $k + 1$ runs of the ordered values of X and k runs of the ordered values of Y or k runs of X's and $k + 1$ runs of Y's. We can form $k + 1$ runs of the n_1 values of X by inserting k dividers into the $n_1 - 1$ spaces between the values of X with no more than one divider per space in

$$\binom{n_1 - 1}{k}$$

ways. Similarly, k runs of n_2 values of Y can be done in

$$\binom{n_2 - 1}{k - 1}$$

ways. These two sets of runs can be placed together to form $2k + 1$ runs in

$$\binom{n_1 - 1}{k}\binom{n_2 - 1}{k - 1}$$

ways. In addition, $k + 1$ runs of the n_2 values of Y and k runs of the n_1 values of X can be placed together to form

$$\binom{n_2 - 1}{k}\binom{n_1 - 1}{k - 1}$$

sets of $2k + 1$ runs. Hence

$$P(R = 2k + 1) = \frac{\binom{n_1 - 1}{k}\binom{n_2 - 1}{k - 1} + \binom{n_1 - 1}{k - 1}\binom{n_2 - 1}{k}}{\binom{n_1 + n_2}{n_2}}, \qquad (10.3)$$

for $2k + 1$ in the space of R.

A test based on the number of runs can be used for testing the hypothesis H_0: $F(z) = G(z)$. The hypothesis is rejected if the observed number of runs r is too small. That is, the critical region is of the form $r \leq c$, where the constant c is determined by using the p.d.f. of R to yield the desired significance level. The run test is sensitive to both differences in location and differences in spread of the two distributions.

Example 10.6.1

Let X and Y equal the percentages of body fat for freshman women and men, respectively, with distribution functions $F(x)$ and $G(y)$. We shall use the run test to test the hypothesis H_0: $F(z) = G(z)$ against the alternative hypothesis H_1: $F(z) < G(z)$. (That is, the alternative hypothesis is that the X distribution is to the right of the Y distribution.) Ten observations of both X and Y that have been ordered are

$$
\begin{array}{llllll}
X: & 16.6 & 16.7 & 18.5 & 19.2 & 21.5 \\
   & 22.4 & 22.6 & 23.2 & 24.2 & 26.3 \\
Y: & 9.4 & 9.7 & 11.3 & 11.8 & 13.3 \\
   & 15.6 & 16.1 & 16.5 & 18.2 & 21.7
\end{array}
$$

The critical region is of the form $r \leq c$. To determine the value of c, we use Formulas 10.2 and 10.3 with $n_1 = n_2 = 10$. Table I is very useful for evaluating these probabilities. We have

$$P(R = 2) = \frac{2}{184,756}; \quad P(R = 3) = \frac{18}{184,756};$$

$$P(R = 4) = \frac{162}{184,756}; \quad P(R = 5) = \frac{648}{184,756};$$

$$P(R = 6) = \frac{2592}{184,756}; \quad P(R = 7) = \frac{6048}{184,756}.$$

The sum of these six probabilities is $9470/184,756 = 0.051$, so we can take for our critical region $C = \{r: r \leq 7\}$ with a significance level of $\alpha = 0.051$. To determine the number of runs, we order the combined samples and underline adjacent x and y values.

$$\underline{9.4 \quad 9.7} \quad \underline{11.3 \quad 11.8 \quad 13.3 \quad 15.6 \quad 16.1 \quad 16.5} \quad \underline{16.6 \quad 16.7}$$
$$\underline{18.2 \quad 18.5 \quad 19.2 \quad 21.5} \quad \underline{21.7 \quad 22.4 \quad 22.6 \quad 23.2 \quad 24.2 \quad 26.3}$$

We see that the number of runs is $r = 6$ so we reject the null hypothesis. Note that the p-value of this test is

$$p\text{-value} = P(R \leq 6) = \frac{3422}{184,756} = 0.0185. \qquad \diamond$$

When n_1 and n_2 are large, say each is at least equal to 10, R can be approximated with a normally distributed random variable. That is, it can be shown that

$$\mu_R = E(R) = \frac{2n_1 n_2}{n_1 + n_2} + 1,$$

$$\text{Var}(R) = \frac{(\mu_R - 1)(\mu_R - 2)}{n_1 + n_2 - 1} = \frac{2n_1 n_2 (2n_1 n_2 - n_1 - n_2)}{(n_1 + n_2)^2 (n_1 + n_2 - 1)}$$

and

$$Z = \frac{R - \mu_R}{\sqrt{\text{Var}(R)}}$$

is approximately $N(0, 1)$. The critical region for testing the null hypothesis H_0: $F(z) = G(z)$ is of the form $z \leq -z_\alpha$, where α is the desired significance level.

Example 10.6.2

We use the normal approximation to calculate the significance level and the p-value for Example 10.6.1. With $n_1 = n_2 = 10$,

$$\mu_R = \frac{2(10)(10)}{10 + 10} + 1 = 11; \quad \sigma_R^2 = \frac{(11 - 1)(11 - 2)}{19} = \frac{90}{19}.$$

With the critical region $C = \{r: r \leq 7\}$, the approximate significance level, using a half unit correction for continuity, is

$$\alpha = P(R \leq 7) = P\left(\frac{R - 11}{\sqrt{90/19}} \leq \frac{7.5 - 11}{\sqrt{90/19}}\right)$$

$$\approx P(Z \leq -1.608) = 0.0539.$$

Note that this value compares very favorably with $\alpha = 0.051$ given in Example 10.6.1. Since $r = 6$, the approximate p-value, using a normal approximation, is

$$p\text{-value} = P(R \leq 6)$$

$$\approx P\left(Z \leq \frac{6.5 - 11}{\sqrt{90/19}}\right) = P(Z \leq -2.068) = 0.0193,$$

which is close to the p-value given in Example 10.6.1.   ◇

Applications of the run test include tests for randomness. Analysis of runs can also be useful in quality-control studies. To illustrate these applications, let $x_1, x_2, \ldots, x_k$ be the observed values of a random variable X, where the subscripts now designate the order in which the outcomes were observed and the observations are not arranged in the order of magnitude. It is possible, in a quality-control situation, that the observations are made systematically every hour, for example. Assume that k is even. The median divides the k numbers into a lower and an upper half. Replace each observation by L if it falls below the median and by U if it falls above the median. Then, for example, a sequence such as

$$\text{U U U L U L L L}$$

might suggest a trend toward decreasing values of X. If trend is the alternative hypothesis to randomness, the critical region would be of the form $r \leq c$. On the other hand, if we have a sequence such as

$$\text{U L U L U L U L,}$$

we would suspect a cyclic effect and would reject the hypothesis of randomness if r were too large. To test both for trend and cyclic effect, the critical region for testing the hypothesis of randomness is of the form $r \leq c_1$ or $r \geq c_2$.

If the sample size k is odd, the number of observations in the "upper half" and "lower half" will differ by one. In this case we will always put the extra observation in the upper group and, of course, $n_2 = n_1 + 1$. If the median is equal to a value that is tied with other values, we will again put the tied values in the upper group and then perform the test in which n_1 and n_2 are not equal to each other.

Example 10.6.3

We shall use a sample of size $k = 14$ to test for both trend and cyclic effect. To determine the critical region for rejecting the hypothesis of randomness, we use the p.d.f. of R with $n_1 = n_2 = 7$. Since

$$P(R = 2) = P(R = 14) = \frac{2}{3432},$$

$$P(R = 3) = P(R = 13) = \frac{12}{3432},$$

$$P(R = 4) = P(R = 12) = \frac{72}{3432},$$

the critical region $\{r: r \le 4 \text{ or } r \ge 12\}$ would yield a test at a significance level of $\alpha = 172/3432 = 0.05$. The 14 observations are

$$\begin{array}{ccccccc} 81.4 & 76.3 & 85.6 & 76.4 & 88.4 & 80.2 & 85.6 \\ 84.6 & 78.3 & 82.8 & 88.1 & 85.4 & 87.7 & 86.6 \end{array}$$

The median of these outcomes is $(84.6 + 85.4)/2 = 85.0$. Replacing each outcome with L if it falls below 85.0 and U if it falls above 85.0 yields the sequence

$$L \ L \ U \ L \ U \ L \ U \ L \ L \ L \ U \ U \ U \ U.$$

Since $r = 8$, the hypothesis of randomness is not rejected.   ◇

EXERCISES

10.6.1 Let the total lengths of the male and female trident lynx spiders be denoted by X and Y, respectively, with corresponding distribution functions $F(x)$ and $G(y)$. Measurement of the lengths, in millimeters, of eight male and eight female spiders yielded the following observations of X:

$$5.40 \quad 5.55 \quad 6.00 \quad 5.00 \quad 5.70 \quad 5.20 \quad 5.45 \quad 4.95$$

and of Y:

$$6.20 \quad 6.25 \quad 5.75 \quad 5.85 \quad 6.55 \quad 6.05 \quad 5.50 \quad 6.65$$

Use these data to test the hypothesis H_0: $F(z) = G(z)$. Let $\alpha = 0.10$, approximately.

10.6.2 Let X and Y denote the times in hours per week that students in two different schools watch television. Let $F(x)$ and $G(y)$ denote the respective distribution functions. To test the hypothesis H_0: $F(z) = G(z)$, a random sample of eight students was selected from each school. Test this hypothesis using the following observations:

$$\begin{array}{lcccccccc} X: & 16.75 & 19.25 & 22.00 & 20.50 & 22.50 & 15.50 & 17.25 & 20.75 \\ Y: & 24.75 & 21.50 & 19.75 & 17.50 & 22.75 & 23.50 & 13.00 & 19.00 \end{array}$$

10.6.3 A parade consists of 6 bands and 12 floats. How many different parade line-ups are possible if the parade begins and ends with a band and there is at least 1 float between each pair of bands?

10.6.4 Given six values of x, list the $\binom{5}{3} = 10$ ways in which three dividers can be inserted between the six values of x, no more than one divider per space.

10.6.5 Use the run test to test H_0: $F(z) = G(z)$ using the observations of X and Y that are given in Exercise 10.5.7.

(a) Show that if $C = \{r: r \leq 11\}$, then the significance level is $\alpha = 0.0457$, approximately.

(b) Calculate the value of r and state your conclusion.

(c) What is the p-value of this test?

10.6.6 Use the run test to test H_0: $F(z) = G(z)$ using the following observations of X and Y that have been ordered for your convenience. Let $\alpha = 0.05$, approximately.

X:	−2.0482	−1.5748	−0.8797	−0.7170	−0.4907
	−0.2051	0.1651	0.2893	0.3186	0.3550
	0.4056	0.6975	0.7113	0.7377	1.7356
Y:	−1.2311	−1.0228	−0.8836	−0.6684	−0.6157
	−0.5755	−0.1019	−0.0297	0.3781	0.7400
	0.8479	1.0901	1.1397	1.1748	1.2921

NOTE: These data were simulated on a computer using the standard normal distribution for X and a $U(-1.5, 1.5)$ for Y.

10.6.7 Use the following sample of size $k = 14$ to test the hypothesis of randomness against the alternative hypothesis of a trend effect at an $\alpha = 0.025$ significance level.

$$12.4 \quad 14.2 \quad 11.7 \quad 14.0 \quad 12.7 \quad 15.7 \quad 12.8$$
$$14.1 \quad 17.9 \quad 18.4 \quad 17.5 \quad 20.2 \quad 20.8 \quad 20.3$$

10.6.8 Use the following sample of size $k = 16$ to test the hypothesis of randomness against the alternative hypothesis of a cyclic effect:

$$12.4 \quad 31.8 \quad 22.2 \quad 24.5 \quad 17.9 \quad 24.6 \quad 15.7 \quad 27.3$$
$$22.7 \quad 26.0 \quad 14.5 \quad 22.0 \quad 21.8 \quad 31.9 \quad 11.5 \quad 28.3$$

What is your conclusion if (a) $\alpha = 0.0317$ and (b) $\alpha = 0.10$?

10.6.9 A powdered soap manufacturer checks periodically throughout a day the weights of soap in the company's "6-pound" boxes. At each of 22 times, four boxes are selected at random, and the average of the weights of the soap in the boxes is recorded. Use the following 22 average weights to test the hypothesis of randomness against the alternative hypothesis of a trend at an approximate significance level of $\alpha = 0.025$.

$$6.050 \quad 6.038 \quad 6.003 \quad 6.015 \quad 6.025 \quad 6.063 \quad 6.033 \quad 6.010$$
$$5.995 \quad 6.020 \quad 6.060 \quad 6.060 \quad 6.065 \quad 6.050 \quad 6.043 \quad 6.040$$
$$6.045 \quad 6.065 \quad 6.055 \quad 6.060 \quad 6.060 \quad 6.070$$

10.6.10 Each hour a manufacturer of mints selects four mints at random from the production line and finds the average weight in grams. For one week the following average weights were observed. Use these weights to test the hypothesis of randomness against the alternative hypothesis of a cyclic effect.

$$
\begin{array}{cccccccccc}
21.2 & 21.7 & 21.3 & 21.4 & 21.8 & 21.9 & 21.6 & 21.7 & 21.3 & 21.9 \\
21.3 & 22.0 & 21.3 & 21.5 & 21.6 & 21.3 & 21.6 & 21.9 & 21.3 & 21.6 \\
21.9 & 22.0 & 21.9 & 21.4 & 21.4 & 21.3 & 21.7 & 21.6 & 21.5 & 21.7 \\
21.3 & 21.7 & 21.0 & 21.3 & 21.3 & 21.6 & 20.9 & 21.4 & &
\end{array}
$$

10.6.11 It is claimed that X is $U(0, 1)$. A random sample of 14 observations of X yielded the following data:

$$
\begin{array}{ccccccc}
0.15 & 0.67 & 0.05 & 0.47 & 0.29 & 0.23 & 0.10 \\
0.01 & 0.96 & 0.92 & 0.51 & 0.73 & 0.91 & 0.82
\end{array}
$$

(a) Test H_0: $m = 0.5$ against the two-sided alternative hypothesis H_1: $m \neq 0.5$ using the sign test with significance level $\alpha = 0.0574$.

(b) Use a run test with these observations of X to test for both trend and cyclic effect at the approximate significance level $\alpha = 0.05$.

10.6.12 On an introductory statistics test, 27 form A and 29 form B tests were used. The order in which the tests were returned is

$$
\begin{array}{ccccccccccccccc}
A & B & B & A & B & A & B & A & A & B & B & A & B & B \\
A & B & B & A & A & A & B & A & B & A & A & A & B & B \\
B & A & B & A & B & A & B & A & A & A & B & A & B & B \\
A & B & A & A & B & A & B & B & A & B & A & B & B & B
\end{array}
$$

(a) Use a run test to test the null hypothesis that this is a random sequence against the alternative hypothesis that there is a cyclic effect. Use the normal approximation and an $\alpha = 0.01$ significance level.

(b) Based on your conclusion, what advice would you give to the professor who gave this test?

10.6.13 In Exercise 1.5.4 the U.S. birthrates are given for 1960–1987. In order, they are

$$
\begin{array}{ccccccc}
23.7 & 23.3 & 22.4 & 21.7 & 21.1 & 19.4 & 18.4 \\
17.8 & 17.6 & 17.9 & 18.4 & 17.2 & 15.6 & 14.8 \\
14.8 & 14.6 & 14.6 & 15.1 & 15.0 & 15.6 & 15.9 \\
15.8 & 15.9 & 15.5 & 15.5 & 15.8 & 15.6 & 15.7
\end{array}
$$

Use the run test to test whether this is a random sequence against the alternative that there is a trend effect. Give the p-value of your test.

10.6.14 A window is manufactured that will be inserted into a car. For attaching the window, five studs are located in the frame in locations A, B, C, D, and E. Periodically, a window is selected randomly from the production line and a standard "stud pull-out test" is performed that measures the force required to pull a stud out of the window. For each window, the tests are always performed in the order A, B, C, D, and E. Seventy observations that resulted from 14 windows were

$$
\begin{array}{cccccccccc}
140 & 159 & 138 & 102 & 84 & 126 & 147 & 126 & 103 & 92 \\
149 & 155 & 135 & 120 & 94 & 149 & 143 & 109 & 101 & 86 \\
144 & 154 & 120 & 105 & 97 & 149 & 151 & 140 & 103 & 99 \\
157 & 140 & 120 & 96 & 87 & 146 & 137 & 120 & 93 & 89 \\
149 & 154 & 139 & 100 & 84 & 148 & 142 & 130 & 98 & 81 \\
112 & 135 & 109 & 84 & 87 & 112 & 135 & 109 & 84 & 87 \\
126 & 118 & 135 & 90 & 77 & 125 & 126 & 131 & 78 & 75 \\
\end{array}
$$

For these data, the median is equal to 120. Use the run test to test whether this is a random sequence against the alternative hypothesis of a cyclic effect. Give the approximate p-value of the test. Interpret your conclusion.

10.7 KOLMOGOROV-SMIRNOV GOODNESS-OF-FIT TEST

In this section we discuss a test that considers the goodness of fit between a hypothesized distribution function and an empirical distribution function. The empirical distribution function is given in terms of the order statistics. Let $y_1 < y_2 < \cdots < y_n$ be the observed values of the order statistics of a random sample $x_1, x_2, \ldots, x_n$ of size n. When no two observations are equal, the empirical distribution function is defined by

$$
F_n(x) = \begin{cases}
0, & x < y_1, \\
k/n, & y_k \leq x < y_{k+1}, \qquad k = 1, 2, \ldots, n-1, \\
1, & y_n \leq x.
\end{cases}
$$

In this case the empirical distribution function has a jump of magnitude $1/n$ occurring at each observation. If n_k observations are equal to x_k, a jump of magnitude n_k/n occurs at x_k.

Suppose that a random sample of size n is taken from a distribution of the continuous type that has the distribution function $F(x)$. How can we measure the "closeness" of $F(x)$ and the empirical distribution function $F_n(x)$? How does the sample size affect this closeness? We give some theoretical results to help answer these questions and then give a test for goodness of fit.

Let $X_1, X_2, \ldots, X_n$ denote a random sample of size n from a distribution of the continuous type with the distribution function $F(x)$. Consider a fixed value of x. Then $W = F_n(x)$, the value of the empirical distribution function at x, can be thought of as a random variable that takes on the values $0, 1/n, 2/n, \ldots, 1$. Now $nW = k$ if, and only if, exactly k observations are less than or equal to x (say success) and $n - k$ observations are greater than x. The probability that an observation is less than or equal to x is given by $F(x)$. That is, the probability of success is $F(x)$. Because of the independence of the random variables $X_1, X_2, \ldots, X_n$, the probability of k successes is given by the binomial distribution, namely

$$P(nW = k) = P\left(W = \frac{k}{n}\right)$$

$$= \binom{n}{k}[F(x)]^k[1 - F(x)]^{n-k}, \qquad k = 0, 1, 2, \ldots, n.$$

Since nW has a binomial distribution with $p = F(x)$, the mean and variance of nW are given by

$$E(nW) = nF(x) \qquad \text{and} \qquad \text{Var}(nW) = n[F(x)][1 - F(x)].$$

Hence the mean and the variance of $W = F_n(x)$ are

$$E[F_n(x)] = E(W) = F(x)$$

and

$$\text{Var}[F_n(x)] = \text{Var}(W) = \frac{F(x)[1 - F(x)]}{n}.$$

Since the variance of $F_n(x)$ gets nearer to zero as n becomes large, $F_n(x)$ and its mean $F(x)$ tend to be closer with large n. As a matter of fact, there is a theorem by Glivenko, the proof of which is beyond the level of this book, which states that with probability 1, $F_n(x)$ converges to $F(x)$ uniformly in x as $n \to x$.

Because of the convergence of the empirical distribution function to the theoretical distribution function, it makes sense to construct a goodness-of-fit test based on the closeness of the empirical and a hypothesized distribution function, say $F_n(x)$ and $F_0(x)$, respectively. We shall use the Kolmogorov-Smirnov statistic defined by

$$D_n = \sup_x[\,|F_n(x) - F_0(x)|\,].$$

That is, D_n is the least upper bound of all pointwise differences $|F_n(x) - F_0(x)|$.

The exact distribution of the statistic D_n can be derived. We shall not derive this distribution but do give some values of the distribution function of D_n, namely $P(D_n \leq d)$, in Table VIII that will be used for goodness-of-fit tests. We would like to point out that the distribution of D_n does not depend on the particular function $F_0(x)$ of the continuous type. [This is essentially due to the fact that $Y = F_0(X)$ has a uniform distribution $U(0, 1)$.] Thus D_n can be thought of as a distribution-free statistic.

We are interested in using the Kolmogorov-Smirnov statistic D_n to test the null hypothesis H_0: $F(x) = F_0(x)$ against all alternatives, H_1: $F(x) \neq F_0(x)$, where $F_0(x)$ is some specified distribution function. Intuitively, we accept H_0 if the empirical distribution function $F_n(x)$ is sufficiently close to $F_0(x)$, that is, if the value of D_n is sufficiently small. The hypothesis H_0 is rejected if the observed value of D_n is greater than the critical value selected from Table VIII, where this critical value depends upon the desired significance level and sample size.

The use of the Kolmogorov-Smirnov statistic is illustrated by two examples.

Example 10.7.1

We shall test the hypothesis H_0: $F(x) = F_0(x)$ against H_1: $F(x) \neq F_0(x)$, where

$$F_0(x) = \begin{cases} 0, & x < 0, \\ x, & 0 \le x < 1, \\ 1, & 1 \le x. \end{cases}$$

That is, the null hypothesis is that X is $U(0, 1)$. If the test is based on a sample of size $n = 10$ and if $a = 0.10$, the critical region is $C = \{d_{10} : d_{10} \ge 0.37\}$, where d_{10} is the observed value of the Kolmogorov-Smirnov statistic D_{10}. Suppose that the observed values of the random sample are 0.62, 0.36, 0.23, 0.76, 0.65, 0.09, 0.55, 0.26, 0.38, and 0.24. In Figure 10.4 we have plotted the empirical and hypothesized distribution functions for $0 \le x \le 1$. We see that $d_{10} = F_{10}(0.65) - F_0(0.65) = 0.25$ and hence H_0 is not rejected. Note that the more carefully you plot the empirical and theoretical distribution functions, the more you can minimize the number of points at which you should calculate the possible value of the Kolmogorov-Smirnov goodness-of-fit statistic.   ◇

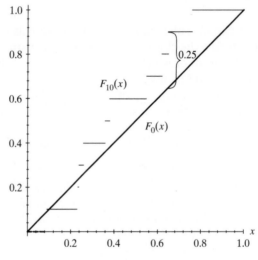

Figure 10.4 H_0: X is $U(0, 1)$

Example 10.7.2

When observing a Poisson process with a mean rate of arrivals $\lambda = 1/\theta$, the random variable W, which denotes the waiting time until the αth arrival, has a gamma distribution. The p.d.f. of W is

$$f(w) = \frac{w^{\alpha-1}e^{-w/\theta}}{\Gamma(\alpha)\theta^\alpha}, \qquad 0 \le w < \infty.$$

A Geiger counter was set up to record the waiting time W in seconds to observe $\alpha = 100$ alpha-particle emissions of barium 133. It is claimed that the number of counts per second has a Poisson distribution with $\lambda = 14.7$ and hence $\theta = 0.068$.

We shall test the hypothesis

$$H_0: \quad F(w) = \int_{-\infty}^{w} f(t)\, dt,$$

where $f(t)$ is the gamma p.d.f. with $\theta = 0.068$ and $\alpha = 100$. Based on 25 observations, H_0 is rejected if $d_{25} \geq 0.24$ for $\alpha = 0.10$. For 25 observations, the empirical and theoretical distribution functions are depicted in Figure 10.5. For these data (Exercise 4.3.6), $d_{25} = 0.117$ and hence H_0 is not rejected.   ◇

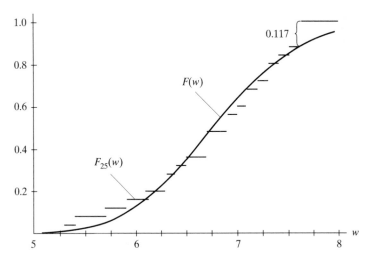

Figure 10.5 H_0: X has a gamma distribution

You will note that we have been assuming that $F(x)$ is a continuous function. That is, we have only considered random variables of the continuous type. This procedure may also be applied in the discrete case. However, in the discrete case, the true significance level will be at most α. That is, the resulting test will be conservative.

Another application of the Kolmogorov-Smirnov statistic is in forming a confidence band for an unknown distribution function $F(x)$. To form a confidence band based on a sample of size n, select a number d such that

$$P(D_n \geq d) = \alpha.$$

Then

$$
\begin{aligned}
1 - \alpha &= P[\sup_x |F_n(x) - F(x)| \leq d] \\
&= P[\,|F_n(x) - F(x)| \leq d \text{ for all } x] \\
&= P[F_n(x) - d \leq F(x) \leq F_n(x) + d \text{ for all } x].
\end{aligned}
$$

Let

$$
F_L(x) = \begin{cases} 0, & F_n(x) - d \leq 0, \\ F_n(x) - d, & F_n(x) - d > 0, \end{cases}
$$

and

$$F_U(x) = \begin{cases} F_n(x) + d, & F_n(x) + d < 1, \\ 1, & F_n(x) + d \geq 0. \end{cases}$$

The two-step functions $F_L(x)$ and $F_U(x)$ yield a $100(1 - \alpha)\%$ confidence band for the unknown distribution function $F(x)$.

Example 10.7.3

A random sample of size $n = 15$ from an unknown distribution yielded the sample values 3.88, 3.97, 4.03, 2.49, 3.18, 3.08, 2.91, 3.43, 2.41, 1.57, 3.78, 3.25, 1.29, 2.57, and 3.40. Now

$$P(D_{15} \geq 0.30) = 0.10.$$

A 90% confidence band for the unknown distribution function $F(x)$ is depicted in Figure 10.6. $\diamond$

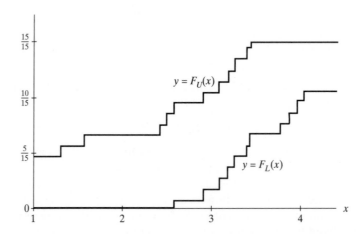

Figure 10.6 A 90% confidence band for an unknown distribution function

EXERCISES

10.7.1 Five observations of X, which have been ordered, are 0.40, 0.51, 0.53, 0.62, and 0.74. Use the Kolmogorov-Smirnov statistic to test the hypothesis that X is $U(0, 1)$. Let $\alpha = 0.20$.

10.7.2 Select 10 sets of 10 random numbers from Table IX. For each set of 10 random numbers, calculate the value of the Kolmogorov-Smirnov statistic d_{10} with $F_0(x) = x, 0 \leq x \leq 1$. Is it true that about 20% of the observations of D_{10} are greater than 0.32? (Compare your results with those of other students, provided that they selected different random numbers.)

10.7.3 A doctor of obstetrics used an ultrasound examination between the 16th and 25th weeks of pregnancy to measure in millimeters the widest diameter of the fetal head. Let X equal this diameter. Use the following 10 observations of X to test the hypothesis that the distribution of X is $N(60, 100)$.

$$56 \quad 65 \quad 47 \quad 57 \quad 62 \quad 48 \quad 68 \quad 75 \quad 79 \quad 49$$

10.7.4 In Exercise 10.4.4 are listed outcomes of 10 simulations of a Cauchy random variable.

(a) Sketch the empirical distribution function and superimpose the Cauchy distribution function.

(b) Use the Kolmogorov-Smirnov statistic to test whether the 10 observations represent a random sample from a Cauchy distribution. Let $\alpha = 0.10$.

10.7.5 Let X equal the distance in feet between bad records on a used computer tape. Test the hypothesis that the distribution of X is exponential with a mean of $\theta = 40$ using the following observations of X:

$$18 \quad 6 \quad 1 \quad 32 \quad 116 \quad 23 \quad 12 \quad 58 \quad 101 \quad 68$$

10.7.6 Construct a 90% confidence band for the unknown distribution function $F(x)$ using the following 15 observations of X:

$$20.2 \quad 85.4 \quad 59.9 \quad 72.7 \quad 88.0 \quad 33.7 \quad 87.1 \quad 99.5 \quad 93.8$$
$$18.4 \quad 60.6 \quad 98.9 \quad 90.9 \quad 86.9 \quad 74.2$$

10.7.7 While testing a computer tape for bad records, the computer operator counted the number of flaws per 100 feet. Let X equal this number and test the hypothesis that the distribution of X is Poisson with a mean of $\lambda = 2.4$ using the following 40 observations of X. Let $\alpha = 0.10$, approximately.

x	Frequency
0	5
1	7
2	12
3	9
4	5
5	1
6	1
	40

10.7.8 Let X equal the number of chocolate chips in a chocolate-chip cookie. Use the Kolmogorov-Smirnov goodness-of-fit statistic and the following data to test the hypothesis that the distribution of X is Poisson with a mean of $\lambda = 5.6$. Let $\alpha = 0.10$, approximately.

x	Frequency
0	0
1	0
2	2
3	8
4	7
5	13
6	13
7	10
8	4
9	4
10	1
	62

10.7.9 In Exercise 10.6.14, 70 observations of a random variable X are listed that give the force required to pull a stud out of a window.

(a) Test the null hypothesis that the distribution of X is $N(117.5, 25^2)$. Use the Kolmogorov-Smirnov statistic with a significance level of $\alpha = 0.20$. What is your conclusion?

(b) Carefully analyzing the graphs of the empirical and theoretical distribution functions or by constructing a histogram, do you agree with the conclusion in part (a)? Why?

(c) Construct and interpret a q-q plot for these data.

10.7.10 In Exercise 10.1.13 the weights in grams of "14-gram" snack packages of peanuts were given as follows:

$$13.9 \quad 14.4 \quad 14.6 \quad 14.7 \quad 14.7 \quad 15.2 \quad 15.2 \quad 15.2$$
$$15.3 \quad 15.4 \quad 15.4 \quad 15.5 \quad 15.6 \quad 15.6 \quad 15.9 \quad 16.4$$

Test the null hypothesis that these weights are observations of a normally distributed random variable with mean $\mu = 15.3$ and standard deviation $\sigma = 0.6$. Use a 10% significance level.

10.7.11 In Exercise 4.2.16 we were asked whether the times that bees spend in high-density flower patches are observations of an exponential random variable. We shall now use the Kolmogorov-Smirnov statistic to help answer this question. Use $\alpha = 0.05$.

(a) Test whether the times are observations of a one-parameter exponential distribution. Estimate θ with $\bar{x}$.

(b) Test whether the times are observations of a two-parameter exponential distribution. Estimate α and θ with their unbiased estimators. That is, letting y_1 equal the minimum, unbiased estimates of θ and α are

$$\widehat{\theta} = [n/(n-1)][\bar{x} - y_1] \quad \text{and} \quad \widehat{\alpha} = y_1 - \widehat{\theta}/n.$$

10.7.12 *Parade* (March 29, 1992) reported the results of a study by Runzheimer International in which data were gathered from around the world on the cost in U.S. dollars per gallon for the least-expensive gasoline available. Each cost in December 1990 (before the Gulf war) was compared with the cost in December 1991. Let X equal the difference, the 1991 cost minus the 1990 cost. Assume that the following $n = 18$ differences represent a random sample of observations of X. How is X distributed? Include estimates of the parameters of X.

$$
\begin{array}{rrrrrr}
-0.28 & -0.40 & 0.06 & 0.04 & -0.29 & 0.38 \\
-0.08 & -0.02 & -0.22 & -0.28 & 0.11 & 0.44 \\
-0.17 & -0.39 & -0.36 & -0.18 & -0.39 & -0.31
\end{array}
$$

11

MULTIVARIATE DISTRIBUTIONS

11.1 MULTIVARIATE DISTRIBUTIONS

In Section 5.1 some properties of multivariate distributions were introduced in a very special case: when the several random variables involved were independent. For illustration, there the joint p.d.f. $f(x, y)$ of X and Y was such that

$$f(x, y) = f_1(x) f_2(y), \qquad x \in R_1, \qquad y \in R_2.$$

In that case it is true that

$$\sum_{x \in R_1} \sum_{y \in R_2} f_1(x) f_2(y) = \sum_{x \in R_1} f_1(x) \sum_{y \in R_2} f_2(y) = (1)(1) = 1.$$

In the continuous case, integrals replace the summations.

However, there are many situations with two or more random variables such that those random variables are not independent; that is, they are dependent. For example, a college admissions department might be interested in the ACT mathematics score X and the ACT verbal score Y of prospective students. Or manufactured items might be classified into three or more categories: Here X might represent the number of good items among n items; Y would be the number of "seconds," and the number of defectives would then be $n - X - Y$. Or to illustrate the Mendelian theory of inheritance in a biology laboratory, 400 kernels of corn could be classified into four categories: smooth and yellow, smooth and purple, wrinkled and yellow, and wrinkled and purple. The numbers in these four categories could be denoted by $X_1, X_2, X_3,$ and $X_4 = 400 - X_1 - X_2 - X_3$, respectively.

In order to deal with situations such as these, it will be necessary to extend certain definitions as well as give new ones. We begin with the discrete case.

Definition 11.1.1. Let X and Y be two random variables defined on a discrete probability space. Let R denote the corresponding two-dimensional space of X and Y, the two random variables of the discrete type. The probability that $X = x$ and $Y = y$ is denoted by $f(x, y) = P(X = x, Y = y)$. The function $f(x, y)$ is called the **joint probability density function** (joint p.d.f.) of X and Y and has the following properties:

(a) $0 \le f(x, y) \le 1$.

(b) $\sum_{(x,y)\in R} f(x, y) = 1$.

(c) $P[(X, Y) \in A] = \sum_{(x,y)\in A} f(x, y)$, where A is a subset of the space R.

The following example will make this definition more meaningful.

Example 11.1.1

Roll a pair of unbiased dice. For each of the 36 sample points with probability 1/36, let X denote the smaller and Y the larger outcome on the dice. For example, if the outcome is $(3, 2)$, then the observed values are $X = 2$, $Y = 3$. The event $\{X = 2, Y = 3\}$ could occur in two ways: $(3, 2)$ or $(2, 3)$; so its probability is

$$\frac{1}{36} + \frac{1}{36} = \frac{2}{36}.$$

If the outcome is $(2, 2)$, then the observed values are $X = 2$, $Y = 2$. Since the event $\{X = 2, Y = 2\}$ can occur in only one way, $P(X = 2, Y = 2) = 1/36$. The joint p.d.f. of X and Y is given by the probabilities

$$f(x, y) = \begin{cases} \dfrac{1}{36}, & 1 \le x = y \le 6, \\ \dfrac{2}{36}, & 1 \le x < y \le 6, \end{cases}$$

when x and y are integers. Figure 11.1 depicts the probabilities of the various points of the space R.   ◇

Notice that certain numbers have been recorded in the bottom and left-hand margins of Figure 11.1. These numbers are the respective column and row totals of the probabilities. The column totals are the respective probabilities that X will assume the values in the x space $R_1 = \{1, 2, 3, 4, 5, 6\}$, and the row totals are the respective probabilities that Y will assume the values in the y space $R_2 = \{1, 2, 3, 4, 5, 6\}$. That is, the totals describe probability density functions of X and Y, respectively. Since each collection of these probabilities is frequently recorded in the margins and satisfies the properties of a p.d.f. of one random variable, each is called a marginal p.d.f.

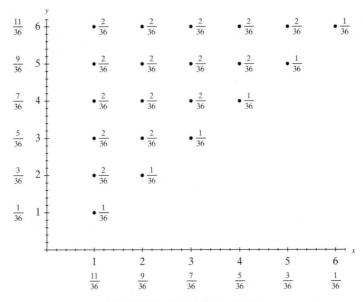

Figure 11.1   Discrete joint p.d.f.

Definition 11.1.2.   Let X and Y have the joint probability density function $f(x, y)$ with space R. The probability density function of X alone, which is called the **marginal probability density function of** X, is defined by

$$f_1(x) = \sum_y f(x, y), \qquad x \in R_1,$$

where the summation is taken over all possible y values for each given x in the x space R_1. That is, the summation is over all (x, y) in R with a given x value. Similarly, the **marginal probability density function of** Y is defined by

$$f_2(y) = \sum_x f(x, y), \qquad y \in R_2,$$

where the summation is taken over all possible x values for each given y in the y space R_2. The random variables X and Y are **independent** if and only if

$$f(x, y) \equiv f_1(x)f_2(y), \qquad x \in R_1, \qquad y \in R_2;$$

otherwise, X and Y are said to be **dependent**.

Example 11.1.2

Let the joint p.d.f. of X and Y be defined by

$$f(x, y) = \frac{x + y}{21}, \qquad x = 1, 2, 3, \qquad y = 1, 2.$$

Then

$$f_1(x) = \sum_y f(x, y) = \sum_{y=1}^{2} \frac{x + y}{21}$$

$$= \frac{x + 1}{21} + \frac{x + 2}{21} = \frac{2x + 3}{21}, \qquad x = 1, 2, 3;$$

and

$$f_2(y) = \sum_x f(x, y) = \sum_{x=1}^{3} \frac{x + y}{21} = \frac{6 + 3y}{21}, \qquad y = 1, 2.$$

Note that both $f_1(x)$ and $f_2(y)$ satisfy the properties of a probability density function. Since $f(x, y) \neq f_1(x) f_2(y)$, X and Y are dependent. $\diamond$

Example 11.1.3

Let the joint p.d.f. of X and Y be

$$f(x, y) = \frac{xy^2}{30}, \qquad x = 1, 2, 3, \qquad y = 1, 2.$$

The marginal probability density functions are

$$f_1(x) = \sum_{y=1}^{2} \frac{xy^2}{30} = \frac{x}{6}, \qquad x = 1, 2, 3,$$

and

$$f_2(y) = \sum_{x=1}^{3} \frac{xy^2}{30} = \frac{y^2}{5}, \qquad y = 1, 2.$$

Then $f(x, y) \equiv f_1(x) f_2(y)$ for $x = 1, 2, 3$ and $y = 1, 2$; thus X and Y are independent. $\diamond$

Example 11.1.4

Let the joint p.d.f. of X and Y be

$$f(x, y) = \frac{xy^2}{13}, \qquad (x, y) = (1, 1), (1, 2), (2, 2).$$

Then the p.d.f. of X is

$$f_1(x) = \begin{cases} \dfrac{5}{13}, & x = 1, \\[2mm] \dfrac{8}{13}, & x = 2, \end{cases}$$

and that of Y is

$$f_2(y) = \begin{cases} \dfrac{1}{13}, & y = 1, \\[2mm] \dfrac{12}{13}, & y = 2. \end{cases}$$

X and Y are dependent because $f(x, y) \neq f_1(x) f_2(y)$ for $x = 1, 2$ and $y = 1, 2$. $\diamond$

Note that in Example 11.1.4 the support R of X and Y is "triangular." Whenever this support R is not "rectangular," the random variables must be dependent because R cannot then equal the product set $\{(x, y): x \in R_1, y \in R_2\}$. That is, if we observe that the support R of X and Y is not a product set, then X and Y must be dependent. For illustration, in Example 11.1.4, X and Y are dependent because $R = \{(1, 1), (1, 2), (2, 2)\}$ is not a product set. On the other hand, if R equals the product set $\{(x, y) : x \in R_1, y \in R_2\}$ and if the formula for $f(x, y)$ is the product of an expression in x alone and an expression in y alone, then X and Y are independent, as illustrated in Example 11.1.3. Example 11.1.2 illustrates the fact that the support can be rectangular but the formula for $f(x, y)$ is not such a product and thus X and Y are dependent.

The notion of a joint p.d.f. of two discrete random variables can be extended to a joint p.d.f. of n random variables of the discrete type. Briefly, the joint p.d.f. of the n random variables $X_1, X_2, \ldots, X_n$ is defined by

$$f(x_1, x_2, \ldots, x_n) = P(X_1 = x_1, X_2 = x_2, \ldots, X_n = x_n)$$

over an appropriate space R. Furthermore, $f(x_1, x_2, \ldots, x_n)$ satisfies properties similar to those given in Definition 11.1.1. In addition, the **marginal probability density function** of one of n discrete random variables, say X_k, is found by summing $f(x_1, x_2, \ldots, x_n)$ over all x_i's except X_k; that is,

$$f_k(x_k) = \sum_{x_1} \cdots \sum_{x_{k-1}} \sum_{x_{k+1}} \cdots \sum_{x_n} f(x_1, x_2, \ldots, x_n), \qquad x_k \in R_k.$$

The random variables $X_1, X_2, \ldots, X_n$ are **mutually independent** if and only if

$$f(x_1, x_2, \ldots, x_n) \equiv f_1(x_1) f_2(x_2) \cdots f_n(x_n), \qquad x_1 \in R_1, x_2 \in R_2, \ldots, x_n \in R_n.$$

If $X_1, X_2, \ldots, X_n$ are not independent, they are said to be **dependent**.

Let $X_1, X_2, \ldots, X_n$ be random variables of the discrete type having a joint distribution. We now consider the mathematical expectation of functions of these random variables. If $u(X_1, X_2, \ldots, X_n)$ is a function of n random variables of the discrete type that have a joint p.d.f. $f(x_1, x_2, \ldots, x_n)$ and space R, then

$$E[u(X_1, X_2, \ldots, X_n)] = \sum_{x_1} \cdots \sum_{x_n} u(x_1, x_2, \ldots, x_n) f(x_1, x_2, \ldots, x_n),$$

if it exists, is called the **mathematical expectation** (or **expected value**) of

$$u(X_1, X_2, \ldots, X_n).$$

Example 11.1.5

There are eight similar chips in a bowl: three marked $(0, 0)$, two marked $(1, 0)$, two marked $(0, 1)$, and one marked $(1, 1)$. A player selects a chip at random and is given the sum of the two coordinates in dollars. If X_1 and X_2 represent those two coordinates, respectively, their joint p.d.f. is

$$f(x_1, x_2) = \frac{3 - x_1 - x_2}{8}, \qquad x_1 = 0, 1 \quad \text{and} \quad x_2 = 0, 1.$$

Thus

$$E(X_1 + X_2) = \sum_{x_2=0}^{1} \sum_{x_1=0}^{1} (x_1 + x_2) \frac{3 - x_1 - x_2}{8}$$

$$= (0)\left(\frac{3}{8}\right) + (1)\left(\frac{2}{8}\right) + (1)\left(\frac{2}{8}\right)(2)\left(\frac{1}{8}\right) = \frac{3}{4}.$$

That is, the expected payoff is 75¢.   ◇

The following mathematical expectations, subject to their existence, have special names:

(a) If $u_1(X_1, X_2, \ldots, X_n) = X_i$, then

$$E[u_1(X_1, X_2, \ldots, X_n)] = E(X_i) = \mu_i$$

is called the **mean** of X_i, $i = 1, 2, \ldots, n$.

(b) If $u_2(X_1, X_2, \ldots, X_n) = (X_i - \mu_i)^2$, then

$$E[u_2(X_1, X_2, \ldots, X_n)] = E[(X_i - \mu_i)^2] = \sigma_i^2 = \text{Var}(X_i)$$

is called the **variance** of X_i, $i = 1, 2, \ldots, n$.

The idea of joint distributions of two or more random variables of the discrete type can be extended to that of two or more random variables of the continuous type. The definitions are really the same except that integrals replace summations. We begin with two random variables of the continuous type.

The **joint probability density function** of two continuous-type random variables is an integrable function $f(x, y)$ with the following properties:

(a) $f(x, y) \geq 0$.

(b) $\int_{-\infty}^{\infty} \int_{-\infty}^{\infty} f(x, y) \, dx \, dy = 1$.

(c) $P[(X, Y) \in A] = \iint_A f(x, y) \, dx \, dy$, where $\{(X, Y) \in A\}$ is an event defined in the plane.

Property (c) implies that $P[(X, Y) \in A]$ is the volume of the solid over the region A in the xy-plane and bounded by the surface $z = f(x, y)$.

Example 11.1.6

Let X and Y have the joint p.d.f. $f(x, y) = e^{-x-y}, 0 < x < \infty, 0 < y < \infty$. The graph of $z = f(x, y)$ is given in Figure 11.2. Let $A = \{(x, y) : 0 < x < \infty, 0 < y < x/3\}$. The probability that (X, Y) falls in A is given by

$$P[(X, Y) \in A] = \int_0^\infty \int_0^{x/3} e^{-x-y} \, dy \, dx = \int_0^\infty e^{-x} [-e^{-y}]_0^{x/3} \, dx$$

$$= \int_0^\infty [e^{-x} - e^{-4x/3}] \, dx = \left[-e^{-x} + \frac{3}{4} e^{-4x/3} \right]_0^\infty$$

$$= \frac{1}{4}. \qquad \diamond$$

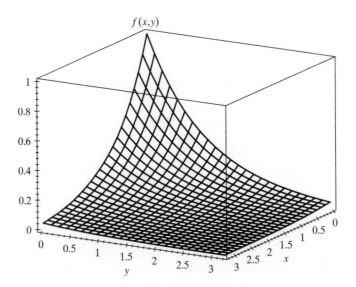

Figure 11.2   Joint p.d.f. $f(x, y) = e^{-x-y}$

The respective **marginal p.d.f.'s** of continuous-type random variables X and Y are given by

$$f_1(x) = \int_{-\infty}^\infty f(x, y) \, dy, \qquad x \in R_1,$$

and

$$f_2(y) = \int_{-\infty}^\infty f(x, y) \, dx, \qquad y \in R_2,$$

where R_1 and R_2 are the spaces of X and Y. The definitions associated with mathematical expectations are the same as those associated with the discrete case after replacing the summations with integrations.

Example 11.1.7

Let X and Y have the joint p.d.f.

$$f(x, y) = 2, \qquad 0 \le x \le y \le 1.$$

Then $R = \{(x, y): 0 \le x \le y \le 1\}$ is the support and, for illustration,

$$P\left(0 \le X \le \frac{1}{2}, 0 \le Y \le \frac{1}{2}\right) = P\left(0 \le X \le Y, 0 \le Y \le \frac{1}{2},\right)$$

$$= \int_0^{1/2} \int_0^y 2\, dx\, dy = \int_0^{1/2} 2y\, dy = \frac{1}{4}.$$

You should draw a figure to illustrate the set of points for which $f(x, y) > 0$ and then shade the region over which the integral is taken. The given probability is the volume above this shaded region under the surface $z = 2$. The marginal p.d.f.'s are given by

$$f_1(x) = \int_x^1 2\, dy = 2(1 - x), \qquad 0 \le x \le 1,$$

and

$$f_2(y) = \int_0^y 2\, dx = 2y, \qquad 0 \le y \le 1.$$

Three illustrations of expected values are

$$E(X) = \int_0^1 \int_x^1 2x\, dy\, dx = \int_0^1 2x(1 - x)\, dx = \frac{1}{3},$$

$$E(Y) = \int_0^1 \int_0^y 2y\, dx\, dy = \int_0^1 2y^2\, dy = \frac{2}{3},$$

$$E(Y^2) = \int_0^1 \int_0^y 2y^2\, dx\, dy = \int_0^1 2y^3\, dy = \frac{1}{2}.$$

From these calculations we see that $E(X)$, $E(Y)$, and $E(Y^2)$ could be calculated using the marginal p.d.f.'s as well as the joint one. ◇

The definition of independent random variables of the continuous type carries over naturally from the discrete case. That is, X and Y are **independent** if and only if the joint p.d.f. factors into the product of their marginal p.d.f.'s; namely,

$$f(x, y) \equiv f_1(x) f_2(y), \qquad x \in R_1, \quad y \in R_2.$$

Thus the random variables X and Y in Example 11.1.6 are independent. In addition, the rules that allow us to easily determine dependent and independent random variables are also valid here. For illustration, X and Y in Example 11.1.7 are dependent because the support R is not a product space, since it is bounded by the diagonal line $y = x$.

If the n random variables, $X_1, X_2, \ldots, X_n$, have joint p.d.f. $f(x_1, x_2, \ldots, x_n)$, the **marginal p.d.f.** of any one of them, say X_k, is given by an $(n-1)$-fold integral in which x_k is fixed; namely,

$$f_k(x_k) = \int_{-\infty}^{\infty} \cdots \int_{-\infty}^{\infty} f(x_1, x_2, \ldots, x_n)dx_1 \cdots dx_{k-1}dx_{k+1} \cdots dx_n, \qquad x_k \in R_k.$$

These n random variables are mutually independent if and only if

$$f(x_1, x_2, \ldots, x_n) \equiv f_1(x_1)f_2(x_2) \cdots f_n(x_n), \qquad x_1 \in R_1, x_2 \in R_2, \cdots, x_n \in R_n.$$

As before, if $X_1, X_2, \ldots, X_n$ are observations of a **random sample of size** n from a distribution with p.d.f. $f(x)$, then the joint p.d.f. of $X_1, X_2, \ldots, X_n$ is

$$f(x_1)f(x_2) \cdots f(x_n).$$

These ideas of marginal distributions are illustrated using extensions of the hypergeometric and the binomial distributions.

Example 11.1.8

Consider a population of 200 students who have just finished a first course in calculus. Of these 200, 40 have earned A's, 60 B's, 70 C's, 20 D's, and 10 F's. A sample of size 25 is taken at random and without replacement from this population so that each possible sample has probability

$$\frac{1}{\binom{200}{25}}$$

of being selected. Within the sample of 25, let X_1 be the number of A students, X_2 the number of B students, X_3 the number of C students, X_4 the number of D students, and $25 - X_1 - X_2 - X_3 - X_4$ the number of F students. The space R of (X_1, X_2, X_3, X_4) is defined by the collection of ordered 4-tuplets of nonnegative integers (x_1, x_2, x_3, x_4) such that $x_1+x_2+x_3+x_4 \leq 25$. The joint p.d.f. of X_1, X_2, X_3, and X_4 is

$$f(x_1, x_2, x_3, x_4) = \frac{\binom{40}{x_1}\binom{60}{x_2}\binom{70}{x_3}\binom{20}{x_4}\binom{10}{25 - x_1 - x_2 - x_3 - x_4}}{\binom{200}{25}},$$

for $(x_1, x_2, x_3, x_4) \in R$, where it is understood that $\binom{k}{j} = 0$ if $j > k$. Without actually summing, we know that the marginal p.d.f. of X_3 is

$$f_3(x_3) = \frac{\binom{70}{x_3}\binom{130}{25 - x_3}}{\binom{200}{25}}, \qquad x_3 = 0, 1, 2, \ldots, 25,$$

since X_3 alone has a hypergeometric distribution.

Similarly, the joint marginal p.d.f. of X_1 and X_2 is

$$f_{12}(x_1, x_2) = \frac{\binom{40}{x_1}\binom{60}{x_2}\binom{100}{25 - x_1 - x_2}}{\binom{200}{25}}, \qquad 0 \le x_1, \quad 0 \le x_2, \quad x_1 + x_2 \le 25.$$

Of course, $f_3(x_3)$ is a hypergeometric p.d.f. and $f_{12}(x_1, x_2)$ and $f(x_1, x_2, x_3, x_4)$ are extensions of that type of p.d.f. Since

$$f(x_1, x_2, x_3, x_4) \ne f_1(x_1) f_2(x_2) f_3(x_3) f_4(x_4),$$

X_1, X_2, X_3, and X_4 are dependent. Note also that the space R is not "rectangular," which would also imply that the random variables are dependent. ◇

The distribution in Example 11.1.8 illustrates an extension of the hypergeometric distribution. In general, instead of two classes, suppose that each of n objects can be placed into one of s disjoint classes so that n_1 objects are in the first class, n_2 in the second, and so on until we find n_s in the sth class. Thus, $n = n_1 + n_2 + \cdots + n_s$. A collection of r objects is selected from these n at random and without replacement. Let the random variable X_i denote the number of observed objects in the sample belonging to the ith class. Find the probability that exactly x_i objects belong to the ith class, $i = 1, 2, \ldots, s$. Here

$$0 \le x_i \le n_i \qquad \text{and} \qquad x_1 + x_2 + \cdots + x_s = r.$$

We can select x_i objects from the ith class in any one of $\binom{n_i}{x_i}$ ways, $i = 1, 2, \ldots, s$. By the multiplication principle, the product

$$\binom{n_1}{x_1}\binom{n_2}{x_2}\cdots\binom{n_s}{x_s}$$

equals the number of ways the joint operation can be performed. If we assume that each of the $\binom{n}{r}$ ways of selecting r objects from $n = n_1 + n_2 + \cdots + n_s$ objects has the same probability, we have that the probability of selecting exactly x_i objects from the ith class, $i = 1, 2, \ldots, s$, is

$$P(X_1 = x_1, X_2 = x_2, \ldots, X_s = x_s) = \frac{\binom{n_1}{x_1}\binom{n_2}{x_2}\cdots\binom{n_s}{x_s}}{\binom{n}{r}},$$

$$0 \le x_i \le n_i \qquad \text{and} \qquad x_1 + x_2 + \cdots + x_s = r.$$

Example 11.1.9

The probability that a 13-card bridge hand (selected at random and without replacement) contains two clubs, four diamonds, three hearts, and four spades is

$$\frac{\binom{13}{2}\binom{13}{4}\binom{13}{3}\binom{13}{4}}{\binom{52}{13}} = \frac{11,404,407,300}{635,013,559,600} = 0.018. \qquad \diamond$$

In Section 7.5 we introduced an extension of the binomial distribution, namely the **multinomial distribution**. In particular, if $X_1, X_2, \ldots, X_{k-1}$ have a multinomial distribution with parameters $n, p_1, p_2, \ldots, p_{k-1}$, (with $X_k = n - X_1 - \cdots - X_{k-1}$ and $p_k = 1 - p_1 - \cdots - p_{k-1}$, for convenience) then the joint p.d.f. is given by

$$P(X_1 = x_1, X_2 = x_2, \ldots, X_k = x_k) = \frac{n!}{x_1! x_2! \cdots x_k!} p_1^{x_1} p_2^{x_2} \cdots p_k^{x_k},$$

where the integer $x_i \geq 0$, $i = 1, 2, \ldots, k$ and $x_1 + x_2 + \cdots + x_k = n$.

When $k = 3$, we often let $X = X_1$ and $Y = X_2$; then $n - X - Y = X_3$. We say that X and Y have a **trinomial distribution**. The joint p.d.f. of X and Y is

$$f(x, y) = \frac{n!}{x! y! (n - x - y)!} p_1^x p_2^y (1 - p_1 - p_2)^{n-x-y},$$

where x and y are nonnegative integers such that $x + y \leq n$. Since the marginal distributions of X and Y are, respectively, $b(n, p_1)$ and $b(n, p_2)$, the product of their probability density functions does not equal $f(x, y)$, and hence they are dependent random variables. Also note that the support of X and Y is triangular, so the random variables must be dependent.

Example 11.1.10

In manufacturing a certain item, it is found that in normal production about 95% of the items are good ones, 4% are "seconds," and 1% are defective. This particular company has a program of quality control by statistical method; and each hour an on-line inspector observes 20 items selected at random, counting the number X of seconds, and the number Y of defectives. If, in fact, the production is normal, we shall find the probability that in this sample of size $n = 20$, at least two seconds or at least two defective items are found. If we let $A = \{(x, y) : x \geq 2 \text{ or } y \geq 2\}$, then

$$P(A) = 1 - P(A')$$

$$= 1 - P(X = 0 \text{ or } 1 \text{ and } Y = 0 \text{ or } 1)$$

$$= 1 - \frac{20!}{0! 0! 20!} (0.04)^0 (0.01)^0 (0.95)^{20} - \frac{20!}{1! 0! 19!} (0.04)^1 (0.01)^0 (0.95)^{19}$$

$$\quad - \frac{20!}{0! 1! 19!} (0.04)^0 (0.01)^1 (0.95)^{19} - \frac{20!}{1! 1! 18!} (0.04)^1 (0.01)^1 (0.95)^{18}$$

$$= 0.204. \qquad \diamond$$

EXERCISES

11.1.1 Let the joint p.d.f. of X and Y be defined by

$$f(x, y) = \frac{x + y}{32}, \quad x = 1, 2, \quad y = 1, 2, 3, 4.$$

Find

(a) $f_1(x)$, the marginal p.d.f. of X.

(b) $f_2(y)$, the marginal p.d.f. of Y.

(c) $P(X > Y)$.

(d) $P(Y = 2X)$.

(e) $P(X + Y = 3)$.

(f) $P(X \le 3 - Y)$.

(g) Are X and Y independent or dependent?

11.1.2 Roll a red and a black four-sided die. Let X equal the outcome on the red die, and let Y equal the outcome on the black die.

(a) On graph paper, show the space of X and Y.

(b) Define the joint p.d.f. on the space (similar to Figure 11.1).

(c) Give the marginal p.d.f. of X in the margin.

(d) Give the marginal p.d.f. of Y in the margin.

(e) Are X and Y dependent or independent? Why?

11.1.3 Roll a red and a black four-sided die. Let X equal the outcome on the red die and let Y equal the sum of the two dice.

(a) On graph paper, describe the space of X and Y.

(b) Define the joint p.d.f. on the space (similar to Figure 11.1).

(c) Give the marginal p.d.f. of X in the margin.

(d) Give the marginal p.d.f. of Y in the margin.

(e) Are X and Y dependent or independent? Why?

11.1.4 Roll a pair of six-sided dice; one red and one black. Let X equal the outcome on the red die and let Y equal the absolute value of the difference of the two outcomes.

(a) On a graph, describe the space of X and Y.

(b) Define the joint p.d.f. on the space (similar to Figure 11.1).

(c) Give the marginal p.d.f.'s in the margins.

(d) Calculate the values of $\mu_Y = E(Y)$ and $\sigma_Y^2 = \text{Var}(Y)$.

11.1.5 Draw 13 cards at random and without replacement from an ordinary deck of playing cards. Among these 13 cards, let X_1 be the number of spades, X_2 the number of hearts, X_3 the number of diamonds, and $13 - X_1 - X_2 - X_3$ the number of clubs.

(a) Give the joint p.d.f. of X_1, X_2, X_3 and describe the support in three-space.

(b) Determine the p.d.f. of X_1 and X_2 and describe the support in two-space.

(c) Find the p.d.f. of X_3, including its support.

11.1.6 In a biology laboratory, corn is used to illustrate the Mendelian theory of inheritance. It is claimed that the four categories for the kernels of corn smooth and yellow, wrinkled and yellow, smooth and purple, and wrinkled and purple, will occur in the ratio 9:3:3:1. Out of 208 kernels of corn, let X_1, X_2, X_3, and $X_4 = 208 - X_1 - X_2 - X_3$ denote the numbers in the four categories, respectively, and assume that the theory is true.

(a) Give the joint p.d.f. of X_1, X_2, X_3, and X_4, and describe the support in 3-space.

(b) Give the marginal p.d.f. of X_1.

11.1.7 A manufactured item is classified as good, a "second," or defective, with probabilities 6/10, 3/10, and 1/10, respectively. Fifteen such items are selected at random from the production line. Let X denote the number of good items, Y the number of seconds, and $15 - X - Y$ the number of defective items.

(a) Give the joint p.d.f. of X and Y, $f(x, y)$.

(b) Sketch the set of points for which $f(x, y) > 0$. From the shape of this region, can X and Y be independent? Why?

(c) Find $P(X = 10, Y = 4)$.

(d) Give the marginal p.d.f. of X.

(e) Find $P(X \leq 11)$.

11.1.8 In a smoking survey among boys between the ages of 12 and 17, 78% prefer to date nonsmokers, 1% prefer to date smokers, and 21% don't care. If seven such boys are selected randomly and X equals the number who prefer to date nonsmokers and Y equals the number who prefer to date smokers,

(a) Determine the joint p.d.f. of X and Y. Be sure to include the support of the p.d.f.

(b) Find the marginal p.d.f. of X. Again include the support.

11.1.9 Let $f(x, y) = 2e^{-x-y}$, $0 \leq x \leq y < \infty$, be the joint p.d.f. of X and Y. Find $f_1(x)$ and $f_2(y)$, the marginal p.d.f.'s of X and Y, respectively. Are X and Y independent?

11.1.10 Let $f(x, y) = 3/2$, $x^2 \leq y \leq 1$, $0 \leq x \leq 1$, be the joint p.d.f. of X and Y. Find

(a) $P(0 \leq X \leq 1/2)$.

(b) $P(1/2 \leq Y \leq 1)$.

(c) $P(1/2 \leq X \leq 1, 1/2 \leq Y \leq 1)$.

(d) $P(X \geq 1/2, Y \geq 1/2)$.

(e) Are X and Y independent?

11.1.11 Let $f(x, y) = 1/4$, $0 \leq x \leq 2$, $0 \leq y \leq 2$, be the joint p.d.f. of X and Y. Find $f_1(x)$ and $f_2(y)$, the marginal probability density functions. Are the two random variables independent?

11.1.12 Let X and Y have the joint p.d.f. $f(x, y) = x + y$, $0 \leq x \leq 1$, $0 \leq y \leq 1$. (a) Find the marginal p.d.f.'s $f_1(x)$ and $f_2(y)$ and show that $f(x, y) \neq f_1(x)f_2(y)$. Thus X and Y are dependent. Compute (b) μ_X, (c) μ_Y, (d) σ_X^2, and (e) σ_Y^2.

11.2 THE CORRELATION COEFFICIENT

In Section 11.1 we introduced the mathematical expectation of a function of two or more random variables, say $X_1, X_2, \ldots, X_n$. We gave the respective special names of mean and variance of X_i to

$$\mu_i = E(X_i) \quad \text{and} \quad \sigma_i^2 = E[(X_i - \mu_i)^2], \quad i = 1, 2, \ldots, n.$$

We introduce two more special names:

(a) If $u_3(X_1, X_2, \ldots, X_n) = (X_i - \mu_i)(X_j - \mu_j), i \neq j$, then

$$E[u_3(X_1, X_2, \ldots, X_n)] = E[(X_i - \mu_i)(X_j - \mu_j)] = \sigma_{ij} = \text{Cov}(X_i, X_j)$$

is called the **covariance** of X_i and X_j.

(b) If the standard deviations σ_i and σ_j are positive, then

$$\rho_{ij} = \frac{\text{Cov}(X_i, X_j)}{\sigma_i \sigma_j} = \frac{\sigma_{ij}}{\sigma_i \sigma_j}$$

is called the **correlation coefficient** of X_i and X_j.

It is convenient that the mean and the variance of X_i can be computed from either the joint p.d.f. or the marginal p.d.f. of X_i. For example, if $n = 2$,

$$\mu_1 = E(X_1) = \sum_{x_1} \sum_{x_2} x_1 f(x_1, x_2)$$

$$= \sum_{x_1} x_1 \left[\sum_{x_2} f(x_1, x_2) \right] = \sum_{x_1} x_1 f_1(x_1).$$

However, to compute the covariance, we need the joint p.d.f.

Before considering the meaning of the covariance and the correlation coefficient, let us note a few simple facts. With $i \neq j$,

$$E[(X_i - \mu_i)(X_j - \mu_j)] = E(X_i X_j - \mu_i X_j - \mu_j X_i + \mu_i \mu_j)$$
$$= E(X_i X_j) - \mu_i E(X_j) - \mu_j E(X_i) + \mu_i \mu_j$$

because it is true that even in the multivariate situation, E is still a linear or distributive operator (see Exercise 11.2.4). Thus

$$\text{Cov}(X_i, X_j) = E(X_i X_j) - \mu_i \mu_j - \mu_j \mu_i + \mu_i \mu_j = E(X_i X_j) - \mu_i \mu_j.$$

Since $\rho_{ij} = \text{Cov}(X_i, X_j)/\sigma_i \sigma_j$, we also have

$$E(X_i X_j) = \mu_i \mu_j + \rho_{ij} \sigma_i \sigma_j.$$

That is, the expected value of the product of two random variables is equal to the product $\mu_i \mu_j$ of their expectations plus their covariance $\rho_{ij} \sigma_i \sigma_j$.

A simple example at this point would be helpful.

Example 11.2.1

Let X_1 and X_2 have the joint p.d.f.

$$f(x_1, x_2) = \frac{x_1 + 2x_2}{18}, \qquad x_1 = 1, 2, \quad x_2 = 1, 2.$$

The marginal probability density functions are, respectively,

$$f_1(x_1) = \sum_{x_2=1}^{2} \frac{x_1 + 2x_2}{18} = \frac{2x_1 + 6}{18}, \qquad x_1 = 1, 2,$$

and

$$f_2(x_2) = \sum_{x_1=1}^{2} \frac{x_1 + 2x_2}{18} = \frac{3 + 4x_2}{18}, \qquad x_2 = 1, 2.$$

Since $f(x_1, x_2) \neq f_1(x_1)f_2(x_2)$, X_1 and X_2 are dependent. The mean and the variance of X_1 are

$$\mu_1 = \sum_{x_1=1}^{2} x_1 \frac{2x_1 + 6}{18} = (1)\left(\frac{8}{18}\right) + (2)\left(\frac{10}{18}\right) = \frac{14}{9}$$

and

$$\sigma_1^2 = \sum_{x_1=1}^{2} x_1^2 \frac{2x_1 + 6}{18} - \left(\frac{14}{9}\right)^2 = \frac{24}{9} - \frac{196}{81} = \frac{20}{81}.$$

The mean and the variance of X_2 are

$$\mu_2 = \sum_{x_2=1}^{2} x_2 \frac{3 + 4x_2}{18} = (1)\left(\frac{7}{18}\right) + (2)\left(\frac{11}{18}\right) = \frac{29}{18}$$

and

$$\sigma_2^2 = \sum_{x_2=1}^{2} x_2^2 \frac{3 + 4x_2}{18} - \left(\frac{29}{18}\right)^2 = \frac{51}{18} - \frac{841}{324} = \frac{77}{324}.$$

The covariance of X_1 and X_2 is

$$\text{Cov}(X_1, X_2) = \sum_{x_2=1}^{2} \sum_{x_1=1}^{2} x_1 x_2 \frac{x_1 + 2x_2}{18} - \left(\frac{14}{9}\right)\left(\frac{29}{18}\right)$$

$$= (1)(1)\left(\frac{3}{18}\right) + (2)(1)\left(\frac{4}{18}\right) + (1)(2)\left(\frac{5}{18}\right)$$

$$+ (2)(2)\left(\frac{6}{18}\right) - \left(\frac{14}{19}\right)\left(\frac{29}{18}\right)$$

$$= \frac{45}{18} - \frac{406}{162} = -\frac{1}{162}.$$

Hence the correlation coefficient is

$$\rho = \frac{-1/162}{\sqrt{(20/81)(77/324)}} = \frac{-1}{\sqrt{1540}} = -0.025. \qquad \diamond$$

Insight into the correlation coefficient ρ of two discrete random variables X and Y may be gained by thoughtfully examining its definition

$$\rho = \frac{\sum_R (x - \mu_X)(y - \mu_Y)f(x, y)}{\sigma_X \sigma_Y},$$

where μ_X, μ_Y, σ_X, and σ_Y denote the respective means and standard deviations. If positive probability is assigned to pairs (x, y) in which both x and y are either simultaneously above or simultaneously below their respective means, the corresponding terms in the summation that defines ρ are positive because both factors $(x - \mu_X)$ and $(y - \mu_Y)$ will be positive or both will be negative. If pairs (x, y), which yield large positive products $(x - \mu_X)(y - \mu_Y)$, contain most of the probability of the distribution, the correlation coefficient will tend to be positive. If, on the other hand, the points (x, y), in which one component is below its mean and the other above its mean, have most of the probability, then the coefficient of correlation will tend to be negative because the products $(x - \mu_X)(y - \mu_Y)$ are negative (see Example 11.1.8). This interpretation of the sign of the correlation coefficient will play an important role in subsequent work.

To gain additional insight into the meaning of the correlation coefficient p, consider the following problem. Think of the points (x, y) in the space R and their corresponding probabilities. Let us consider all possible lines in two-dimensional space, each with finite slope, that pass through the point associated with the means, namely (μ_X, μ_Y). These lines are of the form $y - \mu_Y = b(x - \mu_X)$ or, equivalently, $y = \mu_Y + b(x - \mu_X)$. For each point in R, say (x_0, y_0) so that $f(x_0, y_0) > 0$, consider the vertical distance from that point to one of these lines. Since y_0 is the height of the point above the x axis and $\mu_Y + b(x_0 - \mu_X)$ is the height of the point on the line that is directly above or below the point (x_0, y_0), then the absolute value of the difference of these two heights is the vertical distance from point (x_0, y_0) to the line $y = \mu_Y + b(x - \mu_X)$ That is, the required distance is $|y_0 - \mu_Y - b(x_0 - \mu_X)|$. Let us now square this distance and take the weighted average of all such squares; that is, let us consider the mathematical expectation

$$E\{[(Y - \mu_Y) - b(X - \mu_X)]^2\} = K(b).$$

The problem is to find that line (or that b) which minimizes this expectation of the square $\{Y - \mu_Y - b(X - \mu_X)\}^2$. This is an application of the principle of least squares, and the line is sometimes called the least squares regression line.

The solution of the problem is very easy, since

$$K(b) = E\{(Y - \mu_Y)^2 - 2b(X - \mu_X)(Y - \mu_Y) + b^2(X - \mu_X)^2\}$$
$$= \sigma_Y^2 - 2b\rho\sigma_X\sigma_Y + b^2\sigma_X^2,$$

because E is a linear operator and $E[(X - \mu_X)(Y - \mu_Y)] = \rho\sigma_X\sigma_Y$. Accordingly, the derivative

$$K'(b) = -2\rho\sigma_X\sigma_Y + 2b\sigma_X^2$$

equals zero at $b = \rho\sigma_Y/\sigma_X$; and we see that $K(b)$ obtains its minimum for that b since $K''(b) = 2\sigma_X^2 > 0$. Consequently, the **least squares regression line** (the line of the given form that is the best fit in the foregoing sense) is

$$y = \mu_Y + \rho\frac{\sigma_Y}{\sigma_X}(x - \mu_X).$$

Of course, if $\rho > 0$, the slope of the line is positive; but if $\rho < 0$, the slope is negative.

It is also instructive to note the value of the minimum of

$$K(b) = E\{[(Y - \mu_Y) - b(X - \mu_X)]^2\} = \sigma_Y^2 - 2b\rho\sigma_X\sigma_Y + b^2\sigma_X^2.$$

It is

$$K\left(\rho\frac{\sigma_Y}{\sigma_X}\right) = \sigma_Y^2 - 2\rho\frac{\sigma_Y}{\sigma_X}\rho\sigma_X\sigma_Y + \left(\rho\frac{\sigma_Y}{\sigma_X}\right)^2\sigma_X^2$$
$$= \sigma_Y^2 - 2\rho^2\sigma_Y^2 + \rho^2\sigma_Y^2 = \sigma_Y^2(1 - \rho^2).$$

Since $K(b)$ is the expected value of a square, it must be nonnegative for all b, and we see that $\sigma_Y^2(1 - \rho^2) \geq 0$; that is $\rho^2 \leq 1$, and hence $-1 \leq \rho \leq 1$, which is an important property of the correlation coefficient ρ. If $\rho = 0$, then $K(\rho\sigma_Y/\sigma_X) = \sigma_Y^2$; on the other hand, $K(\rho\sigma_Y/\sigma_X)$ is relatively small if ρ is close to 1 or negative 1. That is, the vertical deviations of the points with positive probability from the line $y = \mu_Y + \rho(\sigma_Y/\sigma_X)(x - \mu_X)$ are small if ρ is close to 1 or negative 1 because $K(\rho\sigma_Y/\sigma_X)$ is the expectation of the square of those deviations. Thus ρ measures, in this sense, the amount of *linearity* in the probability distribution. As a matter of fact, in the discrete case, all the points of positive probability lie on this straight line if and only if ρ is equal to 1 or negative 1.

Remark   More generally, we could have fitted the line $y = a + bx$ by the same application of the principle of least squares. We would then have proved that the "best" line actually passes through the point (μ_X, μ_Y). Recall that in the discussion above we assumed our line to be of that form. Students will find this derivation to be an interesting exercise using partial derivatives (see Exercise 11.2.5).

The following example illustrates a joint discrete distribution for which ρ is negative. In Figure 11.3 the line of best fit or the least squares regression line is also drawn.

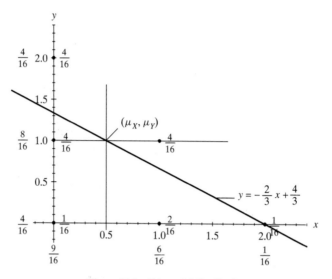

Figure 11.3   Trinomial distribution

Example 11.2.2

Let X equal the number of ones and Y the number of twos and threes when a pair of fair four-sided dice are rolled. Then X and Y have a trinomial distribution with p.d.f.

$$f(x, y) = \frac{2!}{x!y!(2 - x - y)!}\left(\frac{1}{4}\right)^x\left(\frac{2}{4}\right)^y\left(\frac{1}{4}\right)^{2-x-y}, \qquad 0 \le x + y \le 2,$$

where x and y are nonnegative integers. Since the marginal p.d.f. of X is $b(2, 1/4)$ and the marginal p.d.f. of Y is $b(2, 1/2)$, we know that $\mu_X = 1/2$, $\mathrm{Var}(X) = 6/16$, $\mu_Y = 1$, and $\mathrm{Var}(Y) = 1/2$. Since $E(XY) = (1)(1)(4/16) = 4/16$, $\mathrm{Cov}(X, Y) = 4/16 - (1/2)(1) = -4/16$; therefore, the correlation coefficient is $\rho = -1/\sqrt{3}$. Using these values for the parameters, we obtain the line of best fit, namely

$$y = 1 + \left(-\frac{1}{\sqrt{3}}\right)\sqrt{\frac{1/2}{3/8}}\left(x - \frac{1}{2}\right) = -\frac{2}{3}x + \frac{4}{3}.$$

The joint p.d.f. is displayed in Figure 11.3 along with the line of best fit.   ◇

Suppose that X and Y are independent so that $f(x, y) \equiv f_1(x) f_2(y)$ and we want to find the expected value of the product $u(X)v(Y)$. Subject to the existence of the expectations, we know that

$$E[u(X)v(Y)] = \sum_{R_1} \sum_{R_2} u(x)v(y) f(x, y)$$

$$= \sum_{R_1} \sum_{R_2} u(x)v(y) f_1(x) f_2(y)$$

$$= \sum_{R_1} u(x) f_1(x) \sum_{R_2} v(y) f_2(y)$$

$$= E[u(X)]E[v(Y)].$$

This can be used to show that the correlation coefficient of two independent variables is zero. For, in a standard notation, we have

$$\text{Cov}(X, Y) = E[(X - \mu_X)(Y - \mu_Y)]$$

$$= E(X - \mu_X)E(Y - \mu_Y) = 0.$$

The converse of this fact is not necessarily true, however; zero correlation does not in general imply independence. It is most important to keep this straight: independence implies zero correlation, but zero correlation does not necessarily imply independence. The latter is now illustrated.

Example 11.2.3

Let X and Y have the joint p.d.f.

$$f(x, y) = \frac{1}{3}, \qquad (x, y) = (0, 1), (1, 0), (2, 1).$$

Since the support is not "rectangular," X and Y must be dependent. The means of X and Y are $\mu_X = 1$ and $\mu_Y = 2/3$, respectively. Hence

$$\text{Cov}(X, Y) = E(XY) - \mu_X\mu_Y$$

$$= (0)(1)\left(\frac{1}{3}\right) + (1)(0)\left(\frac{1}{3}\right) + (2)(1)\left(\frac{1}{3}\right) - (1)\left(\frac{2}{3}\right) = 0.$$

That is $\rho = 0$, but X and Y are dependent.   ◇

EXERCISES

11.2.1 Let the random variables X and Y have the joint p.d.f

$$f(x, y) = \frac{x + y}{32}, \qquad x = 1, 2, \quad y = 1, 2, 3, 4.$$

Find the means μ_X and μ_Y, the variances σ_X^2 and σ_Y^2, and the correlation coefficient ρ. Are X and Y independent or dependent?

11.2.2 Let X and Y have the joint p.d.f. defined by $f(0, 0) = f(1, 2) = 0.2$, $f(0, 1) = f(1, 1) = 0.3$.

  (a) Depict the points and corresponding probabilities on a graph.

  (b) Give the marginal p.d.f.'s in the "margins."

  (c) Compute μ_X, μ_Y, σ_X^2, σ_Y^2, $\text{Cov}(X, Y)$, and ρ.

  (d) Find the equation of the least squares regression line and draw it on your graph. Does the line make sense to you intuitively?

11.2.3 Roll a fair four-sided die twice. Let X equal the outcome on the first roll and let Y equal the sum of the two rolls.

  (a) Display the joint p.d.f. on a graph along with the marginal probabilities.

  (b) Determine μ_X, μ_Y, σ_X^2, σ_Y^2, $\text{Cov}(X, Y)$, and ρ.

  (c) Find the equation of the least squares regression line and draw it on your graph. Does the line make sense to you intuitively?

11.2.4 In the multivariate situation, show that E is a linear or distributive operator. For convenience, let $n = 2$ and show that

$$E[a_1 u_1(X_1, X_2) + a_2 u_2(X_1, X_2)] = a_1 E[u_1(X_1, X_2)] + a_2 E[u_2(X_1, X_2)].$$

11.2.5 Let X and Y be random variables with respective means μ_X and μ_Y, respective variances σ_X^2 and σ_Y^2, and correlation coefficient ρ. Fit the line $y = a + bx$ by the method of least squares to the probability distribution by minimizing the expectation

$$K(a, b) = E[(Y - a - bX)^2]$$

with respect to a and b. HINT: Consider $\partial K/\partial a = 0$ and $\partial K/\partial b = 0$ and solve simultaneously.

11.2.6 Let X and Y have a trinomial distribution with parameters $n = 3$, $p_1 = 1/6$, and $p_2 = 1/2$. Find

  (a) $E(X)$.

  (b) $E(Y)$.

  (c) $\text{Var}(X)$.

  (d) $\text{Var}(Y)$.

  (e) $\text{Cov}(X, Y)$.

  (f) ρ.

  Note that $\rho = -\sqrt{p_1 p_2/(1 - p_1)(1 - p_2)}$.

11.2.7 Let the joint p.d.f. of X and Y be

$$f(x, y) = 1/4, \qquad (x, y) \in R = \{(0, 0), (1, 1), (1, -1), (2, 0)\}.$$

  (a) Are X and Y independent?

  (b) Calculate $\text{Cov}(X, Y)$ and ρ.

  This also illustrates the fact that dependent random variables can have a correlation coefficient of zero.

11.2.8 The joint p.d.f. of X and Y is $f(x, y) = 1/6, 0 \le x + y \le 2$, where x and y are integers.

(a) Sketch the support of X and Y.

(b) Record the marginal p.d.f.'s $f_1(x)$ and $f_2(y)$ in the "margins."

(c) Compute $\text{Cov}(X, Y)$.

(d) Determine ρ, the correlation coefficient.

(e) Find the best-fitting line and draw it on your figure.

11.2.9 Let X_1, X_2 be a random sample of size $n = 2$ from the distribution with the binomial p.d.f.

$$f(x) = \binom{2}{x}\left(\frac{1}{3}\right)^x\left(\frac{2}{3}\right)^{2-x}, \qquad x = 0, 1, 2.$$

Find the joint p.d.f. of $Y = X_1$ and $W = X_1 + X_2$, determine the marginal p.d.f. of W, and compute the correlation coefficient of Y and W. HINT: Map the nine points (x_1, x_2) in the space of X_1, X_2 into the nine points (y, w) in the space of Y, W along with the corresponding probabilities and proceed as in earlier exercises.

11.2.10 Let X and Y be random variables of the continuous type having the joint p.d.f. $f(x, y) = 2$, $0 \le y \le x \le 1$. Draw a graph that illustrates the domain of this p.d.f.

(a) Find the marginal p.d.f.'s of X and Y.

(b) Compute $\mu_X, \mu_Y, \sigma_X^2, \sigma_Y^2, \text{Cov}(X, Y)$, and ρ.

(c) Determine the equation of the least squares regression line and draw it on your graph. Does the line make sense to you intuitively?

11.3 CONDITIONAL DISTRIBUTIONS

Let X and Y have a joint discrete distribution with p.d.f. $f(x, y)$ on space R. Say the marginal probability density functions are $f_1(x)$ and $f_2(y)$ with spaces R_1 and R_2, respectively. Let event $A = \{X = x\}$ and event $B = \{Y = y\}, (x, y) \in R$. Thus $A \cap B = \{X = x, Y = y\}$. Because

$$P(A \cap B) = P(X = x, Y = y) = f(x, y)$$

and

$$P(B) = P(Y = y) = f_2(y) > 0 \qquad \text{(since } y \in R_2\text{)},$$

we see that the conditional probability of event A given event B is

$$P(A \mid B) = \frac{P(A \cap B)}{P(B)} = \frac{f(x, y)}{f_2(y)}.$$

This leads to the following definition.

Definition 11.3.1.   The **conditional probability density function of** X, given that $Y = y$, is defined by

$$g(x \mid y) = \frac{f(x, y)}{f_2(y)}, \qquad \text{provided that } f_2(y) > 0.$$

Similarly, the **conditional probability density function of** Y, given that $X = x$, is defined by

$$h(y \mid x) = \frac{f(x, y)}{f_1(x)}, \qquad \text{provided that } f_1(x) > 0.$$

Example 11.3.1

Let X and Y have the joint p.d.f.

$$f(x, y) = \frac{x + y}{21}, \qquad x = 1, 2, 3, \quad y = 1, 2.$$

In Example 11.1.2 we showed that

$$f_1(x) = \frac{2x + 3}{21}, \qquad x = 1, 2, 3,$$

and

$$f_2(y) = \frac{3y + 6}{21}, \qquad y = 1, 2.$$

Thus the conditional p.d.f. of X, given $Y = y$, is equal to

$$g(x \mid y) = \frac{(x + y)/21}{(3y + 6)/21} = \frac{x + y}{3y + 6}, \qquad x = 1, 2, 3, \quad \text{when } y = 1 \text{ or } 2.$$

For example,

$$P(X = 2 \mid Y = 2) = g(2 \mid 2) = \frac{4}{12} = \frac{1}{3}.$$

Similarly, the conditional p.d.f. of Y, given $X = x$, is equal to

$$h(y \mid x) = \frac{x + y}{2x + 3}, \qquad y = 1, 2, \quad \text{when } x = 1, 2, \text{ or } 3.$$

The joint p.d.f. $f(x, y)$ is depicted in Figure 11.4 along with the marginal p.d.f.'s. Conditionally, if $y = 2$, we would expect the outcomes of x, namely 1, 2, and 3, to occur in the ratios 3:4:5. This is precisely what $g(x \mid y)$ does:

$$g(1 \mid 2) = \frac{1 + 2}{12}, \qquad g(2 \mid 2) = \frac{2 + 2}{12}, \qquad g(3 \mid 2) = \frac{3 + 2}{12}.$$

Figure 11.5 displays $g(x \mid 1)$ and $g(x \mid 2)$, while Figure 11.6 gives $h(y \mid 1)$, $h(y \mid 2)$, and $h(y \mid 3)$. Compare the probabilities in Figure 11.6 with those in Figure 11.4. They should agree with your intuition as well as with the formula for $h(y \mid x)$.   ◇

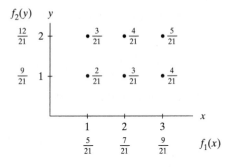

Figure 11.4   Joint and marginal p.d.f.'s

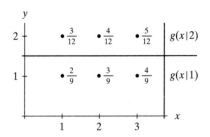

Figure 11.5   Conditional p.d.f.'s of X, given y

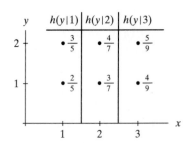

Figure 11.6   Conditional p.d.f.'s of Y, given x

Note that $0 \le h(y \mid x)$. If we sum $h(y \mid x)$ over y for that fixed x, we obtain

$$\sum_{y} h(y \mid x) = \sum_{y} \frac{f(x, y)}{f_1(x)} = \frac{f_1(x)}{f_1(x)} = 1.$$

Thus $h(y \mid x)$ satisfies the conditions of a probability density function, and so we can compute conditional probabilities such as

$$P(a < Y < b \mid X = x) = \sum_{\{y:\, a<y<b\}} h(y \mid x)$$

and conditional expectations such as

$$E[u(Y) \mid X = x] = \sum_y u(y)h(y \mid x)$$

in a manner similar to those associated with probabilities and expectations.

Two special conditional expectations are the **conditional mean** of Y, given $X = x$, defined by

$$\mu_{Y \mid x} = E(Y \mid x) = \sum_y y\, h(y \mid x),$$

and the **conditional variance** of Y, given $X = x$, defined by

$$\sigma^2_{Y \mid x} = E\{[Y - E(Y \mid x)]^2 \mid x\} = \sum_y [y - E(Y \mid x)]^2\, h(y \mid x),$$

which can be computed using

$$\sigma^2_{Y \mid x} = E(Y^2 \mid x) - [E(Y \mid x)]^2.$$

The conditional mean $\mu_{X \mid y}$ and the conditional variance $\sigma^2_{X \mid y}$ are given by similar expressions.

Example 11.3.2

We use the background of Example 11.3.1 and compute $\mu_{Y \mid x}$ and $\sigma^2_{Y \mid x}$, when $x = 3$:

$$\mu_{Y \mid 3} = E(Y \mid X = 3) = \sum_{y=1}^{2} y\, h(y \mid 3)$$

$$= \sum_{y=1}^{2} y\left(\frac{3+y}{9}\right) = 1\left(\frac{4}{9}\right) + 2\left(\frac{5}{9}\right) = \frac{14}{9}$$

and

$$\sigma^2_{Y \mid 3} = E\left[\left(Y - \frac{14}{9}\right)^2 \,\middle|\, X = 3\right] = \sum_{y=1}^{2}\left(y - \frac{14}{9}\right)^2\left(\frac{3+y}{9}\right)$$

$$= \frac{25}{81}\left(\frac{4}{9}\right) + \frac{16}{81}\left(\frac{5}{9}\right) = \frac{20}{81}. \qquad \diamond$$

The conditional mean of X, given $Y = y$, is a function of y alone; the conditional mean of Y, given $X = x$, is a function of x alone. Suppose that the latter conditional mean is a linear function of x; that is, $E(Y \mid x) = a + bx$. Let us find the constants a and b in terms of characteristics μ_X, μ_Y, σ^2_X, σ^2_Y, and ρ. This development will shed additional light on the correlation coefficient ρ; accordingly, we assume that the respective standard deviations σ_X and σ_Y are both positive so that the correlation coefficient will exist.

It is given that

$$\sum_y y\,h(y\,|\,x) = \sum_y y\,\frac{f(x,\,y)}{f_1(x)} = a + bx, \qquad \text{for } x \in R_1,$$

where R_1 is the space of X and R_2 is the space of Y. Hence

$$\sum_y y\,f(x,\,y) = (a + bx)\,f_1(x), \qquad \text{for } x \in R_1 \tag{11.1}$$

and

$$\sum_{x \in R_1} \sum_y y\,f(x,\,y) = \sum_{x \in R_1} (a + bx)\,f_1(x);$$

that is, with μ_X and μ_Y representing the respective means, we have

$$\mu_Y = a + b\mu_X. \tag{11.2}$$

In addition, if we multiply both members of Equation 11.1 by x and sum, we obtain

$$\sum_{x \in R_1} \sum_y xy\,f(x,\,y) = \sum_{x \in R_1} (ax + bx^2)\,f_1(x);$$

that is,

$$E(XY) = aE(X) + bE(X^2)$$

or, equivalently,

$$\mu_X\mu_Y + \rho\sigma_X\sigma_Y = a\mu_X + b(\mu_X^2 + \sigma_X^2). \tag{11.3}$$

The solution of Equations 11.2 and 11.3 is

$$a = \mu_Y - \rho\,\frac{\sigma_Y}{\sigma_X}\,\mu_X \qquad \text{and} \qquad b = \rho\,\frac{\sigma_Y}{\sigma_X},$$

which implies that if $E(Y\,|\,x)$ is linear, it is given by

$$E(Y\,|\,x) = \mu_Y + \rho\,\frac{\sigma_Y}{\sigma_X}\,(x - \mu_X).$$

That is, if the conditional mean of Y, given $X = x$, is linear, it is exactly the same as the best-fitting line (least squares regression line) considered in Section 11.2.

By symmetry, if the conditional mean of X, given $Y = y$, is linear, it is given by

$$E(X\,|\,y) = \mu_X + \rho\,\frac{\sigma_X}{\sigma_Y}\,(y - \mu_Y).$$

We see that the point $[x = \mu_X, E(Y\,|\,x) = \mu_Y]$ satisfies the expression for $E(Y\,|\,x)$; and $[E(X\,|\,y) = \mu_X, y = \mu_Y]$ satisfies the expression for $E(X\,|\,y)$. That is, the point (μ_X, μ_Y) is on each of the two lines. In addition, we note that the product of the coefficient of x in $E(Y\,|\,x)$ and the coefficient of y in $E(X\,|\,y)$ equals ρ^2 and the ratio of these two coefficients equals σ_Y^2/σ_X^2. These observations sometimes prove useful in particular problems.

Example 11.3.3

Let X and Y have the trinomial p.d.f. with parameters n, p_1, p_2, and $1 - p_1 - p_2 = p_3$. That is,

$$f(x, y) = \frac{n!}{x!\, y!\, (n - x - y)!}\, p_1^x\, p_2^y\, p_3^{n-x-y},$$

where x and y are nonnegative integers such that $x + y \le n$. From the development of the trinomial distribution, we note that X and Y have marginal binomial distributions $b(n, p_1)$ and $b(n, p_2)$, respectively. Thus

$$h(y \mid x) = \frac{f(x, y)}{f_1(x)} = \frac{(n - x)!}{y!\, (n - x - y)!} \left(\frac{p_2}{1 - p_1} \right)^y \left(\frac{p_3}{1 - p_1} \right)^{n-x-y},$$

$$y = 0, 1, 2, \ldots, n - x.$$

That is, the conditional p.d.f. of Y, given $X = x$, is binomial

$$b\!\left[n - x, \frac{p_2}{1 - p_1} \right]$$

and thus has conditional mean

$$E(Y \mid x) = (n - x) \frac{p_2}{1 - p_1}.$$

In a similar manner, we obtain

$$E(X \mid y) = (n - y) \frac{p_1}{1 - p_2}.$$

Since each of the conditional means is linear, the product of the respective coefficients of x and y is

$$\rho^2 = \left(\frac{-p_2}{1 - p_1} \right) \left(\frac{-p_1}{1 - p_2} \right) = \frac{p_1 p_2}{(1 - p_1)(1 - p_2)}.$$

However, ρ must be negative because the coefficients of x and y are negative, thus

$$\rho = -\sqrt{\frac{p_1 p_2}{(1 - p_1)(1 - p_2)}}. \qquad \diamond$$

Although we have used random variables of the discrete type to introduce the new definitions, they also hold for random variables of the continuous type. Let X and Y have a distribution of the continuous type with joint p.d.f. $f(x, y)$ and marginal p.d.f.'s $f_1(x)$ and $f_2(y)$, respectively. We have that the conditional p.d.f., mean, and variance of Y, given $X = x$, are, respectively,

$$h(y \mid x) = \frac{f(x, y)}{f_1(x)}, \qquad \text{provided that } f_1(x) > 0,$$

$$E(Y \mid x) = \int_{-\infty}^{\infty} y\, h(y \mid x)\, dy,$$

and

$$\text{Var}(Y \mid x) = E\{[Y - E(Y \mid x)]^2 \mid x\}$$

$$= \int_{-\infty}^{\infty} [y - E(Y \mid x)]^2 \, h(y \mid x) \, dy$$

$$= E[Y^2 \mid x] - [E(Y \mid x)]^2.$$

Similar expressions are associated with the conditional distribution of X, given $Y = y$.

Example 11.3.4

Let X and Y be the random variables of Example 11.1.7. Thus

$$f(x, y) = 2, \qquad 0 \le x \le y \le 1,$$
$$f_1(x) = 2(1 - x), \qquad 0 \le x \le 1,$$
$$f_2(y) = 2y, \qquad 0 \le y \le 1.$$

Before we actually find the conditional p.d.f. of Y, given $X = x$, we shall give an intuitive argument. The joint p.d.f. is constant over the triangular region bounded by $y = x$, $y = 1$, and $x = 0$. If the value of X is known, say $X = x$, then the possible values of Y are between x and 1. Furthermore, we would expect Y to be uniformly distributed on the interval $[x, 1]$. That is, we would anticipate that $h(y \mid x) = 1/(1 - x)$, $x \le y \le 1$.

More formally now, we have by definition that

$$h(y \mid x) = \frac{f(x, y)}{f_1(x)} = \frac{2}{2(1 - x)} = \frac{1}{1 - x}, \qquad x \le y \le 1, \quad 0 \le x \le 1.$$

The conditional mean of Y, given $X = x$, is

$$E(Y \mid x) = \int_x^1 y \, \frac{1}{1 - x} \, dy = \left[\frac{y^2}{2(1 - x)} \right]_x^1 = \frac{1 + x}{2}, \qquad 0 \le x \le 1.$$

Similarly, it can be shown that

$$E(X \mid y) = \frac{y}{2}, \qquad 0 \le y \le 1.$$

The conditional variance of Y, given $X = x$, is

$$E\{[Y - E(Y \mid x)]^2 \mid x\} = \int_x^1 \left(y - \frac{1 + x}{2} \right)^2 \frac{1}{1 - x} \, dy$$

$$= \left[\frac{1}{3(1 - x)} \left(y - \frac{1 + x}{2} \right)^3 \right]_x^1$$

$$= \frac{(1 - x)^2}{12}.$$

Recall that if a random variable W is $U(a, b)$, then $E(W) = (a + b)/2$, and $\text{Var}(W) = (b - a)^2/12$. Since the conditional distribution of Y, given $X = x$, is $U(x, 1)$, we could have written down immediately that $E(Y \mid x) = (x + 1)/2$ and $\text{Var}(Y \mid x) = (1 - x)^2/12$.

An illustration of a computation of a conditional probability is

$$P\left(\frac{3}{4} \leq Y \leq \frac{7}{8} \,\middle|\, X = \frac{1}{4}\right) = \int_{3/4}^{7/8} h\left(y \,\middle|\, \frac{1}{4}\right) dy$$

$$= \int_{3/4}^{7/8} \frac{1}{3/4} \, dy = \frac{1}{6}. \qquad \diamond$$

In general, if $E(Y \mid x)$ is linear, it is equal to

$$E(Y \mid x) = \mu_Y + \rho\left(\frac{\sigma_Y}{\sigma_X}\right)(x - \mu_X).$$

If $E(X \mid y)$ is linear, then

$$E(X \mid y) = \mu_X + \rho\left(\frac{\sigma_X}{\sigma_Y}\right)(y - \mu_Y).$$

Thus, in Example 11.3.4, we see that the product of the coefficients of x in $E(Y \mid x)$ and y in $E(X \mid y)$ is $\rho^2 = 1/4$. Thus $\rho = 1/2$ since each coefficient is positive. Since the ratio of those coefficients is equal to $\sigma_Y^2/\sigma_X^2 = 1$, we have that $\sigma_X^2 = \sigma_Y^2$.

EXERCISES

11.3.1 Let X and Y have the joint p.d.f.

$$f(x, y) = \frac{x + y}{32}, \qquad x = 1, 2, \quad y = 1, 2, 3, 4.$$

(a) Display the joint p.d.f. and the marginal p.d.f.'s on a graph like Figure 11.4.

(b) Find $g(x \mid y)$ and draw a figure like Figure 11.5, depicting the conditional p.d.f.'s for $y = 1, 2, 3$, and 4.

(c) Find $h(y \mid x)$ and draw a figure like Figure 11.6, depicting the conditional p.d.f.'s for $x = 1$ and 2.

(d) Find $P(1 \leq Y \leq 3 \mid X = 1)$, $P(Y \leq 2 \mid X = 2)$, and $P(X = 2 \mid Y = 3)$.

(e) Find $E(Y \mid X = 1)$ and $\text{Var}(Y \mid X = 1)$.

11.3.2 Let the joint p.d.f. $f(x, y)$ of X and Y be given by:

(x, y)	$f(x, y)$
$(1, 1)$	3/8
$(2, 1)$	1/8
$(1, 2)$	1/8
$(2, 2)$	3/8

Find the two conditional probability density functions and the corresponding means and variances.

11.3.3 Let W equal the weight of laundry soap in a "1-kilogram" box that is distributed in Southeast Asia. Suppose that $P(W < 1) = 0.02$ and $P(W > 1.072) = 0.08$. Call a box of soap light, good, or heavy depending on whether $\{W < 1\}$, $\{1 \le W \le 1.072\}$, or $\{W > 1.072\}$, respectively. In a random sample of $n = 50$ boxes, let X equal the number of light boxes and Y the number of good boxes.

(a) What is the joint p.d.f. of X and Y?

(b) Give the name of the distribution of Y along with the values of the parameters of this distribution.

(c) Given that $X = 3$, how is Y distributed conditionally?

(d) Determine $E(Y \mid X = 3)$.

(e) Find ρ, the correlation coefficient of X and Y.

11.3.4 The genes for eye color for a certain male fruit fly are (R, W). The genes for eye color for the mating female fruit fly are (R, W). Their offspring receive one gene for eye color from each parent. If an offspring ends up with either (R, R), (R, W), or (W, R), its eyes will look red. Let X equal the number of offspring having red eyes. Let Y equal the number of red-eyed offspring having (R, W) or (W, R) genes.

(a) If the total number of offspring is $n = 400$, how is X distributed?

(b) Give the values of $E(X)$ and Var(X).

(c) Given that $X = 300$, how is Y distributed?

(d) Give the values of $E(Y \mid X = 300)$ and Var$(Y \mid X = 300)$.

11.3.5 Let X and Y have a trinomial distribution with $n = 2$, $p_1 = 1/4$, and $p_2 = 1/2$.

(a) Give $E(Y \mid x)$.

(b) Compare your answer in part (a) with the equation of the line of best fit in Example 11.2.2. Are they the same? Why?

11.3.6 Roll a pair of four-sided dice for which the outcome is 1, 2, 3, or 4 on each die. Let X denote the smaller and Y the larger outcome on the dice. Then the joint p.d.f. of X and Y is

$$f(x, y) = \begin{cases} \dfrac{1}{16}, & 1 \le x = y \le 4, \\ \dfrac{2}{16}, & 1 \le x < y \le 4. \end{cases}$$

(a) Find $E(Y \mid x)$ for $x = 1, 2, 3, 4$.

(b) Show that the least squares regression line is $y = 0.45x + 2.27$.

(c) Do the points $[x, E(Y \mid x)]$, $x = 1, 2, 3, 4$, lie on the least squares regression line? NOTE: Recall that we discovered that if the conditional mean of Y, given $X = x$, is linear, then it is exactly the same as the least squares regression line.

11.3.7 Using the joint p.d.f. given in Exercise 11.2.3, find the value of $E(Y \mid x)$ for $x = 1, 2, 3, 4$. Is this linear? Do these points lie on the best-fitting line?

11.3.8 An unbiased six-sided die is cast 30 independent times. Let X be the number of one's and Y the number of two's.

(a) What is the joint p.d.f. of X and Y?

(b) Find the conditional p.d.f. of X, given $Y = y$.

(c) Compute $E(X^2 - 4XY + 3Y^2)$.

11.3.9 Let X and Y have a uniform distribution on the set of points with integer coordinates in $R = \{(x, y) : 0 \le x \le 7, x \le y \le x + 2\}$. That is, $f(x, y) = 1/24$, $(x, y) \in R$, and both x and y are integers. Find

(a) $f_1(x)$.

(b) $h(y \mid x)$.

(c) $E(Y \mid x)$.

(d) $\sigma_{Y \mid x}^2$.

(e) $f_2(y)$.

11.3.10 Let $f_1(x) = 1/10$, $x = 0, 1, 2, \ldots, 9$, and $h(y \mid x) = 1/(10 - x)$, $y = x, x + 1, \ldots, 9$. Find

(a) $f(x, y)$.

(b) $f_2(y)$.

(c) $E(Y \mid x)$.

11.3.11 The joint p.d.f. of X and Y is $f(x, y) = 1/6$, $0 \le x + y \le 2$, where x and y are nonnegative integers.

(a) Sketch the domain of $f(x, y)$.

(b) Determine the marginal p.d.f.'s of X and Y.

(c) Find $E(Y \mid x)$ and plot $[x, E(Y \mid x)]$ on the sketch in part (a).

(d) Find $E(X \mid y)$ and plot $[E(X \mid y), y]$ on the sketch in part (a).

(e) Find ρ, the correlation coefficient, using your answers to parts (c) and (d).

11.3.12 For the random variables defined in Example 11.3.4, calculate the correlation coefficient directly from the definition

$$\rho = \frac{\text{Cov}(X, Y)}{\sigma_X \sigma_Y}.$$

11.3.13 Let $f(x, y) = 1/40$, $0 \le x \le 10$, $10 - x \le y \le 14 - x$, be the joint p.d.f. of X and Y.

(a) Sketch the region for which $f(x, y) > 0$.

(b) Find $f_1(x)$, the marginal p.d.f. of X.

(c) Determine $h(y \mid x)$, the conditional p.d.f. of Y, given $X = x$.

(d) Calculate $E(Y \mid x)$, the conditional mean of Y, given $X = x$.

11.3.14 Let $f(x, y) = 1/8, 0 \le y \le 4, y \le x \le y + 2$, be the joint p.d.f. of X and Y.

    (a) Sketch the region for which $f(x, y) > 0$.

    (b) Find $f_1(x)$, the marginal p.d.f. of X.

    (c) Determine $h(y \mid x)$, the conditional p.d.f. of Y, given $X = x$.

    (d) Compute $E(Y \mid x)$, the conditional mean of Y, given $X = x$.

    (e) Graph $y = E(Y \mid x)$ on your sketch in part (a). Is $y = E(Y \mid x)$ linear?

11.3.15 Let X have a uniform distribution $U(0, 2)$, and let the conditional distribution of Y, given $X = x$, be $U(0, x^2)$.

    (a) Determine the joint p.d.f. of X and Y, $f(x, y)$.

    (b) Calculate $f_2(y)$, the marginal p.d.f. of Y.

    (c) Compute $E(X \mid y)$, the conditional mean of X, given $Y = y$.

    (d) Find $E(Y \mid x)$, the conditional mean of Y, given $X = x$.

11.3.16 Let X have a uniform distribution on the interval $(0, 1)$. Given $X = x$, let Y have a uniform distribution on the interval $(0, x)$.

    (a) Define the conditional p.d.f. of Y, given that $X = x$. Be sure to include the domain.

    (b) Find $E(Y \mid x)$.

    (c) Determine the joint p.d.f. of X and Y.

    (d) Find the marginal p.d.f. of Y.

11.3.17 The marginal distribution of X is $U(0, 1)$. The conditional distribution of Y, given $X = x$, is $U(0, e^X)$.

    (a) Determine $h(y \mid x)$, the conditional p.d.f. of Y, given $X = x$.

    (b) Find $E(Y \mid x)$.

    (c) Display the joint p.d.f. of X and Y. Sketch the region where $f(x, y) > 0$.

    (d) Find $f_2(y)$, the marginal p.d.f. of Y.

11.3.18 Let $Y_1 < Y_2 < \cdots < Y_m < \cdots < Y_n$ be the order statistics of a random sample of size n from a continuous distribution that has p.d.f. $f(x)$ and a distribution function $F(x)$, where $0 < F(x) < 1$ for $a < x < b$, $F(a) = 0$, $F(b) = 1$. This exercise outlines a procedure for simulating the first m order statistics in an ordered manner; in other words, first the value of y_1, then y_2, and so on. Prove the following facts.

    (a) The joint marginal p.d.f. of Y_i and Y_{i+1} is

$$g_{i,i+1}(Y_i, Y_{i+1}) =$$

$$\frac{n!}{(i-1)! \, (n-i-1)!} [F(y_i)]^{i-1} [1 - F(y_{i+1})]^{n-i-1} f(y_i) f(y_{i+1}),$$

$$a < y_i < y_{i+1} < b.$$

    (b) The conditional p.d.f. of Y_{i+1} given $Y_i = y_i$, is

$$h(y_{i+1} \mid y_i) = \frac{(n-i)[1 - F(y_{i+1})]^{n-i-1} f(y_{i+1})}{[1 - F(y_i)]^{n-i}}, \qquad a < y_i < y_{i+1} < b.$$

(c) The conditional distribution function of Y_{i+1}, given $Y_i = y_i$, is

$$H(y_{i+1} \mid y_i) = 1 - \left[\frac{1 - F(y_{i+1})}{1 - F(y_i)} \right]^{n-i}, \qquad a < y_i < y_{i+1} < b.$$

(d) The distribution function of Y_1, the first order statistic, is

$$v = 1 - [1 - F(y_1)]^n, \qquad a < y_1 < b.$$

(e) When V is $U(0, 1)$, $y_1 = F^{-1}(1 - [1 - v]^{1/n})$ gives an observed value of Y_1.

(f) Let $v = H(y_{i+1} \mid y_i)$. If V is $U(0, 1)$, then

$$y_{i+1} = H^{-1}(v \mid y_i) = F^{-1}[1 - \{1 - F(y_i)\}\{1 - v\}^{1/(n-i)}]$$

is the observed value of Y_{i+1} for $i = 1, 2, \ldots, n - 1$.

11.3.19 Use the "random numbers" $v_1 = 0.2360$, $v_2 = 0.0914$, $v_3 = 0.8903$, $v_4 = 0.3195$ and the method outlined in Exercise 11.3.18 to simulate $y_1 < y_2 < y_3 < y_4$ in a random sample of size $n = 10$ from

(a) The exponential distribution with mean $\theta = 10$.

(b) The uniform distribution $U(0, 1)$.

11.4 THE BIVARIATE NORMAL DISTRIBUTION

Let X and Y be random variables with joint p.d.f. $f(x, y)$ of the continuous type. Many applications are concerned with the conditional distribution of one of the random variables, say Y, given that $X = x$. For example, X and Y might be a student's grade-point averages from high school and from the first year in college, respectively. Persons in the field of educational testing and measurement are extremely interested in the conditional distribution of Y, given $X = x$, in such situations.

Suppose that we have an application in which we can make the following three assumptions about the conditional distribution of Y, given $X = x$:

(a) It is normal for each real x.

(b) Its mean $E(Y \mid x)$ is a linear function of x.

(c) Its variance is constant; that is, it does not depend upon the given value of x.

Of course, assumption (b), along with a result given in Section 11.3, implies that

$$E(Y \mid x) = \mu_Y + \rho \frac{\sigma_Y}{\sigma_X} (x - \mu_X).$$

Let us consider the implication of assumption (c). The conditional variance is given by

$$\sigma_{Y|x}^2 = \int_{-\infty}^{\infty} \left[y - \mu_Y - \rho \frac{\sigma_Y}{\sigma_X} (x - \mu_X) \right]^2 h(y \mid x)\, dy,$$

where $h(y \mid x)$ is the conditional p.d.f. of Y given $X = x$. Multiply each member of this equation by $f_1(x)$ and integrate on x. Since $\sigma^2_{Y \mid x}$ is a constant, the left-hand member is equal to $\sigma^2_{Y \mid x}$. Thus we have

$$\sigma^2_{Y \mid x} = \int_{-\infty}^{\infty} \int_{-\infty}^{\infty} \left[y - \mu_Y - \rho \frac{\sigma_Y}{\sigma_X} (x - \mu_X) \right]^2 h(y \mid x) f_1(x) \, dy \, dx. \quad (11.4)$$

However, $h(y \mid x) f_1(x) = f(x, y)$; hence the right-hand member is just an expectation and Equation 11.4 can be written as

$$\sigma^2_{Y \mid x} = E \left\{ (Y - \mu_Y)^2 - 2\rho \frac{\sigma_Y}{\sigma_X} (X - \mu_X)(Y - \mu_Y) + \rho^2 \frac{\sigma^2_Y}{\sigma^2_X} (X - \mu_X)^2 \right\}.$$

But using the fact that the expectation E is a linear operator, we have, recalling $E[(X - \mu_X)(Y - \mu_Y)] = \rho \sigma_X \sigma_Y$, that

$$\sigma^2_{Y \mid x} = \sigma^2_Y - 2\rho \frac{\sigma_Y}{\sigma_X} \rho \sigma_X \sigma_Y + \rho^2 \frac{\sigma^2_Y}{\sigma^2_X} \sigma^2_X$$

$$= \sigma^2_Y - 2\rho^2 \sigma^2_Y + \rho^2 \sigma^2_Y = \sigma^2_Y(1 - \rho^2).$$

That is, the conditional variance of Y, for each given x, is $\sigma^2_Y(1 - \rho^2)$. These facts about the conditional mean and variance, along with assumption (a), require that the conditional p.d.f. of Y, given $X = x$, be

$$h(y \mid x) = \frac{1}{\sigma_Y \sqrt{2\pi} \sqrt{1 - \rho^2}} \exp \left[-\frac{[y - \mu_Y - \rho(\sigma_Y/\sigma_X)(x - \mu_X)]^2}{2\sigma^2_Y(1 - \rho^2)} \right],$$

$$-\infty < y < \infty, \text{ for every real } x.$$

Before we make any assumptions about the distribution of X, we give an example and figure to illustrate the implications of our current assumptions.

Example 11.4.1

Let $\mu_X = 10$, $\sigma^2_X = 9$, $\mu_Y = 12$, $\sigma^2_Y = 16$, and $\rho = 0.8$. We have seen that assumptions (a), (b), and (c) imply that the conditional distribution of Y, given $X = x$, is

$$N \left[12 + (0.8) \left(\frac{4}{3} \right) (x - 10), 16(1 - 0.8^2) \right].$$

In Figure 11.7 the conditional mean line

$$E(Y \mid x) = 12 + (0.8) \left(\frac{4}{3} \right) (x - 10) = \left(\frac{3.2}{3} \right) x + \left(\frac{4}{3} \right)$$

has been graphed. For each of $x = 5, 10$, and 15, the p.d.f. of Y, given $X = x$, is given.   ◇

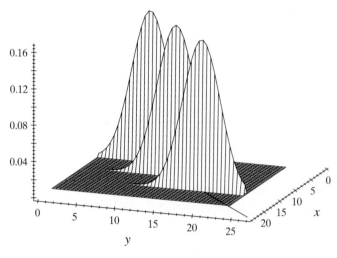

0.16
0.12
0.08
0.04

0   5   10   15   20   25

0   5   10   15   20

y   *x*

Figure 11.7   Conditional p.d.f. of Y, given $x = 5, 10, 15$

Up to this point, nothing has been said about the distribution of X other than that it has mean μ_X and positive variance σ_X^2. Suppose, in addition, we assume that this distribution is also normal; that is, the marginal p.d.f. of X is

$$f_1(x) = \frac{1}{\sigma_X \sqrt{2\pi}} \exp\left[-\frac{(x - \mu_X)^2}{2\sigma_X^2}\right], \qquad -\infty < x < \infty.$$

Hence the joint p.d.f. of X and Y is given by the product

$$f(x, y) = h(y\,|\,x) f_1(x) = \frac{1}{2\pi \sigma_X \sigma_Y \sqrt{1 - \rho^2}} \exp\left[-\frac{q(x, y)}{2}\right], \tag{11.5}$$

where it can be shown (see Exercise 11.4.2) that

$$q(x, y) = \frac{1}{1 - \rho^2}\left[\left(\frac{x - \mu_X}{\sigma_X}\right)^2 - 2\rho\left(\frac{x - \mu_X}{\sigma_X}\right)\left(\frac{y - \mu_Y}{\sigma_Y}\right) + \left(\frac{y - \mu_Y}{\sigma_Y}\right)^2\right].$$

A joint p.d.f. of this form is called a **bivariate normal p.d.f.**

Example 11.4.2

Let us assume that in a certain population of college students, the respective grade-point averages, say X and Y, in high school and the first year in college have an approximate bivariate normal distribution with parameters $\mu_X = 2.9$, $\mu_Y = 2.4$, $\sigma_X = 0.4$, $\sigma_Y = 0.5$, and $\rho = 0.8$.

Then, for illustration,

$$P(2.1 < Y < 3.3) = P\left(\frac{2.1 - 2.4}{0.5} < \frac{Y - 2.4}{0.5} < \frac{3.3 - 2.4}{0.5}\right)$$

$$= \Phi(1.8) - \Phi(-0.6) = 0.6898.$$

Since the conditional p.d.f. of Y, given $X = 3.2$, is normal with mean

$$2.4 + (0.8)\left(\frac{0.5}{0.4}\right)(3.2 - 2.9) = 2.7$$

and standard deviation $(0.5)\sqrt{1 - 0.64} = 0.3$, we have that

$$P(2.1 < Y < 3.3 \mid X = 3.2)$$

$$= P\left(\frac{2.1 - 2.7}{0.3} < \frac{Y - 2.7}{0.3} < \frac{3.3 - 2.7}{0.3} \,\middle|\, X = 3.2\right)$$

$$= \Phi(2) - \Phi(-2) = 0.9544.$$

From a practical point of view, however, the reader should be warned that the correlation coefficient of these grade-point averages is, in many instances, much smaller than 0.8.   ◇

Since x and y enter the bivariate normal p.d.f. in a similar manner, the roles of X and Y could have been interchanged. That is, Y could have been assigned the marginal normal p.d.f. $N(\mu_Y, \sigma_Y^2)$, and the conditional p.d.f. of X, given $Y = y$, would have then been normal, with mean $\mu_X + \rho(\sigma_X/\sigma_Y)(y - \mu_Y)$ and variance $\sigma_X^2(1 - \rho^2)$. Although this is fairly obvious, we do want to make special note of it.

In order to have a better understanding of the geometry of the bivariate normal distribution, consider the graph of $z = f(x, y)$, where $f(x, y)$ is given by Equation 11.5. If we intersect this surface with planes parallel to the yz plane, that is, with $x = x_0$, we have

$$f(x_0, y) = f_1(x_0)h(y \mid x_0).$$

In this equation $f_1(x_0)$ is a constant, and $h(y \mid x_0)$ is a normal p.d.f. Thus $z = f(x_0, y)$ is bell-shaped, that is, has the shape of a normal p.d.f. However, note that it is not necessarily a p.d.f. because of the factor $f_1(x_0)$. Similarly, intersections of the surface $z = f(x, y)$ with planes $y = y_0$, parallel to the xz plane will be bell-shaped.

If

$$0 < z_0 < \frac{1}{2\pi \sigma_X \sigma_Y \sqrt{1 - \rho^2}},$$

then

$$0 < z_0 2\pi \sigma_X \sigma_Y \sqrt{1 - \rho^2} < 1.$$

If we intersect $z = f(x, y)$ with the plane $z = z_0$, which is parallel to the xy plane, we have

$$z_0 2\pi \sigma_X \sigma_Y \sqrt{1 - \rho^2} = \exp\left[\frac{-q(x, y)}{2}\right].$$

Taking the natural logarithm of each side, we obtain

$$\left(\frac{x - \mu_X}{\sigma_X}\right)^2 - 2\rho\left(\frac{x - \mu_X}{\sigma_X}\right)\left(\frac{y - \mu_Y}{\sigma_Y}\right) + \left(\frac{y - \mu_Y}{\sigma_Y}\right)^2$$

$$= -2(1 - \rho^2)\ln(z_0 2\pi \sigma_X \sigma_Y \sqrt{1 - \rho^2}).$$

Thus we see that these intersections are ellipses.

Example 11.4.3

With $\mu_X = 10$, $\sigma_X^2 = 9$, $\mu_Y = 12$, $\sigma_Y^2 = 16$, and $\rho = 0.8$, the bivariate normal p.d.f. has been graphed in Figure 11.8. For $\rho = 0.8$, level curves or contours are given in Figure 11.9. The conditional mean line,

$$E(Y \mid x) = 12 + (0.8)\left(\frac{4}{3}\right)(x - 10) = \frac{3.2}{3}x + \frac{4}{3},$$

is also drawn on Figure 11.9. Note that this line intersects the level curves at points through which vertical tangents $(x = x_0)$ can be drawn to the ellipses.   ◇

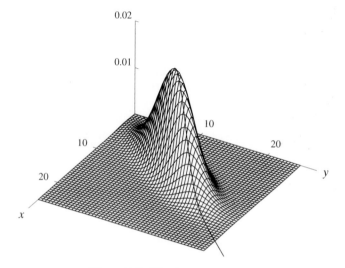

Figure 11.8   Bivariate normal p.d.f.

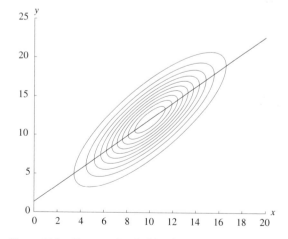

Figure 11.9   Contours for the bivariate normal distribution

We close this section by observing another important property of the correlation coefficient ρ if X and Y have a bivariate normal distribution. In Equation 11.5 of the product $h(y \mid x) f_1(x)$, let us consider the factor $h(y \mid x)$ if $\rho = 0$. We see that this product, which is the joint p.d.f. of X and Y, equals $f_1(x) f_2(y)$ because $h(y \mid x)$ is, when $\rho = 0$, a normal p.d.f. with mean μ_Y and variance σ_Y^2. That is, if $\rho = 0$, the joint p.d.f. factors into the product of the two marginal probability density functions, and hence, X and Y are independent random variables. Of course, if X and Y are any independent random variables (not necessarily normal), we know that ρ, if it exists, is always equal to zero. Thus we have proved the following.

Theorem 11.4.1.   If X and Y have a bivariate normal distribution with correlation coefficient ρ, then X and Y are independent if and only if $\rho = 0$.

Thus, in the bivariate normal case, $\rho = 0$ does imply independence of X and Y.

It should be mentioned here that these characteristics of the bivariate normal distribution can be extended to the trivariate normal distribution or, more generally, the multivariate normal distribution. This is done in more advanced texts assuming some knowledge of matrices; for illustration, see Section 4.10 of R.V. Hogg and A.T. Craig, *Introduction to Mathematical Statistics*, 5th ed. (Upper Saddle River, NJ: Prentice Hall, 1995).

EXERCISES

11.4.1 Let X and Y have a bivariate normal distribution with parameters $\mu_X = -3$, $\mu_Y = 10$, $\sigma_X^2 = 25$, $\sigma_Y^2 = 9$, and $\rho = 3/5$. Compute

(a) $P(-5 < X < 5)$.

(b) $P(-5 < X < 5 \mid Y = 13)$.

(c) $P(7 < Y < 16)$.

(d) $P(7 < Y < 16 \mid X = 2)$.

11.4.2 Show that the expression in the exponent of Equation 11.5 is equal to the function $q(x, y)$ given in the text.

11.4.3 Let X and Y have a bivariate normal distribution with parameters $\mu_X = 2.8$, $\mu_Y = 110$, $\sigma_X^2 = 0.16$, $\sigma_Y^2 = 100$, and $\rho = 0.6$. Compute

(a) $P(106 < Y < 124)$.

(b) $P(106 < Y < 124 \mid X = 3.2)$.

11.4.4 Let X and Y have a bivariate normal distribution with $\mu_X = 70$, $\sigma_X^2 = 100$, $\mu_Y = 80$, $\sigma_Y^2 = 169$, and $\rho = 5/13$. Find

(a) $E(Y \mid X = 72)$.

(b) $\text{Var}(Y \mid X = 72)$.

(c) $P(Y \le 84 \mid X = 72)$.

11.4.5 Let X denote the height in centimeters and Y the weight in kilograms of male college students. Assume that X and Y have a bivariate normal distribution with parameters $\mu_X = 185$, $\sigma_X^2 = 100$, $\mu_Y = 84$, $\sigma_Y^2 = 64$, and $\rho = 3/5$.

(a) Determine the conditional distribution of Y, given that $X = 190$.

(b) Find $P(86.4 < Y < 95.36 \,|\, X = 190)$.

11.4.6 For a freshman taking introductory statistics and majoring in psychology, let X equal the student's ACT mathematics score and Y the student's ACT verbal score. Assume that X and Y have a bivariate normal distribution with $\mu_X = 22.7$, $\sigma_X^2 = 17.64$, $\mu_Y = 22.7$, $\sigma_Y^2 = 12.25$, and $\rho = 0.78$. Find

(a) $P(18.5 < Y < 25.5)$.

(b) $E(Y \,|\, x)$.

(c) $\mathrm{Var}(Y \,|\, x)$.

(d) $P(18.5 < Y < 25.5 \,|\, X = 23)$.

(e) $P(18.5 < Y < 25.5 \,|\, X = 25)$.

(f) For $x = 21, 23$, and 25, draw a graph of $z = h(y \,|\, x)$ similar to Figure 11.7.

11.4.7 For a pair of gallinules, let X equal the weight in grams of the male and Y the weight in grams of the female. Assume that X and Y have a bivariate normal distribution with $\mu_X = 415$, $\sigma_X^2 = 611$, $\mu_Y = 347$, $\sigma_Y^2 = 689$, and $\rho = -0.25$. Find

(a) $P(309.2 < Y < 380.6)$.

(b) $E(Y \,|\, x)$.

(c) $\mathrm{Var}(Y \,|\, x)$.

(d) $P(309.2 < Y < 380.6 \,|\, X = 385.1)$.

11.4.8 Let X and Y have a bivariate normal distribution with parameters $\mu_X = 10$, $\sigma_X^2 = 9$, $\mu_Y = 15$, $\sigma_Y^2 = 16$, and $\rho = 0$. Find

(a) $P(13.6 < Y < 17.2)$.

(b) $E(Y \,|\, x)$.

(c) $\mathrm{Var}(Y \,|\, x)$.

(d) $P(13.6 < Y < 17.2 \,|\, X = 9.1)$.

11.4.9 Let X and Y have a bivariate normal distribution. Find two different lines, $a(x)$ and $b(x)$, parallel to and equidistant from $E(Y \,|\, x)$, such that

$$P[a(x) < Y < b(x) \,|\, X = x] = 0.9544$$

for all real x. Plot $a(x)$, $b(x)$, and $E(Y \,|\, x)$ when $\mu_X = 2$, $\mu_Y = -1$, $\sigma_X = 3$, $\sigma_Y = 5$, and $\rho = 3/5$.

11.4.10 In a college health fitness program, let X denote the weight in kilograms of a male freshman at the beginning of the program and let Y denote his weight change during a semester. Assume that X and Y have a bivariate normal distribution with $\mu_X = 72.30$, $\sigma_X^2 = 110.25$, $\mu_Y = 2.80$, $\sigma_Y^2 = 2.89$, and $\rho = -0.57$. (The lighter students tend to gain weight, while the heavier students tend to lose weight.) Find

(a) $P(2.80 \leq Y \leq 5.35)$.

(b) $P(2.76 \leq y \leq 5.34 \,|\, X = 82.3)$.

11.4.11 For a female freshman in a health fitness program, let X equal her percentage of body fat at the beginning of the program and let Y equal the change in her percentage of body fat measured at the end of the program. Assume that X and Y have a bivariate normal distribution with $\mu_X = 24.5$, $\sigma_X^2 = 4.8^2 = 23.04$, $\mu_Y = -0.2$, $\sigma_Y^2 = 3.0^2 = 9.0$, and $\rho = -0.32$. Find

(a) $P(1.3 \leq Y \leq 5.8)$.

(b) $\mu_{Y|x}$, the conditional mean of Y, given $X = x$.

(c) $\sigma_{Y|x}^2$, the conditional variance of Y, given $X = x$.

(d) $P(1.3 \leq Y \leq 5.8 \,|\, X = 18)$.

11.4.12 For a male freshman in a health fitness program, let X equal his percentage of body fat at the beginning of the program and let Y equal the change in his percentage of body fat measured at the end of the program. Assume that X and Y have a bivariate normal distribution with $\mu_X = 15.00$, $\sigma_X^2 = 4.5^2$, $\mu_Y = -1.55$, $\sigma_Y^2 = 1.5^2$, and $\rho = -0.60$. Find

(a) $P(0.205 \leq Y \leq 0.805)$.

(b) $P(0.21 \leq Y \leq 0.81 \,|\, X = 20)$.

11.5 CORRELATION ANALYSIS

We consider tests about the correlation coefficient ρ of a bivariate normal distribution. Let X and Y have a bivariate normal distribution. We know that if the correlation coefficient ρ is zero, then X and Y are independent random variables. Furthermore, the value of ρ gives a measure of the linear relationship between X and Y. We now give methods for using the sample correlation coefficient to test the hypothesis $H_0: \rho = 0$ and also to form a confidence interval for ρ.

Let $(X_1, Y_1), (X_2, Y_2), \ldots, (X_n, Y_n)$ denote a random sample from a bivariate normal distribution with parameters μ_X, μ_Y, σ_X^2, σ_Y^2, and ρ. Recall that the sample correlation coefficient is

$$R = \frac{\dfrac{1}{n-1} \sum_{i=1}^{n} (X_i - \overline{X})(Y_i - \overline{Y})}{\sqrt{\dfrac{1}{n-1} \sum_{i=1}^{n} (X_i - \overline{X})^2} \sqrt{\dfrac{1}{n-1} \sum_{i=1}^{n} (Y_i - \overline{Y})^2}} = \frac{S_{XY}}{S_X S_Y}.$$

We note that

$$R \frac{S_Y}{S_X} = \frac{S_{XY}}{S_X^2} = \frac{\dfrac{1}{n-1} \sum_{i=1}^{n} (X_i - \overline{X})(Y_i - \overline{Y})}{\dfrac{1}{n-1} \sum_{i=1}^{n} (X_i - \overline{X})^2}.$$

is exactly the solution that we obtained for $\widehat{\beta}$ in Section 8.3 when the X values were fixed at $X_1 = x_1, X_2 = x_2, \ldots, X_n = x_n$. Let us consider these values fixed temporarily so that we are considering conditional distributions, given $X_1 = x_1, \ldots, X_n = x_n$. Moreover, if $H_0: \rho = 0$ is true, then the distributions of $Y_1, Y_2, \ldots, Y_n$ are independent of $x_1, x_2, \ldots, x_n$ and thus $\beta = \rho \sigma_Y / \sigma_X = 0$. Under these conditions the conditional distribution of

$$
\widehat{\beta} = \frac{\displaystyle\sum_{i=1}^{n} (x_i - \overline{x})(Y_i - \overline{Y})}{\displaystyle\sum_{i=1}^{n} (x_i - \overline{x})^2}
$$

is $N[0, \sigma_Y^2/(n-1)s_x^2]$ when $s_x^2 > 0$. Moreover, recall from Section 8.3 that the conditional distribution, given $X_1 = x_1, \ldots, X_n = x_n$, of (see Exercise 11.5.6)

$$
\frac{\sum_{i=1}^{n}[Y_i - \overline{Y} - (S_{xY}/s_x^2)(x_i - \overline{x})]^2}{\sigma_Y^2} = \frac{(n-1)S_Y^2(1-R^2)}{\sigma_Y^2}
$$

is $\chi^2(n-2)$ and is independent of $\widehat{\beta}$. Thus, when $\rho = 0$, the conditional distribution of

$$
T = \frac{(RS_Y/s_x)/(\sigma_Y/\sqrt{n-1}\,s_x)}{\sqrt{[(n-1)\,S_Y^2\,(1-R^2)/\sigma_Y^2][1/(n-2)]}} = \frac{R\sqrt{n-2}}{\sqrt{1-R^2}}
$$

is Student's t with $n-2$ degrees of freedom. However, since the conditional distribution of T, given $X_1 = x_1, \ldots, X_n = x_n$, does not depend on $x_1, x_2, \ldots, x_n$, the unconditional distribution of T must be Student's t with $n-2$ degrees of freedom and T and $(X_1, X_2, \ldots, X_n)$ are independent when $\rho = 0$.

Remark   It is interesting to note that in the discussion about the distribution of T, nothing was said about the distribution of $X_1, X_2, \ldots, X_n$. This means that if X and Y are independent and Y has a normal distribution, then T has a Student's t distribution whatever the distribution of X. Obviously, the roles of X and Y can be reversed in all of this development. In particular, if X and Y are independent, then T and $Y_1, Y_2, \ldots, Y_n$ are also independent.

   Now T can be used to test $H_0: \rho = 0$. If the alternative hypothesis is $H_1: \rho > 0$, we would use the critical region defined by observed $T \geq t_\alpha(n-2)$, since large T implies large R. Obvious modifications would be made for the alternative hypotheses $H_1: \rho < 0$ and $H_1: \rho \neq 0$, the latter leading to a two-sided test.

Using the p.d.f. $h(t)$ of T, we can find the distribution function and p.d.f. of R when $-1 < r < 1$, provided that $\rho = 0$:

$$G(r) = P(R \leq r) = P\left(T \leq \frac{r\sqrt{n-2}}{\sqrt{1-r^2}}\right)$$

$$= \int_{-\infty}^{r\sqrt{n-2}/\sqrt{1-r^2}} h(t)\,dt$$

$$= \int_{-\infty}^{r\sqrt{n-2}/\sqrt{1-r^2}} \frac{\Gamma[(n-1)/2]}{\Gamma(1/2)\,\Gamma[(n-2)/2]} \frac{1}{\sqrt{n-2}}\left(1 + \frac{t^2}{n-2}\right)^{-(n-1)/2} dt.$$

The derivative of $G(r)$, with respect to r, is (see Section A.4)

$$g(r) = h\left(\frac{r\sqrt{n-2}}{\sqrt{1-r^2}}\right)\frac{d(r\sqrt{n-2}/\sqrt{1-r^2})}{dr},$$

which equals

$$g(r) = \frac{\Gamma[(n-1)/2]}{\Gamma(1/2)\,\Gamma[(n-2)/2]}(1 - r^2)^{(n-4)/2}, \qquad -1 < r < 1.$$

Thus, for example, to test the hypothesis $H_0\colon \rho = 0$ against the alternative hypothesis $H_1\colon \rho \neq 0$ at a significance level α, select either a constant $r_{\alpha/2}(n-2)$ or a constant $t_{\alpha/2}(n-2)$ so that

$$\alpha = P(|R| \geq r_{\alpha/2}(n-2); H_0) = P(|T| \geq t_{\alpha/2}(n-2); H_0),$$

depending on the availability of R or T tables.

It is interesting to graph the p.d.f. of R. Note in particular that if $n = 4$, $g(r) = 1/2$, $-1 < r < 1$, and if $n = 6$, $g(r) = (3/4)(1 - r^2)$, $-1 < r < 1$. The graphs of the p.d.f. of R when $n = 8$ and when $n = 14$ are given in Figure 11.10. Recall that this is the p.d.f. of R when $\rho = 0$. As n increases R is more likely to equal values close to 0.

Table XI lists selected values of the distribution function of R when $\rho = 0$. For example, if $n = 8$, the number of degrees of freedom is 6 and $P(R \leq 0.7887) = 0.99$. Also, if $\alpha = 0.10$, then $r_{\alpha/2}(6) = r_{0.05}(6) = 0.6215$ [see Figure 11.10(a)].

It is also possible to obtain an approximate test of size α by using the fact that

$$W = \frac{1}{2}\ln\frac{1+R}{1-R}$$

has an approximate normal distribution with mean $(1/2)\ln[(1+\rho)/(1-\rho)]$ and variance $1/(n-3)$. We accept this statement without proof (see Exercise 11.5.8). Thus

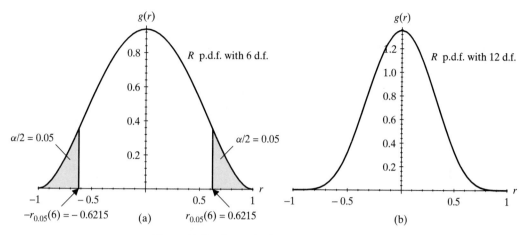

Figure 11.10 R p.d.f.'s when $n = 8$ and $n = 14$

a test of H_0: $p = p_0$ can be based on the statistic

$$Z = \frac{\frac{1}{2}\ln\frac{1+R}{1-R} - \frac{1}{2}\ln\frac{1+\rho_0}{1-\rho_0}}{\sqrt{\frac{1}{n-3}}},$$

which has a distribution that is approximately $N(0, 1)$.

Example 11.5.1

We would like to test the hypothesis H_0: $\rho = 0$ against H_1: $\rho \neq 0$ at an $\alpha = 0.05$ significance level. A random sample of size 18 from a bivariate normal distribution yielded a sample correlation coefficient of $r = 0.35$. Using Table XI in the Appendix, since $0.35 < 0.4683$, H_0 is accepted (not rejected) at an $\alpha = 0.05$ significance level. Using the t distribution, we would reject H_0 if $|t| \geq 2.120 = t_{0.025}(16)$. Since

$$t = \frac{0.35\sqrt{16}}{\sqrt{1 - (0.35)^2}} = 1.495,$$

H_0 is not rejected. If we had used the normal approximation for Z, H_0 would be rejected if $|z| \geq 1.96$. Since

$$z = \frac{(1/2)\ln[(1 + 0.35)/(1 - 0.35)] - 0}{\sqrt{1/(18 - 3)}} = 1.415,$$

H_0 is not rejected. ◇

To develop an approximate $100(1 - \alpha)\%$ confidence interval for ρ, we use the normal approximation for the distribution of Z. Thus we select a constant $c = z_{\alpha/2}$

from Table V so that

$$P\left(-c \le \frac{(1/2) \ln[(1+R)/(1-R)] - (1/2) \ln[(1+\rho)/(1-\rho)]}{\sqrt{1/(n-3)}} \le c\right) \approx 1 - \alpha.$$

After several steps, this becomes

$$P\left(\frac{1 + R - (1 - R)\exp(2c/\sqrt{n-3})}{1 + R + (1 - R)\exp(2c/\sqrt{n-3})} \le \rho \le \right.$$

$$\left.\frac{1 + R - (1 - R)\exp(-2c/\sqrt{n-3})}{1 + R + (1 - R)\exp(-2c/\sqrt{n-3})}\right) \approx 1 - \alpha.$$

Example 11.5.2

Suppose that a random sample of size 12 from a bivariate normal distribution yielded a correlation coefficient of $r = 0.6$. An approximate 95% confidence interval for ρ would be

$$\left[\frac{1 + 0.6 - (1 - 0.6)\exp\left(\frac{2(1.96)}{3}\right)}{1 + 0.6 + (1 - 0.6)\exp\left(\frac{2(1.96)}{3}\right)}, \frac{1 + 0.6 - (1 - 0.6)\exp\left(\frac{-2(1.96)}{3}\right)}{1 + 0.6 + (1 - 0.6)\exp\left(\frac{-2(1.96)}{3}\right)}\right]$$

$$= [0.040, 0.873].$$

If the sample size had been $n = 39$ and $r = 0.6$, the approximate 95% confidence interval would have been $[0.351, 0.770]$. ◇

EXERCISES

11.5.1 A random sample of size $n = 27$ from a bivariate normal distribution yielded a sample correlation coefficient of $r = -0.45$. Would the hypothesis H_0: $\rho = 0$ be rejected in favor of H_1: $\rho \ne 0$ at an $\alpha = 0.05$ significance level?

11.5.2 When bowling it is often possible to score well on the first game and then bowl poorly on the second game, or vice versa. The following six pairs of numbers give the scores of the first and second games bowled by the same person on six consecutive Tuesday evenings. Assume a bivariate normal distribution and use these scores to test the hypothesis H_0: $\rho = 0$ against H_1: $\rho \ne 0$ at $\alpha = 0.10$.

Game 1:	170	190	200	183	187	178
Game 2:	197	178	150	176	205	153

11.5.3 A random sample of size 28 from a bivariate normal distribution yielded a sample correlation coefficient of $r = 0.65$. Find an approximate 90% confidence interval for ρ.

11.5.4 For each part of Example 1.6.2, find an approximate 95% confidence interval for ρ.

11.5.5 To help determine whether gallinules selected their mate on the basis of weight, 14 pairs of gallinules were captured and weighed. Test the null hypothesis $H_0: \rho = 0$ against a two-sided alternative at an $\alpha = 0.01$ significance level. Given that the male and female weights for the $n = 14$ pairs of birds yielded a sample correlation coefficient of $r = -0.252$, would H_0 be rejected?

11.5.6 By squaring the binomial expression $[(Y_i - \overline{Y}) - (S_{xY}/s_x^2)(x_i - \overline{x})]$, show that

$$\sum_{i=1}^{n} [(Y_i - \overline{Y}) - (S_{xY}/s_x^2)(x_i - \overline{x})]^2$$

$$= \sum_{i=1}^{n} (Y_i - \overline{Y})^2 - 2\left(\frac{S_{xY}}{s_x^2}\right)\sum_{i=1}^{n}(x_i - \overline{x})(Y_i - \overline{Y}) + \frac{S_{xY}^2}{s_x^4}\sum_{i=1}^{n}(x_i - \overline{x})^2$$

equals $(n-1)S_Y^2(1 - R^2)$, where $X_1 = x_1, X_2 = x_2, \ldots, X_n = x_n$. HINT: Replace $S_{xY} = \sum_{i=1}^{n}(x_i - \overline{x})(Y_i - \overline{Y})/(n-1)$ by $Rs_x S_Y$.

11.5.7 When $\rho = 0$, show that

(a) The points of inflection for the graph of the p.d.f. of R are at $r = \pm 1/\sqrt{n-5}$ for $n \geq 7$.

(b) $E(R) = 0$.

(c) $\text{Var}(R) = 1/(n-1), n \geq 3$. HINT: Note that $E(R^2) = E[1 - (1 - R^2)]$.

11.5.8 It is true when sampling from a bivariate normal distribution that the correlation coefficient R has an approximate normal distribution $N[\rho, (1-\rho^2)^2/n]$ if the sample size n is large. Since, for large n, R is close to ρ, use two terms of the Taylor's expansion of $u(R)$ about ρ and determine that function $u(R)$ such that it has a variance (essentially) free of ρ. (The solution of this exercise explains why the transformation $(1/2)\ln[(1 + R)/(1 - R)]$ was suggested.)

11.5.9 Let X and Y have a bivariate normal distribution with correlation coefficient ρ. To test $H_0: \rho = 0$ against $H_1: \rho \neq 0$, a random sample of n pairs of observations is selected. Suppose that the sample correlation coefficient is $r = 0.68$. Using a significance level of $\alpha = 0.05$, find the smallest value of the sample size n so that H_0 is rejected.

11.5.10 In a college health fitness program, let X equal the weight in kilograms of a female freshman at the beginning of the program and let Y equal her weight change during the semester. We shall use the following data for $n = 16$ observations of (x, y) to test the null hypothesis $H_0: \rho = 0$ against a two-sided alternative hypothesis:

(61.4, −3.2)	(62.9, 1.4)	(58.7, 1.3)	(49.3, 0.6)
(71.3, 0.2)	(81.5, −2.2)	(60.8, 0.9)	(50.2, 0.2)
(60.3, 2.0)	(54.6, 0.3)	(51.1, 3.7)	(53.3, 0.2)
(81.0, −0.5)	(67.6, −0.8)	(71.4, −0.1)	(72.1, −0.1)

(a) What is the conclusion if $\alpha = 0.10$?

(b) What is the conclusion if $\alpha = 0.05$?

11.5.11 For the data in Exercise 8.3.11, test $H_0: \rho = 0$ against $H_1: \rho > 0$. Use $\alpha = 0.01$.

11.5.12 For the data in Exercise 8.3.12, test $H_0: \rho = 0$ against $H_1: \rho > 0$. In particular, approximate the p-value for this test.

11.6 TRANSFORMATIONS OF RANDOM VARIABLES

In Section 4.5 we considered the transformation of one random variable X with p.d.f. $f(x)$. In particular, in the continuous case, if $Y = u(X)$ was an increasing or decreasing function of X, with inverse $X = v(Y)$, then the p.d.f. of Y is

$$g(y) = |v'(y)| \, f[v(y)], \qquad c < y < d,$$

where the support $c < y < d$ corresponds to the support of X, say $a < x < b$, through the transformation $x = v(y)$.

Recall the discrete case had to be handled a little differently as illustrated by the following example.

Example 11.6.1

Let X be binomial with parameters n and p. Since X has a discrete distribution, $Y = u(X)$ will also have a discrete distribution with the same probabilities as those in the support of X. For illustration, with $n = 3$, $p = 1/4$, and $Y = X^2$, we have

$$g(y) = \binom{3}{\sqrt{y}} \left(\frac{1}{4}\right)^{\sqrt{y}} \left(\frac{3}{4}\right)^{3-\sqrt{y}}, \qquad y = 0, 1, 4, 9. \qquad \diamond$$

For a more interesting problem with the binomial random variable X, suppose that we were to search for a transformation $u(X/n)$ of the relative frequency X/n that would have a variance very little dependent on p itself when n is large. That is, we want the variance of $u(X/n)$ to be essentially a constant.

Consider the function $u(X/n)$ and find, using two terms of Taylor's expansion about p, that

$$u\left(\frac{X}{n}\right) \approx u(p) + u'(p)\left(\frac{X}{n} - p\right).$$

Here terms of higher powers can be disregarded if n is large enough so that X/n is close enough to p. Thus

$$\mathrm{Var}\left[u\left(\frac{X}{n}\right)\right] \approx [u'(p)]^2 \, \mathrm{Var}\left(\frac{X}{n} - p\right) = [u'(p)]^2 \, \frac{p(1-p)}{n}. \qquad (11.6)$$

However, if $\mathrm{Var}[u(X/n)]$ is to be constant with respect to p, then

$$[u'(p)]^2 \, p(1-p) = k \qquad \text{or} \qquad u'(p) = \frac{c}{\sqrt{p(1-p)}}$$

where k and c are constants. We see that

$$u(p) = 2c \arcsin \sqrt{p}$$

is a solution to this differential equation. Thus with $c = 1/2$, we frequently see, in the literature, use of the arcsine transformation, namely

$$Y = \arcsin \sqrt{\frac{X}{n}}$$

which, with large n, has an approximate normal distribution with mean

$$\mu = \arcsin \sqrt{p}$$

and variance (using Equation 11.6)

$$\sigma^2 = [D_p(\arcsin \sqrt{p})]^2 \frac{p(1-p)}{n} = \frac{1}{4n}.$$

There should be one note of warning here: If the function $Y = u(X)$ does not have a single-valued inverse, the determination of the distribution of Y will not be as simple. We did consider one such example in Section 4.4 by finding the distribution of Z^2, where Z is $N(0, 1)$. In this case, there were two inverse functions and special care was exercised. In our examples, we will not consider problems with many inverses; however, we thought that such a warning should be issued here.

When two or more random variables are involved, many interesting problems can result. In the case of a single-valued inverse, the rule is about the same as that in the one-variable case, with the derivative being replaced by the Jacobian. Namely, if X_1 and X_2 are two continuous-type random variables with joint p.d.f. $f(x_1, x_2)$ and if $Y_1 = u_1(X_1, X_2)$, $Y_2 = u_2(X_1, X_2)$ has the single-valued inverse $X_1 = v_1(Y_1, Y_2)$, $X_2 = v_2(Y_1, Y_2)$, then the joint p.d.f. of Y_1 and Y_2 is

$$g(y_1, y_2) = |J| \, f[v_1(y_1, y_2), v_2(y_1, y_2)],$$

where the Jacobian J is the determinant

$$J = \begin{vmatrix} \dfrac{\partial x_1}{\partial y_1} & \dfrac{\partial x_1}{\partial y_2} \\[2mm] \dfrac{\partial x_2}{\partial y_1} & \dfrac{\partial x_2}{\partial y_2} \end{vmatrix}.$$

Of course, we find the support of Y_1, Y_2 by considering the mapping of the support of X_1, X_2 under the transformation $y_1 = u_1(x_1, x_2)$, $y_2 = u_2(x_1, x_2)$.

Example 11.6.2

Let X_1 and X_2 have independent gamma distributions with parameters α, θ and β, θ, respectively. That is, the joint p.d.f. of X_1 and X_2 is

$$f(x_1, x_2) = \frac{1}{\Gamma(\alpha)\Gamma(\beta)\theta^{\alpha+\beta}} x_1^{\alpha-1} x_2^{\beta-1} \exp\left(-\frac{x_1 + x_2}{\theta}\right), \quad 0 < x_1 < \infty, \;\; 0 < x_2 < \infty.$$

Consider

$$Y_1 = \frac{X_1}{X_1 + X_2}, \qquad Y_2 = X_1 + X_2,$$

or, equivalently,

$$X_1 = Y_1 Y_2, \qquad X_2 = Y_2 - Y_1 Y_2.$$

The Jacobian is

$$J = \begin{vmatrix} y_2 & y_1 \\ -y_2 & 1 - y_1 \end{vmatrix} = y_2(1 - y_1) + y_1 y_2 = y_2.$$

Thus the joint p.d.f. $g(y_1, y_2)$ of Y_1 and Y_2 is

$$g(y_1, y_2) = |y_2| \frac{1}{\Gamma(\alpha)\Gamma(\beta)\theta^{\alpha+\beta}} (y_1 y_2)^{\alpha-1} (y_2 - y_1 y_2)^{\beta-1} e^{-y_2/\theta},$$

where the support is $0 < y_1 < 1, 0 < y_2 < \infty$. The marginal p.d.f. of Y_1 is

$$g_1(y_1) = \frac{y_1^{\alpha-1}(1 - y_1)^{\beta-1}}{\Gamma(\alpha)\Gamma(\beta)} \int_0^\infty \frac{y_2^{\alpha+\beta-1}}{\theta^{\alpha+\beta}} e^{-y_2/\theta} \, dy_2.$$

But the integral in this expression is that of a gamma p.d.f. with parameters $\alpha + \beta$ and θ, except for $\Gamma(\alpha + \beta)$ in the denominator; hence the integral equals $\Gamma(\alpha + \beta)$ and

$$g_1(y_1) = \frac{\Gamma(\alpha + \beta)}{\Gamma(\alpha)\Gamma(\beta)} y_1^{\alpha-1}(1 - y_1)^{\beta-1}, \qquad 0 < y_1 < 1.$$

We say that Y_1 has a **beta p.d.f.** with parameters α and β (see Figure 11.11).   ◇

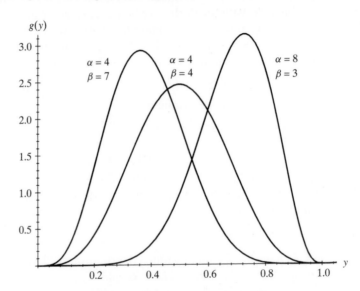

Figure 11.11   Beta distribution p.d.f.'s

Let the distribution of X be $N(\mu, \sigma^2)$. It is not easy to simulate an observation of X using $Y = F(X)$, where Y is uniform, $U(0, 1)$ and F is the distribution function of the desired normal distribution, because we cannot express the normal distribution function $F(x)$ in closed form. The following example provides a method for using "random numbers" to perform this simulation.

Example 11.6.3

(Box-Muller Transformation) Consider the following transformation where X_1 and X_2 are observations of a random sample from $U(0, 1)$:

$$Z_1 = \sqrt{-2 \ln X_1} \, \cos(2\pi X_2), \qquad Z_2 = \sqrt{-2 \ln X_1} \, \sin(2\pi X_2)$$

or, equivalently,

$$X_1 = \exp\left(-\frac{Z_1^2 + Z_2^2}{2}\right) = e^{-q/2}, \qquad X_2 = \frac{1}{2\pi}\arctan\left(\frac{Z_2}{Z_1}\right),$$

which has Jacobian

$$J = \begin{vmatrix} -z_1 e^{-q/2} & -z_2 e^{-q/2} \\[2mm] \dfrac{-z_2}{2\pi(z_1^2 + z_2^2)} & \dfrac{z_1}{2\pi(z_1^2 + z_2^2)} \end{vmatrix} = \frac{-1}{2\pi}e^{-q/2}.$$

Since the joint p.d.f. of X_1 and X_2 is

$$f(x_1, x_2) = 1, \qquad 0 < x_1 < 1, \quad 0 < x_2 < 1,$$

we have that the joint p.d.f. of Z_1 and Z_2 is

$$g(z_1, z_2) = \left| \frac{-1}{2\pi}e^{-q/2} \right| \quad (1)$$

$$= \frac{1}{2\pi}\exp\left(-\frac{z_1^2 + z_2^2}{2}\right), \qquad -\infty < z_1 < \infty, \quad -\infty < z_2 < \infty.$$

(The student should note that there is some difficulty with the definition of the transformation, particularly when $z_1 = 0$. However, these difficulties occur at events with probability zero and hence cause no problems; see Exercise 11.6.11.) Summarizing, note that from two independent $U(0, 1)$ random variables, we have generated two independent $N(0, 1)$ random variables through this **Box-Muller transformation.**   ◇

The techniques described for two random variables can be extended to three or more random variables. We do not give any details here but mention, for illustration, that with three random variables X_1, X_2, X_3 of the continuous type, we need three "new" random variables Z_1, Z_2, Z_3 so that the corresponding Jacobian of the single-valued inverse transformation is the nonzero determinant

$$J = \begin{vmatrix} \dfrac{\partial x_1}{\partial z_1} & \dfrac{\partial x_1}{\partial z_2} & \dfrac{\partial x_1}{\partial z_3} \\[3mm] \dfrac{\partial x_2}{\partial z_1} & \dfrac{\partial x_2}{\partial z_2} & \dfrac{\partial x_2}{\partial z_3} \\[3mm] \dfrac{\partial x_3}{\partial z_1} & \dfrac{\partial x_3}{\partial z_2} & \dfrac{\partial x_3}{\partial z_3} \end{vmatrix}.$$

We are now in a position to derive the distributions of T and F which we have used earlier in the text. In both cases we use the distribution function technique, although we could have used the change-of-variable technique as we did in deriving the beta p.d.f.

Example 11.6.4

We let

$$T = \frac{Z}{\sqrt{U/r}},$$

where Z is a random variable that is $N(0, 1)$, U is a random variable that is $\chi^2(r)$, and Z and U are independent. Thus the joint p.d.f. of Z and U is

$$g(z, u) = \frac{1}{\sqrt{2\pi}} e^{-z^2/2} \frac{1}{\Gamma(r/2)2^{r/2}} u^{r/2-1} e^{-u/2}, \quad -\infty < z < \infty, \quad 0 < u < \infty.$$

The distribution function $F(t) = P(T \le t)$ of T is given by

$$F(t) = P(Z/\sqrt{U/r} \le t)$$

$$= P(Z \le \sqrt{U/r}\, t)$$

$$= \int_0^\infty \int_{-\infty}^{\sqrt{(u/r)}\, t} g(z, u) \, dz \, du.$$

That is,

$$F(t) = \frac{1}{\sqrt{\pi}\, \Gamma(r/2)} \int_0^\infty \left[\int_{-\infty}^{\sqrt{(u/r)}\, t} \frac{e^{-z^2/2}}{2^{(r+1)/2}} dz \right] u^{r/2-1} e^{-u/2} \, du.$$

The p.d.f. of T is the derivative of the distribution function; so applying the fundamental theorem of calculus to the inner integral, we see that

$$f(t) = F'(t) = \frac{1}{\sqrt{\pi}\, \Gamma(r/2)} \int_0^\infty \frac{e^{-(u/2)(t^2/r)}}{2^{(r+1)/2}} \sqrt{\frac{u}{r}}\, u^{r/2-1} e^{-u/2} \, du$$

$$= \frac{1}{\sqrt{\pi r}\, \Gamma(r/2)} \int_0^\infty \frac{u^{(r+1)/2-1}}{2^{(r+1)/2}} e^{-(u/2)(1+t^2/r)} \, du.$$

In the integral make the change of variables

$$y = (1 + t^2/r)u \quad \text{so that} \quad \frac{du}{dy} = \frac{1}{1 + t^2/r}.$$

Thus we find that

$$f(t) = \frac{\Gamma((r + 1)/2)}{\sqrt{\pi r}\, \Gamma(r/2)} \left[\frac{1}{(1 + t^2/r)^{(r+1)/2}} \right] \int_0^\infty \frac{y^{(r+1)/2-1}}{\Gamma((r + 1)/2)\, 2^{(r+1)/2}} e^{-y/2} \, dy.$$

The integral in this last expression for $f(t)$ is equal to 1 because the integrand is like the p.d.f. of a chi-square distribution with $r + 1$ degrees of freedom. Thus the p.d.f. is

$$f(t) = \frac{\Gamma((r + 1)/2)}{\sqrt{\pi r}\, \Gamma(r/2)} \frac{1}{(1 + t^2/r)^{(r+1)/2}}, \quad -\infty < t < \infty. \qquad \diamond$$

Example 11.6.5

We know that

$$F = \frac{U/r_1}{V/r_2},$$

where U and V are independent chi-square variables with r_1 and r_2 degrees of freedom, respectively. Thus the joint p.d.f. of U and V is

$$g(u, v) = \frac{u^{r_1/2-1}e^{-u/2}}{\Gamma(r_1/2)2^{r_1/2}} \frac{v^{r_2/2-1}e^{-v/2}}{\Gamma(r_2/2)2^{r_2/2}}, \qquad 0 < u < \infty, \quad 0 < v < \infty.$$

In this derivation we let $W = F$ to avoid using f as a symbol for a variable. The distribution function $F(w) = P(W \le w)$ of F is

$$F(w) = P\left(\frac{U/r_1}{V/r_2} \le w\right) = P\left(U \le \frac{r_1}{r_2} w V\right)$$

$$= \int_0^\infty \int_0^{(r_1/r_2)wv} g(u, v)\, du\, dv.$$

That is,

$$F(w) = \frac{1}{\Gamma(r_1/2)\Gamma(r_2/2)} \int_0^\infty \left[\int_0^{(r_1/r_2)wv} \frac{u^{r_1/2-1}e^{-u/2}}{2^{(r_1+r_2)/2}}\, du\right] v^{r_2/2-1}e^{-v/2}\, dv.$$

The p.d.f. of $F = W$ is the derivative of the distribution function; so applying the fundamental theorem of calculus to the inner integral, we have

$$f(w) = F'(w)$$

$$= \frac{1}{\Gamma(r_1/2)\Gamma(r_2/2)} \int_0^\infty \frac{[(r_1/r_2)vw]^{r_1/2-1}}{2^{(r_1+r_2)/2}} e^{-(r_1/2r_2)(vw)} \left(\frac{r_1}{r_2} v\right) v^{r_2/2-1}e^{-v/2}\, dv$$

$$= \frac{(r_1/r_2)^{r_1/2}w^{r_1/2-1}}{\Gamma(r_1/2)\Gamma(r_2/2)} \int_0^\infty \frac{v^{(r_1+r_2)/2-1}}{2^{(r_1+r_2)/2}} e^{-(v/2)[1+(r_1/r_2)w]}\, dv.$$

In the integral, make the change of variables

$$y = \left(1 + \frac{r_1}{r_2} w\right) v \qquad \text{so that} \qquad \frac{dv}{dy} = \frac{1}{1 + (r_1/r_2)w}.$$

Thus we see that

$$f(w) = \frac{(r_1/r_2)^{r_1/2}\Gamma[(r_1 + r_2)/2]w^{r_1/2-1}}{\Gamma(r_1/2)\Gamma(r_2/2)[1 + (r_1w/r_2)]^{(r_1+r_2)/2}} \int_0^\infty \frac{y^{(r_1+r_2)/2-1}e^{-y/2}}{\Gamma[(r_1 + r_2)/2]2^{(r_1+r_2)/2}}\, dy$$

$$= \frac{(r_1/r_2)^{r_1/2}\Gamma[(r_1 + r_2)/2]w^{r_1/2-1}}{\Gamma(r_1/2)\Gamma(r_2/2)[1 + (r_1w/r_2)]^{(r_1+r_2)/2}},$$

the p.d.f. of the F distribution. Note that the integral in this last expression for $f(w)$ is equal to 1 because the integrand is like a p.d.f. of a chi-square distribution with $r_1 + r_2$ degrees of freedom. Graphs of p.d.f.'s for the F distribution are given in Figure 11.12.

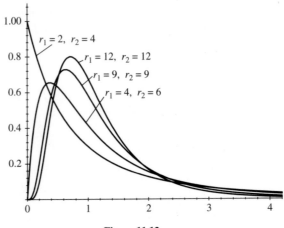

Figure 11.12

EXERCISES

11.6.1 Let X_1, X_2 denote a random sample from a distribution $\chi^2(2)$. Find the joint p.d.f. of $Y_1 = X_1$ and $Y_2 = X_2 + X_1$. Here note that the support of Y_1, Y_2 is $0 < y_1 < y_2 < \infty$. Also find the marginal p.d.f. of each of Y_1 and Y_2. Are Y_1 and Y_2 independent?

11.6.2 Let Z_1, Z_2 be a random sample from the standard normal distribution $N(0, 1)$. Use the transformation defined by polar coordinates $Z_1 = X_1 \cos X_2$, $Z_2 = X_1 \sin X_2$.

   (a) Show that the Jacobian equals x_1. (This explains the factor r of $r\,dr\,d\theta$ in the usual polar coordinate notation.)

   (b) Find the joint p.d.f. of X_1 and X_2.

   (c) Are X_1 and X_2 independent?

11.6.3 Let the independent random variables X_1 and X_2 be $N(0, 1)$ and $\chi^2(r)$, respectively. Let $Y_1 = X_1/\sqrt{X_2/r}$ and $Y_2 = X_2$.

   (a) Find the joint p.d.f. of Y_1 and Y_2.

   (b) Determine the marginal p.d.f. of Y_1 and show that Y_1 has a t distribution. (This is another, but equivalent, way of finding the p.d.f. of T.)

11.6.4 Let T have a t distribution with r degrees of freedom. Show that $E(T) = 0, r \geq 2$, and $\mathrm{Var}(T) = r/(r-2)$, provided that $r \geq 3$, by first finding $E(Z)$, $E(1/\sqrt{U})$, $E(Z^2)$, and $E(1/U)$.

11.6.5 Let T have a t distribution with r degrees of freedom. Show that T^2 has an F distribution with 1 and r degrees of freedom. HINT: Consider $T^2 = U^2/(V/r)$.

11.6.6 Let X_1 and X_2 be independent chi-square random variables with r_1 and r_2 degrees of freedom, respectively. Let $Y_1 = (X_1/r_1)/(X_2/r_2)$ and $Y_2 = X_2$.

   (a) Find the joint p.d.f. of Y_1 and Y_2.

(b) Determine the marginal p.d.f. of Y_1 and show that Y_1 has an F distribution. (This is another, but equivalent, way of finding the p.d.f. of F.)

11.6.7 Find the mean and the variance of an F random variable with r_1 and r_2 degrees of freedom by first finding $E(U)$, $E(1/V)$, $E(U^2)$, and $E(1/V^2)$.

11.6.8 Let X have a Poisson distribution with mean λ. Find a transformation $u(X)$ so that $\text{Var}[u(X)]$ is about free of λ, for large values of λ. HINT: $u(X) \approx u(\lambda) + [u'(\lambda)](X - \lambda)$, provided that $[u''(\lambda)](X-\lambda)^2/2 \approx \text{Var}[u(X)]$ and higher terms can be neglected when λ is large. A solution is $u(X) = \sqrt{X}$.

11.6.9 Generalize Exercise 11.6.8 by assuming that the variance of a distribution is equal to $c\mu^p$, where c is a constant. (Note in Exercise 11.6.8 that this is the case with $p = 1$.) In particular, the transformation $Y = u(X) = X^{1-p/2}$, when $p \neq 2$, or $Y = u(X) = \ln X$, when $p = 2$, seems to produce a random variable Y whose variance is about free of μ. This is the reason transformations like $\sqrt{X}$, $\ln X$, and more generally, X^b are so popular in applications.

11.6.10 Let Z_1 and Z_2 be independent standard normal random variables $N(0, 1)$. Show that $Y_1 = Z_1/Z_2$ has a Cauchy distribution. Here, let $Y_2 = Z_2$. (See Exercise 11.6.18.)

11.6.11 In Example 11.6.3 verify that the given transformation maps $\{(x_1, x_2) : 0 < x_1 < 1, 0 < x_2 < 1\}$ onto $\{(z_1, z_2) : -\infty < z_1 < \infty, -\infty < z_2 < \infty\}$ except for a set of points that has probability 0. HINT: What is the image of vertical line segments? What is the image of horizontal line segments?

11.6.12 Let X_1 and X_2 have independent gamma distributions with parameters α, θ and β, θ, respectively. Let $W = X_1/(X_1 + X_2)$. Use a method which is similar to that given in the derivations of the t and F distributions (Examples 11.6.4 and 11.6.5) to show that the p.d.f. of W is

$$g(w) = \frac{\Gamma(\alpha + \beta)}{\Gamma(\alpha)\Gamma(\beta)} w^{\alpha-1}(1 - w)^{\beta-1}, \qquad 0 < w < 1.$$

We say that W has a **beta distribution** with parameters α and β (see Example 11.6.2).

11.6.13 Let X_1 and X_2 be independent chi-square random variables with r_1 and r_2 degrees of freedom, respectively. Show that

(a) $U = X_1/(X_1 + X_2)$ has a beta distribution with $\alpha = r_1/2$ and $\beta = r_2/2$.

(b) $V = X_2/(X_1 + X_2)$ has a beta distribution with $\alpha = r_2/2$ and $\beta = r_1/2$.

11.6.14 Let X have a beta distribution with parameters α and β (see Example 11.6.2).

(a) Show that the mean and variance of X are

$$\mu = \frac{\alpha}{\alpha + \beta} \quad \text{and} \quad \sigma^2 = \frac{\alpha\beta}{(\alpha + \beta + 1)(\alpha + \beta)^2}.$$

(b) Show that the beta p.d.f. has a mode at $x = (\alpha - 1)/(\alpha + \beta - 2)$.

11.6.15 Determine the constant c such that $f(x) = cx^3(1 - x)^6, 0 < x < 1$, is a p.d.f.

11.6.16 When α and β are integers and $0 < p < 1$, then

$$\int_0^p \frac{\Gamma(\alpha + \beta)}{\Gamma(\alpha)\Gamma(\beta)} y^{\alpha-1}(1 - y)^{\beta-1} \, dy = \sum_{y=\alpha}^{n} \binom{n}{y} p^y (1 - p)^{n-y},$$

where $n = \alpha + \beta - 1$. Verify this formula when $\alpha = 4$ and $\beta = 3$.

11.6.17 Evaluate

$$\int_0^{0.4} \frac{\Gamma(7)}{\Gamma(4)\Gamma(3)} y^3 (1-y)^2 \, dy$$

(a) Using integration.

(b) Using the result of Exercise 11.6.16.

11.6.18 Let W_1, W_2 be a random sample of size $n = 2$ from a Cauchy distribution with p.d.f. $h(\omega) = [\pi(1+\omega^2)]^{-1}$, $-\infty < \omega < \infty$. In this exercise we find the p.d.f. of the sample mean.

(a) Show that the p.d.f. of $X_1 = (1/2)W_1$ is

$$f(x) = \frac{2}{\pi(1+4x^2)}, \qquad -\infty < x < \infty.$$

(b) Let $Y_1 = X_1 + X_2 = \overline{W}$ and $Y_2 = X_1$, where $X_2 = (1/2)W_2$. Show that the joint p.d.f. of Y_1 and Y_2 is

$$g(y_1, y_2) = f(y_1 - y_2)f(y_2), \qquad -\infty < y_1 < \infty, -\infty < y_2 < \infty.$$

(c) Show that the p.d.f. of $Y_1 = \overline{W}$ is given by the **convolution formula**,

$$g_1(y_1) = \int_{-\infty}^{\infty} f(y_1 - y_2)f(y_2)\, dy_2.$$

(d) Show that

$$g_1(y_1) = \frac{1}{\pi(1+y_1^2)}, \qquad -\infty < y_1 < \infty.$$

That is, the p.d.f. of $\overline{W}$ is the same as that of an individual W.

11.6.19 Let Z_1, Z_2, Z_3 be a random sample of size 3 from a standard normal distribution $N(0, 1)$.

(a) How is U distributed if

$$U = \frac{Z_3}{\sqrt{(Z_1^2 + Z_2^2)/2}} \quad ?$$

(b) Let

$$W = \frac{Z_2}{\sqrt{(Z_1^2 + Z_2^2)/2}}.$$

Show that the distribution function of W, for $-\sqrt{2} < w < \sqrt{2}$, can be defined by

$$F(w) = \frac{1}{2} + \frac{1}{\pi}\left(\arcsin \frac{w}{\sqrt{2}}\right) = \frac{1}{2} + \frac{1}{\pi}\left(\arctan \frac{w}{\sqrt{2-w^2}}\right).$$

(c) Show that the p.d.f. of W is

$$f(w) = \frac{1}{\pi\sqrt{2-w^2}}, \qquad -\sqrt{2} < w < \sqrt{2}.$$

Note that this is a U-shaped distribution. Why does it differ so much from that in part (a) when the definitions for U and W are so similar?

(d) Find the means and variances (if they exist) of U and W.

12

THEORY OF STATISTICAL INFERENCE

12.1 SUFFICIENT STATISTICS

This and the next section would have been introduced earlier. However, the notion of conditional distributions is the key concept associated with both sufficient statistics and Bayesian methods. Therefore, we delayed these discussions until that concept was introduced.

We first define a sufficient statistic $Y = u(X_1, X_2, \ldots, X_n)$ for a parameter using a statement that in most books is given as a necessary and sufficient condition for sufficiency, namely the well-known Fisher-Neyman factorization theorem. Using this as a definition, we shall note, by examples, the implications of this definition, one of which is sometimes used as the definition.

Definition 12.1.1. (Factorization Theorem) Let $X_1, X_2, \ldots, X_n$ denote random variables with joint p.d.f. $f(x_1, x_2, \ldots, x_n; \theta)$, which depends on the parameter θ. The statistic $Y = u(X_1, X_2, \ldots, X_n)$ is sufficient for θ if and only if

$$f(x_1, x_2, \ldots, x_n; \theta) = \phi[u(x_1, x_2, \ldots, x_n); \theta]h(x_1, x_2, \ldots, x_n),$$

where ϕ depends on $x_1, x_2, \ldots, x_n$ only through $u(x_1, x_2, \ldots, x_n)$ and the function $h(x_1, x_2, \ldots, x_n)$ does not depend on θ.

Let us consider several important examples and consequences of this definition. We first note, however, that in all instances in this book the random variables $X_1, X_2, \ldots, X_n$ will be of a random sample and hence their joint p.d.f. will be of the form

$$f(x_1; \theta)f(x_2; \theta) \cdots f(x_n; \theta).$$

Example 12.1.1

Let $X_1, X_2, \ldots, X_n$ denote a random sample from a Poisson distribution with parameter $\lambda > 0$. Then

$$f(x_1; \lambda) f(x_2; \lambda) \cdots f(x_n; \lambda) = \frac{\lambda^{\Sigma x_i} e^{-n\lambda}}{x_1! x_2! \cdots x_n!} = (\lambda^{n\bar{x}} e^{-n\lambda}) \left(\frac{1}{x_1! x_2! \cdots x_n!} \right),$$

where $\bar{x} = (1/n) \sum x_i$. Thus, from the factorization theorem (definition), it is clear that the sample mean $\overline{X}$ is a sufficient statistic for λ. It can easily be shown that the maximum likelihood estimator for λ is also $\overline{X}$ (see Exercise 6.1.3).   ◇

In Example 12.1.1, if we replace $n\bar{x}$ by $\sum x_i$, it is quite obvious that the sum $\sum X_i$ is also a sufficient statistic for λ. This certainly agrees with our intuition because if we know one of the statistics, $\overline{X}$ or $\sum X_i$, we can easily find the other. If we generalize this, we see that if Y is sufficient for a parameter θ, then every single-valued function of Y, not involving θ but with a single-valued inverse, is also a sufficient statistic for θ. Again the reason is that knowing either Y or that function of Y, we know the other. More formally, if $W = v(Y) = v[u(X_1, X_2, \ldots, X_n)]$ is that function and $Y = v^{-1}(W)$ is the single-valued inverse, then the display of the factorization theorem can be written as

$$f(x_1, x_2, \ldots, x_n; \theta) = \phi[v^{-1}\{v[u(x_1, x_2, \ldots, x_n)]\}; \theta] h(x_1, x_2, \ldots, x_n).$$

The first factor of the right-hand member of this equation depends on $x_1, x_2, \ldots, x_n$ through $v[u(x_1, x_2, \ldots, x_n)]$, so $W = v[u(X_1, X_2, \ldots, X_n)]$ is a sufficient statistic for θ. We illustrate this fact and the factorization theorem with an underlying distribution of the continuous type.

Example 12.1.2

Let $X_1, X_2, \ldots, X_n$ be a random sample from $N(\mu, 1)$, $-\infty < \mu < \infty$. The joint p.d.f. of these random variables is

$$\frac{1}{(2\pi)^{n/2}} \exp\left[-\frac{1}{2} \sum_{i=1}^{n} (x_i - \mu)^2 \right]$$

$$= \frac{1}{(2\pi)^{n/2}} \exp\left[-\frac{1}{2} \sum_{i=1}^{n} [(x_i - \bar{x}) + (\bar{x} - \mu)]^2 \right]$$

$$= \left\{ \exp\left[-\frac{n}{2} (\bar{x} - \mu)^2 \right] \right\} \left\{ \frac{1}{(2\pi)^{n/2}} \exp\left[-\frac{1}{2} \sum_{i=1}^{n} (x_i - \bar{x})^2 \right] \right\}.$$

From the factorization theorem we see that $\overline{X}$ is sufficient for μ. Now $\overline{X}^3$ is also sufficient for μ because knowing $\overline{X}^3$ is equivalent to having knowledge of the value of $\overline{X}$. However, $\overline{X}^2$ does not have this property, and it is not sufficient for μ.   ◇

One consequence of the sufficiency of a statistic Y is that the conditional probability of any given event A in the support of $X_1, X_2, \ldots, X_n$, given $Y = y$, does not depend on θ. This is sometimes used as the definition of sufficiency and is illustrated by the following example.

Example 12.1.3

Let $X_1, X_2, \ldots, X_n$ be a random sample from a distribution with p.d.f.

$$f(x; p) = p^x (1 - p)^{1-x}, \qquad x = 0, 1,$$

where the parameter p is between zero and one. We know that

$$Y = X_1 + X_2 + \cdots + X_n$$

is $b(n, p)$ and Y is sufficient for p because the joint p.d.f. of $X_1, X_2, \ldots, X_n$ is

$$p^{x_1}(1 - p)^{1-x_1} \cdots p^{x_n}(1 - p)^{1-x_n} = [p^{\Sigma x_i}(1 - p)^{n - \Sigma x_i}](1),$$

where $\phi(y; p) = p^y(1 - p)^{n-y}$ and $h(x_1, x_2, \ldots, x_n) = 1$. What then is the conditional probability $P(X_1 = x_1, \ldots, X_n = x_n \mid Y = y)$, where $y = 0, 1, \ldots, n - 1$, or n? Unless the sum of the nonnegative integers $x_1, x_2, \ldots, x_n$ equals y, this conditional probability is obviously equal to zero, which does not depend on p. Hence it is only interesting to consider the solution when $y = x_1 + \cdots + x_n$. From the definition of conditional probability we have

$$P(X_1 = x_1, \ldots, X_n = x_n \mid Y = y) = \frac{P(X_1 = x_1, \ldots, X_n = x_n)}{P(Y = y)}$$

$$= \frac{p^{x_1}(1 - p)^{1-x_1} \cdots p^{x_n}(1 - p)^{1-x_n}}{\binom{n}{y} p^y (1 - p)^{n-y}}$$

$$= \frac{1}{\binom{n}{y}},$$

where $y = x_1 + \cdots + x_n$. Since y equals the number of ones in the collection $x_1, x_2, \ldots, x_n$, this answer is only the probability of selecting a particular arrangement, namely $x_1, x_2, \ldots, x_n$ of y ones and $n - y$ zeros, and does not depend on the parameter p. That is, given that the sufficient statistic $Y = y$, the conditional probability of $X_1 = x_1, X_2 = x_2, \ldots, X_n = x_n$ does not depend on the parameter p. $\diamond$

It is interesting to observe that the underlying p.d.f. in Examples 12.1.1, 12.1.2, and 12.1.3 can be written in the exponential form

$$f(x; \theta) = \exp[K(x)p(\theta) + S(x) + q(\theta)],$$

where the support is free of θ. That is, we have, respectively,

$$\frac{e^{-\lambda}\lambda^x}{x!} = \exp\{x \ln \lambda - \ln x! - \lambda\}, \qquad x = 0, 1, 2, \ldots,$$

$$\frac{1}{\sqrt{2\pi}}e^{-(x-\mu)^2/2} = \exp\left\{x\mu - \frac{x^2}{2} - \frac{\mu^2}{2} - \frac{1}{2}\ln(2\pi)\right\}, \qquad -\infty < x < \infty,$$

and

$$p^x(1-p)^{1-x} = \exp\left\{x\ln\left(\frac{p}{1-p}\right) + \ln(1-p)\right\}, \qquad x = 0, 1.$$

In each of these examples, the sum $\sum X_i$ of the observations of the random sample was the sufficient statistic for the parameter. This is generalized by Theorem 12.1.1.

Theorem 12.1.1. Let $X_1, X_2, \ldots, X_n$ be a random sample from a distribution with a p.d.f. of the exponential form

$$f(x; \theta) = \exp[K(x)p(\theta) + S(x) + q(\theta)]$$

on a support free of θ. The statistic $\sum_{i=1}^{n} K(X_i)$ is sufficient for θ.

Proof. The joint p.d.f. of $X_1, X_2, \ldots, X_n$ is

$$\exp\left[p(\theta)\sum_{i=1}^{n} K(x_i) + \sum_{i=1}^{n} S(x_i) + nq(\theta)\right]$$

$$= \left\{\exp\left[p(\theta)\sum_{i=1}^{n} K(x_i) + nq(\theta)\right]\right\}\left\{\exp\left[\sum_{i=1}^{n} S(x_i)\right]\right\}$$

In accordance with the factorization theorem, the statistic $\sum_{i=1}^{n} K(X_i)$ is sufficient for θ. $\square$

In many cases Theorem 12.1.1 permits the student to find the sufficient statistic for the parameter with very little effort, as shown by the following example.

Example 12.1.4

Let $X_1, X_2, \ldots, X_n$ be a random sample from an exponential distribution with p.d.f.

$$f(x; \theta) = \frac{1}{\theta}e^{-x/\theta} = \exp\left[x\left(-\frac{1}{\theta}\right) - \ln\theta\right], \qquad 0 < x < \infty,$$

provided $0 < \theta < \infty$. Here $K(x) = x$. Thus $\sum_{i=1}^{n} X_i$ is sufficient for θ; of course, $\bar{X} = \sum_{i=1}^{n} X_i/n$ is also sufficient. $\diamond$

We should also note that if there is a sufficient statistic for the parameter under consideration and if the maximum likelihood estimator of this parameter is unique, then the maximum likelihood estimator is a function of the sufficient statistic. To see this heuristically, consider the following. If a sufficient statistic exists, then the likelihood function is

$$L(\theta) = f(x_1, x_2, \ldots, x_n; \theta) = \phi[u(x_1, x_2, \ldots, x_n); \theta]h(x_1, x_2, \ldots, x_n).$$

Since $h(x_1, x_2, \ldots, x_n)$ does not depend on θ, we maximize $L(\theta)$ by maximizing $\phi[u(x_1, x_2, \ldots, x_n); \theta]$. But ϕ is a function of $x_1, x_2, \ldots, x_n$ only through the statistic $u(x_1, x_2, \ldots, x_n)$. Thus, if there is a unique value of θ that maximizes ϕ, then it must be a function of $u(x_1, x_2, \ldots, x_n)$. That is, $\widehat{\theta}$ is a function of the sufficient statistic $u(X_1, X_2, \ldots, X_n)$. This fact was alluded to in Example 12.1.1, but it could be checked using other examples and exercises.

EXERCISES

12.1.1 Let $X_1, X_2, \ldots, X_n$ be a random sample from the distribution with the p.d.f. $f(x; p) = p(1 - p)^{x-1}$, $x = 1, 2, 3, \ldots$, where $0 < p < 1$.

(a) Show that $Y = \sum_{i=1}^{n} X_i$ is a sufficient statistic for p.

(b) Find a function of $Y = \sum_{i=1}^{n} X_i$ that is an unbiased estimator of $\theta = 1/p$.

12.1.2 Let $X_1, X_2, \ldots, X_n$ be a random sample from a Poisson distribution with mean $\lambda > 0$. Find the conditional probability $P(X_1 = x_1, \ldots, X_n = x_n | Y = y)$, where $Y = X_1 + \cdots + X_n$ and the nonnegative integers $x_1, x_2, \ldots, x_n$ sum to y, showing that this probability does not depend on λ.

12.1.3 Let $X_1, X_2, \ldots, X_n$ be a random sample from $N(0, \sigma^2)$.

(a) Find the sufficient statistic Y for σ^2.

(b) Show that the maximum likelihood estimator for σ^2 is a function of Y.

(c) Is the maximum likelihood estimator for σ^2 unbiased?

12.1.4 Let $X_1, X_2, \ldots, X_n$ be a random sample from a distribution with p.d.f. $f(x; \theta) = \theta x^{\theta-1}$, $0 < x < 1$, where $0 < \theta$.

(a) Find the sufficient statistic Y for θ.

(b) Show that the maximum likelihood estimator $\widehat{\theta}$ is a function of Y.

(c) Argue that $\widehat{\theta}$ is also sufficient for θ.

12.2 BAYESIAN ESTIMATION

We now describe another approach to estimation that is used by a group of statisticians who call themselves Bayesians. To understand fully their approach would require more text than we can allocate to this topic, but let us begin this brief introduction by considering a simple application of the theorem of the Reverend Thomas Bayes (see Section 2.5).

Example 12.2.1

Suppose we know that we are going to select an observation from a Poisson distribution with mean λ equal to 2 or 4. Moreover, we believe prior to performing the experiment that $\lambda = 2$ has about four times as much chance of being the parameter as does $\lambda = 4$; that is, the prior probabilities are $P(\lambda = 2) = 0.8$ and $P(\lambda = 4) = 0.2$. The experiment is now performed and we observe that $x = 6$. At this point, our intuition now tells us that $\lambda = 2$ seems less likely than before as the observation $x = 6$ is much more probable with $\lambda = 4$ than with $\lambda = 2$, because, in an obvious notation,

$$P(X = 6 \mid \lambda = 2) = 0.995 - 0.983 = 0.012$$

and

$$P(X = 6 \mid \lambda = 4) = 0.889 - 0.785 = 0.104,$$

from Table III. Our intuition can be supported by computing the conditional probability of $\lambda = 2$, given $X = 6$;

$$
\begin{aligned}
P(\lambda = 2 \mid X = 6) &= \frac{P(\lambda = 2, X = 6)}{P(X = 6)} \\[2mm]
&= \frac{P(\lambda = 2) P(X = 6 \mid \lambda = 2)}{P(\lambda = 2) P(X = 6 \mid \lambda = 2) + P(\lambda = 4) P(X = 6 \mid \lambda = 4)} \\[2mm]
&= \frac{(0.8)(0.012)}{(0.8)(0.012) + (0.2)(0.104)} = 0.316.
\end{aligned}
$$

This conditional probability is called the posterior probability of $\lambda = 2$, given the data (here $x = 6$). In a similar fashion, the posterior probability of $\lambda = 6$ is found to be 0.684. Thus we see that the probability of $\lambda = 2$ has decreased from 0.8 (the prior probability) to 0.316 (the posterior probability) with the observation of $x = 6$. ◇

In a more practical application the parameter, say θ, can possibly take many more than two values as in Example 12.2.1. Somehow Bayesians must assign prior probabilities to this total parameter space through a prior p.d.f. $h(\theta)$. They have developed procedures for assessing these prior probabilities, and we clearly cannot do justice to these methods here. Somehow, $h(\theta)$ reflects the prior weights that the Bayesian wants to assign to the various possible values of θ. In some instances, if $h(\theta)$ is a constant and thus θ has the uniform prior distribution, we say that the Bayesian has a *noninformative* prior. If, in fact, some knowledge of θ exists in advance of experimentation, noninformative priors should be avoided, if at all possible.

Also, in more practical examples, we usually take several observations, not just one. That is, we take a random sample and there is frequently a good statistic, say Y, for the parameter θ. The p.d.f. of Y, say $g(y; \theta)$, can be thought of as the conditional p.d.f. of Y, given θ; and henceforth in this section we write $g(y; \theta) = g(y \mid \theta)$. Thus we can treat

$$g(y \mid \theta) h(\theta) = k(y, \theta)$$

as the joint p.d.f. of the statistic Y and the parameter. Of course, the marginal p.d.f. of Y is

$$k_1(y) = \int_{-\infty}^{\infty} h(\theta) g(y \mid \theta) \, d\theta.$$

Thus

$$\frac{k(y, \theta)}{k_1(y)} = \frac{g(y \mid \theta)h(\theta)}{k_1(y)} = k(\theta \mid y)$$

would serve as the conditional p.d.f. of the parameter, given $Y = y$. This is essentially Bayes' theorem, and $k(\theta \mid y)$ is called the *posterior p.d.f. of* θ, given $Y = y$.

Bayesians believe that everything that needs to be known about the parameter is summarized in this posterior p.d.f. $k(\theta \mid y)$. Suppose, for illustration, they were pressed into making a point estimate of the parameter θ. They would note that they would be guessing the value of a random variable, here θ, given its p.d.f. $k(\theta \mid y)$. There are many ways that this could be done: for illustrations, the mean, the median, or the mode of that distribution would be reasonable guesses. However, in the final analysis, the best guess would clearly depend upon the penalties for various errors created by incorrect guesses. For illustration, if we were penalized by taking the square of the error between the guess, say $w(y)$, and real value of the parameter θ, clearly we would use the conditional mean

$$w(y) = \int_{-\infty}^{\infty} \theta k(\theta \mid y)\, d\theta$$

as our Bayes estimate of θ. The reason for this is that, in general, if Z is a random variable, then the function of b, $E[(Z - b)^2]$, is minimized by $b = E(Z)$ (see Example 3.2.4). Similarly, if the penalty (loss) function is the absolute value of the error, $|R - w(y)|$, then we use the median of the distribution as, with any random variable Z, $E[|Z - b|]$ is minimized when b equals the median of the distribution of Z (see Exercise 3.2.10).

Example 12.2.2

Suppose that Y has a binomial distribution with parameters n and $p = \theta$. Thus the p.d.f. of Y, given θ, is

$$g(y \mid \theta) = \binom{n}{y} \theta^y (1 - \theta)^{n-y}, \qquad y = 0, 1, 2, \ldots, n.$$

Let us take the prior p.d.f. of the parameter to be the beta p.d.f.

$$h(\theta) = \frac{\Gamma(\alpha + \beta)}{\Gamma(\alpha)\Gamma(\beta)}, \qquad 0 < \theta < 1.$$

Such a prior p.d.f. provides a Bayesian a great deal of flexibility through the selection of the parameters α and β. Thus the joint p.d.f. is a product of a binomial p.d.f. with parameters n and θ and this beta p.d.f., namely,

$$k(y, \theta) = \binom{n}{y} \frac{\Gamma(\alpha + \beta)}{\Gamma(\alpha)\Gamma(\beta)} \theta^{y+\alpha-1} (1 - \theta)^{n-y+\beta-1},$$

on the support given by $y = 0, 1, 2, \ldots, n$ and $0 < \theta < 1$. We find

$$
\begin{aligned}
k_1(y) &= \int_0^1 k(y, \theta) \\
&= \binom{n}{y} \frac{\Gamma(\alpha + \beta)}{\Gamma(\alpha)\Gamma(\beta)} \frac{\Gamma(\alpha + y)\Gamma(n + \beta - y)}{\Gamma(n + \alpha + \beta)}
\end{aligned}
$$

on the support $y = 0, 1, 2, \ldots, n$ by comparing the integral to one involving a beta p.d.f. with parameters $y + \alpha$ and $n - y + \beta$. Therefore,

$$
\begin{aligned}
k(\theta \mid y) &= \frac{k(y, \theta)}{k_1(y)} \\
&= \frac{\Gamma(n + \alpha + \beta)}{\Gamma(\alpha + y)\Gamma(n + \beta - y)} \theta^{y+\alpha-1}(1 - \theta)^{n-y+\beta-1}, \qquad 0 < \theta < 1,
\end{aligned}
$$

which is a beta p.d.f. with parameters $y + \alpha$ and $n - y + \beta$. With the square error loss function we must minimize, with respect to $w(y)$, the integral

$$
\int_0^1 [\theta - w(y)]^2 \, k(\theta \mid y) \, d\theta
$$

to obtain the Bayes solution. But, as noted earlier, if Z is a random variable with a second moment, $E[(Z - b)^2]$ is minimized by $b = E(Z)$. In the preceding display, θ is like the Z with p.d.f. $k(\theta \mid y)$, and $w(y)$ is like the b, so the minimization is accomplished by taking

$$
w(y) = E(\theta \mid y) = \frac{\alpha + y}{\alpha + \beta + n},
$$

which is the mean of the beta distribution with parameters $y + \alpha$ and $n - y + \beta$ (see Exercise 11.6.14). It is very instructive to note that this Bayes solution can be written as

$$
w(y) = \left(\frac{n}{\alpha + \beta + n} \right)\left(\frac{y}{n} \right) + \left(\frac{\alpha + \beta}{\alpha + \beta + n} \right)\left(\frac{\alpha}{\alpha + \beta} \right),
$$

which is a weighted average of the maximum likelihood estimate y/n of θ and the mean $\alpha/(\alpha+\beta)$ of the prior p.d.f. of the parameter. Moreover, the respective weights are $n/(\alpha + \beta + n)$ and $(\alpha + \beta)/(\alpha + \beta + n)$. Thus we see that α and β should be selected so that not only is $\alpha/(\alpha+\beta)$ the desired prior mean, but the sum $\alpha + \beta$ also plays a role corresponding to a sample size. That is, if we want our prior opinion to have as much weight as a sample size of 20, we would take $\alpha + \beta = 20$. So if our prior mean is 3/4, we have that α and β are selected such that $\alpha = 15$ and $\beta = 5$.   ◇

In Example 12.2.2 it is extremely convenient to note that it is not really necessary to determine $k_1(y)$ to find $k(\theta \mid y)$. If we divide $k(y, \theta)$ by $k_1(y)$, we get the product of a factor, which depends on y but does *not* depend on θ, say $c(y)$, and

$$
\theta^{y+\alpha-1}(1 - \theta)^{n-y+\beta-1}.
$$

That is,

$$k(\theta \mid y) = c(y)\,\theta^{y+\alpha-1}(1-\theta)^{n-y+\beta-1}, \qquad 0 < \theta < 1.$$

However, $c(y)$ must be that "constant" needed to make $k(\theta \mid y)$ a p.d.f., namely

$$c(y) = \frac{\Gamma(n+\alpha+\beta)}{\Gamma(y+\alpha)\Gamma(n-y+\beta)}.$$

Accordingly, Bayesians frequently write that $k(\theta \mid y)$ is proportional to $k(y, \theta) = g(y \mid \theta)h(\theta)$; that is,

$$k(\theta \mid y) \propto g(y \mid \theta)\,h(\theta).$$

Then, to actually form the p.d.f. $k(\theta \mid y)$, they simply find a "constant," which is some function of y, so that the expression integrates to one.

Example 12.2.3

Suppose that $Y = \overline{X}$ is the mean of a random sample of size n that arises from the normal distribution $N(\theta, \sigma^2)$, where σ^2 is known. Then $g(y \mid \theta)$ is $N(\theta, \sigma^2/n)$. Further suppose that we are able to assign prior weights to θ through a prior p.d.f. $h(\theta)$, which is $N(\theta_0, \sigma_0^2)$. Then we have that

$$k(\theta \mid y) \propto \frac{1}{\sqrt{2\pi}\,(\sigma/\sqrt{n})}\frac{1}{\sqrt{2\pi}\,\sigma_0}\exp\left[-\frac{(y-\theta)^2}{2(\sigma^2/n)} - \frac{(\theta-\theta_0)^2}{2\sigma_0^2}\right].$$

If we eliminate all constant factors (including factors involving y only), then we have

$$k(\theta \mid y) \propto \exp\left[-\frac{(\sigma_0^2 + \sigma^2/n)\theta^2 - 2(y\sigma_0^2 + \theta_0\sigma^2/n)\theta}{2(\sigma^2/n)\sigma_0^2}\right].$$

This can be simplified, by completing the square, to read (after eliminating factors not involving θ)

$$k(\theta \mid y) \propto \exp\left\{-\frac{[\theta - (y\sigma_0^2 + \theta_0\sigma^2/n)/(\sigma_0^2 + \sigma^2/n)]^2}{[2(\sigma^2/n)\sigma_0^2]/[\sigma_0^2 + (\sigma^2/n)]}\right\}.$$

That is, the posterior p.d.f. of the parameter is obviously normal with mean

$$\frac{y\sigma_0^2 + \theta_0\sigma^2/n}{\sigma_0^2 + \sigma^2/n} = \left(\frac{\sigma_0^2}{\sigma_0^2 + \sigma^2/n}\right)y + \left(\frac{\sigma^2/n}{\sigma_0^2 + \sigma^2/n}\right)\theta_0$$

and variance $(\sigma^2/n)\sigma_0^2/(\sigma_0^2 + \sigma^2/n)$. If the square error loss function is used, then this posterior mean is the Bayes solution. Again note that it is a weighted average of the maximum likelihood estimate $y = \overline{x}$ and the prior mean θ_0. The Bayes solution $w(y)$ will always be a value between the prior judgment and the usual estimate. Also, note here and in Example 12.2.2 that the Bayes solution gets closer to the maximum likelihood estimate as n increases. Thus the Bayesian procedures permit the decision maker to enter his or her prior opinions into the solution in a very formal way so that the influence of those prior notions will be less and less as n increases.   ◇

In Bayesian statistics, all the information is contained in the posterior p.d.f. $k(\theta \mid y)$. In Examples 12.2.2 and 12.2.3 we found Bayesian point estimates using the square error loss function. It should be noted that if the loss function is $|w(y) - \theta|$, the absolute value of the error, then the Bayes solution would be the median of the posterior distribution of the parameter, which is given by $k(\theta \mid y)$. Hence the Bayes solution changes, as it *should*, with different loss functions.

Finally, if an interval estimate of θ is desired, we would find two functions of y, say $u(y)$ and $v(y)$, such that

$$\int_{u(y)}^{v(y)} k(\theta \mid y)\, d\theta = 1 - \alpha,$$

where α is small, say $\alpha = 0.05$. Then the observed interval $u(y)$ to $u(y)$ would serve as an interval estimate for the parameter in the sense that the posterior probability of the parameter being in that interval is $1 - \alpha$. In Example 12.2.3, where the posterior p.d.f. of the parameter was normal, the interval

$$\frac{y\sigma_0^2 + \theta_0 \sigma^2/n}{\sigma_0^2 + \sigma^2/n} \pm 1.96 \sqrt{\frac{(\sigma^2/n)\sigma_0^2}{\sigma_0^2 + \sigma^2/n}}$$

serves as an interval estimate for θ with posterior probability of 0.95.

In closing this short section on Bayesian estimation, it should be noted that we could begin with the sample observations, $X_1, X_2, \ldots, X_n$, rather than some statistic Y. Then, in our discussion, we would replace $g(y \mid \theta)$ by the likelihood function

$$L(\theta) = f(x_1 \mid \theta) f(x_2 \mid \theta) \cdots f(x_n \mid \theta),$$

which is the joint p.d.f. of $X_1, X_2, \ldots, X_n$, given θ. Thus we find that

$$k(\theta \mid x_1, x_2, \ldots, x_n) \propto h(\theta) f(x_1 \mid \theta) f(x_2 \mid \theta) \cdots f(x_n \mid \theta) = h(\theta) L(\theta).$$

Now $k(\theta \mid x_1, x_2, \ldots, x_n)$ contains all the information about θ, given the data. Depending on the loss function we would choose our Bayesian estimate of θ as some characteristic of this posterior distribution, like the mean or the median. It is interesting to observe that if the loss function is zero for some small neighborhood about the true parameter θ and is some large positive constant otherwise, then the Bayesian estimate, $w(x_1, x_2, \ldots, x_n)$, is essentially the mode of this conditional p.d.f., $k(\theta \mid x_1, x_2, \ldots, x_n)$. The reason for this is that we want to take the estimate, $w(x_1, x_2, \ldots, x_n)$, so that it has as much posterior probability as possible in a small neighborhood around it. Finally, note that if $h(\theta)$ if a constant (noninformative prior), then this Bayesian estimate using the mode is exactly the same as the maximum likelihood estimate. More generally, if $h(\theta)$ is not a constant, the Bayesian estimate using the mode can be thought of as a weighted maximum likelihood estimate, in which the weights reflect the Bayesian's prior opinion about θ. That is, that value of θ which maximizes $h(\theta) L(\theta)$ is the mode of the posterior distribution of the parameter given the data and can be used as the Bayesian estimate associated with the appropriate loss function.

EXERCISES

12.2.1 Let $X_1, X_2, \ldots, X_n$ be a random sample from a distribution $b(1, \theta)$. Let the prior p.d.f. of θ be a beta one with parameters α and β. Show that the posterior p.d.f.

$$k(\theta \mid x_1, x_2, \ldots, x_n)$$

is exactly the same as $k(\theta \mid y)$ given in Example 12.2.2. This demonstrates that we often get exactly the same result whether we begin with a good statistic or with the sample observations. HINT: Note that $k(\theta \mid x_1, x_2, \ldots, x_n)$ is proportional to the product of the joint p.d.f. of $X_1, X_2, \ldots, X_n$ and the prior p.d.f. of θ.

12.2.2 Let Y be the sum of the observations of a random sample from a Poisson distribution with mean θ. Let the prior p.d.f. of θ be a gamma one with parameters α and β.

(a) Find the posterior p.d.f. of θ, given $Y = y$.

(b) If the loss function is $[w(y) - \theta]^2$, find the Bayesian point estimate $w(y)$.

(c) Show that this $w(y)$ is a weighted average of the maximum likelihood estimate y/n and the prior mean $\alpha\beta$, with respective weights of $n/(n + 1/\beta)$ and $(1/\beta)/(n + 1/\beta)$.

12.2.3 In Example 12.2.2 take $n = 30$, $\alpha = 15$, and $\beta = 5$.

(a) Using the square error loss, compute the expected loss (risk function) associated with the Bayes solution $w(Y)$.

(b) The risk function associated with the usual estimator Y/n is, of course, $\theta(1 - \theta)/30$. Find those values of θ for which the risk function in part (a) is less than $\theta(1 - \theta)/30$. In particular, if the prior mean $\alpha/(\alpha + \beta) = 3/4$ is a reasonable guess, then the risk function in part (a) is the better of the two for what values of θ (i.e., it is smaller in a neighborhood of $\theta = 3/4$)?

12.2.4 Let the conditional p.d.f. of X, given θ, be $N(0, \sigma^2 = 1/\theta)$. Assume that the unknown θ is a value of a random variable that has a gamma distribution with parameters $\alpha = r/2$ and $\theta = 2/r$ (here this θ is that associated with the gamma distribution, not the θ in $\sigma^2 = 1/\theta$). Show that the marginal p.d.f. of X is that of a t with r degrees of freedom. This procedure is often called compounding and is frequently used by Bayesians in introducing the t-distribution, as well as other distributions.

12.3 ASYMPTOTIC DISTRIBUTIONS OF MAXIMUM LIKELIHOOD ESTIMATORS

Let us consider a distribution with p.d.f. $f(x; \theta)$ such that the parameter θ is not involved in the support of the distribution. Moreover, we want $f(x; \theta)$ to enjoy a number of mathematical properties that we do not list here. However, in particular, we want to be able to find the maximum likelihood estimator $\widehat{\theta}$ by solving

$$\frac{\partial[\ln L(\theta)]}{\partial \theta} = 0,$$

where here we use a partial derivative sign because $L(\theta)$ involves $x_1, x_2, \ldots, x_n$, too.

That is,

$$\frac{\partial[\ln L(\widehat{\theta})]}{\partial\theta} = 0,$$

where now, with $\widehat{\theta}$ in this expression, $L(\widehat{\theta}) = f(X_1; \widehat{\theta})f(X_2; \widehat{\theta})\cdots f(X_n; \widehat{\theta})$. We can approximate the left-hand member of this latter equation by a linear function found from the first two terms of a Taylor's series expanded about θ, namely

$$\frac{\partial[\ln L(\theta)]}{\partial\theta} + (\widehat{\theta} - \theta)\frac{\partial^2[\ln L(\theta)]}{\partial\theta^2} \approx 0,$$

when $L(\theta) = f(X_1; \theta)f(X_2; \theta)\cdots f(X_n; \theta)$.

Obviously, this approximation is good enough only if $\widehat{\theta}$ is close to θ, and an adequate mathematical proof involves those conditions that we have not given here. But a heuristic argument can be made by solving for $\widehat{\theta} - \theta$ to obtain

$$\widehat{\theta} - \theta = \frac{\dfrac{\partial[\ln L(\theta)]}{\partial\theta}}{-\dfrac{\partial^2[\ln L(\theta)]}{\partial\theta^2}}. \tag{12.1}$$

Recall that

$$\ln L(\theta) = \ln f(X_1; \theta) + \ln f(X_2; \theta) + \cdots + \ln f(X_n; \theta)$$

and

$$\frac{\partial \ln L(\theta)}{\partial\theta} = \sum_{i=1}^{n} \frac{\partial[\ln f(X_i; \theta)]}{\partial\theta}, \tag{12.2}$$

which is the numerator of Equation 12.1. However, Expression 12.2 is the sum of the n independent and identically distributed random variables

$$Y_i = \frac{\partial[\ln f(X_i; \theta)]}{\partial\theta}, \qquad i = 1, 2, \ldots, n;$$

and thus, by the central limit theorem, has an approximate normal distribution with mean (in the continuous case) equal to

$$\int_{-\infty}^{\infty} \frac{\partial[\ln f(x; \theta)]}{\partial\theta} f(x; \theta)\, dx = \int_{-\infty}^{\infty} \frac{\partial[f(x; \theta)]}{\partial\theta} \frac{f(x; \theta)}{f(x; \theta)}\, dx$$

$$= \int_{-\infty}^{\infty} \frac{\partial[f(x; \theta)]}{\partial\theta}\, dx$$

$$= \frac{\partial}{\partial\theta}\left[\int_{-\infty}^{\infty} f(x; \theta)\, dx\right]$$

$$= \frac{\partial}{\partial\theta}[1]$$

$$= 0.$$

Clearly, we need a certain mathematical condition so that it is permissible to interchange the operations of integration and differentiation in those last steps. Of course, the integral of $f(x; \theta)$ is equal to one because it is a p.d.f.

Since we now know that the mean of each Y is

$$\int_{-\infty}^{\infty} \frac{\partial[\ln f(x; \theta)]}{\partial \theta} f(x; \theta)\, dx = 0,$$

let us take derivatives of each member of this equation with respect to θ, obtaining

$$\int_{-\infty}^{\infty} \left\{ \frac{\partial^2[\ln f(x; \theta)]}{\partial \theta^2} f(x; \theta) + \frac{\partial[\ln f(x; \theta)]}{\partial \theta} \frac{\partial[f(x; \theta)]}{\partial \theta} \right\} dx = 0.$$

However,

$$\frac{\partial[f(x; \theta)]}{\partial \theta} = \frac{\partial[\ln f(x; \theta)]}{\partial \theta} f(x; \theta);$$

so

$$\int_{-\infty}^{\infty} \left\{ \frac{\partial[\ln f(x; \theta)]}{\partial \theta} \right\}^2 f(x; \theta)\, dx = -\int_{-\infty}^{\infty} \frac{\partial^2[\ln f(x; \theta)]}{\partial \theta^2} f(x; \theta)\, dx.$$

Since $E(Y) = 0$, this last expression provides the variance of $Y = \partial[\ln f(X; \theta)]/\partial \theta$. Then the variance of Expression 12.2 is n times this value, namely

$$-nE\left\{ \frac{\partial^2[\ln f(X; \theta)]}{\partial \theta^2} \right\}.$$

Let us rewrite Equation 12.1 as

$$\frac{\sqrt{n}(\hat{\theta} - \theta)}{\dfrac{1}{\sqrt{-E\{\partial^2[\ln f(X; \theta)]/\partial\theta^2\}}}} = \frac{\dfrac{\partial[\ln L(\theta)]/\partial\theta}{\sqrt{-nE\{\partial^2[\ln f(X; \theta)]/\partial\theta^2\}}}}{\dfrac{\dfrac{1}{n}\dfrac{\partial^2[\ln L(\theta)]}{\partial\theta^2}}{E\{-\partial^2[\ln f(X; \theta)]/\partial\theta^2\}}}. \qquad (12.3)$$

The numerator of Equation 12.3 has an approximate $N(0, 1)$ distribution; and those unstated mathematical conditions require, in some sense, for

$$-\frac{1}{n}\frac{\partial^2[\ln L(\theta)]}{\partial\theta^2} \qquad \text{to converge to} \qquad E\{-\partial^2[\ln f(X; \theta)]/\partial\theta^2\}.$$

Accordingly, the ratios given in Equation 12.3 must be approximately $N(0, 1)$. That is, $\hat{\theta}$ has an approximate normal distribution with mean θ and standard deviation

$$\frac{1}{\sqrt{-nE\{\partial^2[\ln f(X; \theta)]/\partial\theta^2\}}}.$$

Example 12.3.1

(Continuation of Example 6.1.1.) With the underlying exponential p.d.f.

$$f(x; \theta) = \frac{1}{\theta} e^{-x/\theta}, \qquad 0 < x < \infty, \qquad \theta \in \Omega = \{\theta : 0 < \theta < \infty\},$$

$\overline{X}$ is the maximum likelihood estimator. Since

$$\ln f(x; \theta) = -\ln \theta - \frac{x}{\theta}$$

and

$$\frac{\partial[\ln f(x; \theta)]}{\partial \theta} = -\frac{1}{\theta} + \frac{x}{\theta^2} \quad \text{and} \quad \frac{\partial^2[\ln f(x; \theta)]}{\partial \theta} = \frac{1}{\theta^2} - \frac{2x}{\theta^3},$$

we have

$$-E\left[\frac{1}{\theta^2} - \frac{2X}{\theta^3}\right] = -\frac{1}{\theta^2} + \frac{2\theta}{\theta^3} = \frac{1}{\theta^2}$$

because $E(X) = \theta$. That is, $\overline{X}$ has an approximate normal distribution with mean θ and standard deviation $\theta/\sqrt{n}$. Thus the random interval $\overline{X} \pm 1.96(\theta/\sqrt{n})$ has an approximate probability of 0.95 for covering θ. Substituting the observed $\overline{x}$ for θ, as well as for $\overline{X}$, we say that $\overline{x} \pm 1.96\,\overline{x}/\sqrt{n}$ is an approximate 95% confidence interval for θ.   ◇

Example 12.3.2

(Continuation of Exercise 6.1.3.) The maximum likelihood estimator for λ in

$$f(x; \lambda) = \frac{\lambda^x e^{-\lambda}}{x!}, \qquad x = 0, 1, 2, \ldots; \qquad \lambda \in \Omega = \{\lambda : 0 < \lambda < \infty\},$$

is $\widehat{\lambda} = \overline{X}$. Now

$$\ln f(x; \lambda) = x \ln \lambda - \lambda - \ln x!$$

and

$$\frac{\partial[\ln f(x; \lambda)]}{\partial \lambda} = \frac{x}{\lambda} - 1 \quad \text{and} \quad \frac{\partial^2[\ln f(x; \lambda)]}{\partial \lambda^2} = -\frac{x}{\lambda^2}.$$

Thus

$$-E\left(-\frac{X}{\lambda^2}\right) = \frac{\lambda}{\lambda^2} = \frac{1}{\lambda}$$

and $\widehat{\lambda} = \overline{X}$ has an approximate normal distribution with mean λ and standard deviation $\sqrt{\lambda/n}$. Finally $\overline{x} \pm 1.645\sqrt{\overline{x}/n}$ serves as an approximate 90% confidence interval for λ. With the data in Exercise 6.1.3, $\overline{x} = 2.225$ and hence this interval is from 1.887 to 2.563.   ◇

It is interesting that there is another theorem which is somewhat related to the preceding result in that the variance of $\widehat{\theta}$ serves as a lower bound for the variance of every unbiased estimator of θ. Thus we know that if a certain unbiased estimator has a variance equal to that lower bound, we cannot find a better one and hence it is the best in the sense of being the unbiased minimum variance estimator. So, in the limit, the maximum likelihood estimator is this type of best estimator.

We describe the **Rao-Cramér inequality** without proof. Let $X_1, X_2, \ldots, X_n$ be a random sample from a distribution with p.d.f. $f(x; \theta)$, $\theta \in \Omega = \{\theta : c < \theta < d\}$, where the support of X does not depend upon θ so that we can differentiate, with respect to θ, under integral signs like that in the following integral:

$$\int_{-\infty}^{\infty} f(x; \theta)\, dx = 1.$$

If $Y = u(X_1, X_2, \ldots, X_n)$ is an unbiased estimator of θ, then

$$\mathrm{Var}(Y) \geq \frac{1}{n \int_{-\infty}^{\infty} \{[\partial \ln f(x; \theta)/\partial \theta]\}^2 f(x; \theta)\, dx}$$

$$= \frac{-1}{n \int_{-\infty}^{\infty} [\partial^2 \ln f(x; \theta)/\partial \theta^2] f(x; \theta)\, dx}.$$

Note that the two integrals in the respective denominators are the expectations

$$E\left\{ \left[\frac{\partial \ln f(X; \theta)}{\partial \theta} \right]^2 \right\} \quad \text{and} \quad E\left[\frac{\partial^2 \ln f(X; \theta)}{\partial \theta^2} \right];$$

sometimes one is easier to compute than the other.

Note that we have computed this lower bound for each of two distributions: exponential and Poisson. Those respective lower bounds were θ^2/n and λ/n (see Examples 12.3.1 and 12.3.2). Since in each case, the variance of $\overline{X}$ equals the lower bound, then $\overline{X}$ is the unbiased minimum variance estimator.

Let us consider another example.

Example 12.3.3

(Continuation of Exercise 6.1.7.) Let the p.d.f. of X be given by

$$f(x; \theta) = \theta x^{\theta - 1}, \quad 0 < x < 1, \quad \theta \in \Omega = \{\theta : 0 < \theta < \infty\}.$$

We then have

$$\ln f(x; \theta) = \ln \theta + (\theta - 1) \ln x,$$

$$\frac{\partial \ln f(x; \theta)}{\partial \theta} = \frac{1}{\theta} + \ln x,$$

and

$$\frac{\partial^2 \ln f(x; \theta)}{\partial \theta^2} = -\frac{1}{\theta^2}.$$

Since $E(-1/\theta^2) = -1/\theta^2$, the lower bound of the variance of every unbiased estimator of θ is θ^2/n. Moreover, the maximum likelihood estimator $\widehat{\theta} = -n/\ln \prod_{i=1}^n X_i$ has an approximate normal distribution with mean θ and variance θ^2/n. Thus, in a limiting sense, $\widehat{\theta}$ is the unbiased minimum variance estimator of θ.   ◇

To measure the value of estimators, their variances are compared to the Rao-Cramér lower bound. The ratio of the Rao-Cramér lower bound to the actual variance of any unbiased estimator is called the **efficiency** of that estimator. An estimator with efficiency of 50%, say, requires that $1/0.5 = 2$ times as many sample observations are needed to do as well in estimation as can be done with the unbiased minimum variance estimator (the 100% efficient estimator).

EXERCISES

12.3.1 Let $X_1, X_2, \ldots, X_n$ be a random sample from $N(\theta, \sigma^2)$, where σ^2 is known.
  (a) Show that $Y = (X_1 + X_2)/2$ is an unbiased estimator of θ.
  (b) Find the Rao-Cramér lower bound for the variance of an unbiased estimator of θ for a general n.
  (c) What is the efficiency of Y in part (a)?

12.3.2 Let $X_1, X_2, \ldots, X_n$ denote a random sample from $b(1, p)$. We know that $\overline{X}$ is an unbiased estimator of p and that $\mathrm{Var}(\overline{X}) = p(1 - p)/n$ (see Exercise 6.1.12).
  (a) Find the Rao-Cramér lower bound for the variance of every unbiased estimator of p.
  (b) What is the efficiency of $\overline{X}$ as an estimator of p?

12.3.3 (Continuation of Exercise 6.1.2.) When sampling from a normal distribution with known mean μ, $\widehat{\theta} = \sum_{i=1}^n (X_i - \mu)^2/n$ is the maximum likelihood estimator of $\theta = \sigma^2$.
  (a) Determine the Rao-Cramér lower bound.
  (b) What is the approximate distribution of $\widehat{\theta}$?
  (c) What is the exact distribution of $n\widehat{\theta}/\theta$, where $\theta = \sigma^2$?

12.3.4 Find the Rao-Cramér lower bound and thus the asymptotic variance of the maximum likelihood estimator $\widehat{\theta}$ if the random sample $X_1, X_2, \ldots, X_n$ is taken from each of the distributions having the following p.d.f.'s:
  (a) $f(x; \theta) = (1/\theta^2) x\, e^{-x/\theta}$, $0 < x < \infty$, $0 < \theta < \infty$.
  (b) $f(x; \theta) = (1/2\theta^3) x^2\, e^{-x/\theta}$, $0 < x < \infty$, $0 < \theta < \infty$.
  (c) $f(x; \theta) = (1/\theta) x^{(1-\theta)/\theta}$, $0 < x < 1$, $0 < \theta < \infty$.

12.4 CHEBYSHEV'S INEQUALITY AND CONVERGENCE IN PROBABILITY

In this section we use Chebyshev's inequality to show, in another sense, that the sample mean, $\overline{x}$, is a good statistic to use to estimate a population mean μ; the relative frequency of success in n Bernoulli trials, y/n, is a good statistic for estimating p; and the empirical distribution function, $F_n(x)$, can be used to estimate the theoretical distribution function $F(x)$. The effect of the sample size n on these estimates is discussed.

We begin by showing that Chebyshev's inequality gives added significance to the standard deviation in terms of bounding certain probabilities. The inequality is valid for all distributions for which the standard deviation exists. The proof is given for the discrete case, but it holds for the continuous case with integrals replacing summations

Theorem 12.4.1.   (Chebyshev's Inequality) If the random variable X has a mean μ and variance σ^2, then for every $k \geq 1$,

$$P(|X - \mu| \geq k\sigma) \leq \frac{1}{k^2}.$$

Proof.   Let $f(x)$ denote the p.d.f. of X. Then

$$\sigma^2 = E[(X - \mu)^2] = \sum_{x \in R} (x - \mu)^2 f(x)$$

$$= \sum_{x \in A} (x - \mu)^2 f(x) + \sum_{x \in A'} (x - \mu)^2 f(x), \qquad (12.4)$$

where

$$A = \{x : |x - \mu| \geq k\sigma\}.$$

The second term in the right-hand member of Equation 12.4 is the sum of nonnegative numbers and thus is greater than or equal to zero. Hence

$$\sigma^2 \geq \sum_{x \in A} (x - \mu)^2 f(x).$$

However, in A, $|x - \mu| \geq k\sigma$; so

$$\sigma^2 \geq \sum_{x \in A} (k\sigma)^2 f(x) = k^2 \sigma^2 \sum_{x \in A} f(x).$$

But the latter summation equals $P(X \in A)$, and thus

$$\sigma^2 \geq k^2 \sigma^2 P(X \in A) = k^2 \sigma^2 P(|X - \mu| \geq k\sigma).$$

That is,

$$P(|X - \mu| \geq k\sigma) \leq \frac{1}{k^2}. \qquad \square$$

Corollary 12.4.1.   If $\varepsilon = k\sigma$, then

$$P(|X - \mu| \geq \varepsilon) \leq \frac{\sigma^2}{\varepsilon^2}.$$

In words, Chebyshev's inequality states that the probability that X differs from its mean by at least k standard deviations is less than or equal to $1/k^2$. It follows that the probability that X differs from its mean by less than k standard deviations is at least $1 - 1/k^2$. That is,

$$P(|X - \mu| < k\sigma) \geq 1 - \frac{1}{k^2}.$$

From the corollary, it also follows that

$$P(|X - \mu| < \varepsilon) \geq 1 - \frac{\sigma^2}{\varepsilon^2}.$$

Thus Chebyshev's inequality can be used as a bound for certain probabilities. However, in many instances, the bound is not very close to the true probability.

Example 12.4.1

If it is known that X has a mean of 25 and a variance of 16, then, since $\sigma = 4$, a lower bound for $P(17 < X < 33)$ is given by

$$P(17 < X < 33) = P(|X - 25| < 8)$$

$$= P(|X - 25| < 2\sigma) \geq 1 - \frac{1}{4} = 0.75$$

and an upper bound for $P(|X - 25| \geq 12)$ is found to be

$$P(|X - 25| \geq 12) = P(|X - \mu| \geq 3\sigma) \leq \frac{1}{9}. \qquad \diamond$$

Note that the results of the preceding example hold for any distribution with mean 25 and standard deviation 4. But, even stronger, the probability that any random variable X differs from its mean by 3 or more standard deviations is at most $1/9$ by letting $k = 3$ in the theorem. Also the probability that any random variable X differs from its mean by less than 2 standard deviations is at least 3/4 by letting $k = 2$.

The following consideration partially indicates the value of Chebyshev's inequality in theoretical discussions. If Y is the number of successes in n Bernoulli trials with probability p of success on each trial, then Y is $b(n, p)$. Furthermore, Y/n gives the relative frequency of success and, when p is unknown, Y/n can be used as an estimate of p. To gain some insight into the closeness of Y/n to p, we shall use Chebyshev's inequality. With $\varepsilon > 0$, we note that

$$P\left(\left|\frac{Y}{n} - p\right| \geq \varepsilon\right) = P(|Y - np| \geq n\varepsilon)$$

$$= P\left(|Y - np| \geq \frac{\sqrt{n}\varepsilon}{\sqrt{pq}}\sqrt{npq}\right).$$

However, $\mu = np$ and $\sigma = \sqrt{npq}$ are the mean and the standard deviation of Y, so that, with $k = \sqrt{n}\varepsilon/\sqrt{pq}$, we have

$$P\left(\left|\frac{Y}{n} - p\right| \geq \varepsilon\right) = P(|Y - \mu| \geq k\sigma) \leq \frac{1}{k^2} = \frac{pq}{n\varepsilon^2}$$

or, equivalently,

$$P\left(\left|\frac{Y}{n} - p\right| < \varepsilon\right) \geq 1 - \frac{pq}{n\varepsilon^2}. \qquad (12.5)$$

When p is completely unknown, we can use the fact that $pq = p(1 - p)$ is a maximum when $p = 1/2$ in order to find a lower bound for the probability in Equation 12.5. That is,

$$1 - \frac{pq}{n\varepsilon^2} \geq 1 - \frac{(1/2)(1/2)}{n\varepsilon^2}.$$

For illustration, if $\varepsilon = 0.05$ and $n = 400$,

$$P\left(\left|\frac{Y}{400} - p\right| < 0.05\right) \geq 1 - \frac{(1/2)(1/2)}{400(0.0025)} = 0.75.$$

If, on the other hand, it is known that p is close to 1/10, we would have

$$P\left(\left|\frac{Y}{400} - p\right| < 0.05\right) \geq 1 - \frac{(0.1)(0.9)}{400(0.0025)} = 0.91.$$

Note that Chebyshev's inequality is applicable to all distributions with a finite variance, and thus the bound is not always a tight one; that is, the bound is not necessarily close to the true probability.

In general, however, it should be noted that with fixed $\varepsilon > 0$ and $0 < p < 1$, we have that

$$\lim_{n\to\infty} P\left(\left|\frac{Y}{400} - p\right| < \varepsilon\right) \geq \lim_{n\to\infty}\left(1 - \frac{pq}{n\varepsilon^2}\right) = 1.$$

But since the probability of every event is less than or equal to 1, it must be that

$$\lim_{n\to\infty} P\left(\left|\frac{Y}{400} - p\right| < \varepsilon\right) = 1.$$

That is, the probability that the relative frequency Y/n is within ε of p is close to 1 when n is large enough. This is one form of the **law of large numbers**, and we say Y/n **converges in probability** to p.

This theoretical result has something to contribute to our understanding of the properties of the empirical distribution function, $F_n(x)$. For a fixed x, $F_n(x)$ is the proportion of sample observations that are less than or equal to x. That is, if "success" is an observed X being less than or equal to the fixed x, then $F_n(x)$ is the relative frequency of success. The law of large numbers states that this converges, in that probabilistic sense, to the true probability $P(X \leq x)$ of success, namely $p = F(x)$. Thus $F_n(x)$ does approach $F(x)$ in this sense and hence provides an estimate of the distribution function. This property was used in Section 10.7.

A more general form of the law of large numbers is found by considering the mean $\overline{X}$ of a random sample from a distribution with mean μ and variance σ^2. This is more general because the relative frequency Y/n can be thought of as $\overline{X}$ when the sample arises from a Bernoulli distribution. We know that

$$E(\overline{X}) = \mu \quad \text{and} \quad \text{Var}(\overline{X}) = \frac{\sigma^2}{n}.$$

Thus, for every $\varepsilon > 0$,

$$P[\,|\overline{X} - \mu| \geq \varepsilon\,] = P\left[\,|\overline{X} - \mu| \geq \left(\frac{\varepsilon\sqrt{n}}{\sigma}\right)\left(\frac{\sigma}{\sqrt{n}}\right)\right].$$

But the standard deviation of $\overline{X}$ is $\sigma/\sqrt{n}$. Hence, using Chebyshev's inequality with $k = \varepsilon\sqrt{n}/\sigma$, we have

$$P[\,|\overline{X} - \mu| \geq \varepsilon\,] \leq \frac{\sigma^2}{\varepsilon^2 n}.$$

Since probability is nonnegative, it follows that

$$\lim_{n \to \infty} P(|\overline{X} - \mu| \geq \varepsilon) \leq \lim_{n \to \infty} \frac{\sigma^2}{\varepsilon^2 n} = 0.$$

This implies that

$$\lim_{n \to \infty} P(|\overline{X} - \mu| \geq \varepsilon) = 0,$$

or, equivalently,

$$\lim_{n \to \infty} P(|\overline{X} - \mu| < \varepsilon) = 1.$$

The preceding discussion shows that the probability associated with the distribution of $\overline{X}$ becomes concentrated in an arbitrarily small interval centered at μ as n increases. This is a more general form of the law of large numbers, and we say that $\overline{X}$ converges in probability to μ.

Although Chebyshev's inequality is quite useful in theoretical discussion, it also shows that in a collection of numbers, say $x_1, x_2, \ldots, x_n$, a certain proportion of them must be within $k\sqrt{v}$ of $\overline{x}$, where $\overline{x}$ and v are the respective mean and variance of the empirical distribution defined by the numbers. That is,

$$\overline{x} = \frac{1}{n} \sum_{i=1}^{n} x_i \qquad \text{and} \qquad v = \frac{1}{n} \sum_{i=1}^{n} (x_i - \overline{x})^2.$$

To show this, think of the empirical distribution as defining the probability distribution; thus $\overline{x}$ and v are the mean and the variance of this distribution. Hence, for every $k \geq 1$, we have

$$\frac{\mathcal{N}\{x_i : |x_i - \overline{x}| \geq k\sqrt{v}\}}{n} \leq \frac{1}{k^2}$$

or, equivalently,

$$\frac{\mathcal{N}\{x_i : |x_i - \overline{x}| < k\sqrt{v}\}}{n} \geq 1 - \frac{1}{k^2}.$$

In words, the second inequality says: "The proportion of the numbers $x_1, x_2, \ldots, x_n$ that lie within $k\sqrt{v}$ of the mean $\overline{x}$ is at least $1 - 1/k^2$."

EXERCISES

12.4.1 If X is a random variable with mean 33 and variance 16, use Chebyshev's inequality to find

   (a) A lower bound for $P(23 < X < 43)$.

   (b) An upper bound for $P(|X - 33| \geq 14)$.

12.4.2 If $E(X) = 17$ and $E(X^2) = 298$, use Chebyshev's inequality to determine

   (a) A lower bound for $P(10 < X < 24)$.

   (b) An upper bound for $P(|X - 17| \geq 16)$.

12.4.3 Let X denote the outcome when rolling a fair die. Then $\mu = 7/2$ and $\sigma^2 = 35/12$. Note that the maximum deviation of X from μ equals 5/2. Express this deviation in terms of number of standard deviations; that is, find k where $k\sigma = 5/2$. Then determine a lower bound for $P(|X - 3.5| < 2.5)$.

12.4.4 If Y is $b(n, 0.5)$, give a lower bound for $P(|Y/n - 0.5| < 0.08)$ when

   (a) $n = 100$.

   (b) $n = 500$.

   (c) $n = 1000$.

12.4.5 If Y is $b(n, 0.25)$, give a lower bound for $P(|Y/n - 0.25| < 0.05)$ when

   (a) $n = 100$.

   (b) $n = 500$.

   (c) $n = 1000$.

12.4.6 Let $\overline{X}$ be the mean of a random sample of size $n = 15$ from a distribution with mean $\mu = 80$ and variance $\sigma^2 = 60$. Use Chebyshev's inequality to find a lower bound for $P(75 < \overline{X} < 85)$.

12.4.7 The characteristics of the empirical distribution of test scores of 900 students are $\overline{x} = 83$ and $v = 36$, respectively. At least how many students received test scores between 71 and 95?

12.5 BEST CRITICAL REGIONS

In this section we consider the properties a satisfactory test (or critical region) should possess. To introduce our investigation, we begin with a nonstatistical example.

Example 12.5.1

Say that you have α dollars with which to buy books. Suppose further that you are not interested in the books themselves but only in filling as much of your bookshelves as possible. How do you decide which books to buy? Does the following approach seem reasonable? First of all, take all the available free books. Then start choosing those books for which the cost of filling an inch of bookshelf is smallest. That is, choose those books for which the ratio c/w is a minimum, where w is the width of the book in inches and c is the cost of the book. Continue choosing books this way until you have spent the α dollars.

$\diamond$

To see how Example 12.5.1 provides the background for selecting a good critical region of size α, let us consider a test of the simple hypothesis H_0: $\theta = \theta_0$ against a simple alternative hypothesis H_1: $\theta = \theta_1$. In this discussion we assume that the random variables $X_1, X_2, \ldots, X_n$ under consideration have a joint p.d.f. of the discrete type, which we here denote by $L(\theta; x_1, x_2, \ldots, x_n)$. That is,

$$P(X_1 = x_1, X_2 = x_2, \ldots, X_n = x_n) = L(\theta; x_1, x_2, \ldots, x_n).$$

A critical region C of size α is a set of points $(x_1, x_2, \ldots, x_n)$ with the probability of α when $\theta = \theta_0$. For a good test, this set C of points should have a large probability when $\theta = \theta_1$ because, under H_1: $\theta = \theta_1$, we wish to reject H_0: $\theta = \theta_0$. Accordingly, the first point we would place in the critical region C is the one with the smallest ratio

$$\frac{L(\theta_0; x_1, x_2, \ldots, x_n)}{L(\theta_1; x_1, x_2, \ldots, x_n)}.$$

That is, the "cost" in terms of probability under H_0: $\theta = \theta_0$ is small compared to the probability that we can "buy" if $\theta = \theta_1$. The next point to add to C would be the one with the next smallest ratio. We would continue to add points to C in this manner until the probability of C, under H_0: $\theta = \theta_0$, equals α. In this way, we have achieved, for the given significance level α, the region C with the largest probability when H_1: $\theta = \theta_1$ is true. We now formalize this discussion by defining a best critical region and proving the well-known Neyman-Pearson lemma.

Definition 12.5.1.   Consider the test of the simple null hypothesis H_0: $\theta = \theta_0$ against the simple alternative hypothesis H_1: $\theta = \theta_1$. Let C be a critical region of size α; that is, $\alpha = P(C; \theta_0)$. Then C is a **best critical region of size** α if, for every other critical region D of size $\alpha = P(D; \theta_0)$, we have that

$$P(C; \theta_1) \geq P(D; \theta_1).$$

That is, when H_1: $\theta = \theta_1$ is true, the probability of rejecting H_0: $\theta = \theta_0$ using the critical region C is at least as great as the corresponding probability using any other critical region D of size α.

Thus a best critical region of size α is the critical region that has the greatest power among all critical regions of size α. The Neyman-Pearson lemma gives sufficient conditions for a best critical region of size α.

Theorem 12.5.1.   (Neyman-Pearson Lemma) Let $X_1, X_2, \ldots, X_n$ be a random sample of size n from a distribution with p.d.f. $f(x; \theta)$, where θ_0 and θ_1 are two possible values of θ. Denote the joint p.d.f. of $X_1, X_2, \ldots, X_n$ by the likelihood function

$$L(\theta) = L(\theta; x_1, x_2, \ldots, x_n) = f(x_1; \theta)f(x_2; \theta) \cdots f(x_n; \theta).$$

If there exist a positive constant k and a subset C of the sample space such that

(a) $P[(X_1, X_2, \ldots, X_n) \in C; \theta_0] = \alpha$,

(b) $\dfrac{L(\theta_0)}{L(\theta_1)} \le k$ for $(x_1, x_2, \ldots, x_n) \in C$,

(c) $\dfrac{L(\theta_0)}{L(\theta_1)} \ge k$ for $(x_1, x_2, \ldots, x_n) \in C'$,

then C is a best critical region of size α for testing the simple null hypothesis $H_0: \theta = \theta_0$ against the simple alternative hypothesis $H_1: \theta = \theta_1$.

Proof.   We prove the theorem when the random variables are of the continuous type; for discrete-type random variables, replace the integral signs by summation signs. To simplify the exposition, we shall use the following notation:

$$\int_B L(\theta) = \int \cdots \int_B L(\theta; x_1, x_2, \ldots, x_n)\, dx_1\, dx_2 \cdots dx_n.$$

Assume that there exists another critical region of size α, say D such that, in this new notation,

$$\alpha = \int_C L(\theta_0) = \int_D L(\theta_0).$$

So we have

$$0 = \int_C L(\theta_0) - \int_D L(\theta_0)$$

$$= \int_{C \cap D'} L(\theta_0) + \int_{C \cap D} L(\theta_0) - \int_{C \cap D} L(\theta_0) - \int_{C' \cap D} L(\theta_0)$$

and hence

$$0 = \int_{C \cap D'} L(\theta_0) - \int_{C' \cap D} L(\theta_0).$$

By hypothesis (b), $kL(\theta_1) \ge L(\theta_0)$ at each point in C and therefore in $C \cap D'$; thus

$$k \int_{C \cap D'} L(\theta_1) \ge \int_{C \cap D'} L(\theta_0).$$

By hypothesis (c), $kL(\theta_1) \le L(\theta_0)$ at each point in C', and therefore in $C' \cap D$; thus we obtain

$$k \int_{C' \cap D} L(\theta_1) \le \int_{C' \cap D} L(\theta_0).$$

Therefore,

$$0 = \int_{C \cap D'} L(\theta_0) - \int_{C' \cap D} L(\theta_0) \le (k) \left\{ \int_{C \cap D'} L(\theta_1) - \int_{C' \cap D} L(\theta_1) \right\}.$$

That is,

$$0 \leq (k)\left\{\int_{C\cap D'} L(\theta_1) + \int_{C\cap D} L(\theta_1) - \int_{C\cap D} L(\theta_1) - \int_{C'\cap D} L(\theta_1)\right\}$$

or, equivalently,

$$0 \leq (k)\left\{\int_C L(\theta_1) - \int_D L(\theta_1)\right\}.$$

Thus

$$\int_C L(\theta_1) \geq \int_D L(\theta_1);$$

that is, $P(C; \theta_1) \geq P(D; \theta_1)$. Since that is true for every critical region D of size α, C is a best critical region of size α.   □

For a realistic application of the Neyman-Pearson lemma, consider the following, in which the test is based on a random sample from a normal distribution.

Example 12.5.2

Let $X_1, X_2, \ldots, X_n$ be a random sample from the normal distribution $N(\mu, 36)$. We shall find the best critical region for testing the simple hypothesis H_0: $\mu = 50$ against the simple alternative hypothesis H_1: $\mu = 55$. Using the ratio of the likelihood functions, namely $L(50)/L(55)$, we shall find those points in the sample space for which this ratio is less than or equal to some constant k. That is, we shall solve the following inequality:

$$\frac{L(50)}{L(55)} = \frac{(72\pi)^{-n/2} \exp\left[-\left(\frac{1}{72}\right)\sum_{i=1}^n (x_i - 50)^2\right]}{(72\pi)^{-n/2} \exp\left[-\left(\frac{1}{72}\right)\sum_{i=1}^n (x_i - 55)^2\right]}$$

$$= \exp\left[-\left(\frac{1}{72}\right)\left(10\sum_{i=1}^n x_i + n50^2 - n55^2\right)\right] \leq k.$$

If we take the natural logarithm of each member of the inequality, we find that

$$-10\sum_{i=1}^n x_i - n50^2 + n55^2 \leq (72)\ln k.$$

Thus

$$\frac{1}{n}\sum_{i=1}^n x_i \geq -\frac{1}{10n}[n50^2 - n55^2 + (72)\ln k]$$

or equivalently,

$$\bar{x} \geq c,$$

where $c = -(1/10n)[n50^2 - n55^2 + (72) \ln k]$. Thus $L(50)/L(55) \leq k$ is equivalent to $\bar{x} \geq c$. A best critical region is, according to the Neyman-Pearson lemma,

$$C = \{(x_1, x_2, \ldots, x_n) : \bar{x} \geq c\},$$

where c is selected so that the size of the critical region is α. Say $n = 16$ and $c = 53$. Since $\overline{X}$ is $N(50, 36/16)$ under H_0, we have

$$\alpha = P(\overline{X} \geq 53; \mu = 50)$$

$$= P\left(\frac{\overline{X} - 50}{6/4} \geq \frac{3}{6/4}; \mu = 50\right) = 1 - \Phi(2) = 0.0228. \qquad \diamond$$

This example illustrates what is often true, namely, that the inequality

$$L(\theta_0)/L(\theta_1) \leq k$$

can be expressed in terms of a function $u(x_1, x_2, \ldots, x_n)$, say

$$u(x_1, \ldots, x_n) \leq c_1$$

or

$$u(x_1, \ldots, x_n) \geq c_2,$$

where c_1 or c_2 is selected so that the size of the critical region is α. Thus the test can be based on the statistic $u(x_1, \ldots, x_n)$. Also, for illustration, if we want α to be a given value, say 0.05, we could then choose our c_1 or c_2. In Example 12.5.2, with $\alpha = 0.05$, we want

$$0.05 = P(\overline{X} \geq c; \mu = 50)$$

$$= P\left(\frac{\overline{X} - 50}{6/4} \geq \frac{c - 50}{6/4}; \mu = 50\right) = 1 - \Phi\left(\frac{c - 50}{6/4}\right).$$

Hence it must be true that $(c - 50)/(3/2) = 1.645$, or equivalently,

$$c = 50 + \frac{3}{2}(1.645) \approx 52.47.$$

Example 12.5.3

Let $X_1, X_2, \ldots, X_n$ denote a random sample of size n from a Poisson distribution with mean λ. A best critical region for testing H_0: $\lambda = 2$ against H_1: $\lambda = 5$ is given by

$$\frac{L(2)}{L(5)} = \frac{2^{\Sigma x_i} e^{-2n}}{x_1! x_2! \cdots x_n!} \frac{x_1! x_2! \cdots x_n!}{5^{\Sigma x_i} e^{-5n}} \leq k.$$

This inequality is equivalent to

$$\left(\frac{2}{5}\right)^{\Sigma x_i} e^{3n} \leq k \qquad \text{and} \qquad (\Sigma x_i) \ln\left(\frac{2}{5}\right) + 3n \leq \ln k.$$

Since $\ln(2/5) < 0$, this is the same as

$$\sum_{i=1}^{n} x_i \geq \frac{\ln k - 3n}{\ln(2/5)} = c.$$

If $n = 4$ and $c = 13$, then

$$\alpha = P\left(\sum_{i=1}^{4} X_i \geq 13; \lambda = 2\right) = 1 - 0.936 = 0.064,$$

from Table III, since $\sum_{i=1}^{4} X_i$ has a Poisson distribution with mean 8 when $\lambda = 2$. ◇

When H_0: $\theta = \theta_0$ and H_1: $\theta = \theta_1$ are both simple hypotheses, a critical region of size α is a best critical region if the probability of rejecting H_0 when H_1 is true is a maximum when compared with all other critical regions of size α. The test using the best critical region is called a **most powerful test** because it has the greatest value of the power function at $\theta = \theta_1$ when compared with that of other tests of significance level α. If H_1 is a composite hypothesis, the power of a test depends on each simple alternative in H_1.

Definition 12.5.2.   A test, defined by a critical region C of size α, is a **uniformly most powerful test** if it is a most powerful test against each simple alternative in H_1. The critical region C is called a **uniformly most powerful critical region of size** α.

Let us reconsider Example 12.5.2 when the alternative hypothesis is composite.

Example 12.5.4

Let $X_1, \ldots, X_n$ be a random sample from $N(\mu, 36)$. We have seen that when testing H_0: $\mu = 50$ against H_1: $\mu = 55$, a best critical region C is defined by $C = \{(x_1, x_2, \ldots, x_n): \bar{x} \geq c\}$, where c is selected so that the significance level is α. Now consider testing H_0: $\mu = 50$ against the one-sided composite alternative hypothesis H_1: $\mu > 50$. For each simple hypothesis in H_1, say $\mu = \mu_1$, the quotient of the likelihood functions is

$$\frac{L(50)}{L(\mu_1)} = \frac{(72\pi)^{-n/2} \exp\left[-\left(\frac{1}{72}\right) \sum_{i=1}^{n} (x_i - 50)^2\right]}{(72\pi)^{-n/2} \exp\left[-\left(\frac{1}{72}\right) \sum_{i=1}^{n} (x_i - \mu_1)^2\right]}$$

$$= \exp\left[-\frac{1}{72}\left\{2(\mu_1 - 50) \sum_{i=1}^{n} x_i + n(50^2 - \mu_1^2)\right\}\right].$$

Now $L(50)/L(\mu_1) \leq k$ if and only if

$$\overline{x} \geq \frac{(-72)\ln(k)}{2n(\mu_1 - 50)} + \frac{50 + \mu_1}{2} = c.$$

Thus the best critical region of size α for testing H_0: $\mu = 50$ against H_1: $\mu = \mu_1$, where $\mu_1 > 50$, is given by $C = \{(x_1, x_2, \ldots, x_n): \overline{x} \geq c\}$, where c is selected such that $P(\overline{X} \geq c; H_0: \mu = 50) = \alpha$. Note that the same value of c can be used for each $\mu_1 > 50$, but (of course) k does not remain the same. Since the critical region C defines a test that is most powerful against each simple alternative $\mu_1 > 50$, this is a uniformly most powerful test, and C is a uniformly most powerful critical region of size α. Again if $\alpha = 0.05$, then $c \approx 52.47$. $\diamond$

Example 12.5.5

Let Y have the binomial distribution $b(n, p)$. To find a uniformly most powerful test of the simple null hypothesis H_0: $p = p_0$ against the one-sided alternative hypothesis H_1: $p > p_0$, consider, with $p_1 > p_0$,

$$\frac{L(p_0)}{L(p_1)} = \frac{\binom{n}{y} p_0^y (1 - p_0)^{n-y}}{\binom{n}{y} p_1^y (1 - p_1)^{n-y}} \leq k.$$

This is equivalent to

$$\left[\frac{p_0(1 - p_1)}{p_1(1 - p_0)}\right]^y \left[\frac{1 - p_0}{1 - p_1}\right]^n \leq k$$

and

$$y \ln\left[\frac{p_0(1 - p_1)}{p_1(1 - p_0)}\right] \leq \ln k - n \ln\left[\frac{1 - p_0}{1 - p_1}\right].$$

Since $p_0 < p_1$, $p_0(1 - p_1) < p_1(1 - p_0)$; it follows that $\ln[p_0(1 - p_1)/p_1(1 - p_0)] < 0$. Thus we have

$$\frac{y}{n} \geq \frac{\ln k - n \ln[(1 - p_0)/(1 - p_1)]}{n \ln[p_0(1 - p_1)/p_1(1 - p_0)]} = c$$

for each $p_1 > p_0$.

It is interesting to note that if the alternative hypothesis is the one-sided H_1: $p < p_0$, then a uniformly most powerful test is of the form $(y/n) \leq c$. Thus we see that the tests of H_0: $p = p_0$ against the one-sided alternatives given in Table 7.1 are uniformly most powerful. $\diamond$

Exercise 12.5.5 will demonstrate that uniformly most powerful tests do not always exist; in particular, they do not usually exist when the composite alternative hypothesis is two-sided.

We close this section with one easy but important observation. If a sufficient statistic $Y = u(X_1, X_2, \ldots, X_n)$ exists for θ, then, by the factorization theorem,

$$\frac{L(\theta_0)}{L(\theta_1)} = \frac{\phi[u(x_1, x_2, \ldots, x_n); \theta_0] h(x_1, x_2, \ldots, x_n)}{\phi[u(x_1, x_2, \ldots, x_n); \theta_1] h(x_1, x_2, \ldots, x_n)}$$

$$= \frac{\phi[u(x_1, x_2, \ldots, x_n); \theta_0)}{\phi[u(x_1, x_2, \ldots, x_n); \theta_1]} .$$

Thus $L(\theta_0)/L(\theta_1) \leq k$ provides a critical region that is a function of the observations $x_1, x_2, \ldots x_n$ only through the observed value of the sufficient statistic $y = u(x_1, x_2, \ldots, x_n)$. Hence, best critical and uniformly most powerful critical regions are based upon sufficient statistics when they exist.

EXERCISES

12.5.1 Let $X_1, X_2, \ldots, X_n$ be a random sample from a normal distribution $N(\mu, 64)$.

(a) Show that $C = \{(x_1, x_2, \ldots, x_n): \bar{x} \leq c\}$ is a best critical region for testing H_0: $\mu = 80$ against H_1: $\mu = 76$.

(b) Find n and c so that $\alpha = 0.05$ and $\beta = 0.05$, approximately.

12.5.2 Let $X_1, X_2, \ldots, X_n$ be a random sample from $N(0, \sigma^2)$.

(a) Show that $C = \{(x_1, x_2, \ldots, x_n) : \sum_{i=1}^{n} x_i^2 \geq c\}$ is a best critical region for testing H_0: $\sigma^2 = 4$ against H_1: $\sigma^2 = 16$.

(b) If $n = 15$, find the value of c so that $\alpha = 0.05$. HINT: Recall that $\sum_{i=1}^{n} X_i^2/\sigma^2$ is $\chi^2(n)$.

(c) If $n = 15$ and c is the value found in part (b), find the approximate value of $\beta = P(\sum_{i=1}^{n} X_i^2 < c; \sigma^2 = 16)$.

12.5.3 Let X have an exponential distribution with a mean of θ; that is, the p.d.f. of X is $f(x; \theta) = (1/\theta)e^{-x/\theta}, 0 < x < \infty$. Let $X_1, X_2, \ldots, X_n$ be a random sample from this distribution.

(a) Show that a best critical region for testing H_0: $\theta = 3$ against H_1: $\theta = 5$ can be based on the statistic $\sum_{i=1}^{n} X_i$.

(b) If $n = 12$, use the fact that $(2/\theta) \sum_{i=1}^{12} X_i$ is $\chi^2(24)$ to find a best critical region of size $\alpha = 0.10$.

(c) If $n = 12$, find a best critical region of size $\alpha = 0.10$ for testing H_0: $\theta = 3$ against H_1: $\theta = 7$.

(d) If H_1: $\theta > 3$, is the common region found in parts (b) and (c) a uniformly most powerful critical region of size $\alpha = 0.10$?

12.5.4 Let $X_1, X_2, \ldots, X_n$ be a random sample of Bernoulli trials $b(1, p)$.

(a) Show that a best critical region for testing H_0: $p = 0.9$ against H_1: $p = 0.8$ can be based on the statistic $Y = \sum_{i=1}^{n} X_i$, which is $b(n, p)$.

(b) If $C = \{(x_1, x_2, \ldots, x_n): \sum_{i=1}^{n} x_i \leq n(0.85)\}$ and $Y = \sum_{i=1}^{n} X_i$, find the value of n such that $\alpha = 0.10 \doteq P[Y \leq n(0.85); \ p = 0.9]$, approximately. HINT: Use the normal approximation for the binomial distribution.

(c) What is the approximate value of $\beta = P[Y > n(0.85); \ p = 0.8]$ for the test given in part (b)?

(d) Is the test of part (b) a uniformly most powerful test when the alternative hypothesis is H_1: $p < 0.9$?

12.5.5 Let $X_1, X_2, \ldots, X_n$ be a random sample from the normal distribution $N(\mu, 36)$.

(a) Show that a uniformly most powerful critical region for testing H_0: $\mu = 50$ against H_1: $\mu < 50$ is given by $C_2 = \{\overline{x}: \overline{x} \leq c\}$.

(b) With this result and that of Example 12.5.4, argue that a uniformly most powerful test for testing H_0: $\mu = 50$ against H_1: $\mu \neq 50$ does not exist.

12.5.6 Let $X_1, X_2, \ldots, X_n$ be a random sample from the normal distribution $N(\mu, 9)$. To test the hypothesis H_0: $\mu = 80$ against H_1: $\mu \neq 80$, consider the following three critical regions: $C_1 = \{\overline{x}: \overline{x} \geq c_1\}$, $C_2 = \{\overline{x}: \overline{x} \leq c_2\}$, and $C_3 = \{\overline{x}: |\overline{x} - 80| \geq c_3\}$.

(a) If $n = 16$, find the values of c_1, c_2, c_3 such that the size of each critical region is 0.05. That is, find c_1, c_2, c_3 such that

$$0.05 = P(\overline{X} \in C_1; \ \mu = 80) = P(\overline{X} \in C_2; \ \mu = 80)$$
$$= P(\overline{X} \in C_3; \ \mu = 80).$$

(b) Sketch, on the same graph paper, the power functions for these three critical regions.

12.5.7 Let $X_1, X_2, \ldots, X_{10}$ be a random sample of size 10 from a Poisson distribution with mean μ.

(a) Show that a uniformly most powerful critical region for testing H_0: $\mu = 0.5$ against H_1: $\mu > 0.5$ can be defined using the statistic $\sum_{i=1}^{10} X_i$.

(b) What is a uniformly most powerful critical region of size $\alpha = 0.068$? Recall that $\sum_{i=1}^{10} X_i$ has a Poisson distribution with mean 10μ.

(c) Sketch the power function of this test.

12.6 LIKELIHOOD RATIO TESTS

In this section we consider a general test-construction method that is applicable when both the null and alternative hypotheses, say H_0 and H_1, are composite. We continue to assume that the functional form of the p.d.f. is known but that it depends on an unknown parameter or unknown parameters. That is, we assume that the p.d.f. of X is $f(x; \theta)$, where θ represents one or more unknown parameters. We let Ω denote the total parameter space, that is, the set of all possible values of the parameter θ given by either H_0 or H_1. These hypotheses will be stated as follows:

$$H_0: \theta \in \omega, \qquad H_1: \theta \in \omega',$$

where ω is a subset of Ω and ω' is the complement of ω with respect to Ω. The test will be constructed by using a ratio of likelihood functions that have been maximized in ω and Ω, respectively. In a sense, this is a natural generalization of the ratio appearing in the Neyman-Pearson lemma when the two hypotheses were simple.

Definition 12.6.1.   The **likelihood ratio** is the quotient

$$\lambda = \frac{L(\widehat{\omega})}{L(\widehat{\Omega})},$$

where $L(\widehat{\omega})$ is the maximum of the likelihood function with respect to θ when $\theta \in \omega$, and $L(\widehat{\Omega})$ is the maximum of the likelihood function with respect to θ when $\theta \in \Omega$.

Because λ is the quotient of nonnegative functions, $\lambda \geq 0$. In addition, since $\omega \subset \Omega$, we have that $L(\widehat{\omega}) \leq L(\widehat{\Omega})$ and hence $\lambda \leq 1$. Thus $0 \leq \lambda \leq 1$. If the maximum of L in ω is much smaller than that in Ω, it would seem that the data $x_1, x_2, \ldots, x_n$ do not support the hypothesis $H_0: \theta \in \omega$. That is, a small value of the ratio $\lambda = L(\widehat{\omega})/L(\widehat{\Omega})$ would lead to the rejection of H_0. On the other hand, a value of the ratio λ that is close to one would support the null hypothesis H_0. This leads us to the following definition.

Definition 12.6.2.   To test $H_0: \theta \in \omega$ against $H_1: \theta \in \omega'$, the **critical region for the likelihood ratio test** is the set of points in the sample space for which

$$\lambda = \frac{L(\widehat{\omega})}{L(\widehat{\Omega})} \leq k,$$

where $0 < k < 1$ and k is selected so that the test has a desired significance level α.

The following example illustrates these definitions.

Example 12.6.1

Assume that the weight X in ounces of a "10-pound" bag of sugar is $N(\mu, 5)$. We shall test the hypothesis $H_0: \mu = 162$ against the alternative hypothesis $H_1: \mu \neq 162$. Thus $\Omega = \{\mu : -\infty < \mu < \infty\}$ and $\omega = \{162\}$. To find the likelihood ratio, we need $L(\widehat{\omega})$ and $L(\widehat{\Omega})$. When H_0 is true, μ can take on only one value, namely

$\mu = 162$. Thus $L(\widehat{\omega}) = L(162)$. To find $L(\widehat{\Omega})$, we must find the value of μ that maximizes $L(\mu)$. We recall that $\widehat{\mu} = \bar{x}$ is the maximum likelihood estimate of μ. Thus $L(\widehat{\Omega}) = L(\bar{x})$ and the likelihood ratio $\lambda = L(\widehat{\omega})/L(\widehat{\Omega})$ is given by

$$
\lambda = \frac{(10\pi)^{-n/2} \exp\left[-\left(\dfrac{1}{10}\right) \displaystyle\sum_{i=1}^{n} (x_i - 162)^2\right]}{(10\pi)^{-n/2} \exp\left[-\left(\dfrac{1}{10}\right) \displaystyle\sum_{i=1}^{n} (x_i - \bar{x})^2\right]}
$$

$$
= \frac{\exp\left[-\left(\dfrac{1}{10}\right) \displaystyle\sum_{i=1}^{n} (x_i - \bar{x})^2 - \left(\dfrac{n}{10}\right)(\bar{x} - 162)^2\right]}{\exp\left[-\left(\dfrac{1}{10}\right) \displaystyle\sum_{i=1}^{n} (x_i - \bar{x})^2\right]}
$$

$$
= \exp\left[-\frac{n}{10}(\bar{x} - 162)^2\right].
$$

A value of $\bar{x}$ close to 162 would tend to support H_0 and in that case λ is close to 1. On the other hand, an $\bar{x}$ that differs from 162 by too much would tend to support H_1. See Figure 12.1 for the graph of this likelihood ratio when $n = 5$.

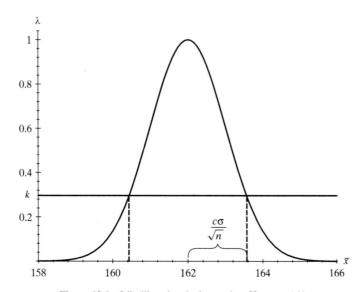

Figure 12.1   Likelihood ratio for testing H_0: $\mu = 160$

A likelihood ratio critical region is given by $\lambda \leq k$, where k is selected so that the significance level of the test is α. Using this criterion and simplifying the inequality as we do using the Neyman-Pearson lemma, we have that $\lambda \leq k$ is equivalent to each of the following inequalities:

$$-\left(\frac{n}{10}\right)(\bar{x} - 162)^2 \leq \ln k,$$

$$(\bar{x} - 162)^2 \geq -\left(\frac{10}{n}\right) \ln k,$$

$$\frac{|\bar{x} - 162|}{\sigma/\sqrt{n}} \geq \frac{\sqrt{-(10/n)\ln k}}{\sigma/\sqrt{n}} = c.$$

Since $Z = (\bar{X} - 162)/(\sigma/\sqrt{n})$ is $N(0, 1)$ when $H_0: \mu = 162$ is true, let $c = z_{\alpha/2}$. Thus the critical region is

$$C = \left\{ \bar{x} : \frac{|\bar{x} - 162|}{\sigma/\sqrt{n}} \geq z_{\alpha/2} \right\}$$

and, for illustration, if $\alpha = 0.05$, we have that $z_{0.025} = 1.96$.   ◇

As illustrated in Example 12.6.1, the inequality $\lambda \leq k$ can often be expressed in terms of a statistic whose distribution is known. Also, note that although the likelihood ratio test is an intuitive test, it leads to the same critical region as that given by the Neyman-Pearson lemma when H_0 and H_1 are both simple hypotheses.

Suppose now that the random sample $X_1, X_2, \ldots, X_n$ arises from the normal population $N(\mu, \sigma^2)$, where both μ and σ^2 are unknown. Let us consider the likelihood ratio test of the null hypothesis $H_0: \mu = \mu_0$ against the two-sided alternative hypothesis $H_1: \mu \neq \mu_0$. For this test

$$\omega = \{(\mu, \sigma^2) : \mu = \mu_0, 0 < \sigma^2 < \infty\}$$

and

$$\Omega = \{(\mu, \sigma^2) : -\infty < \mu < \infty, 0 < \sigma^2 < \infty\}.$$

If $(\mu, \sigma^2) \in \Omega$, the observed maximum likelihood estimates are $\widehat{\mu} = \bar{x}$ and $\widehat{\sigma^2} = (1/n) \sum_{i=1}^{n} (x_i - \bar{x})^2$. Thus

$$L(\widehat{\Omega}) = \left[\frac{1}{2\pi \left(\frac{1}{n}\right) \sum_{i=1}^{n} (x_i - \bar{x})^2} \right]^{n/2} \exp\left[-\frac{\sum_{i=1}^{n} (x_i - \bar{x})^2}{\left(\frac{2}{n}\right) \sum_{i=1}^{n} (x_i - \bar{x})^2} \right]$$

$$= \left[\frac{n e^{-1}}{2\pi \sum_{i=1}^{n} (x_i - \bar{x})^2} \right]^{n/2}.$$

Similarly, if $(\mu, \sigma^2) \in \omega$, the observed maximum likelihood estimates are $\widehat{\mu} = \mu_0$ and $\widehat{\sigma^2} = (1/n) \sum_{i=1}^{n} (x_i - \mu_0)^2$. Thus

$$
L(\widehat{\omega}) = \left[\dfrac{1}{2\pi \left(\dfrac{1}{n}\right) \sum_{i=1}^{n} (x_i - \mu_0)^2} \right]^{n/2} \exp\left[-\dfrac{\sum_{i=1}^{n} (x_i - \mu_0)^2}{\left(\dfrac{2}{n}\right) \sum_{i=1}^{n} (x_i - \mu_0)^2} \right]
$$

$$
= \left[\dfrac{ne^{-1}}{2\pi \sum_{i=1}^{n} (x_i - \mu_0)^2} \right]^{n/2}.
$$

The likelihood ratio $\lambda = L(\widehat{\omega})/L(\widehat{\Omega})$ for this test is

$$
\lambda = \frac{\left[\dfrac{ne^{-1}}{2\pi \sum_{i=1}^{n} (x_i - \mu_0)^2} \right]^{n/2}}{\left[\dfrac{ne^{-1}}{2\pi \sum_{i=1}^{n} (x_i - \bar{x})^2} \right]^{n/2}} = \left[\dfrac{\sum_{i=1}^{n} (x_i - \bar{x})^2}{\sum_{i=1}^{n} (x_i - \mu_0)^2} \right]^{n/2}.
$$

However, note that

$$
\sum_{i=1}^{n} (x_i - \mu_0)^2 = \sum_{i=1}^{n} (x_i - \bar{x} + \bar{x} - \mu_0)^2 = \sum_{i=1}^{n} (x_i - \bar{x})^2 + n(\bar{x} - \mu_0)^2.
$$

If this substitution is made in the denominator of λ we have

$$
\lambda = \left[\dfrac{\sum_{i=1}^{n} (x_i - \bar{x})^2}{\sum_{i=1}^{n} (x_i - \bar{x})^2 + n(\bar{x} - \mu_0)^2} \right]^{n/2}
$$

$$
= \left[\dfrac{1}{1 + \dfrac{n(\bar{x} - \mu_0)^2}{\sum_{i=1}^{n} (x_i - \bar{x})^2}} \right]^{n/2}.
$$

Note that λ is close to 1 when $\bar{x}$ is close to μ_0 and it is small when $\bar{x}$ and μ_0 differ by a great deal. The likelihood ratio test, given by the inequality $\lambda \leq k$, is the same as

$$
\dfrac{1}{1 + \dfrac{n(\bar{x} - \mu_0)^2}{\sum_{i=1}^{n} (x_i - \bar{x})^2}} \leq k^{2/n}
$$

or, equivalently,

$$
\dfrac{n(\bar{x} - \mu_0)^2}{\dfrac{\sum_{i=1}^{n} (x_i - \bar{x})^2}{n - 1}} \geq (n - 1)(k^{-2/n} - 1).
$$

When H_0 is true, $\sqrt{n}(\overline{X} - \mu_0)/\sigma$ is $N(0, 1)$, and $\sum_{i=1}^{n}(X_i - \overline{X})^2/\sigma^2$ has an independent chi-square distribution $\chi^2(n-1)$. Hence, under H_0,

$$T = \frac{\sqrt{n}(\overline{X} - \mu_0)/\sigma}{\sqrt{\sum_{i=1}^{n}(X_i - \overline{X})^2/[\sigma^2(n-1)]}} = \frac{\sqrt{n}(\overline{X} - \mu_0)}{\sqrt{\sum_{i=1}^{n}(X_i - \overline{X})^2/(n-1)}} = \frac{\overline{X} - \mu_0}{S/\sqrt{n}}$$

has a t distribution with $r = n - 1$ degrees of freedom. In accordance with the likelihood ratio test criterion, H_0 is rejected if the observed

$$T^2 \geq (n - 1)(k^{-2/n} - 1).$$

That is, we reject H_0: $\mu = \mu_0$ and accept H_1: $\mu \neq \mu_0$ if the observed $|T| \geq t_{\alpha/2}(n - 1)$.

The reader should observe that this test is exactly the same as that listed in Table 7.5 for testing H_0: $\mu = \mu_0$ against H_1: $\mu \neq \mu_0$. That is, the test listed there is a likelihood ratio test. As a matter of fact, all six of the tests given in Tables 7.5 and 7.7 are likelihood ratio tests. Thus the examples and exercises associated with those tables are illustrations of the use of likelihood ratio tests.

The final development of this section concerns a test about the variance of a normal population. Let $X_1, X_2, \ldots, X_n$ be a random sample from $N(\mu, \sigma^2)$, where μ and σ^2 are unknown. We wish to test H_0: $\sigma^2 = \sigma_0^2$ against H_1: $\sigma^2 \neq \sigma_0^2$. For this, we have

$$\omega = \{(\mu, \sigma^2) : -\infty < \mu < \infty, \sigma^2 = \sigma_0^2\}$$

and

$$\Omega = \{(\mu, \sigma^2) : -\infty < \mu < \infty, 0 < \sigma^2 < \infty\}.$$

As in the test concerning the mean, we obtain

$$L(\widehat{\Omega}) = \left[\frac{ne^{-1}}{2\pi \sum_{i=1}^{n}(x_i - \overline{x})^2}\right]^{n/2}.$$

If $(\mu, \sigma^2) \in \omega$, then $\widehat{\mu} = \overline{x}$ and $\widehat{\sigma^2} = \sigma_0^2$; thus

$$L(\widehat{\omega}) = \left(\frac{1}{2\pi\sigma_0^2}\right)^{n/2} \exp\left[-\frac{\sum_{i=1}^{n}(x_i - \overline{x})^2}{2\sigma_0^2}\right].$$

Accordingly, the likelihood ratio test $\lambda = L(\widehat{\omega})/L(\widehat{\Omega})$ is

$$\lambda = \left(\frac{w}{n}\right)^{n/2} \exp\left(-\frac{w}{2} + \frac{n}{2}\right) \leq k,$$

where $w = \sum_{i=1}^{n}(x_i - \overline{x})^2/\sigma_0^2$. Solving this inequality for w, we obtain a solution of the form $w \leq c_1$ or $w \geq c_2$, where the constants c_1 and c_2 are appropriate functions of the constants k and n so as to achieve the desired significance level α. Since $W = \sum_{i=1}^{n}(X_i - \overline{X})^2/\sigma_0^2$ is $\chi^2(n-1)$ if H_0: $\sigma^2 = \sigma_0^2$ is true, most statisticians modify this test slightly by taking $c_1 = \chi^2_{1-\alpha/2}(n-1)$ and $c_2 = \chi^2_{\alpha/2}(n-1)$. That is, this test and the other tests listed in Table 7.6 are either likelihood ratio tests or slight modifications of likelihood ratio tests. As a matter of fact, most tests involving normal assumptions are likelihood ratio tests or modifications of them involving regression and analysis of variance. These are considered elsewhere in the book.

Note that likelihood ratio tests are based on sufficient statistics when they exist as was also true of best critical and uniformly most powerful critical regions.

EXERCISES

12.6.1 In Example 12.6.1, if $n = 20$ and $\bar{x} = 161.1$, is H_0 accepted at a significance level of size α

(a) If $\alpha = 0.10$?

(b) If $\alpha = 0.05$?

(c) What is the p-value of this test?

12.6.2 Assume that the weight X in ounces of a "10-ounce" box of cornflakes is $N(\mu, 0.03)$. Let $X_1, X_2, \ldots, X_n$ be a random sample from this distribution.

(a) To test the hypothesis H_0: $\mu \geq 10.35$ against the alternative hypothesis H_1: $\mu < 10.35$, what is the critical region of size $\alpha = 0.05$ specified by the likelihood ratio test criterion? HINT: If $\mu \geq 10.35$ and $\bar{x} < 10.35$, note that $\hat{\mu} = 10.35$.

(b) If a random sample of $n = 50$ boxes yielded a sample mean of $\bar{x} = 10.31$, is H_0 rejected? HINT: Find the critical value z_α when H_0 is true by taking $\mu = 10.35$, which is the extreme value in $\mu \geq 10.35$.

(c) What is the p-value of this test?

12.6.3 Let $X_1, X_2, \ldots, X_n$ be a random sample from the normal distribution $N(\mu, 100)$.

(a) To test H_0: $\mu = 230$ against H_1: $\mu > 230$, what is the critical region specified by the likelihood ratio test criterion?

(b) Is this test uniformly most powerful?

(c) If a random sample of $n = 16$ yielded $\bar{x} = 232.6$, is H_0 accepted at a significance level of $\alpha = 0.10$?

(d) What is the p-value of this test?

12.6.4 Let $X_1, X_2, \ldots, X_n$ be a random sample from the normal distribution $N(\mu, 225)$.

(a) To test H_0: $\mu = 59$ against H_1: $\mu \neq 59$, what is the critical region of size $\alpha = 0.05$ specified by the likelihood ratio test criterion?

(b) If a sample of size $n = 100$ yielded $\bar{x} = 56.13$, is H_0 accepted?

(c) What is the p-value of this test?

12.6.5 It is desired to test the hypothesis H_0: $\mu = 30$ against the alternative hypothesis H_1: $\mu \neq 30$, where μ is the mean of a normal distribution and σ^2 is unknown. If a random sample of size $n = 9$ has $\bar{x} = 32.8$ and $s = 4$, is H_0 accepted at an $\alpha = 0.05$ significance level? What is the approximate p-value of this test?

12.6.6 To test H_0: $\mu = 335$ against H_1: $\mu < 335$, under normal assumptions, a random sample of size 17 yielded $\bar{x} = 324.8$ and $s = 40$. Is H_0 accepted at an $\alpha = 0.10$ significance level?

12.6.7 Let X have a normal distribution, in which μ and σ^2 are both unknown. It is desired to test H_0: $\mu = 1.80$ against H_1: $\mu > 1.80$ at an $\alpha = 0.10$ significance level. If a random sample of size $n = 121$ yielded $\bar{x} = 1.84$ and $s = 0.20$, is H_0 accepted or rejected? What is the p-value of this test?

12.6.8 Let $X_1, X_2, \ldots, X_n$ be a random sample from an exponential distribution with mean θ. Show that the likelihood ratio test of H_0: $\theta = \theta_0$ against H_1: $\theta \neq \theta_0$ has a critical region of the form $\sum_{i=1}^{n} x_i \leq c_1$ or $\sum_{i=1}^{n} x_i \geq c_2$. How would you modify this so that chi-square tables can be used easily?

12.6.9 Let independent random samples of sizes n and m be taken respectively from two normal distributions with unknown means μ_X and μ_Y and unknown variances σ_X^2 and σ_Y^2.

(a) Show that the likelihood ratio for testing H_0: $\mu_X = \mu_Y$ against H_1: $\mu_X \neq \mu_Y$ when $\sigma_X^2 = \sigma_Y^2$ is a function of the usual two-sample t-statistic.

(b) Show that the likelihood ratio for testing H_0: $\sigma_X^2 = \sigma_Y^2$ against H_1: $\sigma_X^2 \neq \sigma_Y^2$ is a function of the usual two-sample F-statistic.

The tests described in parts (a) and (b) are listed (the second is modified slightly) in Tables 7.7 and 7.8, respectively.

A

REVIEW OF SELECTED
MATHEMATICAL TECHNIQUES

A.1 ALGEBRA OF SETS

The totality of objects under consideration is called the **universal set** and is denoted by S. Each object in S is called an **element** of S. If a set A is a collection of elements that are also in S, then A is said to be a **subset** of S. In applications of probability, S will usually denote the **sample space**. An **event** A will be a collection of possible outcomes of the experiment and will be a subset of S. We say that event A *has occurred* if the outcome of the experiment is an element of A. The set or event A may be described by listing all of its elements or by defining the properties that its elements must satisfy.

Example A.1.1

A four-sided die, called a tetrahedron, has four faces that are equilateral triangles. These faces are numbered 1, 2, 3, 4. When the tetrahedron is rolled, the outcome of the experiment is the number of the face that is down. If this tetrahedron is rolled twice and we keep track of the first roll and the second roll, then the sample space is that displayed in Figure A.1.

Let A be the event that the first roll is a 1 or a 2. That is,

$$A = \{(x, y): \ x = 1 \text{ or } x = 2\}.$$

Let

$$B = \{(x, y): \ x + y = 6\} = \{(2, 4), (3, 3), (4, 2)\},$$

and let

$$C = \{(x, y): \ x + y \geq 7\} = \{(3, 4), (4, 3), (4, 4)\}.$$

Events A, B, and C are shown in Figure A.1.   ◇

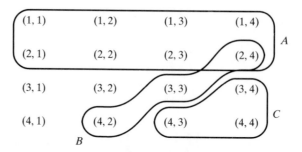

Figure A.1   Sample space for two rolls of a four-sided die

When a is an element in A, we write $a \in A$. When a is not an element in A, we write $a \notin A$. So, in Example A.1.1, we have $(1, 3) \in A$ and $(3, 1) \notin A$. If every element of a set A is also an element in a set B, then A is a **subset** of B. We write $A \subset B$. In probability, if event B occurs whenever event A occurs, then $A \subset B$. The two sets A and B are equal (i.e., $A = B$), if $A \subset B$ and $B \subset A$. Note that it is always true that $A \subset A$ and $A \subset S$, where S is the universal set. We denote the subset that contains no elements by $\emptyset$. This set is called the **null** or **empty** set. For all sets A, $\emptyset \subset A$.

The set of elements in either A or B or possibly both A and B is called the **union** of A and B and is denoted $A \cup B$. The set of elements in both A and B is called the **intersection** of A and B and is denoted $A \cap B$. The **complement** of a set A is the set of elements in the universal set S that are not in the set A and is denoted A'. In probability, if A and B are two events, the event that at least one of the two events has occurred is denoted by $A \cup B$ and the event that both events have occurred is denoted by $A \cap B$. The event that A has not occurred is denoted by A', and the event that A has not occurred but B has occurred is denoted by $A' \cap B$. If $A \cap B = \emptyset$, we say that A and B are **mutually exclusive**. In Example A.1.1, $B \cup C = \{(x, y): x + y \geq 6\}$, $A \cap B = \{(2, 4)\}$, and $A \cap C = \emptyset$. Note that A and C are mutually exclusive. Also, $C' = \{(x, y): x + y \leq 6\}$.

The operations of union and intersection may be extended to more than two sets. Let $A_1, A_2, \ldots, A_n$ be a finite collection of sets. Then the **union**

$$A_1 \cup A_2 \cup \cdots \cup A_n = \bigcup_{k=1}^{n} A_k$$

is the set of elements that belong to at least one A_k, $k = 1, 2, \ldots, n$. The **intersection**

$$A_1 \cap A_2 \cap \cdots \cap A_n = \bigcap_{k=1}^{n} A_k$$

is the set of all elements that belong to every A_k, $k = 1, 2, \ldots, n$. Similarly, let $A_1, A_2, \ldots, A_n, \ldots$ be a countable collection of sets. Then x belongs to the **union**

$$A_1 \cup A_2 \cup A_3 \cup \cdots = \bigcup_{k=1}^{\infty} A_k$$

if x belongs to at least one A_k, $k = 1, 2, 3, \ldots$. Also x belongs to the **intersection**

$$A_1 \cap A_2 \cap A_3 \cap \cdots = \bigcap_{k=1}^{\infty} A_k$$

if x belongs to every A_k, $k = 1, 2, 3, \ldots$

Example A.1.2

Let

$$A_k = \left\{ x : \frac{10}{k+1} \le x \le 10 \right\}, \qquad k = 1, 2, 3, \ldots$$

Then

$$\bigcup_{k=1}^{8} A_k = \left\{ x : \frac{10}{9} \le x \le 10 \right\};$$

$$\bigcup_{k=1}^{\infty} A_k = \{x : 0 < x \le 10\}.$$

Note that the number zero is not in this latter union, since it is not in one of the sets $A_1, A_2, A_3, \ldots$ Also

$$\bigcap_{k=1}^{8} A_k = \{x : 5 \le x \le 10\} = A_1;$$

$$\bigcap_{k=1}^{\infty} A_k = \{x : 5 \le x \le 10\} = A_1,$$

since $A_1 \subset A_k$, $k = 1, 2, 3, \ldots$ ◇

A convenient way to illustrate operations on sets is with a **Venn** diagram. In Figure A.2 the universal set S is represented by the rectangle and its interior and the subsets of S by the points enclosed by the circles. The sets under consideration are the shaded regions.

Set operations satisfy several properties. For example, if A, B, and C are subsets of S, we have the following:

Commutative Laws

$$A \cup B = B \cup A$$
$$A \cap B = B \cap A.$$

Associative Laws

$$(A \cup B) \cup C = A \cup (B \cup C)$$
$$(A \cap B) \cap C = A \cap (B \cap C).$$

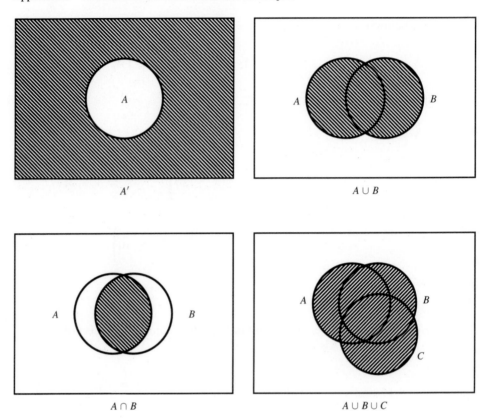

Figure A.2   Venn diagrams

Distributive Laws

$$A \cap (B \cup C) = (A \cap B) \cup (A \cap C)$$
$$A \cup (B \cap C) = (A \cup B) \cap (A \cup C).$$

De Morgan's Laws

$$(A \cup B)' = A' \cap B'$$
$$(A \cap B)' = A' \cup B'.$$

A Venn diagram will be used to justify the first of De Morgan's laws. In Figure A.3(a), $A \cup B$ is represented by horizontal lines, and thus $(A \cup B)'$ is the region represented by vertical lines. In Figure A.3(b), A' is indicated with horizontal lines, and B' is indicated with vertical lines. An element belongs to $A' \cap B'$ if it belongs to both A' and B'. Thus the crosshatched region represents $A' \cap B'$. Clearly, this crosshatched region is the same as that shaded with vertical lines in Figure A.3(a).

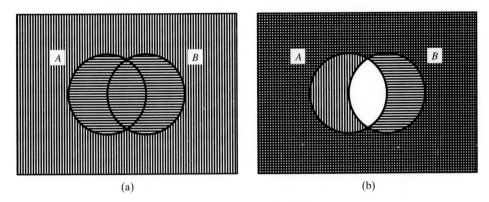

(a)   (b)

Figure A.3   Venn diagrams illustrating De Morgan's Laws

A.2 MATHEMATICAL TOOLS FOR THE HYPERGEOMETRIC DISTRIBUTION

Let X have a hypergeometric distribution. That is, the p.d.f. of X is

$$f(x) = \frac{\binom{N_1}{x}\binom{N_2}{n-x}}{\binom{N_1+N_2}{n}}$$

$$= \frac{\binom{N_1}{x}\binom{N_2}{n-x}}{\binom{N}{n}}, \qquad x \le n,\ x \le N_1,\ n-x \le N_2.$$

To show that $\sum_{x=0}^{n} f(x) = 1$ and to find the mean and variance of X, we use the following theorem.

Theorem A.2.1.

$$\binom{N}{n} = \sum_{x=0}^{n} \binom{N_1}{x}\binom{N_2}{n-x},$$

where $N = N_1 + N_2$ and it is understood that $\binom{k}{j} = 0$ if $j > k$.

Proof.   Because $N = N_1 + N_2$, we have the identity

$$(1+y)^N \equiv (1+y)^{N_1}(1+y)^{N_2}. \tag{A.1}$$

We will expand each of these binomials; and, since the polynomials on each side are identically equal, the coefficients of y^n on each side of Equation A.1 must be equal. Using the binomial expansion, we find that the expansion of the left side of Equation A.1 is

$$(1+y)^N = \sum_{k=0}^{N} \binom{N}{k} y^k$$

$$= \binom{N}{0} + \binom{N}{1}y + \cdots + \binom{N}{n}y^n + \cdots + \binom{N}{N}y^N.$$

The right side of Equation A.1 becomes

$$(1+y)^{N_1}(1+y)^{N_2} = \left[\binom{N_1}{0} + \binom{N_1}{1}y + \cdots + \binom{N_1}{n}y^n + \cdots + \binom{N_1}{N_1}y^{N_1}\right]$$

$$\times \left[\binom{N_2}{0} + \binom{N_2}{1}y + \cdots + \binom{N_2}{n}y^n + \cdots + \binom{N_2}{N_2}y^{N_2}\right].$$

The coefficient of y^n in this product is

$$\binom{N_1}{0}\binom{N_2}{n} + \binom{N_1}{1}\binom{N_2}{n-1} + \cdots + \binom{N_1}{n}\binom{N_2}{0} = \sum_{x=0}^{n}\binom{N_1}{x}\binom{N_2}{n-x}$$

and this sum must be equal to $\binom{N}{n}$, the coefficient of y^n on the left side of Equation A.1.   □

Using Theorem A.2.1, we find that it follows that if X has a hypergeometric distribution with p.d.f. $f(x)$, then

$$\sum_{x=0}^{n} f(x) = \sum_{x=0}^{n} \frac{\binom{N_1}{x}\binom{N_2}{n-x}}{\binom{N}{n}} = 1.$$

To find the mean and variance of a hypergeometric random variable, it is useful to note that, with $n > 0$,

$$\binom{N}{n} = \frac{N!}{n!\,(N-n)!} = \frac{N}{n}\cdot\frac{(N-1)!}{(n-1)!\,(N-n)!} = \frac{N}{n}\binom{N-1}{n-1}.$$

The mean of a hypergeometric random variable X is

$$\mu = \sum_{x=0}^{n} x f(x)$$

$$= \frac{\sum_{x=1}^{n} x \cdot \dfrac{N_1!}{x!\,(N_1 - x)!} \cdot \dfrac{N_2!}{(n - x)!\,(N_2 - n + x)!}}{\dbinom{N}{n}}$$

$$= \frac{N_1 \sum_{x=1}^{n} \dfrac{(N_1 - 1)!}{(x - 1)!\,(N_1 - x)!} \cdot \dfrac{N_2!}{(n - x)!\,(N_2 - n + x)!}}{\dbinom{N}{n}}.$$

If we now make the change of variables $k = x - 1$ in the summation, and replace

$$\dbinom{N}{n} \qquad \text{with} \qquad \dbinom{N}{n}\dbinom{N - 1}{n - 1}$$

in the denominator, this becomes

$$\mu = \frac{N_1}{\dbinom{N}{n}} \frac{\sum_{k=0}^{n-1} \dfrac{(N_1 - 1)!}{k!\,(N_1 - 1 - k)!} \cdot \dfrac{N_2!}{(n - k - 1)!\,(N_2 - n + k + 1)!}}{\dbinom{N - 1}{n - 1}}$$

$$= n\left(\frac{N_1}{N}\right) \frac{\sum_{k=0}^{n-1} \dbinom{N_1 - 1}{k}\dbinom{N_2}{n - 1 - k}}{\dbinom{N - 1}{n - 1}} = n\left(\frac{N_1}{N}\right)$$

because the summation in the expression for μ is equal to $\dbinom{N - 1}{n - 1}$ from Theorem A.2.1.

Note that

$$\text{Var}(X) = \sigma^2 = E[(X - \mu)^2]$$
$$= E[X^2] - \mu^2$$
$$= E[X(X - 1)] + E(X) - \mu^2.$$

So to find the variance of X, we first find $E[X(X-1)]$:

$$E[X(X-1)] = \sum_{x=0}^{n} x(x-1)f(x)$$

$$= \frac{\displaystyle\sum_{x=2}^{n} x(x-1)\,\frac{N_1!}{x!\,(N_1-x)!} \cdot \frac{N_2!}{(n-x)!\,(N_2-n+x)!}}{\dbinom{N}{n}}$$

$$= N_1(N_1-1)\,\frac{\displaystyle\sum_{x=2}^{n} \frac{(N_1-2)!}{(x-2)!\,(N_1-x)!} \cdot \frac{N_2!}{(n-x)!\,(N_2-n+x)!}}{\dbinom{N}{n}}.$$

In the summation, let $k = x - 2$, and in the denominator, note that

$$\binom{N}{n} = \frac{N!}{n!(N-n)!} = \frac{N(N-1)}{n(n-1)} \binom{N-2}{n-2}.$$

Thus

$$E[X(X-1)] = \frac{N_1(N_1-1)}{\dfrac{N(N-1)}{n(n-1)}} \sum_{k=0}^{n-2} \frac{\dbinom{N_1-2}{k}\dbinom{N_2}{n-2-k}}{\dbinom{N-2}{n-2}}$$

$$= \frac{N_1(N_1-1)(n)(n-1)}{N(N-1)}$$

from Theorem A.2.1. Thus the variance of a hypergeometric random variable is, after some algebra,

$$\sigma^2 = \frac{N_1(N_1-1)(n)(n-1)}{N(N-1)} + \frac{nN_1}{N} - \left(\frac{nN_1}{N}\right)^2$$

$$= n\left(\frac{N_1}{N}\right)\left(\frac{N_2}{N}\right)\left(\frac{N-n}{N-1}\right).$$

A.3 LIMITS

We refer the reader to the many fine books on calculus for the definition of a limit and the other concepts used in that subject. Here we simply remind you of some of the techniques we find most useful in probability and statistics.

Early in a calculus course the existence of the following limit is discussed and it is denoted by the letter e:

$$e = \lim_{t \to 0} (1 + t)^{1/t} = \lim_{n \to \infty} \left(1 + \frac{1}{n}\right)^n.$$

Of course, e is an irrational number, which to six significant figures equals 2.71828.

Often it is rather easy to see the value of certain limits. For example, with $-1 < r < 1$, the sum of the geometric progression allows us to write

$$\lim_{n \to \infty} (1 + r + r^2 + \cdots + r^{n-1}) = \lim_{n \to \infty} \left(\frac{1 - r^n}{1 - r}\right) = \frac{1}{1 - r}.$$

That is, the limit of the ratio $(1-r^n)/(1-r)$ is not difficult to find because $\lim_{n \to \infty} r^n = 0$ when $-1 < r < 1$.

However it is not that easy to determine the limit of every ratio; for example, consider

$$\lim_{b \to \infty} (be^{-b}) = \lim_{b \to \infty} \left(\frac{b}{e^b}\right).$$

Since both the numerator and the denominator of the latter ratio are unbounded, we can use **L'Hospital's rule**, taking the limit of the ratio of the derivative of the numerator and the derivative of the denominator. We have

$$\lim_{b \to \infty} \left(\frac{b}{e^b}\right) = \lim_{b \to \infty} \left(\frac{1}{e^b}\right) = 0.$$

This result can be used in the evaluation of the integral

$$\int_0^\infty xe^{-x}\, dx = \lim_{b \to \infty} \int_0^b xe^{-x}\, dx$$

$$= \lim_{b \to \infty} [-xe^{-x} - e^{-x}]_0^b$$

$$= \lim_{b \to \infty} [1 - be^{-b} - e^{-b}] = 1.$$

Note that
$$D_x[-xe^{-x} - e^{-x}] = xe^{-x} - e^{-x} + e^{-x} = xe^{-x};$$

that is, $-xe^{-x} - e^{-x}$ is the antiderivative of xe^{-x}.

Another limit of importance is

$$\lim_{n \to \infty} \left(1 + \frac{b}{n}\right)^n = \lim_{n \to \infty} e^{n \ln(1 + b/n)},$$

where b is a constant.

Since the exponential function is continuous, the limit can be taken to the exponent. That is,

$$\lim_{n \to \infty} \exp[n \ln(1 + b/n)] = \exp[\lim_{n \to \infty} n \ln(1 + b/n)].$$

The limit in the exponent is equal to

$$\lim_{n\to\infty} \frac{\ln(1+b/n)}{1/n} = \lim_{n\to\infty} \frac{\frac{-b/n^2}{1+b/n}}{-1/n^2} = \lim_{n\to\infty} \frac{b}{1+b/n} = b$$

by L'Hospital's rule. Since this limit is equal to b, the original limit is

$$\lim_{n\to\infty} \left(1 + \frac{b}{n}\right)^n = e^b.$$

Applications of this limit in probability occur with $b = -1$ yielding

$$\lim_{n\to\infty} \left(1 - \frac{1}{n}\right)^n = e^{-1}.$$

A.4 INFINITE SERIES

A function $f(x)$ possessing derivatives of all orders at $x = b$ can be expanded in the following **Taylor's series**:

$$f(x) = f(b) + \frac{f'(b)}{1!}(x-b) + \frac{f''(b)}{2!}(x-b)^2 + \frac{f'''(b)}{3!}(x-b)^3 + \cdots.$$

If $b = 0$, we obtain the special case that is often called **Maclaurin's series**;

$$f(x) = f(0) + \frac{f'(0)}{1!}x + \frac{f''(0)}{2!}x^2 + \frac{f'''(0)}{3!}x^3 + \cdots.$$

For example, if $f(x) = e^x$ so that all derivatives of $f(x) = e^x$ are $f^{(r)}(x) = e^x$, then $f^{(r)}(0) = 1$, for $r = 1, 2, 3, \ldots$ Thus the Maclaurin's series expansion of $f(x) = e^x$ is

$$e^x = 1 + \frac{x}{1!} + \frac{x^2}{2!} + \frac{x^3}{3!} + \frac{x^4}{4!} + \cdots.$$

The **ratio test**

$$\lim_{n\to\infty} \left| \frac{x^n/n!}{x^{n-1}/(n-1)!} \right| = \lim_{n\to\infty} \left| \frac{x}{n} \right| = 0$$

shows that this series expansion of e^x converges for all real values of x.

Note, for examples, that

$$e = 1 + \frac{1}{1!} + \frac{1}{2!} + \frac{1}{3!} + \cdots$$

and

$$e^{-1} = 1 - \frac{1}{1!} + \frac{1}{2!} - \frac{1}{3!} + \cdots + \frac{(-1)^n}{n!} + \cdots.$$

For another example, consider

$$h(w) = (1-w)^{-r},$$

where r is a positive integer. Here

$$h'(w) = r(1 - w)^{-(r+1)}$$
$$h''(w) = (r)(r + 1)(1 - w)^{-(r+2)}$$
$$h'''(w) = (r)(r + 1)(r + 2)(1 - w)^{-(r+3)}$$
$$\vdots$$

In general, $h^{(k)}(0) = (r)(r + 1) \cdots (r + k - 1) = (r + k - 1)!/(r - 1)!$. Thus

$$(1 - w)^{-r} = 1 + \frac{(r + 1 - 1)!}{(r - 1)!\,1!} w + \frac{(r + 2 - 1)!}{(r - 1)!\,2!} w^2 + \cdots + \frac{(r + k - 1)!}{(r - 1)!\,k!} w^k + \cdots$$

$$= \sum_{k=0}^{\infty} \binom{r + k - 1}{r - 1} w^k.$$

This is often called the negative binomial series. Using the ratio test,

$$\lim_{n \to \infty} \left| \frac{w^n (r + n - 1)!/[(r - 1)!\,n!]}{w^{n-1}(r + n - 2)!/[(r - 1)!\,(n - 1)!]} \right| = \lim_{n \to \infty} \left| \frac{w(r + n - 1)}{n} \right| = |w|.$$

Thus the series converges when $|w| < 1$ or $-1 < w < 1$.

A negative binomial random variable receives its name from this negative binomial series. Before showing that relationship, we note that, for $-1 < w < 1$,

$$h(w) = \sum_{k=0}^{\infty} \binom{r + k - 1}{r - 1} w^k = (1 - w)^{-r}$$

$$h'(w) = \sum_{k=1}^{\infty} \binom{r + k - 1}{r - 1} k w^{k-1} = r(1 - w)^{-r-1}$$

$$h''(w) = \sum_{k=2}^{\infty} \binom{r + k - 1}{r - 1} k(k - 1) w^{k-2} = r(r + 1)(1 - w)^{-r-2}.$$

The p.d.f. of a negative binomial random variable X is

$$g(x) = \binom{x - 1}{r - 1} p^r q^{x-r}, \qquad x = r, r + 1, r + 2, \cdots.$$

In the series expansion for $h(w) = (1 - w)^{-r}$, let $x = k + r$. Then

$$\sum_{x=r}^{\infty} \binom{x - 1}{r - 1} w^{x-r} = (1 - w)^{-r}.$$

Letting $w = q$ in this equation, we see that

$$\sum_{x=r}^{\infty} g(x) = \sum_{x=r}^{\infty} \binom{x - 1}{r - 1} p^r q^{x-r} = p^r (1 - q)^{-r} = 1.$$

That is, $g(x)$ does satisfy the properties of a p.d.f.

To find the mean of X, we first find

$$E(X - r) = \sum_{x=r}^{\infty} (x - r) \binom{x-1}{r-1} p^r q^{x-r} = \sum_{x=r+1}^{\infty} (x - r) \binom{x-1}{r-1} p^r q^{x-r}.$$

Letting $k = x - r$ in this latter summation and using the expansion of $h'(w)$ gives us

$$E(X - r) = \sum_{k=1}^{\infty} (k) \binom{r+k-1}{r-1} p^r q^k$$

$$= p^r q \sum_{k=1}^{\infty} (k) \binom{r+k-1}{r-1} k q^{k-1}$$

$$= p^r q r (1 - q)^{-r-1} = r\left(\frac{q}{p}\right).$$

Thus

$$E(X) = r + r\left(\frac{q}{p}\right)$$

$$= r\left(1 + \frac{q}{p}\right) = r\left(\frac{1}{p}\right).$$

Similarly, you can show, using $h''(w)$, that

$$E[(X - r)(X - r - 1)] = \left(\frac{q^2}{p^2}\right)(r)(r + 1)$$

and thus

$$\text{Var}(X) = \text{Var}(X - r) = \left(\frac{q^2}{p^2}\right)(r)(r + 1) + r\left(\frac{q}{p}\right) - r^2\left(\frac{q^2}{p^2}\right) = r\left(\frac{q}{p^2}\right).$$

A very special case of the negative binomial series occurs when $r = 1$, obtaining the well known geometric series

$$(1 - w)^{-1} = 1 + w + w^2 + w^3 + \cdots$$

provided that $-1 < w < 1$.

The geometric series gives its name to the geometric probability distribution. Perhaps you recall the geometric series being written, for $-1 < r < 1$,

$$g(r) = \sum_{k=0}^{\infty} ar^k = \frac{a}{1 - r}.$$

To find the mean and the variance of a geometric random variable X, simply let $r = 1$ in the formulas for the mean and the variance of a negative binomial random variable. However, if you want to find the mean and variance directly, you can use

$$g'(r) = \sum_{k=1}^{\infty} akr^{k-1} = \frac{a}{(1-r)^2}$$

and

$$g''(r) = \sum_{k=2}^{\infty} ak(k-1)r^{k-2} = \frac{2a}{(1-r)^3}$$

to find $E(X)$ and $E[X(X-1)]$, respectively.

In applications associated with the geometric random variable, it is also useful to recall that the nth partial sum of a geometric series is

$$s_n = \sum_{k=0}^{n-1} ar^k = \frac{a(1-r^n)}{1-r}.$$

A bonus in this section is the following logarithmic series that produces a useful tool in daily life. Consider

$$f(x) = \ln(1+x)$$
$$f'(x) = (1+x)^{-1}$$
$$f''(x) = (-1)(1+x)^{-2}$$
$$f'''(x) = (-1)(-2)(1+x)^{-3}$$
$$\vdots$$

Thus $f^{(r)}(0) = (-1)^{r-1}(r-1)!$ and

$$\ln(1+x) = \frac{0!}{1!}x - \frac{1!}{2!}x^2 + \frac{2!}{3!}x^3 - \frac{3!}{4!}x^4 + \cdots$$
$$= x - \frac{x^2}{2} + \frac{x^3}{3} - \frac{x^4}{4} + \cdots$$

which converges for $-1 < x \le 1$.

Now consider the following question:"How long does it take money to double in value if the interest rate is i?" Assuming the compounding is on an annual basis and that you begin with \$1, after one year you have \$$(1+i)$ and after two years the number of dollars you have is

$$(1+i) + i(1+i) = (1+i)^2.$$

Continuing this process, the equation that we have to solve is

$$(1+i)^n = 2,$$

the solution of which is

$$n = \frac{\ln 2}{\ln(1+i)}.$$

To approximate the value of n, recall that $\ln 2 \approx 0.693$ and use the series expansion of $f(x) = \ln(1 + x)$ to obtain

$$n \approx \frac{0.693}{i - \dfrac{i^2}{2} + \dfrac{i^3}{3} - \cdots}.$$

Due to the alternating series in the denominator, the denominator is a little less than i. Frequently brokers increase the numerator a little (say to 0.72) and simply divide by i, obtaining the "well known Rule of 72," namely

$$n \approx \frac{72}{100i}.$$

For example, if $i = 0.08$, then $n \approx 72/8 = 9$ provides an excellent approximation (the answer is about 9.006). Many persons find that the Rule of 72 is extremely useful when dealing with money matters.

A.5 INTEGRATION

If $F'(t) = f(t)$, $a \leq t \leq b$, then

$$\int_a^b f(t)\, dt = F(b) - F(a).$$

Thus if $u(x)$ is such that $u'(x)$ exists and $a \leq u(x)$, then

$$\int_a^{u(x)} f(t)\, dt = F[u(x)] - F(b).$$

Taking derivatives of this latter equation, we obtain

$$D_x \left[\int_a^{u(x)} f(t)\, dt \right] = F'[u(x)]u'(x) = f[u(x)]u'(x).$$

For example, with $0 < v$,

$$D_v \left[2 \int_0^{\sqrt{v}} \frac{1}{\sqrt{2\pi}} e^{-z^2/2}\, dz \right] = \left(\frac{2}{\sqrt{2\pi}} e^{-v/2} \right) \frac{1}{2\sqrt{v}} = \frac{v^{(1/2)-1} e^{-v/2}}{\sqrt{\pi}\, 2^{1/2}}.$$

This is needed in proving that if Z is $N(0, 1)$, then Z^2 is $\chi^2(1)$.

The preceding example could be worked by first changing variables in the integral. That is, first using the fact that

$$\int_a^b f(x)\, dx = \int_{u(a)}^{u(b)} f[w(y)]\, w'(y)\, dy$$

where the monotone increasing (decreasing) function $x = w(y)$ has derivative $w'(y)$

and inverse function $y = u(x)$. In that example, $a = 0$, $b = \sqrt{v}$, $z = \sqrt{t}$, $z' = 1/(2\sqrt{t})$, and $t = z^2$ so that

$$2 \int_0^{\sqrt{v}} \frac{1}{\sqrt{2\pi}} e^{-z^2/2} \, dz = 2 \int_0^v \frac{1}{\sqrt{2\pi}} e^{-t/2} \left(\frac{1}{2\sqrt{t}} \right) dt.$$

The derivative of the latter, by one form of the fundamental theorem of calculus, is

$$2 \frac{1}{\sqrt{2\pi}} e^{-v/2} \left(\frac{1}{2\sqrt{v}} \right) = \frac{v^{(1/2)-1} e^{-v/2}}{\sqrt{\pi}\, 2^{1/2}}.$$

Integration by parts is frequently needed. It is based upon the derivative of the product of two functions of x, say $u(x)$ and $v(x)$, namely

$$D_x[u(x)v(x)] = u(x)v'(x) + v(x)u'(x).$$

Thus

$$[u(x)v(x)]_a^b = \int_a^b u(x)v'(x) \, dx + \int_a^b v(x)u'(x) \, dx$$

or, equivalently,

$$\int_a^b u(x)v'(x) \, dx = [u(x)v(x)]_a^b - \int_a^b v(x)u'(x) \, dx.$$

For example, by letting $u(x) = x$ and $v'(x) = e^{-x}$, we have that

$$\int_0^b xe^{-1} \, dx = \left[-xe^{-x} \right]_0^b - \int_0^b (1)(-e^{-x}) \, dx$$

$$= -be^{-b} + \left[-e^{-x} \right]_0^b = -be^{-b} - e^{-b} + 1$$

because $u'(x) = 1$ and $v(x) = -e^{-x}$.

With some thought about the product rule of differentiation, it is not always necessary to assign $u(x)$ and $v'(x)$, however. For illustration, an integral like

$$\int_0^b x^3 e^{-x} \, dx$$

would require integration by parts three times, the first of which would assign $u(x) = x^3$ and $v'(x) = e^{-x}$. But note that

$$D_x(-x^3 e^{-x}) = x^3 e^{-x} - 3x^2 e^{-x}.$$

That is, $-x^3 e^{-x}$ is "almost" the antiderivative of $x^3 e^{-x}$ except for the undesirable term, $-3x^2 e^{-x}$. Clearly,

$$D_x(-x^3 e^{-x} - 3x^2 e^{-x}) = x^3 e^{-x} - 3x^2 e^{-x} + 3x^2 e^{-x} - 6xe^{-x} = x^3 e^{-x} - 6xe^{-x}.$$

So we eliminated that undesirable term $-3x^2 e^{-x}$, but got another one, namely $-6xe^{-x}$. However,

$$D_x(-x^3 e^{-x} - 3x^2 e^{-x} - 6xe^{-x}) = x^3 e^{-x} - 6e^{-x}$$

and finally

$$D_x(-x^3e^{-x} - 3x^2e^{-x} - 6xe^{-x} - 6e^{-x}) = x^3e^{-x}.$$

That is,

$$-x^3e^{-x} - 3x^2e^{-x} - 6xe^{-x} - 6e^{-x}$$

is the antiderivative of x^3e^{-x} and can be written down without ever assigning u and v.

As practice in this technique, consider

$$\int_0^{\pi/2} x^2 \cos x \, dx = \left[x^2 \sin x + 2x \cos x - 2 \sin x\right]_0^{\pi/2}.$$

Now $x^2 \sin x$ is our first guess because we obtain $x^2 \cos x$ when we differentiate the $\sin x$ factor. But we get the undesirable term $2x \sin x$. That is why we add $2x \cos x$ as the derivative of the $\cos x$ is $-\sin x$ and $-2x \sin x$ eliminates $2x \sin x$. But the second term of the derivative of $2x \cos x$ is $2 \cos x$, which we get rid of by taking the derivative of the next term, $-2 \sin x$.

Possibly the best advice is to take the derivative of the right-hand member, here

$$x^2 \sin x + 2x \cos x - 2 \sin x,$$

and note how the terms cancel leaving only $x^2 \cos x$. Then practice on integrals like

$$\int x^4 e^{-x} \, dx, \qquad \int x^3 \sin x \, dx, \qquad \int x^5 e^x \, dx.$$

A.6 MULTIVARIATE CALCULUS

We really only make some suggestions about functions of two variables, say

$$z = f(x, y).$$

But these remarks can be extended to more than two variables. The two first *partial derivatives* with respect to x and y, denoted by $\dfrac{\partial z}{\partial x}$ and $\dfrac{\partial z}{\partial y}$, can be found in the usual manner of differentiating by treating the "other" variable as a constant. For illustration,

$$\frac{\partial(x^2y + \sin x)}{\partial x} = 2xy + \cos x$$

and

$$\frac{\partial(e^{xy^2})}{\partial y} = (e^{xy^2})(2xy).$$

The second partial derivatives are simply first partial derivatives of the first partial derivatives. If $z = e^{xy^2}$, then

$$\frac{\partial}{\partial x}\left(\frac{\partial z}{\partial y}\right) = \frac{\partial}{\partial x}(2xye^{xy^2}) = 2xye^{xy^2}(y^2) + 2ye^{xy^2}.$$

For notation we use

$$\frac{\partial}{\partial x}\left(\frac{\partial z}{\partial x}\right) = \frac{\partial^2 z}{\partial x^2}, \qquad \frac{\partial}{\partial x}\left(\frac{\partial z}{\partial y}\right) = \frac{\partial^2 z}{\partial x \partial y},$$

$$\frac{\partial}{\partial y}\left(\frac{\partial z}{\partial x}\right) = \frac{\partial^2 z}{\partial y \partial x}, \qquad \frac{\partial}{\partial y}\left(\frac{\partial z}{\partial y}\right) = \frac{\partial^2 z}{\partial y^2},$$

In general,

$$\frac{\partial^2 z}{\partial x \partial y} = \frac{\partial^2 z}{\partial y \partial x},$$

provided the partial derivatives involved are continuous functions.

As you might guess, at a relative maximum or minimum of $z = f(x, y)$, we have

$$\frac{\partial z}{\partial x} = 0 \qquad \text{and} \qquad \frac{\partial z}{\partial y} = 0,$$

provided that the derivatives exist. To assure us that we have a maximum or minimum, we need

$$\left(\frac{\partial^2 z}{\partial x \partial y}\right)^2 - \left(\frac{\partial^2 z}{\partial x^2}\right)\left(\frac{\partial^2 z}{\partial y^2}\right) < 0.$$

Moreover, it is a relative minimum if $\frac{\partial^2 z}{\partial x^2} > 0$ and a relative maximum if $\frac{\partial^2 z}{\partial x^2} < 0$.

A major problem in statistics, called least squares, is to find a and b to minimize

$$K(a, b) = \sum_{i=1}^{n} (y_i - a - bx_i)^2.$$

Thus the solution of the two equations

$$\frac{\partial K}{\partial a} = \sum_{i=1}^{n} 2(y_i - a - bx_i)(-1) = 0$$

$$\frac{\partial K}{\partial b} = \sum_{i=1}^{n} 2(y_i - a - bx_i)(-x_i) = 0$$

would possibly give us a point (a, b) that minimizes $K(a, b)$. Taking second partial derivatives,

$$\frac{\partial^2 K}{\partial a^2} = \sum_{i=1}^{n} 2(-1)(-1) = 2n > 0$$

$$\frac{\partial^2 K}{\partial b^2} = \sum_{i=1}^{n} 2(-x_i)(-x_i) = 2\sum_{i=1}^{n} x_i^2 > 0$$

$$\frac{\partial^2 K}{\partial a \partial b} = \sum_{i=1}^{n} 2(-1)(-x_i) = 2\sum_{i=1}^{n} x_i$$

we note that

$$\left(2 \sum_{i=1}^{n} x_i \right)^2 - (2n) \left(2 \sum_{i=1}^{n} x_i^2 \right) < 0$$

because $(\sum_{i=1}^{n} x_i)^2 < n \sum_{i=1}^{n} x_i^2$ provided all x_i are not equal. Noting that $\dfrac{\partial^2 z}{\partial x^2} > 0$, we see that the solution of the two equations, $\dfrac{\partial K}{\partial a} = 0$ and $\dfrac{\partial K}{\partial b} = 0$, provides the only minimizing solution.

The value of the *double integral*

$$\iint_A f(x, y) \, dx \, dy$$

can usually be evaluated by an iterated process; that is, by evaluating two successive single integrals. For illustration, say $A = \{(x, y): 0 \le x \le 1, 0 \le y \le x\}$ as given in Figure A.4.

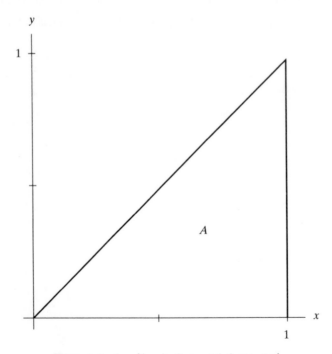

Figure A.4 $A = \{(x, y): 0 \le x \le 1, 0 \le y \le x\}$

Then

$$\iint_A (x + x^3 y^2)\, dx\, dy = \int_0^1 \left[\int_0^x (x + x^3 y^2)\, dy \right] dx$$

$$= \int_0^1 \left[xy + \frac{x^3 y^3}{3} \right]_0^x dx$$

$$= \int_0^1 \left(x^2 + \frac{x^6}{3} \right) dx = \left[\frac{x^3}{3} + \frac{x^7}{3 \cdot 7} \right]_0^1$$

$$= \frac{1}{3} + \frac{1}{21} = \frac{8}{21}.$$

When placing the limits on the iterated integral, note that for each fixed x between zero and one, y is restricted to the interval zero to x. Also in the inner integral on y, x is treated as a constant.

In evaluating this double integral we could have restricted y to the interval zero to one, then x would be between y and one. That is, we would have evaluated the iterated integral

$$\int_0^1 \left[\int_y^1 (x + x^3 y^2)\, dx \right] dy = \int_0^1 \left[\frac{x^2}{2} + \frac{x^4 y^2}{4} \right]_y^1 dy$$

$$= \int_0^1 \left[\frac{1}{2} + \frac{y^2}{4} - \frac{y^2}{2} - \frac{y^6}{4} \right] dy$$

$$= \left[\frac{y}{2} - \frac{y^3}{3 \cdot 4} - \frac{y^7}{7 \cdot 4} \right]_0^1$$

$$= \frac{1}{2} - \frac{1}{12} - \frac{1}{28} = \frac{8}{21}.$$

Finally, we can *change variables* in a double integral

$$\iint_A f(x, y)\, dx\, dy.$$

If $f(x, y)$ is a joint p.d.f. of random variables X and Y of the continuous type, then the double integral represents $P[(X, Y) \in A]$. Consider only one-to-one transformations, say $z = u_1(x, y)$ and $w = u_2(x, y)$ with inverse transformation given by $x = v_1(z, w)$ and $y = v_2(z, w)$. The determinant of order two

$$J = \begin{vmatrix} \dfrac{\partial x}{\partial z} & \dfrac{\partial x}{\partial w} \\[2mm] \dfrac{\partial y}{\partial z} & \dfrac{\partial y}{\partial w} \end{vmatrix}$$

is called the *Jacobian* of the inverse transformation. Moreover, say the region A maps onto B in the (z, w) space.

Since we are usually dealing with probabilities in this book, we fixed the *sign* of the integral so that it is positive (by using the absolute value of the Jacobian). Then it is true that

$$\iint_A f(x, y)\, dx\, dy = \iint_B f[v_1(z, w), v_2(z, w)]\,|J|\, dz\, dw.$$

For illustration, let

$$f(x, y) = \frac{1}{2\pi} e^{-(x^2+y^2)/2}, \qquad -\infty < x < \infty, \qquad -\infty < y < \infty,$$

which is the joint p.d.f. of two independent normal variables, each with mean zero and variance one. Say $A = \{(x, y): 0 \le x^2 + y^2 \le 1\}$ and consider

$$P(A) = \iint_A f(x, y)\, dx\, dy.$$

This is impossible to deal with directly in the x, y variables. However, consider the inverse transformation to polar coordinates, namely

$$x = r\cos\theta, \qquad y = r\sin\theta$$

with Jacobian

$$J = \begin{vmatrix} \cos\theta & -r\sin\theta \\ \sin\theta & r\cos\theta \end{vmatrix} = r(\cos^2\theta + \sin^2\theta) = r.$$

Since A maps onto $B = \{(r, \theta): 0 \le r \le 1,\ 0 \le \theta < 2\pi\}$, we have that

$$P(A) = \int_0^{2\pi} \left(\int_0^1 \frac{1}{2\pi} e^{-r^2/2}\, r\, dr \right) d\theta$$

$$= \int_0^{2\pi} \left[-\frac{1}{2\pi} e^{-r^2/2} \right]_0^1 d\theta$$

$$= \int_0^{2\pi} \frac{1}{2\pi}(1 - e^{-1/2})\, d\theta$$

$$= \frac{1}{2\pi}(1 - e^{-1/2})\, 2\pi = 1 - e^{-1/2}.$$

B

REFERENCES

Aspin, A. A., "Tables for Use in Comparisons Whose Accuracy Involves Two Variances, Separately Estimated," *Biometrika*, **36** (1949), pp. 290–296.

Box, G. E. P., W. G. Hunter, and J. S. Hunter, *Statistics for Experimenters*, John Wiley & Sons, Inc., New York, 1978.

Box, G. E. P., and M. E. Muller, "A Note on the Generation of Random Normal Deviates," *Ann. Math. Statist.*," **29** (1958), p. 610.

Crisman, R., "Shortest Confidence Interval for the Standard Deviation of a Normal Distribution," *J. Undergrad. Math.*, **7**, 2 (1975), p. 57.

Guenther, William C., "Shortest Confidence Intervals," *Amer. Statist.*, **23**, 1(1969), p. 22.

Hogg, R. V., and J. Ledolter, *Applied Statistics for Engineers and Physical Scientists*, 2nd ed., Macmillan Publishing Company, New York, 1992.

Hogg, R. V., and A. T. Craig, *Introduction to Mathematical Statistics*, 5th ed., Prentice Hall, NJ, 1995.

Hogg, R. V., and A. T. Craig, "On the Decomposition of Certain Chi-Square Variables," *Ann. Math. Statist.*, **29** (1958), p. 608.

Karian, Z. A., and E. A. Tanis, *Probability & Statistics Explorations with MAPLE*, Prentice Hall, NJ, 1995.

Montgomery, D. C., *Design and Analysis of Experiments*, 2nd ed., John Wiley & Sons, Inc., New York, 1984.

Natrella, M. G., *Experimental Statistics, National Bureau of Standards Handbook 91*, U.S. Government Printing Office, Washington DC, 1963.

Pearson, K., "On the Criterion That a Given System of Deviations from the Probable in the Case of a Correlated System of Variables Is Such That It Can Be Reasonably Supposed to Have Arisen from Random Sampling," *Phil. Mag.*, Series 5, **50** (1900), p. 157.

Snee, R. D., L. B. Hare, and J. R. Trout, *Experiments in Industry*, American Society of Quality Conrol, Milwaukee, Wis., 1985.

Tate, R. F., and G. W. Klett, "Optimum Confidence Intervals for the Variance of a Normal Distribution," *J. Am. Statist. Assoc.*, **54** (1959), p. 674.

Tukey, J. W., *Exploratory Data Analysis*, Addison-Wesley Publishing Company, Reading, Mass., 1977.

Velleman, P. F., D. C. Hoaglin, *Applications, Basics, and Computing of Exploratory Data Analysis*, Duxbury Press, Boston, 1981.

Wilcoxon, F., "Individual Comparisons by Ranking Methods," *Biometrics Bull.*, **1** (1945), p. 80.

C

TABLES

Table I	Binomial Coefficients	664
Table II	The Binomial Distribution	665–71
Table III	The Poisson Distribution	672–674
Table IV	The Chi-Square Distribution	675
Table Va	The Standard Normal Distribution Function	676
Table Vb	The Standard Normal Right Tail Probabilities	677
Table VI	The t Distribution	678
Table VII	The F Distribution	679–83
Table VIII	Kolmogrov-Smirnov Acceptance Limits	684
Table IX	Random Numbers on the Interval (0, 1)	685
Table X	Divisors for the Confidence Interval for σ of Minimum Length	686
Table XI	Distribution Function of the Correlation Coefficient R	687

TABLE I
Binomial Coefficients

$$\binom{n}{r} = \frac{n!}{r!(n-r)!} = \binom{n}{n-r}$$

n	$\binom{n}{0}$	$\binom{n}{1}$	$\binom{n}{2}$	$\binom{n}{3}$	$\binom{n}{4}$	$\binom{n}{5}$	$\binom{n}{6}$	$\binom{n}{7}$	$\binom{n}{8}$	$\binom{n}{9}$	$\binom{n}{10}$	$\binom{n}{11}$	$\binom{n}{12}$	$\binom{n}{13}$
0	1													
1	1	1												
2	1	2	1											
3	1	3	3	1										
4	1	4	6	4	1									
5	1	5	10	10	5	1								
6	1	6	15	20	15	6	1							
7	1	7	21	35	35	21	7	1						
8	1	8	28	56	70	56	28	8	1					
9	1	9	36	84	126	126	84	36	9	1				
10	1	10	45	120	210	252	210	120	45	10	1			
11	1	11	55	165	330	462	462	330	165	55	11	1		
12	1	12	66	220	495	792	924	792	495	220	66	12	1	
13	1	13	78	286	715	1,287	1,716	1,716	1,287	715	286	78	13	1
14	1	14	91	364	1,001	2,002	3,003	3,432	3,003	2,002	1,001	364	91	14
15	1	15	105	455	1,365	3,003	5,005	6,435	6,435	5,005	3,003	1,365	455	105
16	1	16	120	560	1,820	4,368	8,008	11,440	12,870	11,440	8,008	4,368	1,820	560
17	1	17	136	680	2,380	6,188	12,376	19,448	24,310	24,310	19,448	12,376	6,188	2,380
18	1	18	153	816	3,060	8,568	18,564	31,824	43,758	48,620	43,758	31,824	18,564	8,568
19	1	19	171	969	3,876	11,628	27,132	50,388	75,582	92,378	92,378	75,582	50,388	27,132
20	1	20	190	1,140	4,845	15,504	38,760	77,520	125,970	167,960	184,756	167,960	125,970	77,520
21	1	21	210	1,330	5,985	20,349	54,264	116,280	203,490	293,930	352,716	352,716	293,930	203,490
22	1	22	231	1,540	7,315	26,334	74,613	170,544	319,770	497,420	646,646	705,432	646,646	497,420
23	1	23	253	1,771	8,855	33,649	100,947	245,157	490,314	817,190	1,144,066	1,352,078	1,352,078	1,144,066
24	1	24	276	2,024	10,626	42,504	134,596	346,104	735,471	1,307,504	1,961,256	2,496,144	2,704,156	2,496,144
25	1	25	300	2,300	12,650	53,130	177,100	480,700	1,081,575	2,042,975	3,268,760	4,457,400	5,200,300	5,200,300
26	1	26	325	2,600	14,950	65,780	230,230	657,800	1,562,275	3,124,550	5,311,735	7,726,160	9,657,700	10,400,600

For $r > 13$ you may use the identity $\binom{n}{r} = \binom{n}{n-r}$.

TABLE II
The Binomial Distribution

$$F(x) = P(X \leq x) = \sum_{k=0}^{x} \frac{n!}{k!(n-k)!} p^k (1-p)^{n-k}$$

								p					
n	x	0.05	0.10	0.15	0.20	0.25	0.30	0.35	0.40	0.45	0.50
2	0	0.9025	0.8100	0.7225	0.6400	0.5625	0.4900	0.4225	0.3600	0.3025	0.2500
	1	0.9975	0.9900	0.9775	0.9600	0.9375	0.9100	0.8775	0.8400	0.7975	0.7500
	2	1.0000	1.0000	1.0000	1.0000	1.0000	1.0000	1.0000	1.0000	1.0000	1.0000
3	0	0.8574	0.7290	0.6141	0.5120	0.4219	0.3430	0.2746	0.2160	0.1664	0.1250
	1	0.9928	0.9720	0.9392	0.8960	0.8438	0.7840	0.7182	0.6480	0.5748	0.5000
	2	0.9999	0.9990	0.9966	0.9920	0.9844	0.9730	0.9571	0.9360	0.9089	0.8750
	3	1.0000	1.0000	1.0000	1.0000	1.0000	1.0000	1.0000	1.0000	1.0000	1.0000
4	0	0.8145	0.6561	0.5220	0.4096	0.3164	0.2401	0.1785	0.1296	0.0915	0.0625
	1	0.9860	0.9477	0.8905	0.8192	0.7383	0.6517	0.5630	0.4752	0.3910	0.3125
	2	0.9995	0.9963	0.9880	0.9728	0.9492	0.9163	0.8735	0.8208	0.7585	0.6875
	3	1.0000	0.9999	0.9995	0.9984	0.9961	0.9919	0.9850	0.9744	0.9590	0.9375
	4	1.0000	1.0000	1.0000	1.0000	1.0000	1.0000	1.0000	1.0000	1.0000	1.0000

TABLE II (continued)

$$F(x) = P(X \le x) = \sum_{k=0}^{x} \frac{n!}{k!(n-k)!} p^k (1-p)^{n-k}$$

n	x	\multicolumn{10}{c}{p}
		0.05	0.10	0.15	0.20	0.25	0.30	0.35	0.40	0.45	0.50
5	0	0.7738	0.5905	0.4437	0.3277	0.2373	0.1681	0.1160	0.0778	0.0503	0.0312
	1	0.9774	0.9185	0.8352	0.7373	0.6328	0.5282	0.4284	0.3370	0.2562	0.1875
	2	0.9988	0.9914	0.9734	0.9421	0.8965	0.8369	0.7648	0.6826	0.5931	0.5000
	3	1.0000	0.9995	0.9978	0.9933	0.9844	0.9692	0.9460	0.9130	0.8688	0.8125
	4	1.0000	1.0000	0.9999	0.9997	0.9990	0.9976	0.9947	0.9898	0.9815	0.9688
	5	1.0000	1.0000	1.0000	1.0000	1.0000	1.0000	1.0000	1.0000	1.0000	1.0000
6	0	0.7351	0.5314	0.3771	0.2621	0.1780	0.1176	0.0754	0.0467	0.0277	0.0156
	1	0.9672	0.8857	0.7765	0.6553	0.5339	0.4202	0.3191	0.2333	0.1636	0.1094
	2	0.9978	0.9842	0.9527	0.9011	0.8306	0.7443	0.6471	0.5443	0.4415	0.3438
	3	0.9999	0.9987	0.9941	0.9830	0.9624	0.9295	0.8826	0.8208	0.7447	0.6562
	4	1.0000	0.9999	0.9996	0.9984	0.9954	0.9891	0.9777	0.9590	0.9308	0.8906
	5	1.0000	1.0000	1.0000	0.9999	0.9998	0.9993	0.9982	0.9959	0.9917	0.9844
	6	1.0000	1.0000	1.0000	1.0000	1.0000	1.0000	1.0000	1.0000	1.0000	1.0000
7	0	0.6983	0.4783	0.3206	0.2097	0.1335	0.0824	0.0490	0.0280	0.0152	0.0078
	1	0.9556	0.8503	0.7166	0.5767	0.4449	0.3294	0.2338	0.1586	0.1024	0.0625
	2	0.9962	0.9743	0.9262	0.8520	0.7564	0.6471	0.5323	0.4199	0.3164	0.2266
	3	0.9998	0.9973	0.9879	0.9667	0.9294	0.8740	0.8002	0.7102	0.6083	0.5000
	4	1.0000	0.9998	0.9988	0.9953	0.9871	0.9712	0.9444	0.9037	0.8471	0.7734
	5	1.0000	1.0000	0.9999	0.9996	0.9987	0.9962	0.9910	0.9812	0.9643	0.9375
	6	1.0000	1.0000	1.0000	1.0000	0.9999	0.9998	0.9994	0.9984	0.9963	0.9922
	7	1.0000	1.0000	1.0000	1.0000	1.0000	1.0000	1.0000	1.0000	1.0000	1.0000
8	0	0.6634	0.4305	0.2725	0.1678	0.1001	0.0576	0.0319	0.0168	0.0084	0.0039
	1	0.9428	0.8131	0.6572	0.5033	0.3671	0.2553	0.1691	0.1064	0.0632	0.0352
	2	0.9942	0.9619	0.8948	0.7969	0.6785	0.5518	0.4278	0.3154	0.2201	0.1445
	3	0.9996	0.9950	0.9786	0.9437	0.8862	0.8059	0.7064	0.5941	0.4770	0.3633

n	x										
	4	1.0000	0.9996	0.9971	0.9896	0.9727	0.9420	0.8939	0.8263	0.7396	0.6367
	5	1.0000	1.0000	0.9998	0.9988	0.9958	0.9887	0.9747	0.9502	0.9115	0.8555
	6	1.0000	1.0000	1.0000	0.9999	0.9996	0.9987	0.9964	0.9915	0.9819	0.9648
	7	1.0000	1.0000	1.0000	1.0000	1.0000	0.9999	0.9998	0.9993	0.9983	0.9961
	8	1.0000	1.0000	1.0000	1.0000	1.0000	1.0000	1.0000	1.0000	1.0000	1.0000
9	0	0.6302	0.3874	0.2316	0.1342	0.0751	0.0404	0.0207	0.0101	0.0046	0.0020
	1	0.9288	0.7748	0.5995	0.4362	0.3003	0.1960	0.1211	0.0705	0.0385	0.0195
	2	0.9916	0.9470	0.8591	0.7382	0.6007	0.4628	0.3373	0.2318	0.1495	0.0898
	3	0.9994	0.9917	0.9661	0.9144	0.8343	0.7297	0.6089	0.4826	0.3614	0.2539
	4	1.0000	0.9991	0.9944	0.9804	0.9511	0.9012	0.8283	0.7334	0.6214	0.5000
	5	1.0000	0.9999	0.9994	0.9969	0.9900	0.9747	0.9464	0.9006	0.8342	0.7461
	6	1.0000	1.0000	1.0000	0.9997	0.9987	0.9957	0.9888	0.9750	0.9502	0.9102
	7	1.0000	1.0000	1.0000	1.0000	0.9999	0.9996	0.9986	0.9962	0.9909	0.9805
	8	1.0000	1.0000	1.0000	1.0000	1.0000	1.0000	0.9999	0.9997	0.9992	0.9980
	9	1.0000	1.0000	1.0000	1.0000	1.0000	1.0000	1.0000	1.0000	1.0000	1.0000
10	0	0.5987	0.3487	0.1969	0.1074	0.0563	0.0282	0.0135	0.0060	0.0025	0.0010
	1	0.9139	0.7361	0.5443	0.3758	0.2440	0.1493	0.0860	0.0464	0.0233	0.0107
	2	0.9885	0.9298	0.8202	0.6778	0.5256	0.3828	0.2616	0.1673	0.0996	0.0547
	3	0.9990	0.9872	0.9500	0.8791	0.7759	0.6496	0.5138	0.3823	0.2660	0.1719
	4	0.9999	0.9984	0.9901	0.9672	0.9219	0.8497	0.7515	0.6331	0.5044	0.3770
	5	1.0000	0.9999	0.9986	0.9936	0.9803	0.9527	0.9051	0.8338	0.7384	0.6230
	6	1.0000	1.0000	0.9999	0.9991	0.9965	0.9894	0.9740	0.9452	0.8980	0.8281
	7	1.0000	1.0000	1.0000	0.9999	0.9996	0.9984	0.9952	0.9877	0.9726	0.9453
	8	1.0000	1.0000	1.0000	1.0000	1.0000	0.9999	0.9995	0.9983	0.9955	0.9893
	9	1.0000	1.0000	1.0000	1.0000	1.0000	1.0000	1.0000	0.9999	0.9997	0.9990
	10	1.0000	1.0000	1.0000	1.0000	1.0000	1.0000	1.0000	1.0000	1.0000	1.0000
11	0	0.5688	0.3138	0.1673	0.0859	0.0422	0.0198	0.0088	0.0036	0.0014	0.0005
	1	0.8981	0.6974	0.4922	0.3221	0.1971	0.1130	0.0606	0.0302	0.0139	0.0059
	2	0.9848	0.9104	0.7788	0.6174	0.4552	0.3127	0.2001	0.1189	0.0652	0.0327
	3	0.9984	0.9815	0.9306	0.8389	0.7133	0.5696	0.4256	0.2963	0.1911	0.1133
	4	0.9999	0.9972	0.9841	0.9496	0.8854	0.7897	0.6683	0.5328	0.3971	0.2744
	5	1.0000	0.9997	0.9973	0.9883	0.9657	0.9218	0.8513	0.7535	0.6331	0.5000
	6	1.0000	1.0000	0.9997	0.9980	0.9924	0.9784	0.9499	0.9006	0.8262	0.7256
	7	1.0000	1.0000	1.0000	0.9998	0.9988	0.9957	0.9878	0.9707	0.9390	0.8867

TABLE II (continued)

$$F(x) = P(X \le x) = \sum_{k=0}^{x} \frac{n!}{k!(n-k)!} p^k (1-p)^{n-k}$$

n	x	0.05	0.10	0.15	0.20	0.25	0.30	0.35	0.40	0.45	0.50
	8	1.0000	1.0000	1.0000	1.0000	0.9999	0.9994	0.9980	0.9941	0.9852	0.9673
	9	1.0000	1.0000	1.0000	1.0000	1.0000	1.0000	0.9998	0.9993	0.9978	0.9941
	10	1.0000	1.0000	1.0000	1.0000	1.0000	1.0000	1.0000	1.0000	0.9998	0.9995
	11	1.0000	1.0000	1.0000	1.0000	1.0000	1.0000	1.0000	1.0000	1.0000	1.0000
12	0	0.5404	0.2824	0.1422	0.0687	0.0317	0.0138	0.0057	0.0022	0.0008	0.0002
	1	0.8816	0.6590	0.4435	0.2749	0.1584	0.0850	0.0424	0.0196	0.0083	0.0032
	2	0.9804	0.8891	0.7358	0.5583	0.3907	0.2528	0.1513	0.0834	0.0421	0.0193
	3	0.9978	0.9744	0.9078	0.7946	0.6488	0.4925	0.3467	0.2253	0.1345	0.0730
	4	0.9998	0.9957	0.9761	0.9274	0.8424	0.7237	0.5833	0.4382	0.3044	0.1938
	5	1.0000	0.9995	0.9954	0.9806	0.9456	0.8822	0.7873	0.6652	0.5269	0.3872
	6	1.0000	0.9999	0.9993	0.9961	0.9857	0.9614	0.9154	0.8418	0.7393	0.6128
	7	1.0000	1.0000	0.9999	0.9994	0.9972	0.9905	0.9745	0.9427	0.8883	0.8062
	8	1.0000	1.0000	1.0000	0.9999	0.9996	0.9983	0.9944	0.9847	0.9644	0.9270
	9	1.0000	1.0000	1.0000	1.0000	1.0000	0.9998	0.9992	0.9972	0.9921	0.9807
	10	1.0000	1.0000	1.0000	1.0000	1.0000	1.0000	0.9999	0.9997	0.9989	0.9968
	11	1.0000	1.0000	1.0000	1.0000	1.0000	1.0000	1.0000	1.0000	0.9999	0.9998
	12	1.0000	1.0000	1.0000	1.0000	1.0000	1.0000	1.0000	1.0000	1.0000	1.0000
13	0	0.5133	0.2542	0.1209	0.0550	0.0238	0.0097	0.0037	0.0013	0.0004	0.0001
	1	0.8646	0.6213	0.3983	0.2336	0.1267	0.0637	0.0296	0.0126	0.0049	0.0017
	2	0.9755	0.8661	0.6920	0.5017	0.3326	0.2025	0.1132	0.0579	0.0269	0.0112
	3	0.9969	0.9658	0.8820	0.7473	0.5843	0.4206	0.2783	0.1686	0.0929	0.0461
	4	0.9997	0.9935	0.9658	0.9009	0.7940	0.6543	0.5005	0.3530	0.2279	0.1334
	5	1.0000	0.9991	0.9924	0.9700	0.9198	0.8346	0.7159	0.5744	0.4268	0.2905
	6	1.0000	0.9999	0.9987	0.9930	0.9757	0.9376	0.8705	0.7712	0.6437	0.5000
	7	1.0000	1.0000	0.9998	0.9988	0.9944	0.9818	0.9538	0.9023	0.8212	0.7095
	8	1.0000	1.0000	1.0000	0.9998	0.9990	0.9960	0.9874	0.9679	0.9302	0.8666

n	k										
	9	0.9539	0.9797	0.9922	0.9975	0.9993	0.9999	1.0000	1.0000	1.0000	1.0000
	10	0.9888	0.9959	0.9987	0.9997	0.9999	1.0000	1.0000	1.0000	1.0000	1.0000
	11	0.9983	0.9995	0.9999	1.0000	1.0000	1.0000	1.0000	1.0000	1.0000	1.0000
	12	0.9999	1.0000	1.0000	1.0000	1.0000	1.0000	1.0000	1.0000	1.0000	1.0000
	13	1.0000	1.0000	1.0000	1.0000	1.0000	1.0000	1.0000	1.0000	1.0000	1.0000
14	0	0.0001	0.0002	0.0008	0.0024	0.0068	0.0178	0.0440	0.1028	0.2288	0.4877
	1	0.0009	0.0029	0.0081	0.0205	0.0475	0.1010	0.1979	0.3567	0.5846	0.8470
	2	0.0065	0.0170	0.0398	0.0839	0.1608	0.2811	0.4481	0.6479	0.8416	0.9699
	3	0.0287	0.0632	0.1243	0.2205	0.3552	0.5213	0.6982	0.8535	0.9559	0.9958
	4	0.0898	0.1672	0.2793	0.4227	0.5842	0.7415	0.8702	0.9533	0.9908	0.9996
	5	0.2120	0.3373	0.4859	0.6405	0.7805	0.8883	0.9561	0.9885	0.9985	1.0000
	6	0.3953	0.5461	0.6925	0.8164	0.9067	0.9617	0.9884	0.9978	0.9998	1.0000
	7	0.6047	0.7414	0.8499	0.9247	0.9685	0.9897	0.9976	0.9997	1.0000	1.0000
	8	0.7880	0.8811	0.9417	0.9757	0.9917	0.9978	0.9996	1.0000	1.0000	1.0000
	9	0.9102	0.9574	0.9825	0.9940	0.9983	0.9997	1.0000	1.0000	1.0000	1.0000
	10	0.9713	0.9886	0.9961	0.9989	0.9998	1.0000	1.0000	1.0000	1.0000	1.0000
	11	0.9935	0.9978	0.9994	0.9999	1.0000	1.0000	1.0000	1.0000	1.0000	1.0000
	12	0.9991	0.9997	0.9999	1.0000	1.0000	1.0000	1.0000	1.0000	1.0000	1.0000
	13	0.9999	1.0000	1.0000	1.0000	1.0000	1.0000	1.0000	1.0000	1.0000	1.0000
	14	1.0000	1.0000	1.0000	1.0000	1.0000	1.0000	1.0000	1.0000	1.0000	1.0000
15	0	0.0000	0.0001	0.0005	0.0016	0.0047	0.0134	0.0352	0.0874	0.2059	0.4633
	1	0.0005	0.0017	0.0052	0.0142	0.0353	0.0802	0.1671	0.3186	0.5490	0.8290
	2	0.0037	0.0107	0.0271	0.0617	0.1268	0.2361	0.3980	0.6042	0.8159	0.9638
	3	0.0176	0.0424	0.0905	0.1727	0.2969	0.4613	0.6482	0.8227	0.9444	0.9945
	4	0.0592	0.1204	0.2173	0.3519	0.5155	0.6865	0.8358	0.9383	0.9873	0.9994
	5	0.1509	0.2608	0.4032	0.5643	0.7216	0.8516	0.9389	0.9832	0.9978	0.9999
	6	0.3036	0.4522	0.6098	0.7548	0.8689	0.9434	0.9819	0.9964	0.9997	1.0000
	7	0.5000	0.6535	0.7869	0.8868	0.9500	0.9827	0.9958	0.9994	1.0000	1.0000
	8	0.6964	0.8182	0.9050	0.9578	0.9848	0.9958	0.9992	0.9999	1.0000	1.0000
	9	0.8491	0.9231	0.9662	0.9876	0.9963	0.9992	0.9999	1.0000	1.0000	1.0000
	10	0.9408	0.9745	0.9907	0.9972	0.9993	0.9999	1.0000	1.0000	1.0000	1.0000
	11	0.9824	0.9937	0.9981	0.9995	0.9999	1.0000	1.0000	1.0000	1.0000	1.0000
	12	0.9963	0.9989	0.9997	0.9999	1.0000	1.0000	1.0000	1.0000	1.0000	1.0000
	13	0.9995	0.9999	1.0000	1.0000	1.0000	1.0000	1.0000	1.0000	1.0000	1.0000
	14	1.0000	1.0000	1.0000	1.0000	1.0000	1.0000	1.0000	1.0000	1.0000	1.0000
	15	1.0000	1.0000	1.0000	1.0000	1.0000	1.0000	1.0000	1.0000	1.0000	1.0000

TABLE II (continued)

$$F(x) = P(X \le x) = \sum_{k=0}^{x} \frac{n!}{k!(n-k)!} p^k (1-p)^{n-k}$$

n	x	p 0.05	0.10	0.15	0.20	0.25	0.30	0.35	0.40	0.45	0.50
16	0	0.4401	0.1853	0.0743	0.0281	0.0100	0.0033	0.0010	0.0003	0.0001	0.0000
	1	0.8108	0.5147	0.2839	0.1407	0.0635	0.0261	0.0098	0.0033	0.0010	0.0003
	2	0.9571	0.7892	0.5614	0.3518	0.1971	0.0994	0.0451	0.0183	0.0066	0.0021
	3	0.9930	0.9316	0.7899	0.5981	0.4050	0.2459	0.1339	0.0651	0.0281	0.0106
	4	0.9991	0.9830	0.9209	0.7982	0.6302	0.4499	0.2892	0.1666	0.0853	0.0384
	5	0.9999	0.9967	0.9765	0.9183	0.8103	0.6598	0.4900	0.3288	0.1976	0.1051
	6	1.0000	0.9995	0.9944	0.9733	0.9204	0.8247	0.6881	0.5272	0.3660	0.2272
	7	1.0000	0.9999	0.9989	0.9930	0.9729	0.9256	0.8406	0.7161	0.5629	0.4018
	8	1.0000	1.0000	0.9998	0.9985	0.9925	0.9743	0.9329	0.8577	0.7441	0.5982
	9	1.0000	1.0000	1.0000	0.9998	0.9984	0.9929	0.9771	0.9417	0.8759	0.7728
	10	1.0000	1.0000	1.0000	1.0000	0.9997	0.9984	0.9938	0.9809	0.9514	0.8949
	11	1.0000	1.0000	1.0000	1.0000	1.0000	0.9997	0.9987	0.9951	0.9851	0.9616
	12	1.0000	1.0000	1.0000	1.0000	1.0000	1.0000	0.9998	0.9991	0.9965	0.9894
	13	1.0000	1.0000	1.0000	1.0000	1.0000	1.0000	1.0000	0.9999	0.9994	0.9979
	14	1.0000	1.0000	1.0000	1.0000	1.0000	1.0000	1.0000	1.0000	0.9999	0.9997
	15	1.0000	1.0000	1.0000	1.0000	1.0000	1.0000	1.0000	1.0000	1.0000	1.0000
	16	1.0000	1.0000	1.0000	1.0000	1.0000	1.0000	1.0000	1.0000	1.0000	1.0000
20	0	0.3585	0.1216	0.0388	0.0115	0.0032	0.0008	0.0002	0.0000	0.0000	0.0000
	1	0.7358	0.3917	0.1756	0.0692	0.0243	0.0076	0.0021	0.0005	0.0001	0.0000
	2	0.9245	0.6769	0.4049	0.2061	0.0913	0.0355	0.0121	0.0036	0.0009	0.0002
	3	0.9841	0.8670	0.6477	0.4114	0.2252	0.1071	0.0444	0.0160	0.0049	0.0013
	4	0.9974	0.9568	0.8298	0.6296	0.4148	0.2375	0.1182	0.0510	0.0189	0.0059
	5	0.9997	0.9887	0.9327	0.8042	0.6172	0.4164	0.2454	0.1256	0.0553	0.0207
	6	1.0000	0.9976	0.9781	0.9133	0.7858	0.6080	0.4166	0.2500	0.1299	0.0577
	7	1.0000	0.9996	0.9941	0.9679	0.8982	0.7723	0.6010	0.4159	0.2520	0.1316
	8	1.0000	0.9999	0.9987	0.9900	0.9591	0.8867	0.7624	0.5956	0.4143	0.2517
	9	1.0000	1.0000	0.9998	0.9974	0.9861	0.9520	0.8782	0.7553	0.5914	0.4119

10	0.5881	0.7507	0.8725	0.9468	0.9829	0.9961	0.9994	1.0000	1.0000
11	0.7483	0.8692	0.9435	0.9804	0.9949	0.9991	0.9999	1.0000	1.0000
12	0.8684	0.9420	0.9790	0.9940	0.9987	0.9998	1.0000	1.0000	1.0000
13	0.9423	0.9786	0.9935	0.9985	0.9997	1.0000	1.0000	1.0000	1.0000
14	0.9793	0.9936	0.9984	0.9997	1.0000	1.0000	1.0000	1.0000	1.0000
15	0.9941	0.9985	0.9997	1.0000	1.0000	1.0000	1.0000	1.0000	1.0000
16	0.9987	0.9997	1.0000	1.0000	1.0000	1.0000	1.0000	1.0000	1.0000
17	0.9998	1.0000	1.0000	1.0000	1.0000	1.0000	1.0000	1.0000	1.0000
18	1.0000	1.0000	1.0000	1.0000	1.0000	1.0000	1.0000	1.0000	1.0000
19	1.0000	1.0000	1.0000	1.0000	1.0000	1.0000	1.0000	1.0000	1.0000
20	1.0000	1.0000	1.0000	1.0000	1.0000	1.0000	1.0000	1.0000	1.0000
0	0.0000	0.0000	0.0000	0.0000	0.0001	0.0008	0.0038	0.0172	0.0718
1	0.0000	0.0000	0.0001	0.0003	0.0016	0.0070	0.0274	0.0931	0.2712
2	0.0000	0.0001	0.0004	0.0021	0.0090	0.0321	0.0982	0.2537	0.5371
3	0.0001	0.0005	0.0024	0.0097	0.0332	0.0962	0.2340	0.4711	0.7636
4	0.0005	0.0023	0.0095	0.0320	0.0905	0.2137	0.4207	0.6821	0.9020
5	0.0020	0.0086	0.0294	0.0826	0.1935	0.3783	0.6167	0.8385	0.9666
6	0.0073	0.0258	0.0736	0.1734	0.3407	0.5611	0.7800	0.9305	0.9905
7	0.0216	0.0639	0.1536	0.3061	0.5118	0.7265	0.8909	0.9745	0.9977
8	0.0539	0.1340	0.2735	0.4668	0.6769	0.8506	0.9532	0.9920	0.9995
9	0.1148	0.2424	0.4246	0.6303	0.8106	0.9287	0.9827	0.9979	0.9999
10	0.2122	0.3843	0.5858	0.7712	0.9022	0.9703	0.9944	0.9995	1.0000
11	0.3450	0.5426	0.7323	0.8746	0.9558	0.9893	0.9985	0.9999	1.0000
12	0.5000	0.6937	0.8462	0.9396	0.9825	0.9966	0.9996	1.0000	1.0000
13	0.6550	0.8173	0.9222	0.9745	0.9940	0.9991	0.9999	1.0000	1.0000
14	0.7878	0.9040	0.9656	0.9907	0.9982	0.9998	1.0000	1.0000	1.0000
15	0.8852	0.9560	0.9868	0.9971	0.9995	1.0000	1.0000	1.0000	1.0000
16	0.9461	0.9826	0.9957	0.9992	0.9999	1.0000	1.0000	1.0000	1.0000
17	0.9784	0.9942	0.9988	0.9998	1.0000	1.0000	1.0000	1.0000	1.0000
18	0.9927	0.9984	0.9997	1.0000	1.0000	1.0000	1.0000	1.0000	1.0000
19	0.9980	0.9996	0.9999	1.0000	1.0000	1.0000	1.0000	1.0000	1.0000
20	0.9995	0.9999	1.0000	1.0000	1.0000	1.0000	1.0000	1.0000	1.0000
21	0.9999	1.0000	1.0000	1.0000	1.0000	1.0000	1.0000	1.0000	1.0000
22	1.0000	1.0000	1.0000	1.0000	1.0000	1.0000	1.0000	1.0000	1.0000
23	1.0000	1.0000	1.0000	1.0000	1.0000	1.0000	1.0000	1.0000	1.0000
24	1.0000	1.0000	1.0000	1.0000	1.0000	1.0000	1.0000	1.0000	1.0000
25	1.0000	1.0000	1.0000	1.0000	1.0000	1.0000	1.0000	1.0000	1.0000

25

TABLE III
The Poisson Distribution

Poisson, λ= 3.8

Poisson, λ = 3.8

$$F(x) = P(X \le x) = \sum_{k=0}^{x} \frac{\lambda^k e^{-\lambda}}{k!}$$

	$\lambda = E(X)$									
x	0.1	0.2	0.3	0.4	0.5	0.6	0.7	0.8	0.9	1.0
0	0.905	0.819	0.741	0.670	0.607	0.549	0.497	0.449	0.407	0.368
1	0.995	0.982	0.963	0.938	0.910	0.878	0.844	0.809	0.772	0.736
2	1.000	0.999	0.996	0.992	0.986	0.977	0.966	0.953	0.937	0.920
3	1.000	1.000	1.000	0.999	0.998	0.997	0.994	0.991	0.987	0.981
4	1.000	1.000	1.000	1.000	1.000	1.000	0.999	0.999	0.998	0.996
5	1.000	1.000	1.000	1.000	1.000	1.000	1.000	1.000	1.000	0.999
6	1.000	1.000	1.000	1.000	1.000	1.000	1.000	1.000	1.000	1.000

x	1.1	1.2	1.3	1.4	1.5	1.6	1.7	1.8	1.9	2.0
0	0.333	0.301	0.273	0.247	0.223	0.202	0.183	0.165	0.150	0.135
1	0.699	0.663	0.627	0.592	0.558	0.525	0.493	0.463	0.434	0.406
2	0.900	0.879	0.857	0.833	0.809	0.783	0.757	0.731	0.704	0.677
3	0.974	0.966	0.957	0.946	0.934	0.921	0.907	0.891	0.875	0.857
4	0.995	0.992	0.989	0.986	0.981	0.976	0.970	0.964	0.956	0.947
5	0.999	0.998	0.998	0.997	0.996	0.994	0.992	0.990	0.987	0.983
6	1.000	1.000	1.000	0.999	0.999	0.999	0.998	0.997	0.997	0.995
7	1.000	1.000	1.000	1.000	1.000	1.000	1.000	0.999	0.999	0.999
8	1.000	1.000	1.000	1.000	1.000	1.000	1.000	1.000	1.000	1.000

x	2.2	2.4	2.6	2.8	3.0	3.2	3.4	3.6	3.8	4.0
0	0.111	0.091	0.074	0.061	0.050	0.041	0.033	0.027	0.022	0.018
1	0.355	0.308	0.267	0.231	0.199	0.171	0.147	0.126	0.107	0.092
2	0.623	0.570	0.518	0.469	0.423	0.380	0.340	0.303	0.269	0.238
3	0.819	0.779	0.736	0.692	0.647	0.603	0.558	0.515	0.473	0.433
4	0.928	0.904	0.877	0.848	0.815	0.781	0.744	0.706	0.668	0.629
5	0.975	0.964	0.951	0.935	0.916	0.895	0.871	0.844	0.816	0.785
6	0.993	0.988	0.983	0.976	0.966	0.955	0.942	0.927	0.909	0.889
7	0.998	0.997	0.995	0.992	0.988	0.983	0.977	0.969	0.960	0.949
8	1.000	0.999	0.999	0.998	0.996	0.994	0.992	0.988	0.984	0.979
9	1.000	1.000	1.000	0.999	0.999	0.998	0.997	0.996	0.994	0.992

TABLE III *(continued)*

10	1.000	1.000	1.000	1.000	1.000	1.000	0.999	0.999	0.998	0.997
11	1.000	1.000	1.000	1.000	1.000	1.000	1.000	1.000	0.999	0.999
12	1.000	1.000	1.000	1.000	1.000	1.000	1.000	1.000	1.000	1.000

x	4.2	4.4	4.6	4.8	5.0	5.2	5.4	5.6	5.8	6.0
0	0.015	0.012	0.010	0.008	0.007	0.006	0.005	0.004	0.003	0.002
1	0.078	0.066	0.056	0.048	0.040	0.034	0.029	0.024	0.021	0.017
2	0.210	0.185	0.163	0.143	0.125	0.109	0.095	0.082	0.072	0.062
3	0.395	0.359	0.326	0.294	0.265	0.238	0.213	0.191	0.170	0.151
4	0.590	0.551	0.513	0.476	0.440	0.406	0.373	0.342	0.313	0.285
5	0.753	0.720	0.686	0.651	0.616	0.581	0.546	0.512	0.478	0.446
6	0.867	0.844	0.818	0.791	0.762	0.732	0.702	0.670	0.638	0.606
7	0.936	0.921	0.905	0.887	0.867	0.845	0.822	0.797	0.771	0.744
8	0.972	0.964	0.955	0.944	0.932	0.918	0.903	0.886	0.867	0.847
9	0.989	0.985	0.980	0.975	0.968	0.960	0.951	0.941	0.929	0.916
10	0.996	0.994	0.992	0.990	0.986	0.982	0.977	0.972	0.965	0.957
11	0.999	0.998	0.997	0.996	0.995	0.993	0.990	0.988	0.984	0.980
12	1.000	0.999	0.999	0.999	0.998	0.997	0.996	0.995	0.993	0.991
13	1.000	1.000	1.000	1.000	0.999	0.999	0.999	0.998	0.997	0.996
14	1.000	1.000	1.000	1.000	1.000	1.000	0.999	0.999	0.999	0.999
15	1.000	1.000	1.000	1.000	1.000	1.000	1.000	1.000	1.000	0.999
16	1.000	1.000	1.000	1.000	1.000	1.000	1.000	1.000	1.000	1.000

x	6.5	7.0	7.5	8.0	8.5	9.0	9.5	10.0	10.5	11.0
0	0.002	0.001	0.001	0.000	0.000	0.000	0.000	0.000	0.000	0.000
1	0.011	0.007	0.005	0.003	0.002	0.001	0.001	0.000	0.000	0.000
2	0.043	0.030	0.020	0.014	0.009	0.006	0.004	0.003	0.002	0.001
3	0.112	0.082	0.059	0.042	0.030	0.021	0.015	0.010	0.007	0.005
4	0.224	0.173	0.132	0.100	0.074	0.055	0.040	0.029	0.021	0.015
5	0.369	0.301	0.241	0.191	0.150	0.116	0.089	0.067	0.050	0.038
6	0.527	0.450	0.378	0.313	0.256	0.207	0.165	0.130	0.102	0.079
7	0.673	0.599	0.525	0.453	0.386	0.324	0.269	0.220	0.179	0.143
8	0.792	0.729	0.662	0.593	0.523	0.456	0.392	0.333	0.279	0.232
9	0.877	0.830	0.776	0.717	0.653	0.587	0.522	0.458	0.397	0.341
10	0.933	0.901	0.862	0.816	0.763	0.706	0.645	0.583	0.521	0.460
11	0.966	0.947	0.921	0.888	0.849	0.803	0.752	0.697	0.639	0.579
12	0.984	0.973	0.957	0.936	0.909	0.876	0.836	0.792	0.742	0.689
13	0.993	0.987	0.978	0.966	0.949	0.926	0.898	0.864	0.825	0.781
14	0.997	0.994	0.990	0.983	0.973	0.959	0.940	0.917	0.888	0.854
15	0.999	0.998	0.995	0.992	0.986	0.978	0.967	0.951	0.932	0.907
16	1.000	0.999	0.998	0.996	0.993	0.989	0.982	0.973	0.960	0.944
17	1.000	1.000	0.999	0.998	0.997	0.995	0.991	0.986	0.978	0.968
18	1.000	1.000	1.000	0.999	0.999	0.998	0.996	0.993	0.988	0.982
19	1.000	1.000	1.000	1.000	0.999	0.999	0.998	0.997	0.994	0.991
20	1.000	1.000	1.000	1.000	1.000	1.000	0.999	0.998	0.997	0.995
21	1.000	1.000	1.000	1.000	1.000	1.000	1.000	0.999	0.999	0.998
22	1.000	1.000	1.000	1.000	1.000	1.000	1.000	1.000	0.999	0.999
23	1.000	1.000	1.000	1.000	1.000	1.000	1.000	1.000	1.000	1.000

TABLE III (*continued*)

x	11.5	12.0	12.5	13.0	13.5	14.0	14.5	15.0	15.5	16.0
0	0.000	0.000	0.000	0.000	0.000	0.000	0.000	0.000	0.000	0.000
1	0.000	0.000	0.000	0.000	0.000	0.000	0.000	0.000	0.000	0.000
2	0.001	0.001	0.000	0.000	0.000	0.000	0.000	0.000	0.000	0.000
3	0.003	0.002	0.002	0.001	0.001	0.000	0.000	0.000	0.000	0.000
4	0.011	0.008	0.005	0.004	0.003	0.002	0.001	0.001	0.001	0.000
5	0.028	0.020	0.015	0.011	0.008	0.006	0.004	0.003	0.002	0.001
6	0.060	0.046	0.035	0.026	0.019	0.014	0.010	0.008	0.006	0.004
7	0.114	0.090	0.070	0.054	0.041	0.032	0.024	0.018	0.013	0.010
8	0.191	0.155	0.125	0.100	0.079	0.062	0.048	0.037	0.029	0.022
9	0.289	0.242	0.201	0.166	0.135	0.109	0.088	0.070	0.055	0.043
10	0.402	0.347	0.297	0.252	0.211	0.176	0.145	0.118	0.096	0.077
11	0.520	0.462	0.406	0.353	0.304	0.260	0.220	0.185	0.154	0.127
12	0.633	0.576	0.519	0.463	0.409	0.358	0.311	0.268	0.228	0.193
13	0.733	0.682	0.628	0.573	0.518	0.464	0.413	0.363	0.317	0.275
14	0.815	0.772	0.725	0.675	0.623	0.570	0.518	0.466	0.415	0.368
15	0.878	0.844	0.806	0.764	0.718	0.669	0.619	0.568	0.517	0.467
16	0.924	0.899	0.869	0.835	0.798	0.756	0.711	0.664	0.615	0.566
17	0.954	0.937	0.916	0.890	0.861	0.827	0.790	0.749	0.705	0.659
18	0.974	0.963	0.948	0.930	0.908	0.883	0.853	0.819	0.782	0.742
19	0.986	0.979	0.969	0.957	0.942	0.923	0.901	0.875	0.846	0.812
20	0.992	0.988	0.983	0.975	0.965	0.952	0.936	0.917	0.894	0.868
21	0.996	0.994	0.991	0.986	0.980	0.971	0.960	0.947	0.930	0.911
22	0.998	0.997	0.995	0.992	0.989	0.983	0.976	0.967	0.956	0.942
23	0.999	0.999	0.998	0.996	0.994	0.991	0.986	0.981	0.973	0.963
24	1.000	0.999	0.999	0.998	0.997	0.995	0.992	0.989	0.984	0.978
25	1.000	1.000	0.999	0.999	0.998	0.997	0.996	0.994	0.991	0.987
26	1.000	1.000	1.000	1.000	0.999	0.999	0.998	0.997	0.995	0.993
27	1.000	1.000	1.000	1.000	1.000	0.999	0.999	0.998	0.997	0.996
28	1.000	1.000	1.000	1.000	1.000	1.000	0.999	0.999	0.999	0.998
29	1.000	1.000	1.000	1.000	1.000	1.000	1.000	1.000	0.999	0.999
30	1.000	1.000	1.000	1.000	1.000	1.000	1.000	1.000	1.000	0.999
31	1.000	1.000	1.000	1.000	1.000	1.000	1.000	1.000	1.000	1.000
32	1.000	1.000	1.000	1.000	1.000	1.000	1.000	1.000	1.000	1.000
33	1.000	1.000	1.000	1.000	1.000	1.000	1.000	1.000	1.000	1.000
34	1.000	1.000	1.000	1.000	1.000	1.000	1.000	1.000	1.000	1.000
35	1.000	1.000	1.000	1.000	1.000	1.000	1.000	1.000	1.000	1.000

TABLE IV

The Chi-Square Distribution

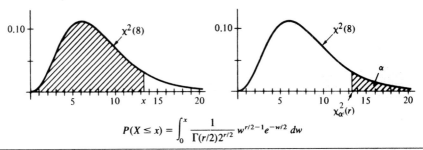

$$P(X \le x) = \int_0^x \frac{1}{\Gamma(r/2)2^{r/2}} w^{r/2-1} e^{-w/2} \, dw$$

				$P(X \le x)$				
	0.010	0.025	0.050	0.100	0.900	0.950	0.975	0.990
r	$\chi^2_{0.99}(r)$	$\chi^2_{0.975}(r)$	$\chi^2_{0.95}(r)$	$\chi^2_{0.90}(r)$	$\chi^2_{0.10}(r)$	$\chi^2_{0.05}(r)$	$\chi^2_{0.025}(r)$	$\chi^2_{0.01}(r)$
1	0.000	0.001	0.004	0.016	2.706	3.841	5.024	6.635
2	0.020	0.051	0.103	0.211	4.605	5.991	7.378	9.210
3	0.115	0.216	0.352	0.584	6.251	7.815	9.348	11.34
4	0.297	0.484	0.711	1.064	7.779	9.488	11.14	13.28
5	0.554	0.831	1.145	1.610	9.236	11.07	12.83	15.09
6	0.872	1.237	1.635	2.204	10.64	12.59	14.45	16.81
7	1.239	1.690	2.167	2.833	12.02	14.07	16.01	18.48
8	1.646	2.180	2.733	3.490	13.36	15.51	17.54	20.09
9	2.088	2.700	3.325	4.168	14.68	16.92	19.02	21.67
10	2.558	3.247	3.940	4.865	15.99	18.31	20.48	23.21
11	3.053	3.816	4.575	5.578	17.28	19.68	21.92	24.72
12	3.571	4.404	5.226	6.304	18.55	21.03	23.34	26.22
13	4.107	5.009	5.892	7.042	19.81	22.36	24.74	27.69
14	4.660	5.629	6.571	7.790	21.06	23.68	26.12	29.14
15	5.229	6.262	7.261	8.547	22.31	25.00	27.49	30.58
16	5.812	6.908	7.962	9.312	23.54	26.30	28.84	32.00
17	6.408	7.564	8.672	10.08	24.77	27.59	30.19	33.41
18	7.015	8.231	9.390	10.86	25.99	28.87	31.53	34.80
19	7.633	8.907	10.12	11.65	27.20	30.14	32.85	36.19
20	8.260	9.591	10.85	12.44	28.41	31.41	34.17	37.57
21	8.897	10.28	11.59	13.24	29.62	32.67	35.48	38.93
22	9.542	10.98	12.34	14.04	30.81	33.92	36.78	40.29
23	10.20	11.69	13.09	14.85	32.01	35.17	38.08	41.64
24	10.86	12.40	13.85	15.66	33.20	36.42	39.36	42.98
25	11.52	13.12	14.61	16.47	34.38	37.65	40.65	44.31
26	12.20	13.84	15.38	17.29	35.56	38.88	41.92	45.64
27	12.88	14.57	16.15	18.11	36.74	40.11	43.19	46.96
28	13.56	15.31	16.93	18.94	37.92	41.34	44.46	48.28
29	14.26	16.05	17.71	19.77	39.09	42.56	45.72	49.59
30	14.95	16.79	18.49	20.60	40.26	43.77	46.98	50.89
40	22.16	24.43	26.51	29.05	51.80	55.76	59.34	63.69
50	29.71	32.36	34.76	37.69	63.17	67.50	71.42	76.15
60	37.48	40.48	43.19	46.46	74.40	79.08	83.30	88.38
70	45.44	48.76	51.74	55.33	85.53	90.53	95.02	100.4
80	53.34	57.15	60.39	64.28	96.58	101.9	106.6	112.3

This table is abridged and adapted from Table III in *Biometrika Tables for Statisticians*, edited by E. S. Pearson and H. O. Hartley. It is published here with the kind permission of the *Biometrika* Trustees.

TABLE Va
The Normal Distribution

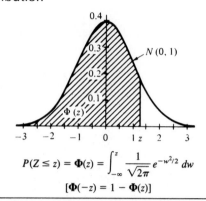

$$P(Z \le z) = \Phi(z) = \int_{-\infty}^{z} \frac{1}{\sqrt{2\pi}} e^{-w^2/2} \, dw$$

$$[\Phi(-z) = 1 - \Phi(z)]$$

z	0.00	0.01	0.02	0.03	0.04	0.05	0.06	0.07	0.08	0.09
0.0	0.5000	0.5040	0.5080	0.5120	0.5160	0.5199	0.5239	0.5279	0.5319	0.5359
0.1	0.5398	0.5438	0.5478	0.5517	0.5557	0.5596	0.5636	0.5675	0.5714	0.5753
0.2	0.5793	0.5832	0.5871	0.5910	0.5948	0.5987	0.6026	0.6064	0.6103	0.6141
0.3	0.6179	0.6217	0.6255	0.6293	0.6331	0.6368	0.6406	0.6443	0.6480	0.6517
0.4	0.6554	0.6591	0.6628	0.6664	0.6700	0.6736	0.6772	0.6808	0.6844	0.6879
0.5	0.6915	0.6950	0.6985	0.7019	0.7054	0.7088	0.7123	0.7157	0.7190	0.7224
0.6	0.7257	0.7291	0.7324	0.7357	0.7389	0.7422	0.7454	0.7486	0.7517	0.7549
0.7	0.7580	0.7611	0.7642	0.7673	0.7703	0.7734	0.7764	0.7794	0.7823	0.7852
0.8	0.7881	0.7910	0.7939	0.7967	0.7995	0.8023	0.8051	0.8078	0.8106	0.8133
0.9	0.8159	0.8186	0.8212	0.8238	0.8264	0.8289	0.8315	0.8340	0.8365	0.8389
1.0	0.8413	0.8438	0.8461	0.8485	0.8508	0.8531	0.8554	0.8577	0.8599	0.8621
1.1	0.8643	0.8665	0.8686	0.8708	0.8729	0.8749	0.8770	0.8790	0.8810	0.8830
1.2	0.8849	0.8869	0.8888	0.8907	0.8925	0.8944	0.8962	0.8980	0.8997	0.9015
1.3	0.9032	0.9049	0.9066	0.9082	0.9099	0.9115	0.9131	0.9147	0.9162	0.9177
1.4	0.9192	0.9207	0.9222	0.9236	0.9251	0.9265	0.9279	0.9292	0.9306	0.9319
1.5	0.9332	0.9345	0.9357	0.9370	0.9382	0.9394	0.9406	0.9418	0.9429	0.9441
1.6	0.9452	0.9463	0.9474	0.9484	0.9495	0.9505	0.9515	0.9525	0.9535	0.9545
1.7	0.9554	0.9564	0.9573	0.9582	0.9591	0.9599	0.9608	0.9616	0.9625	0.9633
1.8	0.9641	0.9649	0.9656	0.9664	0.9671	0.9678	0.9686	0.9693	0.9699	0.9706
1.9	0.9713	0.9719	0.9726	0.9732	0.9738	0.9744	0.9750	0.9756	0.9761	0.9767
2.0	0.9772	0.9778	0.9783	0.9788	0.9793	0.9798	0.9803	0.9808	0.9812	0.9817
2.1	0.9821	0.9826	0.9830	0.9834	0.9838	0.9842	0.9846	0.9850	0.9854	0.9857
2.2	0.9861	0.9864	0.9868	0.9871	0.9875	0.9878	0.9881	0.9884	0.9887	0.9890
2.3	0.9893	0.9896	0.9898	0.9901	0.9904	0.9906	0.9909	0.9911	0.9913	0.9916
2.4	0.9918	0.9920	0.9922	0.9925	0.9927	0.9929	0.9931	0.9932	0.9934	0.9936
2.5	0.9938	0.9940	0.9941	0.9943	0.9945	0.9946	0.9948	0.9949	0.9951	0.9952
2.6	0.9953	0.9955	0.9956	0.9957	0.9959	0.9960	0.9961	0.9962	0.9963	0.9964
2.7	0.9965	0.9966	0.9967	0.9968	0.9969	0.9970	0.9971	0.9972	0.9973	0.9974
2.8	0.9974	0.9975	0.9976	0.9977	0.9977	0.9978	0.9979	0.9979	0.9980	0.9981
2.9	0.9981	0.9982	0.9982	0.9983	0.9984	0.9984	0.9985	0.9985	0.9986	0.9986
3.0	0.9987	0.9987	0.9987	0.9988	0.9988	0.9989	0.9989	0.9989	0.9990	0.9990

α	0.400	0.300	0.200	0.100	0.050	0.025	0.010	0.005	0.001
z_α	0.253	0.524	0.842	1.282	1.645	1.960	2.326	2.576	3.090
$z_{\alpha/2}$	0.842	1.036	1.282	1.645	1.960	2.240	2.576	2.807	3.291

TABLE **Vb**

The Normal Distribution

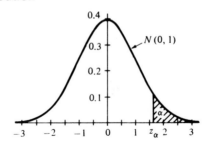

$$P(Z > z_\alpha) = \alpha$$
$$P(Z > z) = 1 - \Phi(z) = \Phi(-z)$$

z_α	0.00	0.01	0.02	0.03	0.04	0.05	0.06	0.07	0.08	0.09
0.0	0.5000	0.4960	0.4920	0.4880	0.4840	0.4801	0.4761	0.4721	0.4681	0.4641
0.1	0.4602	0.4562	0.4522	0.4483	0.4443	0.4404	0.4364	0.4325	0.4286	0.4247
0.2	0.4207	0.4168	0.4129	0.4090	0.4052	0.4013	0.3974	0.3936	0.3897	0.3859
0.3	0.3821	0.3783	0.3745	0.3707	0.3669	0.3632	0.3594	0.3557	0.3520	0.3483
0.4	0.3446	0.3409	0.3372	0.3336	0.3300	0.3264	0.3228	0.3192	0.3156	0.3121
0.5	0.3085	0.3050	0.3015	0.2981	0.2946	0.2912	0.2877	0.2843	0.2810	0.2776
0.6	0.2743	0.2709	0.2676	0.2643	0.2611	0.2578	0.2546	0.2514	0.2483	0.2451
0.7	0.2420	0.2389	0.2358	0.2327	0.2296	0.2266	0.2236	0.2206	0.2177	0.2148
0.8	0.2119	0.2090	0.2061	0.2033	0.2005	0.1977	0.1949	0.1922	0.1894	0.1867
0.9	0.1841	0.1814	0.1788	0.1762	0.1736	0.1711	0.1685	0.1660	0.1635	0.1611
1.0	0.1587	0.1562	0.1539	0.1515	0.1492	0.1469	0.1446	0.1423	0.1401	0.1379
1.1	0.1357	0.1335	0.1314	0.1292	0.1271	0.1251	0.1230	0.1210	0.1190	0.1170
1.2	0.1151	0.1131	0.1112	0.1093	0.1075	0.1056	0.1038	0.1020	0.1003	0.0985
1.3	0.0968	0.0951	0.0934	0.0918	0.0901	0.0885	0.0869	0.0853	0.0838	0.0823
1.4	0.0808	0.0793	0.0778	0.0764	0.0749	0.0735	0.0721	0.0708	0.0694	0.0681
1.5	0.0668	0.0655	0.0643	0.0630	0.0618	0.0606	0.0594	0.0582	0.0571	0.0559
1.6	0.0548	0.0537	0.0526	0.0516	0.0505	0.0495	0.0485	0.0475	0.0465	0.0455
1.7	0.0446	0.0436	0.0427	0.0418	0.0409	0.0401	0.0392	0.0384	0.0375	0.0367
1.8	0.0359	0.0351	0.0344	0.0336	0.0329	0.0322	0.0314	0.0307	0.0301	0.0294
1.9	0.0287	0.0281	0.0274	0.0268	0.0262	0.0256	0.0250	0.0244	0.0239	0.0233
2.0	0.0228	0.0222	0.0217	0.0212	0.0207	0.0202	0.0197	0.0192	0.0188	0.0183
2.1	0.0179	0.0174	0.0170	0.0166	0.0162	0.0158	0.0154	0.0150	0.0146	0.0143
2.2	0.0139	0.0136	0.0132	0.0129	0.0125	0.0122	0.0119	0.0116	0.0113	0.0110
2.3	0.0107	0.0104	0.0102	0.0099	0.0096	0.0094	0.0091	0.0089	0.0087	0.0084
2.4	0.0082	0.0080	0.0078	0.0075	0.0073	0.0071	0.0069	0.0068	0.0066	0.0064
2.5	0.0062	0.0060	0.0059	0.0057	0.0055	0.0054	0.0052	0.0051	0.0049	0.0048
2.6	0.0047	0.0045	0.0044	0.0043	0.0041	0.0040	0.0039	0.0038	0.0037	0.0036
2.7	0.0035	0.0034	0.0033	0.0032	0.0031	0.0030	0.0029	0.0028	0.0027	0.0026
2.8	0.0026	0.0025	0.0024	0.0023	0.0023	0.0022	0.0021	0.0021	0.0020	0.0019
2.9	0.0019	0.0018	0.0018	0.0017	0.0016	0.0016	0.0015	0.0015	0.0014	0.0014
3.0	0.0013	0.0013	0.0013	0.0012	0.0012	0.0011	0.0011	0.0011	0.0010	0.0010
3.1	0.0010	0.0009	0.0009	0.0009	0.0008	0.0008	0.0008	0.0008	0.0007	0.0007
3.2	0.0007	0.0007	0.0006	0.0006	0.0006	0.0006	0.0006	0.0005	0.0005	0.0005
3.3	0.0005	0.0005	0.0005	0.0004	0.0004	0.0004	0.0004	0.0004	0.0004	0.0003
3.4	0.0003	0.0003	0.0003	0.0003	0.0003	0.0003	0.0003	0.0003	0.0003	0.0002

TABLE VI
The t Distribution

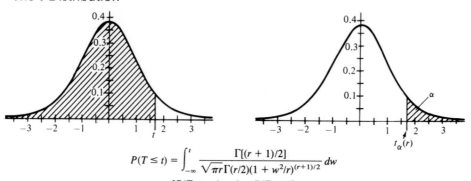

$$P(T \le t) = \int_{-\infty}^{t} \frac{\Gamma[(r + 1)/2]}{\sqrt{\pi r}\,\Gamma(r/2)(1 + w^2/r)^{(r+1)/2}}\, dw$$

$$[P(T \le -t) = 1 - P(T \le t)]$$

				$P(T \le t)$			
	0.60	0.75	0.90	0.95	0.975	0.99	0.995
r	$t_{0.40}(r)$	$t_{0.25}(r)$	$t_{0.10}(r)$	$t_{0.05}(r)$	$t_{0.025}(r)$	$t_{0.01}(r)$	$t_{0.005}(r)$
1	0.325	1.000	3.078	6.314	12.706	31.821	63.657
2	0.289	0.816	1.886	2.920	4.303	6.965	9.925
3	0.277	0.765	1.638	2.353	3.182	4.541	5.841
4	0.271	0.741	1.533	2.132	2.776	3.747	4.604
5	0.267	0.727	1.476	2.015	2.571	3.365	4.032
6	0.265	0.718	1.440	1.943	2.447	3.143	3.707
7	0.263	0.711	1.415	1.895	2.365	2.998	3.499
8	0.262	0.706	1.397	1.860	2.306	2.896	3.355
9	0.261	0.703	1.383	1.833	2.262	2.821	3.250
10	0.260	0.700	1.372	1.812	2.228	2.764	3.169
11	0.260	0.697	1.363	1.796	2.201	2.718	3.106
12	0.259	0.695	1.356	1.782	2.179	2.681	3.055
13	0.259	0.694	1.350	1.771	2.160	2.650	3.012
14	0.258	0.692	1.345	1.761	2.145	2.624	2.997
15	0.258	0.691	1.341	1.753	2.131	2.602	2.947
16	0.258	0.690	1.337	1.746	2.120	2.583	2.921
17	0.257	0.689	1.333	1.740	2.110	2.567	2.898
18	0.257	0.688	1.330	1.734	2.101	2.552	2.878
19	0.257	0.688	1.328	1.729	2.093	2.539	2.861
20	0.257	0.687	1.325	1.725	2.086	2.528	2.845
21	0.257	0.686	1.323	1.721	2.080	2.518	2.831
22	0.256	0.686	1.321	1.717	2.074	2.508	2.819
23	0.256	0.685	1.319	1.714	2.069	2.500	2.807
24	0.256	0.685	1.318	1.711	2.064	2.492	2.797
25	0.256	0.684	1.316	1.708	2.060	2.485	2.787
26	0.256	0.684	1.315	1.706	2.056	2.479	2.779
27	0.256	0.684	1.314	1.703	2.052	2.473	2.771
28	0.256	0.683	1.313	1.701	2.048	2.467	2.763
29	0.256	0.683	1.311	1.699	2.045	2.462	2.756
30	0.256	0.683	1.310	1.697	2.042	2.457	2.750
∞	0.253	0.674	1.282	1.645	1.960	2.326	2.576

This table is taken from Table III of Fisher and Yates: *Statistical Tables for Biological, Agricultural, and Medical Research,* published by Longman Group Ltd., London (previously published by Oliver and Boyd, Edinburgh), by permission of the authors and publishers.

Table VII
The F Distribution

$$P(F \leq f) = \int_0^f \frac{\Gamma[(r_1 + r_2)/2](r_1/r_2)^{r_1/2} w^{r_1/2-1}}{\Gamma(r_1/2)\Gamma(r_2/2)(1 + r_1 w/r_2)^{(r_1+r_2)/2}}\, dw$$

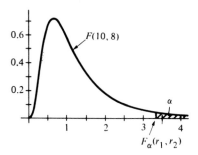

TABLE VII (continued)

$$P(F \le f) = \int_0^f \frac{\Gamma[(r_1+r_2)/2](r_1/r_2)^{r_1/2}w^{r_1/2-1}}{\Gamma(r_1/2)\Gamma(r_2/2)(1+r_1w/r_2)^{(r_1+r_2)/2}}\,dw$$

α	$P(F \le f)$	Den. d.f. r_2	1	2	3	4	5	6	7	8	9	10
							Numerator Degrees of Freedom, r_1					
0.05	0.95	1	161.4	199.5	215.7	224.6	230.2	234.0	236.8	238.9	240.5	241.9
0.025	0.975		647.79	799.50	864.16	899.58	921.85	937.11	948.22	956.66	963.28	968.63
0.01	0.99		4052	4999.5	5403	5625	5764	5859	5928	5981	6022	6056
0.05	0.95	2	18.51	19.00	19.16	19.25	19.30	19.33	19.35	19.37	19.38	19.40
0.025	0.975		38.51	39.00	39.17	39.25	39.30	39.33	39.36	39.37	39.39	39.40
0.01	0.99		98.50	99.00	99.17	99.25	99.30	99.33	99.36	99.37	99.39	99.40
0.05	0.95	3	10.13	9.55	9.28	9.12	9.01	8.94	8.89	8.85	8.81	8.79
0.025	0.975		17.44	16.04	15.44	15.10	14.88	14.73	14.62	14.54	14.47	14.42
0.01	0.99		34.12	30.82	29.46	28.71	28.24	27.91	27.67	27.49	27.35	27.23
0.05	0.95	4	7.71	6.94	6.59	6.39	6.26	6.16	6.09	6.04	6.00	5.96
0.025	0.975		12.22	10.65	9.98	9.60	9.36	9.20	9.07	8.98	8.90	8.84
0.01	0.99		21.20	18.00	16.69	15.98	15.52	15.21	14.98	14.80	14.66	14.55
0.05	0.95	5	6.61	5.79	5.41	5.19	5.05	4.95	4.88	4.82	4.77	4.74
0.025	0.975		10.01	8.43	7.76	7.39	7.15	6.98	6.85	6.76	6.68	6.62
0.01	0.99		16.26	13.27	12.06	11.39	10.97	10.67	10.46	10.29	10.16	10.05
0.05	0.95	6	5.99	5.14	4.76	4.53	4.39	4.28	4.21	4.15	4.10	4.06
0.025	0.975		8.81	7.26	6.60	6.23	5.99	5.82	5.70	5.60	5.52	5.46
0.01	0.99		13.75	10.92	9.78	9.15	8.75	8.47	8.26	8.10	7.98	7.87
0.05	0.95	7	5.59	4.74	4.35	4.12	3.97	3.87	3.79	3.73	3.68	3.64
0.025	0.975		8.07	6.54	5.89	5.52	5.29	5.12	4.99	4.90	4.82	4.76
0.01	0.99		12.25	9.55	8.45	7.85	7.46	7.19	6.99	6.84	6.72	6.62
0.05	0.95	8	5.32	4.46	4.07	3.84	3.69	3.58	3.50	3.44	3.39	3.35
0.025	0.975		7.57	6.06	5.42	5.05	4.82	4.65	4.53	4.43	4.36	4.30
0.01	0.99		11.26	8.65	7.59	7.01	6.63	6.37	6.18	6.03	5.91	5.81
0.05	0.95	9	5.12	4.26	3.86	3.63	3.48	3.37	3.29	3.23	3.18	3.14
0.025	0.975		7.21	5.71	5.08	4.72	4.48	4.32	4.20	4.10	4.03	3.96
0.01	0.99		10.56	8.02	6.99	6.42	6.06	5.80	5.61	5.47	5.35	5.26
0.05	0.95	10	4.96	4.10	3.71	3.48	3.33	3.22	3.14	3.07	3.02	2.98
0.025	0.975		6.94	5.46	4.83	4.47	4.24	4.07	3.95	3.85	3.78	3.72
0.01	0.99		10.04	7.56	6.55	5.99	5.64	5.39	5.20	5.06	4.94	4.85

ν_2			$\nu_1=1$	2	3	4	5	6	7	8	9	10
12	0.95	0.05	4.75	3.89	3.49	3.26	3.11	3.00	2.91	2.85	2.80	2.75
	0.975	0.025	6.55	5.10	4.47	4.12	3.89	3.73	3.61	3.51	3.44	3.37
	0.99	0.01	9.33	6.93	5.95	5.41	5.06	4.82	4.64	4.50	4.39	4.30
15	0.95	0.05	4.54	3.68	3.29	3.06	2.90	2.79	2.71	2.64	2.59	2.54
	0.975	0.025	6.20	4.77	4.15	3.80	3.58	3.41	3.29	3.20	3.12	3.06
	0.99	0.01	8.68	6.36	5.42	4.89	4.56	4.32	4.14	4.00	3.89	3.80
20	0.95	0.05	4.35	3.49	3.10	2.87	2.71	2.60	2.51	2.45	2.39	2.35
	0.975	0.025	5.87	4.46	3.86	3.51	3.29	3.13	3.01	2.91	2.84	2.77
	0.99	0.01	8.10	5.85	4.94	4.43	4.10	3.87	3.70	3.56	3.46	3.37
24	0.95	0.05	4.26	3.40	3.01	2.78	2.62	2.51	2.42	2.36	2.30	2.25
	0.975	0.025	5.72	4.32	3.72	3.38	3.15	2.99	2.87	2.78	2.70	2.64
	0.99	0.01	7.82	5.61	4.72	4.22	3.90	3.67	3.50	3.36	3.26	3.17
30	0.95	0.05	4.17	3.32	2.92	2.69	2.53	2.42	2.33	2.27	2.21	2.16
	0.975	0.025	5.57	4.18	3.59	3.25	3.03	2.87	2.75	2.65	2.57	2.51
	0.99	0.01	7.56	5.39	4.51	4.02	3.70	3.47	3.30	3.17	3.07	2.98
40	0.95	0.05	4.08	3.23	2.84	2.61	2.45	2.34	2.25	2.18	2.12	2.08
	0.975	0.025	5.42	4.05	3.46	3.13	2.90	2.74	2.62	2.53	2.45	2.39
	0.99	0.01	7.31	5.18	4.31	3.83	3.51	3.29	3.12	2.99	2.89	2.80
60	0.95	0.05	4.00	3.15	2.76	2.53	2.37	2.25	2.17	2.10	2.04	1.99
	0.975	0.025	5.29	3.93	3.34	3.01	2.79	2.63	2.51	2.41	2.33	2.27
	0.99	0.01	7.08	4.98	4.13	3.65	3.34	3.12	2.95	2.82	2.72	2.63
120	0.95	0.05	3.92	3.07	2.68	2.45	2.29	2.17	2.09	2.02	1.96	1.91
	0.975	0.025	5.15	3.80	3.23	2.89	2.67	2.52	2.39	2.30	2.22	2.16
	0.99	0.01	6.85	4.79	3.95	3.48	3.17	2.96	2.79	2.66	2.56	2.47
∞	0.95	0.05	3.84	3.00	2.60	2.37	2.21	2.10	2.01	1.94	1.88	1.83
	0.975	0.025	5.02	3.69	3.12	2.79	2.57	2.41	2.29	2.19	2.11	2.05
	0.99	0.01	6.63	4.61	3.78	3.32	3.02	2.80	2.64	2.51	2.41	2.32

TABLE VII (continued)

$$P(F \le f) = \int_0^f \frac{\Gamma[(r_1 + r_2)/2](r_1/r_2)^{r_1/2} w^{r_1/2-1}}{\Gamma(r_1/2)\Gamma(r_2/2)(1 + r_1 w/r_2)^{(r_1+r_2)/2}}\, dw$$

α	$P(F \le f)$	Den. d.f. r_2	\multicolumn{9}{c}{Numerator Degrees of Freedom, r_1}								
			12	15	20	24	30	40	60	120	∞
0.05	0.95	1	243.9	245.9	248.0	249.1	250.1	251.1	252.2	253.3	254.3
0.025	0.975		976.71	984.87	993.10	997.25	1001.4	1005.6	1009.8	1014.0	1018.3
0.01	0.99		6106	6157	6209	6235	6261	6287	6313	6339	6366
0.05	0.95	2	19.41	19.43	19.45	19.45	19.46	19.47	19.48	19.49	19.50
0.025	0.975		39.42	39.43	39.45	39.46	39.47	39.47	39.48	39.49	39.50
0.01	0.99		99.42	99.43	99.45	99.46	99.47	99.47	99.48	99.49	99.50
0.05	0.95	3	8.74	8.70	8.66	8.64	8.62	8.59	8.57	8.55	8.53
0.025	0.975		14.34	14.25	14.17	14.12	14.08	14.04	13.99	13.95	13.90
0.01	0.99		27.05	26.87	26.69	26.60	26.50	26.41	26.32	26.22	26.13
0.05	0.95	4	5.91	5.86	5.80	5.77	5.75	5.72	5.69	5.66	5.63
0.025	0.975		8.75	8.66	8.56	8.51	8.46	8.41	8.36	8.31	8.26
0.01	0.99		14.37	14.20	14.02	13.93	13.84	13.75	13.65	13.56	13.46
0.05	0.95	5	4.68	4.62	4.56	4.53	4.50	4.46	4.43	4.40	4.36
0.025	0.975		6.52	6.43	6.33	6.28	6.23	6.18	6.12	6.07	6.02
0.01	0.99		9.89	9.72	9.55	9.47	9.38	9.29	9.20	9.11	9.02
0.05	0.95	6	4.00	3.94	3.87	3.84	3.81	3.77	3.74	3.70	3.67
0.025	0.975		5.37	5.27	5.17	5.12	5.07	5.01	4.96	4.90	4.85
0.01	0.99		7.72	7.56	7.40	7.31	7.23	7.14	7.06	6.97	6.88
0.05	0.95	7	3.57	3.51	3.41	3.41	3.38	3.34	3.30	3.27	3.23
0.025	0.975		4.67	4.57	4.47	4.42	4.36	4.31	4.25	4.20	4.14
0.01	0.99		6.47	6.31	6.16	6.07	5.99	5.91	5.82	5.74	5.65
0.05	0.95	8	3.28	3.22	3.15	3.12	3.08	3.04	3.01	2.97	2.93
0.025	0.975		4.20	4.10	4.00	3.95	3.89	3.84	3.78	3.73	3.67
0.01	0.99		5.67	5.52	5.36	5.28	5.20	5.12	5.03	4.95	4.86
0.05	0.95	9	3.07	3.01	2.94	2.90	2.86	2.83	2.79	2.75	2.71
0.025	0.975		3.87	3.77	3.67	3.61	3.56	3.51	3.45	3.39	3.33
0.01	0.99		5.11	4.96	4.81	4.73	4.65	4.57	4.48	4.40	4.31

ν_2											
10	0.05	0.95	2.91	2.85	2.77	2.74	2.70	2.66	2.62	2.58	2.54
	0.025	0.975	3.62	3.52	3.42	3.37	3.31	3.26	3.20	3.14	3.08
	0.01	0.99	4.71	4.56	4.41	4.33	4.25	4.17	4.08	4.00	3.91
12	0.05	0.95	2.69	2.62	2.54	2.51	2.47	2.43	2.38	2.34	2.30
	0.025	0.975	3.28	3.18	3.07	3.02	2.96	2.91	2.85	2.79	2.72
	0.01	0.99	4.16	4.01	3.86	3.78	3.70	3.62	3.54	3.45	3.36
15	0.05	0.95	2.48	2.40	2.33	2.29	2.25	2.20	2.16	2.11	2.07
	0.025	0.975	2.96	2.86	2.76	2.70	2.64	2.59	2.52	2.46	2.40
	0.01	0.99	3.67	3.52	3.37	3.29	3.21	3.13	3.05	2.96	2.87
20	0.05	0.95	2.28	2.20	2.12	2.08	2.04	1.99	1.95	1.90	1.84
	0.025	0.975	2.68	2.57	2.46	2.41	2.35	2.29	2.22	2.16	2.09
	0.01	0.99	3.23	3.09	2.94	2.86	2.78	2.69	2.61	2.52	2.42
24	0.05	0.95	2.18	2.11	2.03	1.98	1.94	1.89	1.84	1.79	1.73
	0.025	0.975	2.54	2.44	2.33	2.27	2.21	2.15	2.08	2.01	1.94
	0.01	0.99	3.03	2.89	2.74	2.66	2.58	2.49	2.40	2.31	2.21
30	0.05	0.95	2.09	2.01	1.93	1.89	1.84	1.79	1.74	1.68	1.62
	0.025	0.975	2.41	2.31	2.20	2.14	2.07	2.01	1.94	1.87	1.79
	0.01	0.99	2.84	2.70	2.55	2.47	2.39	2.30	2.21	2.11	2.01
40	0.05	0.95	2.00	1.92	1.84	1.79	1.74	1.69	1.64	1.58	1.51
	0.025	0.975	2.29	2.18	2.07	2.01	1.94	1.88	1.80	1.72	1.64
	0.01	0.99	2.66	2.52	2.37	2.29	2.20	2.11	2.02	1.92	1.80
60	0.05	0.95	1.92	1.84	1.75	1.70	1.65	1.59	1.53	1.47	1.39
	0.025	0.975	2.17	2.06	1.94	1.88	1.82	1.74	1.67	1.58	1.48
	0.01	0.99	2.50	2.35	2.20	2.12	2.03	1.94	1.84	1.73	1.60
120	0.05	0.95	1.83	1.75	1.66	1.61	1.55	1.50	1.43	1.35	1.25
	0.025	0.975	2.05	1.95	1.82	1.76	1.69	1.61	1.53	1.43	1.31
	0.01	0.99	2.34	2.19	2.03	1.95	1.86	1.76	1.66	1.53	1.38
∞	0.05	0.95	1.75	1.67	1.57	1.52	1.46	1.39	1.32	1.22	1.00
	0.025	0.975	1.94	1.83	1.71	1.64	1.57	1.48	1.39	1.27	1.00
	0.01	0.99	2.18	2.04	1.88	1.79	1.70	1.59	1.47	1.32	1.00

TABLE **VIII**
Kolmogorov–Smirnov Acceptance Limits

$$D_n = \sup_x [|F_n(x) - F_0(x)|]$$
$$\alpha = 1 - P(D_n \leq d)$$

	α			
n	0.20	0.10	0.05	0.01
1	0.90	0.95	0.98	0.99
2	0.68	0.78	0.84	0.93
3	0.56	0.64	0.71	0.83
4	0.49	0.56	0.62	0.73
5	0.45	0.51	0.56	0.67
6	0.41	0.47	0.52	0.62
7	0.38	0.44	0.49	0.58
8	0.36	0.41	0.46	0.54
9	0.34	0.39	0.43	0.51
10	0.32	0.37	0.41	0.49
11	0.31	0.35	0.39	0.47
12	0.30	0.34	0.38	0.45
13	0.28	0.32	0.36	0.43
14	0.27	0.31	0.35	0.42
15	0.27	0.30	0.34	0.40
16	0.26	0.30	0.33	0.39
17	0.25	0.29	0.32	0.38
18	0.24	0.28	0.31	0.37
19	0.24	0.27	0.30	0.36
20	0.23	0.26	0.29	0.35
25	0.21	0.24	0.26	0.32
30	0.19	0.22	0.24	0.29
35	0.18	0.21	0.23	0.27
40	0.17	0.19	0.21	0.25
45	0.16	0.18	0.20	0.24
Large n	$\dfrac{1.07}{\sqrt{n}}$	$\dfrac{1.22}{\sqrt{n}}$	$\dfrac{1.36}{\sqrt{n}}$	$\dfrac{1.63}{\sqrt{n}}$

TABLE IX
Random Numbers on the Interval (0, 1)

3407	1440	6960	8675	5649	5793	1514
5044	9859	4658	7779	7986	0520	6697
0045	4999	4930	7408	7551	3124	0527
7536	1448	7843	4801	3147	3071	4749
7653	4231	1233	4409	0609	6448	2900
6157	1144	4779	0951	3757	9562	2354
6593	8668	4871	0946	3155	3941	9662
3187	7434	0315	4418	1569	1101	0043
4780	1071	6814	2733	7968	8541	1003
9414	6170	2581	1398	2429	4763	9192
1948	2360	7244	9682	5418	0596	4971
1843	0914	9705	7861	6861	7865	7293
4944	8903	0460	0188	0530	7790	9118
3882	3195	8287	3298	9532	9066	8225
6596	9009	2055	4081	4842	7852	5915
4793	2503	2906	6807	2028	1075	7175
2112	0232	5334	1443	7306	6418	9639
0743	1083	8071	9779	5973	1141	4393
8856	5352	3384	8891	9189	1680	3192
8027	4975	2346	5786	0693	5615	2047
3134	1688	4071	3766	0570	2142	3492
0633	9002	1305	2256	5956	9256	8979
8771	6069	1598	4275	6017	5946	8189
2672	1304	2186	8279	2430	4896	3698
3136	1916	8886	8617	9312	5070	2720
6490	7491	6562	5355	3794	3555	7510
8628	0501	4618	3364	6709	1289	0543
9270	0504	5018	7013	4423	2147	4089
5723	3807	4997	4699	2231	3193	8130
6228	8874	7271	2621	5746	6333	0345
7645	3379	8376	3030	0351	8290	3640
6842	5836	6203	6171	2698	4086	5469
6126	7792	9337	7773	7286	4236	1788
4956	0215	3468	8038	6144	9753	3131
1327	4736	6229	8965	7215	6458	3937
9188	1516	5279	5433	2254	5768	8718
0271	9627	9442	9217	4656	7603	8826
2127	1847	1331	5122	8332	8195	3322
2102	9201	2911	7318	7670	6079	2676
1706	6011	5280	5552	5180	4630	4747
7501	7635	2301	0889	6955	8113	4364
5705	1900	7144	8707	9065	8163	9846
3234	2599	3295	9160	8441	0085	9317
5641	4935	7971	8917	1978	5649	5799
2127	1868	3664	9376	1984	6315	8396

TABLE X

Divisors for the Confidence Interval for σ of Minimum Length

Let $X_1, X_2, \ldots, X_n$ be a random sample of size n from $N(\mu, \sigma^2)$; let $\bar{x} = (1/n) \Sigma_{i=1}^n x_i$ and $s^2 = (1/[n-1]) \Sigma_{i=1}^n (x_i - \bar{x})^2$. A $(100\gamma)\%$ confidence interval for σ which has minimum length is given by $[\sqrt{(n-1)s^2/b}, \sqrt{(n-1)s^2/a}]$. In the table, a is the upper number, b the lower number; $r = n - 1$, the number of degrees of freedom; and $\gamma = 1 - \alpha$, the confidence level.

	$\gamma = 1 - \alpha$				$\gamma = 1 - \alpha$		
r	0.90	0.95	0.99	r	0.90	0.95	0.99
2	0.206	0.101	0.020	16	8.774	7.604	5.649
	12.521	15.111	20.865		29.233	32.072	38.097
3	0.565	0.345	0.114	17	9.505	8.282	6.226
	13.153	15.589	20.973		30.480	33.362	39.469
4	1.020	0.692	0.294	18	10.242	8.969	6.814
	14.180	16.573	21.838		31.721	34.647	40.835
5	1.535	1.109	0.546	19	10.986	9.663	7.413
	15.350	17.743	22.985		32.959	35.927	42.195
6	2.093	1.578	0.837	20	11.736	10.365	8.021
	16.581	18.996	24.262		34.192	37.202	43.550
7	2.683	2.085	1.214	21	12.492	11.073	8.638
	17.839	20.286	25.602		35.420	38.472	44.899
8	3.298	2.623	1.611	22	13.253	11.788	9.264
	19.110	21.595	26.975		36.646	39.738	46.243
9	3.934	3.187	2.039	23	14.019	12.509	9.898
	20.385	22.912	28.364		37.867	41.000	47.586
10	4.588	3.773	2.496	24	14.790	13.236	10.539
	21.660	24.230	29.760		39.084	42.257	48.914
11	5.257	4.377	2.976	25	15.565	13.968	11.186
	22.933	25.548	31.158		40.299	43.510	50.243
12	5.940	4.997	3.477	26	16.344	14.704	11.841
	24.202	26.862	32.554		41.509	44.760	51.566
13	6.634	5.631	3.997	27	17.127	15.446	12.501
	25.467	28.172	33.947		42.717	46.006	52.886
14	7.338	6.278	4.533	28	17.914	16.192	13.168
	26.727	29.477	35.336		43.922	47.248	54.200
15	8.052	6.936	5.084	29	18.705	16.942	13.840
	27.982	30.777	36.719		45.123	48.487	55.511

TABLE XI

Distribution Function of the Correlation Coefficient R

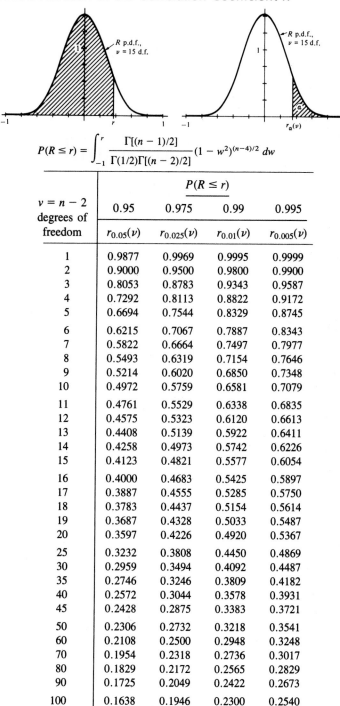

$$P(R \leq r) = \int_{-1}^{r} \frac{\Gamma[(n-1)/2]}{\Gamma(1/2)\Gamma[(n-2)/2]} (1 - w^2)^{(n-4)/2} \, dw$$

$v = n - 2$ degrees of freedom	$P(R \leq r)$			
	0.95	0.975	0.99	0.995
	$r_{0.05}(v)$	$r_{0.025}(v)$	$r_{0.01}(v)$	$r_{0.005}(v)$
1	0.9877	0.9969	0.9995	0.9999
2	0.9000	0.9500	0.9800	0.9900
3	0.8053	0.8783	0.9343	0.9587
4	0.7292	0.8113	0.8822	0.9172
5	0.6694	0.7544	0.8329	0.8745
6	0.6215	0.7067	0.7887	0.8343
7	0.5822	0.6664	0.7497	0.7977
8	0.5493	0.6319	0.7154	0.7646
9	0.5214	0.6020	0.6850	0.7348
10	0.4972	0.5759	0.6581	0.7079
11	0.4761	0.5529	0.6338	0.6835
12	0.4575	0.5323	0.6120	0.6613
13	0.4408	0.5139	0.5922	0.6411
14	0.4258	0.4973	0.5742	0.6226
15	0.4123	0.4821	0.5577	0.6054
16	0.4000	0.4683	0.5425	0.5897
17	0.3887	0.4555	0.5285	0.5750
18	0.3783	0.4437	0.5154	0.5614
19	0.3687	0.4328	0.5033	0.5487
20	0.3597	0.4226	0.4920	0.5367
25	0.3232	0.3808	0.4450	0.4869
30	0.2959	0.3494	0.4092	0.4487
35	0.2746	0.3246	0.3809	0.4182
40	0.2572	0.3044	0.3578	0.3931
45	0.2428	0.2875	0.3383	0.3721
50	0.2306	0.2732	0.3218	0.3541
60	0.2108	0.2500	0.2948	0.3248
70	0.1954	0.2318	0.2736	0.3017
80	0.1829	0.2172	0.2565	0.2829
90	0.1725	0.2049	0.2422	0.2673
100	0.1638	0.1946	0.2300	0.2540

Answers to Odd-numbered Exercises

CHAPTER 1

1.1.1   (a) $S = \{1, 2, 3, \ldots, 36\}$;

       (b) $S = \{w : 19 \leq w \leq 23\}$, where w is the weight in grams;

       (c) $S = \{HHH, HHT, HTH, THH, HTT, THT, TTH, TTT\}$;

1.1.3   (a) Litter size: 1   2   3   4

             Frequency: 5   10   4   1

       (d) Yes, 2.

1.1.5   (a) Clutch size: 4   5   6   7   8   9   10

             Frequency: 4   5   6   4   6   7   1

       (c) No.

1.1.7   (a) $[8, 6, 3, 2, 1, 3, 5]$.

1.1.9   (a) $[1, 1, 3, 4, 20, 23, 16, 10, 3, 0, 1]$;

       (c) Fairly symmetrical.

1.2.1   (a) 1.1;

       (b) 0.035;

       (c) 0.1871.

1.2.3   (a) 2;

       (b) 2.05, 259/380, $\sqrt{24{,}605}/190$.

1.2.5   8.18.

1.2.7   2.6, 5.2, 6.6, 9.0, 3.6.

1.2.9
    (a) [3, 15, 8, 1, 0, 5, 8, 2];

    (b) $\bar{x} = 17.30, s_x = 2.34$.

1.2.11
    (a) $\bar{x} = 112.12, s = 231.3576$.

    (d) Half of the observations are less than 48.

1.3.1
    (a)

Stem	Leaf	Frequency
127	8	1
128	8	1
129	5 8 9	3
130	8	1
131	2 3 4 4 5 5 7	7
132	2 7 7 8	4
133	7 9	2
134	8	1

(Multiply numbers by 10^{-1}.)

    (b) 131.3, 7.0, 2.575, 131.45, 131.47, 3.034.

1.3.3
    (a)

Stem	Leaf	Frequency
0•	612	1
1∗	450	1
1•	560 889 961 994	4
2∗	065 142 151 172 195 290	6
2•	510 545 788 817 880 921 938	7
3∗	011 041 051 060 062 080 090	7

(Multiply numbers by 10^{-2}.)

    (b) 6.12, 20.4725, 25.2750, 30.1850, 30.90

    (d) Skewed to the left.

1.3.5   (a)

Stem	Leaf	Frequency
0*f*	5 5 5 5 5 5 5 5 5 5 5 5 5 5 5 5 5 5 5	19
0*s*	6 6 6 6 6 6 6 6 6 6 6 6 6 7 7 7 7 7 7 7	21
0●	8 8 8 8 8 8 8 9 9 9 9 9 9	13
1*	0 0 0 0 0 0 1 1 1 1 1 1 1	13
1*t*	2 2 2 3 3	5
1*f*	4 5 5 5 5	5
1*s*	6 6 6 6 7 7	6
1●	8 8 9	3
2*	0 1 1 1 1	5
2*t*	3 3	2
2*f*	4 4 4 5	4
2*s*		0
2●		0
3*		0
3*t*		0
3*f*	4 5	2
3*s*		0
3●		0
4*		0
4*t*		0
4*f*	5	1
4*s*		0
4●		0
5*		0
5*t*		0
5*f*	5	1

(b) 5, 6, 9, 15, 55;

(c) 9, 28.5, 42;

(e) 22.8.

1.3.7   (a) 1, 6.75, 32, 90.75, 527;

(b) 84, 216.75, 342.75.

1.4.1

Frequency	Old Method Leaves	Stem	New Method Leaves	Frequency
0		0s	78	1
0		0•		0
1	08	1*	12 15	2
0		1t	24	1
0		1f		0
1	76	1s	67 68	2
5	99 96 92 83 80	1•	89	1
1	14	2*	15	1
3	37 32 27	2t	37	1
0		2f	56	1
0		2s	69	1
0		2•	81	1
0		3*		0
0		3t		0
0		3f		0
0		3s		0
1	94	3•		0

(Multiply numbers by 10^{-1}.)

1.4.3 (a)

Frequency	Leaves for Females	Stem	Leaves for Males	Frequency
1	7	2s		0
2	9 8	2•		0
3	1 1 1	3*		0
1	2	3t		0
8	5 5 5 4 4 4 4 4	3f		0
4	7 7 7 6	3s		0
2	8 8	3•		0
2	1 0	4*	1 1	2
2	2 2	4t	3 3	2
0		4f	5	1
0		4s	6 7	2
0		4•	8 8	2
0		5*	1	1
0		5t	3 3 3	3
0		5f	4 4 5	3
0		5s	6 7	2
0		5•	8 8	2
0		6*	0 1	2
0		6t	3	1
0		6f		0
0		6s	7 7	2

(Multiply numbers by 10^{-1}.)

(b) 2.7, 3.15, 3.5, 3.75, 4.2; 4.1, 4.65, 5.3, 5.8, 6.7;
(e) Male FVC is greater than female FVC and also has a greater standard deviation.

1.4.5   (a)

Freq	New Brand	Stem	Competing Brand	Freq
1	36	6t		0
0		6f	43 54 56	3
0		6s	68 68	2
5	97 92 90 85 80	6•	80 81 88 95 99 99	6
4	14 09 05 03	7*	00 03 03 06 13 14 14 15 17	9
3	39 32 24	7t	20 21 21 26 28 28 28 29 39	9
9	54 53 51 46 46 42 40 40 40	7f	41 41 46 47	4
6	75 73 73 66 64 64	7s	60 65 66 69 74	5
8	96 89 87 86 85 83 82 81	7•	80 82 83 83 83 84 88 98	8
6	16 10 08 07 07 04	8*	03 04 05 17	4
6	35 30 28 23 21 21	8t		0
1	45	8f		0
1	66	8s		0

(b) 636, 737.25, 769.5, 807, 866; 643, 702.25, 728, 775.5, 817;
(e) 764.86, 49.67; 733.50, 44.36.

1.4.7   (a) 2.70, 4.25, 6.10, 7.15, 11.40; 2.10, 3.10, 4.80, 5.50, 6.45;
(c)

Frequency	American Beech	Stem	Sugar Maple	Frequency
3	85 70 70	2	10 35 35	3
0		3	10 10 15 90 90	5
4	50 50 25 25	4	00 80	2
2	00 00	5	00 00 15 35 50	5
5	50 50 50 15 10	6	00 00 25 45	4
1	15	7		0
1	80	8		0
2	70 35	9		0
0		10		0
1	40	11		0

(Multiply numbers by 10^{-2}.)

1.6.1 (a) $r = 0.6859$;
       (b) $\widehat{y} = 0.2566 + 1.4013x$.

1.6.3 (b) $\widehat{y} = 0.0298 + 0.8822x$.

1.6.5 (b) $\widehat{y} = 73.7717 + 0.6938x$.

1.6.7 (b) $\widehat{y} = 1.2615 - 0.1641x$.

1.6.9 (b) $r = 0.9842$;
       (c) $\widehat{y} = 209.8168 - 21.3099x + 16.2631x^2 - 0.8323x^3$.

CHAPTER 2

2.1.1   (a) 12/52; (b) 2/52; (c) 16/52; (d) 1; (e) 0.

2.1.3   27/64.

2.1.5   (a) (i) 4/8, (ii) 2/8, (iii) 3/8, (iv) 4/8;
        (b) (i) 1/8, (ii) 0, (iii) 2/8;
        (c) (i) 5/8, (ii) 5/8, (iii) 5/8.

2.1.7   0.6.

2.1.9   (a) $3(1/3) - 3(1/3)^2 + (1/3)^3$;
        (b) $P(A_1 \cup A_2 \cup A_3) = 1 - [1 - 3(1/3) + 3(1/3)^2 - (1/3)^3] = 1 - (1 - 1/3)^3$.

2.1.11 (a) $1 - (1/2)^{10}$; (b) $1 - (1/2)^{20}$; (c) $1 - (1/2)^{20}$;
        (d) $1 - (1/2)^{10}$; (e) $(1/2)^{10} - (1/2)^{20}$; (f) $(1/2)^{20}$.

2.1.13 (a) $S = \{00, 0, 1, 2, 3, \ldots, 36\}$; (b) $P(A) = 2/38$;
        (c) $P(B) = 4/38$; (d) $P(D) = 18/38$.

2.1.15 2/3.

2.2.1   4096.

2.2.3   (a) 6,760,000; (b) 17,576,000.

2.2.5   (a) 24; (b) 256.

2.2.7   (a) 0.0024; (b) 0.0012; (c) 0.0006; (d) 0.0004.

2.2.9   (a) 2; (b) 8; (c) 20; (d) 40.

2.2.11 (a) 96; (b) 393,216; (c) 2,973,696.

2.2.13 (a) 362,880; (b) 84; (c) 512.

2.2.15 (a) 0.00024; (b) 0.00144; (c) 0.02113; (d) 0.04754; (e) 0.42257.

2.2.17 14,702,688.

2.2.19 (a) 560; (b) 3,688.

2.3.1   (a) 5000/1,000,000; (b) 78,515/1,000,000;
        (c) 73,630/995,000; (d) 4,885/78,515.

2.3.3   (a) 5/35; (b) 26/35; (c) 5/19; (d) 9/23; (e) left.

2.3.5   (a) 3/10; (b) 3/5; (c) 3/7.

2.3.7   (a) $S = \{(R, R), (R, W), (W, R), (W, W)\}$; (b) 1/3.

2.3.9   (a) 1/33; (b) 5/66.

2.3.11   (a) 1/220; (b) 3/110; (c) 9/55.

2.3.13   (b) 8/36; (c) 5/11;
          (e) $8/36 + 2[(5/36)(5/11) + (4/36)(4/10) + (3/36)(3/9)] = 0.49293$.

2.3.15   (a) 365^r; (b) $_{365}P_r$; (c) $1 - {_{365}P_r}/365^r$; (d) 23.

2.3.17   (a) 49/153; (b) 28/49

2.3.19   (f) $1 - 1/e$.

2.4.1   (a) 0.14; (b) 0.76; (c) 0.86.

2.4.3   (a) 1/6; (b) 1/12; (c) 1/4; (d) 1/4; (e) 1/2.

2.4.5   Yes; $0.9 = 0.8 + 0.5 - (0.8)(0.5)$.

2.4.7   0.398.

2.4.9   (a) 0.36; (b) 0.49; (c) 0.01.

2.4.11   (a) No; (b) Only if $P(A) = 0$ or $P(B) = 1$.

2.4.13 $(2/3)^3(1/3)^2$; $(2/3)^3(1/3)^2$.

2.4.15   (a) $10(1/2)^6$; (b) $[(120)(45)/15,504][7/15]$;
          (b) neither model is very good.

2.4.17   (a) 9000/1,000,000; (b) 900/1,000,000; (c) 90/1,000,000;
          (d) 9/1,000,000; (e) 1/1,000,000.

2.4.19

n	3	6	9	12	15
(a)	0.7037	0.6651	0.6536	0.6480	0.6447
(b)	0.6667	0.6319	0.6321	0.6321	0.6321

(c) Very little when $n > 20$ sampling with replacement, and when $n > 8$ sampling without replacement.

(d) Convergence is faster when sampling without replacement.

2.5.1   (a) 21/32; (b) 16/21.

2.5.3   (a) 11/24; (b) 2/11.

2.5.5   (a) 220/276; (b) 512/944.

2.5.7   5/14.

CHAPTER 3

3.1.3   (a) 10; (b) 1/55; (c) 3; (d) 1/30; (e) $n(n + 1)/2$.

3.1.5   (b)

x	Frequency	Relative Frequency	$f(x)$
1	38	0.38	0.40
2	27	0.27	0.30
3	21	0.21	0.20
4	14	0.14	0.10

3.1.7   (a) $f(x) = \dfrac{13 - 2x}{36}$, $x = 1, 2, 3, 4, 5, 6$;

3.1.9   (c)

x	Frequency	Relative Frequency	$f(x)$
0	13	0.325	0.253
1	16	0.400	0.422
2	9	0.225	0.251
3	2	0.050	0.066

3.2.1 $5.

3.2.3   (a) 3; (b) 7; (c) 4/3; (d) 7/3; (e) $(2n + 1)/3$.

3.2.5 $-91/300$ or -30.33 cents.

3.2.7 $E(X) = -17/216 = -0.0787$.

3.2.9 $E(X) = \sum_{x=1}^{n} \frac{1}{x}\left(\frac{1}{n}\right); \frac{137}{300} = 0.4567$;

$$\frac{1}{100}\left[\frac{\ln(101) + 1 + \ln(100)}{2}\right] = 0.05110 \approx 0.05187.$$

3.2.11   (a) $0.46; (b) lose $0.54.

3.2.13 $-$0.01414.

3.3.1   1.4; 1.44; 1.2.

3.3.3   (a) 15; 50; (b) 5; 0; (c) 5/3; 5/9.

3.3.5   (a) 3; 28/11; (b) 1; 2/3.

3.3.7   (a) 16; (b) 6; (c) 16.

3.3.9   (a) (ii) 127/64; (iii) $\sqrt{7359}/64$; (iv) 1.6635;
(b) (ii) 385/64; (iii) $\sqrt{7359}/64$; (iv) -1.6635.

3.3.11   (a) $25/8 = 3.125$, $55/64 = 0.8594$, $\sqrt{55}/8 = 0.9270$;
(b) $79/25 = 3.16$, $662/825 = 0.8024$, $\sqrt{21{,}846}/165 = 0.8958$.

3.3.13   (b) 75, (f) (i) $\bar{x} = 72.78$, (ii) $s^2 = 8.7188$, (iii) $s = 2.9528$, (iv) 5378.34.

3.4.1   (a) $f(x) = 0.001$, $x = 000, 001, \ldots, 999$;
(b) 499.5;
(c) 83,333.25; (d) 288.675.

3.4.3   (a) $\mu = 6.5$, $\bar{x} = 6.567$;
(b) $\sigma^2 = 143/12 = 11.9167$, $s^2 = 12.1141$,
(c) $\sigma = 3.4521$, $s = 3.4805$; (e) The fit is quite good.

3.4.5   (a)
$$g(y) = \frac{1}{b - a + 1}, a \leq y \leq b, y \text{ an integer};$$

(b) $\dfrac{a + b}{2}$; $\sigma^2 = \dfrac{(b - a + 1)^2 - 1}{12}$, $\sigma = \sqrt{\sigma^2}$.

3.4.7 (a) 39/98; (b) 221/245;
       (c) $\mu = 3/5, \sigma^2 = 564/1{,}225$.

3.4.9 0.416.

3.4.11 (a)
$$P(X = x) = \frac{\binom{22}{x}\binom{58}{10 - x}}{\binom{80}{10}}, x = 0, 1, 2, \ldots, 10,$$
       (b) $\mu = 11/4, \sigma^2 = 2{,}233/1{,}264, \sigma = 1.329$;
       (c) 2.45, 3.20, 3.95, 4.70, and 5.46;
       (d) 2.07; (e) $-\$\,0.5316$; (f) $-\$\,0.4511$.

3.4.13 (a) $\mu = 16/5, \sigma^2 = 512/325, \sigma = 16\sqrt{26}/65$;
       (b) $\mu = 8, \sigma^2 = 64/31, \sigma = 8\sqrt{31}/31$;
       (c) $\mu = 16/5, \sigma^2 = 4{,}096/1{,}975, \sigma = 64\sqrt{79}/395$;
       (d) $\mu = 8, \sigma^2 = 256/79, \sigma = 16\sqrt{79}/79$.

3.4.15 1.000, 0.900, 0.809, 0.0006.

3.5.1 (a) $f(x) = (7/18)^x (11/18)^{1-x}, x = 0, 1; \mu = 7/18; \sigma^2 = 77/324$.

3.5.3 (a) $(1/5)^2(4/5)^4 = 0.0164$;
       (b) $\dfrac{6!}{2!4!}(1/5)^2(4/5)^4 = 0.2458$.

3.5.5 (a) 0.2459; (b) 0.5501; (c) 0.2040; (d) 11.2, 3.36, 1.833.

3.5.7 (a) $b(2000, \pi/4)$; (b) 1570.796, 337.096, 18.360; (c) π;
       (d) $V_n = \pi^{n/2}/\Gamma(n/2+1)$ is the volume of a ball of radius one in n-space.

3.5.9 (a) $b(9, 0.10)$; (b) 0.9, 0.81, 0.9; (c) (i) 0.1722, (ii) 0.2252.

3.5.11 (a) 0.8725; (b) 0.5841; (c) 0.1597.

3.5.13 0.1268.

3.5.15 (a) $\mu = 7.2, \sigma^2 = 1.44, \sigma = 1.2$;
       (c) $\bar{x} = 7.32, s^2 = 1.4117, s = 1.1882$.

3.5.17 (a) 0.6513; (b) 0.7941.

3.5.19

x	$b(8, 0.2)$	$N_1 = 8, N_2 = 32$	$N_1 = 16, N_2 = 64$
0	0.1678	0 1368	0.1527
1	0.3355	0.3501	0.3429
2	0.2936	0.3299	0.3104
3	0.1468	0.1466	0.1473
4	0.0459	0.0327	0.0399
5	0.0092	0.0036	0.0063
6	0.0011	0.0002	0.0006
7	0.0001	0.0000	0.0000
8	0.0000	0.0000	0.0000

3.6.1   (a) 1/8; (b) 7/8.

3.6.3   (a) $(4/5)^3(1/5)$.

3.6.5   (a) 0.4604; (b) 0.5580; (c) 0.0184.

3.6.7   (a) 3000; (b) 0.3679; (c) 0.4866.

3.6.9   1/4.

3.6.11   (a) $f(x) = (1/2)^{x-1}$, $x = 2, 3, 4, \ldots$;
         (b) 3, 2, 1.414;
         (c) (i) 3/4, (ii) 1/8, (iii) 1/4.

3.6.13   (a) $f(x) = (x-1)/2^x$, $x = 2, 3, \ldots$;
         (b) 4, 4, 2;
         (c) (i) 1/2, (ii) 5/16, (iii) 1/4.

3.6.15   25/3.

3.6.17   33.2187.

3.6.19   (a) 0.0670; (b) 0.4823; (c) 0.1319.

3.6.23   (a) 16.667, 11.111, 3.333; (b) 0.124.

3.7.1   (a) 0.693; (b) 0.762; (c) 0.433.

3.7.3   0.540.

3.7.5   0.558.

3.7.7   0.947.

3.7.9   (a) 0.217; (b) 0.971.

3.7.11   (a) $\bar{x} = 7.94$, $s^2 = 7.73$, yes.

3.7.13   (a) $\bar{x} = 4.269$, $s^2 = 4.045$, yes;
         (b) 0.395, 0.346; (c) 0.247, 0.269; (d) yes.

3.7.15   (b) $\bar{x} = 5.597$, possibly.

3.7.17   (a) 2.967; (b) 2.792; (c) 1.671.

3.7.19   If $n = 80$, $Ac = 3$, $Oc(0.02) \approx 0.92$, $Oc(0.08) \approx 0.12$.

3.8.1   (a) $\eta(t) = (1/3)t + (1/3)t^2 + (1/3)t^3$;
         (b) $\eta(t) = t^5$;
         (c) $\eta(t) = 0.4t + 0.3t^2 + 0.2t^3 + 0.1t^4$;
         (d) $\eta(t) = 0.7 + 0.3t$.

3.8.5 $\mu = 2$, $\sigma^2 = 4/5$,

$$f(x) = \begin{cases} 2/5, & x = 1, \\ 1/5, & x = 2, \\ 2/5, & x = 3. \end{cases}$$

$\eta(t) = (2/5)t + (1/5)t^2 + (2/5)t^3$.

3.8.7   (a) 0.1366; (b) 0.109; (c) 0.35; (d) $(0.65)^2(0.35)$ (e) 1/15.

CHAPTER 4

4.1.1   (b)
$$F(x) = \begin{cases} 0, & x < 0, \\ x(2-x), & 0 \le x < 1, \\ 1, & 1 \le x. \end{cases}$$
(d) (i) 3/4, (ii) 1/2, (iii) 0, (iv) 1/16.

4.1.3   (a) (i) 3; (ii) $F(x) = x^4, 0 \le x \le 1$;
(b) (i) 3/16; (ii) $F(x) = (1/18)x^{3/2}, 0 \le x \le 4$;
(c) (i) 1/4; (ii) $F(x) = x^{1/4}, 0 \le x \le 1$.

4.1.5   (a) $4/5, 2/75, \sqrt{6}/15$;
(b) $12/5, 192/175; 8\sqrt{21}/35$;
(c) $1/5, 16/225, 4/15$.

4.1.7   (a) $\pi/2, (\pi^2 - 8)/4$;
(c) $F(x) = (1 - \cos x)/2, 0 \le x \le \pi$.

4.1.9   (a) d = 2; (b) $E(Y) = 2$; (c) $E(Y^2)$ is unbounded.

4.1.11 (a) $\mu = 0, \sigma^2 = 3/5$;
(b) $\mu = 0, \sigma^2 = 1/3$;
(c) $\mu = 0, \sigma^2 = 1/6$.

4.1.13 (a) $\mu = 2$; (b) $\sigma^2 = 2$.

4.1.15 $f(x) = \dfrac{e^{-x}}{(1+e^{-x})^2} = \dfrac{e^{-x}}{(1+e^{-x})^2} \dfrac{e^{2x}}{e^{2x}} = \dfrac{e^x}{(e^x+1)^2}$.

4.1.17 (a) $\pi_{0.5} = 0$; (b) $\pi_{0.25} = -0.50$; (c) $\pi_{0.90} = 0.80$.

4.1.19 (a) $F_n(x) = nx, 0 \le x \le 1/n$.

4.2.3   (a) $f(x) = 1/10, 0 < x < 10$; (b) 0.2; (c) 0.6;
(d) $\mu = 5$; (e) $\sigma^2 = 25/3$.

4.2.5   (a) $G(w) = (w-a)/(b-a), a \le w \le b$;
(b) $U(a, b)$.

4.2.7   (a) 2; 4; $M(t) = (1 - 2t)^{-1}, t < 1/2$;
(b) $e^{-3/2}$; (c) $e^{-3/2}$.

4.2.9   (a) $f(x) = (1/3)e^{-x/3}, 0 < x < \infty; 3; 9$;
(b) $f(x) = 3e^{-3x}, 0 < x < \infty; 1/3; 1/9$.

4.2.11 $P(X > x + y \mid X > x) = \dfrac{P(X > x + y)}{P(X > x)} = \dfrac{e^{-(x+y)/\theta}}{e^{-x/\theta}} = P(X > y)$.

4.2.13 (a) $\lambda = 0.025$; (b) exponential distribution, $\theta = 40$;
(c) $\mu = 40, \sigma^2 = 1600$; (d) (i) $1 - e^{-1/2}$, (ii) e^{-1}, (iii) e^{-1}.

4.2.15 (a) 10.524; 9.320; yes.

4.2.17 (a) $F(x) = 1 - e^{-(x-\alpha)/\theta}, \alpha \le x < \infty$;

(b) $\theta + \alpha; \theta^2$.

4.2.19 (a) $-\theta \ln 0.75 = 0.2877 \theta$;

(b) $(\ln 0.75 + 1)\theta = 0.7123 \theta$;

(c) $-\theta \ln 0.25 = 1.3863 \theta$;

(d) $(-\ln 0.25 - 1) \theta = 0.3863 \theta$.

4.3.1 (a) $f(x) = \dfrac{1}{\Gamma(10)(3/2)^{10}} x^9 e^{-2x/3}, 0 \le x < \infty$;

(b) $M(t) = (1 - 3t/2)^{-10}, t < 2/3; \mu = 15; \sigma^2 = 45/2$.

4.3.5 $f(x) = \dfrac{1}{\Gamma(20)7^{20}} x^{19} e^{-x/7}, 0 \le x < \infty; \mu = 140; \sigma^2 = 980$.

4.3.7 (a) Gamma distribution with $\alpha = 3, \theta = 40$.

(b) $\mu = 120, \sigma^2 = 4800$; (c) 0.4562.

4.3.9 (a) 0.025; (b) 0.05; (c) 0.94; (d) 8.672; (e) 30.19.

4.3.11 (a) 0.80; (b) $a = 11.69, b = 38.08$;

(c) $\mu = 23, \sigma^2 = 46$; (d) 35.17, 13.09.

4.3.13 $r - 2$.

4.3.15 0.9444.

4.3.17 (a) With class boundaries 0.005, 2.005, 4.005, and so forth, the class frequencies are 3, 8, 14, 19, 15, 13, 12, 7, 5, 3, 0, 1.

(c) $\bar{x} = 9.362, s^2 = 19.6410$;

(d) Compare 0.06 with 0.05; 0.12 with 0.10.

4.4.1 (a) (a) 0.2784; (b) 0.7209; (c) 0.9616; (d) 0.0019;

(e) (e) 0.9500; (f) 0.6826; (g) 0.9544; (h) 0.9974.

4.4.3 (a) 2.326; (b) -2.576; (c) 1.67; (d) -2.17.

4.4.5 (a) 1.96; (b) 1.96; (c) 1.645; (d) 1.645.

4.4.7 (a) 0.3849; (b) 0.5403; (c) 0.0603; (d) 0.0013;

(e) 0.6826; (f) 0.9544; (g) 0.9974; (h) 0.99.

4.4.9 (a) 0.6326; (b) 50.

4.4.11 (a) The frequencies of 450, 451, 452, and so on are 2, 3, 3, 5, 6, 15, 19, 22, 11, 5, 5, 3, 0, 1;

(b) yes.

4.4.13 X could be normal because of the linearity of the q-q plot.

4.4.15 (a) 0.8673; (b) 0.6026; (c) 0.3103; (d) 0.6821.

4.4.17   (a)

Stem	Leaf	Frequency
7•	938	1
8*	032 089	2
8t	222 268 383	3
8f	442 490 528 572	4
8s	674 734 786	3
8•	850 873 920	3
9*	069 150	2
9t	243	1

(Multiply numbers by 10^{-4}.)

(b) Normal quantiles: $-1.645, -1.282, -1.036, -0.842, -0.674, -0.524,$ $-0.385, -0.253, -0.126, 0, 0.126, 0.253, 0.385, 0.524, 0.674, 0.842,$ $1.036, 1.282, 1.645.$

(c) Yes, because of the linearity of the q-q plot.

4.4.19   0.025.

4.4.21   (a) $N(0, 1)$; (b) $N(-1, 1)$; (c) $N(2, 1)$.

4.5.1 $g(y) = 2y, 0 < y < 1.$

4.5.3 $g(y) = (1/8)y^5 e^{-y^2/2}, 0 < y < \infty.$

4.5.5   (d) (i) 0.250, (ii) 0.0628, (iii) 0.0317.

4.5.9 $g(y) = \dfrac{2}{\sqrt{2\pi}} e^{-y^2/2}, 0 < y < \infty.$

4.5.11 $g(y) = \dfrac{1}{\pi(1 + y)\sqrt{y}}, 0 < y < \infty.$

4.5.13 $M(t) = (1 - 2t)^{-1}, t < 1/2$; exponential distribution with $\theta = 2$.

4.6.1   (a) 0.5; (b) 0; (c) 0.25; (d) 0.75; (e) 0.625; (f) 0.75.

4.6.3   (b) $\mu = 31/24, \sigma^2 = 167/567$;
         (c) 15/64; 1/4; 0; 11/16.

4.6.5   (a)
$$F(x) = \begin{cases} 0, & x < 0, \\ x/2, & 0 \le x < 1, \\ 1/2, & 1 \le x < 2, \\ 4/6, & 2 \le x < 4, \\ 5/6, & 4 \le x < 6, \\ 1, & 6 \le x. \end{cases}$$

(b) \$2.25.

4.6.7   (b) 1/4; (c) 3/8; (d) 19/24; (e) 191/576.

CHAPTER 5

5.1.1   (a) 0.0182; (b) 0.0337.

5.1.3   (a) 36/125; (b) 2/7.

5.1.5

$$f(x) = \begin{cases} 1/9, & x = 2, 6, \\ 2/9, & x = 3, 5, \\ 3/9, & x = 4. \end{cases}$$

$\mu_Y = 4, \sigma_Y^2 = 4/3.$

5.1.7   (a) 729/4096; (b) $\mu = 3/2$; $\sigma^2 = 3/40$.

5.1.9   (a) 0.0035; (b) 8; (c) $\mu = 4, \sigma^2 = 7/3$.

5.1.11   (a) $\mu = 10, \sigma^2 = 20$;
        (b) $(1 - 2t)^{-10/2}$, $t < 1/2$;
        (c) $\mu = 10, \sigma^2 = 20$, equal.

5.1.13   (c) $\mu = 13{,}315{,}424/3{,}011{,}805 = 4.4211$.

5.2.1

$$g(y) = \begin{cases} 1/36, & y = 2, \\ 4/36, & y = 3, \\ 10/36, & y = 4, \\ 12/36, & y = 5, \\ 9/36, & y = 6; \end{cases}$$

$\mu = 14/3, \sigma^2 = 10/9.$

5.2.3   (a)

$$g(y) = \begin{cases} 1/64, & y = 3, 12, \\ 3/64, & y = 4, 11, \\ 6/64, & y = 5, 10, \\ 10/64, & y = 6, 9, \\ 12/64, & y = 7, 8. \end{cases}$$

5.2.5 $\mu_Y = 1, \sigma_Y^2 = 61$.

5.2.7   (a) $M(t) = e^{7(e^t - 1)}$; (b) Poisson, $\lambda = 7$; (c) 0.800.

5.2.9   (a) $M(t) = (pe^t)^5/(1 - qe^t)^5$, $t < -\ln q$;
        (b) negative binomial with parameters $r = 5$ and p.

5.2.11   (a) $M(t) = 1/(1 - 5t)^{21}$, $t < 1/5$;
        (b) gamma distribution, $\alpha = 21, \theta = 5$.

5.2.13   (a) $(1/24)(e^{2t} + 2e^{3t} + 3e^{4t} + 4e^{5t} + 4e^{6t} + 4e^{7t} + 3e^{8t} + 2e^{9t} + e^{10t})$;
        (b)

$$g(w) = \begin{cases} 1/24, & w = 2, 10, \\ 2/24, & w = 3, 9, \\ 3/24, & w = 4, 8, \\ 4/24, & w = 5, 6, 7. \end{cases}$$

5.2.15 (a)
$$h_1(w_1) = \begin{cases} 1/36, & w_1 = 0, \\ 4/36, & w_1 = 1, \\ 10/36, & w_1 = 2, \\ 12/36, & w_1 = 3, \\ 9/36, & w_1 = 4. \end{cases}$$

(b) $h_2(w) = h_1(w)$;

(c)
$$h(w) = \begin{cases} 1/1296, & w = 0 \\ 8/1296, & w = 1, \\ 36/1296, & w = 2, \\ 104/1296, & w = 3, \\ 214/1296, & w = 4, \\ 312/1296, & w = 5, \\ 324/1296, & w = 6, \\ 216/1296, & w = 7, \\ 81/1296, & w = 8. \end{cases}$$

5.2.17 (b) $\mu_Y = 25/3$, $\sigma_Y^2 = 130/9$;

(c)
$$P(Y = y) = \begin{cases} 96/1024, & y = 4, \\ 144/1024, & y = 5, \\ 150/1024, & y = 6, \\ 135/1024, & y = 7. \end{cases}$$

5.3.1   (a) 0.4772; (b) 0.8561.

5.3.3   (a) 46.58, 2.56; (b) 0.8447.

5.3.5   0.925.

5.3.7   (b) 0.05466; 0.3102.

5.3.9   (a) 0.8962; (b) 0.8962; (c) 87.31; 229.58.

5.3.11 0.8997.

5.3.13 (a) 0.3085; (b) 0.2267.

5.3.15 25.

5.4.1   0.4772.

5.4.3   0.8185.

5.4.5   (a) $\chi^2(18)$; (b) 0.0756, 0.9974.

5.4.7   0.6247.

5.4.9   0.95.

5.4.11 (a) The frequencies are 5, 6, 12, 18, 31, 31, 20, 17, 6, 4; the fifth and sixth classes; (c) yes.

5.5.1   (a) 0.2878, 0.2881; (b) 0.4428, 0.4435; (c) 0.1550, 0.1554.

5.5.3   0.9258.

5.5.5   0.6915.

5.5.7   0.3085.

5.5.9   0.6247.

5.5.11   (a) 0.5548; (b) 0.3823.

5.5.13   (a) 0.5641; (b) 0.0111.

5.5.15   (a) 0.3802; (b) 0.7571.

5.5.17   0.2779 (exact is 0.2791).

5.6.1   (a) 0.9984; (b) 0.998.

5.6.3 $M(t) = \left[1 - \dfrac{2t\sigma^2}{n-1} \right]^{-(n-1)/2} \rightarrow e^{\sigma^2 t}.$

CHAPTER 6

6.1.3   (b) $\overline{x} = 89/40 = 2.225$.

6.1.5   (a) $\widehat{\theta} = \overline{X}/2$; (b) $\widehat{\theta} = \overline{X}/3$; (c) $\widehat{\theta}$ equals the sample median.

6.1.7   (c) (i) $\widehat{\theta} = 0.5493, \widetilde{\theta} = 0.5975$, (ii) $\widehat{\theta} = 2.2101, \widetilde{\theta} = 2.4004$,
            (iii) $\widehat{\theta} = 0.9588, \widetilde{\theta} = 0.8646$.

6.1.9 $1/9.5 = 2/19 = 0.1053$.

6.1.11   (a) $\widetilde{p} = 1/\overline{X} = n/\sum_{i=1}^{n} X_i$;
            (b) $\widetilde{p}$ equals the number of successes divided by the number of Bernoulli
                trials;
            (c) $5/63 = 0.0794$.

6.1.15   (a) $\widetilde{\theta} = \overline{X}$; (b) yes; (c) 7.382; (d) 7.485.

6.1.19   7.2.

6.2.1 $[71.35, 76.25]$.

6.2.3 $[2.03, 2.15]$.

6.2.5   (a) $[48.467, 72.266]$ or $[48.076, 72.657]$.

6.2.7 $[19.47, 22.33]$.

6.2.9   (a) 3.24; (b) $s^2 = 0.2372, s = 0.487$; (c) $[2.99, 3.50]$.

6.2.11   (a) 14.72, 1.381; (b) $[14.292, 15.148]$.

6.2.13 $[22.74, 25.48]$.

6.2.15 $[0.757, \infty)$.

6.2.17   (a) 29.49, 3.41; (b) $[0, 31.259]$;
            (c) yes, because of the linearity of the q-q plot of the data and the corre-
                sponding normal quantiles.

6.3.1 $[-59.725, -43.275]$.

6.3.3 $[-5.845, 0.845]$.

6.3.5 $[-2.723, -1.750]$.

6.3.7   (a) $\overline{d} = 0.45$; (b) $[-0.49, 1.39]$.

6.3.9 $[150.76, 231.24]$.

6.4.1   (a) $s = 6.144$; (b) [4.406, 10.142] or [4.107, 9.521].

6.4.3   (a) $\bar{x} = 273.04$, $s^2 = 3,155.54$; (b) [2,079.43, 5,468.08];
(c) [45.60, 73.95]; (d) [44.02, 71.56]; (e) yes.

6.4.5   (a) $\left[\dfrac{\sum_{i=1}^{n} (X_i - \mu)^2}{\chi^2_{\alpha/2}(n)} , \dfrac{\sum_{i=1}^{n} (X_i - \mu)^2}{\chi^2_{1-\alpha/2}(n)} \right]$.

6.4.7   (a) [130.268, 364.503]; (b) $\bar{x} = 204$, $s = 195.45$, yes.

6.4.9   (a) 0.5688; (b) [0.1124, 3.3045].

6.4.11 (a) 0.499; (b) [0.119, 2.404].

6.4.13 $g(a) = g(b)$ and $a = -b$; so $a = -t_{\alpha/2}(n-1) = -b$.

6.4.15 (a) [0.041, 0.599];
(b) No. The compact disks look like they come from two distributions, single length and double length.

6.5.1   (a) 0.0374; (b) [0.0227, 0.0521]; (c) [0.0252, 0.0550]; (d) [0, 0.0497].

6.5.3   (a) 0.69; (b) [0.6613, 0.7187].

6.5.5   (a) 0.506; (b) [0.461, 0.551]; (c) not necessarily.

6.5.7   [0.074, 0.286].

6.5.9   (a) 0.680; (b) [0.646, 0.714].

6.5.11 (a) [0.445, 0.515]; (b) 91.04%.

6.5.13 [0.207, 0.253].

6.5.15 (a) 0.2115; (b) [0.155, 0.268].

6.5.17 (a) 0.2183; (b) [0.164, 0.273].

6.6.1   117.

6.6.3   (a) 1,083; (b) [6.047, 6.049]; (c) $58,800; (d) 0.0145.

6.6.5   (a) 257; (b) yes.

6.6.7   (a) 1,068; (b) 2,401; (c) 752.

6.6.9   2,305.

6.6.11 451.

6.6.13 (a) 38; (b) [0.621, 0.845].

6.6.15 601.

6.6.17 235.

CHAPTER 7

7.1.1 $\alpha = 7/27$; $\beta = 7/27$.

7.1.3   (a) 0.3032 using $b(100, 0.08)$, 0.313 using Poisson approximation, 0.2902 using normal approximation;
(b) 0.1064 using $b(100, 0.04)$, 0.111 using Poisson approximation, 0.1010 using normal approximation.

7.1.5   (a) $\alpha = 0.1056$; (b) $\beta = 0.3524$.

7.1.7   (a) $z = 2.266 > 1.645$, reject H_0;

          (b) $z = 2.266 < 2.326$, do not reject H_0.

7.1.9   (a) $z = 1.752 > 1.645$, reject H_0;

          (b) $z = 1.752 < 1.96$, do not reject H_0.

7.1.11   (a) $z = \dfrac{y/n - 0.40}{\sqrt{(0.40)(0.60)/n}} \geq 1.645$;

          (b) $z = 2.215 > 1.645$, reject H_0.

7.1.13   (a) H_0: $p = 0.40$, H_1: $p > 0.40$, $z \geq 2.326$;

          (b) $2.236 < 2.326$, do not reject H_0.

7.1.15   (a) H_0: $p = 0.037$, H_1: $p > 0.037$;

          (b) $z \geq 2.326$; (c) $z = 2.722 > 2.326$, reject H_0.

7.1.17 $|z| \geq 1.960$; (b) $1.726 < 1.960$, do not reject H_0.

7.1.19 $[0.007, 0.071]$, yes.

7.2.1   (a)
$$K(\mu) = \Phi\left(\frac{22.5 - \mu}{3/2}\right); \alpha = 0.0478;$$

          (b) $\bar{x} = 24.1225 > 22.5$, do not reject H_0;

          (c) 0.2793.

7.2.3   (a)
$$K(\mu) = \Phi\left(\frac{510.77 - \mu}{15}\right);$$

          (b) $\alpha = 0.10$;

          (c) 0.5000;

          (e) (i) 0.0655, (ii) 0.0150.

7.2.5 $n = 25, c = 1.6$.

7.2.7 $n = 40, c = 678.38$.

7.2.9   (a)
$$K(p) = \sum_{y=14}^{25} \binom{25}{y} p^y (1 - p)^{25-y}, 0.40 \leq p \leq 1.0;$$

          (b) $\alpha = 0.0778$;

          (c) $0.1827, 0.3450, 0.7323, 0.9558, 0.9985, 1.0000$;

          (e) yes;

          (f) 0.0344.

7.2.11   With $n = 130$, $c = 8.5$, $\alpha \approx 0.055$, $\beta \approx 0.094$.

7.3.1   (a) $1.4 < 1.645$, do not reject H_0; (b) $1.4 > 1.282$, reject H_0.

          (c) p-value $= 0.0808$.

7.3.3   (a) $z = (\bar{x} - 170)/2$, $z \geq 1.645$;
         (b) $1.260 < 1.645$, do not reject H_0;
         (c) 0.1038.

7.3.5   (a) $t = (\bar{x} - 3{,}315)/(s/\sqrt{30}) \leq -1.699$;
         (b) $-1.414 > -1.699$, do not reject H_0;
         (c) $0.05 < p$-value < 0.10 or p-value ≈ 0.08.

7.3.7   (a) $t = (\bar{x} - 47)/(s/\sqrt{20}) \leq -1.729$;
         (b) $-1.789 < -1.729$, reject H_0;
         (c) $0.025 < p$-value < 0.05, p-value ≈ 0.045.

7.3.9   (a) $t \geq 2.764$; (b) $4.028 > 2.764$, reject H_0; (c) p-value < 0.005;
         (d) $\chi^2 \leq 3.940$; (e) $4.223 > 3.940$, do not reject H_0;
         (f) $0.05 < p$-value < 0.10.

7.3.11   (a) $\chi^2 = 24s^2/140^2$, $\chi^2 \leq 36.42$;
         (b) $29.18 < 36.42$, do not reject H_0.

7.3.13   (a) $59.53 < 64.28$, reject H_0;
         (b) $0.025 < p$-value < 0.05.

7.3.15   (a) $0.015 < 0.058 < 0.078$, do not reject H_0;
         (b) $[0.030, 0.158]$, yes.

7.3.17 $1.607 < 1.645$, do not reject H_0.

7.3.19 $1.477 < 1.833$, do not reject H_0.

7.3.21 $\bar{d} = 0.357$, $s_d = 0.561$, $t = 2.985$, p-value < 0.005.

7.4.1   (a) $t \leq -1.734$; (b) $t = -2.221 < -1.734$, reject H_0.

7.4.3   (a) $2.759 < 4.20$, do not reject H_0 at $\alpha = 0.05$;
         (b) $|t| = 0.374 < 2.086$, do not reject H_0 at $\alpha = 0.05$.

7.4.5   (a) $t < -1.706$; (b) $-1.714 < -1.706$, reject H_0;
         (c) $0.025 < p$-value < 0.05; (e) $1.836 < 3.28$, do not reject.

7.4.7   (a) $t < -2.552$; (b) $t = -3.638 < -2.552$, reject H_0.

7.4.9   (a) $z > 1.645$; (b) $1.957 > 1.645$, reject H_0; (c) p-value $= 0.0252$.

7.4.11   (a) $z = 2.245 > 1.645$, reject H_0; (b) p-value $= 0.0124$.

7.4.13 $F = 3.30 > 3.07$, reject $\sigma_X^2 = \sigma_Y^2$;
         Welch's test: $t = -8.119 < -t_{0.05}(12) = -1.782$.

7.4.15 $F = 1.895 < 3.37$, do not reject H_0.

7.5.1 $6.25 < 7.815$, do not reject if $\alpha = 0.05$; p-value ≈ 0.10.

7.5.3 $7.60 < 16.92$, do not reject H_0.

7.5.5 $3.65 < 7.815$, do not reject H_0.

7.5.7   (a) $q_3 \geq 7.815$;
         (b) $q_3 = 1.744 < 7.815$; do not reject H_0.

7.5.9 $6.113 < 16.92$, do not reject H_0.

7.5.11 $2.010 < 11.07 = \chi^2_{0.05}(5)$, do not reject null hypothesis.

7.5.13 $11.378 < \chi^2_{0.05}(8) = 15.51$, do not reject null hypothesis.

7.5.15   (a) Frequencies are 2, 4, 21, 33, 20, 14, 4, 1, 1;
         (b) $q = 4.653 < 9.488$, do not reject H_0.

7.5.17   (a) $608.70 > 28.87 = \chi^2_{0.05}(18)$, reject H_0;
         (b) $32.31 > 28.87 = \chi^2_{0.05}(18)$, reject H_0;
         (c) $12.79 < 28.87 = \chi^2_{0.05}(18)$, do not reject H_0;
         (e) $n \geq 10$.

7.6.1 $3.23 < 11.07$, do not reject H_0.

7.6.3 $2.40 < 5.991$, do not reject H_0.

7.6.5 $5.975 < \chi^2_{0.05}(2) = 5.991$, do not reject the null hypothesis;
         however, p-value ≈ 0.05.

7.6.7 $8.449 < \chi^2_{0.05}(4) = 9.488$, do not reject the null hypothesis;
         however, $0.05 < p$-value < 0.10; p-value $= 0.076$.

7.6.9 $4.149 > \chi^2_{0.05}(1) = 3.841$, reject the null hypothesis;
         $0.025 < p$-value < 0.05; p-value ≈ 0.042.

7.6.11 $23.78 > 21.03$, reject hypothesis of independence.

7.6.13   (a) $39.591 > 9.488$, reject hypothesis of independence;
         (b) $7.117 > 5.991$, reject hypothesis of independence;
         (c) $11.398 > 9.488$, reject hypothesis of independence.

7.6.15 $9.925 < 12.59 = \chi^2_{0.05}(6)$, do not reject independence.

CHAPTER 8

8.1.1 $7.875 > 4.26$, reject H_0.

8.1.3 $13.773 > 4.07$, reject H_0.

8.1.5 $5.171 > 3.25$, reject H_0.

8.1.7 $14.757 > 4.43$, reject H_0.

8.1.9   (a) $F \geq 4.07$;
         (b) $4.106 > 4.07$, reject H_0;
         (c) $4.106 < 5.42$, do not reject H_0;
         (d) $0.025 < p$-value < 0.05, p-value ≈ 0.05.

8.1.11 $10.224 > 4.26$, reject H_0.

8.1.13   (a) $F \geq 5.61$;
         (b) $6.337 > 5.61$; reject H_0.

8.2.1 $18.00 > 5.14$, reject H_A.

8.2.3   (a) $7.624 > 4.46$, reject H_A;
        (b) $15.538 > 3.84$, reject H_B.

8.2.5   (a) $1.723 < 2.90$, accept H_{AB};
        (b) $5.533 > 4.15$, reject H_A;
        (c) $28.645 > 2.90$, reject H_B.

8.2.7 $1.727 < 2.37$, accept H_{AB};
        $2.238 < 3.27$, do not reject H_A;
        $2.063 < 2.87$, do not reject H_B.

8.2.9 $0.111 < 3.89$, accept H_{AB};
        $9.692 > 4.75$, reject H_A;
        $2.606 < 3.89$, do not reject H_B.

8.3.1   (a) $\widehat{y} = 86.8 + (842/829)(x - 74.5)$;
        (c) $\widehat{\sigma^2} = 17.9998$.

8.3.3   (a) $\widehat{y} = 6.478 + 4.483x$.

8.3.5   Solve for α: $-t_{\alpha/2}(n-2) \le \dfrac{\widehat{\alpha} - \alpha}{\sqrt{\widehat{\sigma^2}/(n-1)}} \le t_{\alpha/2}(n-2)$.

8.3.7 $[83.341, 90.259]$, $[0.478, 1.553]$, $[10.265, 82.578]$.

8.3.9   (a) $\widehat{y} = 15.695 + 0.620x$;
        (c) $\widehat{\sigma^2} = 0.2595$;
        (d) $[0.4416, 0.7984]$.

8.3.11 (a) $\widehat{y} = 1.896 + 0.538x$;
        (c) $\widehat{\alpha} = 6.9667$, $\widehat{\beta} = 0.5385$, $\widehat{\sigma^2} = 0.0549$;
        (d) $[6.8263, 7.1070]$, $[0.2081, 0.8688]$, $[0.0333, 0.1643]$.

8.4.1   (a) $4.359 > 2.306$, reject H_0;
        (b) $19.005 > 5.32$, reject H_0;
        (c) $[75.283, 85.113]$, $[83.838, 90.777]$, $[89.107, 99.728]$;
        (d) $[68.206, 92.190]$, $[75.833, 98.783]$, $[82.258, 106.577]$.

8.4.3   (a) $2.148 < 2.160$, do not reject H_0;
        (b) $4.615 < 4.68$, do not reject H_0;
        (c) $[62.793, 129.495]$, $[101.854, 135.266]$, $[119.483, 162.470]$;
        (d) $[28.926, 163.361]$, $[57.856, 179.265]$, $[78.784, 203.169]$.

8.4.5   (a) $[4.897, 8.444]$, $[9.464, 12.068]$, $[12.718, 17.004]$;
        (b) $[1.899, 11.442]$, $[6.149, 15.383]$, $[9.940, 19.782]$.

8.4.7 (a) $2.369 > 1.771$, reject H_0;

(b) $[19.669, 26.856]$, $[22.122, 27.441]$, $[24.048, 28.551]$, $[25.191, 30.445]$, $[25.791, 32.882]$;

(c) $[15.530, 30.996]$, $[17.306, 32.256]$, $[18.915, 33.684]$, $[20.351, 35.285]$, $[21.618, 37.055]$.

8.4.9 $\hat{y} = 1.1037 + 2.0327x - 0.2974x^2 + 0.6204x^3$.

CHAPTER 9

9.1.1 (a) 158.97, 12.1525, 30.55; (f) yes.

9.1.3 (a) 335.176, 0.5214, 1.29; (f) no.

9.1.5 $\bar{p} = 0.0254$, $UCL = 0.0726$, yes.

9.1.7 (a) 0, 5.82; (b) yes.

9.2.1

2^2 Design			
Run	A	B	AB	Observations
1	$-$	$-$	$+$	X_1
2	$+$	$-$	$-$	X_2
3	$-$	$+$	$-$	X_3
4	$+$	$+$	$+$	X_4

(a) $[A] = (-X_1 + X_2 - X_3 + X_4)/4$,
$[B] = (-X_1 - X_2 + X_3 + X_4)/4$,
$[AB] = (X_1 - X_2 - X_3 + X_4)/4$.

(b) It is sufficient to compare the coefficients on both sides of the equations of X_1^2, X_1X_2, X_1X_3, and X_1X_4 which are $3/4$, $-1/2$, $-1/2$, respectively.

(c) Each is $\chi^2(1)$.

9.2.3 $[A]$ is $N(0, \sigma^2/2)$ so $E[(X_2 - X_1)^2/4] = \sigma^2/2$ or $E[(X_2 - X_1)^2/2] = \sigma^2$.

9.2.5 (a) $[A] = -4$, $[B] = 12$, $[C] = -1.125$, $[D] = -2.75$, $[AB] = 0.5$, $[AC] = 0.375$, $[AD] = 0$, $[BC] = -0.625$, $[BD] = 2.25$, $[CD] = -0.125$, $[ABC] = -0.375$, $[ABD] = 0.25$, $[ACD] = -0.125$, $[BCD] = -0.375$, $[ABCD] = -0.125$.

(b) There is clearly a temperature (B) effect. There is also a catalyst charge (A) effect and probably a concentration (D) and a temperature-concentration (BD) effect.

CHAPTER 10

10.1.1   (a) $y_1 = 5$, $y_2 = 6$, $y_3 = 7$, $y_4 = 9, \ldots, y_{10} = 78$;

      (b) $\tilde{m} = 31.5$, $\tilde{\pi}_{0.80} = 72.6$;

      (c) $\tilde{q}_1 = 6.75$, $\tilde{q}_3 = 68.75$.

10.1.3   (a) $g_3(y) = 10(1 - e^{-y/3})^2 e^{-y}, 0 < y < \infty$;

      (b) $5(1 - e^{-5/3})^4 e^{-5/3} = 0.7599$;

      (c) $e^{-5/3} = 0.1889$.

10.1.5   (a) 0.2553; (b) 0.7483.

10.1.7   (a) 0.8697; (b) 0.8634.

10.1.9   (a) $g_1(y) = 19(e^{-y/\theta})^{18} \dfrac{1}{\theta} e^{-y/\theta}, 0 < y < \infty$;

      (b) 1/20.

10.1.11 (a) $E(W_1) = E(W_2) = E(W_3) = \theta$; (b) W_3;

      (d) (i) 0.5098, 0.0276; (ii) 0.5242, 0.0433; (iii) 0.5026, 0.0249;

      (e) W_3, unbiased and smallest variance.

10.1.13 (a) 14.75; (b) 0.8995.

10.2.1   (a) 0.7812; (b) 0.7844; (c) 0.4528.

10.2.3   (a) (6.31, 7.40); (b) (6.58, 7.22), 0.8204.

10.2.5   (2.72, 2.84), 93.46%.

10.2.7   (15.40, 17.05).

10.2.9   (a)

Stem	Leaf	Frequency
0	09 11 13 24 34 51 57 75 79 98	10
1	36 52 61 98	4
2	15 38 44 72 88	5
3	15 15 50	3
4	45	1
5		0
6	20	1
7	23	1

      (b) $y_3 = 13$, $y_9 = 79$, $y_{10} = 98$, $y_{13} = 161$, $y_{17} = 244$;

      (c) $\tilde{\pi}_{0.25} = 54$, $\tilde{m} = 161$, $\tilde{\pi}_{0.75} = 301.5$;

      (d) (i) (13, 98), 89.66%; (ii) (79, 244), 89.22%; (iii) (238, 455), 89.66%.

10.2.11 (a)

Stem	Leaf	Frequency
8●	789	3
9*	00122234	8
9●	555788889	9
10*	0012	4
10●	6	1

(Multiply numbers by 10^{-2}.)

(b) $\tilde{m} = 0.95$; (c) (0.92, 0.98), 95.68%;

(d) $\tilde{q}_1 = 0.915$; (e) (0.89, 0.93), 89.66%;

(f) $\tilde{q}_3 = 0.985$; (g) (0.97, 1.02), 96.33%.

10.3.1 (a) $9 < 10$, do not reject H_0; (b) p-value $= 0.1334$.

10.3.3 (a) $C = \{w: \ w \le 7 \text{ or } w \ge 17\}$, $\alpha \approx 0.0662$, ($\alpha = 0.0639$), $w = 13$, do not reject.

10.3.5 Let Y equal the number of drivers going 60 mph or less. Then $C = \{y: \ y \ge 87\}$, $\alpha \approx 0.0521$, reject H_0.

10.3.7 $t = 2.162 > 2.120$, reject H_0.

10.3.9 (a) $7 < 8$, do not reject H_0 at $\alpha = 0.0547$;

(b) $1.477 < 1.833$, do not reject H_0.

10.3.11 (a) $C = \{w: \ w \le 3 \text{ or } x \ge 11\}$, $\alpha = 0.0574$;

(b) $w = 3$, reject H_0.

10.4.1 (a) $-55 < -47.08$, reject H_0; (b) 0.0296.

10.4.3 $w = 132$, p-value $= 0.0306$ for a one-sided alternative, p-value $= 0.0613$ for a two-sided alternative.

10.4.5 $74 > 63.6$, reject H_0; p-value $= 0.0296$.

10.4.7 $w = 54$, $z = 1.533$, p-value $= 0.0661$, do not reject H_0.

10.4.9 (a) Let Y equal the number of weights less that 1.14 pounds. Then $C = \{y: \ y \le 4\}$, $\alpha = 0.0898$.

(b) $y = 5$, fail to reject H_0;

(c) p-value $= 0.2120$;

(d) $z = 2.072 > 1.282$, reject H_0; (e) p-value $= 0.0207$.

10.5.1 (a) $v = 8 > 7$, reject H_0; p-value $= 0.0115$;

(b) $w = 145 > 126$ or $z = 3.024 > 1.645$, reject H_0; p-value ≈ 0.0014;

(c) $v = 4$, p-value $= 0.1517$, do not reject.

10.5.3 (a) $C = \{v: \ v \le 2 \text{ or } v \ge 8\}$, $\alpha = 0.0230$, $v = 4$, do not reject H_0;

(b) $C = \{w: \ w \le 79 \text{ or } w \ge 131\}$, $\alpha \approx 0.0539$, $w = 95$, do not reject H_0.

10.5.5 (a) $C=\{v: v \le 2 \text{ or } v \ge 8\}, \alpha=0.0230, v=6$, do not reject H_0;
        (b) $C=\{w: w \le 79 \text{ or } w \ge 131\}, \alpha \approx 0.0539, w=107.5$, do not reject H_0.

10.5.7 (a) $C=\{w: w \le 184 \text{ or } w \ge 280\}, \alpha \approx 0.0489, w=241$, do not reject H_0.

10.5.9 (a) $C=\{v: v \ge 7\}, \alpha=0.0402, v=7, p$-value $= 0.0402$, reject H_0;
        (b) $C=\{w: w \ge 174\}, \alpha \approx 0.0545, w=187, p$-value ≈ 0.0067, reject H_0.

10.6.1 $C = \{r: r \le 6\}, \alpha = 43/429, r = 6$, reject H_0.

10.6.3 $(6!)(12!)\binom{11}{4} = 113{,}810{,}780{,}160{,}000.$

10.6.5 (b) $r = 11$, reject H_0; (c) p-value $= 0.0457$.

10.6.7 $r = 4, p$-value $= 43/1{,}716 = 0.025$, reject the hypothesis of randomness. There appears to be a trend effect.

10.6.9 $r = 7, p$-value $= 1{,}331/58{,}786 = 0.0226$, reject the hypothesis of randomness.

10.6.11 (a) $C = \{y: y \le 3 \text{ or } y \ge 11\}, y = 7$, do not reject H_0;
        (b) $C = \{r: r \le 4 \text{ or } r \ge 12\}, \alpha = 43/858 = 0.05, r = 4$, reject the hypothesis of randomness. There appears to be a trend effect.

10.6.13 $C = \{r: r \le 10\}, \alpha=919/22{,}287 = 0.041, r = 6,$
        p-value $= 1{,}133/1{,}002{,}915 = 0.0011$, reject H_0.

10.7.1 $d_5 = 0.40 < 0.45$, do not reject H_0.

10.7.3 $d_{10} = 0.1643$ at $x = 49$, do not reject H_0.

10.7.5 $d_{10} = 0.1654$ at $x = 58$, do not reject H_0.

10.7.7 $d_{40} = 0.04559$, do not reject H_0.

10.7.9 (a) $d_{70} = 0.1295 > 1.07/\sqrt{70} = 0.1279$, reject H_0;
        (b) the histogram is bimodal;
        (c) the data appear to be more uniform than normal.

10.7.11 (a) $\widehat{\theta} = 208.487, d_{39} = 0.1352$ at $x = 95$;
        (b) $\widehat{\alpha} < 0$ which is not compatible with the experiment.

CHAPTER 11

11.1.1 (a) $f_1(x) = (2x + 5)/16, x = 1, 2$;
        (b) $f_2(y) = (2y + 3)/32, y = 1, 2, 3, 4$;
        (c) 3/32; (d) 9/32; (e) 3/16; (f) 1/4; (g) dependent.

11.1.3 (b) $f(x, y) = 1/16, x = 1, 2, 3, 4; y = x + 1, x + 2, x + 3, x + 4$;
        (c) $f_1(x) = 1/4, x = 1, 2, 3, 4$;
        (d) $f_2(y) = (4 - |y - 5|)/16, y = 2, 3, 4, 5, 6, 7, 8$;
        (e) dependent because the space is not rectangular.

11.1.5 (a) $f(x_1, x_2, x_3) = \binom{13}{x_1}\binom{13}{x_2}\binom{13}{x_3}\binom{13}{13 - x_1 - x_2 - x_3}\Big/\binom{52}{13}$,

$0 \le x_1 + x_2 + x_3 \le 13$;

(b) $f_{12}(x_1, x_2) = \binom{13}{x_1}\binom{13}{x_2}\binom{26}{13 - x_1 - x_2}\Big/\binom{52}{13}$, $0 \le x_1 + x_2 \le 13$;

(c) $f_3(x_3) = \binom{13}{x_3}\binom{39}{13 - x_3}\Big/\binom{52}{13}$, $0 \le x_3 \le 13$;

11.1.7 (a) $f(x, y) = \dfrac{15!}{x!\, y!\, (15 - x - y)!}\left(\dfrac{6}{10}\right)^x\left(\dfrac{3}{10}\right)^y\left(\dfrac{1}{10}\right)^{15 - x - y}$,

$0 \le x + y \le 15$;

(b) no, because the space is not rectangular;

(c) 0.0735;

(d) X is $b(15, 0.6)$;

(e) 0.9095.

11.1.9 $f_1(x) = 2e^{-2x}, 0 < x < \infty$; $f_2(y) = 2e^{-y}(1 - e^{-y}), 0 < y < \infty$; no.

11.1.11 $f_1(x) = 1/2, 0 \le x \le 2$; $f_2(y) = 1/2, 0 \le y \le 2$; yes.

11.2.1 $\mu_X = 25/16$; $\mu_Y = 45/16$; $\sigma_X^2 = 63/256$; $\sigma_Y^2 = 295/256$; $\text{Cov}(X, Y) = -5/256$; $\rho = -0.0367$; dependent.

11.2.3 (a) $f(x, y) = 1/16, x = 1, 2, 3, 4$; $y = x + 1, x + 2, x + 3, x + 4$;

$f_1(x) = 1/4, x = 1, 2, 3, 4$;

$f_2(y) = (4 - |y - 5|)/16, y = 2, 3, \ldots, 8$;

(b) $\mu_X = 5/2$; $\mu_Y = 5$; $\sigma_X^2 = 5/4$; $\sigma_Y^2 = 5/2$; $\text{Cov}(X, Y) = 5/4$;

$\rho = \sqrt{2}/2$.

(c) $y = x + 2.5$.

11.2.5 $a = \mu_Y - \mu_X b, b = \text{Cov}(X, Y)/\sigma_X^2$.

11.2.7 (a) No; (b) $\text{Cov}(X, Y) = 0, \rho = 0$.

11.2.9 W is $b(4, 1/3)$; $\rho = \sqrt{2}/2$.

11.3.1 (d) 9/14, 7/18, 5/9; (e) 20/7, 55/49.

11.3.3 (a) $f(x, y) = \dfrac{50!}{x!\, y!\, (50 - x - y)!}(0.02)^x (0.90)^y (0.08)^{50 - x - y}$,

$0 \le x + y \le 50$;

(b) Y is $b(50, 0.90)$;

(c) $b(47, 0.90/0.98)$;

(d) 2115/49; (e) $\rho = -3/7$.

11.3.5 (a) $E(Y \mid x) = 2(2/3) - (2/3)x, x = 1, 2$; (b) yes.

11.3.7 $E(Y \mid x) = x + 5/2, x = 1, 2, 3, 4$; it is linear; yes.

11.3.9   (a) $f_1(x) = 1/8, x = 0, 1, \ldots, 7$;
        (b) $h(y \mid x) = 1/3, y = x, x+1, x+2$, for $x = 0, 1, \ldots, 7$;
        (c) $E(Y \mid x) = x+1$, for $x = 0, 1, \ldots, 7$;
        (d) $\sigma_Y^2 = 2/3$;
        (e)
$$f_2(y) = \begin{cases} 1/24, & y = 0, 9, \\ 2/24, & y = 1, 8, \\ 3/24, & y = 2, 3, 4, 5, 6, 7. \end{cases}$$

11.3.11 (b) $f_1(x) = (3-x)/6, x = 0, 1, 2$; $f_2(y) = (3-y)/6, y = 0, 1, 2$;
        (c) $E(Y \mid x) = 1 - x/2$;
        (d) $E(X \mid y) = 1 - y/2$;
        (e) $\rho = -1/2$.

11.3.13 (b) $f_1(x) = 1/10, 0 \le x \le 10$;
        (c) $h(y \mid x) = 1/4, 10 - x \le y \le 14 - x$ for $0 \le x \le 10$;
        (d) $E(Y \mid x) = 12 - x$.

11.3.15 (a) $f(x, y) = 1/(2x^2), 0 < x < 2, 0 < y < x^2$;
        (b) $f_2(y) = (2 - \sqrt{y})/(4\sqrt{y}), 0 < y < 4$;
        (c) $E(X \mid y) = [2\sqrt{y} \ln(2/\sqrt{y})]/[2 - \sqrt{y}]$;
        (d) $E(Y \mid x) = x^2/2$.

11.3.17 (a) $h(y \mid x) = e^{-x}, 0 < y < e^x$ for $0 < x < 1$;
        (b) $E(Y \mid x) = e^x/2, 0 < x < 1$;
        (c) $f(x, y) = e^{-x}, 0 < x < 1, 0 < y < e^x$;
        (d) $f_2(y) = \begin{cases} 1 - 1/e, & 0 < y < 1, \\ 1/y - 1/e, & 1 \le y < e. \end{cases}$

11.3.19 (a) 0.2692, 0.3757, 3.1382, 3.6881;
        (b) 0.0266, 0.0369, 0.2694, 0.3085.

11.4.1   (a) 0.6006; (b) 0.7888; (c) 0.8185; (d) 0.9371.

11.4.3   (a) 0.5746; (b) 0.7357.

11.4.5   (a) $N(86.4, 40.96)$; (b) 0.4192.

11.4.7   (a) 0.8248;
        (b) $E(Y \mid x) = 457.1735 - 0.2655x$;
        (c) $\text{Var}(Y \mid x) = 645.9375$;
        (d) 0.8079.

11.4.9 $a(x) = x - 11, b(x) = x + 5$.

11.4.11 (a) 0.2857; (b) $\mu_{Y|x} = -0.2x + 4.7$;
        (c) $\sigma_{Y|x}^2 = 8.0784$; (d) 0.4230.

11.5.1 $-0.45 < -0.3808$, reject H_0.

11.5.3 $[0.419, 0.802]$.

11.5.5 $|r| = 0.252 < 0.6613$, do not reject H_0.

11.5.9 $n = 9$.

11.5.11 $r = 0.699 > 0.5922$, so reject H_0.

11.6.1 $g(y_1, y_2) = (1/4)e^{-y_2/2}, 0 < y_1 < y_2 < \infty; g_1(y_1) = (1/2)e^{-y_1/2},$
$0 < y_1 < \infty; g_2(y_2) = (y_2/4)e^{-y_2/2}, 0 < y_2 < \infty;$ no.

11.6.3   (a)
$$g(y_1, y_2) = \frac{1}{\sqrt{2\pi}\,\Gamma(r/2)2^{r/2}}\, y_2^{r/2-1}\, e^{-(y_2/2)(1+y_1^2/r)}\left(\frac{\sqrt{y_2}}{\sqrt{r}}\right),$$
$-\infty < y_1 < \infty, 0 < y_2 < \infty;$

(b)
$$g_1(y_1) = \frac{\Gamma[(r+1)/2]}{\sqrt{\pi r}\,\Gamma(r/2)(1+y_1^2/r)^{(r+1)/2}}, \quad -\infty < y_1 < \infty.$$

11.6.7 $\mu = \dfrac{r_2}{r_2 - 2},\ r_2 > 2; \sigma^2 = \dfrac{2r_2^2(r_1 + r_2 - 2)}{r_1(r_2 - 2)^2(r_2 - 4)},\ r_2 > 4.$

11.6.15 840.

11.6.17 (a) 0.1792; (b) 0.1792.

11.6.19 (a) t with 2 degrees of freedom.
        (d) $\mu_U = 0, \mu_W = 0, \sigma_W^2 = 1.$

CHAPTER 12

12.1.1   (a) $f(x; p) = \exp\{x \ln(1-p) + \ln[p/(1-p)]\};$
        $K(x) = x; \sum\limits_{i=1}^{n} X_i$ is sufficient.

        (b) $\overline{X}.$

12.1.3   (a) $\sum\limits_{i=1}^{n} X_i^2$; (b) $\widehat{\sigma^2} = \left(\dfrac{1}{n}\right)\sum\limits_{i=1}^{n} X_i^2$; (c) yes.

12.2.1 $k(\theta \mid x_1, x_2, \ldots, x_n) \quad \propto \quad \theta^{\Sigma x_i}(1-\theta)^{n-\Sigma x_i}\left[\dfrac{\Gamma(\alpha+\beta)}{\Gamma(\alpha)\Gamma(\beta)}\right]\theta^{\alpha-1}(1-\theta)^{\beta-1}$
        $\propto \quad \theta^{\Sigma x_i + \alpha - 1}(1-\theta)^{n-\Sigma x_i + \beta - 1}.$

12.2.3   (a) $E[\{w(Y) - \theta\}^2] \quad = \quad \{E[w(Y) - \theta]\}^2 + \text{Var}[w(Y)]$
        $= \quad (74\theta^2 - 114\theta + 45)/500;$
        (b) $\theta = 0.569$ to $\theta = 0.872.$

12.3.1   (b) σ^2/n; (c) 1/5.

12.3.3   (a) $2\theta^2/n$; (b) $N(\theta, 2\theta^2/n)$; (c) $\chi^2(n).$

12.4.1   (a) 0.84; (b) 0.082.

12.4.3 $k = 1.464; 8/15.$

12.4.5   (a) 0.25; (b) 0.85; (c) 0.925.

12.4.7   675.

12.5.1 (a)
$$\frac{L(80)}{L(76)} = \exp\left[\frac{6}{128}\sum_{i=1}^{n} x_i - \frac{624n}{128}\right] \le k \text{ or } \bar{x} \le c;$$
(b) $n = 43, c = 78.$

12.5.3 (a)
$$\frac{L(3)}{L(5)} \le k \text{ if and only if } \sum_{i=1}^{n} x_i \ge (-15/2)[\ln(k) - \ln(5/3)^n] = c;$$
(b) $\bar{x} \ge 4.15;$
(c) $\bar{x} \ge 4.15;$
(d) yes.

12.5.5 (a)
$$\frac{L(50)}{L(\mu_1)} \le k \text{ if and only if } \bar{x} \le \frac{(-72)\ln(k)}{2n(\mu_1 - 50)} + \frac{50 + \mu_1}{2} = c;$$

12.5.7 (a)
$$\frac{L(0.5)}{L(\mu)} \le k \text{ if and only if } \sum_{i=1}^{n} x_i \ge \frac{\ln(k) + n(0.05 - \mu)}{\ln(0.5/\mu)} = c;$$
(b) $\sum_{i=1}^{10} x_i \ge 9.$

12.6.1 (a) $|-1.80| > 1.645$, reject H_0;
(b) $|-1.80| < 1.96$, do not reject H_0;
(c) p-value $= 0.0718.$

12.6.3 (a) $\bar{x} \ge 230 + 10z_\alpha/\sqrt{n}$ or $\dfrac{\bar{x} - 230}{10/\sqrt{n}} \ge z_\alpha;$
(b) yes; (c) $1.04 < 1.282$, do not reject H_0; (d) p-value $= 0.1492.$

12.6.5 (a) $|2.10| < 2.306$, do not reject H_0; $0.05 < p$-value $< 0.10.$

12.6.7 $2.20 > 1.282$, reject H_0; p-value $= 0.0139.$

Index

Acceptance number, 164
Acceptance sampling plan, 130
Additive model, 426
Algebra of sets, 62, 641
Alternative hypothesis, 337
Analysis of variance, 418
ANOVA table, 418, 428
Approximate chi-square distribution, 387
Approximate distributions, 163, 263, 270, 277, 593, 617
Arcsine transformation, 596
Assessing prior probabilities, 610
Associative laws, 643
Asymptotic distributions, 615

Back-to-back stem-and-leaf display, 32
Bar graph, 108
Bayes' formula, 102
Bayes' solution, 611
Bayes' Theorem, 102, 610
Bayesian methods, 609
Bernoulli distribution, 136
Bernoulli experiment, 135
Bernoulli trials, 135
Best critical region, 626
Best-fitting line, 51

Beta distribution, 598, 603
Biased estimator, 287
Binomial coefficients, 75
Binomial distribution, 137, 158, 163, 170, 175, 320
Birthday problem, 92
Bivariate hypergeometric distribution, 561
Bivariate normal distribution, 583
Blocking, 486
Box-and-whisker diagram, 26, 34
Box plot, 26
Box-Muller transformation, 598

c-chart, 471
Cauchy distribution, 603, 604
Censoring, 233
Central Limit Theorem, 263, 279, 616
Change-of-variable technique, 223, 597, 659
Chebyshev's Inequality, 621
Chi-square distribution, 204, 215, 256, 387, 417
Chi-square goodness of fit tests, 385
Chi-square statistic, 387, 400
Chi-square tests, 387, 390
Class boundaries, 7
Class intervals, 7

Class limits, 7
Class marks, 7, 18
Column effect, 426
Combinations, 75
Commutative laws, 643
Complement of a set, 62, 642
Composite hypothesis, 346
Compounding, 615
Conditional distribution, 573
Conditional expectation, 575
Conditional mean, 575
Conditional probability, 84
Conditional probability density function,
    573
Conditional variance, 575
Confidence band for distribution function,
    547
Confidence coefficient, 295
Confidence intervals, 295, 301, 302
    for correlation coefficient, 593
    for regression line, 453
    for means, 295, 300, 305
    for proportions, 320
    for percentiles, 504
    for standard deviations, 312
    for variances, 312
Contingency tables, 399
Continuous-type data, 7
Continuous-type random variables, 180
Control charts, 463
    for flaws, 471
    for the mean, 463
    for p, 469
    for the range, 467
    for the standard deviation, 467
Converges in probability, 623
Convolution formula, 246, 253, 604
Correlation analysis, 590
Correlation coefficient, 51, 565, 576, 588, 590
    of sample, 51
Counting data, 5
Covariance, 51, 565
Craps, 91
Critical region, 339, 626, 634
Cumulative distribution function, 139

Data, 2,5,7
    continuous-type, 7
    counting, 5
    discrete, 5
Deciles, 26
Degrees of freedom, 204, 390
De Morgan's laws, 644
Density histogram, 6
Dependent events, 94
Dependent random variables, 554
Design of experiments, 477, 483
Discrete data, 5
Discrete-type random variables, 107
Discrete uniform distribution, 126
Distinguishable permutations, 77
Distribution, 4, 173, 222
    Bernoulli, 136
    beta, 598, 603
    binomial, 137, 158, 163, 170, 175, 320
    bivariate normal, 583
    Cauchy, 603, 604
    chi-square, 204, 215, 256, 387, 417
    discrete uniform, 126
    empirical, 122, 193, 624
    exponential, 194
    F, 314, 433, 601, 602
    gamma, 202
    geometric, 147, 172, 652
    half normal, 230
    hypergeometric, 128, 332, 560, 645
    limiting, 278
    logistic, 200, 229
    lognormal, 222
    mixed, 230
    multinomial, 387, 562
    negative binomial, 152, 177, 651
    normal, 209, 254, 259, 294, 417
    of linear combination, 259
    of order statistics, 496
    of sample mean, 254
    of sum of chi-square variables, 256
    Poisson, 158
    posterior, 611
    prior, 610
    rectangular, 191
    shifted exponential, 200, 503

Distribution, (*continued*)
  standard normal, 211
  t, 298, 361, 600, 602, 638
  trinomial, 562, 577
  uniform, 126, 191, 224, 227
Distribution-free confidence intervals, 504
Distribution-free inferences, 504
Distribution function, 139, 182, 224, 227
Distribution function technique, 222, 600
Distributive laws, 644
Distributive operator, 115

Efficiency, 620
Element, 641
Empirical distribution, 122, 193, 624
Empirical distribution function, 193, 544, 624
Empirical rule, 19
Empty set, 62, 642
Equality of means, 414
Equally likely events, 68
Error structure, 358
Estimate(s), 4, 282
Estimator, 282
  Bayesian, 611
  biased, 287
  efficiency of, 620
  maximum likelihood, 283, 608, 610, 634
  method of moments, 288
  minimum chi-square, 390
  point, 282
  sufficient, 605, 608, 632
  unbiased, 287
Event(s), 62, 641
  equally likely, 68
  exhaustive, 62
  independent, 94, 404
  mutually exclusive, 62, 642
  occurred, 62
Exhaustive events, 62
Expected value, 113, 184, 239, 247, 556
Exploratory data analysis, 22
Exponential distribution, 194
Exponential form, 607

F-distribution, 314, 433, 601, 602
Factorial designs, 476
Factorial moment, 172
Factorial moment-generating function, 172
Factorization Theorem, 605
Factors, 426, 476
Fibonacci numbers, 155
First quartile, 26, 186
Fisher-Neyman Factorization Theorem, 605
Five-number summary, 26
Frequency, 5
  histogram, 5, 8
  table, 5

Gamma distribution, 202
Gamma function, 201
General factorial designs, 476
Generating function, 168
Geometric distribution, 147, 172, 652
Goodness of fit, 385
Graphical methods, 32

Half normal distribution, 230
Histogram, 5
  density, 6
Hypergeometric distribution, 128, 332, 560, 645
Hypothesis, 336

Independent events, 93, 94, 404
Independent random variables, 237, 242, 247, 554
Independent trials, 135
Infinite series, 649
Inner fences, 28
Integration, 654
Interaction, 430, 476
Interquartile range, 26, 29
Intersection of sets, 62, 642
Interval estimate, 295, 614

Jacobian, 597, 659
Joint probability density function, 237, 553, 557

Kolmogorov-Smirnov test, 544

Law of large numbers, 278, 623
Least squares, 49, 657
Least squares regression line, 51, 568, 576
L'Hospital's rule, 649
Likelihood function, 283, 634
Likelihood ratio, 634
Likelihood ratio test, 634
Limiting distribution, 278
Limits, 648
Linear association, 55, 568
Linear model, 439, 458
Linear operator, 115
Logistic distribution, 200, 229
Lognormal distribution, 222
Lognormal model, 222
Loss function, 611

Maclaurin's series, 177
Mann-Whitney test, 531
Marginal probability density function, 554, 558
Matching, 92
Mathematical expectation, 113, 239, 556
Maximum error of estimate, 328
Maximum likelihood estimates, 285, 608, 614
Maximum likelihood estimators, 284, 608, 610, 634
Mean, 118, 184, 249, 557
  of distribution, 118, 184, 557
  of empirical distribution, 122
  of linear combination, 249
  of sample, 15
  of X, 118, 184, 557
Mean square due to error, 452
Mean square due to regression, 452
Mean square error, 452
Median, 26, 186
Median test, 527
Method of least squares, 49, 440
Method of moments estimator, 288
Methods of enumeration, 71
Midrange, 28

Minimum chi-square estimator, 390
Mixed distribution, 230
Modal class, 10
Modal interval, 10
Mode, 5, 10, 614
Model, 391, 395
Moment, 121, 174
Moment-generating function, 173, 185, 250, 277
Most powerful test, 630
Multinomial coefficient, 78, 386
Multinomial distribution, 387, 562
Multiple regression, 458
Multiplication principle, 71
Multiplication rule, 86
Multivariate calculus, 656
Multivariate distributions, 552
Multivariate hypergeometric distribution, 560
Mutually exclusive events, 62, 642
Mutually independent events, 95
Mutually independent random variables, 556

Negative binomial distribution, 152, 177, 651
Negative binomial expansion, 153, 651
Neyman-Pearson Lemma, 626
Noncentral F, 433
Noninformative prior, 610
Nonparametric methods, 493
Normal distribution, 209, 254, 259, 294, 417
Normal equations, 458
Null hypothesis, 337
Null set, 62, 642

One-factor experiment, 414
One-sided confidence intervals, 302, 341
One-sided test, 339
Operating characteristic curve, 130, 165
Order statistics, 24, 37, 493
Ordered sample, 73, 79
Ordered stem-and-leaf display, 23
Outer fences, 28
Outcome, 3
Outliers, 28

p-chart, 469
Paired *t*-test, 361
Pairwise independent events, 95
Parameter, 137, 282
Parameter space, 282
Partition, 102
Pascal's equation, 81
Pascal's triangle, 81
Pearson, Karl, 385
Penalty function, 611
Percentiles, 25, 186, 499
Permutation, 73, 77
Plan-do-study-act, 2
Point estimate, 282
Poisson distribution, 158
Poisson process, 157, 176, 201
Population, 4
Posterior distribution, 611
Posterior probability, 102
Power function, 348
Power of test, 348
Prediction interval, 453
Prior distribution, 610
Prior probability, 102
Probability, 5, 64, 65, 85
Probability density function, 107, 180, 237
Probability-generating function, 169
Probability histogram, 108
Probability value, 352
Pseudo-random numbers, 192
P-value, 352, 365

Quadratic expression, 58
Quadratic form, 417
Quality improvement, 462
Quantiles, 34
Quantile-quantile plot, 35
Quartiles, 26, 186

R-chart, 467
Random experiments, 3
Random interval, 295
Random number generator, 192
Random numbers, 192

Random sample, 136, 240, 242, 560
Random variable, 4, 105, 237
   of the continuous type, 180
   of the discrete type, 107
Range, 7, 29
Rank, 519
Rao-Cramér Inequality, 619
Ratio test, 650
Real quadratic form, 417
Rectangular distribution, 191
Regression, 439
Relative frequency, 5, 63, 109
Relative frequency histogram, 6, 8
Residual, 442
*r*th factorial moment, 172
*r*th moment, 121, 174, 185
Row effect, 426
Rule of, 72, 653
Run, 536
Run test, 536

s-chart, 467
Sample(s), 4, 15
Sample correlation coefficient, 51
Sample covariance, 51
Sample mean, 15, 254, 257
Sample percentiles, 25, 499
Sample size, 326, 352
   of a random sample, 326, 352
   to estimate μ, 326
   to estimate p, 328
Sample space, 641
Sample standard deviation, 15
Sample variance, 15, 257
Sampling distribution theory, 245
Sampling with replacement, 74
Sampling without replacement, 74
Scatter plot, 48
Scientific method, 2
Second quartile, 26
Set function, 65
Shewhart control charts, 462
Shifted exponential distribution, 200, 503
Shortest confidence interval for σ, 313, 319
Sign test, 513

Significance level, 337, 346
Simple hypothesis, 337
Simulation, 225
Skewness, 9, 27, 123
Space
  of X, 105
  outcome, 3
Standard deviation of distribution, 119, 185
Standard deviation of sample, 15
Standard deviation of X, 119, 185
Standard error of mean, 358
Standard normal distribution, 211
Standard order, 477
Standardize, 51
Statistics, 2, 245, 282
Statistical hypothesis, 336, 346
Statistical inference, 4
Statistical process control, 463
Statistical quality control, 463
Statistically independent events, 94
Stem-and-leaf display, 22
Stochastically independent events, 94
Subset, 62, 642
Sufficient statistics, 605, 608, 632
Support, 107
Suspected outliers, 28

Tail-end probability, 353
Taylor's series, 596
T-distribution, 298, 361, 600, 602, 638
Tests of statistical hypotheses, 336
  for correlation coefficient, 590
  for equality of multinomial distributions, 400
  for means, 358, 372
  for percentiles, 512
  for proportions, 336, 340
  for randomness, 536
  for regression coefficients, 450
  for variance(s), 358, 379
Test statistic, 339, 358

Testing probabilistic models, 391, 395
Third quartile, 26, 186
Time sequence, 41
Transformations of random variables, 596
Tree diagram, 4, 72
Trinomial distribution, 562, 577
Tukey, John, 22
Two factor ANOVA, 426
Two-sided confidence intervals, 301, 341
Two-sided stem-and-leaf display, 32
Two-sided test, 339
Type I error, 336, 346
Type II error, 337, 346

Unbiased estimator, 287
Uniform distribution, 126, 191, 224, 227
Uniformly most powerful critical region, 630
Uniformly most powerful test, 630
Union of sets, 62, 642
Universal set, 62, 641
Upper percent point, 205, 212, 299, 314

Variability, 2, 488
Variance, 15, 119, 249, 557
  of distribution, 119, 184, 557
  of empirical distribution, 122
  of linear combination, 249
  of sample, 15
  of X, 119, 184, 557
Venn diagram, 643

Waiting time, 194, 201
Welch's T, 308, 378
Wilcoxon test, 519, 531

$\bar{x}$-chart, 463

Confidence Intervals

Parameter	Assumptions	Endpoints
μ	$N(\mu, \sigma^2)$ or n large, σ^2 known	$\bar{x} \pm z_{\alpha/2} \dfrac{\sigma}{\sqrt{n}}$
μ	$N(\mu, \sigma^2)$, σ^2 unknown	$\bar{x} \pm t_{\alpha/2}(n-1) \dfrac{s}{\sqrt{n}}$
$\mu_X - \mu_Y$	$N(\mu_X, \sigma_X^2)$, $N(\mu_Y, \sigma_Y^2)$, σ_X^2, σ_Y^2 known	$\bar{x} - \bar{y} \pm z_{\alpha/2} \sqrt{\dfrac{\sigma_X^2}{n} + \dfrac{\sigma_Y^2}{m}}$
$\mu_X - \mu_Y$	Variances unknown, large samples	$\bar{x} - \bar{y} \pm z_{\alpha/2} \sqrt{\dfrac{s_x^2}{n} + \dfrac{s_y^2}{m}}$
$\mu_X - \mu_Y$	$N(\mu_X, \sigma_X^2)$, $N(\mu_Y, \sigma_Y^2)$, $\sigma_X^2 = \sigma_Y^2$, unknown	$\bar{x} - \bar{y} \pm t_{\alpha/2}(n+m-2) s_p \sqrt{\dfrac{1}{n} + \dfrac{1}{m}}$, $\quad s_p = \sqrt{\dfrac{(n-1)s_x^2 + (m-1)s_y^2}{n+m-2}}$
$\mu_D = \mu_X - \mu_Y$	X and Y normal, but dependent	$\bar{d} \pm t_{\alpha/2}(n-1) \dfrac{s_d}{\sqrt{n}}$
σ^2	$N(\mu, \sigma^2)$	$\dfrac{(n-1)s^2}{\chi_{\alpha/2}^2(n-1)}, \dfrac{(n-1)s^2}{\chi_{1-\alpha/2}^2(n-1)}$
$\dfrac{\sigma_X^2}{\sigma_Y^2}$	$N(\mu_X, \sigma_X^2)$, $N(\mu_Y, \sigma_Y^2)$	$\dfrac{s_x^2/s_y^2}{F_{\alpha/2}(n-1, m-1)}, \quad F_{\alpha/2}(m-1, n-1) \dfrac{s_x^2}{s_y^2}$
p	$b(n, p)$, n is large	$\dfrac{y}{n} \pm z_{\alpha/2} \sqrt{\dfrac{(y/n)(1 - y/n)}{n}}$
$p_1 - p_2$	$b(n_1, p_1)$, $b(n_2, p_2)$	$\hat{p}_1 - \hat{p}_2 \pm z_{\alpha/2} \sqrt{\dfrac{\hat{p}_1(1 - \hat{p}_1)}{n_1} + \dfrac{\hat{p}_2(1 - \hat{p}_2)}{n_2}}$, $\quad \hat{p}_1 = y_1/n_1, \ \hat{p}_2 = y_2/n_2$